Compendium of Organic Synthetic Methods

Compendium of Organic Synthetic Methods

Volume 10

MICHAEL B. SMITH

DEPARTMENT OF CHEMISTRY
THE UNIVERSITY OF CONNECTICUT
STORRS, CONNECTICUT

WILEY-INTERSCIENCE

A JOHN WILEY & SONS, INC., PUBLICATION

Cover illustration was adapted from "Disconnect By the Numbers: A Beginner's Guide to Synthesis" by M. B. Smith. *Journal of Chemical Education,* **1990,** 67, 848–856.

This text is printed on acid-free paper. ♾

Published simultaneously in Canada.

For ordering and customer service, call 1-800-CALL WILEY.

Library of Congress Cataloging Card Number: 71-162800

ISBN 0-471-20201-0

Printed in the United States of America

10 9 8 7 6 5 4 3 2 1

CONTENTS

PREFACE

Since the original volume in this series by Ian and Shuyen Harrison, the goal of the *Compendium of Organic Synthetic Methods* was to facilitate the search for functional group transformations in the original literature of Organic Chemistry. In Volume 2, difunctional compounds were added and this compilation was continued by Louis Hegedus and Leroy Wade for Volume 3 of the series. Wade became the author for Volume 4 and continued with Volume 5. I began editing the series with Volume 6, where I introduced an author index for the first time and added a new chapter (Chapter 15, Oxides). Volume 7 introduced Sections 378 (Oxides - Alkynes) through Section 390 (Oxides - Oxides). The *Compendium* is a handy desktop reference that will remain a valuable tool to the working Organic chemist, allowing a "quick check" of the literature. It also allows one to "browse" for new reactions and transformations that may be of interest. The body of Organic literature is very large and the *Compendium* is a focused and highly representative review of the literature and is offered in that context.

Compendium of Organic Synthetic Methods, Volume 10 contains both functional group transformations and carbon-carbon bond forming reactions from the literature appearing in the years 1996, 1997 and 1998. The classification schemes used for volumes 6-9 have been continued. Difunctional compounds appear in Chapter 16. The experienced user of the *Compendium* will require no special instructions for the use of Volume 10. Author citations and the Author Index have been continued as in Volumes 6-9.

Every effort has been made to keep the manuscript error free. Where there are errors, I take full responsibility. If there are questions or comments, the reader is encouraged to contact me directly at the address, phone, fax, or Email addresses given below.

As I have throughout my writing career, I thank my wife Sarah and my son Steven who have shown unfailing patience and devotion during this work. I also thank Dr. Darla Henderson, the editor for this volume.

Michael B. Smith

Department of Chemistry
University of Connecticut
55 N. Eagleville Road
Storrs, Connecticut 06269-3060

Voice phone: (860)-486-2881
Fax: (860)-486-2981
Email: smith@nucleus.chem.uconn.edu

Storrs, Connecticut

ABBREVIATIONS

Ac	Acetyl	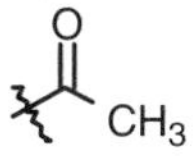
acac	Acetylacetonate	
AIBN	*azo-bis*-isobutyronitrile	
aq.	Aqueous	
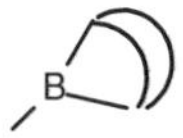	9-Borabicyclo[3.3.1]nonylboryl	
9-BBN	9-Borabicyclo[3.3.1]nonane	
BER	Borohydride exchange resin	
BINAP	*2R,3S*-2,2'-*bis*-(diphenylphosphino)-1,1'-binaphthyl	
Bn	benzyl	
Bz	benzoyl	
BOC	*t*-Butoxycarbonyl	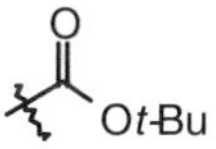
bpy (Bipy)	2,2'-Bipyridyl	
Bu	*n*-Butyl	$-CH_2CH_2CH_2CH_3$
CAM	Carboxamidomethyl	
CAN	Ceric ammonium nitrate	$(NH_4)_2Ce(NO_3)_6$
c-	cyclo-	
cat.	Catalytic	
Cbz	Carbobenzyloxy	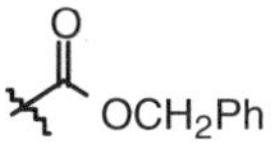
Chirald	2S,3R-(+)-4-dimethylamino-1,2-diphenyl-3-methylbutan-2-ol	
COD	1,5-Cyclooctadienyl	
COT	1,3,5-cyclooctatrienyl	
Cp	Cyclopentadienyl .	
CSA	Camphorsulfonic acid	
CTAB	cetyltrimethylammonium bromide	$C_{16}H_{33}NMe_3{}^+Br^-$
Cy (c-C_6H_{11})	Cyclohexyl	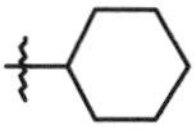
°C	Temperature in Degrees Centigrade	
DABCO	1,4-Diazabicylco[2.2.2]octane	
dba	dibenzylidene acetone	
DBE	1,2-Dibromoethane	$BrCH_2CH_2Br$
DBN	1,8-Diazabicyclo[5.4.0]undec-7-ene	
DBU	1,5-Diazabicyclo[4.3.0}non-5-ene	
DCC	1,3-Dicyclohexylcarbodiimide	c-C_6H_{13}-N=C=N-c-C_6H_{13}
DCE	1,2-Dichloroethane	$ClCH_2CH_2Cl$
DDQ	2,3-Dichloro-5,6-dicyano-1,4-benzoquinone	
% de	% Diasteromeric excess	
DEA	Diethylamine	$HN(CH_2CH_3)_2$

DEAD	Diethylazodicarboxylate	$EtO_2C\text{-}N{=}NCO_2Et$
Dibal-H	Diisobutylaluminum hydride	$(Me_2CHCH_2)_2AlH$
Diphos (dppe)	1,2-*bis*-(Diphenylphosphino)ethane	$Ph_2PCH_2CH_2PPh_2$
Diphos-4 (dppb)	1,4-*bis*-(Diphenylphosphino)butane	$Ph_2P(CH_2)_4PPh_2$
DMAP	4-Dimethylaminopyridine	
DMA	Dimethylacetamide	
DME	Dimethoxyethane	$MeOCH_2CH_2OMe$
DMF	*N,N'*-Dimethylformamide	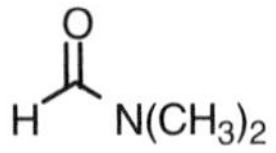
dmp	*bis*-[1,3-Di(p-methoxyphenyl)-1,3-propanedionato]	
dpm	dipivaloylmethanato	
dppb	1,4-*bis*-(Diphenylphosphino)butane	$Ph_2P(CH_2)_4PPh_2$
dppe	1,2-*bis*-(Diphenylphosphino)ethane	$Ph_2PCH_2CH_2PPh_2$
dppf	*bis*-(Diphenylphosphino)ferrocene	
dppp	1,3-*bis*-(Diphenylphosphino)propane	$Ph_2P(CH_2)_3PPh_2$
dvb	Divinylbenzene	
e^-	Electrolysis	
% ee	% Enantiomeric excess	
EE	1-Ethoxyethoxy	EtO(Me)CH-
Et	Ethyl	$\text{-}CH_2CH_3$
EDA	Ethylenediamine	$H_2NCH_2CH_2NH_2$
EDTA	Ethylenediaminetetraacetic acid	
FMN	Flavin mononucleotide	
fod	*tris*-(6,6,7,7,8,8,8)-Heptafluoro-2,2-dimethyl-3,5-octanedionate	
Fp	Cyclopentadienyl-*bis*-carbonyl iron	
FVP	Flash Vacuum Pyrolysis	
h	hour (hours)	
hν	Irradiation with light	
1,5-HD	1,5-Hexadienyl	
HMPA	Hexamethylphosphoramide	$(Me_2N)_3P{=}O$
HMPT	Hexamethylphosphorus triamide	$(Me_2N)_3P$
iPr	Isopropyl	$\text{-}CH(CH_3)_2$
LICA (LIPCA)	Lithium cyclohexylisopropylamide	
LDA	Lithium diisopropylamide	$LiN(iPr)_2$
LHMDS	Lithium hexamethyl disilazide	$LiN(SiMe_3)_2$
LTMP	Lithium 2,2,6,6-tetramethylpiperidide	
MABR	Methylaluminum *bis*-(4-bromo-2,6-di-*tert*-butylphenoxide)	
MAD	*bis*-(2,6-di-*t*-butyl-4-methylphenoxy)methyl aluminum	
mCPBA	*meta*-Chloroperoxybenzoic acid	
Me	Methyl	$\text{-}CH_3$
MEM	β-Methoxyethoxymethyl	$MeOCH_2CH_2OCH_2\text{-}$
Mes	Mesityl	2,4,6-tri-Me-C_6H_2
MOM	Methoxymethyl	$MeOCH_2\text{-}$
Ms	Methanesulfonyl	$CH_3SO_2\text{-}$
MS	Molecular Sieves (3Å or 4Å)	
MTM	Methylthiomethyl	$CH_3SCH_2\text{-}$
NAD	Nicotinamide adenine dinucleotide	
NADP	Sodium triphosphopyridine nucleotide	

Napth	Naphthyl ($C_{10}H_8$)	
NBD	Norbornadiene	
NBS	*N*-Bromosuccinimide	
NCS	*N*-Chlorosuccinimide	
NIS	*N*-Iodosuccinimide	
Ni(R)	Raney nickel	
NMP	*N*-Methyl-2-pyrrolidinone	
Oxone	2 $KHSO_5 \cdot KHSO_4 \cdot K_2SO_4$	
	Polymeric backbone	
PCC	Pyridinium chlorochromate	
PDC	Pyridinium dichromate	
PEG	Polyethylene glycol	
Ph	Phenyl	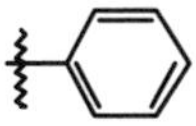
PhH	Benzene	
PhMe	Toluene	
Phth	Phthaloyl	
pic	2-Pyridinecarboxylate	
Pip	Piperidino	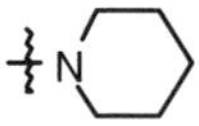N
PMP	4-methoxyphenyl	
Pr	*n*-Propyl	$-CH_2CH_2CH_3$
Py	Pyridine	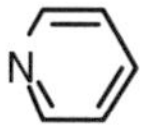N
quant.	Quantitative yield	
Red-Al	$[(MeOCH_2CH_2O)_2AlH_2]Na$	
sBu	*sec*-Butyl	$CH_3CH_2CH(CH_3)$
sBuLi	*sec*-Butyllithium	$CH_3CH_2CH(Li)CH_3$
Siamyl	Diisoamyl	$(CH_3)_2CHCH(CH_3)$-
TADDOL	α,α,α',α'-tetraaryl-4,5-dimethoxy-1,3-dioxolane	
TASF	*tris*-(Diethylamino)sulfonium difluorotrimethyl silicate	
TBAF	Tetrabutylammonium fluoride	n-$Bu_4N^+F^-$
TBDMS	*t*-Butyldimethylsilyl	t-$BuMe_2Si$
TBHP (*t*-BuOOH)	*t*-Butylhydroperoxide	Me_3COOH
t-Bu	*tert*-Butyl	$-C(CH_3)_3$
TEBA	Triethylbenzylammonium	$Bn(Et)_3N^+$
TEMPO	Tetramethylpiperdinyloxy free radical	
TFA	Trifluoroacetic acid	CF_3COOH
TFAA	Trifluoroacetic anhydride	$(CF_3CO)_2O$
Tf (OTf)	Triflate	$-SO_2CF_3$ ($-OSO_2CF_3$)
THF	Tetrahydrofuran	
THP	Tetrahydropyran	
TMEDA	Tetramethylethylenediamine	$Me_2NCH_2CH_2NMe_2$
TMG	1,1,3,3-Tetramethylguanidine	
TMS	Trimethylsilyl	$-Si(CH_3)_3$
TMP	2,2,6,6-Tetramethylpiperidine	
TPAP	tetra-*n*-Propylammonium perruthenate	

Tol	Tolyl	4-$C_6H_4CH_3$
Tr	Trityl	-CPh_3
TRIS	Triisopropylphenylsulfonyl	
Ts(Tos)	Tosyl = *p*-Toluenesulfonyl	4-$MeC_6H_4SO_2$
X_c	Chiral auxiliary	

INDEX, MONOFUNCTIONAL COMPOUNDS

Sections—**heavy type**
Pages—light type

PROTECTION

	Sect.	Pg.
Carboxylic acids	30A	11
Alcohols, phenols	45A	40
Aldehydes	60A	54
Amides	90A	117
Amines	105A	138
Ketones	180A	203

Blanks in the table correspond to sections for which no additional examples were found in the literature

PREPARATION OF →

FROM ↓

	Alkynes	Carboxylic acid derivatives	Alcohols, phenols	Aldehydes	Alkyls, methylenes, aryls	Amides	Amines	Esters	Ethers, epxoides	Halides, sulfonates	Hydrides (RH)	Ketones	Nitriles	Alkenes	Miscellaneous
Alkynes	**1** 1				**61** 57	**76** 100	**91** 118	**106** 139	**121** 157			**166** 183		**196** 212	
Carboxylic acid derivatives		**17** 8	**32** 12	**47** 45		**77** 100		**107** 139			**152** 177	**167** 183	**182** 207		**212** 229
Alcohols, phenols		**18** 8	**33** 12	**48** 46		**78** 101		**108** 141	**123** 157	**138** 170	**153** 177	**168** 185		**198** 213	**213** 229
Aldehydes		**19** 8	**34** 13		**64** 60	**79** 102	**94** 118	**109** 145	**124** 158			**169** 187	**184** 207	**199** 214	
Alkyls, methylenes, aryls	**6** 3	**20** 8	**35** 26	**50** 50	**65** 60	**80** 102				**140** 171		**170** 188			**215** 230
Amides			**36** 27	**51** 50		**81** 103	**96** 120	**111** 146				**171** 190	**186** 207		
Amines			**37** 27			**82** 108	**97** 121			**142** 173	**157** 178	**172** 190	**187** 207		**217** 231
Esters	**8** 4		**38** 28		**68** 63	**83** 111	**98** 127	**113** 147			**158** 178	**173** 190		**203** 218	
Ethers, epxoides			**39** 28	**54** 51	**69** 64		**99** 127	**114** 149	**129** 160		**159** 179	**174** 192		**204** 218	**219** 231
Halides, sulfonates	**10** 4	**25** 9	**40** 32	**55** 52	**70** 65	**85** 111	**100** 128	**115** 150	**130** 161	**145** 174	**160** 180	**175** 192	**190** 208	**205** 219	**220** 236
Hydrides (RH)	**11** 5		**41** 33	**56** 52		**86** 112		**116** 152		**146** 174		**176** 194			
Ketones	**12** 5	**27** 10	**42** 34		**72** 71	**87** 113	**102** 131	**117** 153	**132** 163	**147** 175	**162** 181	**177** 194		**207** 222	
Nitriles					**73** 72	**88** 113	**103** 131				**163** 182				
Alkenes		**29** 10	**44** 39	**59** 53	**74** 72	**89** 114	**104** 132	**119** 155	**134** 163	**149** 175		**179** 198	**194** 209	**209** 223	
Miscellaneous				**60** 54	**75** 97	**90** 115	**105** 133	**120** 156	**135** 169	**150** 176	**165** 182	**180** 199	**195** 209	**210** 227	**225** 237

INDEX, DIFUNCTIONAL COMPOUNDS

Sections—**heavy type**
Pages—light type

Blanks in the table correspond to sections for which additional examples were found in the literature.

Sections—**heavy type**
Pages—light type

Alkyne	Carboxylic acid	Alcohol, Phenol	Aldehyde	Amide	Amine	Ester	Ether, Epoxide	Halide	Ketone	Nitrile	Alkene	Oxide
300 238												
	312 243											
302 239	**313** 243	**323** 247										
303 240		**324** 253										
	315 244	**325** 253	**334** 278	**342** 280								
	316 245	**326** 255		**343** 280	**350** 290							
	317 245	**327** 257		**344** 281	**351** 291	**357** 302						
307 241		**328** 260	**337** 278	**345** 282		**358** 303	**363** 312					
	319 246	**329** 264		**346** 282	**353** 293	**359** 304	**364** 312	**368** 323				
309 241	**320** 246	**330** 266		**347** 283	**354** 293	**360** 305	**365** 313	**369** 357324	**372** 326			
		331 272		**348** 284	**355** 295		**366** 350316	**370** 325	**373** 328			
311 242	**322** 246	**332** 273	**341** 279	**349** 285	**356** 296	**362** 307	**367** 317	**371** 326	**374** 328	**376** 335	**377** 337	
378 341		**380** 341		**382** 343	**383** 343	**384** 344	**385** 345		**387** 345		**389** 347	

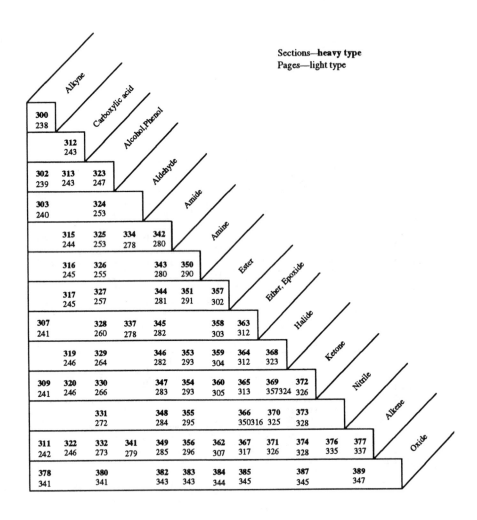

INTRODUCTION

Relationship between Volume 10 and Previous Volumes. *Compendium of Organic Synthetic Methods, Volume 10* presents about 1200 examples of published reactions for the preparation of monofunctional compounds, up-dating the 11850 in Volumes 1-9. Volume 10 contains about 550 examples of reactions that prepare difunctional compounds with various functional groups. Reviews have long been a feature of this series, but Volume 10 adds only 13 pertinent reviews in the various sections.

Chapters 1-14 continue as in Volumes 1-9, as does Chapter 15, introduced in Volume 6. Difunctional compounds appear in Chapter 16, as in Volumes 6-9. The sections on oxides as part of difunctional compounds, introduced in volume 7, continues in Chapter 16 of Volumes 8-10 with Sections 378 (Oxides-Alkynes) through Section 390 (Oxides-Oxides).

Following Chapter 16 is a complete alphabetical listing of all authors (last name, initials). The authors for each citation appear below the reaction. The principle author is indicated by underlining (i.e., Kwon, T.W.; Smith, M.B.), as done previously in Volumes 7-9.

Classification and Organization of Reactions Forming Monofunctional Compounds. Chemical transformations are classified according to the reacting functional group of the starting material and the functional group formed. Those reactions that give products with the same functional group form a chapter. The reactions in each chapter are further classified into sections on the basis of the functional group of the starting material. Within each section, reactions are loosely arranged in descending order of year cited (1998-1996), although an effort has been made to put similar reactions together when possible. Review articles are collected at the end of each appropriate section.

The classification is unaffected by allylic, vinylic, or acetylenic unsaturation appearing in both starting material and product, or by increases or decreases in the length of carbon chains; for example, the reactions *t*-BuOH → *t*-BuCOOH, $PhCH_2OH$ → PhCOOH, and $PhCH{=}CHCH_2OH$ → PhCH=CHCOOH would all be considered as preparations of carboxylic acids from alcohols. Conjugate reduction and alkylation of unsaturated ketones, aldehydes, esters, acids, and nitriles have been placed in Sections 74D and 74E (Alkyls from Alkenes), respectively.

The terms hydrides, alkyls, and aryls classify compounds containing reacting hydrogens, alkyl groups, and aryl groups, respectively; for example, RCH_2-H → RCH_2COOH (carboxylic acids from hydrides), RMe → RCOOH (carboxylic acids from alkyls), RPh → RCOOH (carboxylic acids from aryls). Note the distinction between R_2CO → R_2CH_2 (methylenes from ketones) and RCOR' → RH (hydrides from ketones). Alkylations involving additions across double bonds are found in Section 74 (alkyls, methylenes, and aryls from alkenes).

The following examples illustrate the classification of some potentially confusing cases:

RCH=CHCOOH	→	$RCH{=}CH_2$	Hydrides from carboxylic acids
$RCH{=}CH_2$	→	RCH=CHCOOH	Carboxylic acids from hydrides
ArH	→	ArCOOH	Carboxylic acids from hydrides
ArH	→	ArOAc	Esters from hydrides
RCHO	→	RH	Hydrides from aldehydes
RCH=CHCHO	→	$RCH{=}CH_2$	Hydrides from aldehydes
RCHO	→	RCH_3	Alkyls from aldehydes
R_2CH_2	→	R_2CO	Ketones from methylenes
RCH_2COR	→	R_2CHCOR	Ketones from ketones
$RCH{=}CH_2$	→	RCH_2CH_3	Alkyls from alkenes (Hydrogenation of Alkenes)
RBr + HC+CH	→	RC+CR	Acetylenes from halides; also acetylenes from acetylenes
ROH + RCOOH	→	RCOOR	Esters from alcohols; also esters from carboxylic acids
RCH=CHCHO	→	RCH_2CH_2CHO	Alkyls from alkenes (Conjugate Reduction)
RCH=CHCN	→	RCH_2CH_2CN	Alkyls from alkenes (Conjugate Reduction)

How to Use the Book to Locate Examples of the Preparation of Protection of Monofunctional Compounds. Examples of the preparation of one functional group from another are found in the monofunctional index on p x, which lists the corresponding section and page. Sections that contain examples of the reactions of a functional group are found in the horizontal rows of this index. Section 1 gives examples of the reactions of acetylenes that form new acetylenes; Section 16 gives reactions of acetylenes that form carboxylic acids; and Section 31 gives reactions of acetylenes that form alcohols.

Examples of alkylation, dealkylation, homologation, isomerization, and transposition are found in Sections 1, 17, 33, and so on, lying close to a diagonal of the index. These sections correspond to such topics as the preparation of acetylenes from acetylenes; carboxylic acids from carboxylic acids; and alcohols, thiols, and phenols from alcohols, thiols, and phenols. Alkylations that involve conjugate additions across a double bond are found in Section 74E (Alkyls, Methylenes, and Aryls from Alkenes).

Examples of name reactions can be found by first considering the nature of the starting material and product. The Wittig reaction, for instance, is in Section 199 (Alkenes from Aldehydes) and Section 207 (Alkenes from Ketones). The aldol condensation can be found in the chapters on difunctional compounds in Section 324 (Alcohol, Thiol-Aldehyde) and in Section 330

(Alcohol, Thiol-Ketone). Examples of the synthetically important alkene metathesis reaction are mostly found in Section 209 (Alkenes from Alkenes).

Examples of the protection of acetylenes, carboxylic acids, alcohols, phenols, aldehydes, amides, amines, esters, ketones, and alkenes are also presented. Sections (designated with an A: 15A, 30A, etc.) are labeled "protecting group: reactions" and are located at the end of pertinent chapters.

Some pairs of functional groups such as alcohol, ester; carboxylic acid, ester; amine, amide; and carboxylic acid, amide can be interconverted by simple reactions. When a member of these groups is the desired product or starting material, the other member should also be consulted in the text.

The original literature must be used to determine the generality of reactions, although this is occasionally stated in the citation. This is only done in cases where such generality is stated clearly in the original citation. A reaction given in this book for a primary aliphatic substrate may also be applicable to tertiary or aromatic compounds. This book provides very limited experimental conditions or precautions and the reader is referred to the original literature before attempting a reaction. **In no instance should a citation in this book be taken as a complete experimental procedure. Failure to refer to the original literature prior to beginning laboratory work could be hazardous.** The original papers usually yield a further set of references to previous work. Papers that appear after those publications can usually be found by consulting *Chemical Abstracts and the Science Citation Index.*

Classification and Organization of Reactions Forming Difunctional Compounds. This chapter considers all possible difunctional compounds formed from the groups acetylene, carboxylic acid, alcohol, thiol, aldehyde, amide, amine, ester, ether, epoxide, thioether, halide, ketone, nitrile, and alkene. Reactions that form difunctional compounds are classified into sections on the basis of two functional groups in the product that are pertinent to the reaction. The relative positions of the groups do not affect the classification. Thus preparations of 1,2-amino-alcohols, 1,3-amino-alcohols, and 1,4-amino-alcohols are included in a single section (Section 326, Alcohol-Amine). Difunctional compounds that have an oxide as the second group are found in the appropriate section (Sections 278 - 290). The nitroketone product of oxidation of a nitroalcohol is found in Section 386 (Ketone-Oxide). Conversion of an oxide (such as nitro or a sulfone moiety) to another functional group is generally found in the "Miscellaneous" section of the sections concerning monofunctional compounds. Conversion of a nitroalkane to an amine, for example, is found in Section 105 (Amines from Miscellaneous Compounds). The following examples illustrate applications of this classification system:

Difunctional Product	*Section Title*
RC≡C-C≡CR	Acetylene-Acetylene
RCH(OH)COOH	Carboxylic acid-Alcohol
RCH=CHOMe	Ether-Alkene
$RCHF_2$	Halide-Halide
$RCH(Br)CH_2F$	Halide-Halide
$RCH(OAc)CH_2OH$	Alcohol-Ester
$RCH(OH)CO_2Me$	Alcohol-Ester
$RCH{=}CHCH_2CO_2Me$	Ester-Alkene
RCH=CHOAc	Ester-Alkene
$RCH(OMe)CH_2SO_2CH_2CH_2OH$	Alcohol-Ether
$RSO_2CH_2CH_2OH$	Alcohol-Oxide

How to Use the Book to Locate Examples of the Preparation of Difunctional Compounds. The difunctional index on p xi gives the section and page corresponding to each difunctional product. Thus Section 327 (Alcohol, Thiol-Ester) contains examples of the preparation of hydroxyesters; Section 323 (Alcohol, Thiol-Alcohol, Thiol) contains examples of the preparation of diols.

Some preparations of alkene and acetylenic compounds from alkene and acetylenic starting materials can, in principle, be classified in either the monofunctional or difunctional sections; for example, the transformation RCH=CHBr → RCH=CHCOOH could be consider as preparing carboxylic acids from halides (Section 25, monofunctional compounds) or preparing a carboxylic acid-alkene (Section 322, difunctional compounds). The choice usually depends on the focus of the particular paper where this reaction was found. In such cases both sections should be consulted.

Reactions applicable to both aldehyde and ketone starting materials are in many cases illustrated by an example that uses only one of them. Likewise, many citations for reactions found in the Aldehyde-X sections, will include examples that could be placed in the Ketone-X section. Again, the choice is dictated by the paper where the reaction was found.

Many literature preparations of difunctional compounds are extensions of the methods applicable to monofunctional compounds. As an example, the reaction RCl → ROH might be used for the preparation of diols from an appropriate dichloro compound. Such methods are difficult to categorize and may be found in either the monofunctional or difunctional sections, depending on the focus of the original paper.

The user should bear in mind that the pairs of functional groups alcohol, ester; carboxylic acids, ester; amine, amide; and carboxylic acid, amide can be interconverted by simple reactions. Compounds of the type $RCH(OAc)CH_2OAc$ (ester-ester) would thus be of interest to anyone preparing the diol $RCH(OH)CH_2OH$ (alcohol-alcohol).

Sources of Literature Citations. I thought it would be useful for a reader of this *Compendium* to see those journals that contain the most new synthetic methodology). The accompanying graph shows that *Tetrahedron Letters* and *Journal of Organic Chemistry* account for roughly 50% of all the citations in Volume 10. This book was not edited to favor one journal, section or type of article over another. Undoubtedly, my own personal preferences are part of the selection, but I believe that this compilation is an accurate representation of new synthetic methods that appear in the literature for this period. Therefore, I believe the accompanying graph reflects those journals where new synthetic methodology is located. I should point out that the category "15 other journals" includes: *Accts. Chem. Res.; Acta Chem. Scand.; Angew. Chem. Int. Ed. Engl.; Bull. Chim. Soc. Fr.; Bull. Chim. Soc. Belg.; Bull. Chem. Soc. Jpn.; J. Chem. Res. (S); Can. J. Chem.; Heterocycles; J. Heterocyclic Chem.; J. Indian Chem. Soc.; Liebigs Ann. Chem.; Org. Prep. Proceed Int.;* and, *European J. Org. Chem.* In addition, nine more journals were examined but no references were recorded.

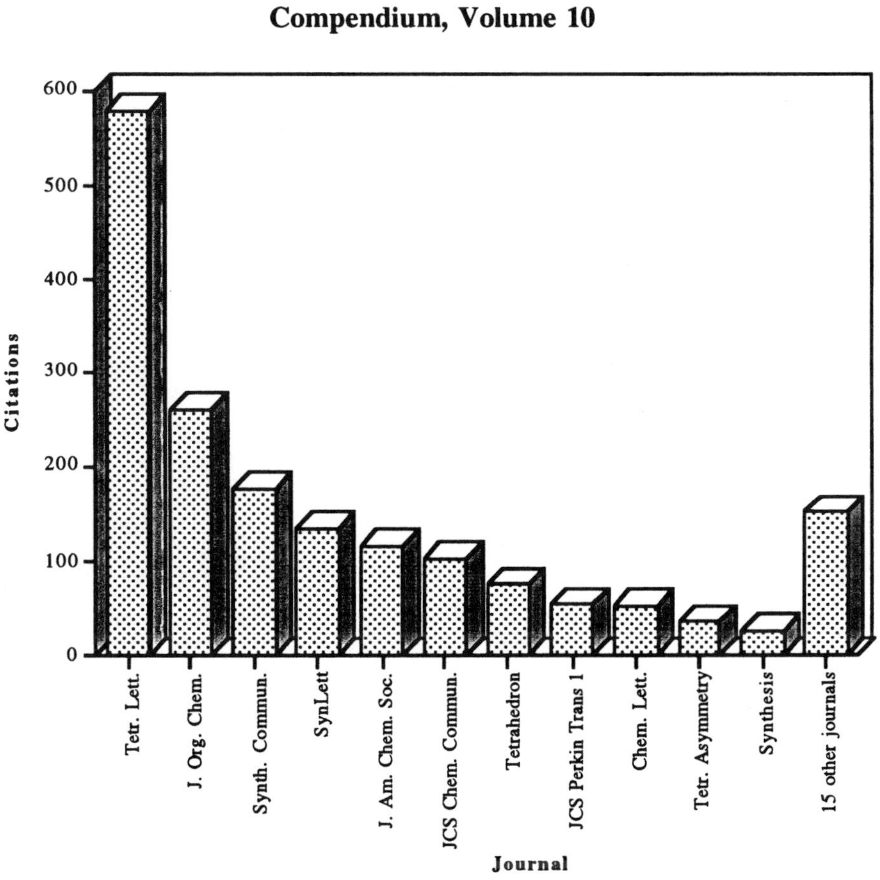
Compendium, Volume 10
Citations
600
500
400
300
200
100
0
Tetr. Lett.
J. Org. Chem.
Synth. Commun.
SynLett
J. Am. Chem. Soc.
JCS Chem. Commun.
Tetrahedron
JCS Perkin Trans 1
Chem. Lett.
Tetr. Asymmetry
Synthesis
15 other journals
Journal

Compendium of Organic Synthetic Methods

CHAPTER 1

PREPARATION OF ALKYNES

SECTION 1: ALKYNES FROM ALKYNES

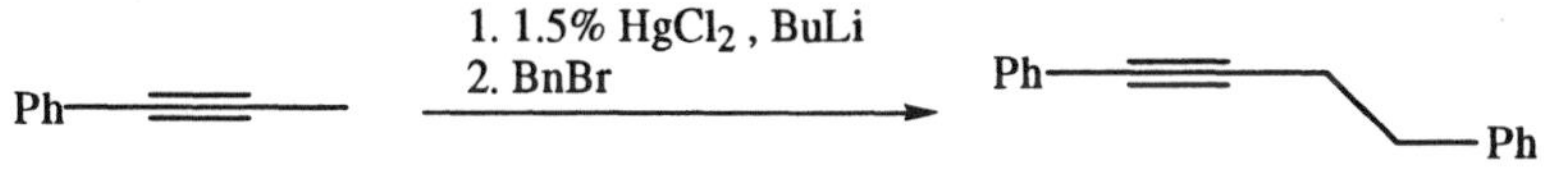

1. 1.5% HgCl2 , BuLi
2. BnBr
Ph
Ph
Ph

65%

Ma, S.; Wang, L. *J. Org. Chem.*, ***1998***, *63*, 3497.

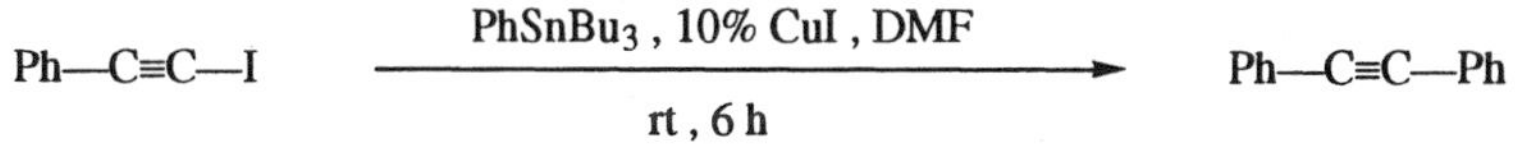

Ph—C≡C—I
PhSnBu3 , 10% CuI , DMF
rt , 6 h
Ph—C≡C—Ph

71%

Kang, S.–K.; Kim, W.–Y.; Jiao, X. *Synthesis*, ***1998***, 1252.

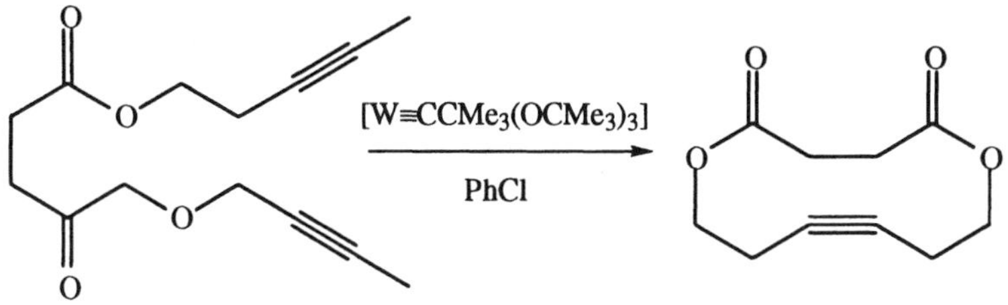

[W≡CCMe3(OCMe3)3]
PhCl

73%

Fürstner, A.; Seidel, G. *Angew. Chem. Int. Ed.*, ***1998***, *37*, 1734.

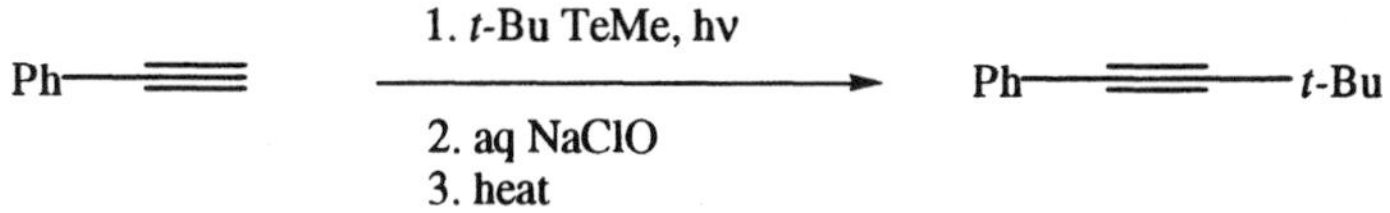

Ph
1. t-Bu TeMe, hν
2. aq NaClO
3. heat
Ph
t-Bu

92x88%

Terao, J.; Kambe, N.; Sonoda, N. *Tetrahedron Lett.*, ***1998***, *39*, 5511.

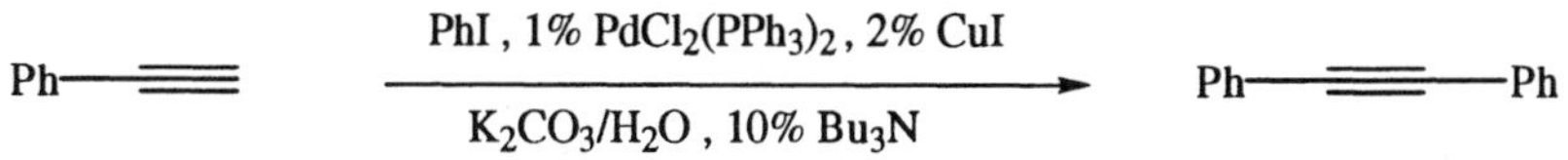

Ph
PhI , 1% PdCl2(PPh3)2 , 2% CuI
K2CO3/H2O , 10% Bu3N
Ph
Ph

98%

Bumagin, N.A.; Sukhomlinova, L.I.; Luzikova, E.V.; Tolstaya, T.P.; Belestskaya, I.P. *Tetrahedron Lett.*. ***1996***. *37*. 897.

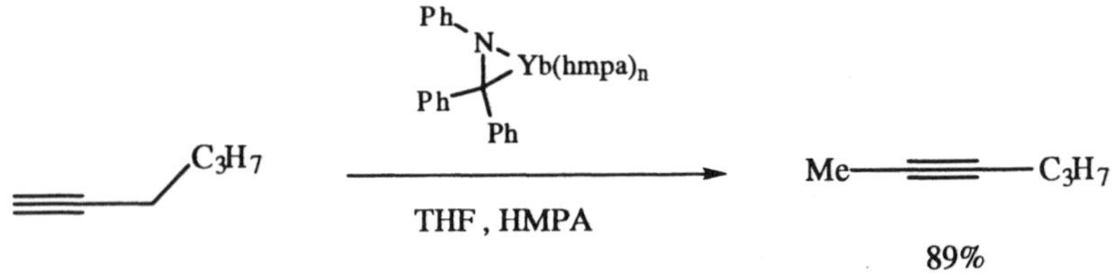
Yb(hmpa)n
THF , HMPA
89%

Makioka, Y.; Saiki, A.; Takaki, K.; Taniguchi, Y.; Kitamura, T.; Fujiwara, Y. *Chem. Lett.*, ***1997***, 27.

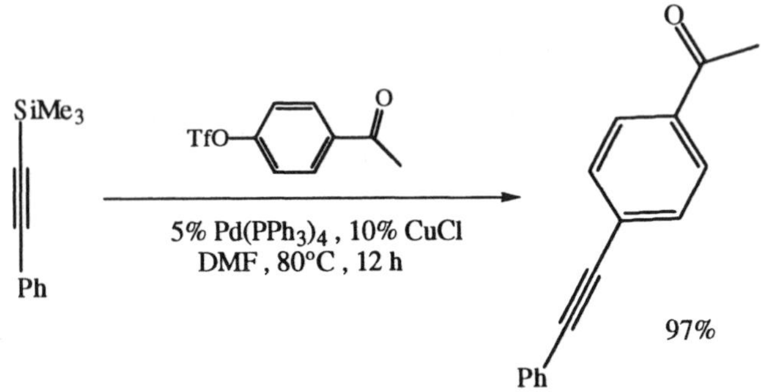
SiMe3
TfO
5% Pd(PPh3)4 , 10% CuCl
DMF , 80°C , 12 h
97%

Nishihara, Y.; Ikegashira, K.; Mori, A.; Hiyama, T. *Chem. Lett.*, ***1997***, 1233.

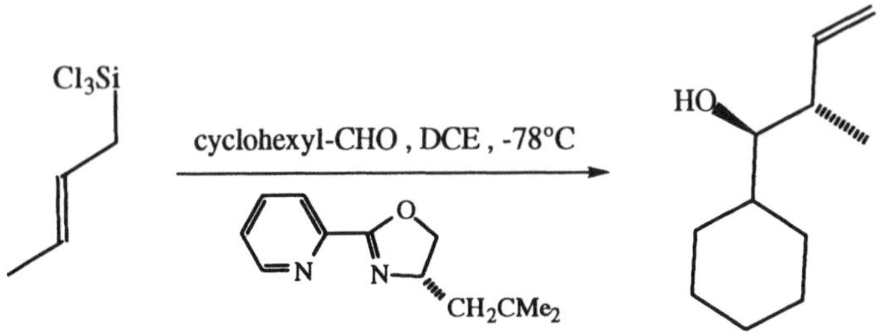
Cl3Si
cyclohexyl-CHO , DCE , -78°C
CH2CMe2
HO

72% (74% ee)

Angell, R.M.; Barrett, A.G.M.; Braddock, D.C.; Swallow, S.; Vickery, B.D. *Chem. Commun.*, ***1997***, 919.

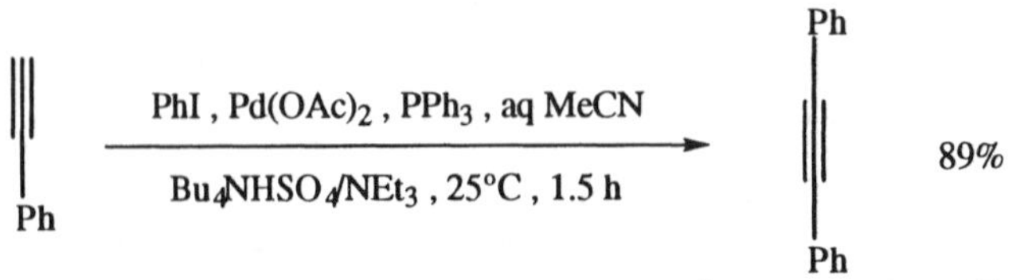
PhI , Pd(OAc)2 , PPh3 , aq MeCN
Bu4NHSO4/NEt3 , 25°C , 1.5 h
89%

Nguefack, J.–F.; Bolitt, V.; Sinou, D. *Tetrahedron Lett.*, ***1996***, *37*, 5527.

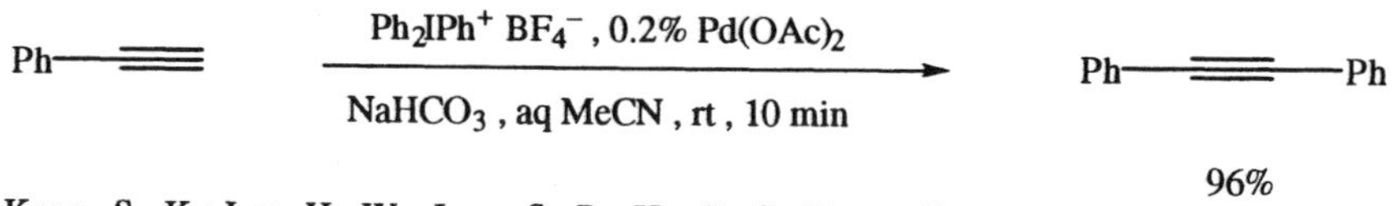
Ph
Ph₂IPh⁺ BF₄⁻ , 0.2% Pd(OAc)₂
NaHCO₃ , aq MeCN , rt , 10 min
Ph
Ph
96%

Kang, S.-K.; Lee, H.-W.; Jang, S.-B.; Ho, P.-S. *Chem. Commun.*, ***1996***, 835.

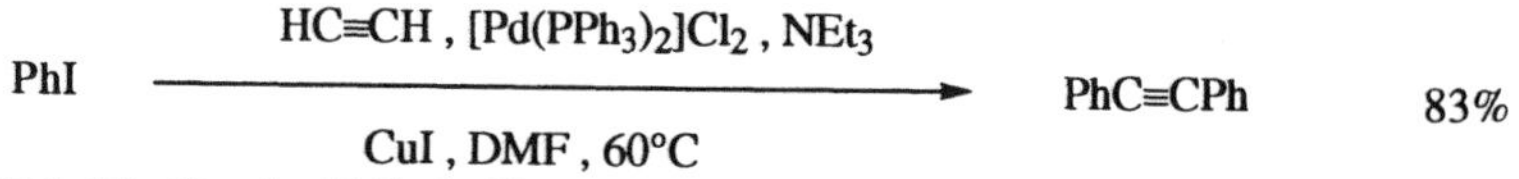
PhI
HC≡CH , [Pd(PPh₃)₂]Cl₂ , NEt₃
CuI , DMF , 60°C
PhC≡CPh
83%

Pal, M.; Kundu, N.G. *J. Chem. Soc., Perkin Trans. 1*, ***1996***, 449.

SECTION 2: ALKYNES FROM ACID DERIVATIVES

NO ADDITIONAL EXAMPLES

SECTION 3: ALKYNES FROM ALCOHOLS AND THIOLS

NO ADDITIONAL EXAMPLES

SECTION 4: ALKYNES FROM ALDEHYDES

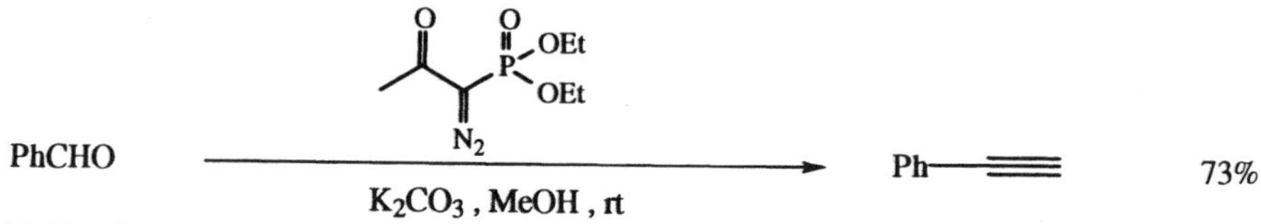
O
O
P
OEt
OEt
N₂
PhCHO
K₂CO₃ , MeOH , rt
Ph
73%

Müller, S.; Liepold, B.; Roth, G.J.; Bestmann, H.J. *SynLett.*, ***1996***, 521.

SECTION 5: ALKYNES FROM ALKYLS, METHYLENES AND ARYLS

NO ADDITIONAL EXAMPLES

SECTION 6: ALKYNES FROM AMIDES

NO ADDITIONAL EXAMPLES

SECTION 7: ALKYNES FROM AMINES

NO ADDITIONAL EXAMPLES

SECTION 8: ALKYNES FROM ESTERS

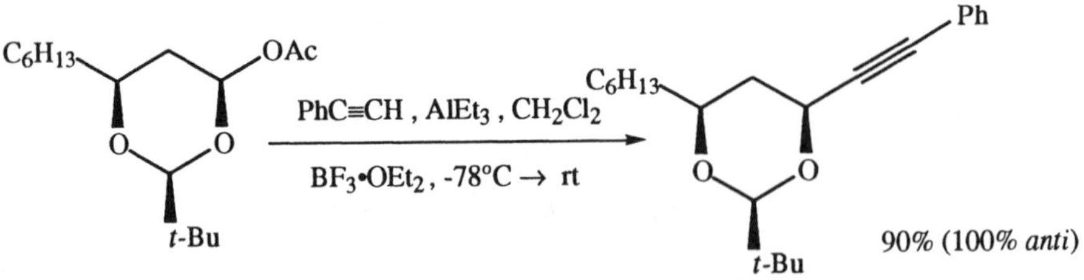
C6H13
OAc
O
O
t-Bu
PhC≡CH , AlEt3 , CH2Cl2
BF3•OEt2 , -78°C → rt
C6H13
Ph
O
O
t-Bu
90% (100% anti)

Powell, N.A.; Rychnovsky, S.D. *Tetrahedron Lett.*, ***1998***, *39*, 3103.

SECTION 9: ALKYNES FROM ETHERS, EPOXIDES AND THIOETHERS

NO ADDITIONAL EXAMPLES

SECTION 10: ALKYNES FROM HALIDES AND SULFONATES

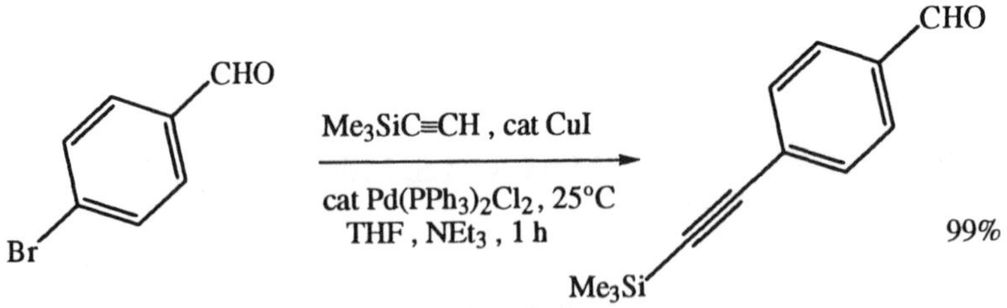
CHO
Br
Me3SiC≡CH , cat CuI
cat Pd(PPh3)2Cl2 , 25°C
THF , NEt3 , 1 h
CHO
Me3Si
99%

Thorand, S.; Krause, N. *J. Org. Chem.*, ***1998***, *63*, 8551.

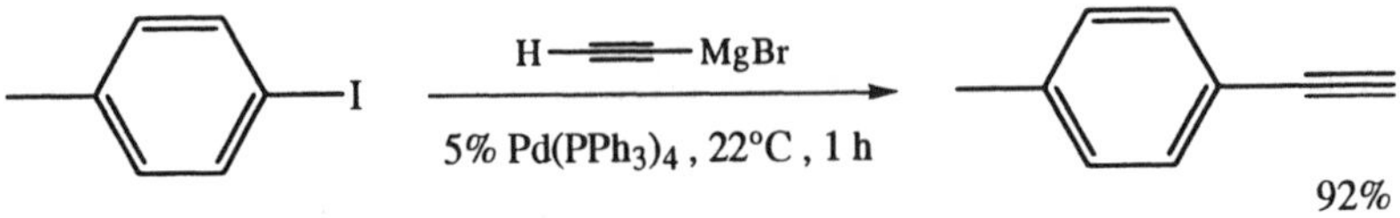
I
H—≡—MgBr
5% Pd(PPh3)4 , 22°C , 1 h
92%

Negishi, E.; Kotora, M.; Xu, C. *J. Org. Chem.*, ***1997***, *62*, 8957.

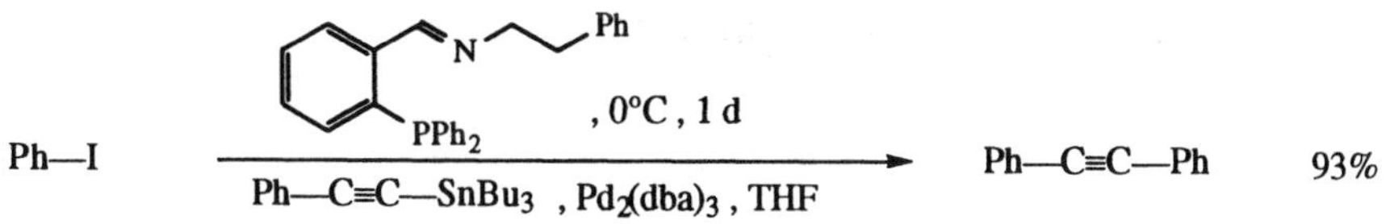
Ph—I
Ph
N
PPh2
, 0°C , 1 d
Ph—C≡C—SnBu3 , Pd2(dba)3 , THF
Ph—C≡C—Ph
93%

Shirakawa, E.; Yoshida, H.; Takaya, H. *Tetrahedron Lett.*, ***1997***, *38*, 3759.

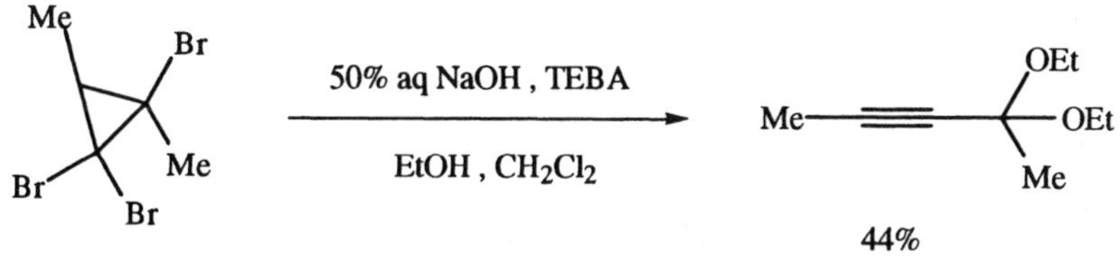
Me
Br
Me
Br
Br
50% aq NaOH , TEBA
EtOH , CH2Cl2
Me
OEt
OEt
Me
44%

Snydes, L.K.; Bakstad, E. *Acta Chem. Scand. B*, ***1996***, *50*, 446.

SECTION 11: ALKYNES FROM HYDRIDES

For examples of the reaction RC≡CH → RC≡C-C≡CR[1], see section 300 (Alkyne-Alkyne).

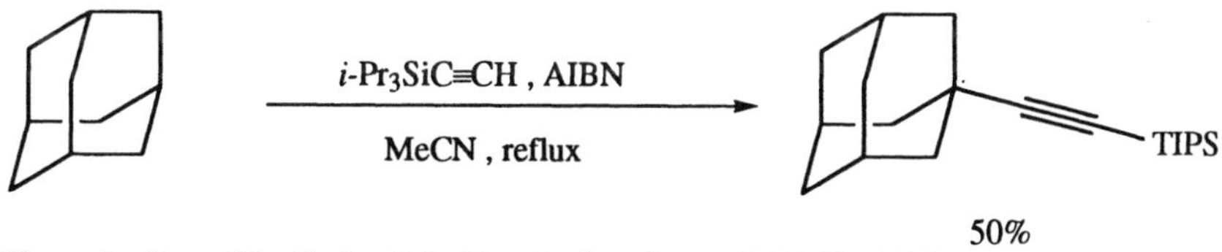
i-Pr3SiC≡CH , AIBN
MeCN , reflux
TIPS
50%

Xiang, J.; Jiang, W.; Fuchs, P.L. *Tetrahedron Lett.*, ***1997***, *38*, 6635.

SECTION 12: ALKYNES FROM KETONES

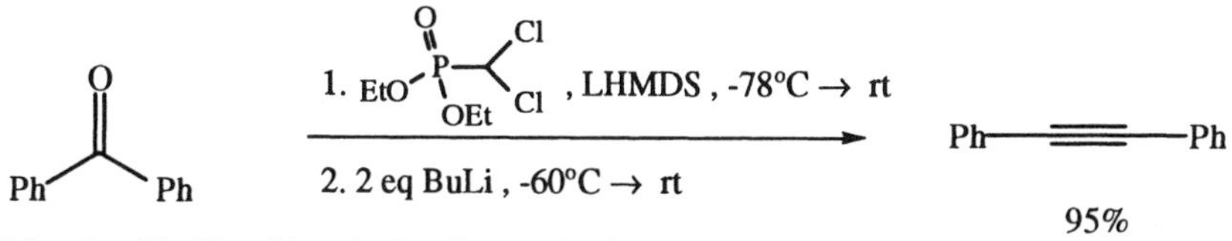
O
Ph
Ph
1. EtO P Cl Cl OEt O , LHMDS , -78°C → rt
2. 2 eq BuLi , -60°C → rt
Ph
Ph
95%

Mouriès, V.; Waschbüsch, R.; Carran, J.; Savignac, P. *Synthesis*, ***1998***, 271.

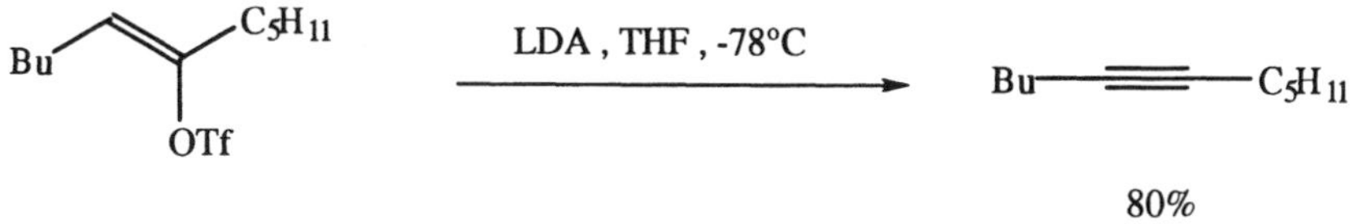
Bu
C5H11
OTf
LDA , THF , -78°C
Bu
C5H11
80%

Brummond, K.M.; Gesenberg, K.D.; Kent, J.L.; Kerekes, A.D. *Tetrahedron Lett.*, ***1998***, *39*, 8613.

SECTION 13: ALKYNES FROM NITRILES

NO ADDITIONAL EXAMPLES

SECTION 14: ALKYNES FROM ALKENES

NO ADDITIONAL EXAMPLES

SECTION 15: ALKYNES FROM MISCELLANEOUS COMPOUNDS

NO ADDITIONAL EXAMPLES

SECTION 15A: PROTECTION OF ALKYNES

NO ADDITIONAL EXAMPLES

CHAPTER 2

PREPARATION OF ACID DERIVATIVES AND ANHYDRIDES

SECTION 16: ACID DERIVATIVES FROM ALKYNES

NO ADDITIONAL EXAMPLES

SECTION 17: ACID DERIVATIVES FROM ACID DERIVATIVES

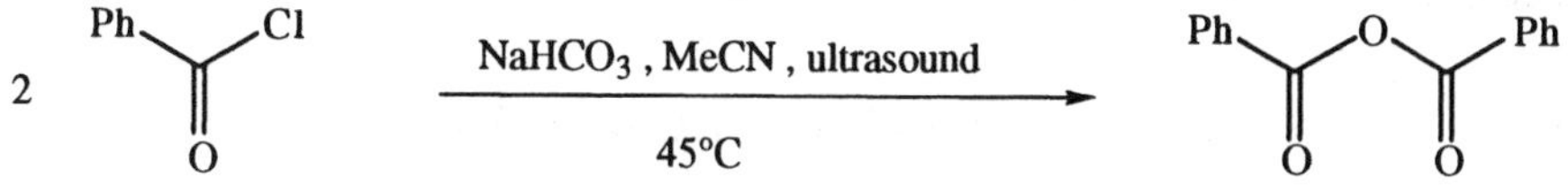

2
Ph
Cl
O
NaHCO3 , MeCN , ultrasound
45°C
Ph
O
Ph
O
O

98%

Hu, Y.; Wang, J.-X.; Li, S. *Synth. Commun.*, ***1997***, *27*, 243.

SECTION 18: ACID DERIVATIVES FROM ALCOHOLS AND THIOLS

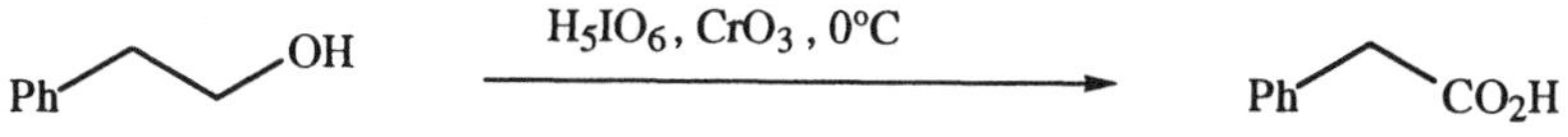

Ph
OH
H5IO6, CrO3 , 0°C
Ph
CO2H

96%

Zhao, M.; Li, J.; Song, Z.; Desmond, R.; Tschaen, D.M.; Grabowski, E.J.J.; Reider, P.J. *Tetrahedron Lett.*, ***1998***, *39*, 5323.

SECTION 19: ACID DERIVATIVES FROM ALDEHYDES

PhCHO $\xrightarrow{\text{Oxone , aq acetone , 25°C , 1 h}}$ PhCOOH 71%

Webb, K.S.; Ruszkay, S.J. *Tetrahedron*, ***1998***, *54*, 401.

PhCHO —— H_2O_2 , aq HCl, MeOH ——→ $PhCO_2Me$ 95%

also works with acetals

Takeda, T.; Watanabe, H.; Kitahara, T. *SynLett.*, ***1997***, 1149.

SECTION 20: ACID DERIVATIVES FROM ALKYLS, METHYLENES AND ARYLS

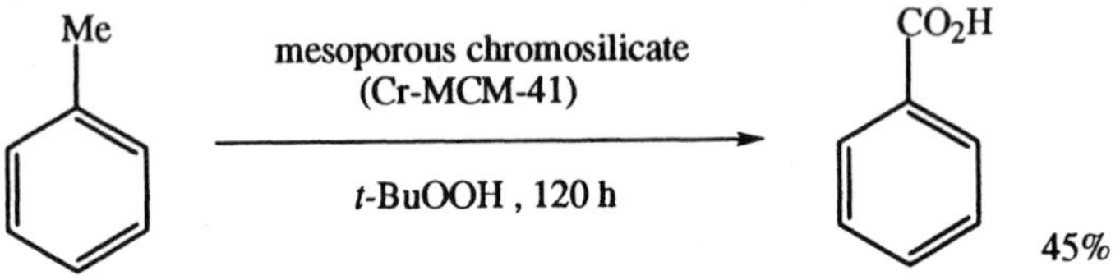
Me
mesoporous chromosilicate
(Cr-MCM-41)
t-BuOOH , 120 h
CO2H
45%

forms ketones from other arenes

Das, T.K.; Chaudhari, K.; Nandanan, E.; Chandwadkar, A.J.H.; Sudalai, A.; Ravidranathan, T Sivasanker, S. *Tetrahedron Lett.*, ***1997***, *38*, 3631.

REVIEW:

"Exploitation of Synthetic Reactions via C—H Bond Activation by Transition Metal Catalysts. Carboxylation and Aminomethylation of Alkanes or Arenes."
Fujiwara, Y.; Takaki, K.; Tankguchi, Y. *SynLett.*, ***1996***, 591.

SECTION 21: ACID DERIVATIVES FROM AMIDES

NO ADDITIONAL EXAMPLES

SECTION 22: ACID DERIVATIVES FROM AMINES

NO ADDITIONAL EXAMPLES

SECTION 23: ACID DERIVATIVES FROM ESTERS

NO ADDITIONAL EXAMPLES

Other reactions useful for the hydrolysis of esters may be found in Section 30A (Protection of Carboxylic Acids).

SECTION 24: ACID DERIVATIVES FROM ETHERS, EPOXIDES AND THIOETHERS

NO ADDITIONAL EXAMPLES

SECTION 25: ACID DERIVATIVES FROM HALIDES AND SULFONATES

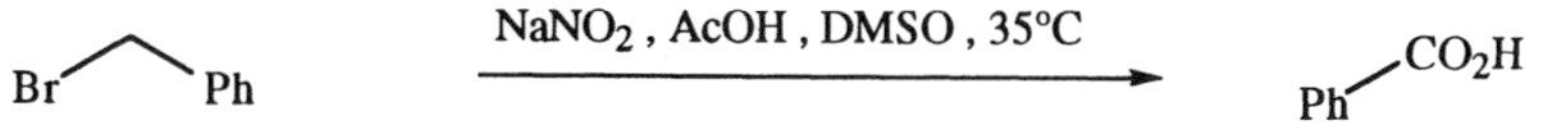
Br
Ph
NaNO2 , AcOH , DMSO , 35°C
CO2H
Ph

80%

Matt, C.; Wagner, A.; Mioskowski, C. *J. Org. Chem.*, ***1997***, *62*, 234.

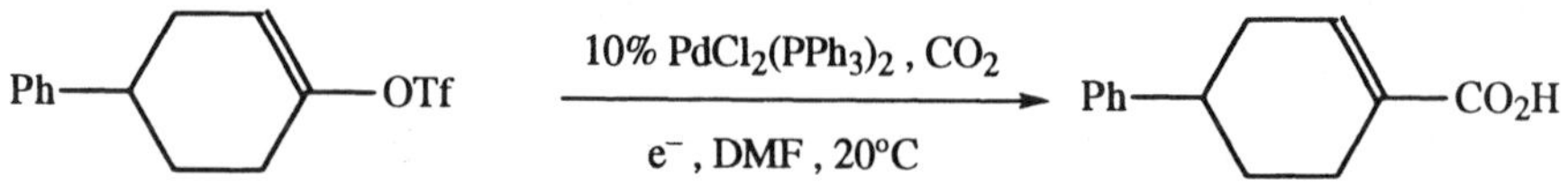
Ph
OTf
10% PdCl2(PPh3)2 , CO2
e⁻ , DMF , 20°C
Ph
CO2H

Jutand, A.; Négri, S. *SynLett.*, ***1997***, 719.

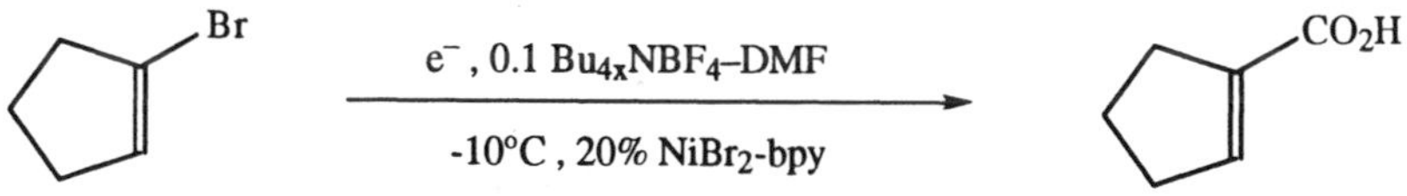
Br
e⁻ , 0.1 Bu4xNBF4–DMF
-10°C , 20% NiBr2-bpy
CO2H

73%

Kamekawa, H.; Kudoh, H.; Senboku, H.; Tokuda, M. *Chem. Lett.*, ***1997***, 917.

SECTION 26: ACID DERIVATIVES FROM HYDRIDES

NO ADDITIONAL EXAMPLES

SECTION 27: ACID DERIVATIVES FROM KETONES

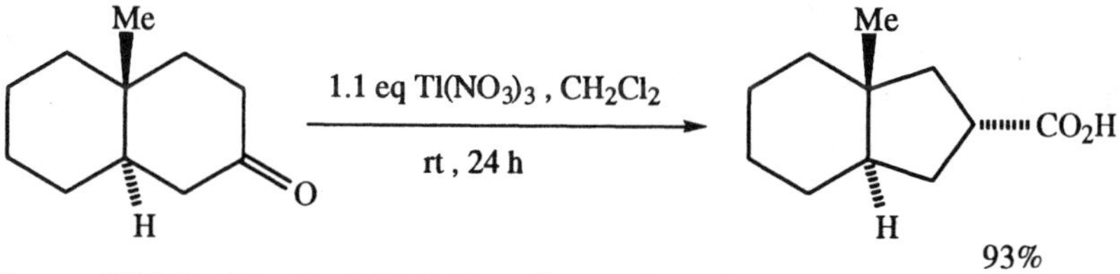
Me
H
O
1.1 eq Tl(NO3)3 , CH2Cl2
rt , 24 h
Me
H
CO2H

93%

Ferraz, H.M.C.; Silva Jr., L.F. *J. Org. Chem.*, ***1998***, *63*, 1716.

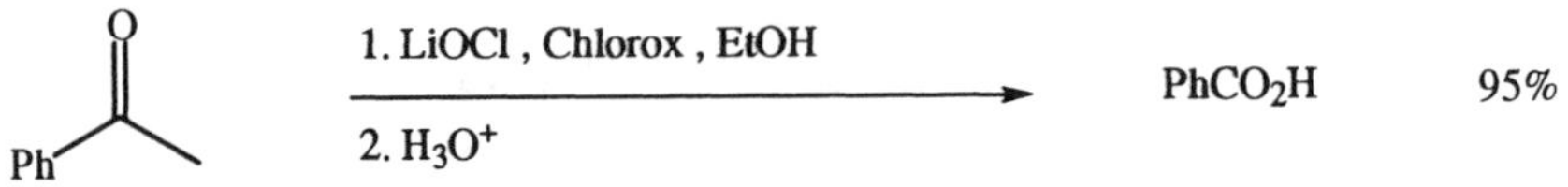
O
Ph
1. LiOCl , Chlorox , EtOH
2. H_3O^+
$PhCO_2H$
95%

Madler, M.M.; Klucik, J.; Soell, P.S.; Brown, C.W.; Liu, S.; Berlin, K.D.; Benbrook, D.M.; Birckbichler, P.J.; Nelson, E.C. *Org. Prep. Proceed. Int.*, ***1998***, *30*, 230.

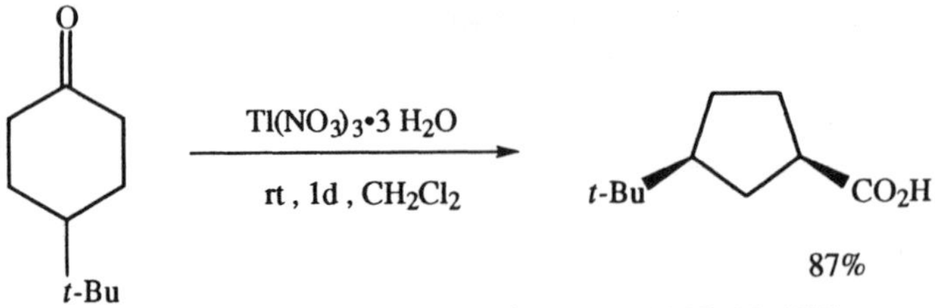
O
$Tl(NO_3)_3 \cdot 3\ H_2O$
rt , 1d , CH_2Cl_2
t-Bu
CO_2H
87%
t-Bu

Ferraz, H.M.C.; Silva Jr., L.F. *Tetrahedron Lett.*, ***1997***, *38*, 1899.

SECTION 28: ACID DERIVATIVES FROM NITRILES

NO ADDITIONAL EXAMPLES

SECTION 29: ACID DERIVATIVES FROM ALKENES

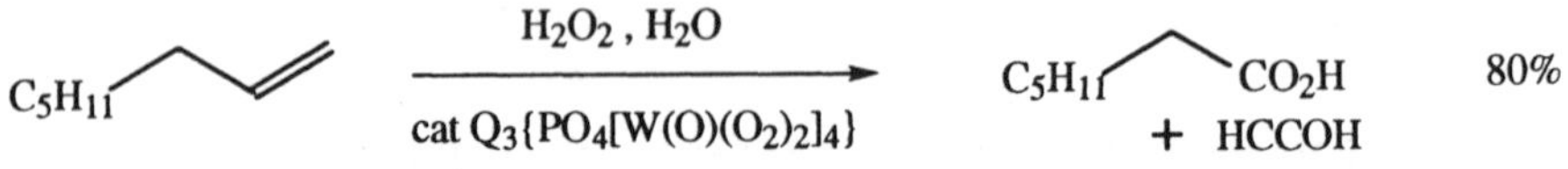
C_5H_{11}
H_2O_2 , H_2O
cat $Q_3\{PO_4[W(O)(O_2)_2]_4\}$
C_5H_{11}
CO_2H
+ HCCOH
80%

$Q = [(C_8H_{17})_3NMe]^+$

Antonelli, E.; D'Aloisio, R.; Gambaro, M.; Fiorani, T.; Venturello, C. *J. Org. Chem.*, ***1998***, *63*, 7190.

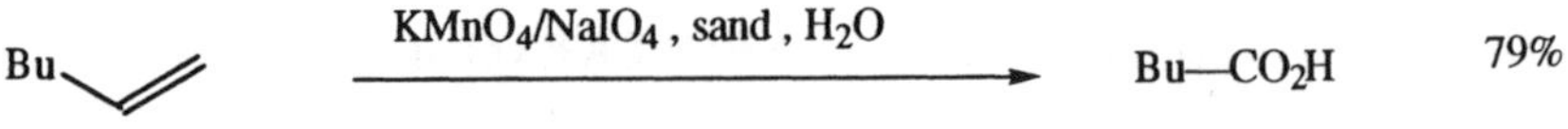
Bu
$KMnO_4/NaIO_4$, sand , H_2O
Bu—CO_2H
79%

Juang, B.; Gupton, J.T.; Hansen, K.C.; Idoux, J.P. *Synth. Commun.*, ***1996***, *26*, 165.

SECTION 30: ACID DERIVATIVES FROM MISCELLANEOUS COMPOUNDS

NO ADDITIONAL EXAMPLES

SECTION 30A: PROTECTION OF CARBOXYLIC ACID DERIVATIVES

Other reactions useful for the protection of carboxylic acids are included in Section 107 (Esters from Carboxylic Acids and Acid Halides) and Section 23 (Carboxylic Acids from Esters).

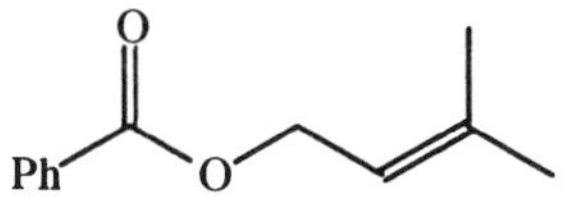

$S\text{-}SnO_2$, PhOMe, toluene / heat, 7h → PhCOOH 82%

Chavan, S.P.; Zubaidha, P.K.; Dantale, S.W.; Keshavaraja, A.; Ramaswamy, A.V.; Ravindranathan, T. *Tetrahedron Lett.*, ***1996***, *37*, 237.

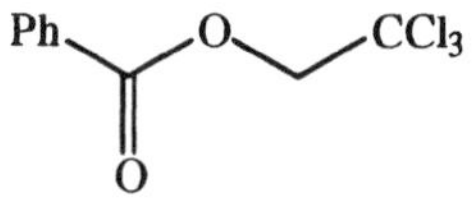

1. Te/$NaBH_4$, DMF, 80°C / 2. 2M HCl → PhCOOH

Blay, G.; Cardona, L.; García, B.; García, C.L.; Pedro, J.R. *Synth. Commun.*, ***1998***, *28*, 1405

CHAPTER 3

PREPARATION OF ALCOHOLS

NO ADDITIONAL EXAMPLES

SECTION 31: ALCOHOLS AND THIOLS FROM ALKYNES

NO ADDITIONAL EXAMPLES

SECTION 32: ALCOHOLS AND THIOLS FROM ACID DERIVATIVES

$C_{17}H_{35}$—COOH —(BOP, $NaBH_4$, *i*-Pr_2NEt, THF)→ $C_{17}H_{35}$—CH_2OH 99%

BOP = benzotriazol-1-yloxytris(dimethylamino)phosphonium hexafluorophosphate

McGeary, R.P. *Tetrahedron Lett.*, ***1998***, *39*, 3319.

SECTION 33: ALCOHOLS AND THIOLS FROM ALCOHOL AND THIOLS

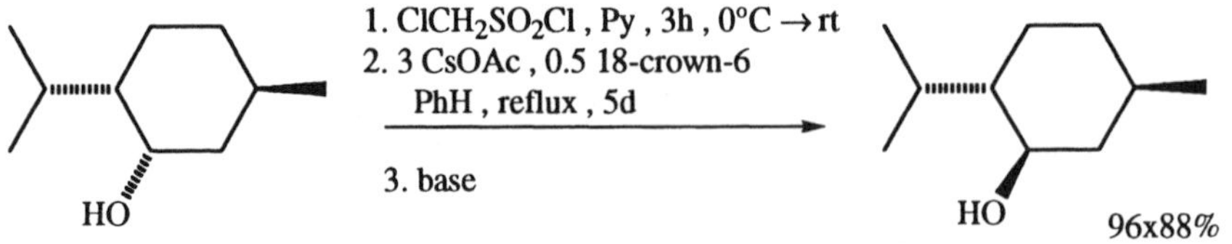
1. ClCH2SO2Cl , Py , 3h , 0°C → rt
2. 3 CsOAc , 0.5 18-crown-6
PhH , reflux , 5d
3. base
HO
HO
96x88%

Shimizu, T.; Hiranuma, S.; Nakata, T. *Tetrahedron Lett.*, ***1996***, *37*, 6145.

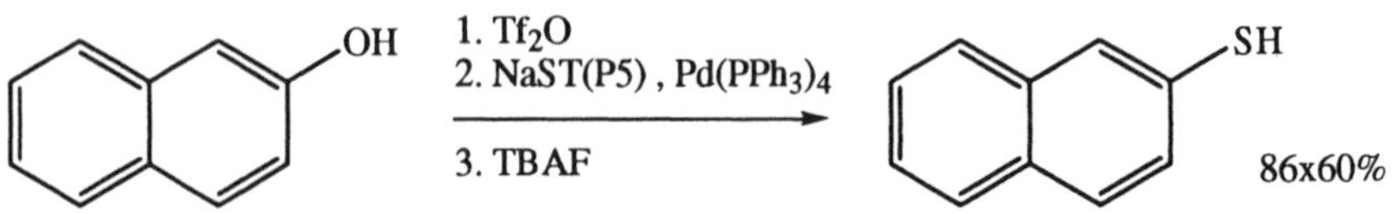
OH
1. Tf2O
2. NaST(P5) , Pd(PPh3)4
3. TBAF
SH
86x60%

Arnould, J.C.; Didelot, M.; Cadilhac, C.; Pasquet, M.J. *Tetrahedron Lett.*, ***1996***, *37*, 4523.

SECTION 34: ALCOHOLS AND THIOLS FROM ALDEHYDES

The following reaction types are included in this section:

A. Reductions of Aldehydes to Alcohols

B. Alkylation of Aldehydes, forming Alcohols.

Coupling of Aldehydes to form Diols is found in Section 323 (Alcohol-Alcohol).

SECTION 34A: REDUCTIONS OF ALDEHYDES TO ALCOHOLS

PhCHO $\xrightarrow[-78°C \to rt]{[InCl_3/Bu_3SnH/THF]}$ $PhCH_2OH$

Miyai, T.; Inoue, K.; Yasuda, M.; Shibata, I.; Baba, A. *Tetrahedron Lett.*, ***1998***, *39*, 1929.

PhCHO $\xrightarrow{FeCl_3/Zn\text{ , DMF/}H_2O\text{ , 1 h}}$ $PhCH_2OH$ 81%

also reduces ketones

Sadavaarte, V.S.; Swami, S.S.; Desai, D.G. *Synth. Commun.*, ***1998***, *28*, 1139.

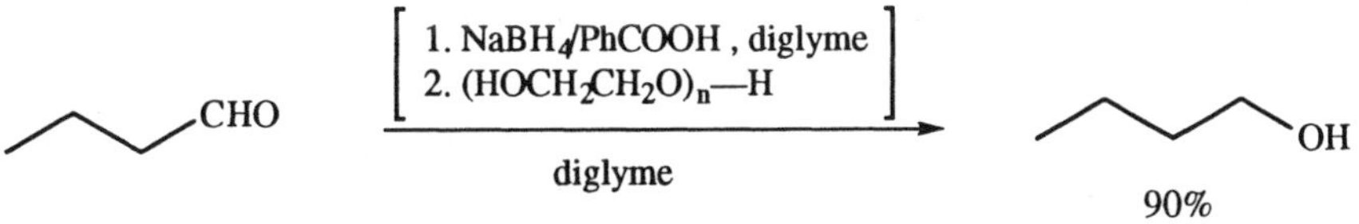
CHO
1. NaBH$_4$/PhCOOH , diglyme
2. (HOCH$_2$CH$_2$O)$_n$—H
diglyme
OH
90%

these conditions are selective for aldehydes

Blanton, J.R. *Synth. Commun.*, ***1997***, *27*, 2093.

PhCHO $\xrightarrow{PMHS\text{ , }ZnCl_2\text{ , 1 d}}$ $PhCH_2OH$ 72%

PMHS = polymethylhydrosiloxane

Chandrasekhar, S.; Reddy, Y.R.; Tamarao, C. *Synth. Commun.*, ***1997***, *27*, 2251.

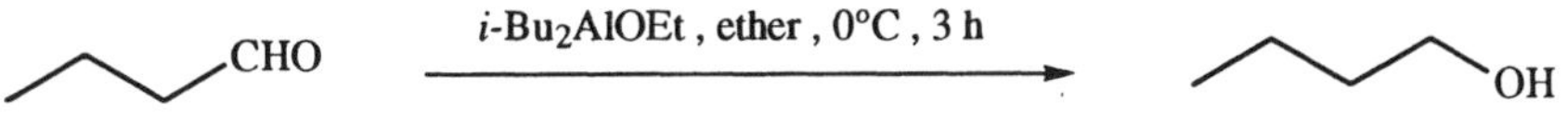
CHO
i-Bu$_2$AlOEt , ether , 0°C , 3 h
OH

Cha, J.S.; Kwon, O.O.; Kwon, S.Y. *Org. Prep. Proceed. Int.*, ***1996***, *28*, 355.

SECTION 34B: ALKYLATION OF ALDEHYDES, FORMING ALCOHOLS

ASYMMETRIC ALKYLATIONS

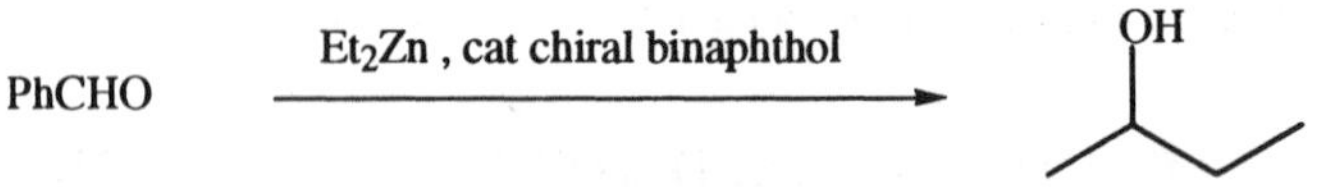
PhCHO
Et_2Zn , cat chiral binaphthol
OH

95% (>99% ee)

Huang, W.–S.; Hu, Q.–S.; Pu, L. *J. Org. Chem.*, ***1998***, *63*, 1364.

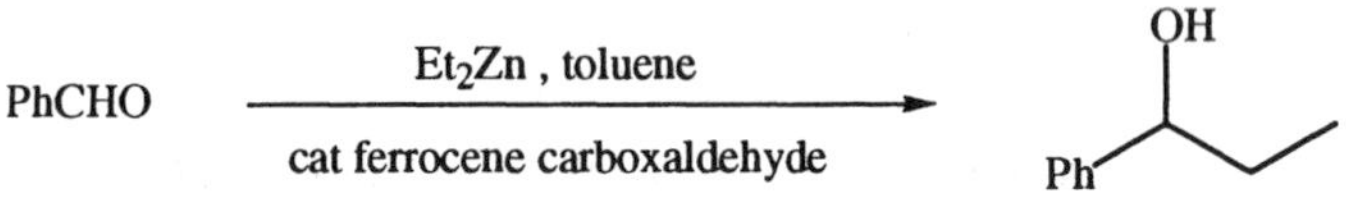
PhCHO
Et_2Zn , toluene
cat ferrocene carboxaldehyde
OH
Ph

94% (91% ee , *R*)

Fukuzawa, S.–i.; Kato, H. *SynLett.*, ***1998***, 727.

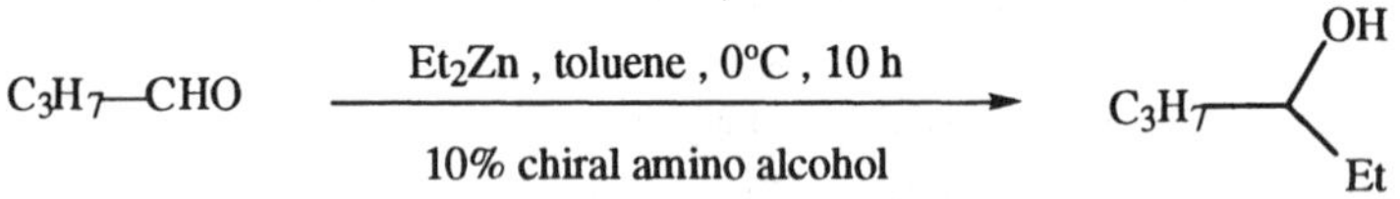
C_3H_7—CHO
Et_2Zn , toluene , 0°C , 10 h
10% chiral amino alcohol
OH
C_3H_7
Et

88% (82% ee , *R*)

Cho, B.T.; Chun, Y.S. *Tetrahedron Asymmetry*, ***1998***, *9*, 1489.

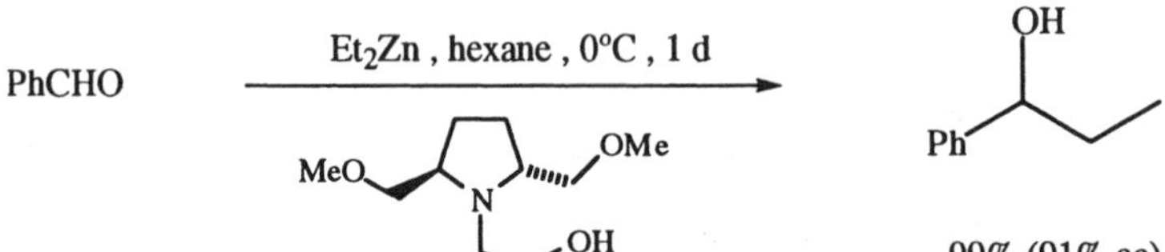
PhCHO
Et_2Zn , hexane , 0°C , 1 d
MeO
OMe
N
OH
OH
Ph

99% (91% ee)

Shi, M.; Satoh, Y.; Masaki, Y. *J. Chem. Soc., Perkin Trans. 1*, ***1998***, 2547.

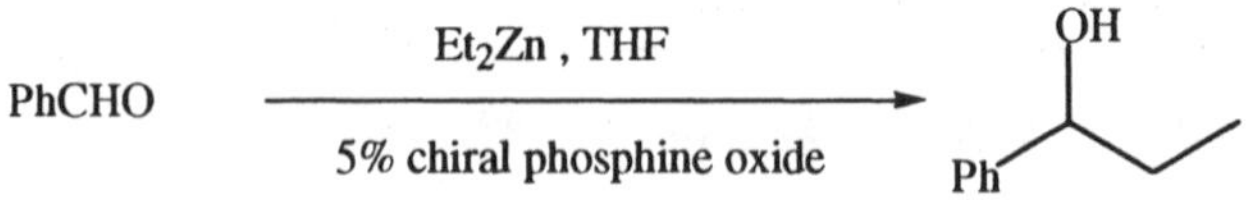
PhCHO
Et_2Zn , THF
5% chiral phosphine oxide
OH
Ph

quant (82% ee)

Legrand, O.; Brunel, J.–M.; Buono, G. *Tetrahedron Lett.*, ***1998***, *39*, 9419.

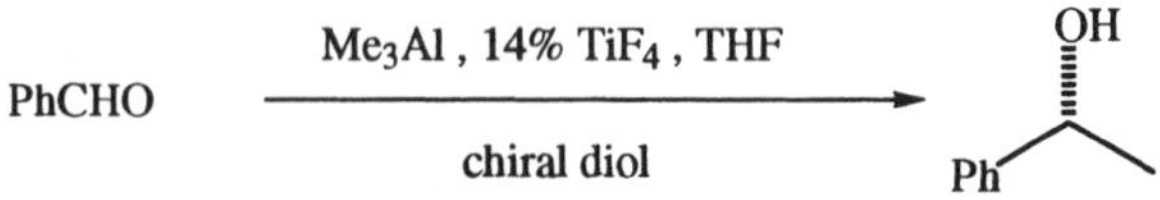
PhCHO
Me_3Al , 14% TiF_4 , THF
chiral diol
OH
Ph

80% (80% ee)

Pagenkopf, B.L.; Carreira, E.M. *Tetrahedron Lett.*, ***1998***, *39*, 9593.

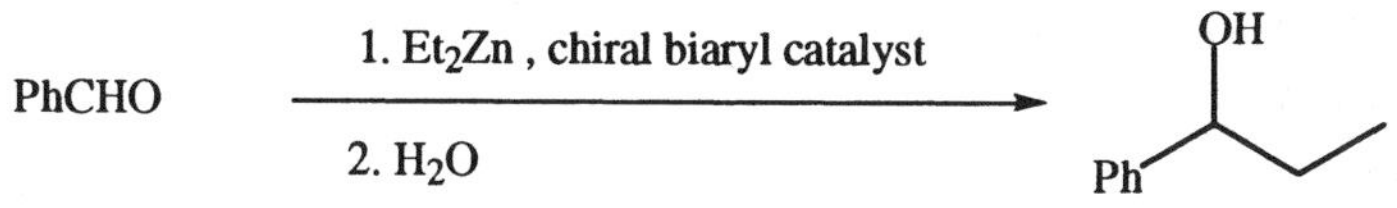
PhCHO
1. Et2Zn , chiral biaryl catalyst
2. H2O
OH
Ph

quant (% er 1:99)

<u>Bringmann, G.</u>; Breuning, M. *Tetrahedron Asymmetry,* ***1998***, *9*, 667.

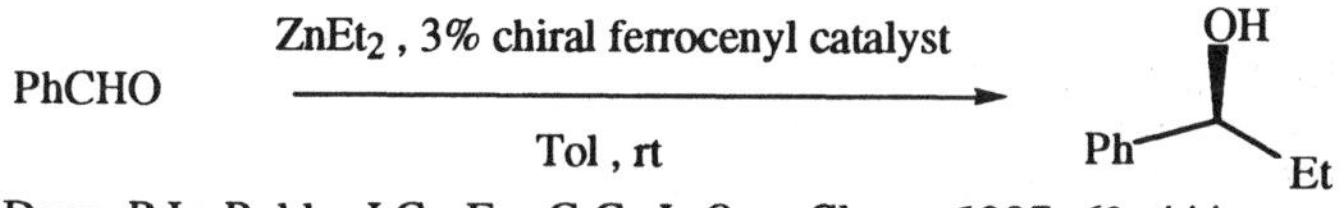
PhCHO
ZnEt2 , 3% chiral ferrocenyl catalyst
Tol , rt
OH
Ph
Et

88% (90% ee)

Dosa, P.I.; Ruble, J.C.; <u>Fu, G.C.</u> *J. Org. Chem.,* ***1997***, *62*, 444.

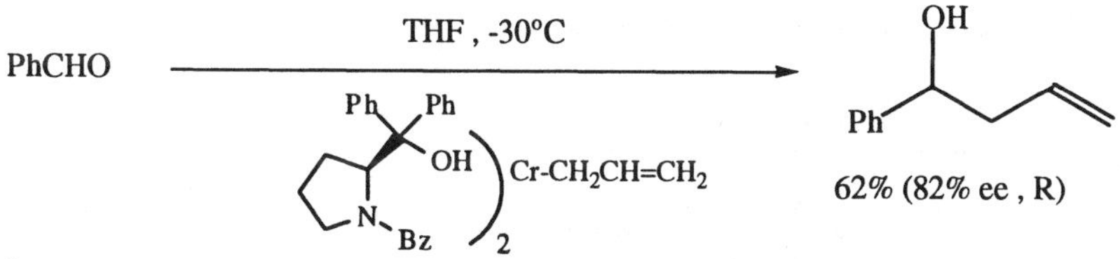
PhCHO
THF , -30°C
Ph
Ph
OH
Cr-CH2CH=CH2
N
Bz
2
OH
Ph

62% (82% ee , R)

Sugimoto, K.; Aoyagi, S.; <u>Kibayashi, C.</u> *J. Org. Chem.,* ***1997***, *62*, 2322.

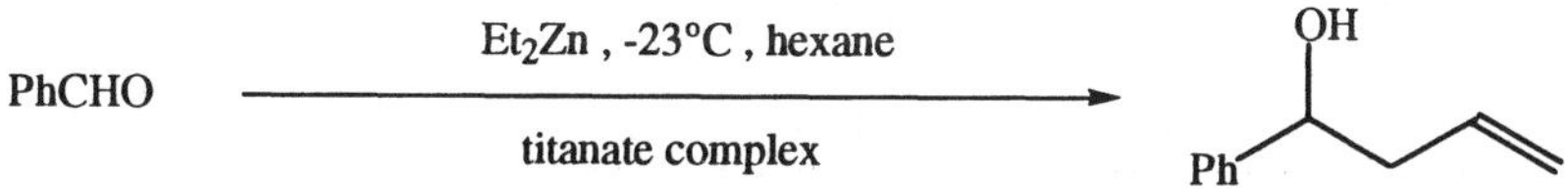
PhCHO
Et2Zn , -23°C , hexane
titanate complex
OH
Ph

99% (94% ee)

Qiu, J.; Guo, C.; <u>Zhang, X.</u> *J. Org. Chem.,* ***1997***, *62*, 2665.

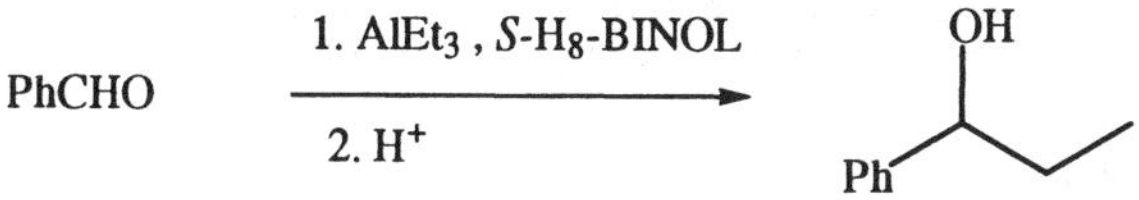
PhCHO
1. AlEt3 , S-H8-BINOL
2. H+
OH
Ph

quant (96% ee , *S*)

<u>Chan, A.S.C.</u>; Zhang, F.–y.; Yip, C.–W. *J. Am. Chem. Soc.,* ***1997***, *119*, 4080.

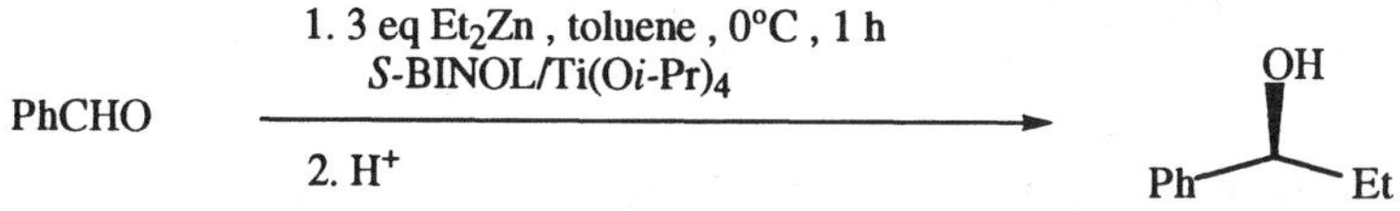
PhCHO
1. 3 eq Et2Zn , toluene , 0°C , 1 h
S-BINOL/Ti(Oi-Pr)4
2. H+
OH
Ph
Et

>98% (84% ee , *S*)

Mori, M.; <u>Nakai, T.</u> *Tetrahedron Lett.,* ***1997***, *38*, 6233.

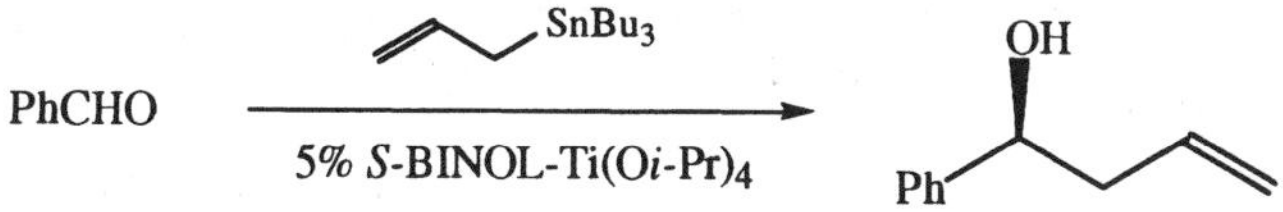
PhCHO
SnBu3
5% S-BINOL-Ti(Oi-Pr)4
OH
Ph

83% (93 ee)

<u>Yu, C.–M.</u>; Cho, H.–S.; Yoon, S.–K.; Jung, W.–H. *SynLett.,* ***1997***, 889.

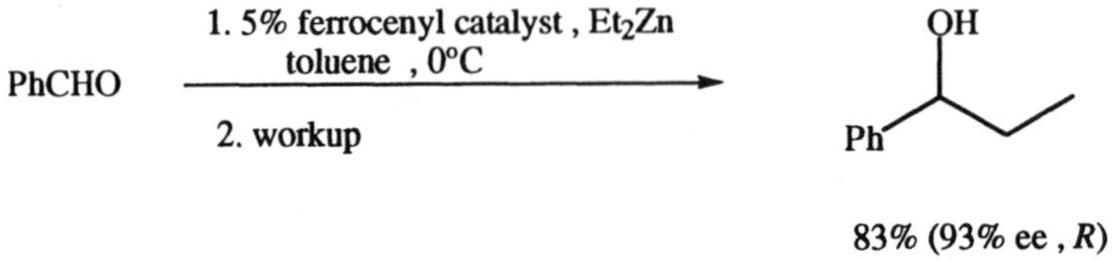
PhCHO
1. 5% ferrocenyl catalyst , Et_2Zn
toluene , 0°C
2. workup
OH
Ph
83% (93% ee , *R*)

Bolm, C.; Fernández, K.M.; Seger, A.; Raabe, G. *SynLett.*, ***1997***, 1051.

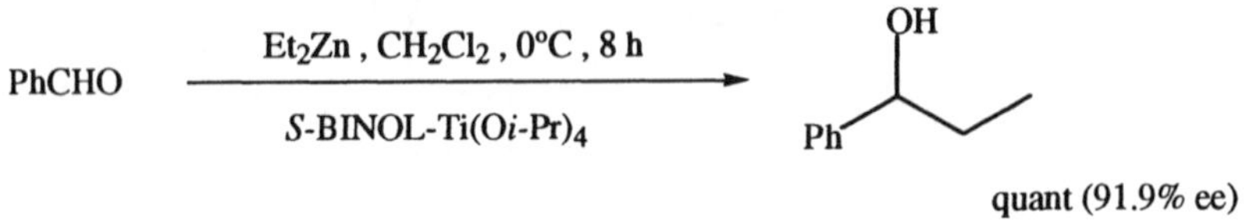
PhCHO
Et_2Zn , CH_2Cl_2 , 0°C , 8 h
S-BINOL-Ti(O*i*-Pr)$_4$
OH
Ph
quant (91.9% ee)

Zhang, F.-Y.; Yip, C.-W.; Cao, R.; Chan, A.S.C. *Tetrahedron Asymmetry*, ***1997***, *8*, 585.

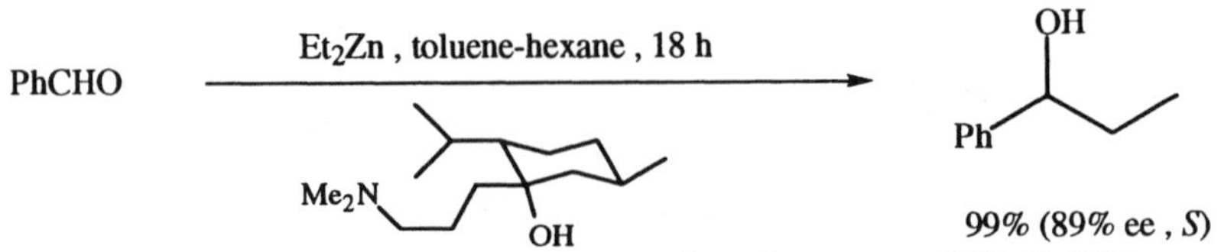
PhCHO
Et_2Zn , toluene-hexane , 18 h
Me_2N
OH
OH
Ph
99% (89% ee , *S*)

Genov, M.; Dimitrov, V.; Ivanova, V. *Tetrahedron Asymmetry*, ***1997***, *8*, 3703.

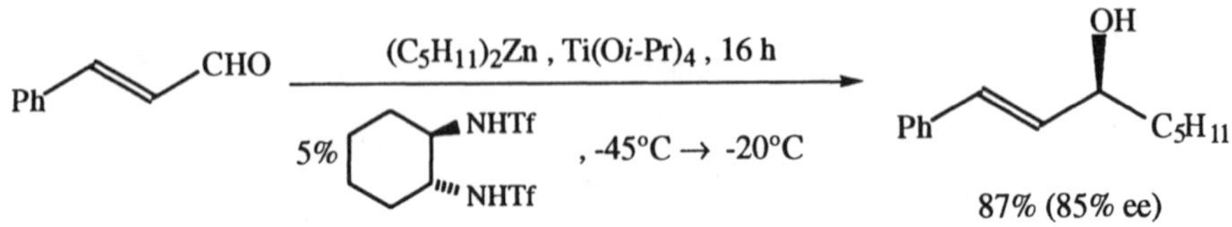
Ph
CHO
$(C_5H_{11})_2Zn$, Ti(O*i*-Pr)$_4$, 16 h
5%
NHTf
NHTf
, -45°C → -20°C
OH
Ph
C_5H_{11}
87% (85% ee)

Vettel, S.; Lutz, C.; Diefenbach, A.; Haderlein, G.; Hammerschmidt, S.; Kühling, K.; Mofid, M.-R.; Zimmerman, T.; Knochel, P. *Tetrahedron Asymmetry*, ***1997***, *8*, 779.

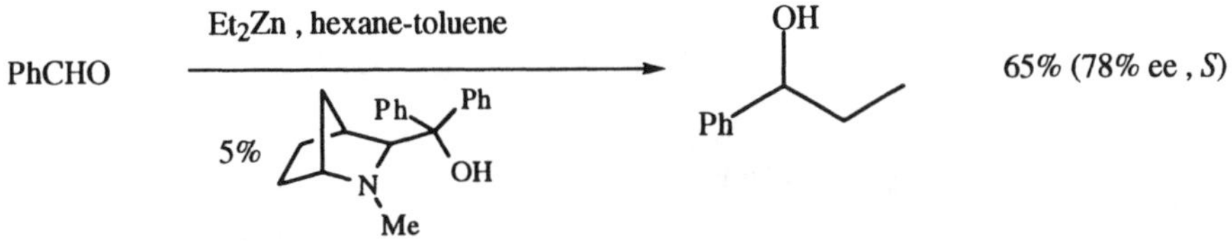
PhCHO
Et_2Zn , hexane-toluene
5%
Ph
Ph
OH
N
Me
OH
Ph
65% (78% ee , *S*)

Nakano, H.; Kumagai, N.; Matsuzaki, H.; Kubato, C.; Hongo, H. *Tetrahedron Asymmetry*, ***1997***, *8*, 1391.

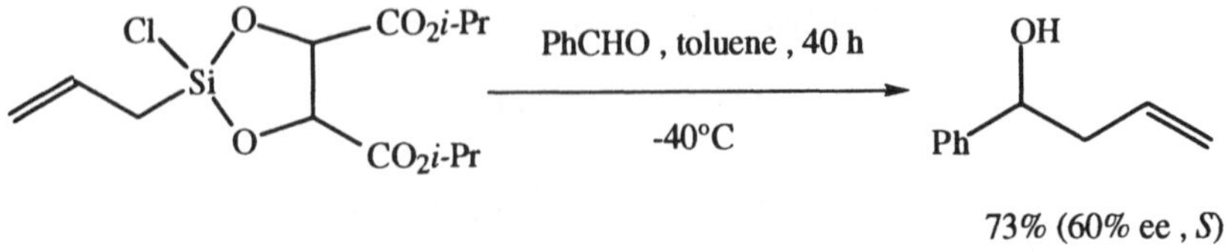
Cl
O
Si
O
CO_2*i*-Pr
CO_2*i*-Pr
PhCHO , toluene , 40 h
-40°C
OH
Ph
73% (60% ee , *S*)

Zhang, L.C.; Sakurai, H.; Kira, M. *Chem. Lett.*, ***1997***, 129.

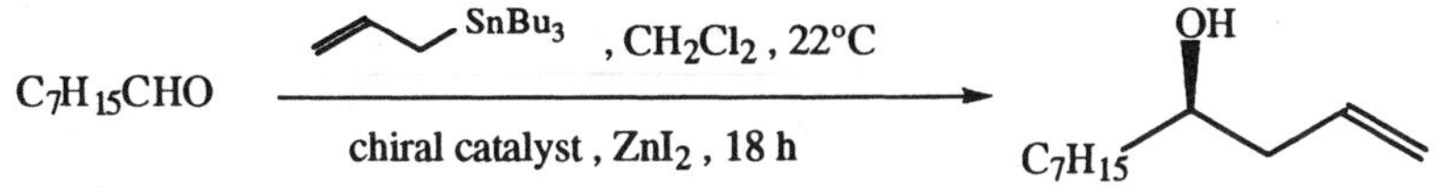
C7H15CHO
SnBu3 , CH2Cl2 , 22°C
chiral catalyst , ZnI2 , 18 h
OH
C7H15

68% (23% ee)

Coi, P.G.; Orioli, P.; Tagliavini, E.; Umani-Ronchi, A. *Tetrahedron Lett.*, ***1997***, *38*, 145.

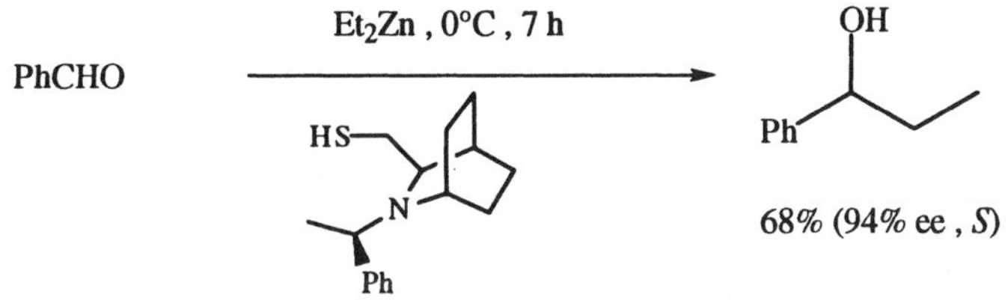
PhCHO
Et2Zn , 0°C , 7 h
HS
N
Ph
OH
Ph
68% (94% ee , S)

Nakano, H.; Iwasa, K.; Hongo, H. *Heterocycles*, ***1997***, *46*, 267.

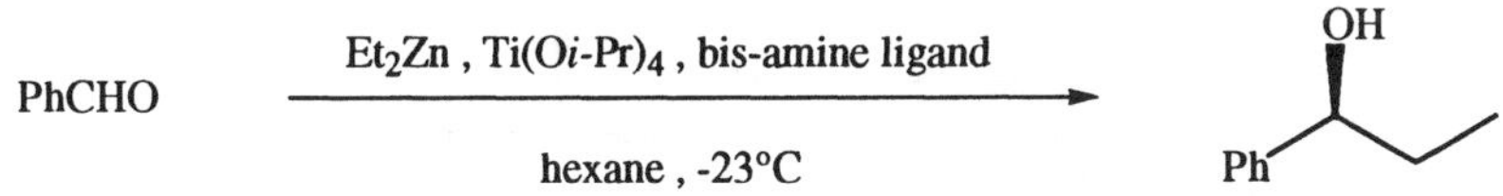
PhCHO
Et2Zn , Ti(Oi-Pr)4 , bis-amine ligand
hexane , -23°C
OH
Ph

98% (99% ee , *S*)

Guo, C.; Qiu, J.; Zhang, X.; Verdugo, D.; Larter, M.L.; Christie, R.; Kenney, P.; Walsh, P.J. *Tetrahedron*, ***1997***, *53*, 4145.

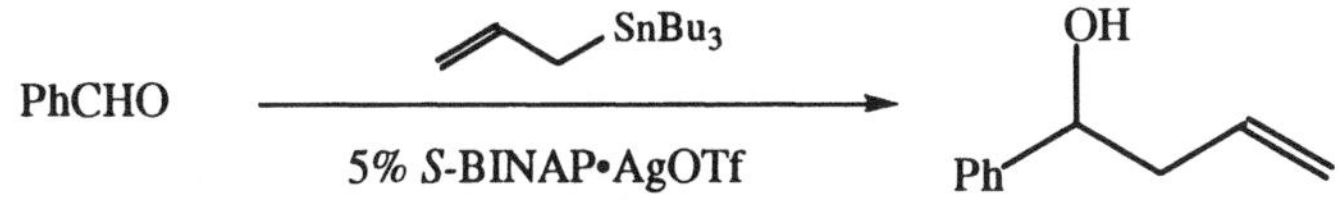
PhCHO
SnBu3
5% S-BINAP•AgOTf
OH
Ph

88% (96% ee , *S*)

Yanagisawa, A.; Nakashima, H.; Ishiba, A.; Yamamoto, H. *J. Am. Chem. Soc.*, ***1996***, *118*, 4723.

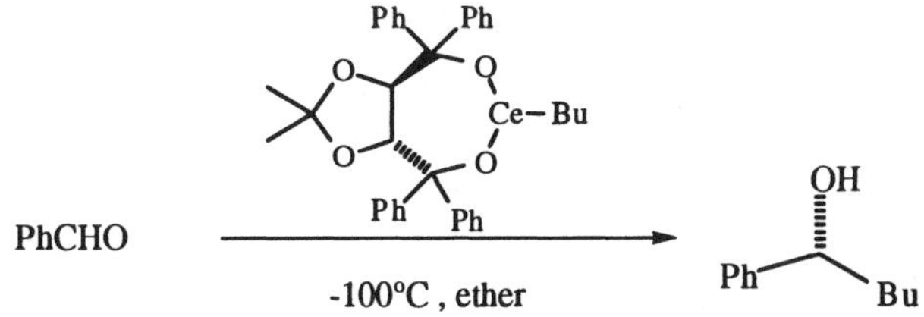
PhCHO
Ph Ph
O O
Ce—Bu
O O
Ph Ph
-100°C , ether
OH
Ph
Bu

66% (66% ee)

GreeVes, N.; Pease, J.E.; Bowden, M.C.; Brown, S.M. *Tetrahedron Lett.*, ***1996***, *37*, 2675.

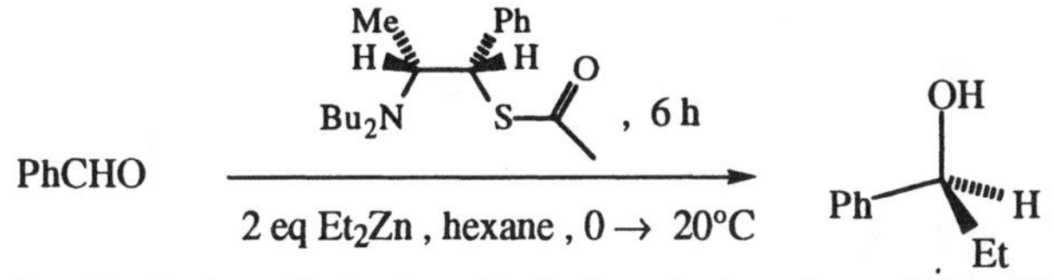
PhCHO
Me Ph
H H
O
Bu2N S
, 6 h
2 eq Et2Zn , hexane , 0 → 20°C
OH
Ph H
Et

quant (>99% ee , *S*)

Jin, M.-J.; Ahn, S.-J.; Lee, K.-S. *Tetrahedron Lett.*, ***1996***, *37*, 8767.

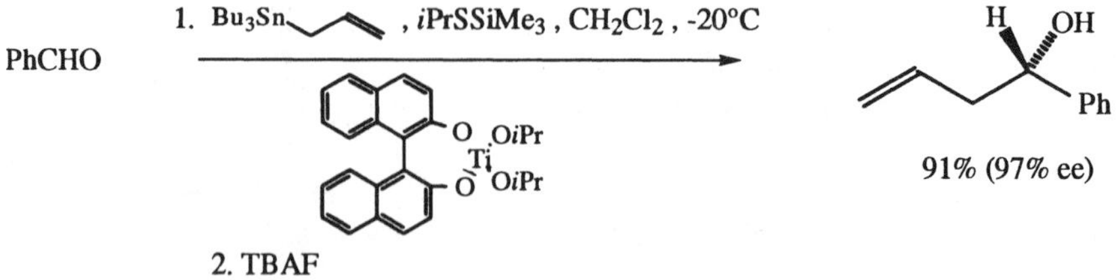
PhCHO
1. Bu3Sn, iPrSSiMe3, CH2Cl2, -20°C
O OiPr
Ti
O OiPr
2. TBAF
H OH
Ph
91% (97% ee)

<u>Yu, C.-M.</u>; Choi, H.-S.; Jung, W.-H; Lee, S.S. *Tetrahedron Lett.*, ***1996***, *37*, 7095.

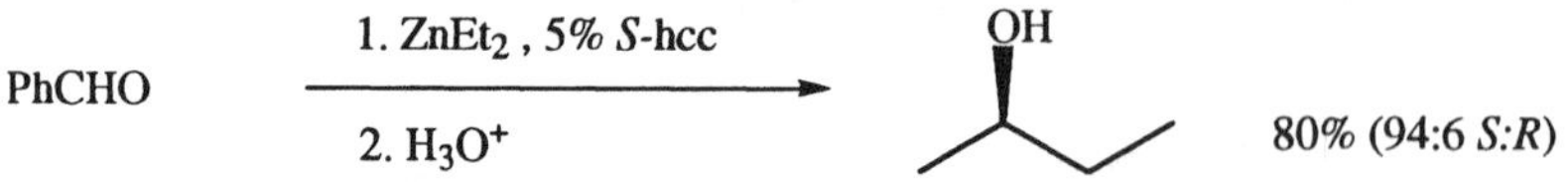
PhCHO
1. ZnEt2, 5% S-hcc
2. H3O+
OH
80% (94:6 S:R)

hcc = hyperbranched chiral catalysts

<u>Bolm, C.</u>; Derrien, N.; Seger, A. *SynLett.*, ***1996***, 387.

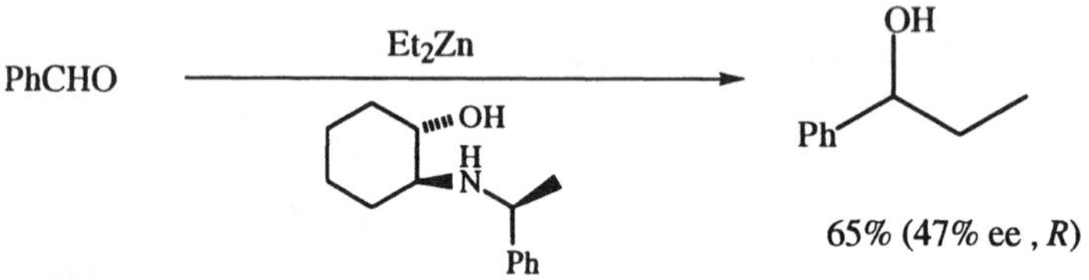
PhCHO
Et2Zn
OH
H
N
Ph
OH
Ph
65% (47% ee, R)

de Parrodi, C.A.; <u>Juaristi, E.</u>; Quintero-Cortés, L.; Amador, P.
Tetrahedron Asymmetry, ***1996***, *7*, 1915.

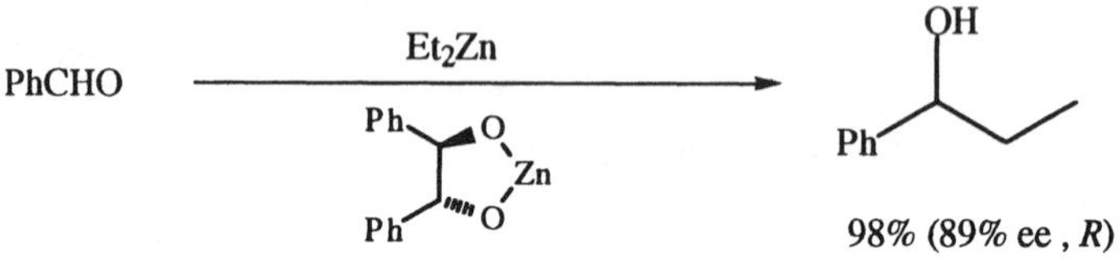
PhCHO
Et2Zn
Ph
O
Zn
Ph
O
OH
Ph
98% (89% ee, R)

Prasad, K.R.K.; <u>Joshi, N.N</u>. *Tetrahedron Asymmetry*, ***1996***, *7*, 1957.

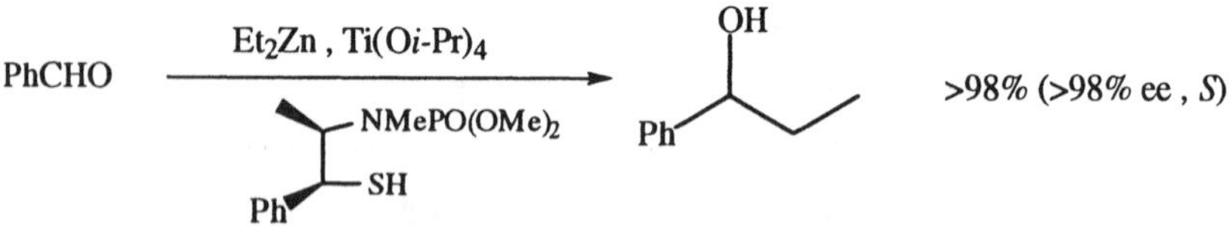
PhCHO
Et2Zn, Ti(Oi-Pr)4
NMePO(OMe)2
SH
Ph
OH
Ph
>98% (>98% ee, S)

Hulst, R.; Heres, H.; Fitzpatrick, K.; Pepper, N.C.M.W.; Kellogg, R.M.
Tetrahedron Asymmetry, ***1996***, *7*, 2755.

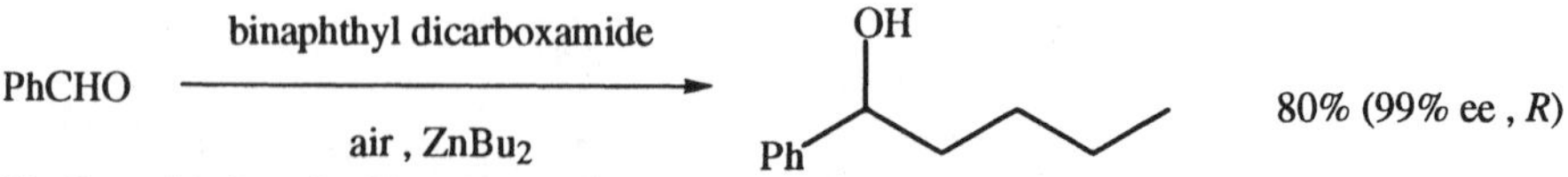
PhCHO
binaphthyl dicarboxamide
air, ZnBu2
OH
Ph
80% (99% ee, R)

Kitajima, H.; Ito, K.; <u>Katsuki, T.</u> *Chem. Lett.*, ***1996***, 343.

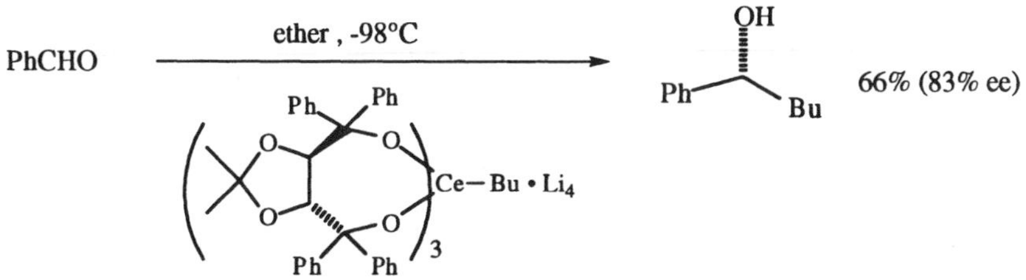
PhCHO
ether , -98°C
OH
Ph
Bu
66% (83% ee)
Ph
Ph
O
O
O
O
Ce—Bu • Li4
Ph
Ph
3

Greeves, N.; Pease, J.E. *Tetrahedron Lett.*, ***1996***, *37*, 5821.

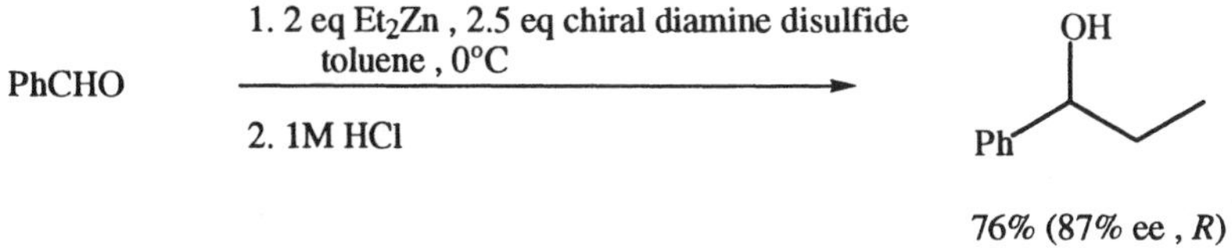
PhCHO
1. 2 eq Et2Zn , 2.5 eq chiral diamine disulfide
toluene , 0°C
2. 1M HCl
OH
Ph
76% (87% ee , R)

Bibson, C.L. *Chem. Commun.*, ***1996***, 645.

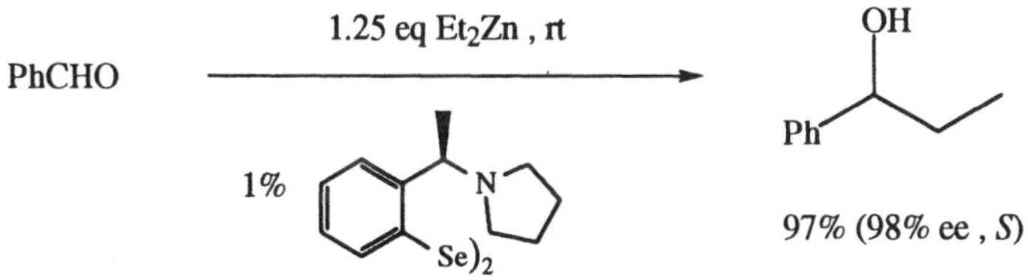
PhCHO
1.25 eq Et2Zn , rt
1%
N
Se)2
OH
Ph
97% (98% ee , S)

Wirth, T.; Kulicke, K.J.; Fragale, G. *Helv. Chim. Acta*, ***1996***, *79*, 1957.

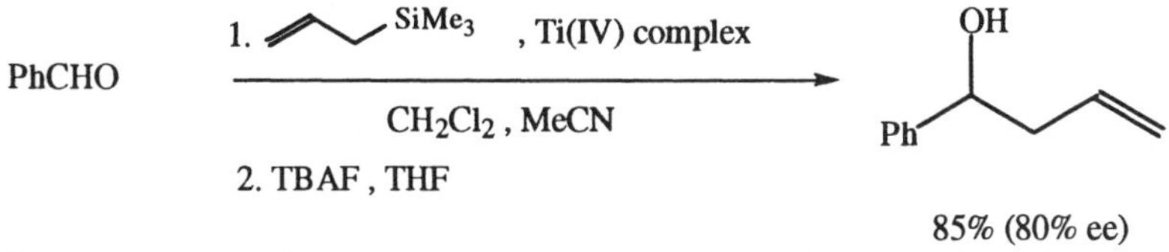
PhCHO
1.
SiMe3
, Ti(IV) complex
CH2Cl2 , MeCN
2. TBAF , THF
OH
Ph
85% (80% ee)

Gauthier Jr., D.R.; Carreira, E.M. *Angew. Chem. Int. Ed.*, ***1996***, *35*, 2363.

NON-ASYMMETRIC ALKYLATIONS

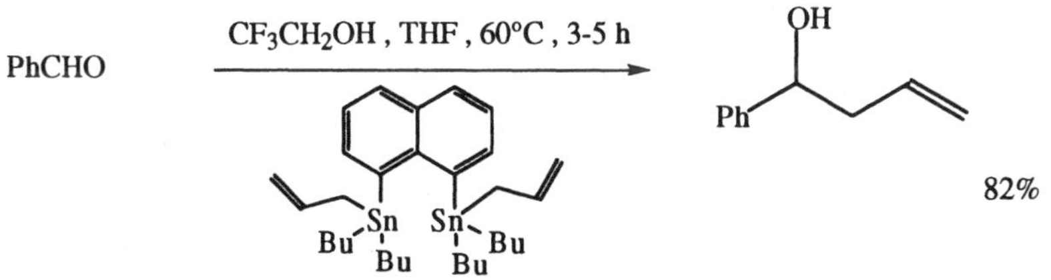
PhCHO
CF3CH2OH , THF , 60°C , 3-5 h
Sn
Sn
Bu
Bu
Bu
Bu
OH
Ph
82%

Asao, N.; Abe, N.; Tan, Z.; Maruoka, K. *SynLett.*, ***1998***, 377.

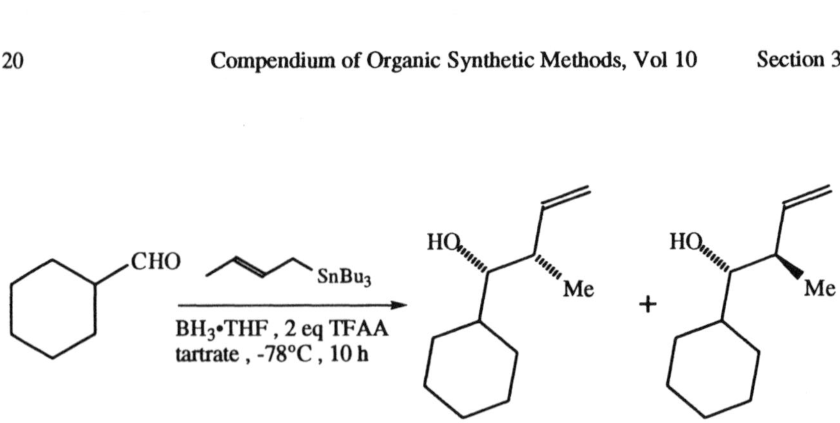
CHO
SnBu3
BH3•THF, 2 eq TFAA
tartrate, -78°C, 10 h
HO
Me
+
HO
Me

(93 : 7) 70%

Marshall, J.A.; Palovich, M.R. *J. Org. Chem.*, ***1998***, *63*, 4381.

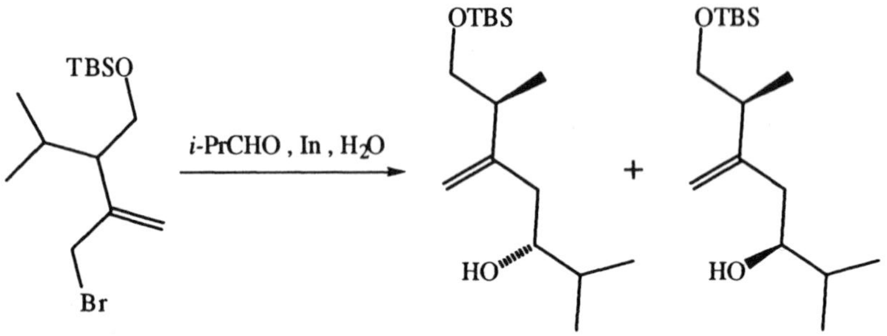
TBSO
Br
i-PrCHO, In, H2O
OTBS
HO
+
OTBS
HO

(65 : 35) 52%

Paquette, L.A.; Bennett, G.D.; Isaac, M.B.; Chhatriwalla, A. *J. Org. Chem.*, ***1998***, *63*, 1836

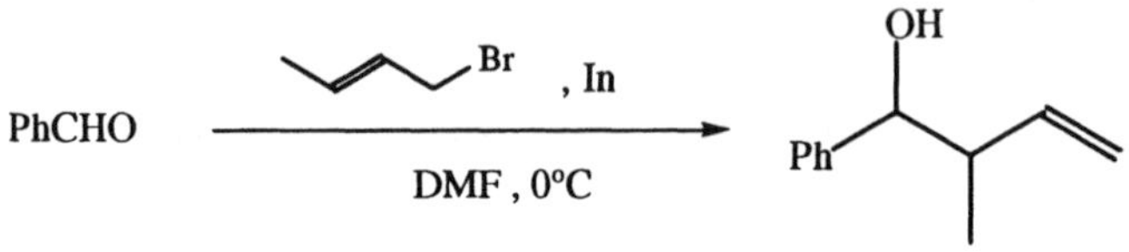
PhCHO
Br
, In
DMF, 0°C
OH
Ph

>95% (65:35 *syn:anti*)

Lloyd-Jones, G.C.; Russell, T. *SynLett.*, ***1998***, 903.

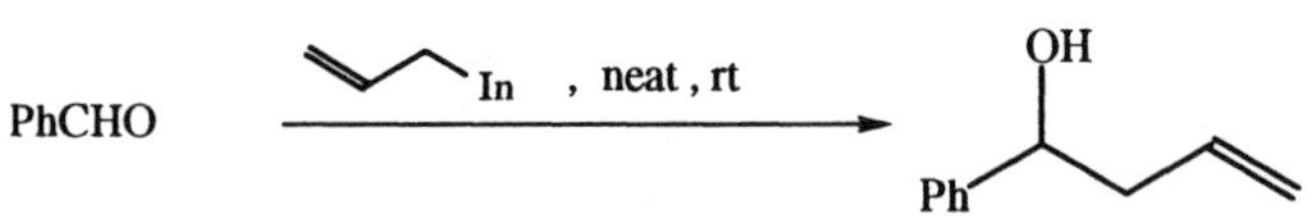
PhCHO
In
, neat, rt
OH
Ph

88%

Yi, X.–H.; Haberman, J.X.; Li, C.–J. *Synth. Commun.*, ***1998***, *28*, 2999.

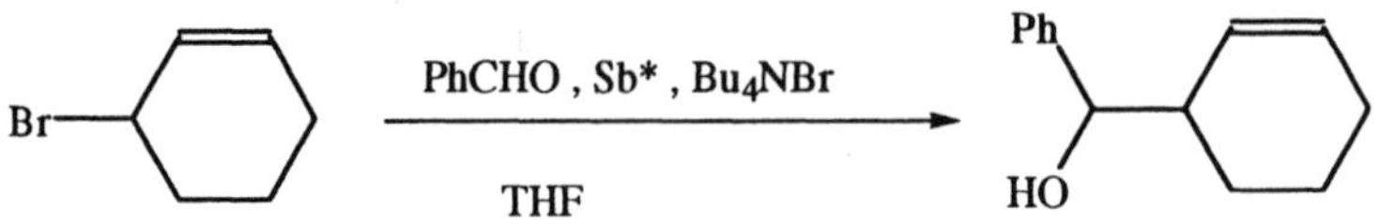
Br
PhCHO, Sb*, Bu4NBr
THF
Ph
HO

51% (92:8 *erythro:threo*)

Jim, Q.–H.; Ren, P.–D.; Li, Y.–Q.; Yao, Z.–P. *Synth. Commun.*, ***1998***, *28*, 4151.

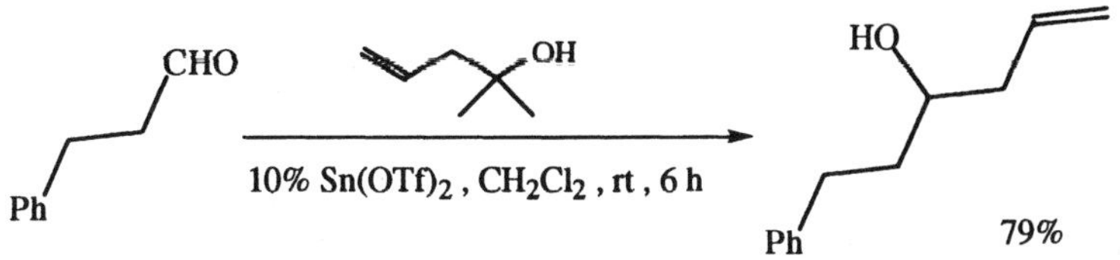
CHO
Ph
OH
10% Sn(OTf)2 , CH2Cl2 , rt , 6 h
HO
Ph
79%

Nokami, J.; Yoshizane, K.; Matsuura, H.; Sumida, S.–i. *J. Am. Chem. Soc.*, ***1998***, *120*, 6609

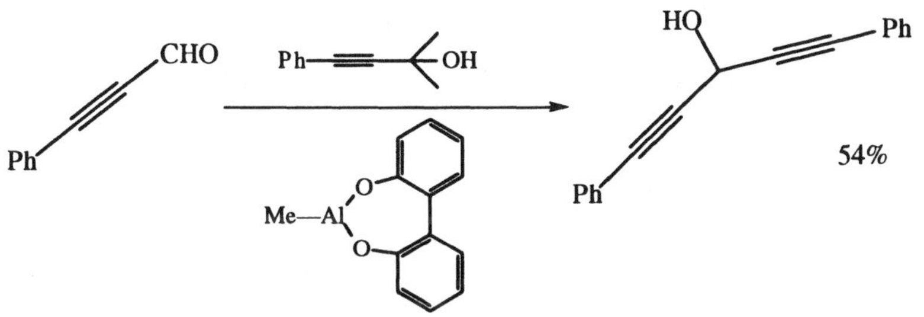
CHO
Ph
Ph
OH
Me—Al
O
O
HO
Ph
Ph
54%

Ooi, T.; Miura, T.; Maruoka, K. *J. Am. Chem. Soc.*, ***1998***, *120*, 10790.

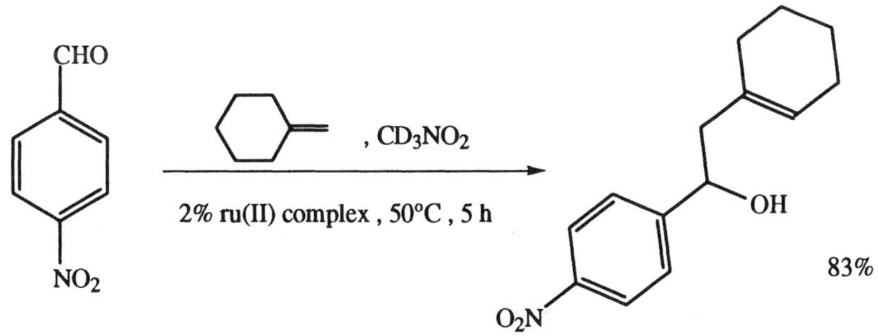
CHO
NO2
, CD3NO2
2% ru(II) complex , 50°C , 5 h
OH
O2N
83%

Ellis, W.W.; Odenkirk, W.; Bosnich, B. *Chem. Commun.*, ***1998***, 1311.

PhCHO
PhB(OH)2 , Rh(acac)dppf
aq DME
OH
Ph
Ph
92%

Sakai, M.; Ueda, M.; Miyaura, N. *Angew. Chem. Int. Ed.*, ***1998***, *37*, 3279.

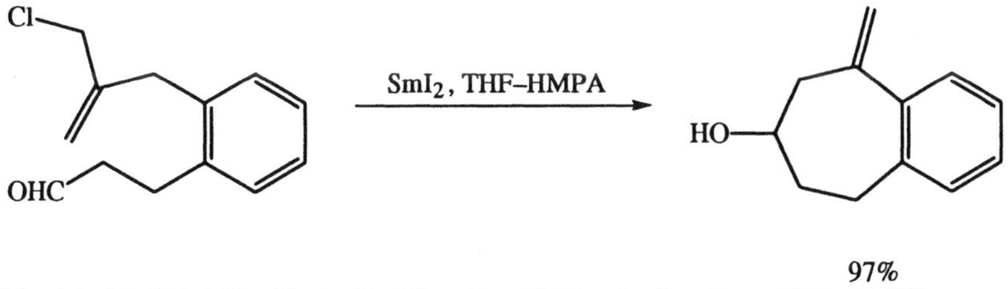
Cl
OHC
SmI2 , THF–HMPA
HO
97%

Matsuda, F.; Sakai, T.; Okada, N.; Miyashita, M. *Tetrahedron Lett.*, ***1998***, *39*, 863

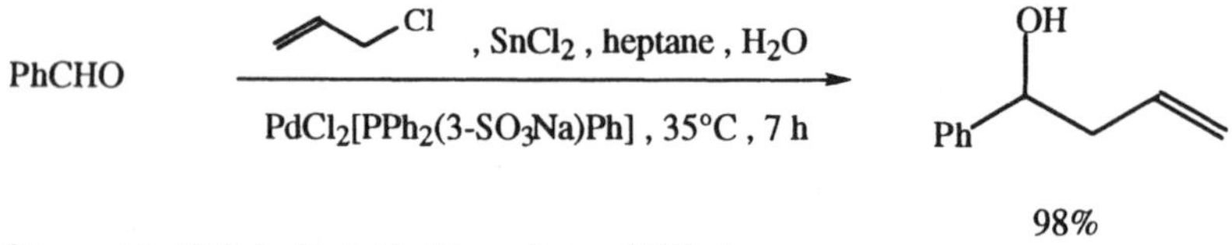
Cl
PhCHO
, SnCl2 , heptane , H2O
PdCl2[PPh2(3-SO3Na)Ph] , 35°C , 7 h
OH
Ph
98%

Okano, T.; Kiji, J.; Doi, T. *Chem. Lett.*, ***1998***, 5.

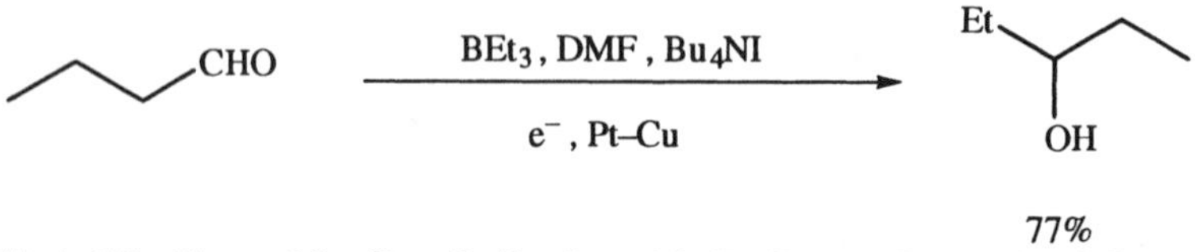
CHO
BEt3 , DMF , Bu4NI
e⁻ , Pt–Cu
Et
OH
77%

Choi, J.H.; Youm, J.S.; Cho, C.–G.; Czae, M.–Z.; Hwang, B.K.; Kim, J.S. *Tetrahedron Lett.*, ***1998***, *39*, 4835.

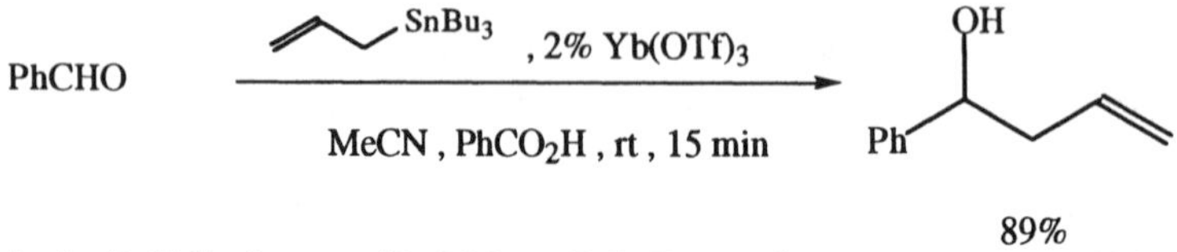
SnBu3
PhCHO
, 2% Yb(OTf)3
MeCN , PhCO2H , rt , 15 min
OH
Ph
89%

Aspinall, H.C.; Greeves, N.; McIver, E.G. *Tetrahedron Lett.*, ***1998***, *39*, 9283.

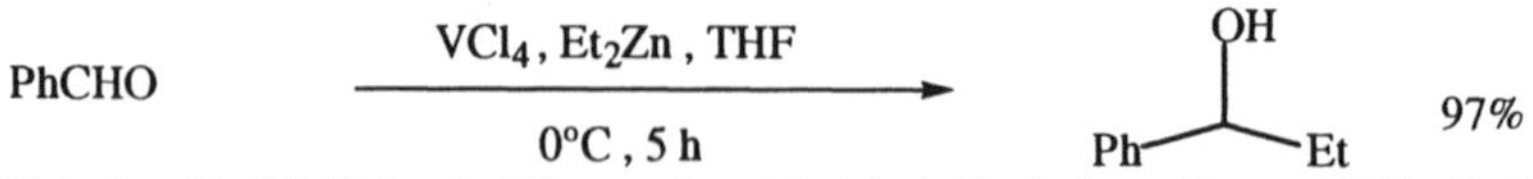
PhCHO
VCl4 , Et2Zn , THF
0°C , 5 h
OH
Ph
Et
97%

Kataoka, Y.; Makihira, I.; Utsunomiya, M.; Tani, K. *J. Org. Chem.*, ***1997***, *62*, 8540.

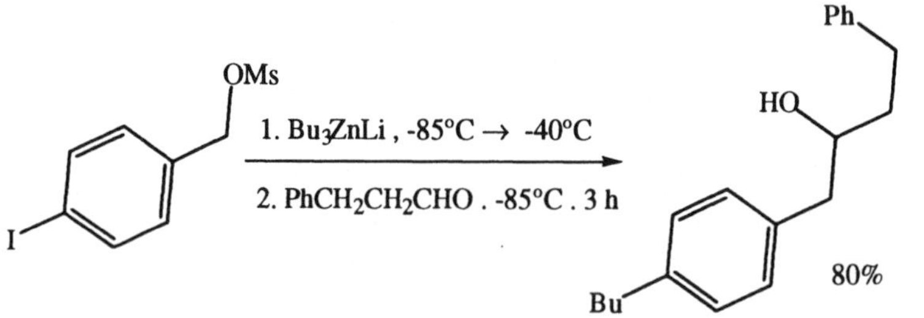
OMs
I
1. Bu3ZnLi , -85°C → -40°C
2. PhCH2CH2CHO . -85°C . 3 h
Ph
HO
Bu
80%

Harada, T.; Kaneko, T.; Fujiwra, R.; Oku, A. *J. Org. Chem.*, ***1997***, *62*, 8966.

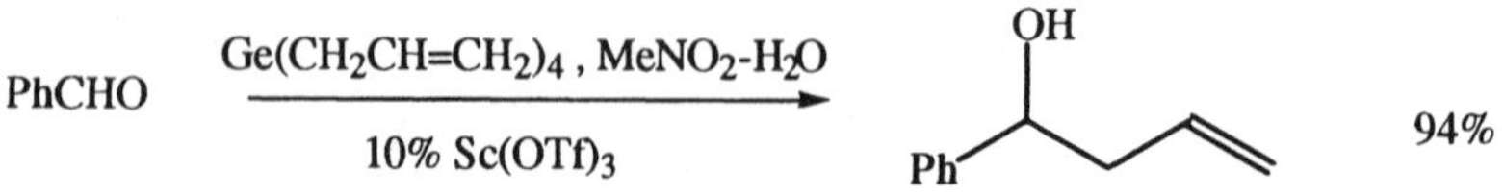
PhCHO
Ge(CH2CH=CH2)4 , MeNO2-H2O
10% Sc(OTf)3
OH
Ph
94%

Akiyama, T.; Iwai, J. *Tetrahedron Lett.*, ***1997***, *38*, 853.

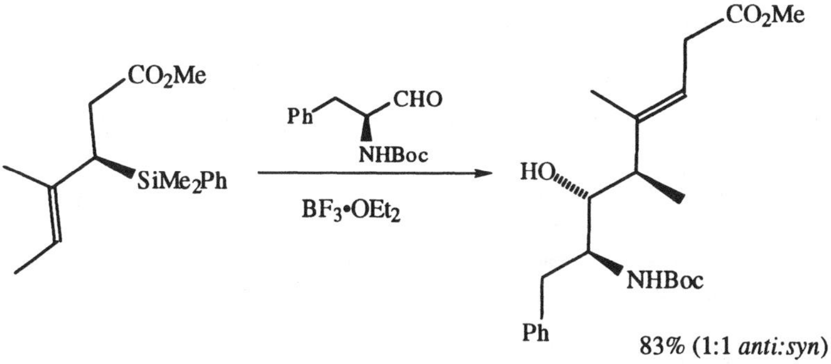

83% (1:1 *anti:syn*)

Panek, J.S.; Liu, P. *Tetrahedron Lett.*, ***1997***, *38*, 5127.

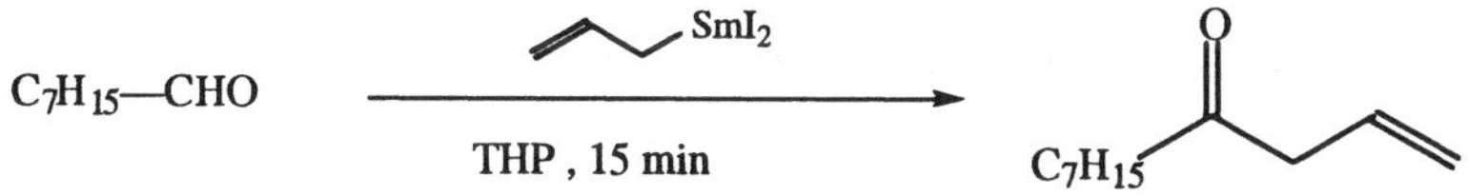

75%

Hamann-Gaudinet, B.; Namy, J.-L.; Kagan, H.B. *Tetrahedron Lett.*, ***1997***, *38*, 6585.

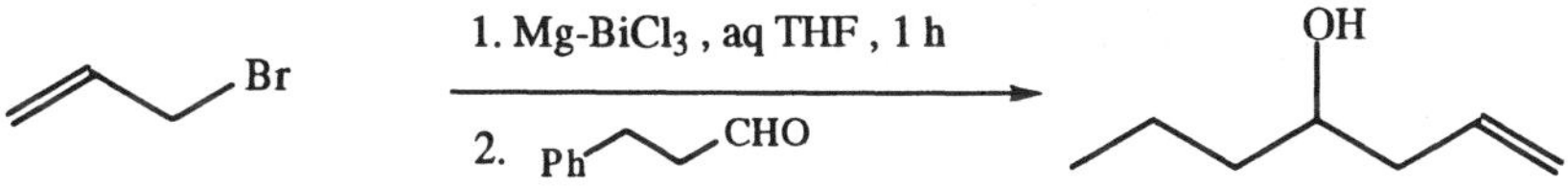

84%

Wada, M.; Fukuma, T.; Morioka, M.; Takahashi, T.; Miyoshi, N. *Tetrahedron Lett.*, ***1997***, *38*, 8045.

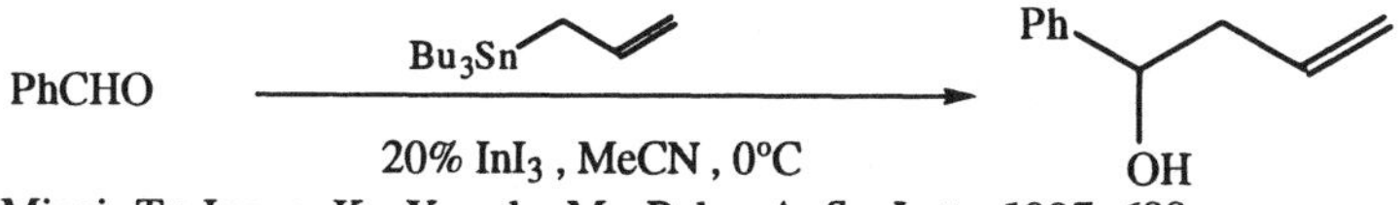

99%

Miyai, T.; Inoue, K.; Yasuda, M.; Baba, A. *SynLett.*, ***1997***, 699.

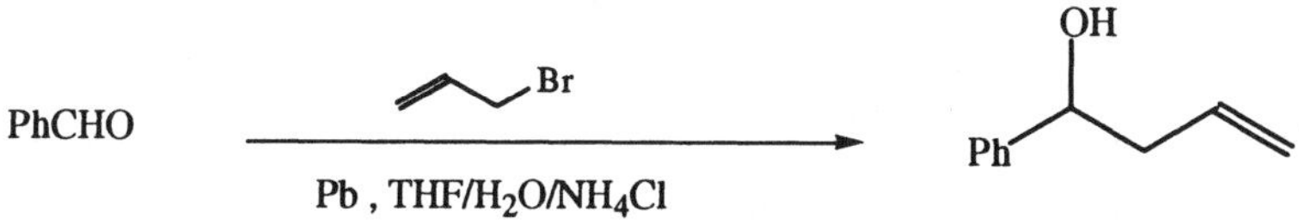

99%

Zhou, J.-Y.; Jia, Y.; Sung, G.-F.; Wu, S.-H. *Synth. Commun.*, ***1997***, *27*, 1899.

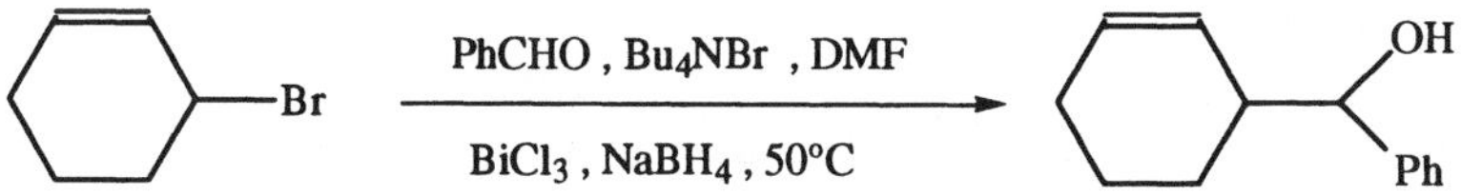

86% (*erythro* only)

Ren, P.-D.; Shao, D.; Dong, T.-W. *Synth. Commun.*, ***1997***, *27*, 2569.

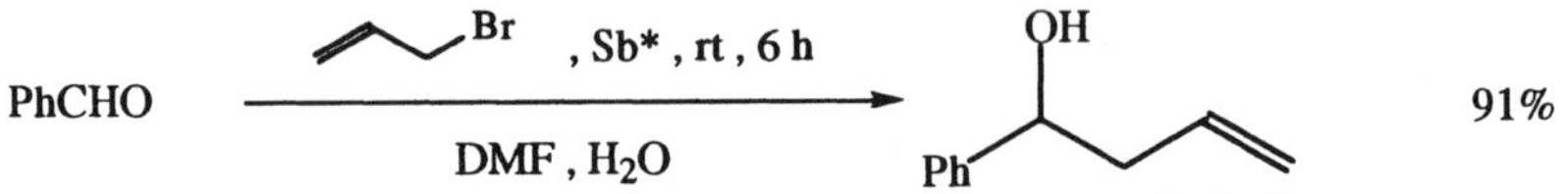
Br
, Sb* , rt , 6 h
PhCHO
DMF , H_2O
OH
Ph
91%

Ren, P.-D.; Jin, Q.-H.; Yao, Z.P. *Synth. Commun.*, ***1997***, *27*, 2761.

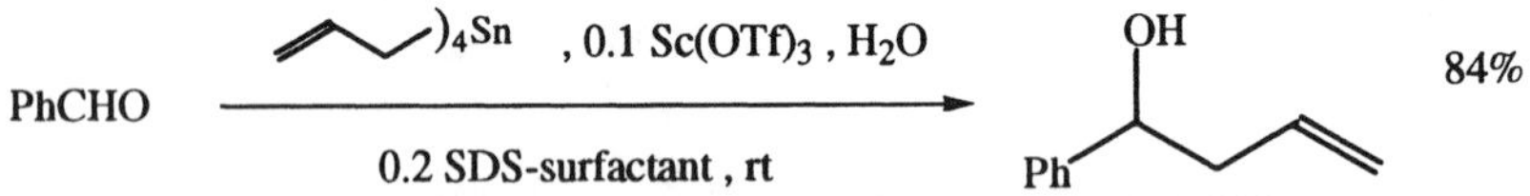
)$_4$Sn
, 0.1 $Sc(OTf)_3$, H_2O
PhCHO
0.2 SDS-surfactant , rt
OH
Ph
84%

Kobayashi, S.; Wakabayashi, T.; Oyamada, H. *Chem. Lett.*, ***1997***, 831.

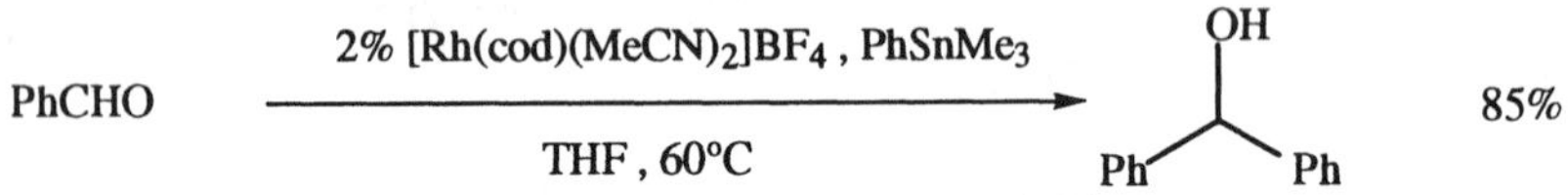
2% $[Rh(cod)(MeCN)_2]BF_4$, $PhSnMe_3$
PhCHO
THF , 60°C
OH
Ph
Ph
85%

Oi, S.; Moro, M.; Inoue, Y. *Chem. Commun.*, ***1997***, 1621.

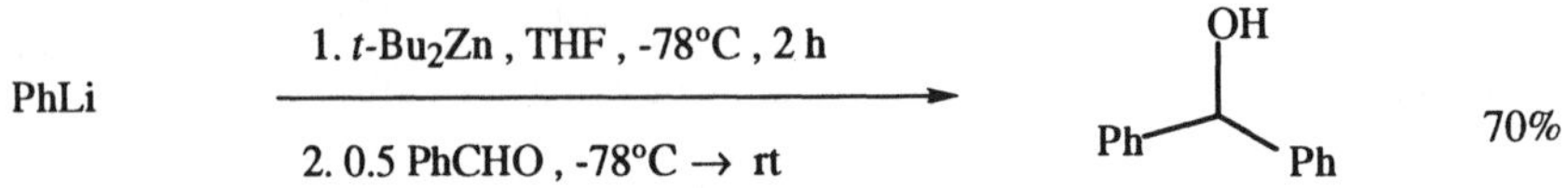
1. t-Bu_2Zn , THF , -78°C , 2 h
PhLi
2. 0.5 PhCHO , -78°C → rt
OH
Ph
Ph
70%

Kondo, Y.; Fujinami, M.; Uchiyama, M.; Sakamoto, T. *J. Chem. Soc., Perkin Trans. 1*, ***1997***, 799.

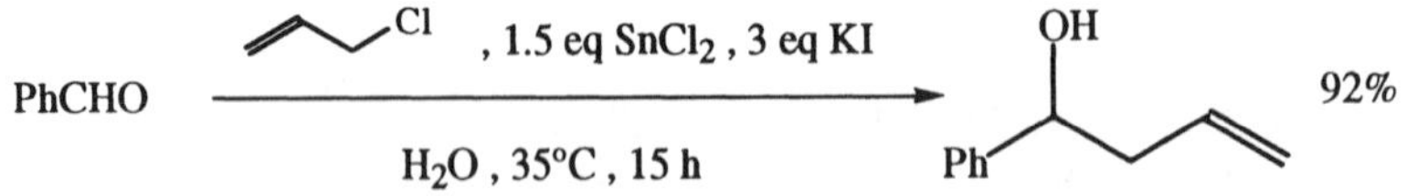
Cl
, 1.5 eq $SnCl_2$, 3 eq KI
PhCHO
H_2O , 35°C , 15 h
OH
Ph
92%

Houllemare, D.; Outurquin, F.; Paulmier, C. *J. Chem. Soc., Perkin Trans. 1*, ***1997***, 1629.

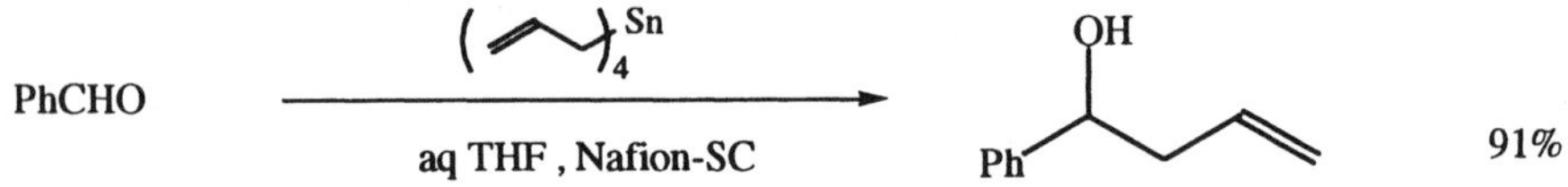
Sn
4
PhCHO
aq THF , Nafion-SC
OH
Ph
91%

Kobayashi, S.; Nagayama, S. *J. Org. Chem.*, ***1996***, *61*, 2256.

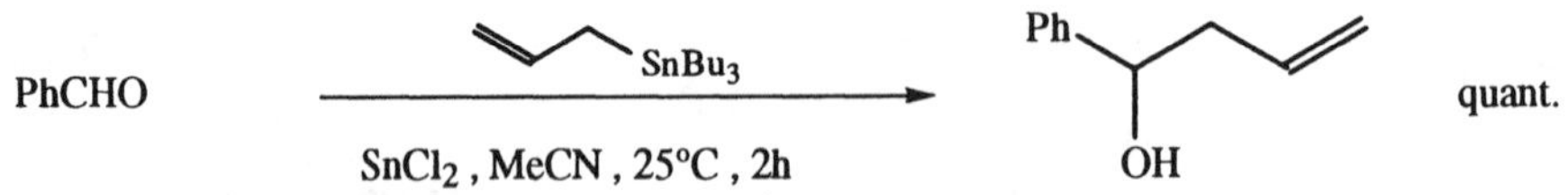
$SnBu_3$
PhCHO
$SnCl_2$, MeCN , 25°C , 2h
Ph
OH
quant.

Riguet, E.; Klement, I.; Reddy, Ch.K.; Cahiez, G.; Knochel, P. *Tetrahedron Lett.*, ***1996***, *37*, 5865.

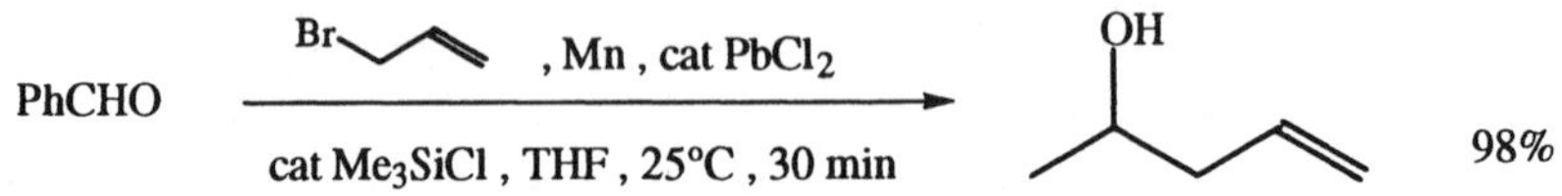
Br
, Mn , cat $PbCl_2$
PhCHO
cat Me_3SiCl , THF , 25°C , 30 min
OH
98%

Takai, K.; Ueda, T.; Hayashi, T.; Moriwake, T. *Tetrahedron Lett.*, ***1996***, *37*, 7049.

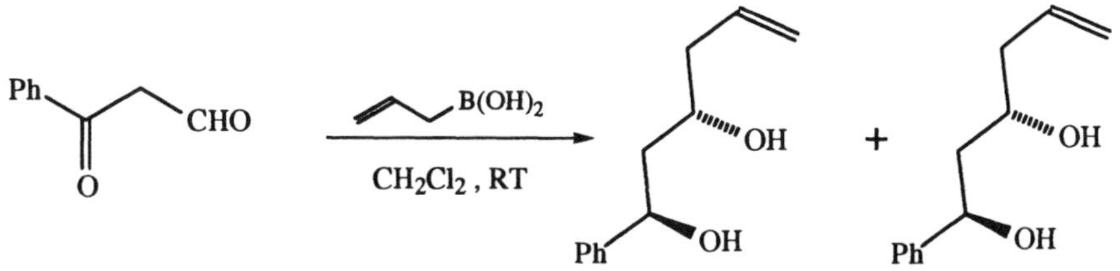

(72 : 28) 85%

Kabalka, G.W.; Narayana, C.; Reddy, N.K. *Tetrahedron Lett.*, ***1996***, *37*, 2181.

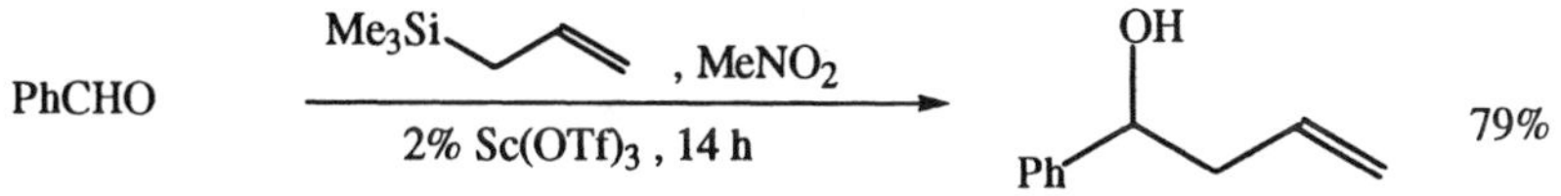

(0% with no $Sc(OTf)_3$ after 48 h)

Aggarwal, V.K.; Vennall, G.P. *Tetrahedron Lett.*, ***1996***, *37*, 3745.

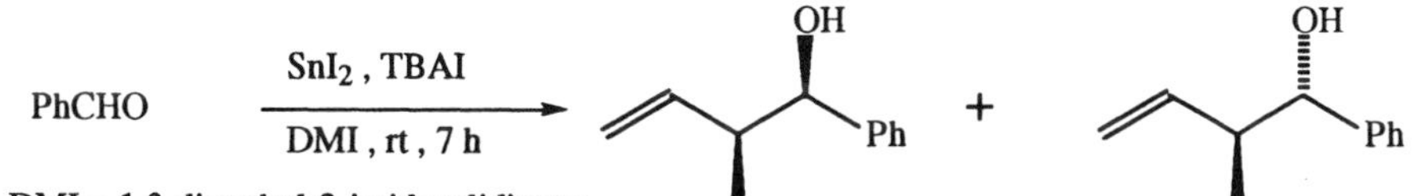

DMI = 1,3-dimethyl-2-imidazolidinone

(87 : 13) 88%

Masuyama, Y.; Kishida, M.; Kurusu, Y. *Tetrahedron Lett.*, ***1996***, *37*, 7103.

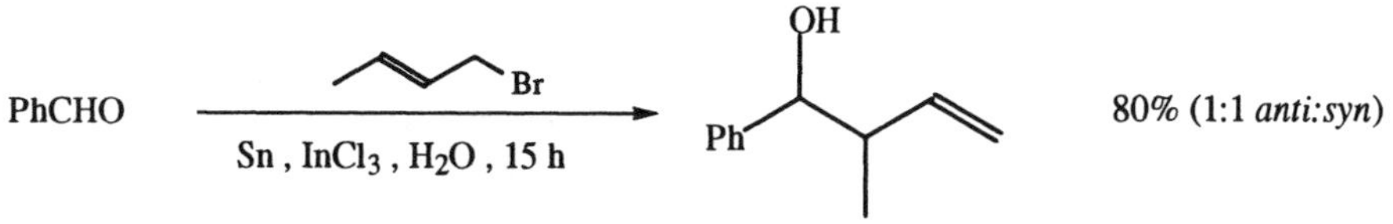

Li, X.-R.; Loh, T.-P. *Tetrahedron Asymmetry*, ***1996***, *7*, 1535.

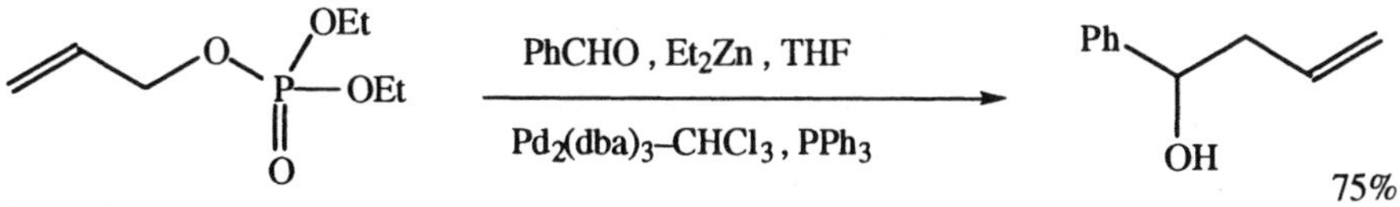

75%

Kang, S.-K.; Kim, D.-Y.; Hong, R.-K.; Ho, P.-S. *Synth. Commun.*, ***1996***, *26*, 1493.

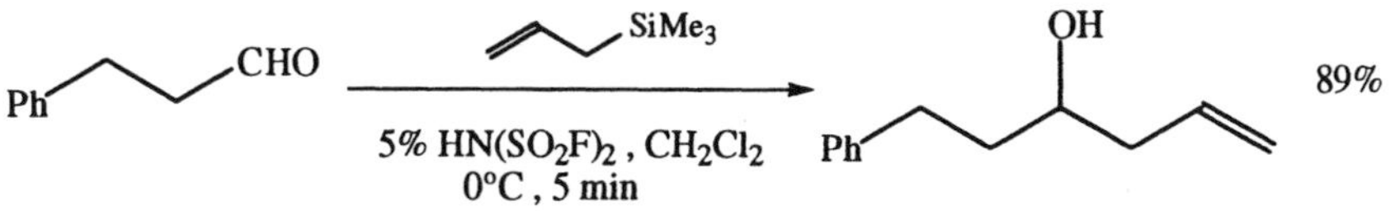

Kaur, G.; Manju, K.; Trehan, S. *Chem. Commun.*, ***1996***, 581.

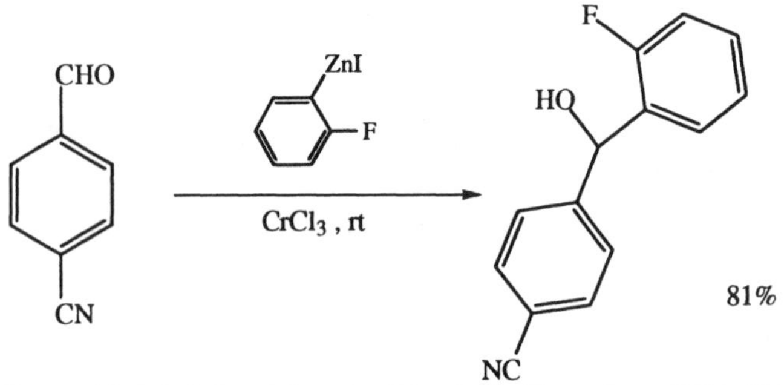
CHO
CN
ZnI
F
CrCl3 , rt
HO
F
NC
81%

Ogawa, Y.; Mori, M.; Saiga, A.; Takagi, K. *Chem. Lett.*, ***1996***, 1069.

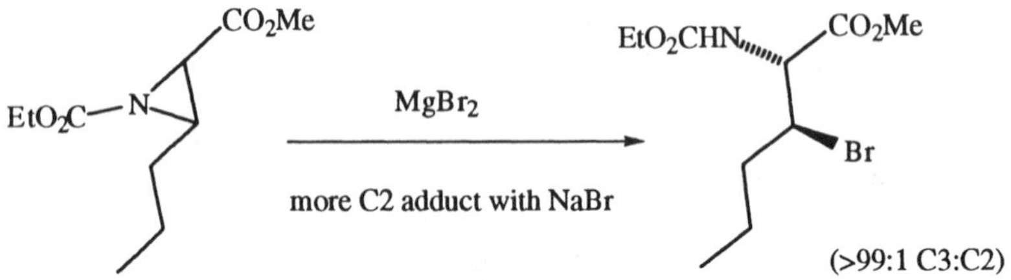
CO2Me
EtO2C—N
MgBr2
more C2 adduct with NaBr
EtO2CHN
CO2Me
Br
(>99:1 C3:C2)

Righi, G.; D'Achille, R.; Bonini, C. *Tetrahedron Lett.*, ***1996***, *37*, 6893.

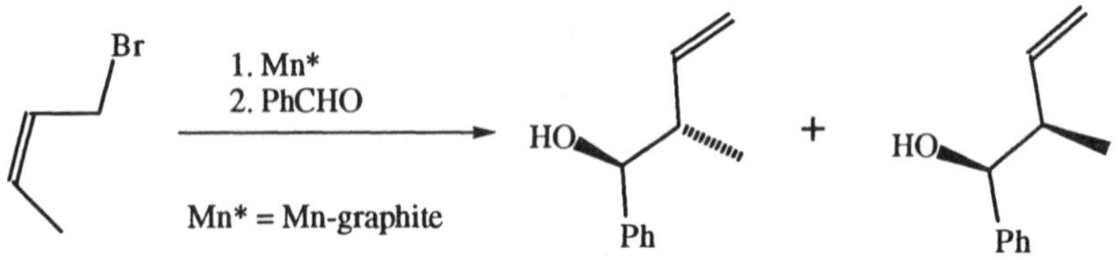
Br
1. Mn*
2. PhCHO
Mn* = Mn-graphite
HO
Ph
+
HO
Ph

(41:59 *threo:erythro*) 65%

Furstner, A.; Brunner, H. *Tetrahedron Lett.*, ***1996***, *37*, 7009.

SECTION 35: ALCOHOLS AND THIOLS FROM ALKYLS, METHYLENES AND ARYLS

No examples of the reaction $RR^1 \rightarrow ROH$ (R^1 = alkyl, aryl, etc.) occur in the literature. For reactions of the type RH → ROH (R = alkyl or aryl) see Section 41 (Alcohols and Phenols from Hydrides).

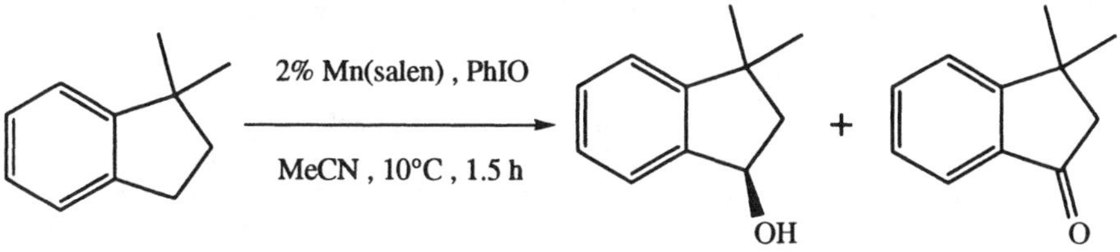
2% Mn(salen) , PhIO
MeCN , 10°C , 1.5 h
+
OH
O

(26 : 1) 18%
39% ee

Hamada, T.; Irie, R.; Mihara, J.; Hamachi, K.; Katsuki, T. *Tetrahedron*, ***1998***, *54*, 10017.

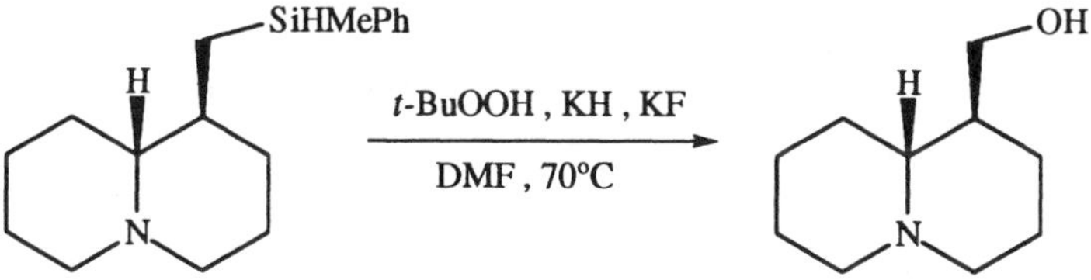
SiHMePh
H
N
t-BuOOH , KH , KF
DMF , 70°C
OH
H
N

57%

Smitrovich, J.H.; Woerpel, K.A. *J. Org. Chem.*, ***1996***, *61*, 6044.

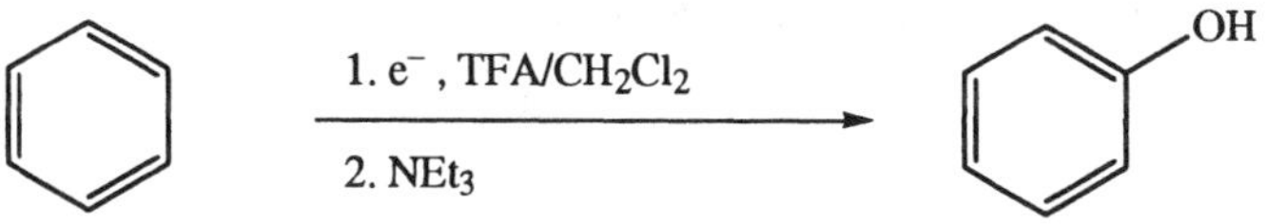
1. e⁻ , TFA/CH2Cl2
2. NEt3
OH

72%

Fujimoto, K.; Toduda, Y.; Maekawa, H.; Matsubara, Y.; Mizuno, T.; Nishiguchi, I. *Tetrahedron*, ***1996***, *52*, 3889.

SECTION 36: ALCOHOLS AND THIOLS FROM AMIDES

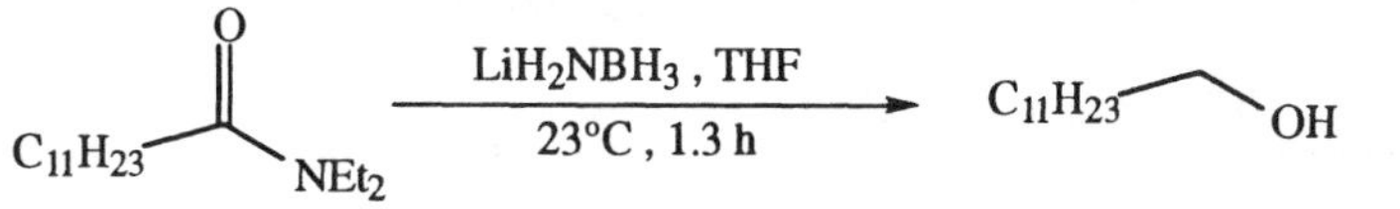
O
C11H23
NEt2
LiH2NBH3 , THF
23°C , 1.3 h
C11H23
OH

94%

Myers, A.G.; Yang, B.H.; Kopecky, D.J. *Tetrahedron Lett.* ***1996***, *37*, 3623.

SECTION 37: ALCOHOLS AND THIOLS FROM AMINES

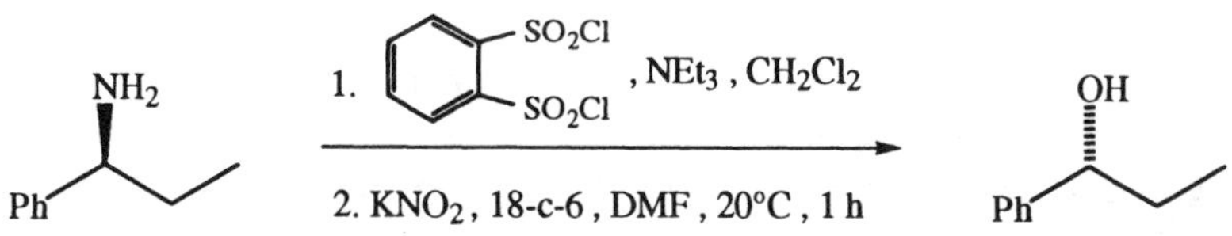
NH2
Ph
1.
SO2Cl
SO2Cl
, NEt3 , CH2Cl2
2. KNO2 , 18-c-6 , DMF , 20°C , 1 h
OH
Ph

70% (40% ee)

Sørbye, K.; Tautermann, C.; Carlsen, P.; Fiksdahl, A. *Tetrahedron Asymmetry*, ***1998***, *9*, 681

SECTION 38: ALCOHOLS AND THIOLS FROM ESTERS

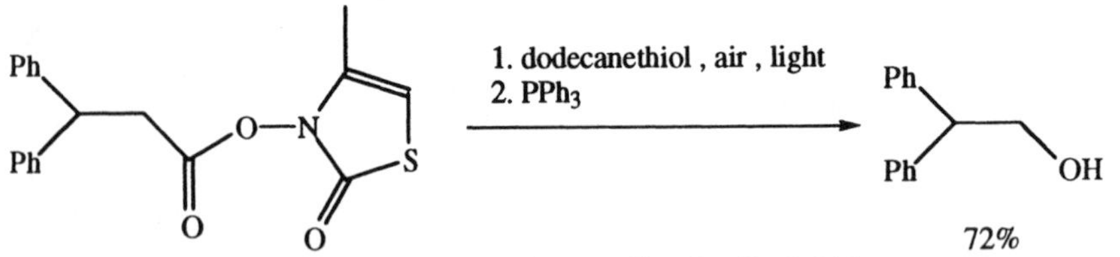
Ph
Ph
O
N
S
O
O
1. dodecanethiol , air , light
2. PPh3
Ph
Ph
OH
72%

Barton, D.H.R.; Géro, S.D.; Holliday, P.; Quiclet-Sire, B.; Zard, S.Z. *Tetrahedron*, ***1998***, *54*, 6751.

$$C_{15}H_{31}CO_2Me \xrightarrow{H_2\text{ , Ru(acac)}_3\text{ , NEt}_3\text{ , FIPA}} C_{15}H_{31}CH_2OH \quad 94\%$$

FIPA = 1,1,1,3,3,3-hexafluoropropan-2-ol

Teunissen, H.T.; Elsevier, C.J. *Chem. Lett.*, ***1998***, 1367.

NaSMe , MeOH , 27°C
C7H15
SAc
C7H15
SH
91%

Wallace, O.B.; Springer, D.M. *Tetrahedron Lett.*, ***1998***, *39*, 2693.

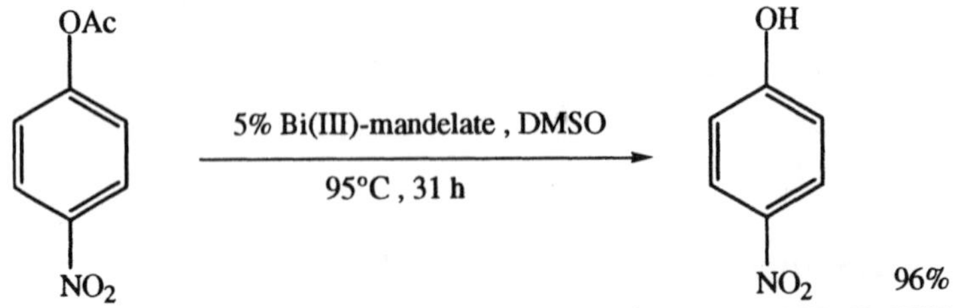
OAc
NO2
5% Bi(III)-mandelate , DMSO
95°C , 31 h
OH
NO2
96%

Le Boisselier, V.; Postel, M.; Duñach, E. *Tetrahedron Lett.*, ***1997***, *38*, 2981.

$$PhCO_2Me \xrightarrow[5\text{ h}]{10\%\text{ cyclohexene , Zn(BH}_4)_2} PhCH_2OH \quad \text{quant}$$

Narasimhan, S.; Madhavan, S.; Prasad, K.G. *Synth. Commun.*, ***1997***, *27*, 385.

SECTION 39: ALCOHOLS AND THIOLS FROM ETHERS, EPOXIDES AND THIOETHERS

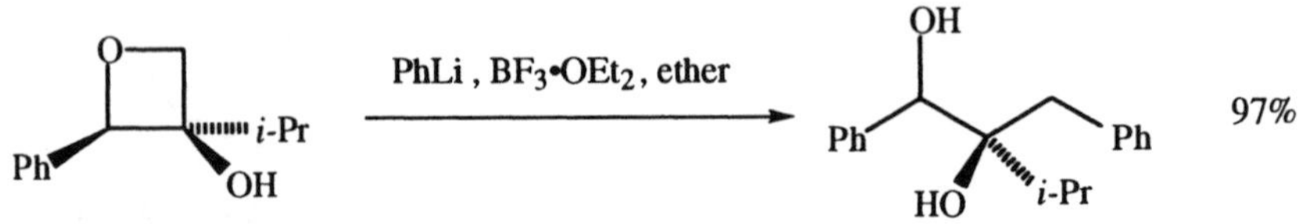
O
i-Pr
Ph
OH
PhLi , BF3•OEt2 , ether
OH
Ph
Ph
HO
i-Pr
97%

Bach, T.; Eilers, F. *Eur. J. Org. Chem.*, ***1998***, 2161.

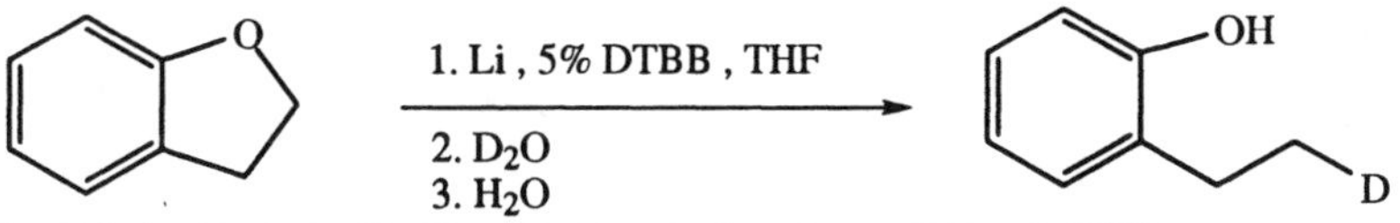
1. Li , 5% DTBB , THF
2. D_2O
3. H_2O
OH
D

Bachki, A.; Foubelo, F.; Yus, M. *Tetrahedron Lett.*, ***1998***, *39*, 7759.

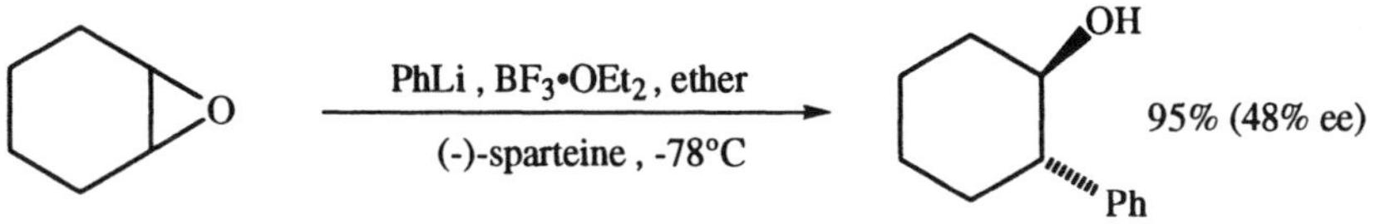
PhLi , BF_3•OEt_2 , ether
(-)-sparteine , -78°C
OH
Ph
95% (48% ee)

Alexakis, A.; Vrancken, E.; Mangeney, P. *SynLett.*, ***1998***, 1165.

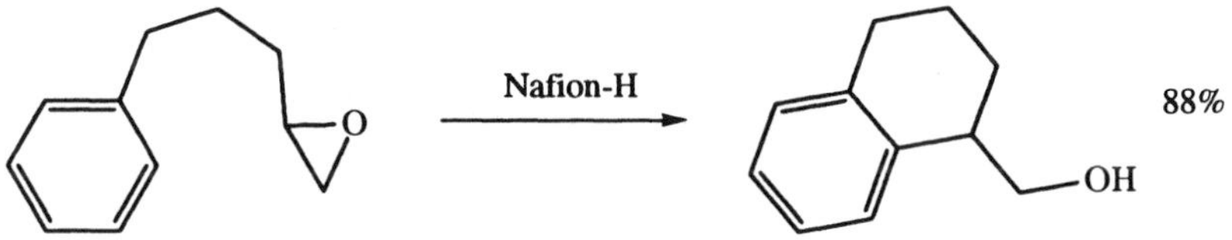
Nafion-H
OH
88%

Taylor, S.K.; Dickinson, M.G.; May. S.A.; Pickering, D.A.; Sadek, P.C. *Synthesis*, ***1998***, 1133.

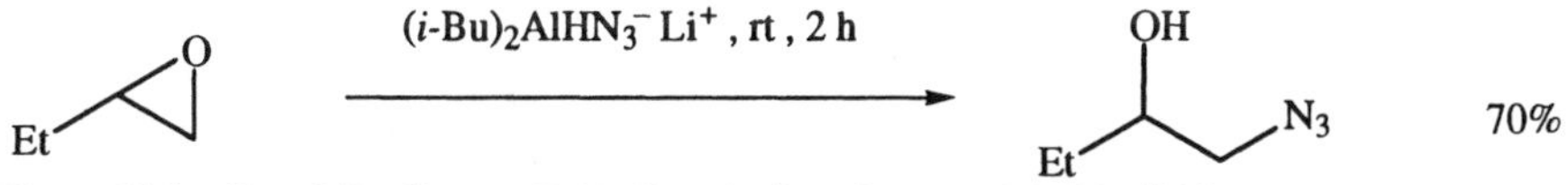
$(i\text{-Bu})_2AlHN_3^-\ Li^+$, rt , 2 h
Et
OH
N_3
70%

Youn, Y.S.; Cho, I.S.; Chung, B.Y. *Tetrahedron Lett.*, ***1998***, *39*, 4337.

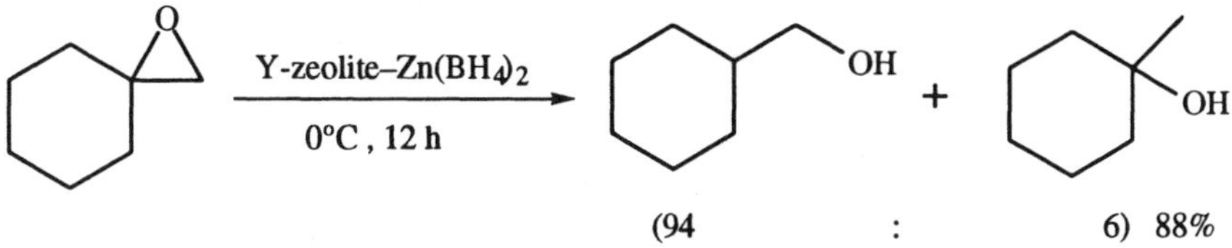
Y-zeolite–$Zn(BH_4)_2$
0°C , 12 h
OH
+
OH
(94 : 6) 88%

Sreekumar, R.; Padmakumar, R.; Rugmini, P. *Tetrahedron Lett.*, ***1998***, *39*, 5151.

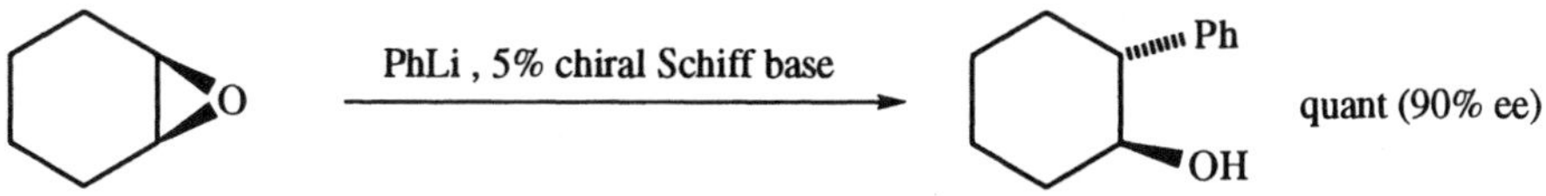
PhLi , 5% chiral Schiff base
Ph
OH
quant (90% ee)

Oguni, N.; Miyagi, Y.; Itoh, K. *Tetrahedron Lett.*, ***1998***, *39*, 9023.

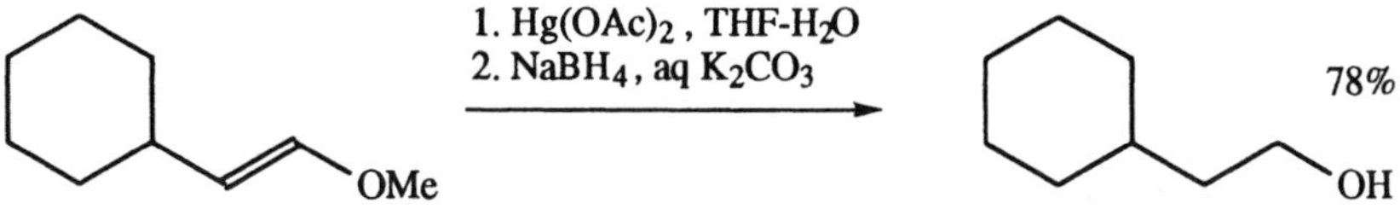
1. $Hg(OAc)_2$, THF-H_2O
2. $NaBH_4$, aq K_2CO_3
OMe
OH
78%

Crouch, R.D.; Mitten, J.V.; Span, A.R.; Dai, H.G. *Tetrahedron Lett.*, ***1997***, *38*, 791.

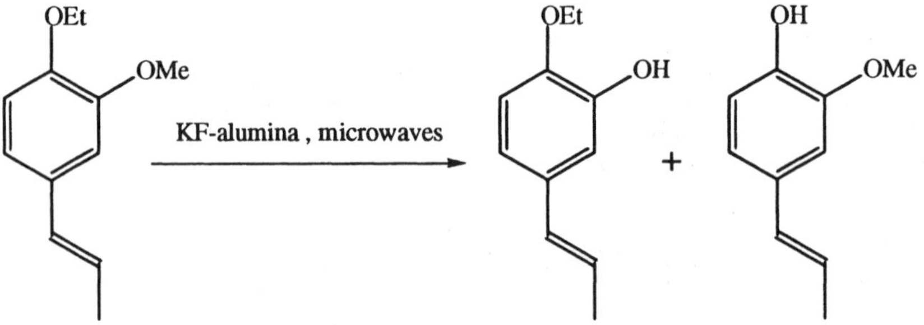
OEt
OMe
KF-alumina , microwaves
OEt
OH
+
OH
OMe

20 min , 139°C	-	82%
+ ethylene glycol 1 h , 228°C	56%	3%

Ousaïd, A.; Thach, L.N.; Loupy, A. *Tetrahedron Lett.*, ***1997***, *38*, 2451.

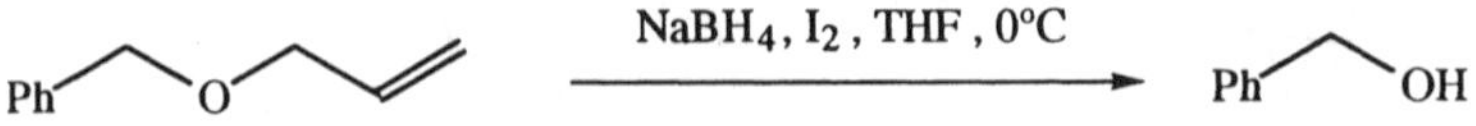
Ph
O
NaBH4 , I2 , THF , 0°C
Ph
OH

95%

Thomas, R.M.; Mohan, G.H.; Iyengar, D.S. *Tetrahedron Lett.*, ***1997***, *38*, 4721.

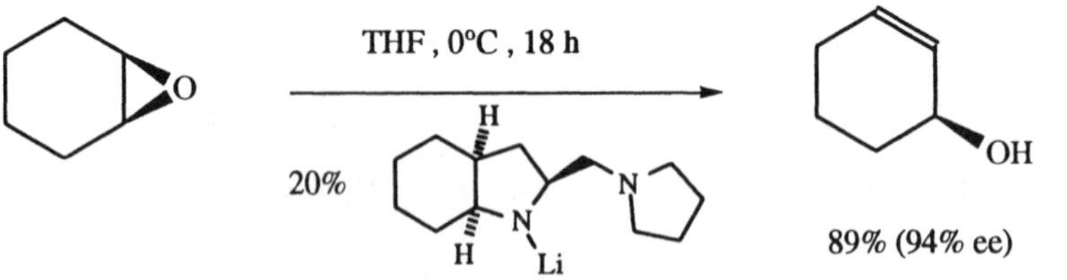
O
THF , 0°C , 18 h
20%
H
N
H
Li
N
OH

89% (94% ee)

Asami, M.;. Suga, T.; Honda, K.; Inoue, S. *Tetrahedron Lett.*, ***1997***, *38*, 6425.

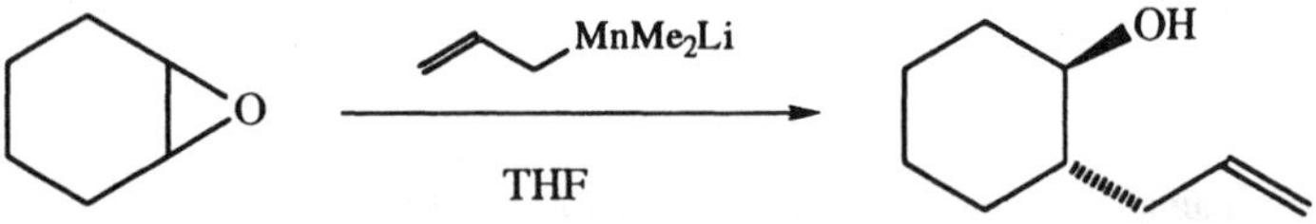
O
MnMe2Li
THF
OH

80%

Tang, J.; Yorimitsu, H.; Kakiya, H.; Inoue, R.; Shinokubo, H.; Oshima, K. *Tetrahedron Lett.*, ***1997***, *38*, 901

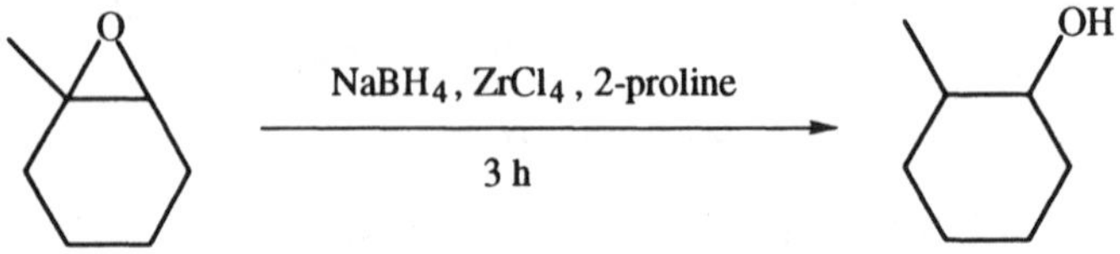
O
NaBH4 , ZrCl4 , 2-proline
3 h
OH

60% (44% ee , *RS*)

Laxmi, Y.R.S.; Iyengar, D.S. *Synth. Commun.*, ***1997***, *27*, 1731.

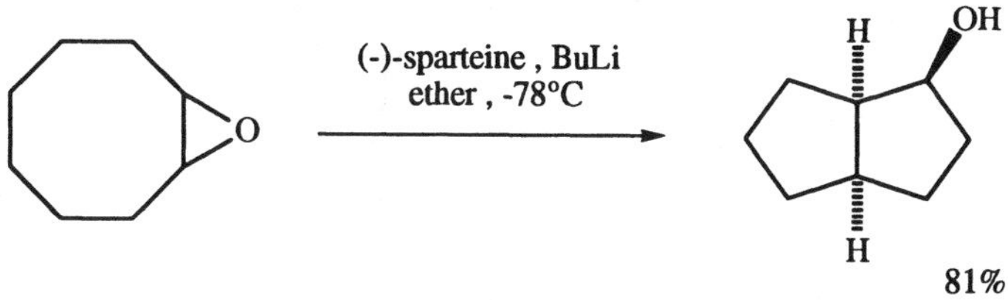
(-)-sparteine , BuLi
ether , -78°C
O
H
OH
H

81% (70% ee)

Hodgson, D.M.; Lee, C.P. *Tetrahedron Asymmetry*, ***1997***, *8*, 2307.

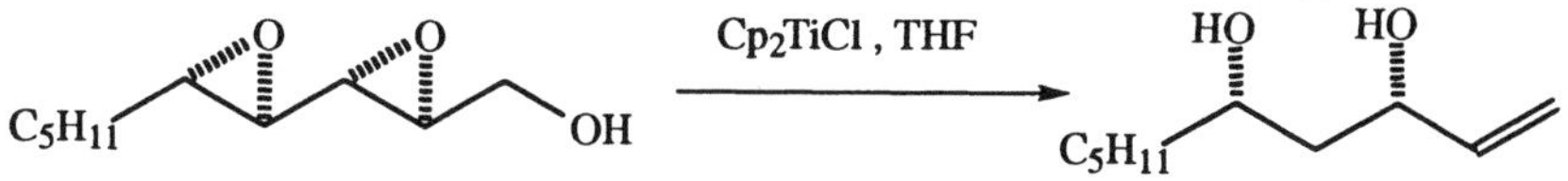
O
O
C_5H_{11}
OH
Cp_2TiCl , THF
HO
HO
C_5H_{11}

78%

Yadav, J.S.; Srinivas, D. *Chem. Lett.*, ***1997***, 905.

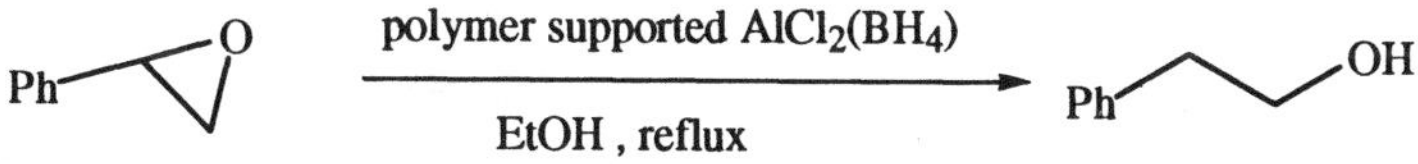
Ph
O
polymer supported $AlCl_2(BH_4)$
EtOH , reflux
Ph
OH

98%

Tamami, B.; Lakouraj, M.M.; Yeganeh, H. *J. Chem. Res. (S)*, ***1997***, 330.

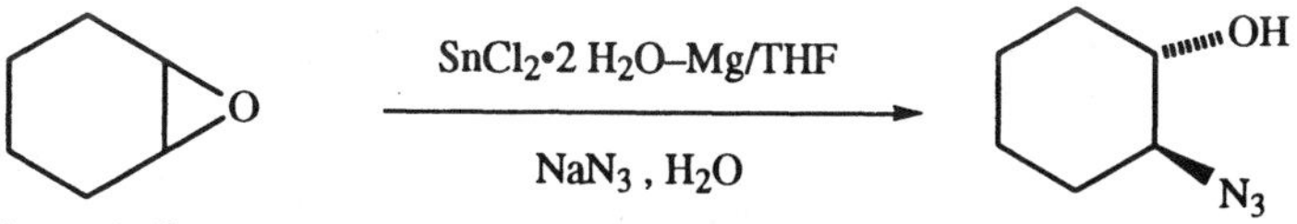
O
$SnCl_2$•2 H_2O–Mg/THF
NaN_3 , H_2O
OH
N_3

85%

Sarangi, C.; Das, N.B.; Nanda, B.; Nayak, A.; Sharma, R.P. *J. Chem. Res., (S)*, ***1997***, 378.

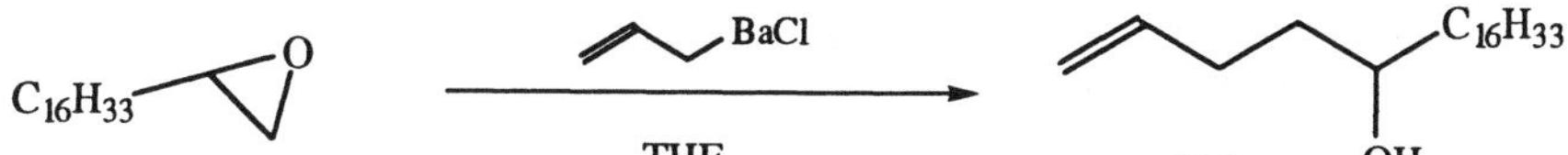
$C_{16}H_{33}$
O
BaCl
THF
$C_{16}H_{33}$
OH

81%

Yasue, K.; Yanagisawa, A.; Yamamoto, H. *Bull. Chem. Soc. Jpn.*, ***1997***, *70*, 493.

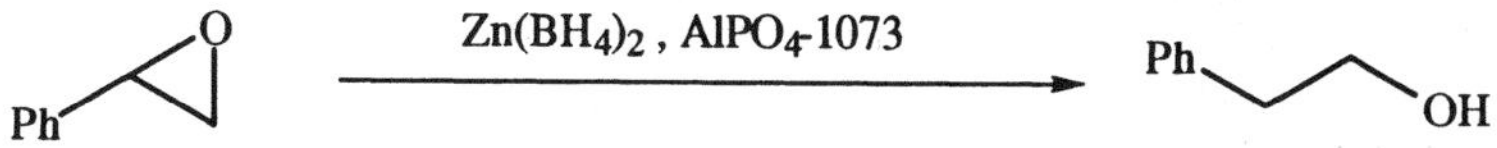
O
Ph
$Zn(BH_4)_2$, $AlPO_4$-1073
Ph
OH

97%

Campelo, J.M.; Chakraborty, R.; Marinas, J.M. *Synth. Commun.*, ***1996***, *26*, 415.

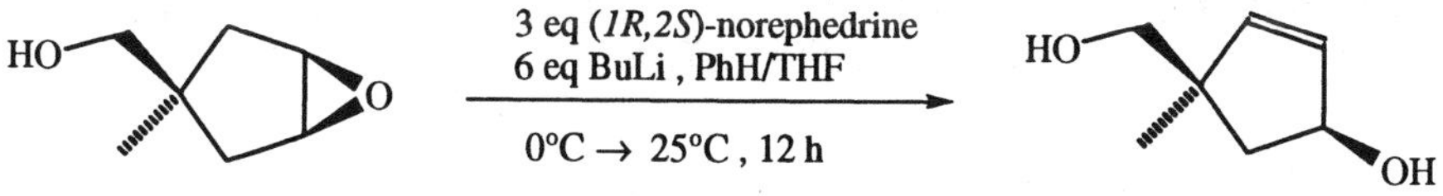
HO
O
3 eq (*1R,2S*)-norephedrine
6 eq BuLi , PhH/THF
0°C → 25°C , 12 h
HO
OH

63% (99% ee)

Hodgson, D.M.; Gibbs, A.R. *Tetrahedron Asymmetry*, ***1996***, *7*, 407.

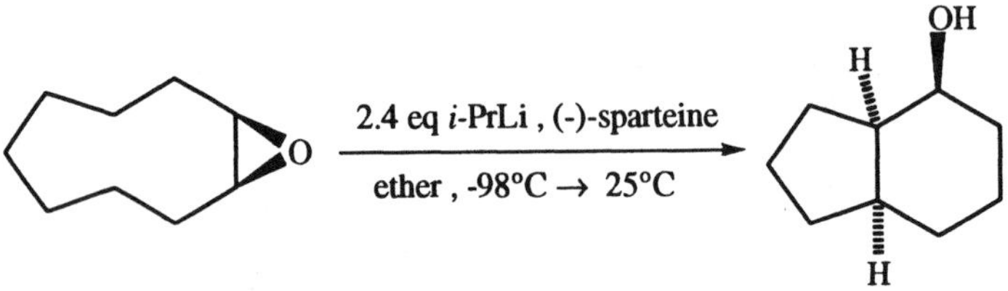
2.4 eq *i*-PrLi , (-)-sparteine
ether , -98°C → 25°C
O
OH
H
H

77% (83% ee)

Hodgson, D.M.; Lee, G.–P. *Chem. Commun.*, ***1996***, 1015.

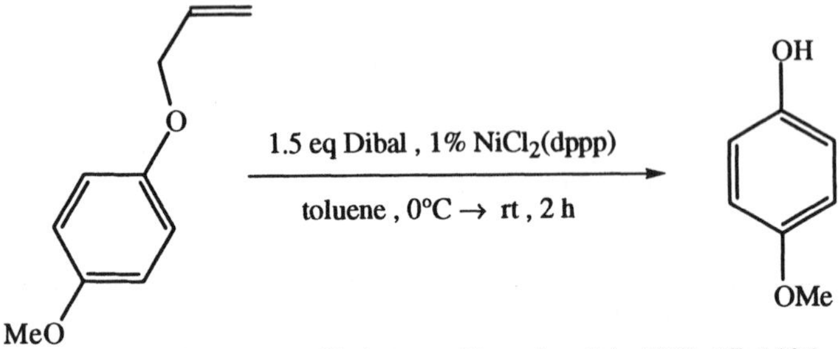
O
MeO
1.5 eq Dibal , 1% NiCl2(dppp)
toluene , 0°C → rt , 2 h
OH
OMe

90%

Taniguchi, T.; Ogasawara, K. *Angew. Chem. Int. Ed.*, ***1998***, *37*, 1136.

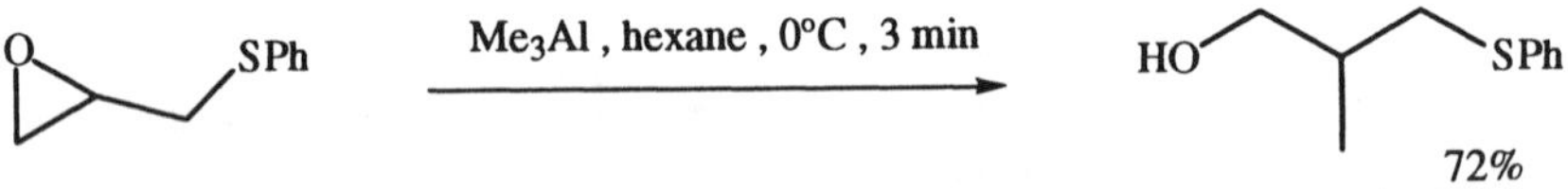
O
SPh
Me3Al , hexane , 0°C , 3 min
HO
SPh

72%

Liu, C.; Hashimoto, Y.; Kudo, K.; Saigo, K. *Bull. Chem. Soc. Jpn.*, ***1996***, *69*, 2095.

Additional examples of ether cleavages may be found in Section 45A (Protection of Alcohols and Thiols).

REVIEWS:

"Lewis Acid Induced Rearrangement of 1-Hetero-2,3-epoxides. Synthesis, Reactivity and Synthetic Applications of Homochiral Thiiranium and Aziridinium Ion Intermediates."
Rayner, C.M. *SynLett.*, ***1997***, *11.*

"Dealkylation of Ethers. A Review."
Ranu, B.C.; Bhar, S. *Org. Prep. Proceed. Int.*, ***1996***, *28*, 372.

SECTION 40: ALCOHOLS AND THIOLS FROM HALIDES AND SULFONATES

$C_{12}H_{25}Br \xrightarrow[\text{AIBN , NaBH}_3\text{CN}]{2\%\ (C_6F_{13}CH_2CH_2)_3SnH\text{ , CO , 90°C , 3 h}} C_{13}H_{27}OH$ 42%

Ryu, I.; Niguma, T.; Minakata, S.; Komatsu, M.; Hadida, S.; Curran, D.P. *Tetrahedron Lett.*. ***1997***. *38*. 7883.

見計品御届書

見計期限 2002/11/09
(200301)

B1037059

有限会社 緑 書 房

担当 橘 亮太

本社 〒560-0082 大阪府豊中市新千里東町1丁目4番2号 千里ライフサイエンスセンタービル12階
電話 06−6831−2828(代)
FAX 06−6831−2847

東京支店 〒113-0033 東京都文京区本郷5丁目29−13 (赤門アビタシオン)
電話 03−3816−5913(代)
FAX 03−3816−5914

大阪大学 基礎工学部 (326005500)
金田研究室 (1260730 基化 3)様
お客様IDはC1203400でございます。

価額欄は消費税抜きの価格です。
消費税は別枠のとおりになっております。

日付

	点数	価額	消費税額
P0020551 J WILEY			
SMITH,M.B.- COMPENDIUM OF ORGANIC SYNTHETIC METHODS, VOL.10.	1	¥16,999	¥849
(2002) ISBN 0471202010 LC 71162800	請求合計	¥17,848	
Ref. ZAIKO			
20021009 B1037059 H6012058 C1203400 0000017848 0010110000			

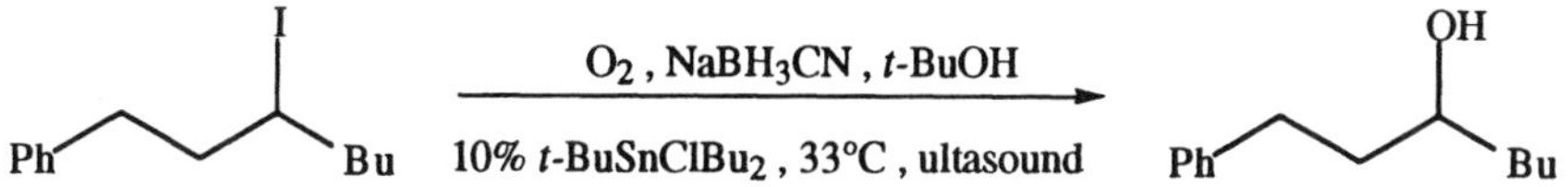
I
Ph
Bu
O_2 , $NaBH_3CN$, t-BuOH
10% t-$BuSnClBu_2$, 33°C , ultasound
OH
Ph
Bu

64%

Sawamura, M.; Kawaguchi, Y.; Sato, K.; Nakamura, E. *Chem. Lett.*, ***1997***, 705.

SECTION 41: ALCOHOLS AND THIOLS FROM HYDRIDES

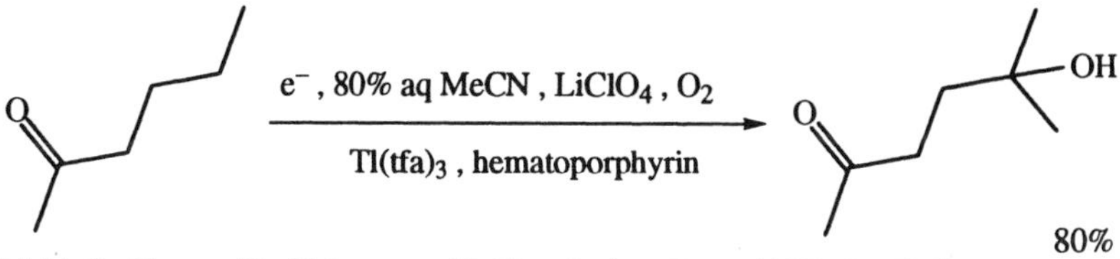
O
e^- , 80% aq MeCN , $LiClO_4$, O_2
$Tl(tfa)_3$, hematoporphyrin
O
OH

80%

Maki, S.; Konno, K.; Takayama, H. *Tetrahedron Lett.*, ***1997***, *38*, 7067.

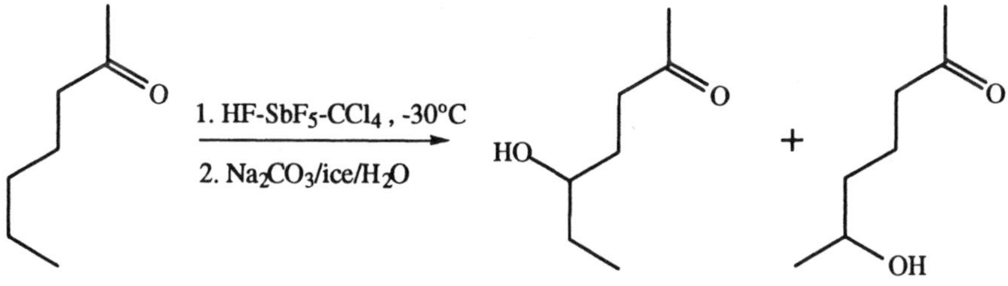
O
1. HF-SbF_5-CCl_4 , -30°C
2. Na_2CO_3/ice/H_2O
HO
O
+
O
OH

(3 : 1) 67%

Martin, A.; Jouannetaud, M.-P.; Jacquesy, J.-C. *Tetrahedron Lett.*, ***1996***, *37*, 2967.

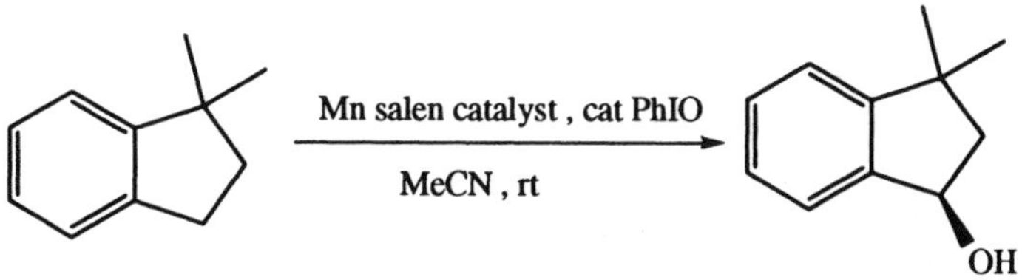
Mn salen catalyst , cat PhIO
MeCN , rt
OH

25%
(36% ee , R)

Hamachi, K.; Irie, R.; Katsuki, T. *Tetrahedron Lett.* ***1996***, *37*, 4979.

SECTION 42: ALCOHOLS AND THIOLS FROM KETONES

The following reaction types are included in this section:

A. Reductions of Ketones to Alcohols
B. Alkylations of Ketones, forming Alcohols

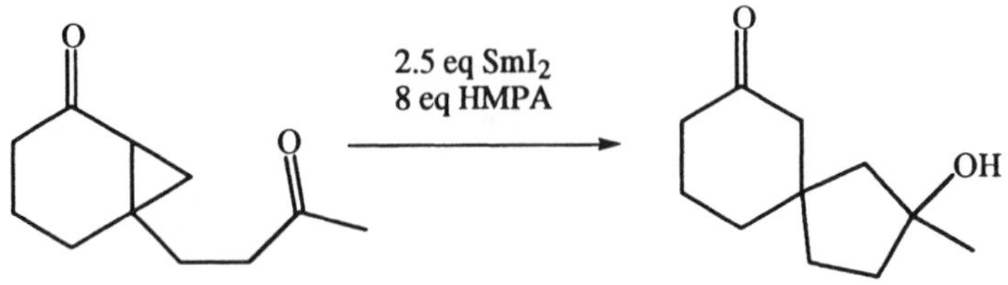
2.5 eq SmI2
8 eq HMPA
OH

79%

Molander, G.A.; Alonso-Alija, C. *Tetrahedron*, ***1997***, *53*, 8067.

Coupling of ketones to give diols is found in Section 323 (Alcohol → Alcohol).

SECTION 42A: REDUCTION OF KETONES TO ALCOHOLS

ASYMMETRIC REDUCTION

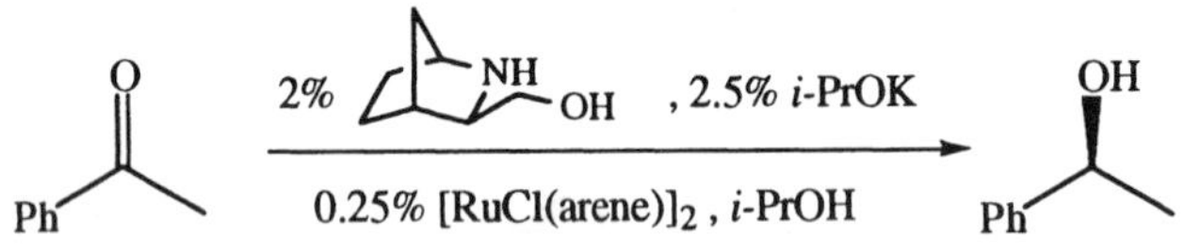
2%
NH
OH
, 2.5% i-PrOK
0.25% [RuCl(arene)]2 , i-PrOH
Ph
OH
Ph

quant (94% ee , *S*)

arene = hexamethylbenzene

Alonso, D.A.; Guijarro, D.; Pinho, P.; Temme, O.; Andersson, P.G.
J. Org. Chem., ***1998***, *63*, 2749.
For an identical reaction using $B(OMe)_3$ and BH_3•SMe_2, see
Pinho, P.; Guijarro, D.; Andersson, P.G. *Tetrahedron*, ***1998***, *54*, 7897.

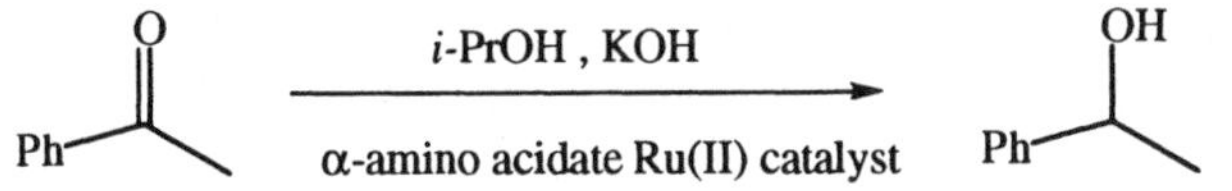
i-PrOH , KOH
α-amino acidate Ru(II) catalyst
Ph
OH
Ph

72% (81% ee , *R*)

Ohta, T.; Nakahara, S.–i.; Shigemura, Y.; Hattori, K.; Furukawa, I. *Chem. Lett.*, ***1998***, 491.

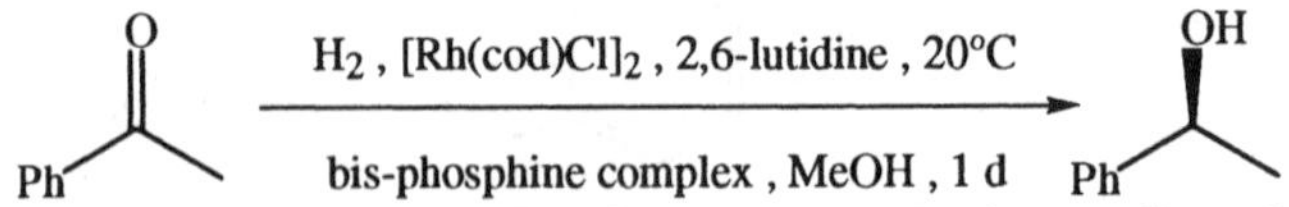
H2 , [Rh(cod)Cl]2 , 2,6-lutidine , 20°C
bis-phosphine complex , MeOH , 1 d
Ph
OH
Ph

93% (95% ee)

Jiang, Q.; Jiang, Y.; Xiao, D.; Cao, P.; Zhang, X. *Angew. Chem. Int. Ed.*, ***1998***, *37*, 1100.

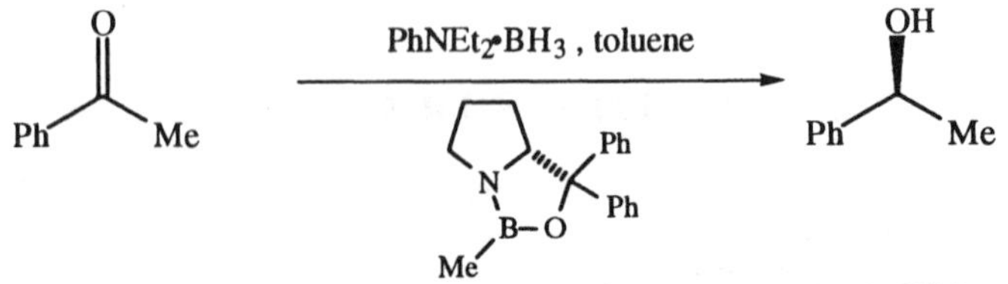
PhNEt2•BH3 , toluene
Ph
Me
Ph
Ph
N
B–O
Me
OH
Ph
Me

85% (97% ee)

Salunkhe, A.M.; Burkhardt, E.R. *Tetrahedron Lett.*, ***1997***, *38*, 1523.

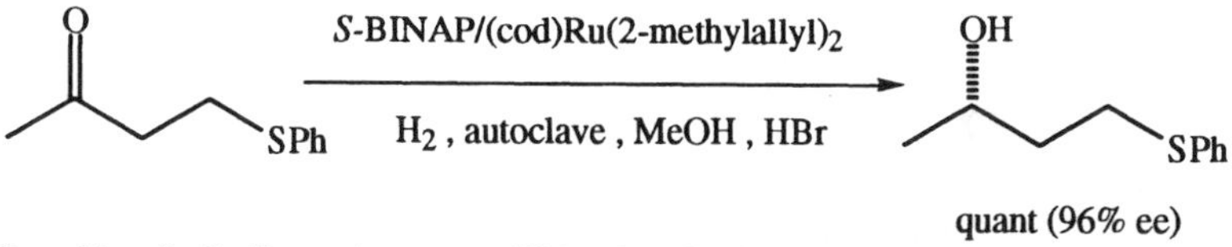
O
S-BINAP/(cod)Ru(2-methylallyl)2
H2 , autoclave , MeOH , HBr
SPh
OH
SPh
quant (96% ee)

Tranchier, J.–P.; Ratovelomanana-Vidal, V.; Genêt, J.–P.; Tong, S.; Cohen, T. *Tetrahedron Lett.*, ***1997***, *38*, 2951.

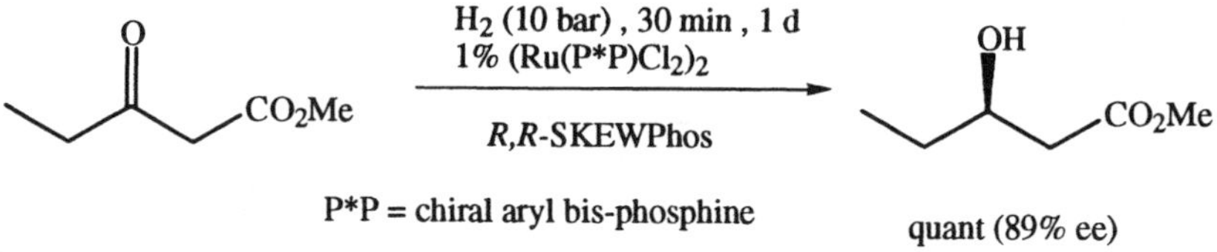
O
CO2Me
H2 (10 bar) , 30 min , 1 d
1% (Ru(P*P)Cl2)2
R,R-SKEWPhos
OH
CO2Me
P*P = chiral aryl bis-phosphine
quant (89% ee)

Blanc, D.; Henry, J.–C.; Ratovelomanana-Vidal, V.; Genêt, J.–P. *Tetrahedron Lett.*, ***1997***, *38*, 6603.

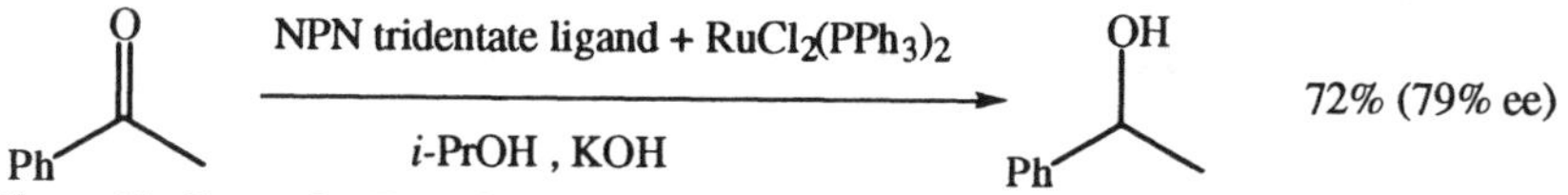
O
Ph
NPN tridentate ligand + RuCl2(PPh3)2
i-PrOH , KOH
OH
Ph
72% (79% ee)

Jiang, Y.; Jiang, Q.; Zhu, G.; Zhang, X. *Tetrahedron Lett.*, ***1997***, *38*, 215.

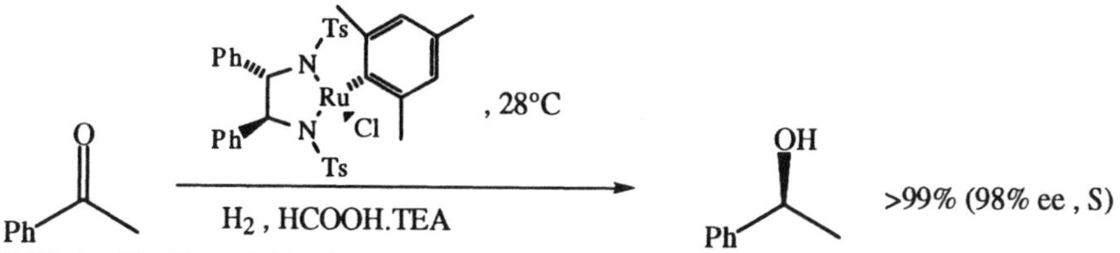
Ts
Ph
N
Ru
Cl
N
Ts
, 28°C
O
Ph
H2 , HCOOH.TEA
OH
Ph
>99% (98% ee , S)

Fujii, A.; Hashiguchi, S.; Uematsu, N.; Ikariya, T.; Noyori, R. *J. Am. Chem. Soc.*, ***1996***, *118*, 2521.

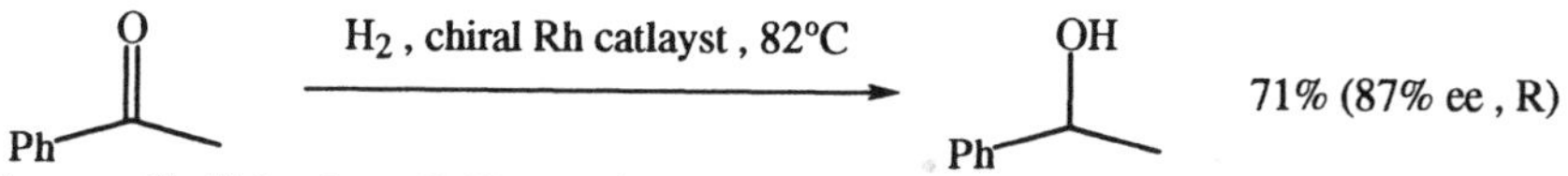
O
Ph
H2 , chiral Rh catlayst , 82°C
OH
Ph
71% (87% ee , R)

Langer, T.; Helmchen, G. *Tetrahedron Lett.*, ***1996***, *37*, 1381.

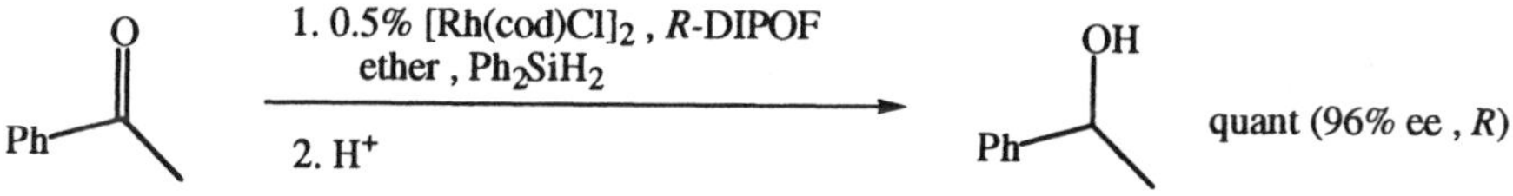
O
Ph
1. 0.5% [Rh(cod)Cl]2 , R-DIPOF
ether , Ph2SiH2
2. H+
OH
Ph
quant (96% ee , R)

DIPOF = (*SSS*)-[2-(4,5-diphenyl-4,5-dihydro-1,3-oxazol-2-yl)ferrocenyl] diphenylphosphine

Nishibayashi, Y.; Segawa, K.; Takada, H.; Ohe, K.; Uemura, S. *Chem. Commun.*, ***1996***, 847

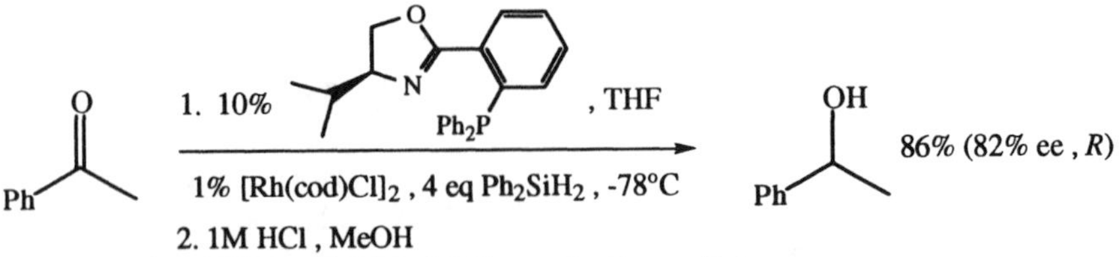
O
Ph
1. 10%
O
N
Ph2P
, THF
1% [Rh(cod)Cl]2 , 4 eq Ph2SiH2 , -78°C
2. 1M HCl , MeOH
OH
Ph
86% (82% ee , R)

Newman, L.M.; Williams, J.M.J.; McCague, R.; Potter, G.A. *Tetrahedron Asymmetry*, ***1996***, *7*, 1597.

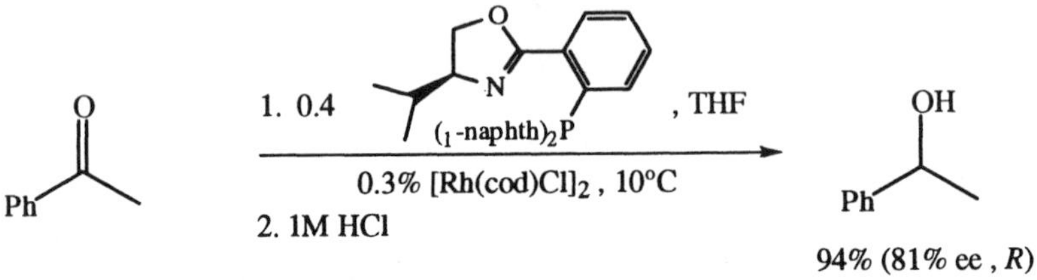
O
Ph
1. 0.4
O
N
(1-naphth)2P
, THF
0.3% [Rh(cod)Cl]2 , 10°C
2. 1M HCl
OH
Ph
94% (81% ee , R)

Langer, T.; Janssen, J.; Helmchen, G. *Tetrahedron Asymmetry*, ***1996***, *7*, 1599.

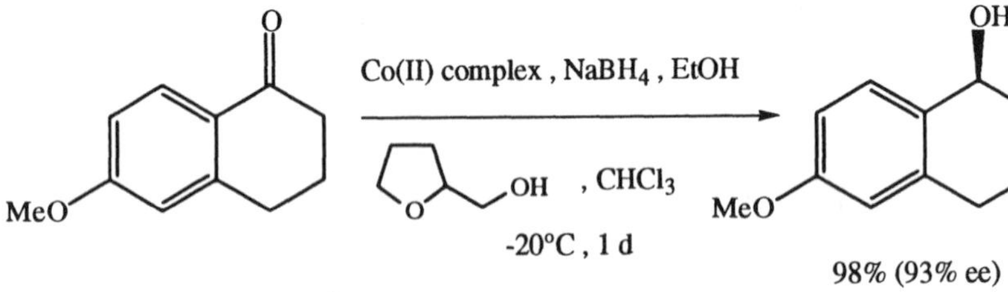
O
MeO
Co(II) complex , NaBH4 , EtOH
O
OH
, CHCl3
-20°C , 1 d
OH
MeO
98% (93% ee)

Sugi, K.D.; Nagata, T.; Yamada, T.; Mukaiyama, T. *Chem. Lett.*, ***1996***, 737.

NON-ASYMMETRIC REDUCTION

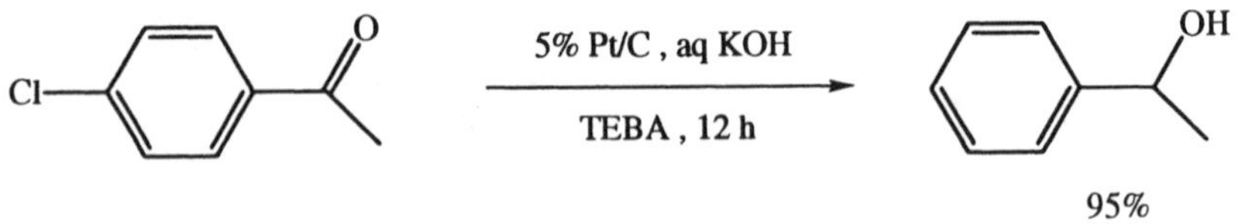
Cl
O
5% Pt/C , aq KOH
TEBA , 12 h
OH
95%

Selva, M.; Tundo, P.; Perosa, A. *J. Org. Chem.*, ***1998***, *63*, 3266.

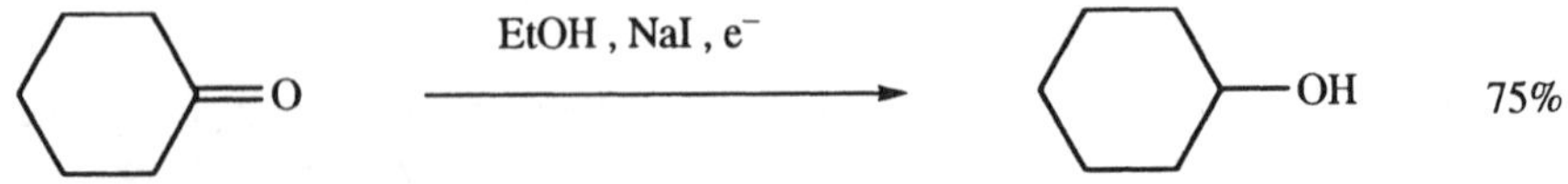
O
EtOH , NaI , e⁻
OH
75%

Wei-Dong, Y.; Cli, Y.; Ar-Xing, W. *Synth. Commun.*, ***1998***, *28*, 2827.

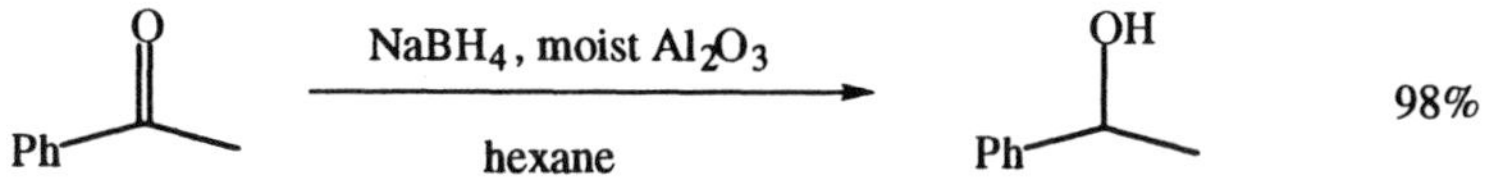
O
NaBH4, moist Al2O3
hexane
OH
Ph
Ph
98%

Yakabe, S.; Hirano, M.; Morimoto, T. *Can. J. Chem.*, ***1998***, *76*, 1916.

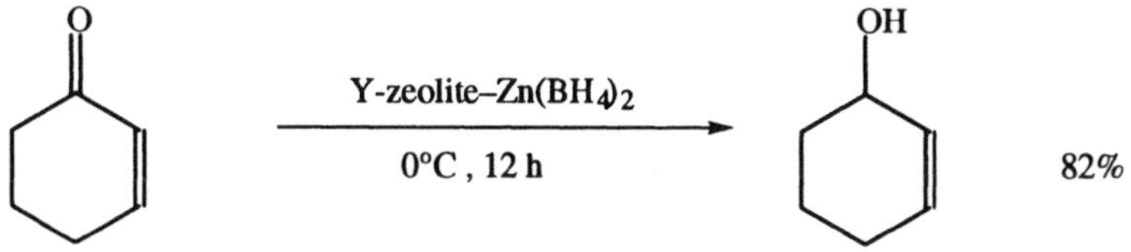
O
Y-zeolite–Zn(BH4)2
0°C, 12 h
OH
82%

Sreekumar, R.; Padmakumar, R.; Rugmini, P. *Tetrahedron Lett.*, ***1998***, *39*, 5151.

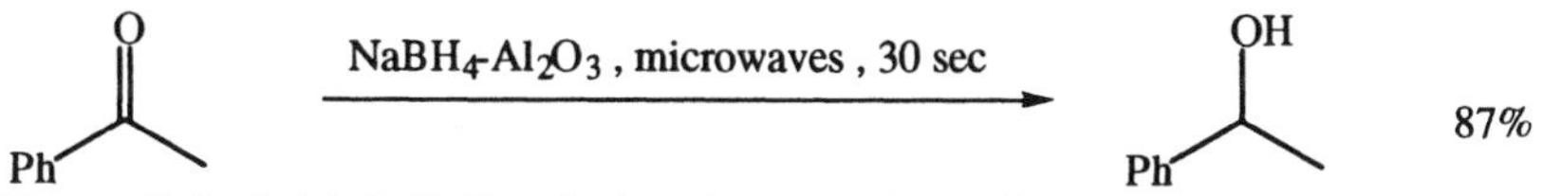
O
NaBH4-Al2O3, microwaves, 30 sec
OH
Ph
Ph
87%

Varma, R.S.; Saini, R.K. *Tetrahedron Lett.*, ***1997***, *38*, 4337.

O
NaBH4/Amberlyst-15, THF
10 min
OH
Ph
Ph
Ph
Ph
98%

Caycho, J.R.; Tellado, F.G.; de Marma, P.; Tellado, J.J.M. *Tetrahedron Lett.*, ***1997***, *38*, 277.

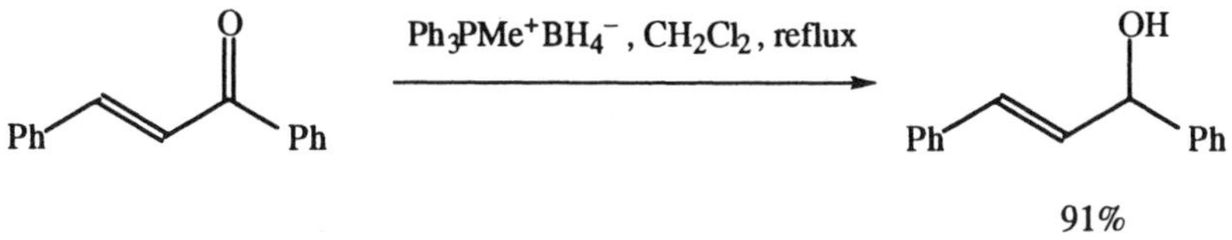
O
Ph3PMe+BH4−, CH2Cl2, reflux
OH
Ph
Ph
Ph
Ph
91%

Firouzabadi, H.; Adibi, M. *Synth. Commun.*, ***1996***, *26*, 2429.

SECTION 42B: ALKYLATION OF KETONES, FORMING ALCOHOLS

Aldol reactions are listed in Section 330 (Ketone-Alcohol)

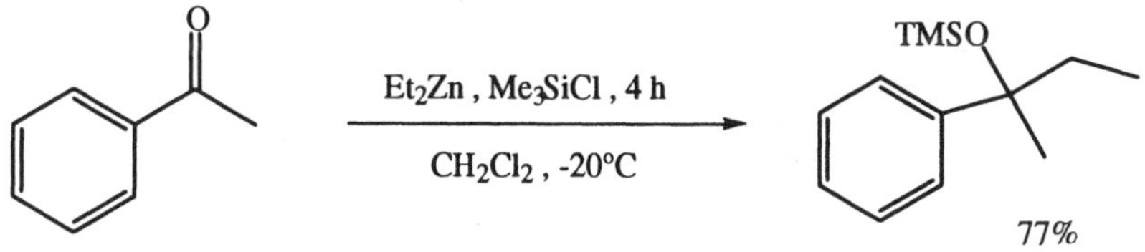
O
Et2Zn, Me3SiCl, 4 h
CH2Cl2, -20°C
TMSO
77%

Alvisi, C.; Casolari, S.; Costa, A.L.; Ritiani, M.; Tagliavini, E. *J. Org. Chem.* ***1998***, *63*, 1330.

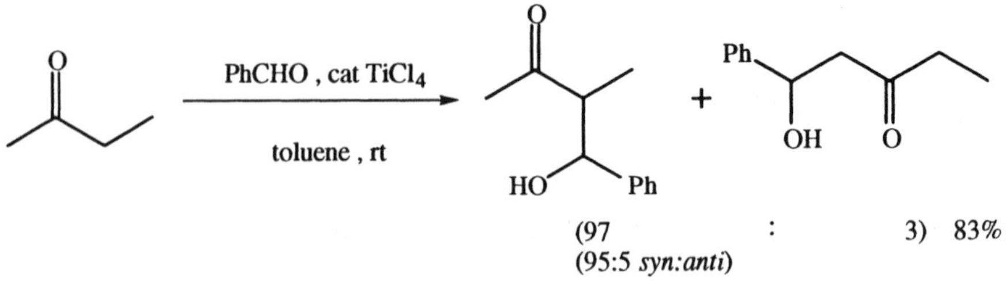
PhCHO , cat $TiCl_4$
toluene , rt
O
HO
Ph
+
Ph
OH
O
(97 : 3) 83%
(95:5 *syn:anti*)

Doas, P.I.; Fu, G.C. *J. Am. Chem. Soc.*, ***1998***, *120*, 445.

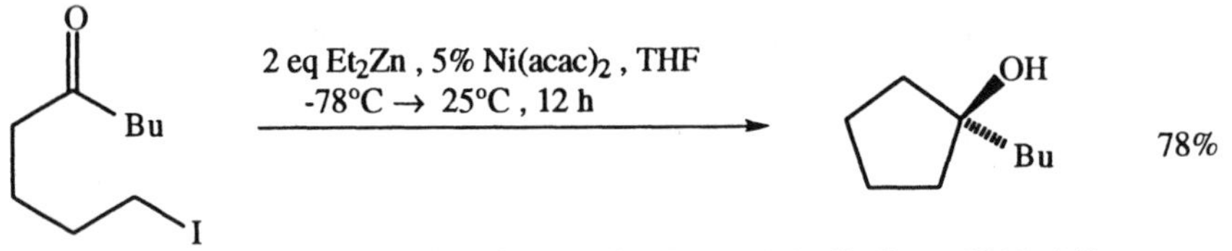
O
Bu
I
2 eq Et_2Zn , 5% $Ni(acac)_2$, THF
-78°C → 25°C , 12 h
OH
Bu
78%

Stüdemann, T.; Ibrahim-Ouali, M.; Cahiez, G.; Knochel, P. *SynLett.*, ***1998***, 143.

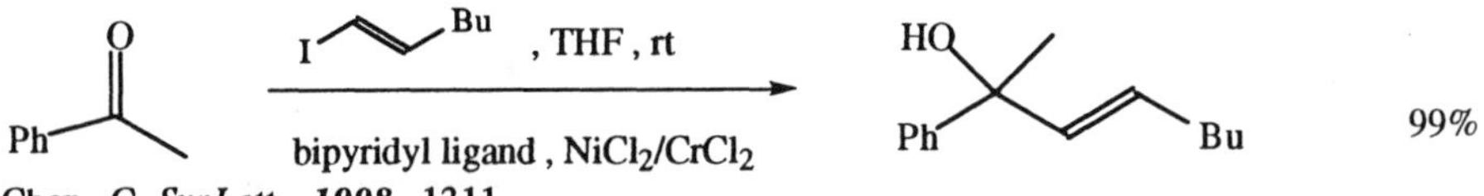
O
Ph
I
Bu
, THF , rt
bipyridyl ligand , $NiCl_2/CrCl_2$
HO
Ph
Bu
99%

Chen, C. *SynLett.*, ***1998***, 1311.

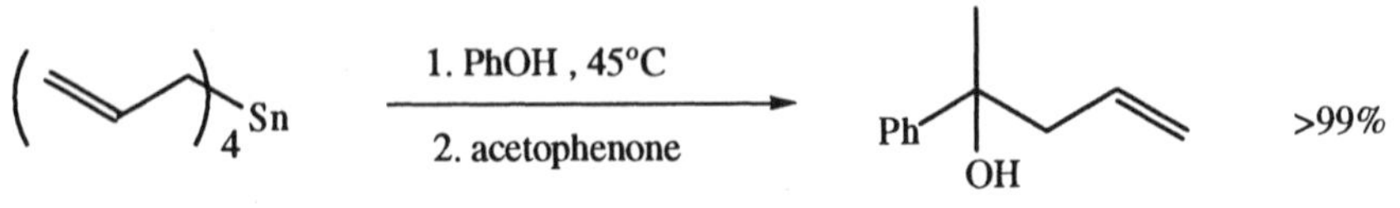
Sn
4
1. PhOH , 45°C
2. acetophenone
Ph
OH
>99%

without phenol, yield is <5%

Yasuda, M.; Kitahara, N.; Fujibayashi, T.; Baba, A. *Chem. Lett.*, ***1998***, 743.

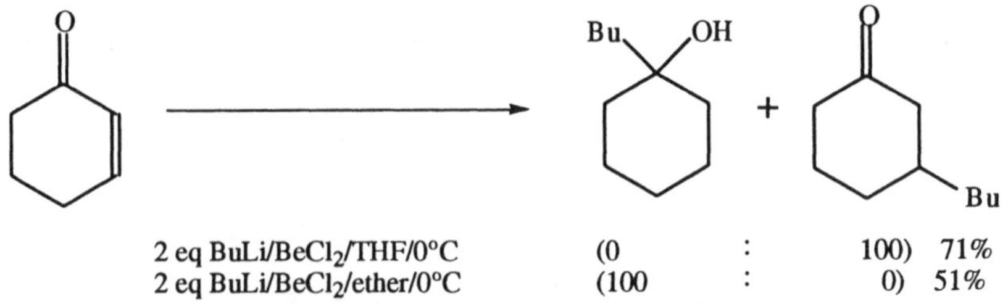
O
Bu
OH
+
O
Bu
2 eq BuLi/$BeCl_2$/THF/0°C (0 : 100) 71%
2 eq BuLi/$BeCl_2$/ether/0°C (100 : 0) 51%

Krief, A.; de Vos, M.J.; De Lombart, S.; Bosref, J.; Couty, F. *Tetrahedron Lett.*. ***1997***. *38*. 6295.

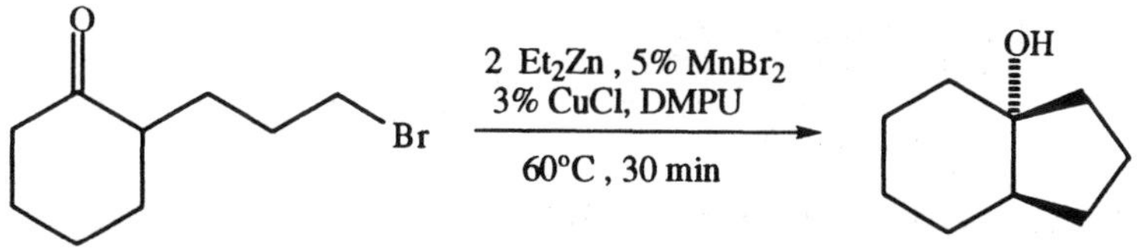
O
Br
2 Et_2Zn , 5% $MnBr_2$
3% CuCl, DMPU
60°C , 30 min
OH

82%

Riguet, E.; Klement, I.; Reddy, Ch.K.; Cahiez, G.; Knochel, P.
Tetrahedron Lett., ***1996***, *37*, 5865.

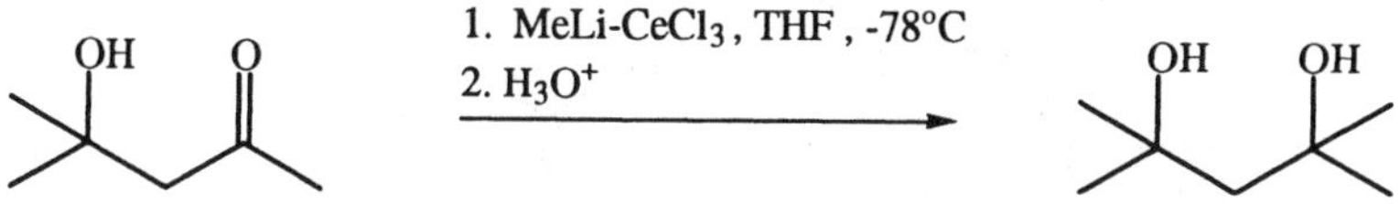
OH
O
1. MeLi-$CeCl_3$, THF , -78°C
2. H_3O^+
OH
OH

97%

Bartoli, G.; Bosco, M.; Van Beek, J.; Sambri, L.; Marcantoni, E.
Tetrahedron Lett., ***1996***, *37*, 2293.

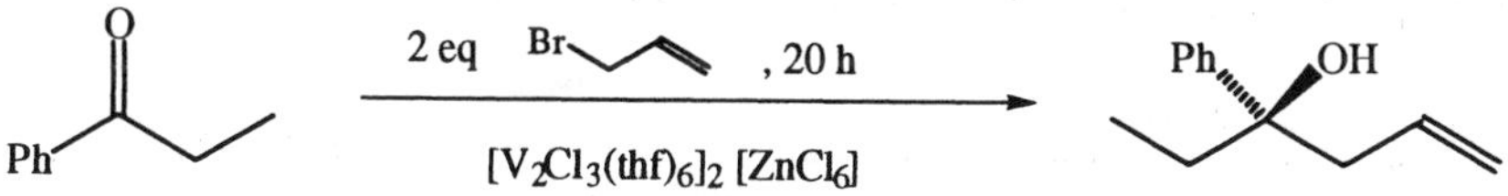
O
Ph
2 eq
Br
, 20 h
$[V_2Cl_3(thf)_6]_2$ $[ZnCl_6]$
Ph
OH

97%

Kataoka, Y.; Makihara, I.; Tani, K. *Tetrahedron Lett.*, ***1996***, *37*, 7083.

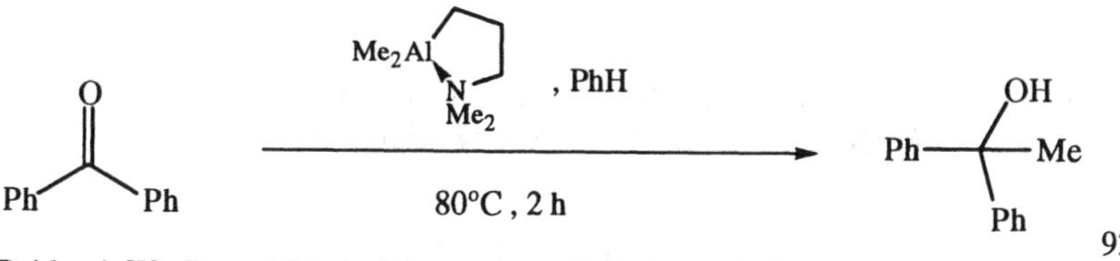
O
Ph
Ph
Me_2Al
N
Me_2
, PhH
80°C , 2 h
OH
Ph
Me
Ph

93%

Baidossi, W.; Resenfeld, A.; Wassermann, B.C.; Schulte, S.; Schumann, H.; Blum, J.
Synthesis, ***1996***, 1127.

SECTION 43: ALCOHOLS AND THIOLS FROM NITRILES

NO ADDITIONAL EXAMPLES

SECTION 44: ALCOHOLS AND THIOLS FROM ALKENES

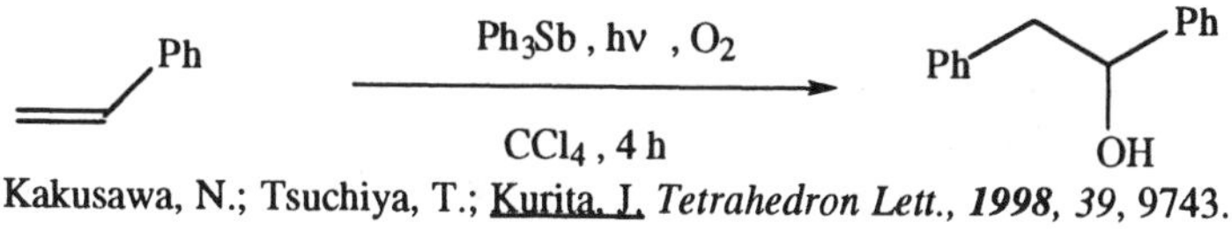
Ph
Ph_3Sb , hν , O_2
CCl_4 , 4 h
Ph
Ph
OH

48%

Kakusawa, N.; Tsuchiya, T.; Kurita, J. *Tetrahedron Lett.*, ***1998***, *39*, 9743.

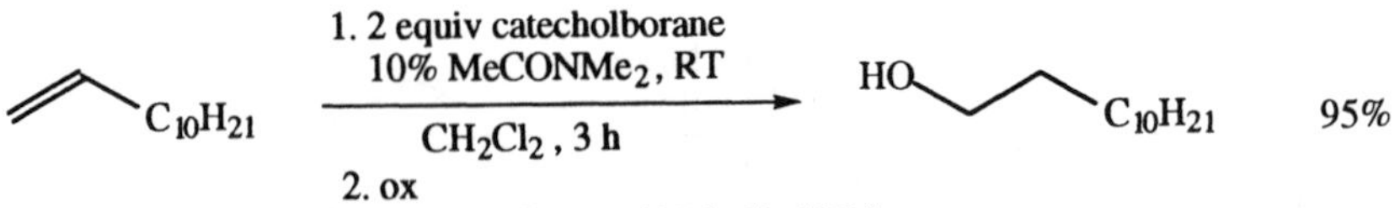
1. 2 equiv catecholborane
10% MeCONMe2, RT
CH2Cl2, 3 h
2. ox
C10H21
HO
C10H21
95%

Garrett, C.E.; Fu, G.C. *J. Org. Chem.*, ***1996***, *61*, 3224.

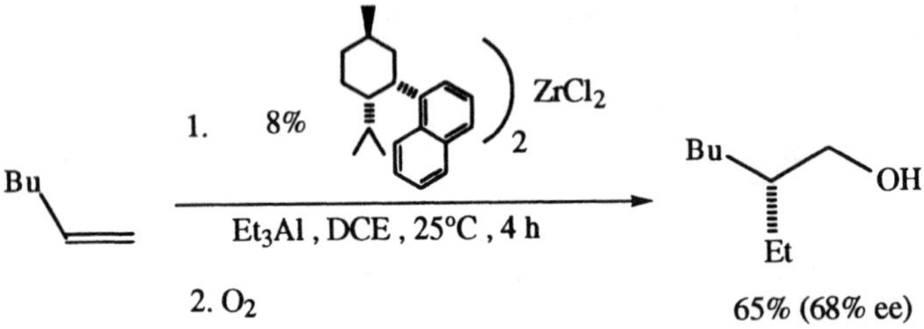
1. 8%
ZrCl2
2
Bu
Et3Al, DCE, 25°C, 4 h
2. O2
Bu
OH
Et
65% (68% ee)

Kondakov, D.Y.; Negishi, E. *J. Am. Chem. Soc.*, ***1996***, *118*, 1577.

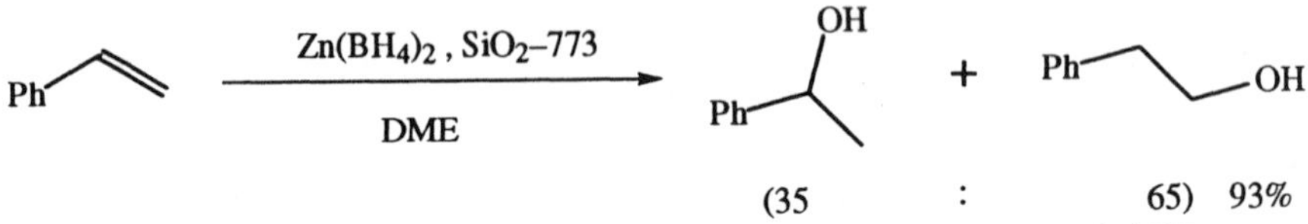
Ph
Zn(BH4)2, SiO2–773
DME
OH
Ph
+
Ph
OH
(35 : 65) 93%

Campelo, J.M.; Chakraborty, R.; Marinas, J.M. *Synth. Commun.*, ***1996***, *26*, 1639.

SECTION 45: ALCOHOLS AND THIOLS FROM MISCELLANEOUS COMPOUNDS

NO ADDITIONAL EXAMPLES

SECTION 45A: PROTECTION OF ALCOHOLS AND THIOLS

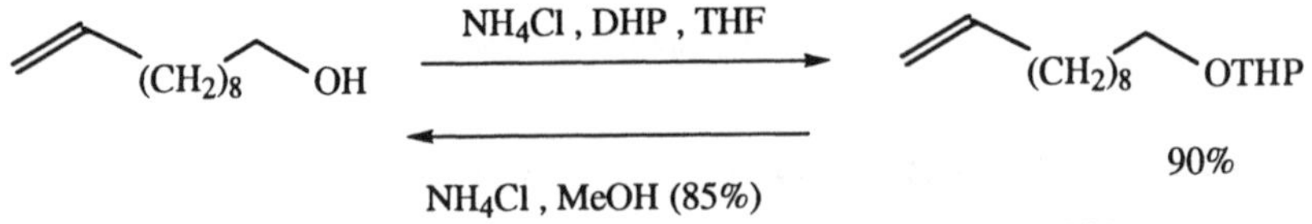
(CH2)8
OH
NH4Cl, DHP, THF
NH4Cl, MeOH (85%)
(CH2)8
OTHP
90%

Yadav, J.S.; Srinivas, D.; Reddy, G.S. *Synth. Commun.*, ***1998***, *28*, 1399.

$C_8H_{17}OTHP$ —(expansive graphite, MeOH; 50°C, 0.7 h)→ $C_8H_{17}OH$ 94%

Zhang, Z.-H.; Li, T.-S.; Jin, T.-S.; Wang, J.-X. *J. Chem. Soc. (S)*, ***1998***, 152.

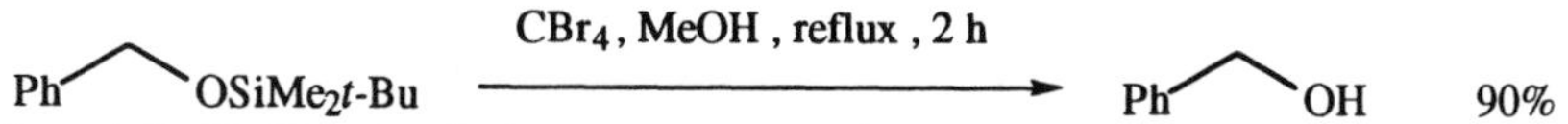
Ph OSiMe2t-Bu
CBr4, MeOH, reflux, 2 h
Ph OH
90%

Lee, A.S.-Y.; Yeh, H.-C.; Shie, J.-J. *Tetrahedron Lett.*, ***1998***, *39*, 5249.

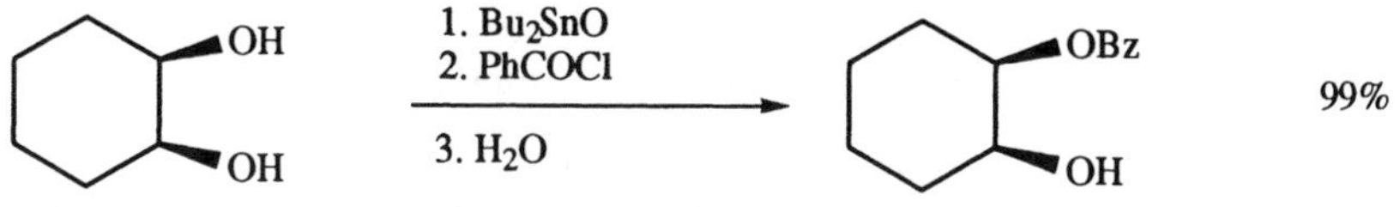
OH
OH
1. Bu2SnO
2. PhCOCl
3. H2O
OBz
OH
99%

Maki, T.; Iwasaki, F.; Matsumura, Y. *Tetrahedron Lett.*, ***1998***, *39*, 5601.

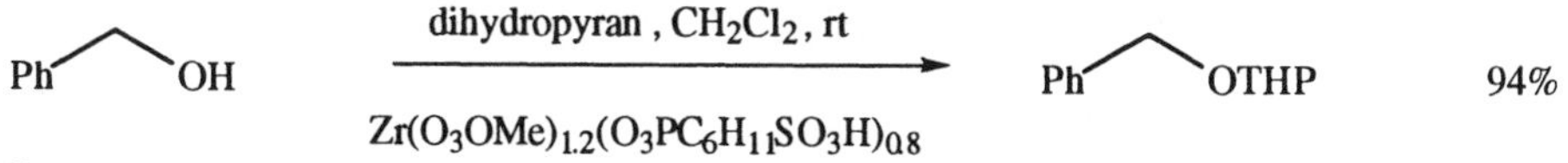
Ph OH
dihydropyran, CH2Cl2, rt
Zr(O3OMe)1.2(O3PC6H11SO3H)0.8
Ph OTHP
94%

Curini, M.; Epifano, F.; Marcotullio, M.C.; Rosati, O.; Costantino, U. *Tetrahedron Lett.*, ***1998***, *39*, 8159.

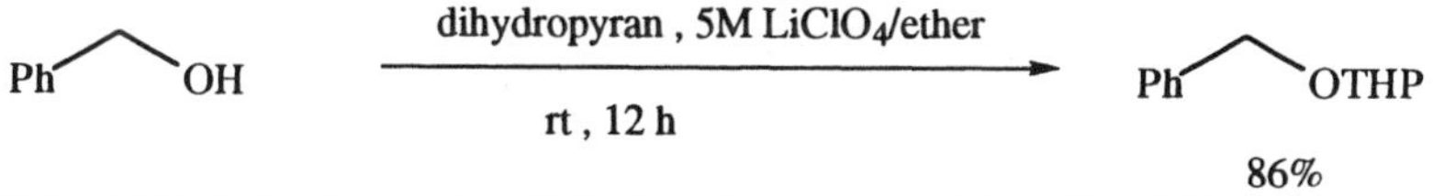
Ph OH
dihydropyran, 5M LiClO4/ether
rt, 12 h
Ph OTHP
86%

Babu, B.S.; Balasubramanian, K.K. *Tetrahedron Lett.*, ***1998***, *39*, 9287.

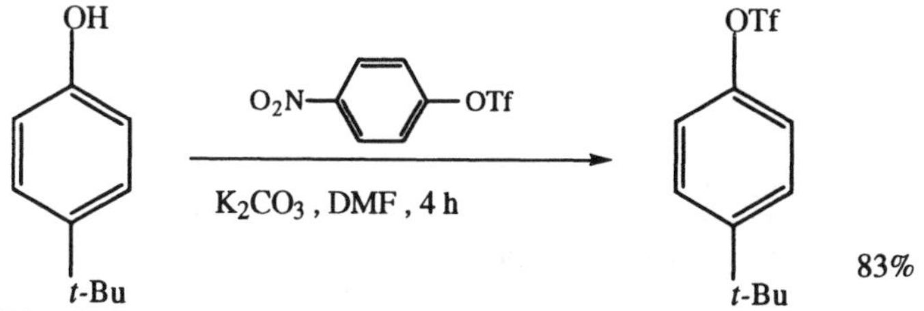
OH
t-Bu
O2N OTf
K2CO3, DMF, 4 h
OTf
t-Bu
83%

Zhu, J.; Bigot, A.; Dau, M.E.T.H. *Tetrahedron Lett.*, ***1997***, *38*, 1181.

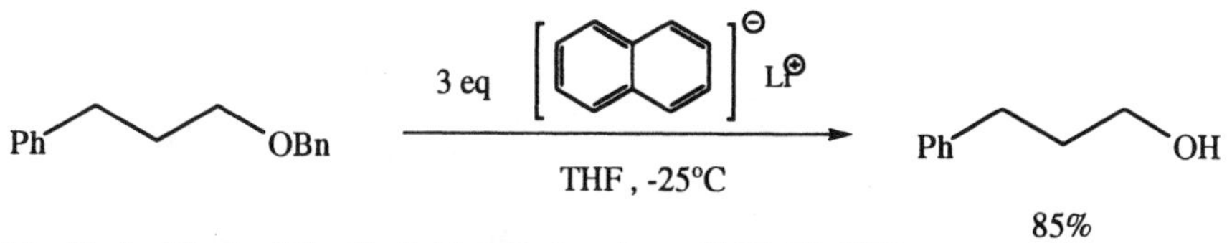
Ph OBn
3 eq
Li
THF, -25°C
Ph OH
85%

Liu, H.-J.; Yip, J.; Shia, K.-S. *Tetrahedron Lett.*, ***1997***, *38*, 2253.

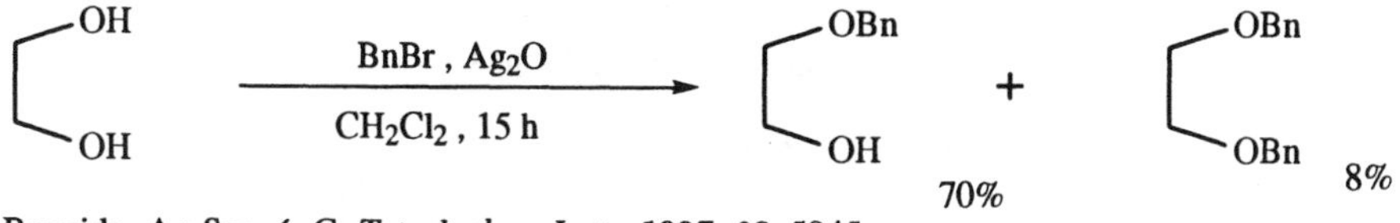
OH
OH
BnBr, Ag2O
CH2Cl2, 15 h
OBn
OH
70%
+
OBn
OBn
8%

Bouzide, A.; Sauvé, G. *Tetrahedron Lett.*, ***1997***, *38*, 5945.

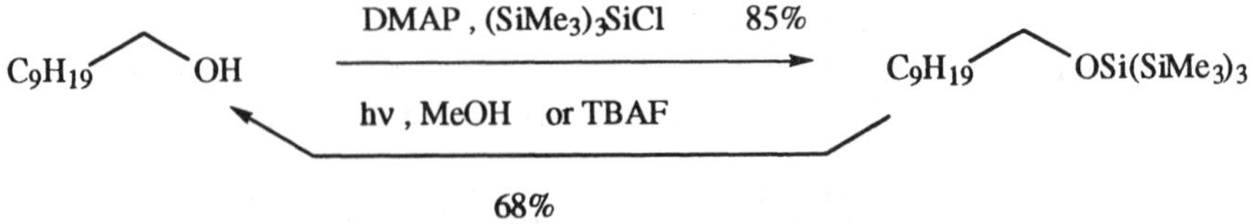
DMAP , (SiMe3)3SiCl 85%
C9H19 OH
C9H19 OSi(SiMe3)3
hv , MeOH or TBAF
68%

Brook, M.A.; Gottardo, C.; Balduzzi, S.; Mohamed, M. *Tetrahedron Lett.*, ***1997***, *38*, 6997.

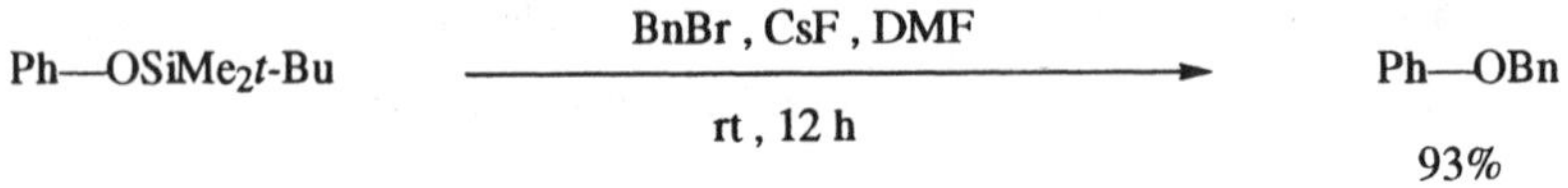
Ph—OSiMe2t-Bu
BnBr , CsF , DMF
rt , 12 h
Ph—OBn
93%

Oriyama, T.; Noda, K.; Yatabe, K. *SynLett.*, ***1997***, 701.

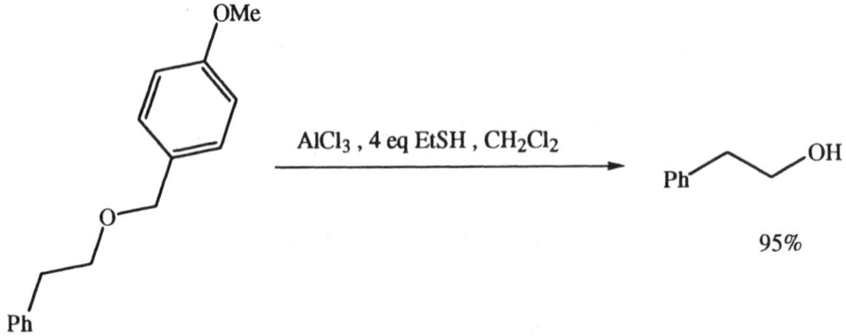
OMe
O
Ph
AlCl3 , 4 eq EtSH , CH2Cl2
Ph
OH
95%

Bouzide, A.; Sauvé, G. *SynLett.*, ***1997***, 1153.

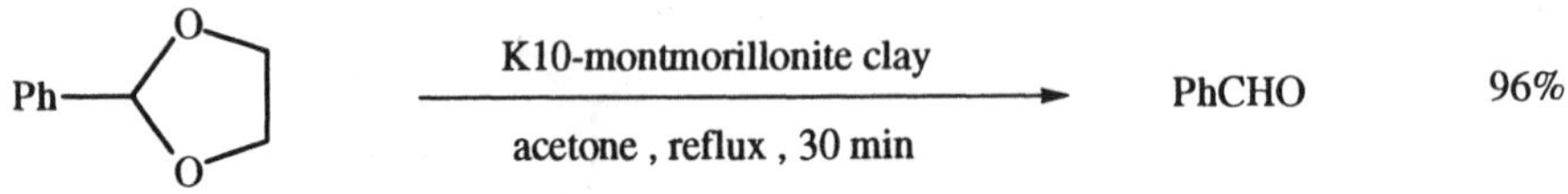
O
Ph
O
K10-montmorillonite clay
acetone , reflux , 30 min
PhCHO
96%

also works with ketals

Li, T.-S.; Li, S.-H. *Synth. Commun.*, ***1997***, *27*, 2299.

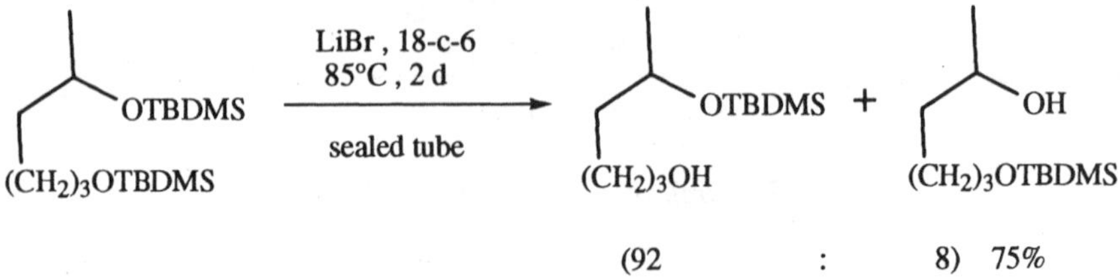
OTBDMS
(CH2)3OTBDMS
LiBr , 18-c-6
85°C , 2 d
sealed tube
OTBDMS
(CH2)3OH
+
OH
(CH2)3OTBDMS
(92 : 8) 75%

Tandon, M.; Begley, T.P. *Synth. Commun.*, ***1997***, *27*, 2953.

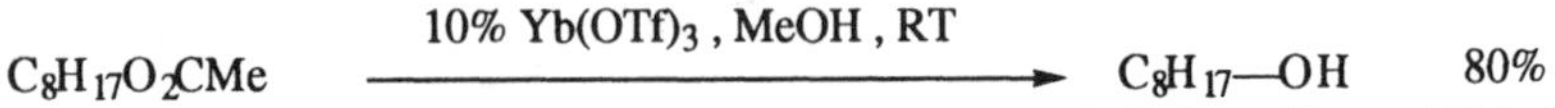
10% Yb(OTf)3 , MeOH , RT
C8H17O2CMe
C8H17—OH
80%

Hanamoto, T.; Sugiomoto, Y.; Yokoyama, Y.; Inanaga, J. *J. Org. Chem.*, ***1996***, *61*, 4491.

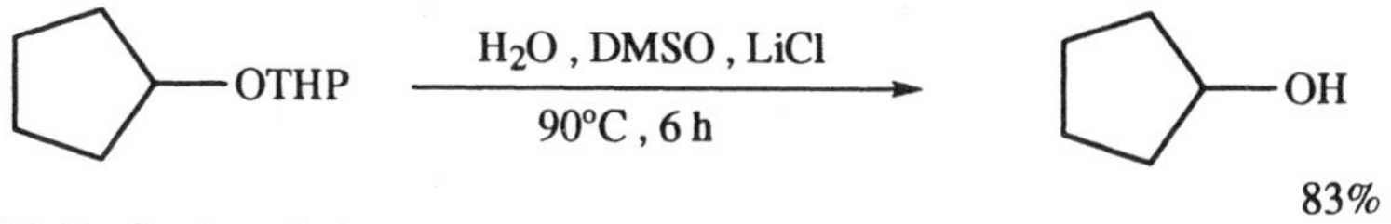
OTHP
H_2O , DMSO , LiCl
90°C , 6 h
OH
83%

Maiti, G.; Roy, S.C. *J. Org. Chem.*, ***1996***, *61*, 6038.

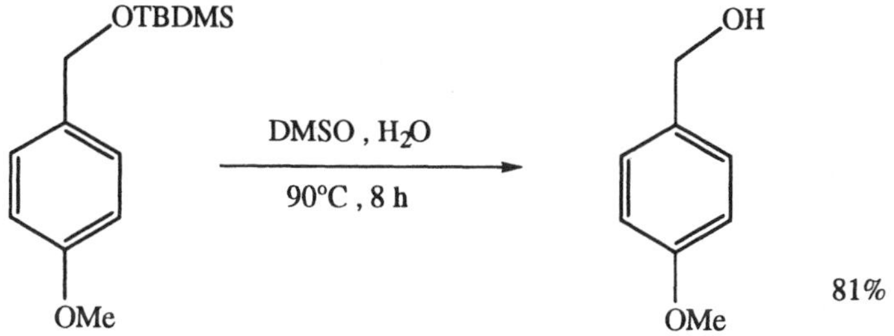
OTBDMS
DMSO , H_2O
90°C , 8 h
OH
81%
OMe
OMe

Maiti, G.; Roy, S.C. *Tetrahedron Lett.*, ***1997***, *38*, 495.

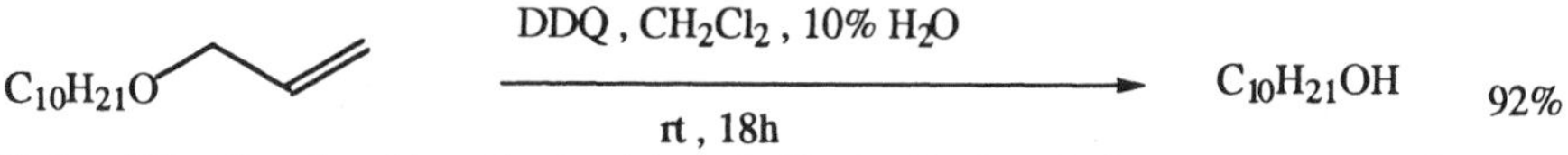
$C_{10}H_{21}O$
DDQ , CH_2Cl_2 , 10% H_2O
rt , 18h
$C_{10}H_{21}OH$
92%

Yadav, J.S.; Chandrasekhar, S.; Sumithra, G.; Kache, R. *Tetrahedron Lett.*, ***1996***, *37*, 6603.

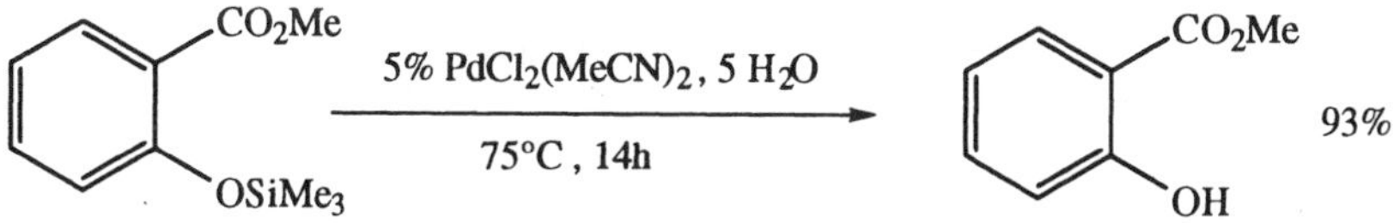
CO_2Me
$OSiMe_3$
5% $PdCl_2(MeCN)_2$, 5 H_2O
75°C , 14h
CO_2Me
OH
93%

Wilson, N.S.; Keay, B.A. *Tetrahedron Lett.*, ***1996***, *37*, 153.

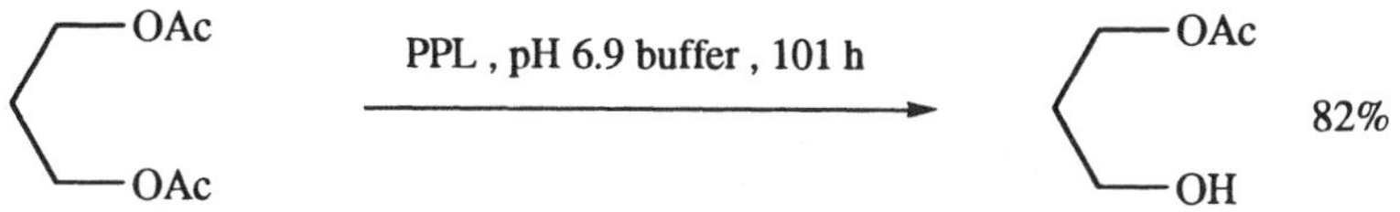
OAc
OAc
PPL , pH 6.9 buffer , 101 h
OAc
OH
82%

Houille, O.; Schmittberger, T.; Uguen, D. *Tetrahedron Lett.*, ***1996***, *37*, 625.

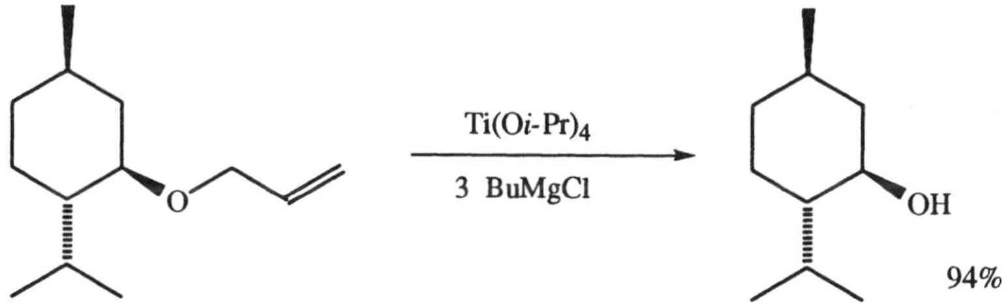
O
Ti(O*i*-Pr)$_4$
3 BuMgCl
OH
94%

Lee, J.; Cha, J.K. *Tetrahedron Lett.*, ***1996***, *37*, 3663.

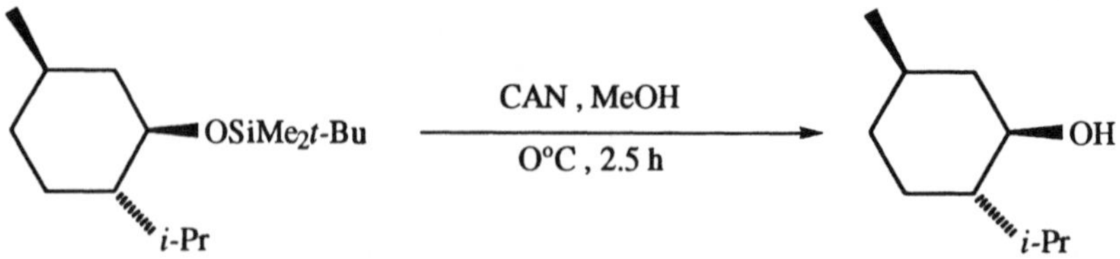

OSiMe$_2$$t$-Bu
i-Pr
CAN , MeOH
0°C , 2.5 h
OH
i-Pr

95%

Datta Gupta, A.; Singh, R.; Singh, V.K. *SynLett.*, ***1996***, 69.

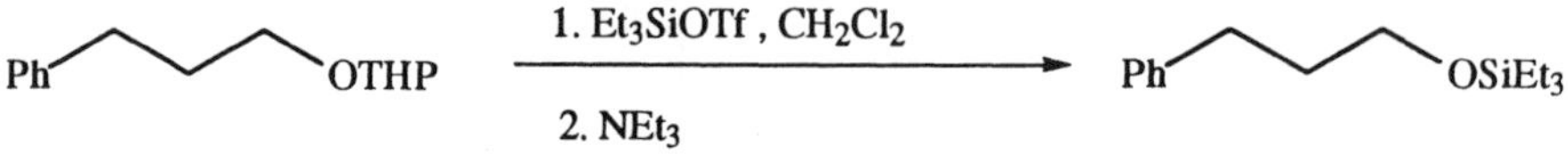

Ph
OTHP
1. Et_3SiOTf , CH_2Cl_2
2. NEt_3
Ph
$OSiEt_3$

77%

Oriyama, T.; Yatabe, K.; Sugawara, S.; Machiguchi, Y.; Koga, G. *SynLett.*, ***1996***, 523.

$PhCH_2OH$ —— dihydropyran , $CuCl_2$ / CH_2Cl_2 ——→ $PhCH_2OTHP$ 83%

Bhalerao, U.T.; Davis, K.J.; Rao, B.V. *Synth. Commun.*, ***1996***, *26*, 3081.

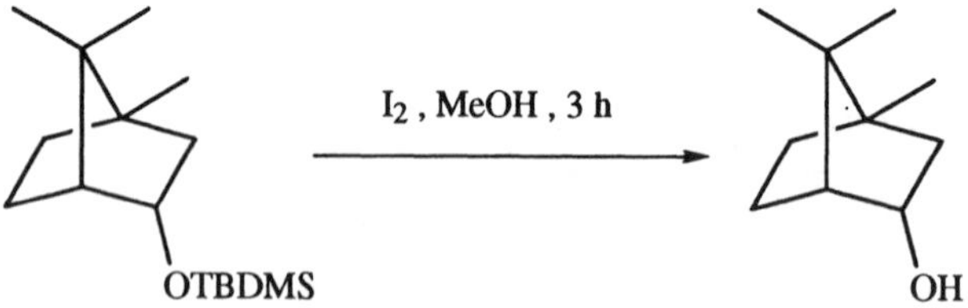

I_2 , MeOH , 3 h
OTBDMS
OH

94%

Vaino, A.R.; Szarek, W. *Chem. Commun.*, ***1996***, 2351.

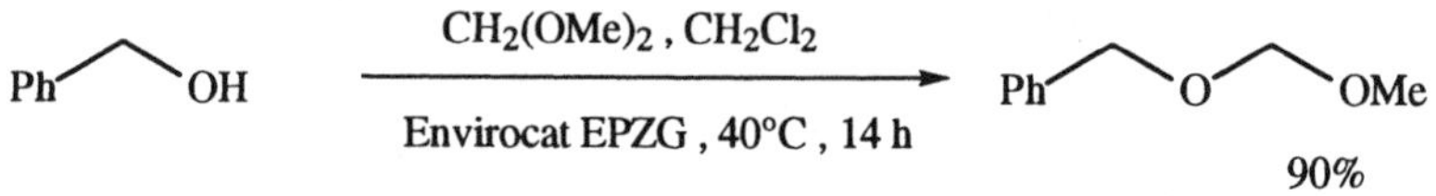

Ph
OH
$CH_2(OMe)_2$, CH_2Cl_2
Envirocat EPZG , 40°C , 14 h
Ph
O
OMe

90%

Bandgar, B.P.; Hajare, C.T.; Wadgaonkar, P.P. *J. Chem. Res. (S)*, ***1996***, 90.

REVIEW:

"Selective Deprotection of Silyl Ethers."

Nelson, T.D.; Crouch, R.D. *Synthesis*, ***1996***, 1031.

CHAPTER 4
PREPARATION OF ALDEHYDES

SECTION 46: ALDEHYDES FROM ALKYNES

NO ADDITIONAL EXAMPLES

SECTION 47: ALDEHYDES FROM ACID DERIVATIVES

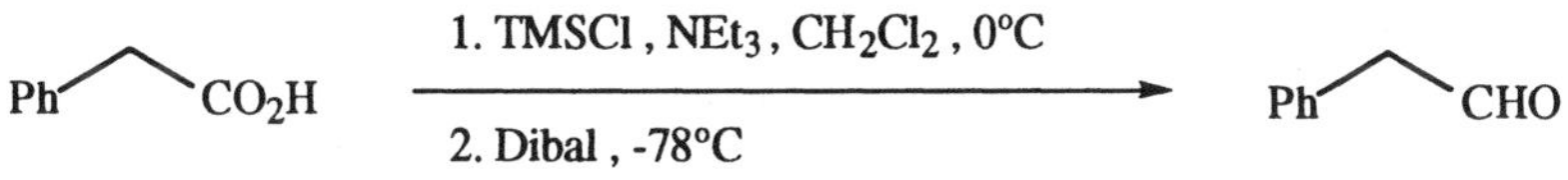

Ph
CO_2H
1. TMSCl , NEt_3 , CH_2Cl_2 , 0°C
2. Dibal , -78°C
Ph
CHO

83%

Chandrasekhar, S.; Kumar, M.S.; Muralidhar, B. *Tetrahedron Lett*, ***1998***, *39*, 909.

$PhCH_2CO_2H$ —— Claycop–H_2O_2 , microwaves , 1 min ——→ PhCHO 83%

Varma, R.S.; Dahiya, R. *Tetrahedron Lett*, ***1998***, *39*, 1307.

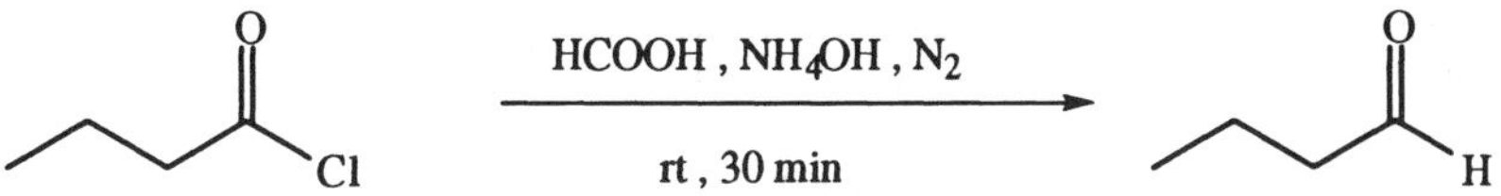

O
Cl
HCOOH , NH_4OH , N_2
rt , 30 min
O
H

90%

Shamsuddin, K.M.; Zubairi, Md.O.; Musharraf, M.A. *Tetrahedron Lett.*, ***1998***, *39*, 8153.

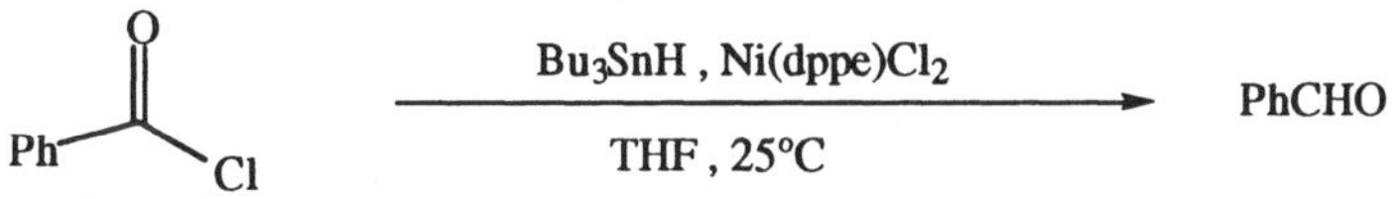

O
Ph
Cl
Bu_3SnH , Ni(dppe)Cl_2
THF , 25°C
PhCHO

90%

Malanga, C.; Mannucci, S.; Lardicci, L. *Tetrahedron Lett*, ***1997***, *38*, 8093.

C_9H_{19}—CO_2H —— 1. heat; 2. Na , EtOH; 3. H^+ ; H_2O ——→ C_9H_{19}—CHO 79%

Shi, Z.; Gu, H. *Synth. Commun.*, ***1997***, *27*, 2701.

SECTION 48: ALDEHYDES FROM ALCOHOLS AND THIOLS

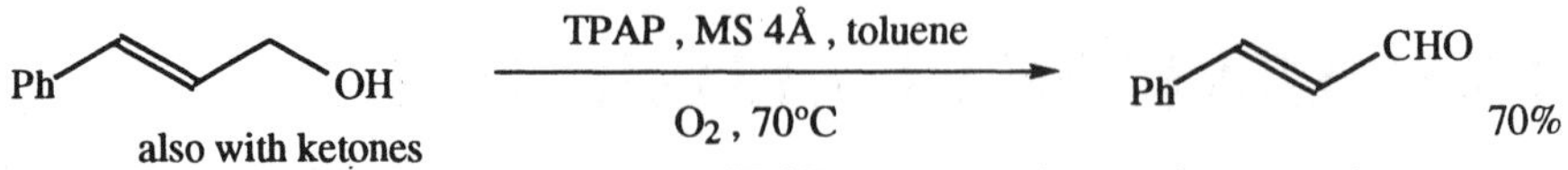

Ph
OH
TPAP , MS 4Å , toluene
O_2 , 70°C
Ph
CHO
70%

also with ketones

TPAP = tetrapropylammonium perruthenate

Mahrwald, R.; Gündogan, B. *J. Am. Chem. Soc.*, ***1998***, *120*, 413.

$PhCH_2OH$ —— 1% [Cn*Ru(CF_3CO_2)$_3$•H_2O], CH_2Cl_2 , TBHP/DCE , rt , 12h ——→ PhCHO

Cn* = N,N',N"-trimethyl-1,4,7-triazacyclononane

88% (70% conversion)

also, with 2° alcohols to ketones

Fung, W.–H.; Yu, W.–Y.; Che, C.–M. *J. Org. Chem.*, ***1998***, *63*, 2873.

$PhCH_2OH$ —— DMCC/alumina , cyclohexane / 30°C , 12 h ——→ PhCHO 93%

DMCC = dimethylammonium dichlorochromate

Zhang, G.–S.; Shi, Q.–Z.; Chen, M.–F.; Cai, K. *Org. Prep. Proceed. Int.*, ***1998***, *30*, 215.

$C_9H_{19}CH_2OH$ —— O_2 , K_2CO_3 , PhF , 5% CuCl / Phen , 5% DBAD , 80°C ——→ $C_9H_{19}CHO$ 65%

Markó, I.E.; Gautier, A.; Chellé-Regnaut, I.; Giles, P.R.; Tsukazaki, M.; Urch, C.J.; Brown, S.M. *J. Org. Chem.*, ***1998***, *63*, 7576.

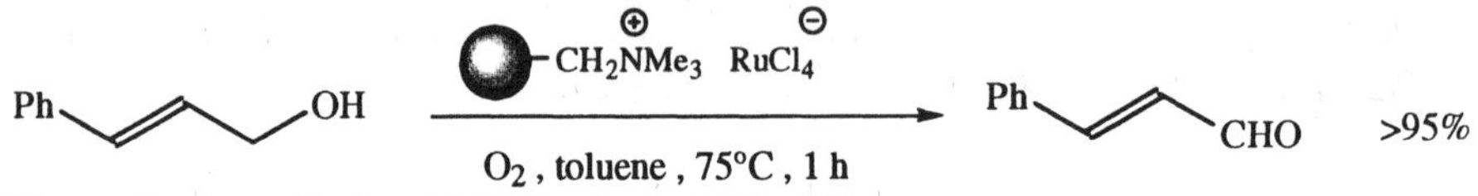

Ph
OH
CH2NMe3 RuCl4
O_2 , toluene , 75°C , 1 h
Ph
CHO
>95%

Hinzen, B.; Lenz, R.; Ley, S.V. *Synthesis*, ***1998***, 977.

$C_7H_{15}OH$ —— (isoquinoline–H)$^+$ $Cr_2O_7^{-2}$ / AcOH , CH_2Cl_2 ——→ $C_6H_{13}CHO$ 64%

Srinivasan, R.; Akila, S.; Caroline, J.; Balasubramanian, K.K.
Synth. Commun., ***1998***, *28*, 2245.

$PhCH_2OH$ —— CeO_2/$RuCl_3$, O_2 , toluene ——→ PhCHO 80%

Vocanson, F.; Guo, Y.P.; Namy, J.–L.; Kagan, H.B. *Synth. Commun.*, ***1998***, *28*, 2577.

$PhCH_2OH$ —— PhI(OAc)$_2$-alumina , microwaves / 2 min ——→ PhCHO 94%

Varma, R.S.; Saini, R.K.; Dahiya, R. *J. Chem. Res. (S)*, ***1998***, 120.

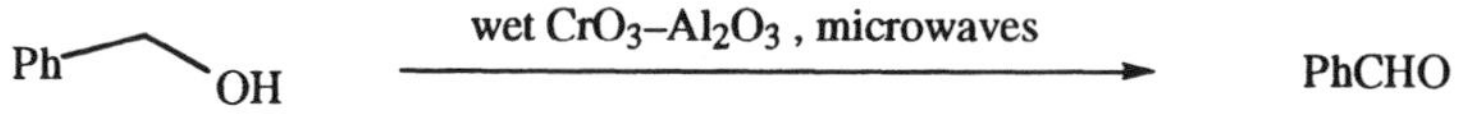
Ph
OH
wet CrO_3–Al_2O_3 , microwaves
PhCHO

76%

Varma, R.S.; Saini, R.K. *Tetrahedron Lett.*, ***1998***, *39*, 1481.

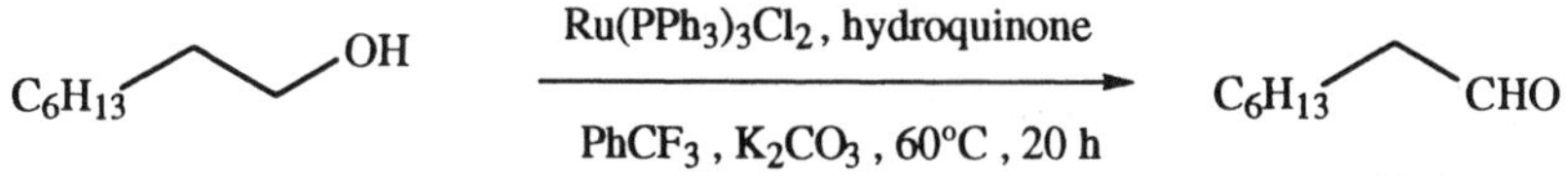
C_6H_{13}
OH
$Ru(PPh_3)_3Cl_2$, hydroquinone
$PhCF_3$, K_2CO_3 , 60°C , 20 h
C_6H_{13}
CHO

90%

Hanyu, A.; Takezawa, E.; Sakaguchi, S.; Ishii, Y. *Tetrahedron Lett.*, ***1998***, *39*, 5557.

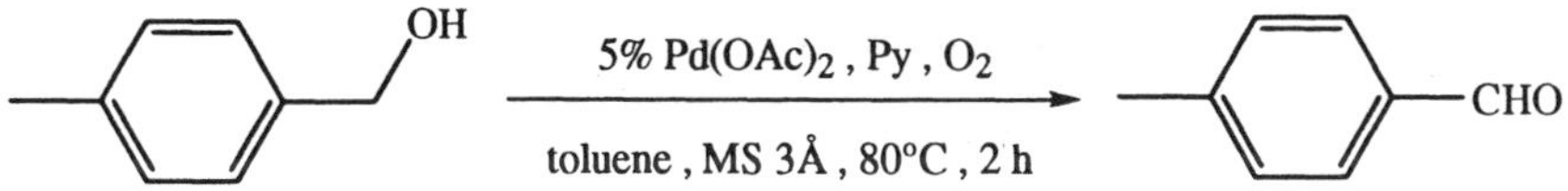
OH
5% $Pd(OAc)_2$, Py , O_2
toluene , MS 3Å , 80°C , 2 h
CHO

can also make ketones

95%

Nishimura, T.; Onoue, T.; Ohe, K.; Uemura, S. *Tetrahedron Lett.*, ***1998***, *39*, 6011.

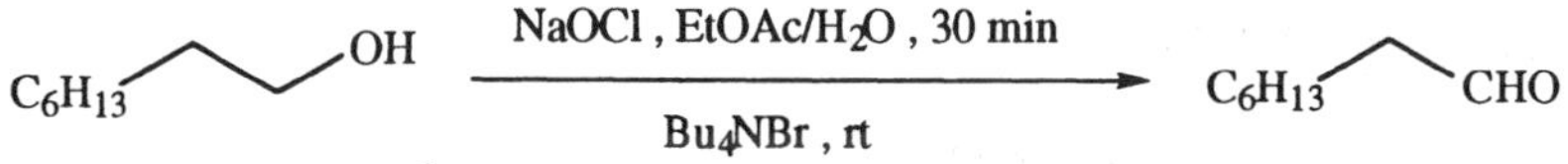
C_6H_{13}
OH
NaOCl , EtOAc/H_2O , 30 min
Bu_4NBr , rt
C_6H_{13}
CHO

can also make ketones

86%

Mirafzal, G.A.; Lozeva, A.M. *Tetrahedron Lett.*, ***1998***, *39*, 7263.

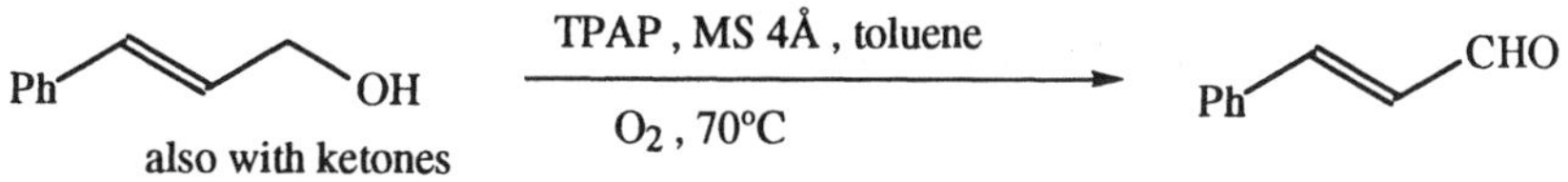
Ph
OH
TPAP , MS 4Å , toluene
O_2 , 70°C
Ph
CHO

70%

also with ketones

TPAP = tetrapropylammonium perruthenate

Markó, I.E.; Giles, P.R.; Tsukazaki, M.; Chellé–Regnaut, I.; Urch, C.J.; Brown, S.M. *J. Am. Chem. Soc.*, ***1997***, *119*, 12661.

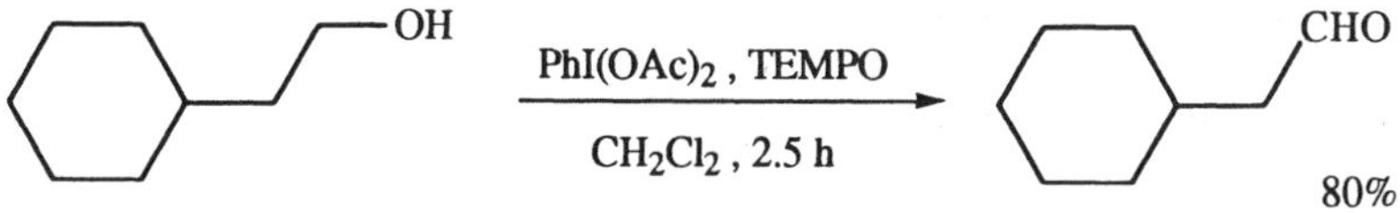
OH
$PhI(OAc)_2$, TEMPO
CH_2Cl_2 , 2.5 h
CHO

80%

DeMico, A.; Margarita, R.; Parlanti, L.; Vescovi, A.; Piancatelli, G. *J. Org. Chem.*, ***1997***, *62*, 6974.

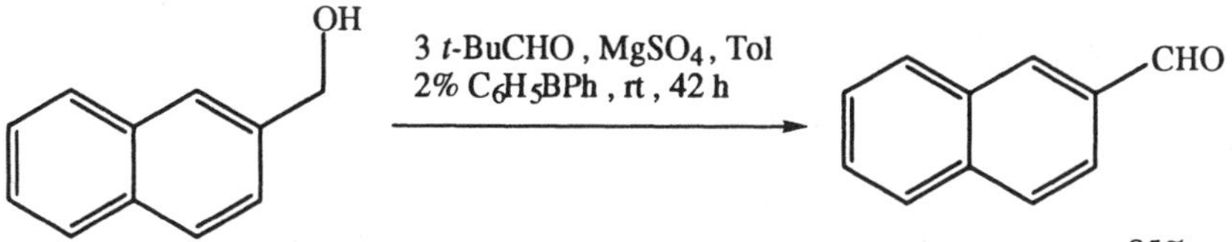
OH
3 *t*-BuCHO , $MgSO_4$, Tol
2% C_6H_5BPh , rt , 42 h
CHO

85%

Ishihara, K.; Kurihara, H.; Yamamoto, H. *J. Org. Chem.*, ***1997***, *62*, 5664.

$PhCH_2OH \xrightarrow{\text{Clayfen, microwaves, 5 sec}} PhCHO$ 92%

also with 2° alcohols

Varma, R.S.; Dahiya, R. *Tetrahedron Lett.*, ***1997***, *38*, 2043.

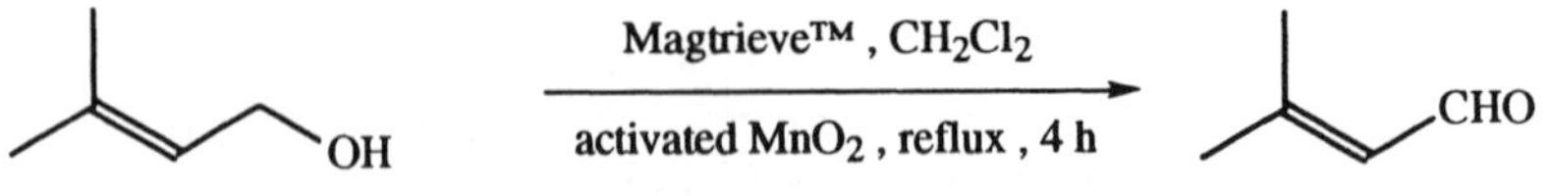

OH
Magtrieve™ , CH2Cl2
activated MnO2 , reflux , 4 h
CHO

Magtrieve = magnetically retrievable Cr oxidant 90%

Lee, R.A.; Donald, D.S. *Tetrahedron Lett.*, ***1997***, *38*, 3857.

$PhCH_2OH \xrightarrow[\text{1 min (no solvent)}]{PhI(OAc)_2\text{–alumina, microwaves}} PhCHO$ 94%

Varma, R.S.; Dahiya, R.; Saini, R.K. *Tetrahedron Lett.*, ***1997***, *38*, 7029.

$PhCH_2OH \xrightarrow{MnO_2{\bullet}SiO_2\text{, microwaves, 20 sec}} PhCHO$ 88%

Varma, R.S.; Saini, R.K.; Dahiya, R. *Tetrahedron Lett.*, ***1997***, *38*, 7823.

$PhCH_2OH \xrightarrow[30°C\text{, 1 h}]{Al_2O_3\text{, }NH_4^+CrO_3Cl^-} PhCHO$ 96%

Zhang, G.-S.; Shi, Q.-Z.; Chen, M.-F.; Cai, K. *Synth. Commun.*, ***1997***, *27*, 953.

$PhCH_2OH \xrightarrow{NH_4^+\ CrO_3Cl^-\text{, }SiO_2\text{, }30°C} PhCHO$ 95%

can also form ketones

Zhang, G.-S.; Shi, Q.-Z.; Chen, M.-F.; Cai, K. *Synth. Commun.*, ***1997***, *27*, 3691.

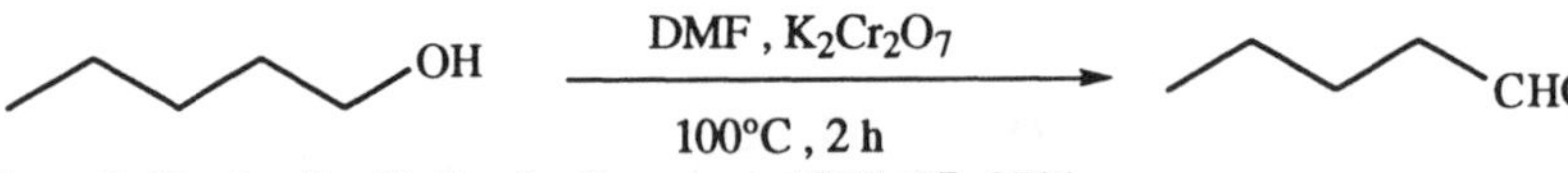

OH
DMF , K2Cr2O7
100°C , 2 h
CHO

Lou, J.-D.; Lu, L.-H. *Synth. Commun.*, ***1997***, *27*, 3701.

$PhCH_2OH \xrightarrow{\text{10\% TPAP, }O_2\text{, }CH_2Cl_2\text{, MS 4Å, rt}} PhCHO$ 98%

also for ketones

Lenz, R.; Ley, S.V. *J. Chem. Soc., Perkin Trans. 1*, ***1997***, 3291.

$PhCH_2OH \xrightarrow[Bu_4NOH]{NbCl_5\text{, PPA, }H_2O_2} PhCHO$ 86%

de Souza Batista, C.M.; de Souza Melo, S.C.; Gelbard, G.; Lachter, E.R. *J. Chem. Res. (S)*, ***1997***, 92.

$PhCH_2OH \xrightarrow[75°C,\ 8\ h]{K_2Cr_2O_7,\ PhH,\ H_2O} PhCHO$ 86%

also for ketones

Lou, J.-D. *J. Chem. Res. (S)*, ***1997***, 206.

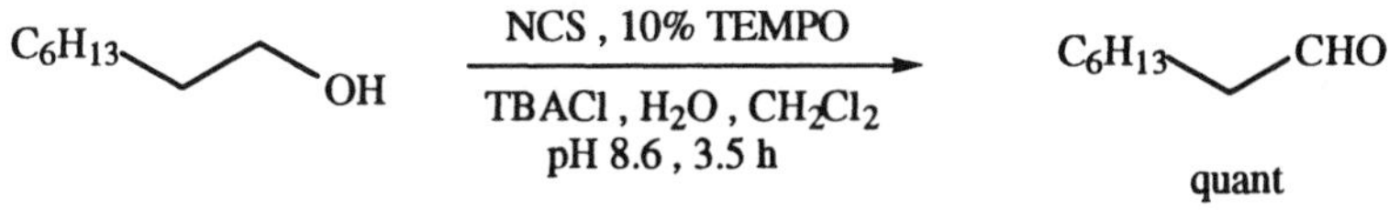
C6H13
OH
NCS , 10% TEMPO
TBACl , H2O , CH2Cl2
pH 8.6 , 3.5 h
C6H13
CHO
quant

Einhorn, J.; Einhorn, C.; Ratajczak, F.; Pierre, J.-L. *J. Org. Chem.*, ***1996***, *61*, 7452.

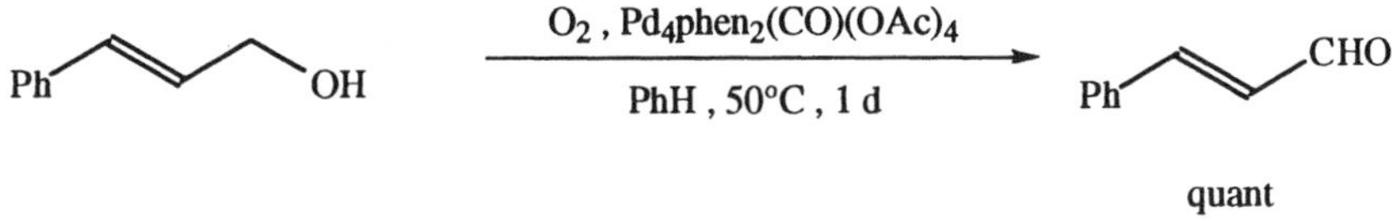
Ph
OH
O2 , Pd4phen2(CO)(OAc)4
PhH , 50°C , 1 d
Ph
CHO
quant

Kaneda, K.; Fujii, M.; Morioka, K. *J. Org. Chem.*, ***1996***, *61*, 4502.

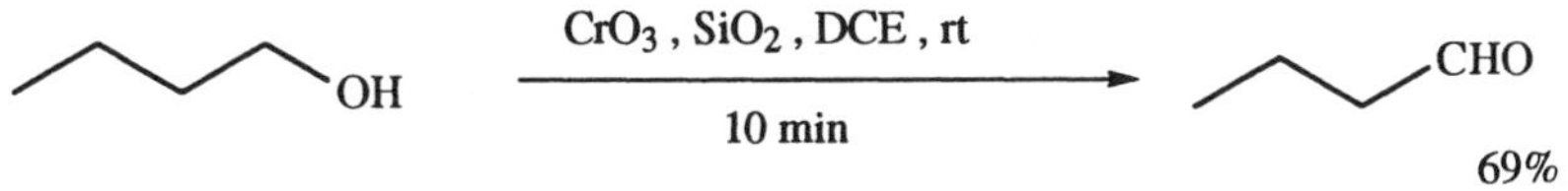
OH
CrO3 , SiO2 , DCE , rt
10 min
CHO
69%

Khadilkar, B.; Chitnavis, A.; Khare, A. *Synth. Commun.*, ***1996***, *26*, 205.

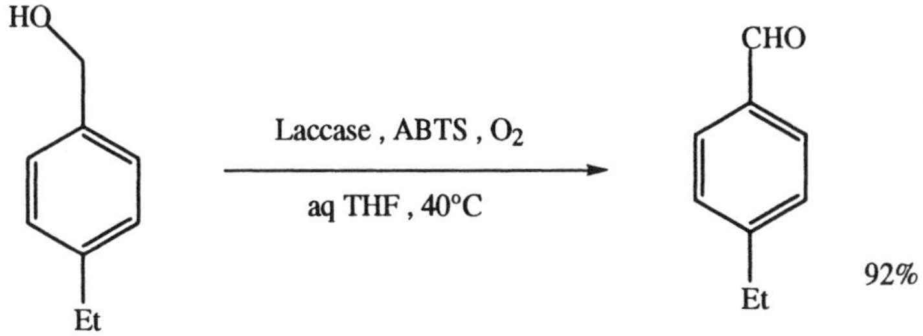
HO
Et
Laccase , ABTS , O2
aq THF , 40°C
CHO
Et
92%

ABTS = 2,2'-azino'*bis*-(3-ethylbenzothiazoline)-6-sulfonic acid

Rosenau, T.; Potthast, A.; Chen, C.L.; Gratzl, J.S. *Synth. Commun.*, ***1996***, *26*, 315.

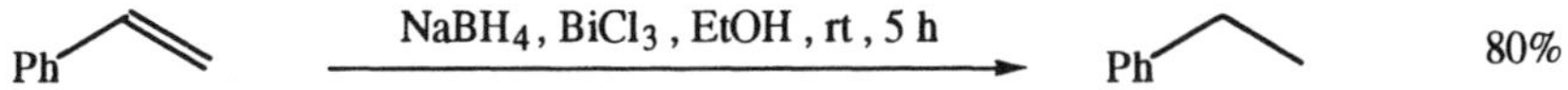
Ph
NaBH4 , BiCl3 , EtOH , rt , 5 h
Ph
80%

Lee, J.G.; Lee, J.A.; Sohn, S.Y. *Synth. Commun.*, ***1996***, *26*, 543.

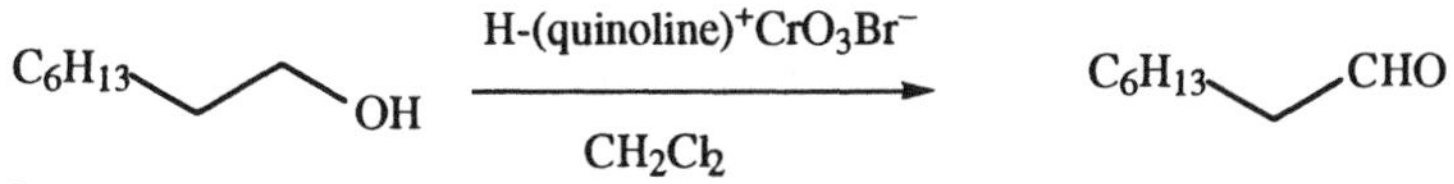
C6H13
OH
H-(quinoline)+CrO3Br−
CH2Cl2
C6H13
CHO

Özgün, B.; Değirmenbaşi, N. *Synth. Commun.*, ***1996***, *26*, 3601.

REVIEW:

"On the Use of Stable Organic Nitroxyl Radicals for the Oxidation of Primary and Secondary Alcohols."

deNooy, A.E.J.; Besemer, A.C.; Bekkum, H. *Synthesis*, ***1996***, 1153.

SECTION 49: ALDEHYDES FROM ALKYNES

Conjugate reductions and Michael Alkylations of conjugated aldehydes are listed in Section 74 (Alkyls from Alkenes).

NO ADDITIONAL EXAMPLES

Related Methods: Aldehydes from Ketones (Section 57)
Ketones from Ketones (Section 177)
Also via: Alkenyl aldehydes (Section 341)

SECTION 50: ALDEHYDES FROM ALKYLS, METHYLENES AND ARYLS

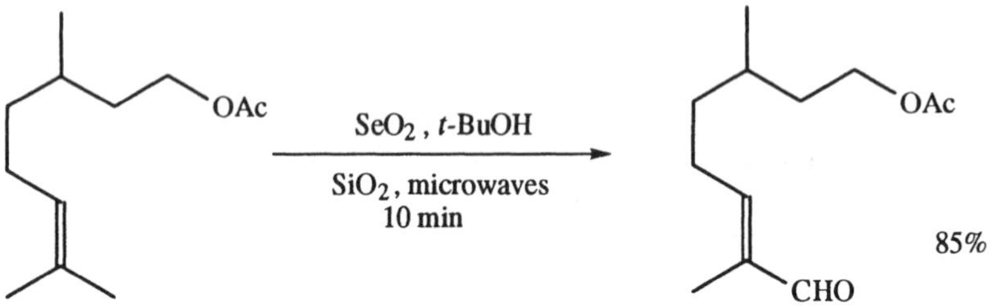
OAc
SeO2 , t-BuOH
SiO2, microwaves
10 min
OAc
85%
CHO

Singh, J.; Sharma, M.; Kad, G.L.; Chhabra, B.R. *J. Chem. Res. (S)*, ***1997***, 264.

SECTION 51: ALDEHYDES FROM AMIDES

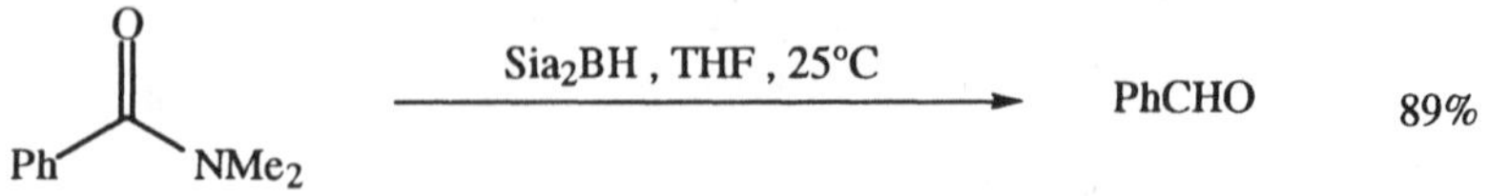
O
Ph
NMe2
Sia2BH , THF , 25°C
PhCHO
89%

Godjoian, G.; Singaram, B. *Tetrahedron Lett.*, ***1997***, *38*, 1717.

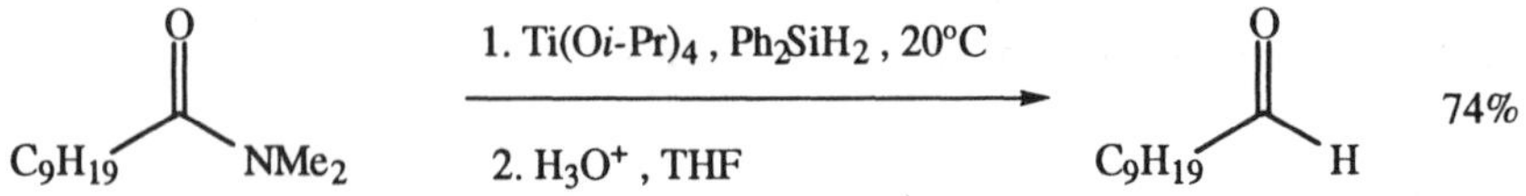
O
C9H19
NMe2
1. Ti(Oi-Pr)4 , Ph2SiH2 , 20°C
2. H3O+ , THF
O
C9H19
H
74%

Bower, S.; Kreutzer, K.A.; Buchwald, S.L. *Angew. Chem. Int., Ed.*, ***1996***, *35*, 1515.

SECTION 52: ALDEHYDES FROM AMINES

NO ADDITIONAL EXAMPLES

Related Methods: Ketones from Amines (Section 172)

SECTION 53: ALDEHYDES FROM ESTERS

NO ADDITIONAL EXAMPLES

SECTION 54: ALDEHYDES FROM ETHERS, EPOXIDES AND THIOETHERS

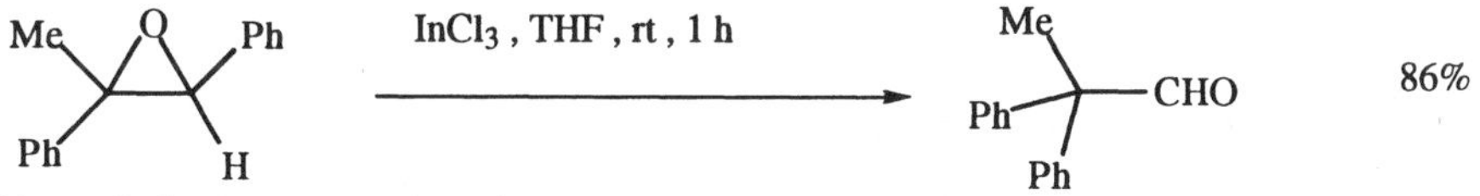
Me
O
Ph
Ph
H
InCl3 , THF , rt , 1 h
Me
Ph
CHO
Ph
86%

Ranu, B.C.; Jana, U. *J. Org. Chem.*, ***1998***, *63*, 8212.

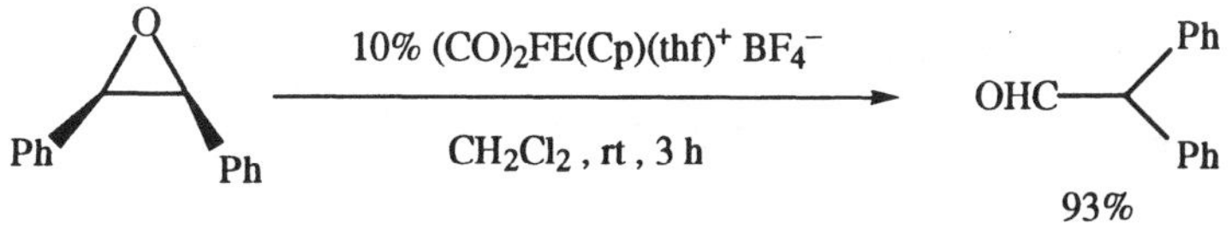
O
Ph
Ph
10% (CO)2FE(Cp)(thf)+ BF4−
CH2Cl2 , rt , 3 h
Ph
OHC
Ph
93%

Picione, J.; Mahmood, S.J.; Gill, A.; Hilliard, M.; Hossain, M.M. *Tetrahedron Lett.*, ***1998***, *39*, 2681.

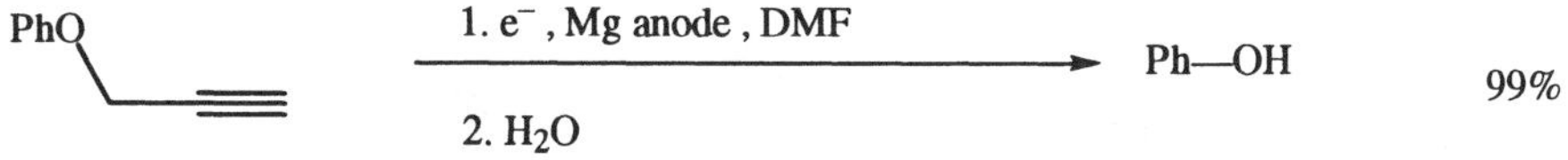
PhO
1. e− , Mg anode , DMF
2. H2O
Ph—OH
99%

Olivero, S.; Duñach, E. *Tetrahedron Lett.*, ***1997***, *38*. 6193.

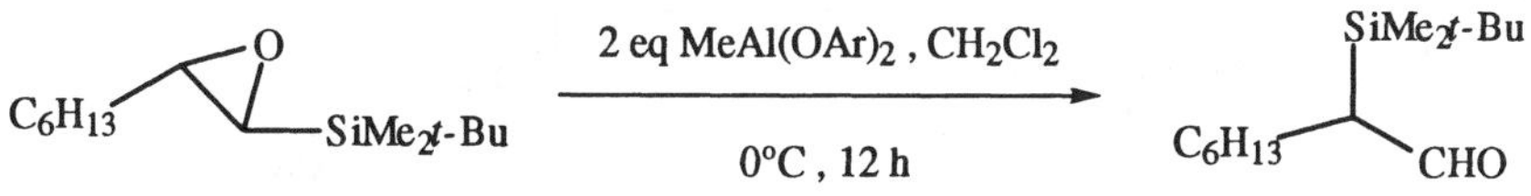
O
C6H13
SiMe2t-Bu
2 eq MeAl(OAr)2 , CH2Cl2
0°C , 12 h
SiMe2t-Bu
C6H13
CHO

does not work with -$SiEt_3$ and yields are poor with $SiPh_3$

79%

Ooi, T.; Kiba, T.; Maruoka, K. *Chem. Lett.*, ***1997***, 519.

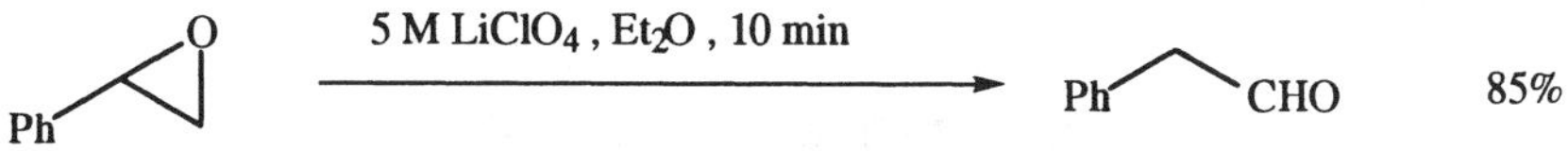
O
Ph
5 M LiClO4 , Et2O , 10 min
Ph
CHO
85%

Sudha, R.; Narasimhan, K.M.; Saraswathy, V.G.; Sankararaman, S. *J. Org. Chem.*, ***1996***, *61*, 1877.

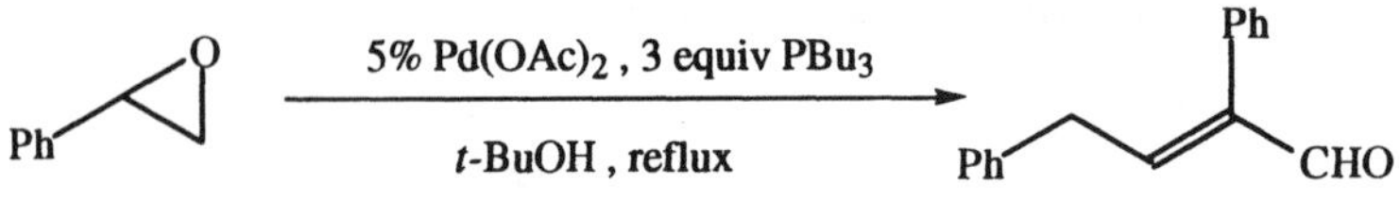
Ph
O
5% Pd(OAc)2 , 3 equiv PBu3
t-BuOH , reflux
Ph
Ph
CHO

79%

Kim, J.–H.; Kulawiec, R.J. *J. Org. Chem.*, ***1996***, *61*, 7656.

Related Methods: Ketones from Ethers and Epoxides (Section 174)

SECTION 55: ALDEHYDES FROM HALIDES AND SULFONATES

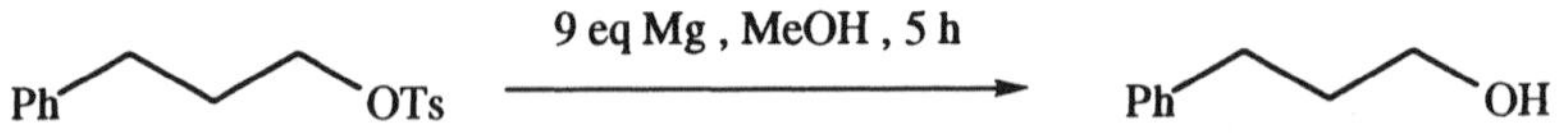
Ph
OTs
9 eq Mg , MeOH , 5 h
Ph
OH

90%

Sridhar, M.; Kumar, B.A.; Narender, R. *Tetrahedron Lett.*, ***1998***, *39*, 2847.

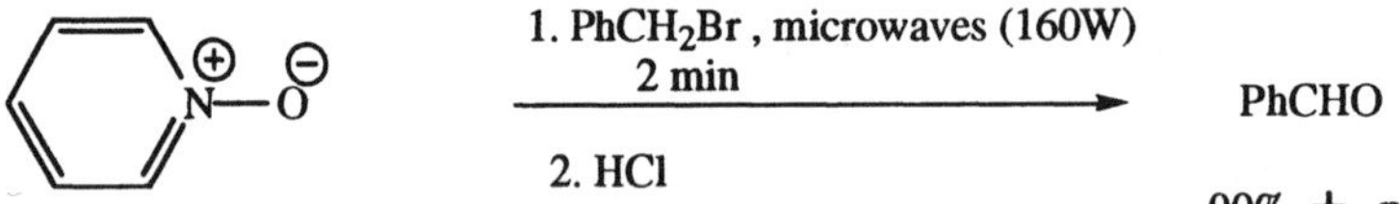
N
O
1. PhCH2Br , microwaves (160W)
2 min
2. HCl
PhCHO

90% + pyridine

Barbry, D.; Champagne, P. *Tetrahedron Lett*, ***1996***, *37*, 7725.

SECTION 56: ALDEHYDES FROM HYDRIDES

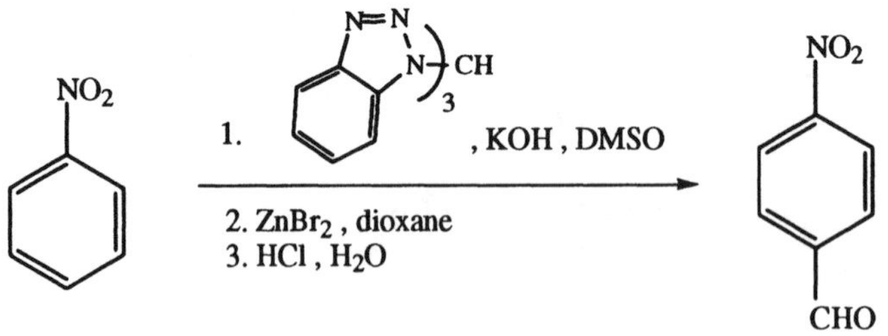
NO2
N=N
N
CH
3
1.
, KOH , DMSO
2. ZnBr2 , dioxane
3. HCl , H2O
NO2
CHO

85x86%

Katritzky, A.R.; Xie, L. *Tetrahedron Lett.* ***1996***, *37*, 347.

SECTION 57: ALDEHYDES FROM KETONES

NO ADDITIONAL EXAMPLES

SECTION 58: ALDEHYDES FROM NITRILES

NO ADDITIONAL EXAMPLES

SECTION 59: ALDEHYDES FROM ALKENES

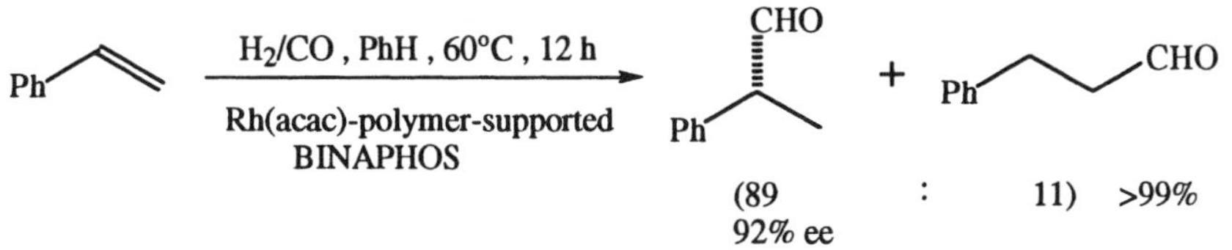
H_2/CO, PhH, 60°C, 12 h
Rh(acac)-polymer-supported BINAPHOS
CHO
Ph
+
Ph
CHO
(89 : 11) >99%
92% ee

Nozaki, K.; Itoi, Y.; Shibahara, F.; Shirakawa, E.; Ohta, T.; Takaya, H.; Hiyama, T. *J. Am. Chem. Soc.*, ***1998***, *120*, 4051.

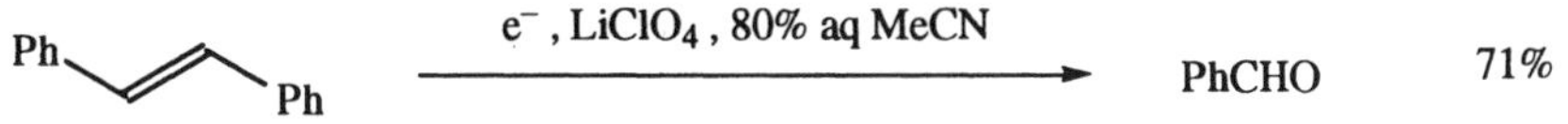
Ph
Ph
e^-, $LiClO_4$, 80% aq MeCN
PhCHO
71%

Maki, S.; Niwa, H.; Hirano, T. *SynLett*, ***1997***, 1385.

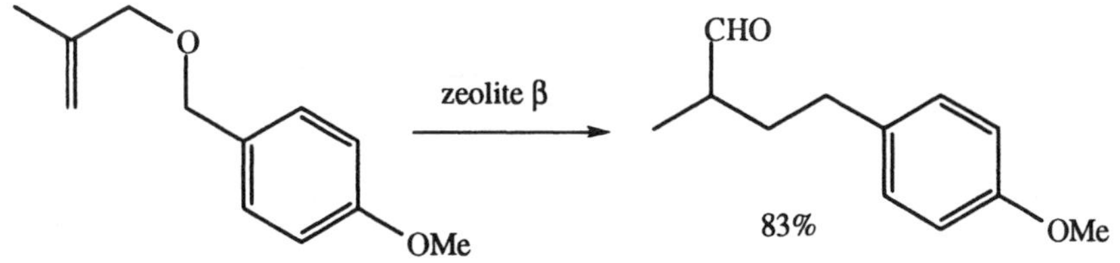
O
OMe
zeolite β
CHO
83%
OMe

Wennerberg, J.; Eklund, L.; Polla, M.; Frejd, T. *Chem. Commun.*, ***1997***, 445.

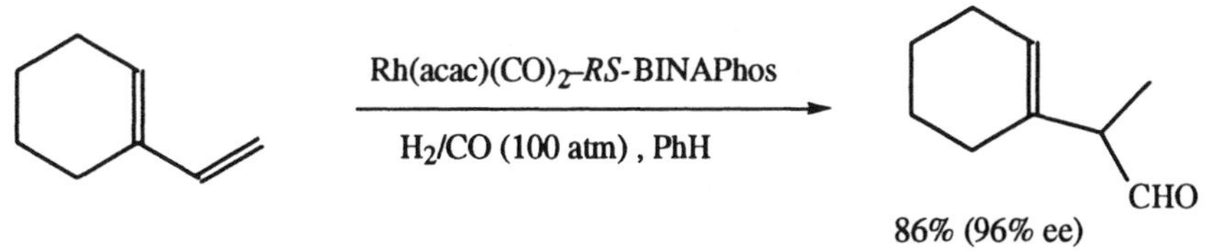
$Rh(acac)(CO)_2$-*RS*-BINAPhos
H_2/CO (100 atm), PhH
CHO
86% (96% ee)

Horiuchi, T.; Ohta, T.; Nozaki, K.; Takaya, H. *Chem. Commun.*, ***1996***, 155.

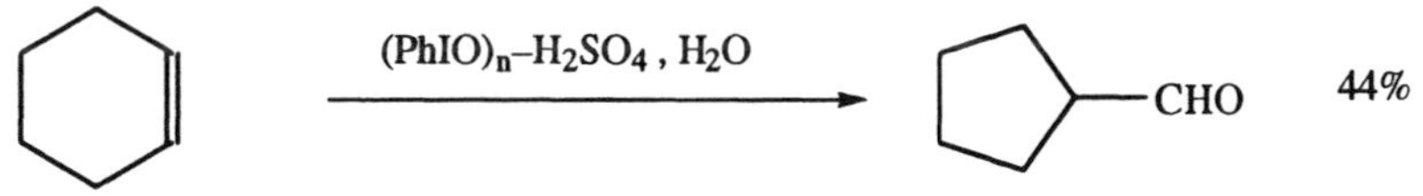
$(PhIO)_n$–H_2SO_4, H_2O
CHO
44%

Moriarty, R.M.; Prakash, O.; Duncan, M.P.; Vaid, R.K.; Rani, N. *J. Chem. Res. (S)*, ***1996***, 432.

Related Methods: Ketones from Alkenes (Section 179)

SECTION 60: ALDEHYDES FROM MISCELLANEOUS COMPOUNDS

$PhCH_2ONO$ —(BF_3, ether)→ PhCHO 97%

Pérez G., C. S.; Pérez, G S.; Zavalas, M.A.; Pérez G., R.M.; Guadarrama M, F.O. *Synth. Commun.*, ***1998***, *28*, 3011.

$PhCH{=}NNMe_2$ —($BiCl_3$, H_2O/THF, 2 min; microwaves)→ PhCHO 98%

Boruah, A.; Baruah, B.; Prajapati, D.; Sandhu, J.S. *SynLett*, ***1997***, 1251.

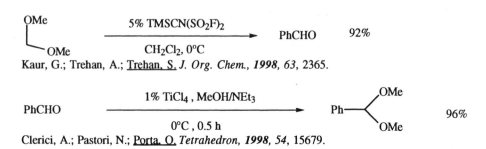

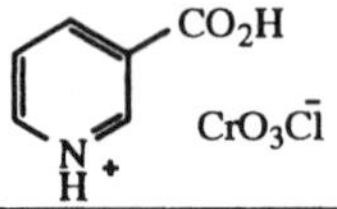

CO_2H
CrO_3Cl^-
N
H

PhCH=N—OH ——→ PhCHO 85%

Baltork, I.M.; Pouranshirvani, Sh. *Synth. Commun.*, ***1996***, *26*, 1.

PhCH=C—NHTs —($Tl(NO_3)_3$, MeOH, rt)→ PhCHO 90%

Wang, J.; Lin, J.; Zheng, Y.; Huang, J. *Synth. Commun.*, ***1997***, *27*, 2583.

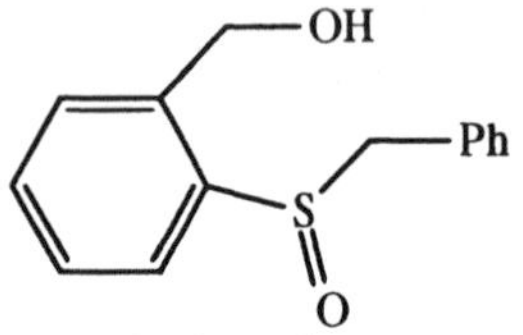

OH
Ph
S
O

—(0.1 TsOH•H_2O, MeCN, 90°C)→ PhCHO 65%

Naka, H.; Sato, S.; Horn, E.; Furukawa, N. *Heterocycles*, ***1997***, *46*, 177.

SECTION 60A: PROTECTION OF ALDEHYDES

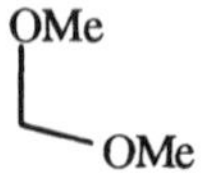

OMe
OMe

—(5% $TMSCN(SO_2F)_2$; CH_2Cl_2, 0°C)→ PhCHO 92%

Kaur, G.; Trehan, A.; Trehan, S. *J. Org. Chem.*, ***1998***, *63*, 2365.

PhCHO —(1% $TiCl_4$, MeOH/NEt_3; 0°C, 0.5 h)→

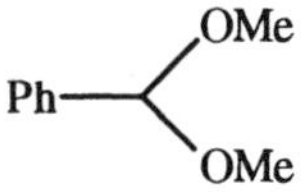

Ph
OMe
OMe

96%

Clerici, A.; Pastori, N.; Porta, O. *Tetrahedron*, ***1998***, *54*, 15679.

$PhCHO(OAc)_2$ —[$FeSO_4$, toluene, heat, 10min]→ PhCHO 98%

Jim. T–S.; Ma, Y.–R.; Zhang, Z.–H.; Li, T.–S. *Org. Prep. Proceed. Int.*, ***1998***, *30*, 463.

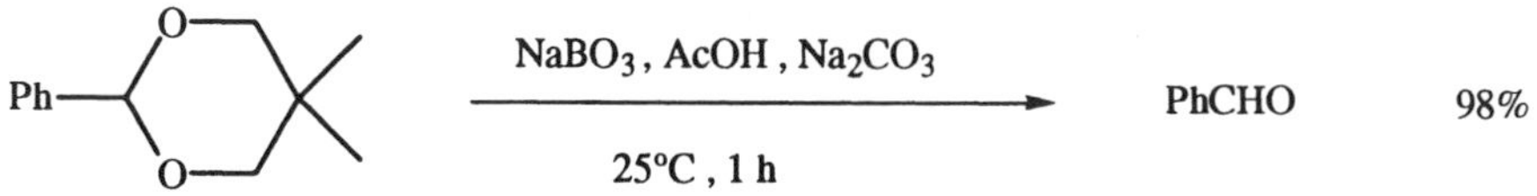
Ph
O
O
$NaBO_3$, AcOH, Na_2CO_3
25°C, 1 h
PhCHO
98%

Bandgar, B.P.; Kulkarni, S.A.; Nigal, J.N. *Org. Prep. Proceed. Int.*, ***1998***, *30*, 706.

$PhCH_2OTHP$ —[bipy-$HCrO_3Cl$, MeCN, reflux]→ PhCHO 93%

Mohammadpoor-Baltork, I.; Kharamesh, B. *J. Chem. Soc. (S)*, ***1998***, 146.

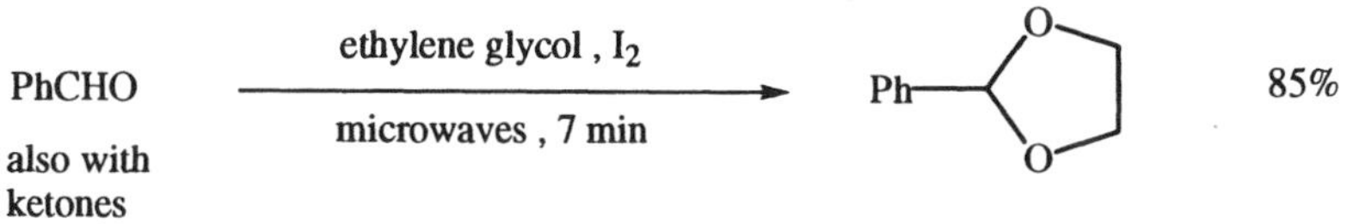
PhCHO
also with
ketones
ethylene glycol, I_2
microwaves, 7 min
Ph
O
O
85%

Kalita, D.J.; Borah, R.; Sarma, J.C. *Tetrahedron Lett.*, ***1998***, *39*, 4573.

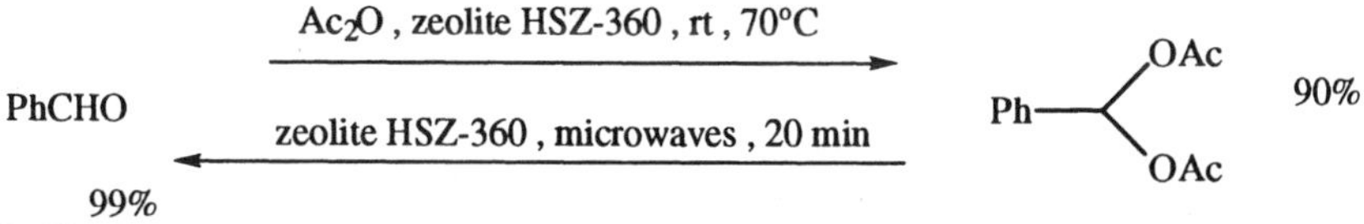
PhCHO
Ac_2O, zeolite HSZ-360, rt, 70°C
zeolite HSZ-360, microwaves, 20 min
99%
Ph
OAc
OAc
90%

Ballini, R.; Bordoni, M.; Bosica, G.; Maggi, R.; Sartori, G. *Tetrahedron Lett*, ***1998***, *39*, 7587

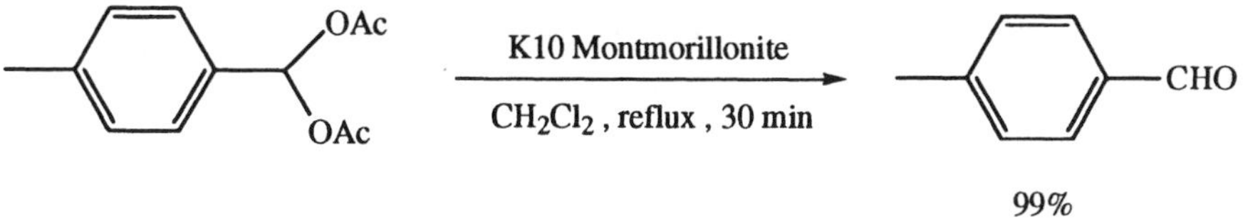
OAc
OAc
K10 Montmorillonite
CH_2Cl_2, reflux, 30 min
CHO
99%

Li, T.S.; Zhang, Z.H.; Fu, C.–G. *Tetrahedron Lett.*, ***1997***, *38*, 3285.

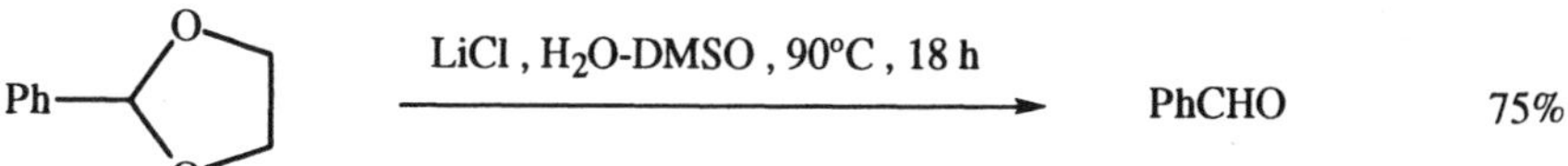
Ph
O
O
LiCl, H_2O-DMSO, 90°C, 18 h
PhCHO
75%

Mandal, P.K.; Dutta, P.; Roy, S.C. *Tetrahedron Lett.*, ***1997***, *38*, 7271.

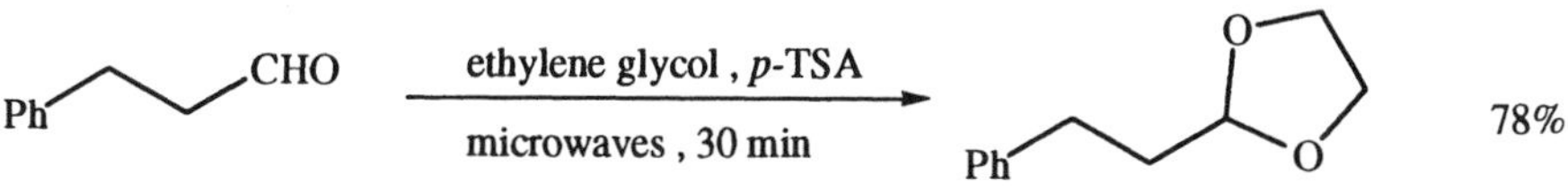
Ph
CHO
ethylene glycol, *p*-TSA
microwaves, 30 min
Ph
O
O
78%

Pério, B.; Dozias, M.–J.; Jacqualt, P.; Hamelin, J. *Tetrahedron Lett.*, ***1997***, *38*, 7867.

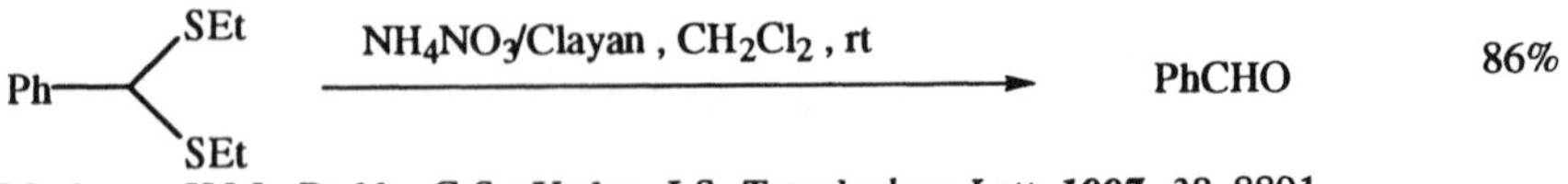
SEt
Ph
SEt
NH4NO3/Clayan , CH2Cl2 , rt
PhCHO
86%

Meshram, H.M.; Reddy, G.S.; Yadav, J.S. *Tetrahedron Lett*, ***1997***, *38*, 8891.

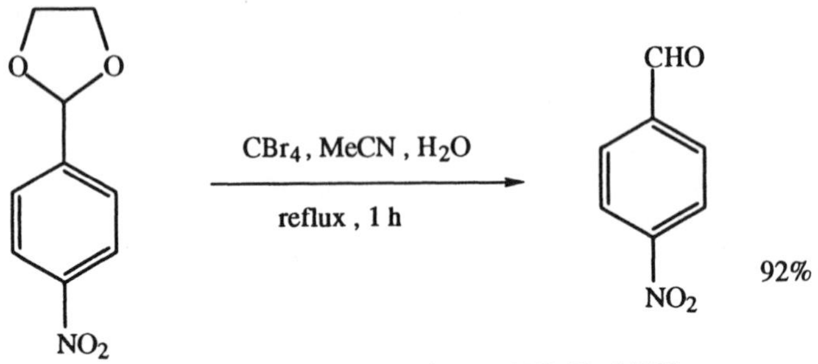
O
O
NO2
CBr4, MeCN , H2O
reflux , 1 h
CHO
NO2
92%

Lee, A.S.-Y.; Cheng, C.-L. *Tetrahedron*, ***1997***, *53*, 14255.

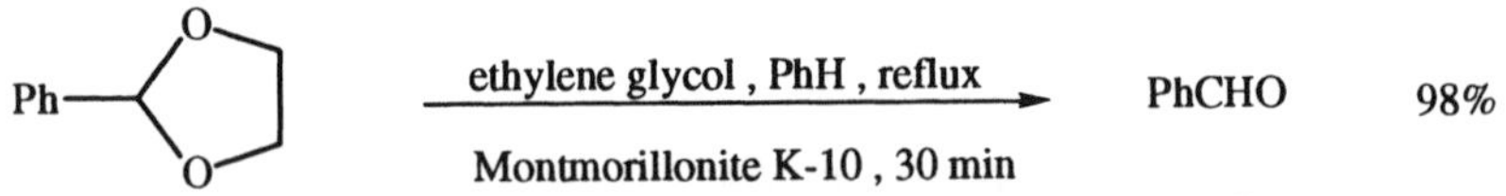
Ph
O
O
ethylene glycol , PhH , reflux
Montmorillonite K-10 , 30 min
PhCHO
98%

Li, T.-S.; Li, S.-H.; Li, J.-T.; Li, H.-Z. *J. Chem. Res. (S)*, ***1997***, 26.

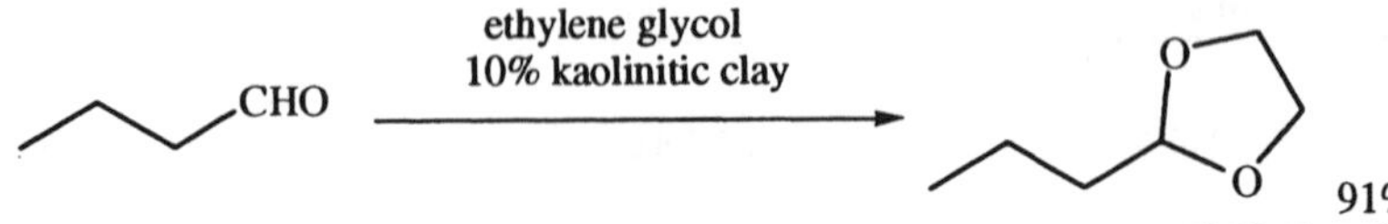
CHO
ethylene glycol
10% kaolinitic clay
O
O
91%

Ponde, D.; Borate, H.B.; Sudalai, A.; Ravindranathan, T.; Deshpande, V.H. *Tetrahedron Lett.*, ***1996***, *37*, 4605.

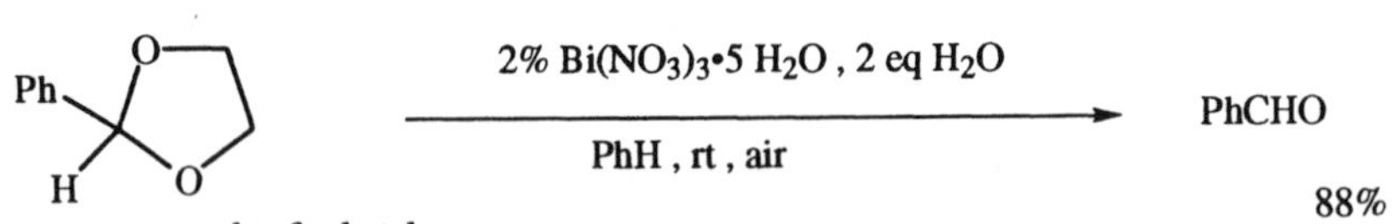
Ph
H
O
O
2% Bi(NO3)3•5 H2O , 2 eq H2O
PhH , rt , air
PhCHO
88%

also for ketals

Komatsu, N.; Tankguchi, A.; Uda, M.; Suzuki, H. *Chem. Commun*, ***1996***, 1847.

CHAPTER 5

PREPARATION OF ALKYLS, METHYLENES AND ARYLS

This chapter lists the conversion of functional groups into methyl, ethyl, propyl, etc. as well as methylene (CH_2), phenyl, etc.

SECTION 61: ALKYLS, METHYLENES AND ARYLS FROM ALKYNES

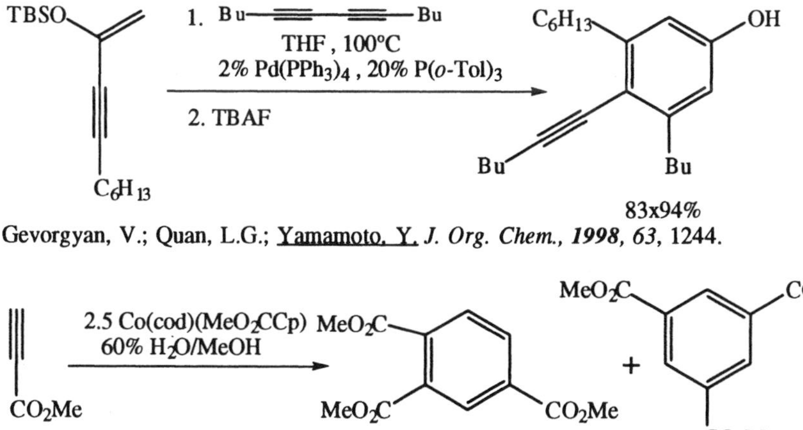

TBSO
C_6H_{13}
1. Bu—≡—≡—Bu
THF, 100°C
2% $Pd(PPh_3)_4$, 20% P(o-Tol)$_3$
2. TBAF
C_6H_{13}
OH
Bu
Bu

83x94%

Gevorgyan, V.; Quan, L.G.; Yamamoto, Y. *J. Org. Chem.*, ***1998***, *63*, 1244.

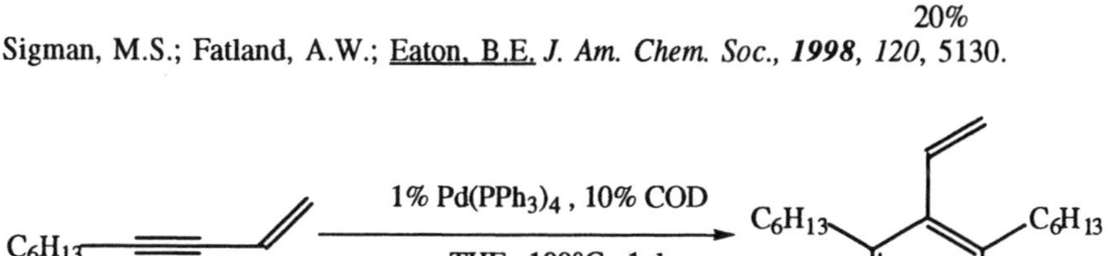

CO_2Me
2.5 Co(cod)(MeO_2CCp)
60% H_2O/MeOH
MeO_2C
MeO_2C
CO_2Me
+
MeO_2C
CO_2Me
CO_2Me

47%

20%

Sigman, M.S.; Fatland, A.W.; Eaton, B.E. *J. Am. Chem. Soc.*, ***1998***, *120*, 5130.

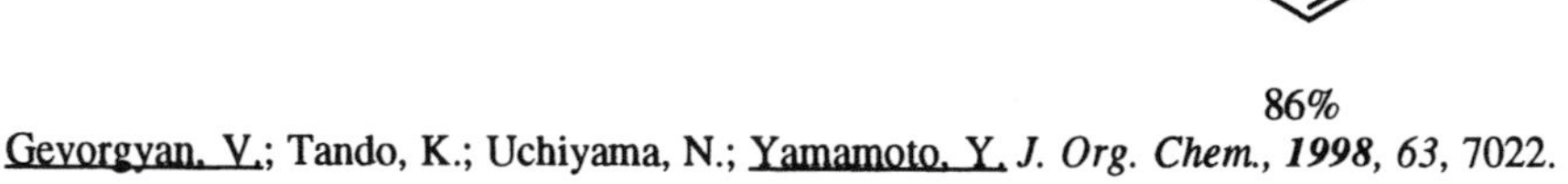

C_6H_{13}
1% $Pd(PPh_3)_4$, 10% COD
THF, 100°C, 1 d
C_6H_{13}
C_6H_{13}

86%

Gevorgyan, V.; Tando, K.; Uchiyama, N.; Yamamoto, Y. *J. Org. Chem.*, ***1998***, *63*, 7022.

2.1 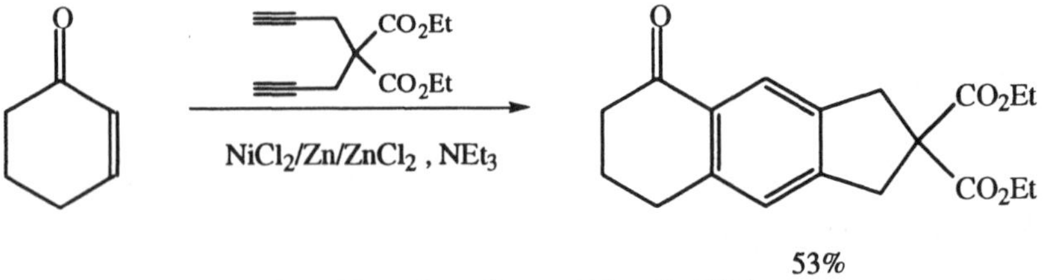
O
CO2Et
CO2Et
NiCl2/Zn/ZnCl2 , NEt3
O
CO2Et
CO2Et
53%

Ikeda, S.-i.; Watanabe, H.; Sato, Y. *J. Org. Chem.*, ***1998***, *63*, 7026.

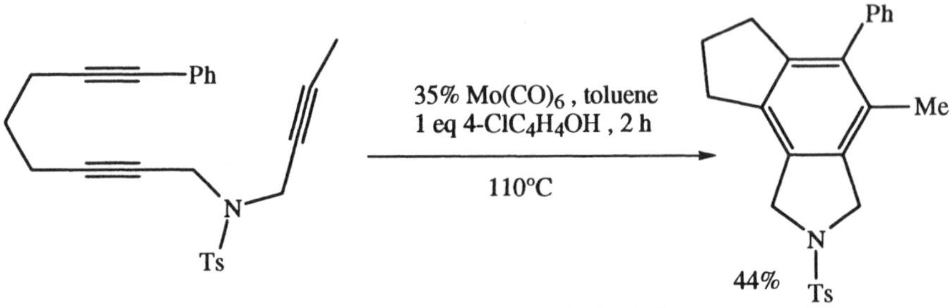
Ph
N
Ts
35% Mo(CO)6 , toluene
1 eq 4-ClC4H4OH , 2 h
110°C
Ph
Me
N
Ts
44%

Nishida, M.; Shiga, H.; Mori, M. *J. Org. Chem.*, ***1998***, *63*, 8606.

3 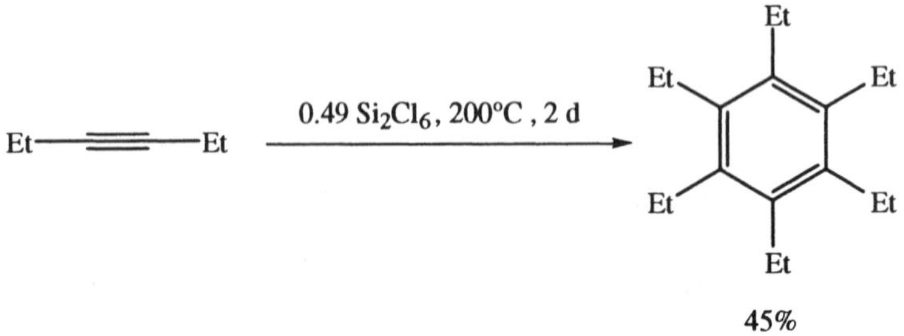
Et
Et
0.49 Si2Cl6, 200°C , 2 d
Et
Et
Et
Et
Et
Et
45%

Yang, J.; Verkade, J.G. *J. Am. Chem. Soc.*, ***1998***, *120*, 6834.

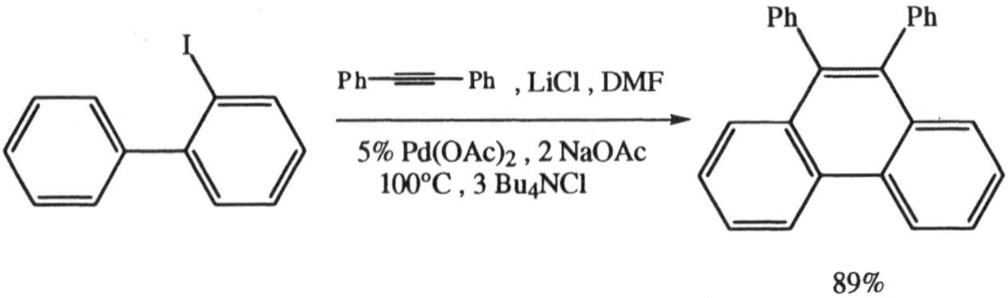
I
Ph
Ph , LiCl , DMF
5% Pd(OAc)2 , 2 NaOAc
100°C , 3 Bu4NCl
Ph
Ph
89%

Larock, R.C.; Doty, M.J.; Tian, Q.; Zenner, J.M. *J. Org. Chem.*, ***1997***, *62*, 7536.

Ph
NiCl2•2 H2O , Li , cat naphthalene
THF , rt
Ph
70%

Alonso, F.; Yus, M. *Tetrahedron Lett.*, ***1997***, *38*, 149.

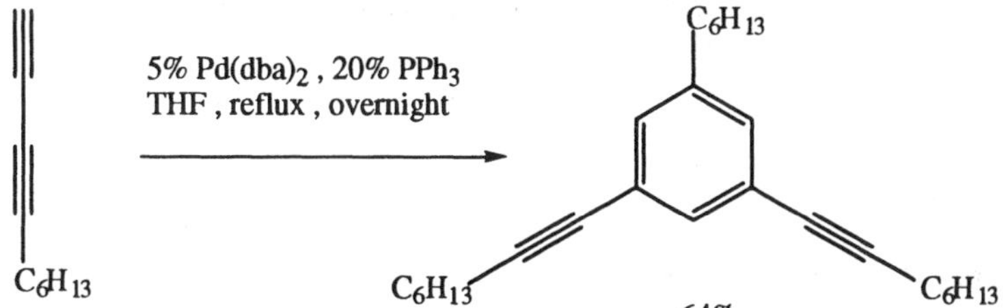

5% Pd(dba)$_2$, 20% PPh$_3$
THF , reflux , overnight
C$_6$H$_{13}$
C$_6$H$_{13}$
C$_6$H$_{13}$
C$_6$H$_{13}$

64%

Takeda, A.; Ohno, A.; Kadota, I.; Gevorgyan, V.; Yamamoto, Y.
J. Am. Chem. Soc., ***1997***, *119*, 4547.

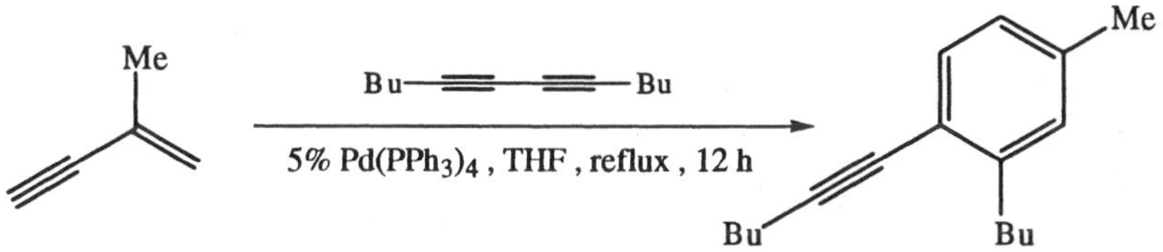

Me
Bu——≡——≡——Bu
5% Pd(PPh$_3$)$_4$, THF , reflux , 12 h
Me
Bu
Bu

89%

Gevorgyan, V.; Takeda, A.; Yamamoto, Y. *J. Am. Chem. Soc.*, ***1997***, *118*, 11313.

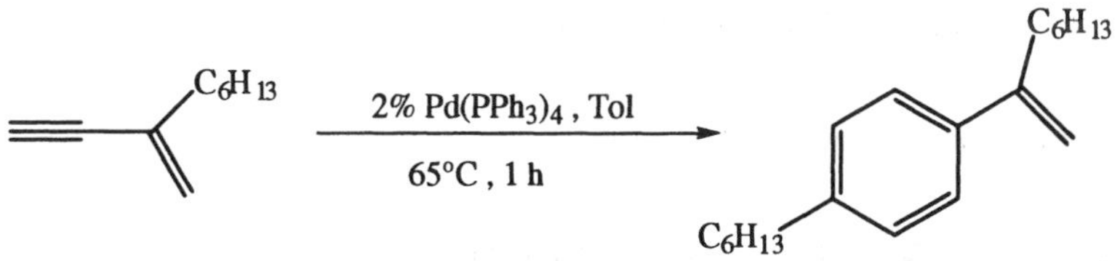

C$_6$H$_{13}$
2% Pd(PPh$_3$)$_4$, Tol
65°C , 1 h
C$_6$H$_{13}$
C$_6$H$_{13}$

77%

Saito, S.; Salter, M.M.; Gevorgyan, V.; Tsuboya, N.; Tando, K.; Yamamoto, Y.
J. Am. Chem. Soc., ***1996***, *118*, 3970.

SECTION 62: ALKYLS, METHYLENES AND ARYLS FROM ACID DERIVATIVES

NO ADDITIONAL EXAMPLES

SECTION 63: ALKYLS, METHYLENES AND ARYLS FROM ALCOHOLS AND THIOLS

NO ADDITIONAL EXAMPLES

SECTION 64: ALKYLS, METHYLENES AND ARYLS FROM ALDEHYDES

PhCHO $\xrightarrow[25°C, 14 h]{PhH, Me_2SiClH, 5\% InCl_3}$ $PhCH_2Ph$ 79%

Miyami, T.; Onishi, Y.; Baba, A. *Tetrahedron Lett.*, ***1998***, *39*, 6291.

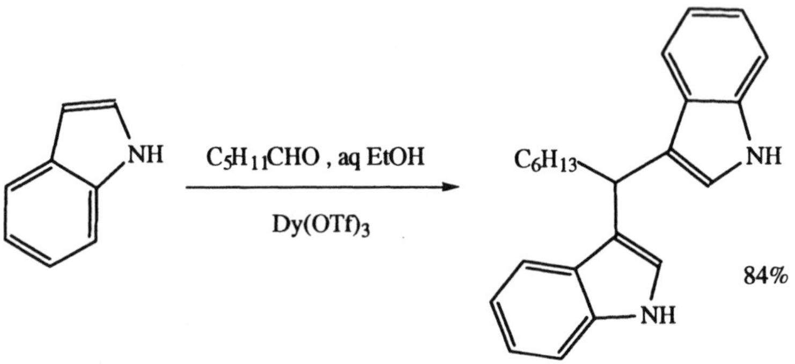
NH
$C_5H_{11}CHO$, aq EtOH
$Dy(OTf)_3$
C_6H_{13}
NH
NH
84%

Chen, D.; Yu, L.; Wang, P.G. *Tetrahedron Lett.*, ***1996***, *37*, 4467.

Related Methods: Alkyls, Methylenes and Aryls from Ketones (Section 72)

SECTION 65: ALKYLS, METHYLENES AND ARYLS FROM ALKYLS, METHYLENES AND ARYLS

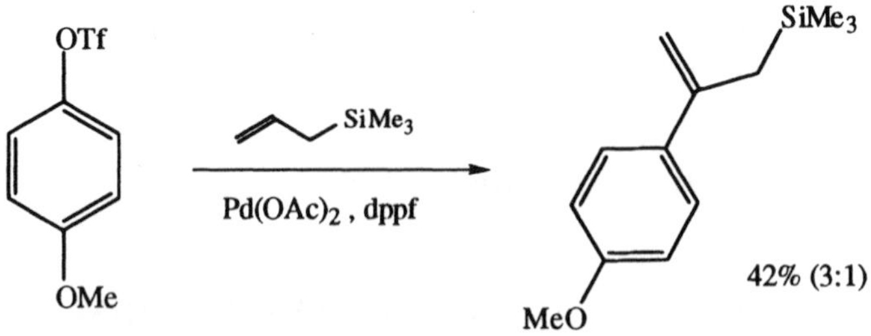
OTf
OMe
$SiMe_3$
$Pd(OAc)_2$, dppf
$SiMe_3$
MeO
42% (3:1)

Olofsson, K.; Larhed, M.; Hallberg, A. *J. Org. Chem.*, ***1998***, *63*, 5076.

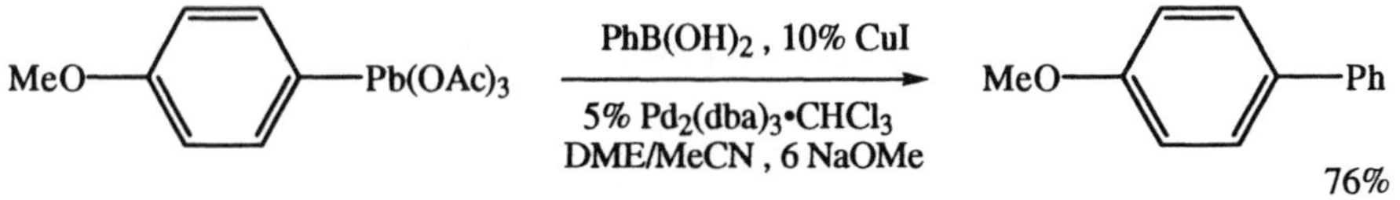
MeO
$Pb(OAc)_3$
$PhB(OH)_2$, 10% CuI
5% $Pd_2(dba)_3$•$CHCl_3$
DME/MeCN, 6 NaOMe
MeO
Ph
76%

Kang, S.-K.; Ryu, H.-C.; Son, H.-J. *SynLett*, ***1998***, 771.

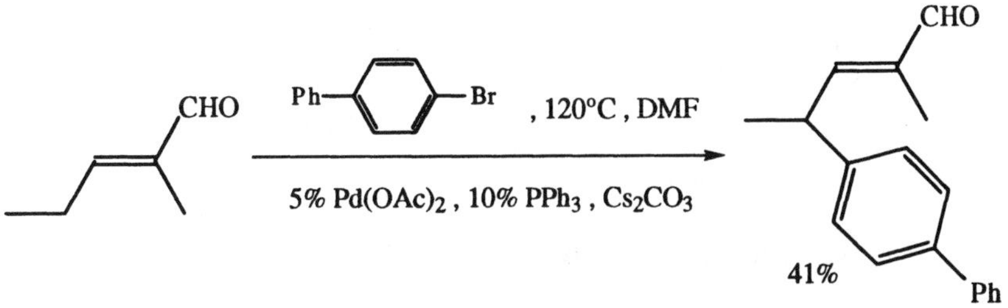

Terao, Y.; Satoh, T.; Miura, M.; Nomura, N.M. *Tetrahedron Lett.*, ***1998***, *39*, 6203.

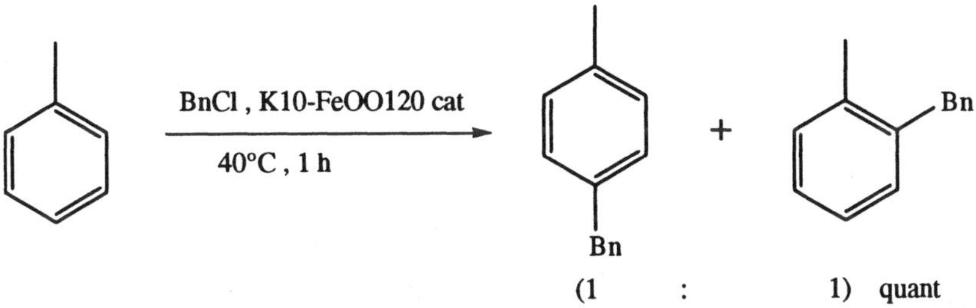

Pai, S.G.; Bajpai, A.R.; Deshpande, A.B.; Samant, S.D. *Synth. Commun.*, ***1997***, *27*, 2267.

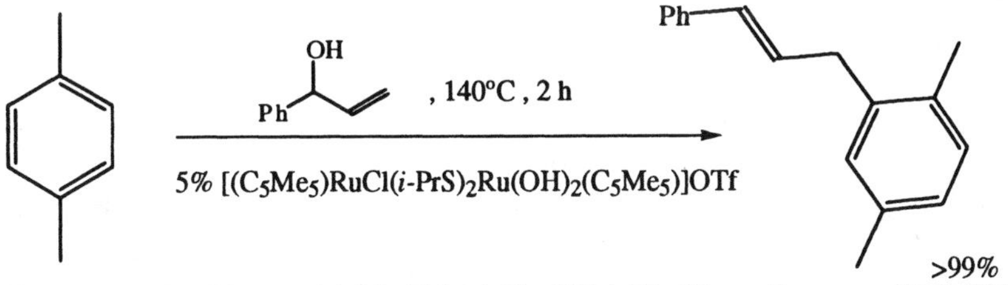

Nishibayashi, Y.; Yamanashi, M.; Takagi, Y.; Hidai, M. *Chem. Commun.*, ***1997***, 859.

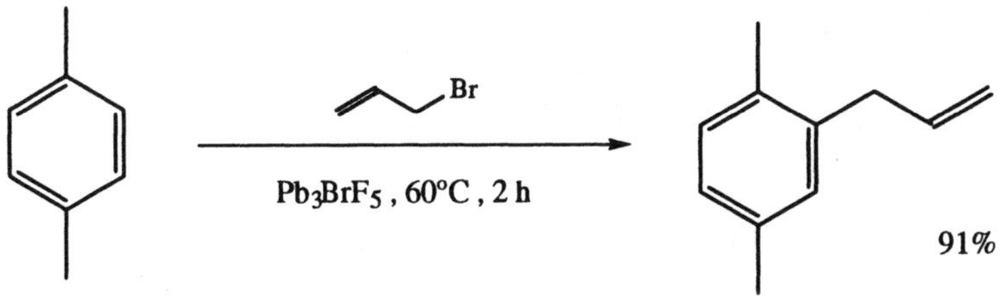

Ichihara, J. *Chem. Commun.*, ***1997***, 1921.

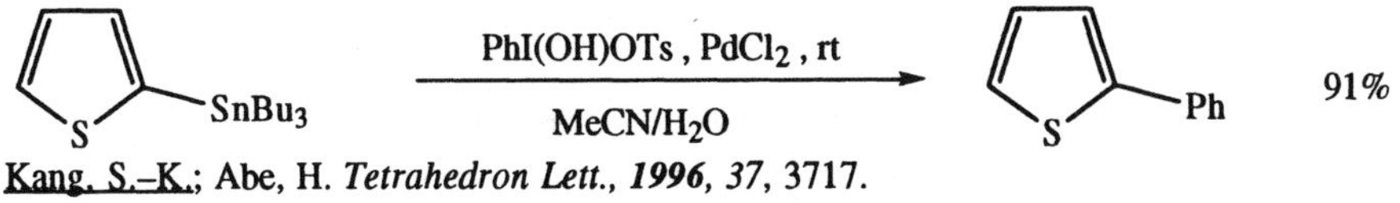

Kang, S.-K.; Abe, H. *Tetrahedron Lett.*, ***1996***, *37*, 3717.

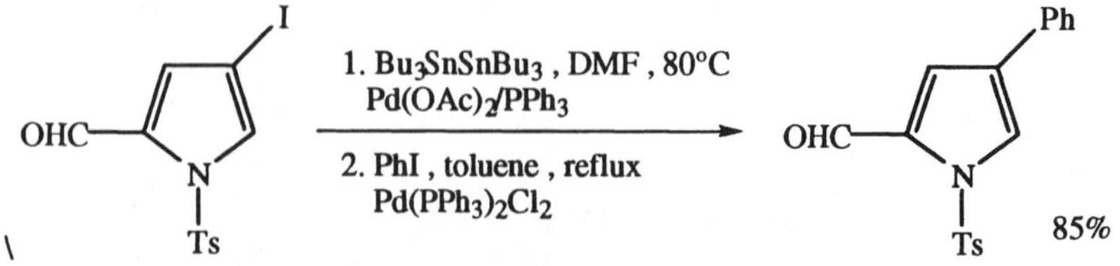
I
OHC
N
Ts
1. $Bu_3SnSnBu_3$, DMF , 80°C
$Pd(OAc)_2/PPh_3$
2. PhI , toluene , reflux
$Pd(PPh_3)_2Cl_2$
Ph
OHC
N
Ts
85%

Wang, J.; Scott, A.I. *Tetrahedron Lett.*, ***1996***, *37*, 3247.

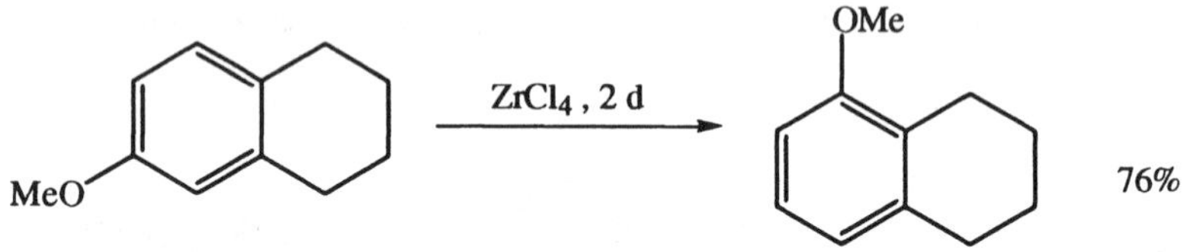
MeO
$ZrCl_4$, 2 d
OMe
76%

Harrowven, D.C.; Dainty, R.F. *Tetrahedron Lett.* ***1996***, *37*, 3607.

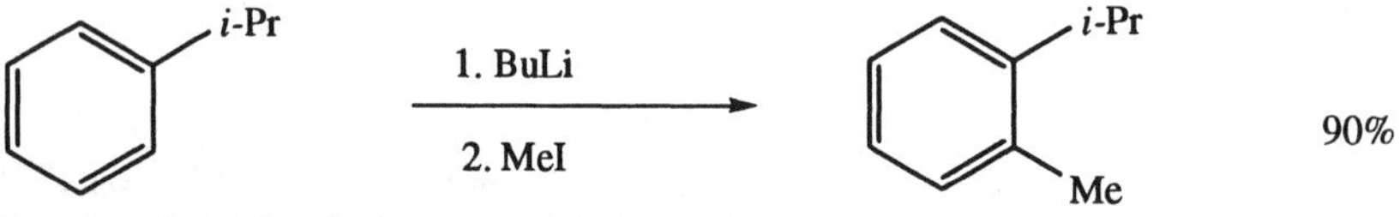
i-Pr
1. BuLi
2. MeI
i-Pr
Me
90%

Spangler, L.A. *Tetrahedron Lett.* ***1996***, *37*, 3639.

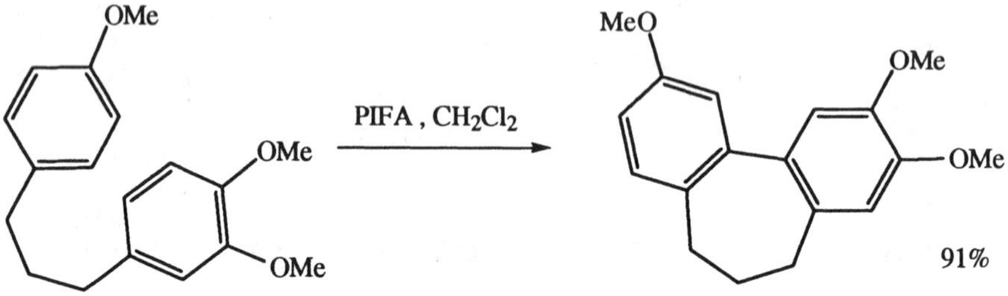
OMe
OMe
OMe
PIFA , CH_2Cl_2
MeO
OMe
OMe
91%

Kita, Y.; Gyoten, M.; Ohtsube, M.; Tohma, H.; Takada, T. *Chem. Commun.*, ***1996***, 1481.

SECTION 66: ALKYLS, METHYLENES AND ARYLS FROM AMIDES

NO ADDITIONAL EXAMPLES

SECTION 67: ALKYLS, METHYLENES AND ARYLS FROM AMINES

NO ADDITIONAL EXAMPLES

SECTION 68: ALKYLS, METHYLENES AND ARYLS FROM ESTERS

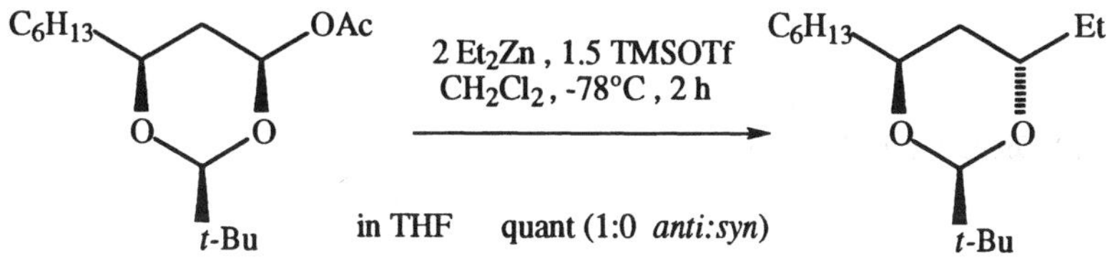

C_6H_{13}
OAc
O
O
t-Bu
2 Et_2Zn , 1.5 TMSOTf
CH_2Cl_2 , -78°C , 2 h
in THF quant (1:0 *anti:syn*)
C_6H_{13}
Et
O
O
t-Bu

91%
(51:1 *antisyn*)

Rychnovsky, S.D.; Powell, N.A. *J. Org. Chem.*, ***1997***, *62*, 6460.

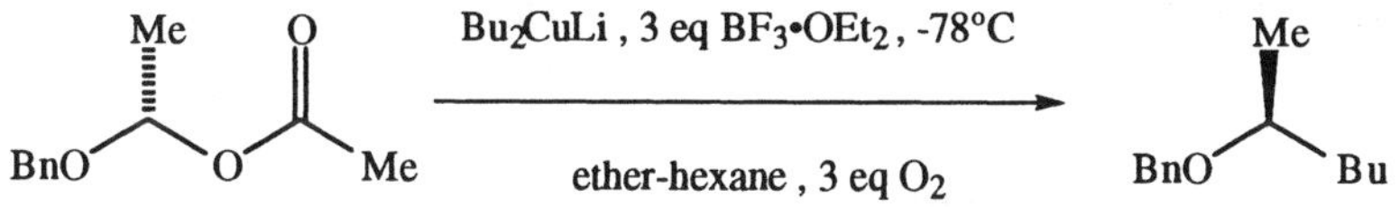

Me
O
BnO
O
Me
Bu_2CuLi , 3 eq $BF_3•OEt_2$, -78°C
ether-hexane , 3 eq O_2
Me
BnO
Bu

53% (86% ee - inversion)

Matsutani, H.; Ichikawa, S.; Yarava, J.; Kusumoto, T.; Hiyama, T.
J. Am. Chem. Soc., ***1997***, *119*, 4541.

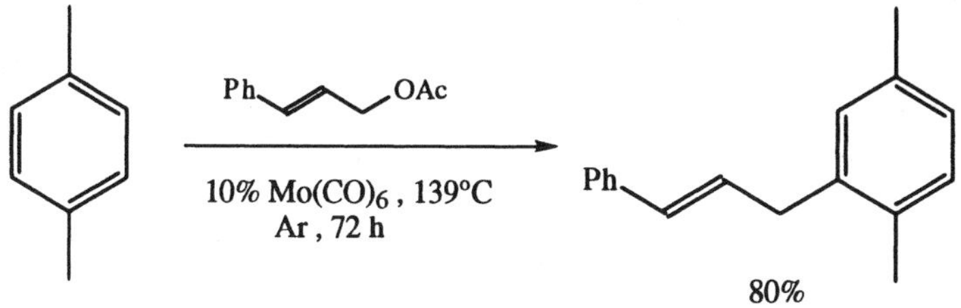

Ph
OAc
10% $Mo(CO)_6$, 139°C
Ar , 72 h
Ph

80%

Shimizu, I.; Sakamoto, T.; Kawaragi, S.; Maruyama, Y.; Yamamoto, A.
Chem. Lett., ***1997***, 137.

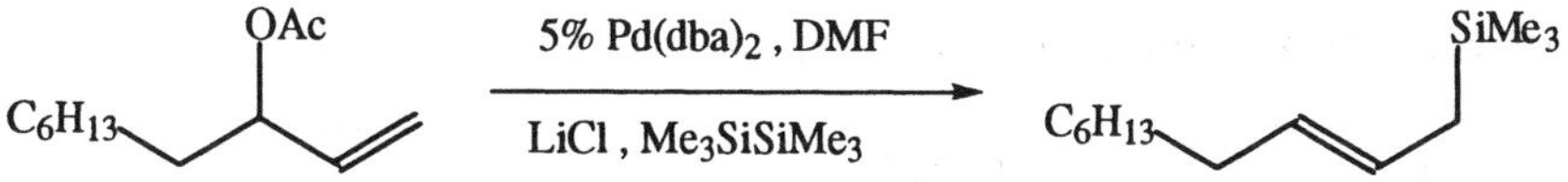

OAc
C_6H_{13}
5% $Pd(dba)_2$, DMF
LiCl , $Me_3SiSiMe_3$
$SiMe_3$
C_6H_{13}

81% (95:5 *E:Z*)

Tsuji, Y.; Funato, M.; Ozawa, M.; Ogiyama, H.; Kajita, S.; Kawamura, T.
J. Org. Chem., ***1996***, *61*, 5779.

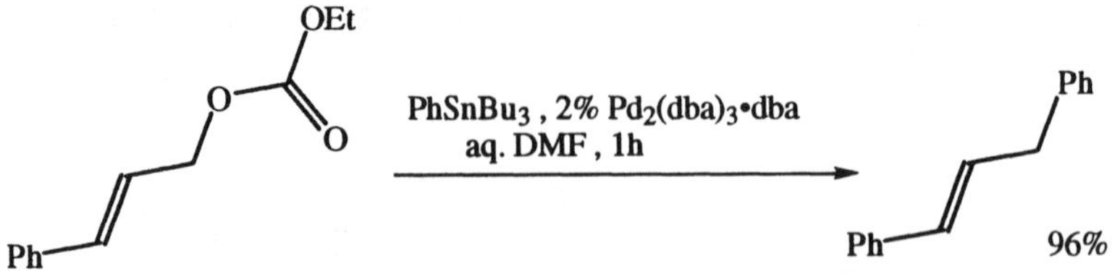
OEt
O
O
PhSnBu3 , 2% Pd2(dba)3•dba
aq. DMF , 1h
Ph
Ph
Ph
96%

Castaño, A.M.; Echavarren, A.M. *Tetrahedron Lett.*, ***1996***, *37*, 6587.

SECTION 69: ALKYLS, METHYLENES AND ARYLS FROM ETHERS, EPOXIDES AND THIOETHERS

The conversion ROR → RR' (R' = alkyl, aryl) is included in this section.

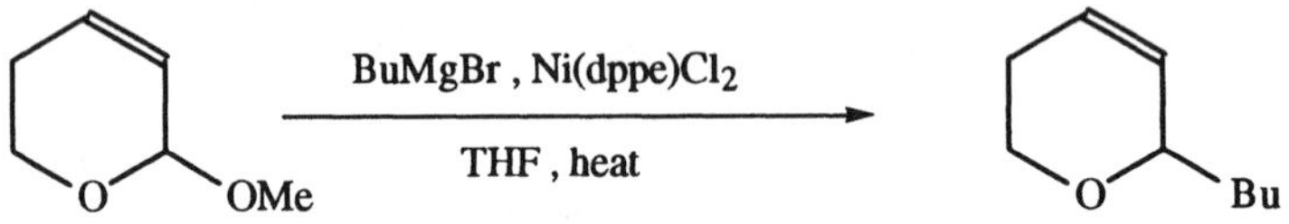
O
OMe
BuMgBr , Ni(dppe)Cl2
THF , heat
O
Bu

95%

Guagnano, V.; Lardicci, L.; Malanga, C.; Menicagli, R. *Tetrahedron Lett.*, ***1998***, *39*, 2025.

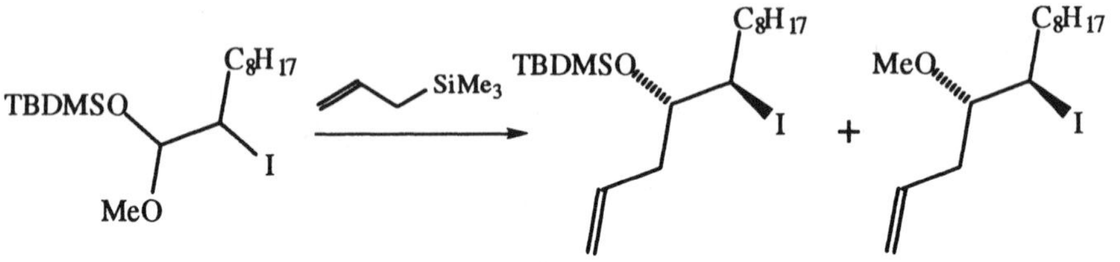
C8H17
TBDMSO
I
MeO
SiMe3
TBDMSO
C8H17
I
+
MeO
C8H17
I

$TiCl_4$	85% (>99:1 *anti:syn*)	—
TMSOTf	—	89% (96:4 *anti:syn*)

Maeda, K.; Shinokubo, H.; Oshima, K. *J. Org. Chem.*, ***1997***, *62*, 6425.

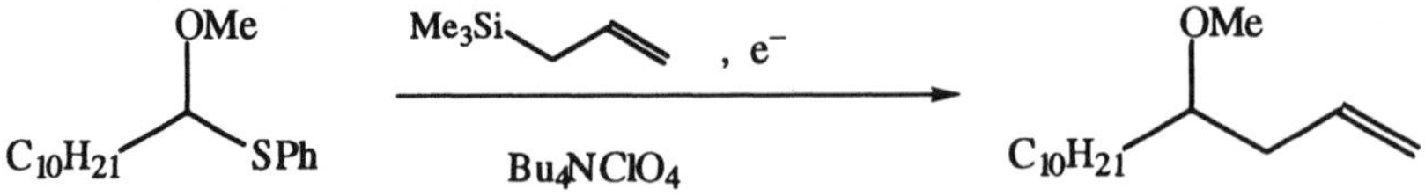
OMe
C10H21
SPh
Me3Si
, e⁻
Bu4NClO4
OMe
C10H21

61%

Yoshida, J.; Sugawara, M.; Kise, N. *Tetrahedron Lett.*, ***1996***, *37*, 3157.

SECTION 70: ALKYLS, METHYLENES AND ARYLS FROM HALIDES AND SULFONATES

The replacement of halogen by alkyl or aryl groups is included in this section. For the conversion of RX → RH (X = halogen) see Section 160 (Hydrides from Halides and Sulfonates).

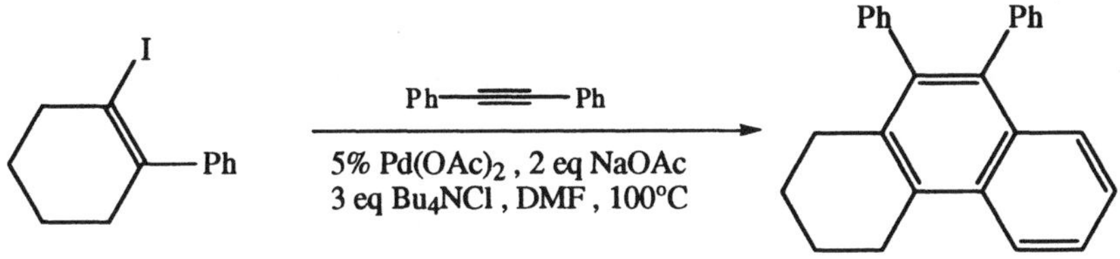
I
Ph
Ph—≡—Ph
5% $Pd(OAc)_2$, 2 eq NaOAc
3 eq Bu_4NCl , DMF , 100°C
Ph
Ph

Larock, R.C.; Tian, Q. *J. Org. Chem.*, ***1998***, *63*, 2002.

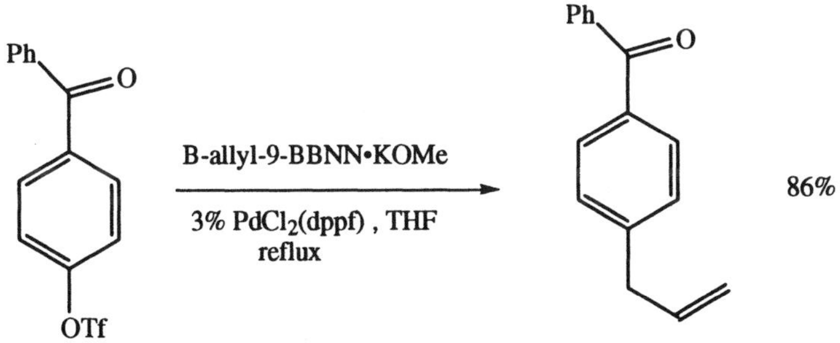
Ph
O
OTf
B-allyl-9-BBNN•KOMe
3% $PdCl_2$(dppf) , THF
reflux
Ph
O
86%

Fürstner, A.; Seidel, G. *SynLett*, ***1998***, 161.

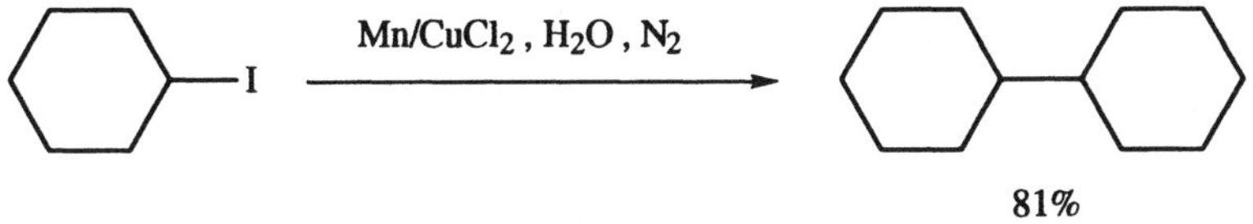
I
Mn/$CuCl_2$, H_2O , N_2
81%

Ma, J.; Chan, T.–H. *Tetrahedron Lett.*, ***1998***, *39*, 2499.

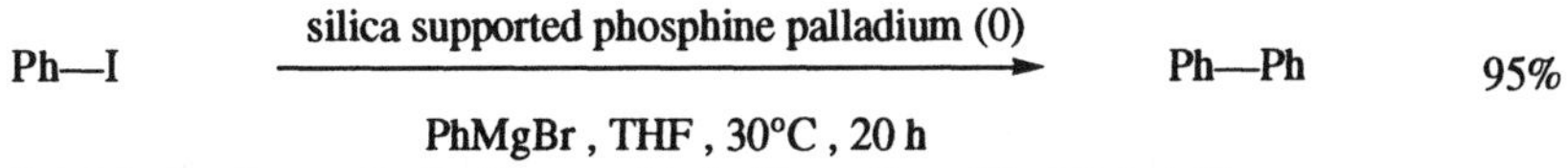
Ph—I
silica supported phosphine palladium (0)
PhMgBr , THF , 30°C , 20 h
Ph—Ph
95%

Cai, M.–Z.; Song, C.–S.; Huang, X. *J. Chem. Res. (S)*, ***1998***, 264.

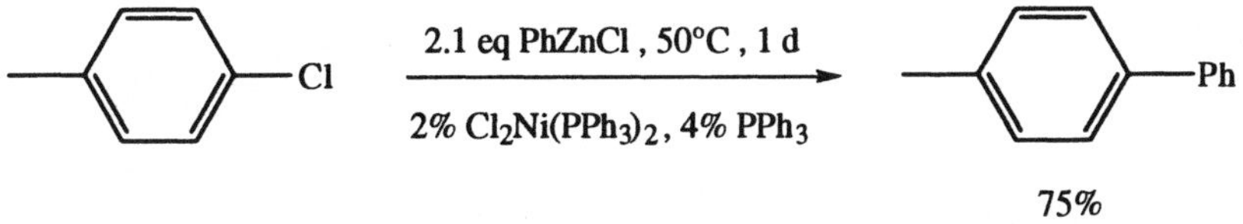
Cl
2.1 eq PhZnCl , 50°C , 1 d
2% $Cl_2Ni(PPh_3)_2$, 4% PPh_3
Ph
75%

Miller, J.A.; Farrell, R.P. *Tetrahedron Lett.*, ***1998***, *39*, 6441.

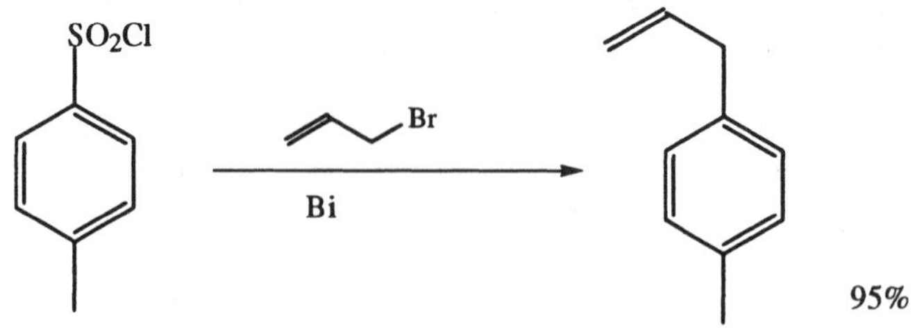
SO2Cl
Br
Bi
95%

Baruah, M.; Boruah, A.; Prajapati, D.; Sandhu, J.S. *SynLett*, ***1998***, 1083.

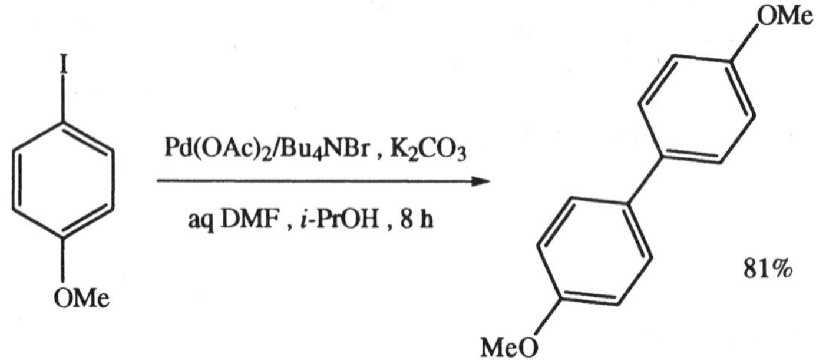
I
OMe
Pd(OAc)2/Bu4NBr , K2CO3
aq DMF , i-PrOH , 8 h
OMe
MeO
81%

Penalva, V.; Hassan, J.; Lavenot, L.; Gozzi, C.; LeMaire, M. *Tetrahedron Lett.*, ***1998***, *39*, 2559

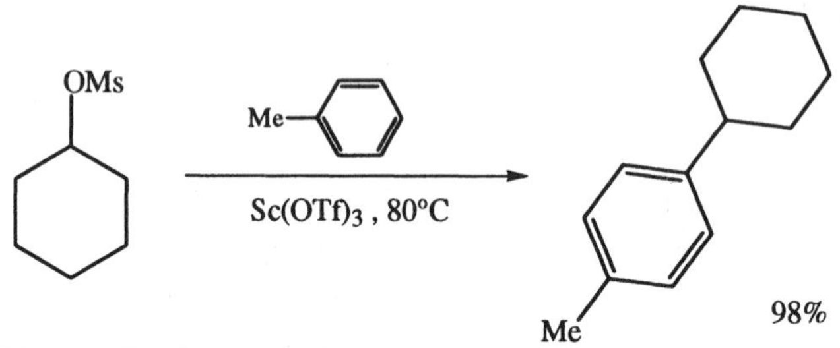
OMs
Me
Sc(OTf)3 , 80°C
Me
98%

Kotsuki, H.; Oshisi, T.; Inoue, M. *SynLett*, ***1998***, 255.

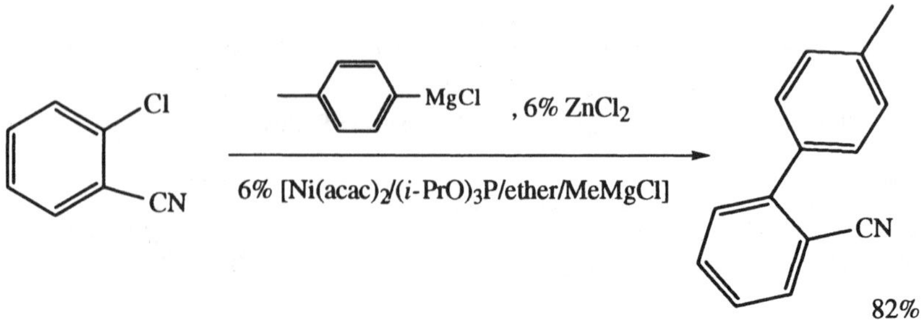
Cl
CN
MgCl
, 6% ZnCl2
6% [Ni(acac)2/(i-PrO)3P/ether/MeMgCl]
CN
82%

Miller, J.A.; Farrell, R.P. *Tetrahedron Lett*, ***1998***, *39*, 7275.

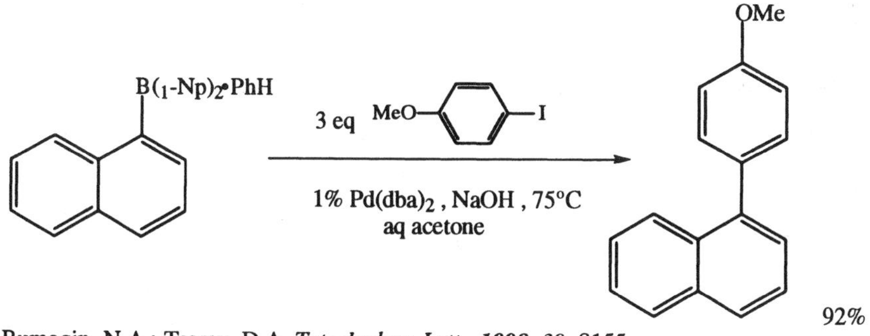
B(1-Np)2•PhH
3 eq MeO—C6H4—I
1% Pd(dba)2 , NaOH , 75°C
aq acetone
OMe
92%

Bumagin, N.A.; Tsarev, D.A. *Tetrahedron Lett.*, ***1998***, *39*, 8155.

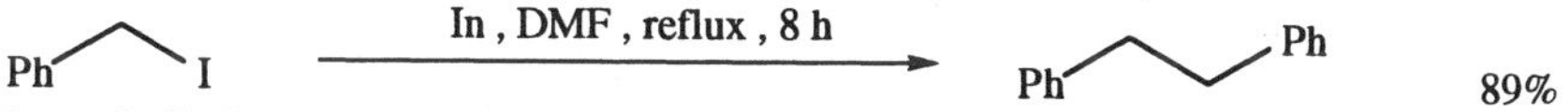
Ph
I
In , DMF , reflux , 8 h
Ph
Ph
89%

Ranu, B.C.; Dutta, P.; Sarkar, A. *Tetrahedron Lett.*, ***1998***, *39*, 9557.

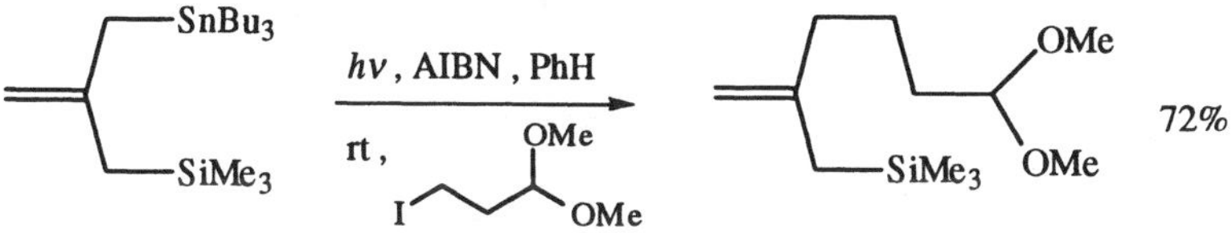
SnBu3
SiMe3
hν , AIBN , PhH
rt ,
OMe
I
OMe
OMe
SiMe3
OMe
72%

Clive, D.L.J.; Paul, C.C.; Wang, Z. *J. Org. Chem.*, ***1997***, *62*, 7028.

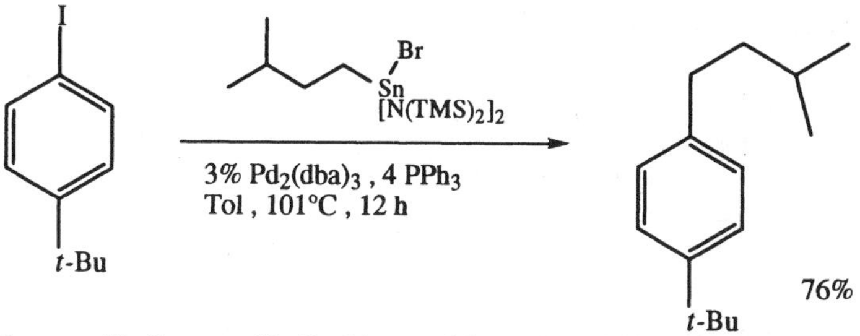
I
t-Bu
Br
Sn
[N(TMS)2]2
3% Pd2(dba)3 , 4 PPh3
Tol , 101°C , 12 h
76%
t-Bu

Fouquet, E.; Pereyre, M.; Rodriguez, A.L. *J. Org. Chem.*, ***1997***, *62*, 5242.

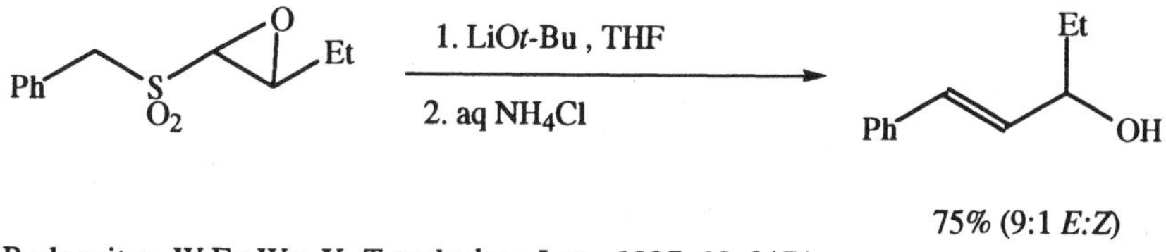
Ph
S
O2
O
Et
1. LiOt-Bu , THF
2. aq NH4Cl
Et
Ph
OH
75% (9:1 E:Z)

Berkowitza, W.F.; Wu, Y. *Tetrahedron Lett.*, ***1997***, *38*, 3171.

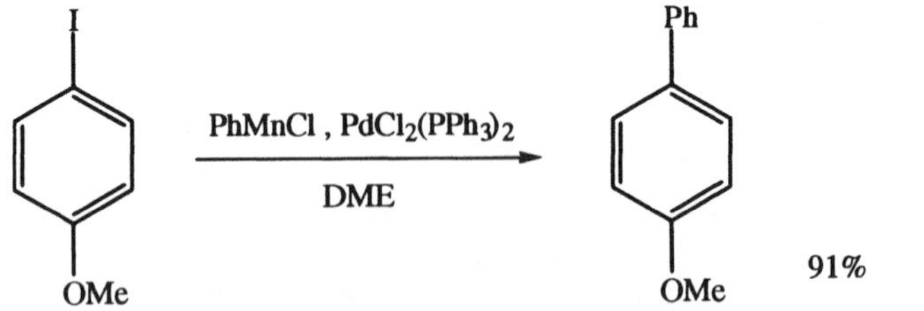
I
Ph
PhMnCl , PdCl2(PPh3)2
DME
OMe
OMe
91%

Riguet, E.; Alami, M.; Cahiez, G. *Tetrahedron Lett.*, ***1997***, *38*, 4397.

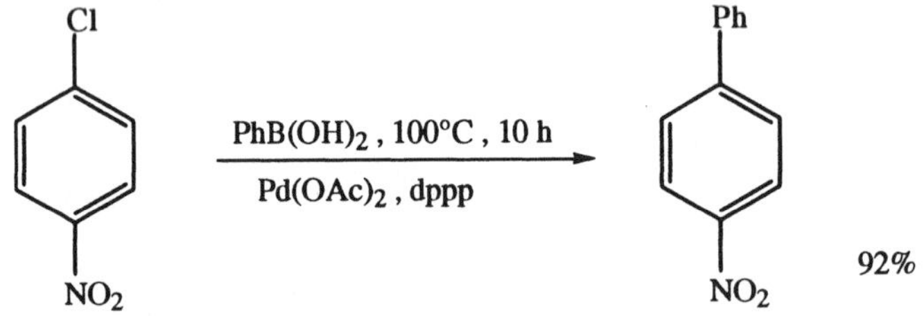
Cl
Ph
PhB(OH)2 , 100°C , 10 h
Pd(OAc)2 , dppp
NO2
NO2
92%

Shen, W. *Tetrahedron Lett.*, ***1997***, *38*, 5575.

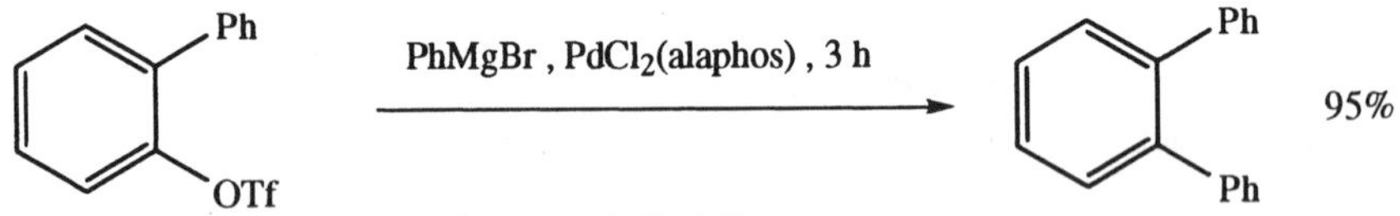
Ph
OTf
PhMgBr , PdCl2(alaphos) , 3 h
Ph
Ph
95%

Kamikawa, T.; Hayashi, T. *SynLett*, ***1997***, 163.

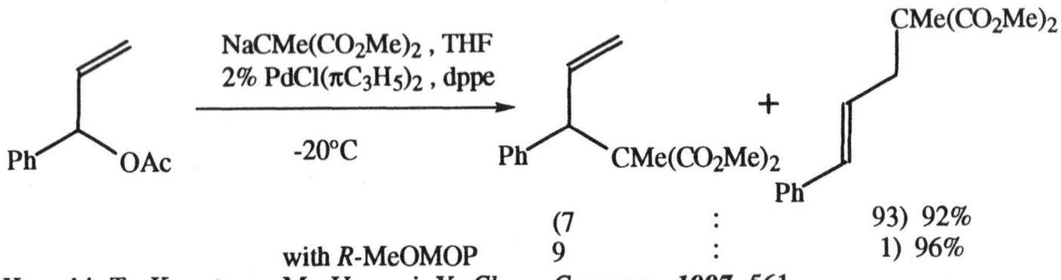
NaCMe(CO2Me)2 , THF
2% PdCl(πC3H5)2 , dppe
-20°C
Ph
OAc
Ph
CMe(CO2Me)2
+
CMe(CO2Me)2
Ph
(7 : 93) 92%
with R-MeOMOP 9 : 1) 96%

Hayashi, T.; Kawatsura, M.; Uozumi, Y. *Chem. Commun.*, ***1997***, 561.

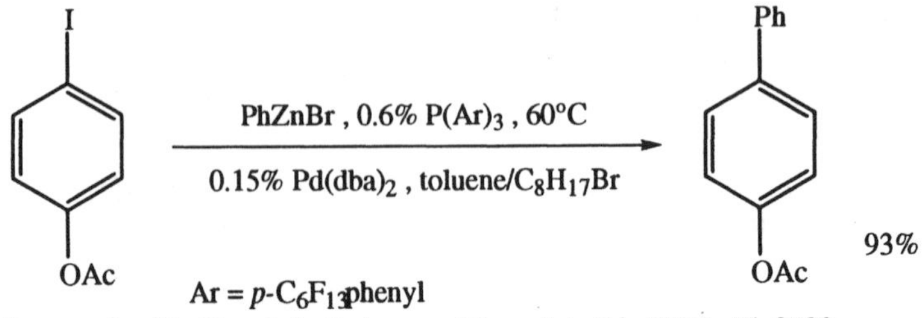
I
Ph
PhZnBr , 0.6% P(Ar)3 , 60°C
0.15% Pd(dba)2 , toluene/C8H17Br
OAc
OAc
93%
Ar = p-C6F13phenyl

Betzemeier, B.; Knochel, P. *Angew. Chem. Int. Ed.*, ***1997***, *36*, 2623.

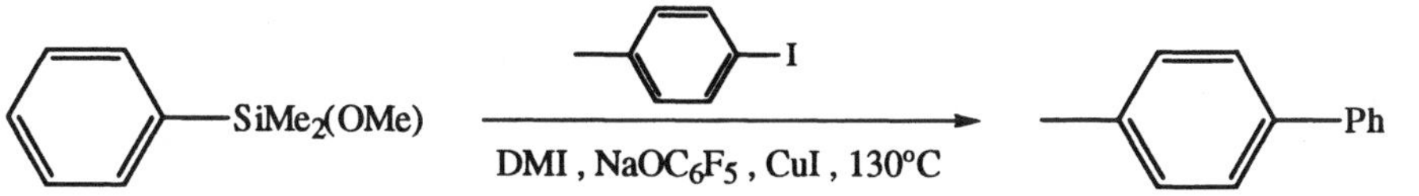
SiMe2(OMe)
I
DMI, NaOC6F5, CuI, 130°C
Ph

does not work well with all SiR_3 derivatives

72%

Ito, H.; Sensui, H.; Arimoto, K.; Miura, K.; Hosomi, A. *Chem. Lett.*, ***1997***, 679.

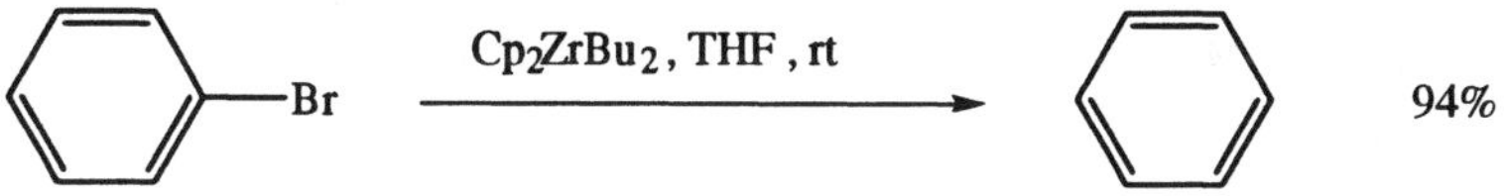
Br
Cp2ZrBu2, THF, rt
94%

Hara, R.; Sun, W.–H.; Nishihara, Y.; Takahashi, T. *Chem. Lett.*, ***1997***, 1251.

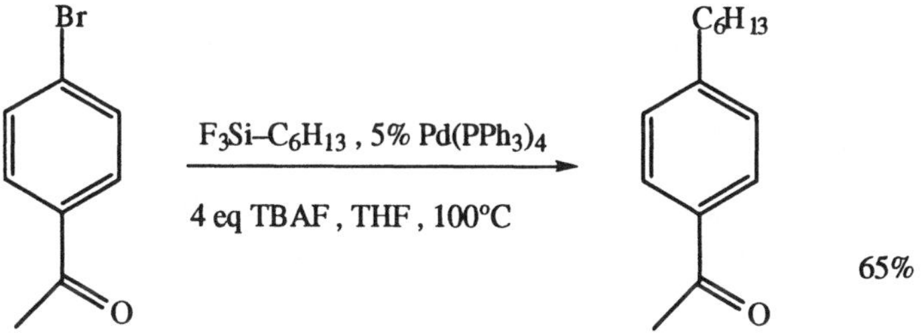
Br
O
F3Si–C6H13, 5% Pd(PPh3)4
4 eq TBAF, THF, 100°C
C6H13
O
65%

Matsuhashi, H.; Asai, S.; Hirabayashi, K.; Hatanaka, Y.; Mori, A.; Hiyama, T. *Bull. Chem. Soc. Jpn.*, ***1997***, *70*, 437.

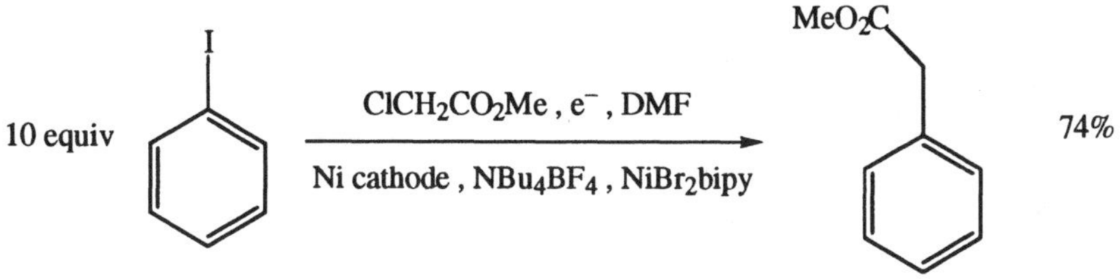
10 equiv
I
ClCH2CO2Me, e⁻, DMF
Ni cathode, NBu4BF4, NiBr2bipy
MeO2C
74%

Durandetti, M.; Nédékec, J.–Y.; Périchon, J. *J. Org. Chem.*, ***1996***, *61*, 1748.

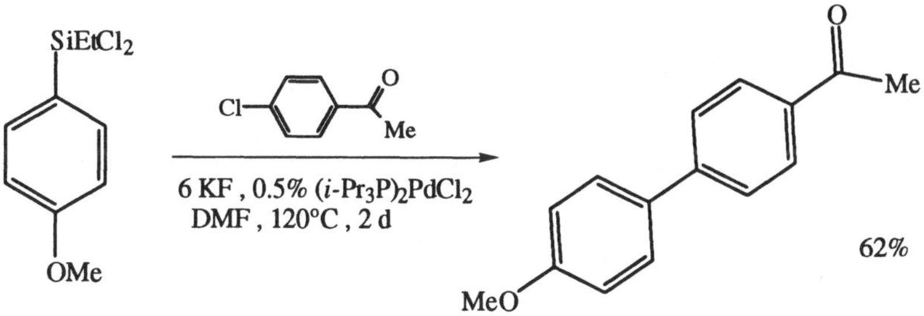
SiEtCl2
OMe
Cl
O
Me
6 KF, 0.5% (i-Pr3P)2PdCl2
DMF, 120°C, 2 d
O
Me
MeO
62%

Gouda, K.; Hagiwara, E.; Hatanaka, Y.; Hiyama, T. *J. Org. Chem.*, ***1996***, *61*, 7232.

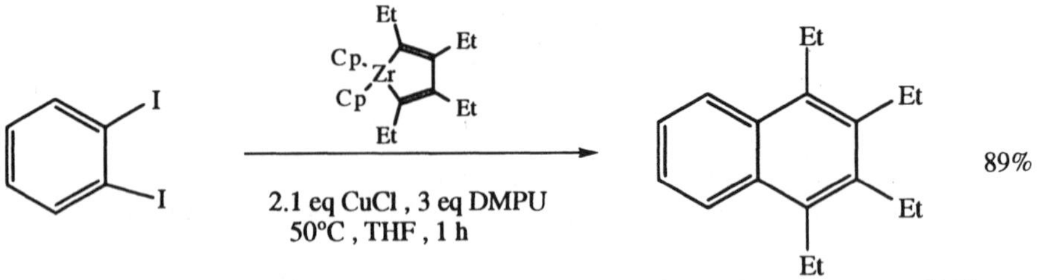

I
I
Cp
Zr
Cp
Et
Et
Et
Et
2.1 eq CuCl , 3 eq DMPU
50°C , THF , 1 h
Et
Et
Et
Et
89%

Takahashi, T.; Hara, R.; Nishihara, Y.; Kotora, M. *J. Am. Chem. Soc.*, ***1996***, *118*, 5154.

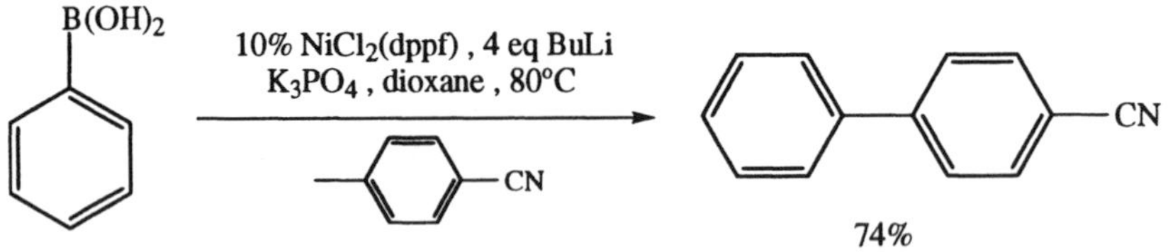

B(OH)2
10% NiCl2(dppf) , 4 eq BuLi
K3PO4 , dioxane , 80°C
CN
CN
74%

Saito, S.; Sakai, M.; Miyaura, N. *Tetrahedron Lett*, ***1996***, *37*, 2993.

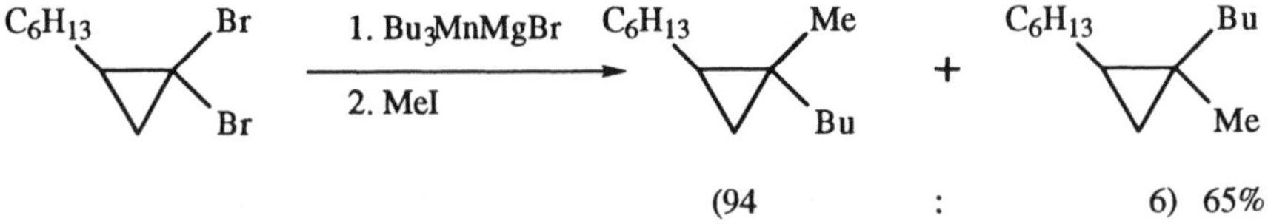

C6H13
Br
Br
1. Bu3MnMgBr
2. MeI
C6H13
Me
Bu
+
C6H13
Bu
Me
(94 : 6) 65%

Inoeu, R.; Shinokubo, H.; Oshima, K. *Tetrahedron Lett*, ***1996***, *37*, 5377.

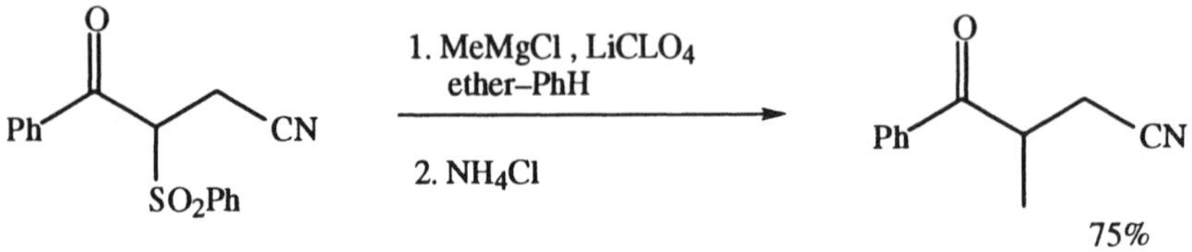

O
Ph
CN
SO2Ph
1. MeMgCl , LiCLO4
ether–PhH
2. NH4Cl
O
Ph
CN
75%

Giovannini, R.; Petrini, M. *SynLett*, ***1996***, 1001.

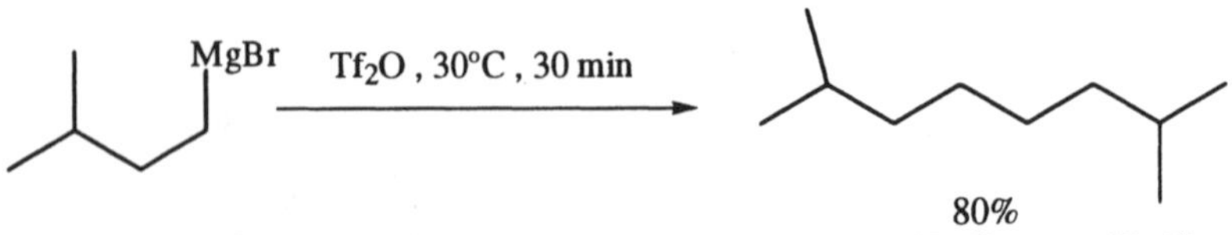

MgBr
Tf2O , 30°C , 30 min
80%

Nishiyama, T.; Seshita, T.; Shodai, H.; Aoki, K.; Kameyama, H.; Komura, K. *Chem. Lett.*, ***1996***, 549.

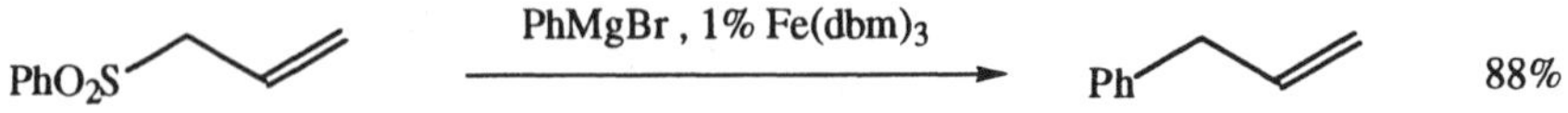

PhO2S
PhMgBr , 1% Fe(dbm)3
Ph
88%

Gai, Y.; Julia, M.; Verpeaux, J.-N. *Bull. Soc. Chim. Fr.*, ***1996***, *133*, 805.

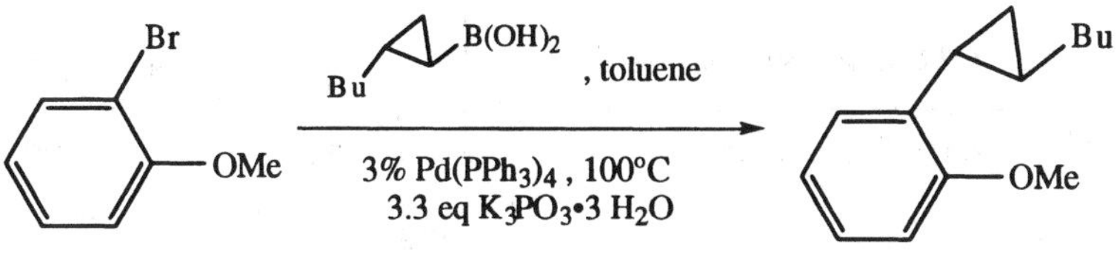

89%

Wang, X.–Z.; Deng, M.–Z. *J. Chem. Soc., Perkin Trans. 1*, ***1996***, 2663.

SECTION 71: ALKYLS, METHYLENES AND ARYLS FROM HYDRIDES

This section lists examples of the reaction of RH → RR' (R,R' = alkyl or aryl). For the reaction C=CH → C=C-R (R = alkyl or aryl), see Section 209 (Alkenes from Alkenes). For alkylations of ketones and esters, see Section 177 (Ketones from Ketones) and Section 113 (Esters from Esters).

NO ADDITIONAL EXAMPLES

SECTION 72: ALKYLS, METHYLENES AND ARYLS FROM KETONES

The conversions $R_2C=O \rightarrow R\text{-}R, R_2CH_2, R_2CHR'$, etc. are listed in this section.

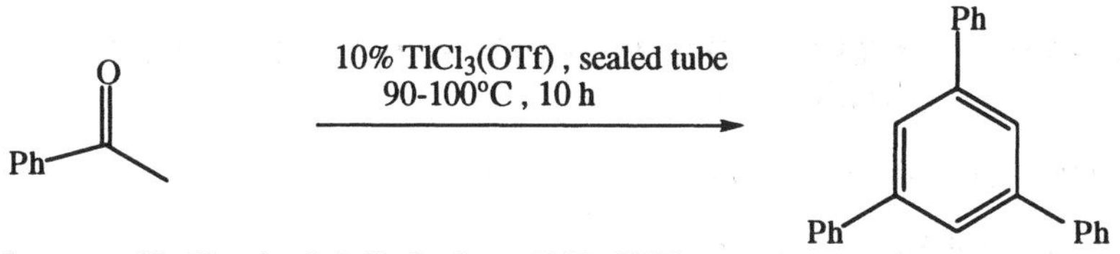

99%

Iranpoor, N.; Zeynizaded, B. *SynLett*, ***1998***, 1079.

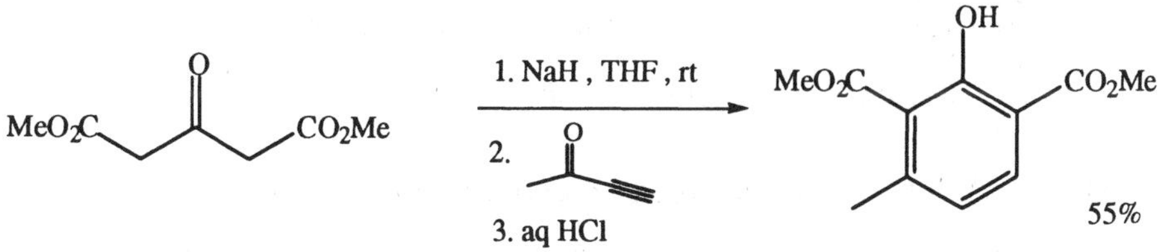

55%

Covarrubias–Zúñiga, A. *Synth. Commun.*, ***1998***, *28*, 1525.

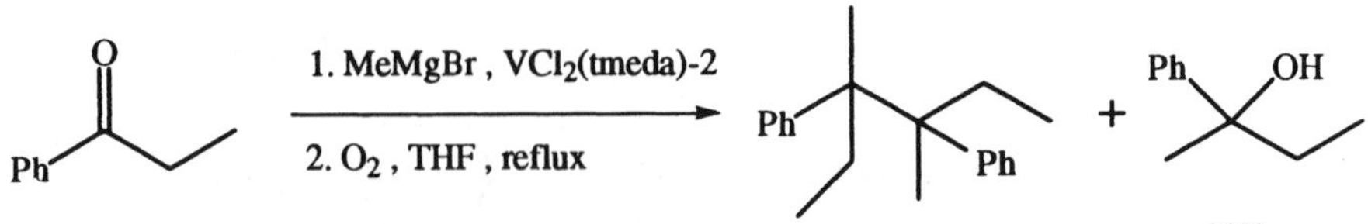

89% 11%

Kataoka, Y.; Akiyama, H.; Makihira, I.; Tani, K. *J. Org. Chem.*, ***1997***, *62*, 8109.

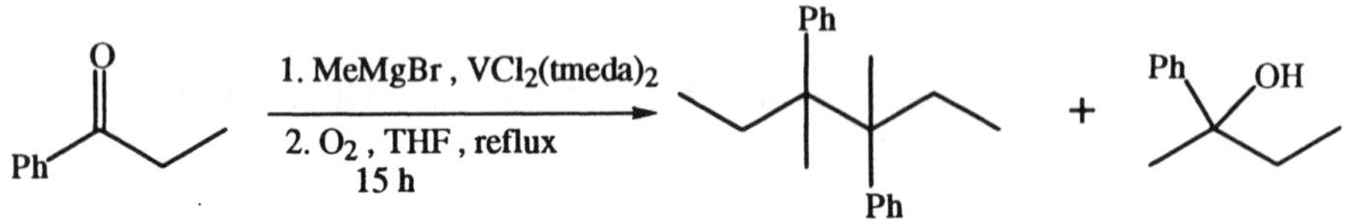

89% 11%

Kataoka, Y.; Akiyama, H.; Makihira, I.; Tani, K. *J. Org. Chem.*, ***1996***, *61*, 6094.

SECTION 73 ALKYLS, METHYLENES AND ARYLS FROM NITRILES

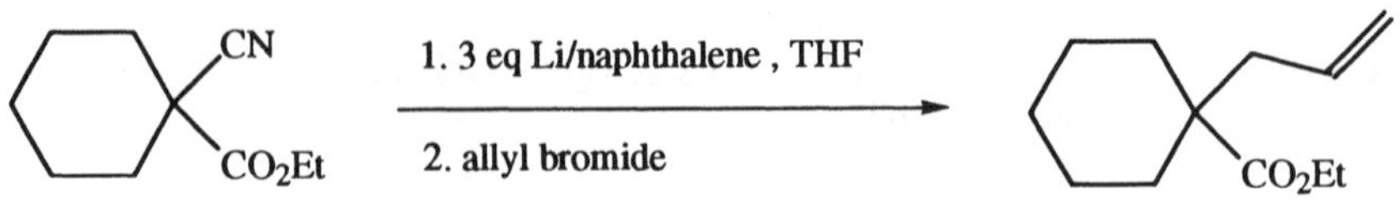

82%

Shia, K.-S.; Chang, N.-Y.; Yie, J.; Liu, H.-J. *Tetrahedron Lett*, ***1997***, *38*, 7713.

SECTION 74: ALKYLS, METHYLENES AND ARYLS FROM ALKENES

The following reaction types are included in this section:

A. Hydrogenation of Alkenes (and Aryls)
B. Formation of Aryls
C. Alkylations and Arylations of Alkenes
D. Conjugate Reduction of Conjugated Aldehydes, Ketones, Acids, Esters and Nitriles
E. Conjugate Alkylations
F. Cyclopropanations, including halocyclopropanations

SECTION 74A: Hydrogenation of Alkenes (and Aryls)

Reduction of aryls to dienes are listed in Section 377 (Alkene-Alkene).

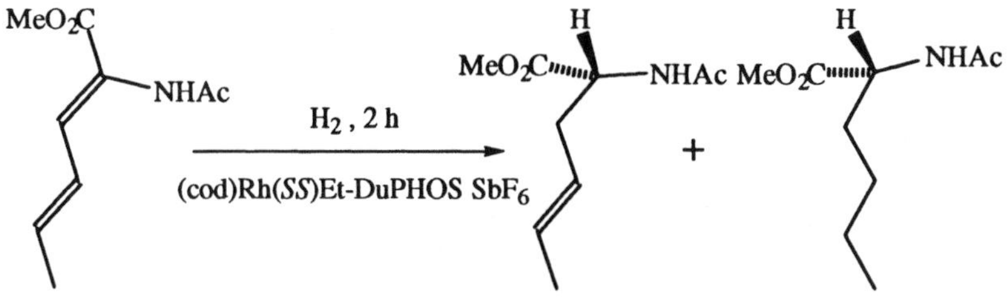
MeO2C
NHAc
H2 , 2 h
(cod)Rh(SS)Et-DuPHOS SbF6
H
MeO2C
NHAc
+
H
MeO2C
NHAc

(200
(99.2% ee , *S*) : 1) quant

Burk, M.J.; Allen, J.G.; Kiesman, W.F. *J. Am. Chem. Soc.*, ***1998***, *120*, 657.

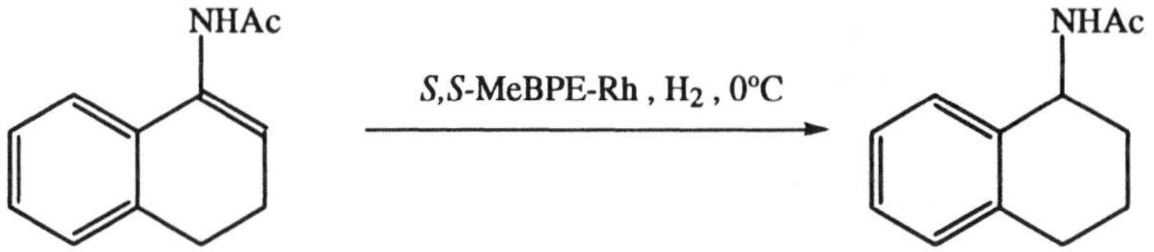
NHAc
S,S-MeBPE-Rh , H2 , 0°C
NHAc

92% ee

Burk, M.J.; Casy, G.; Johnson, N.B. *J. Org. Chem.*, ***1998***, *63*, 6084.

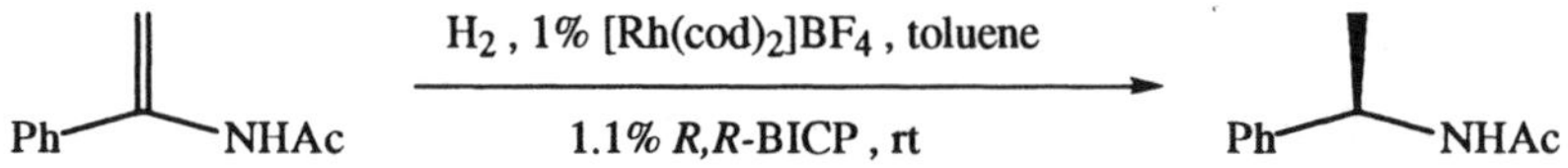
Ph
NHAc
H2 , 1% [Rh(cod)2]BF4 , toluene
1.1% *R,R*-BICP , rt
Ph
NHAc

86% ee

Zhu, G.; Zhang, X. *J. Org. Chem.*, ***1998***, *63*, 9590.

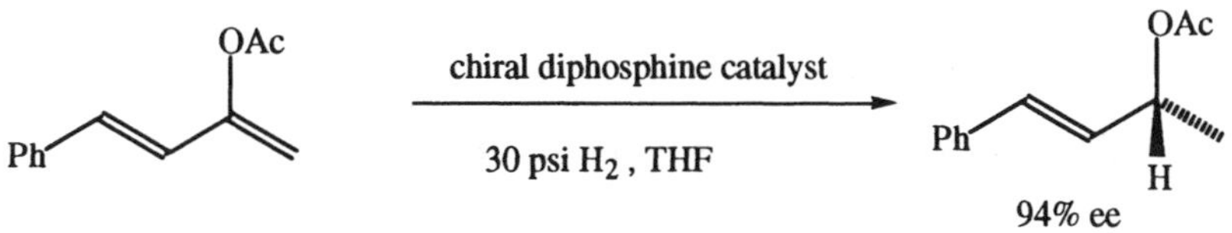
OAc
Ph
chiral diphosphine catalyst
30 psi H2 , THF
OAc
Ph
H

94% ee

Boaz, N.W. *Tetrahedron Lett.*, ***1998***, *39*, 5505.

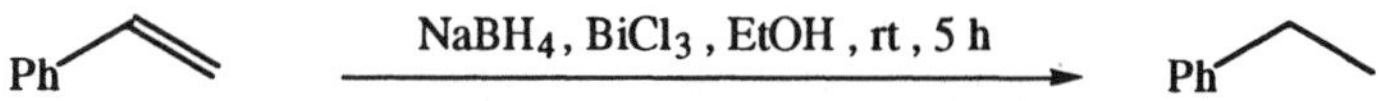
Ph
NaBH4 , BiCl3 , EtOH , rt , 5 h
Ph

80%

Ren, P.–D.; Pan, S.–F.; Dong, T.–W.; Wu, S.–H. *Synth. Commun.*, ***1996***, *26*, 763.

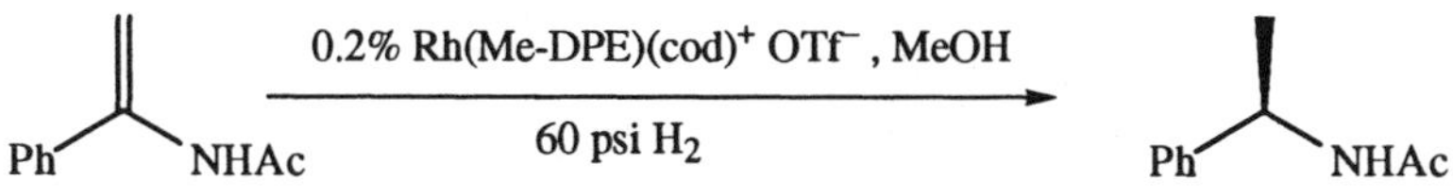
Ph
NHAc
0.2% Rh(Me-DPE)(cod)+ OTf− , MeOH
60 psi H2
Ph
NHAc

Me-DPE = 1,2-*bis*(*trans*-2,5-dimethylphopholanoethane 95% ee , *R*

Burk, M.J.; Wang, Y.M.; Lee, J.R. *J. Am. Chem. Soc.*, ***1996***, *118*, 5142.

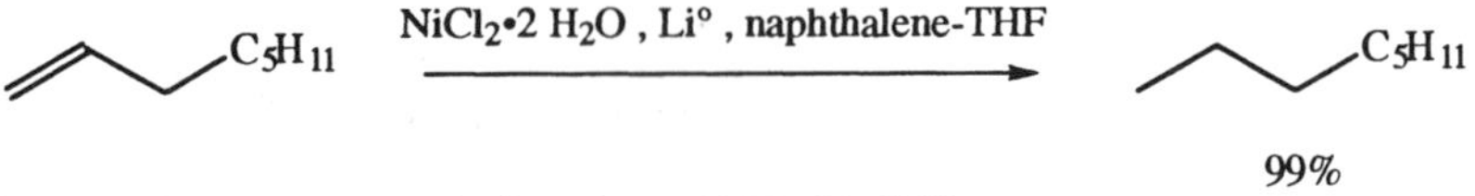

C5H11
NiCl2•2 H2O , Li° , naphthalene-THF
C5H11
99%

Alonso, F.; Yus, M. *Tetrahedron Lett.*, ***1996***, *37*, 6925.

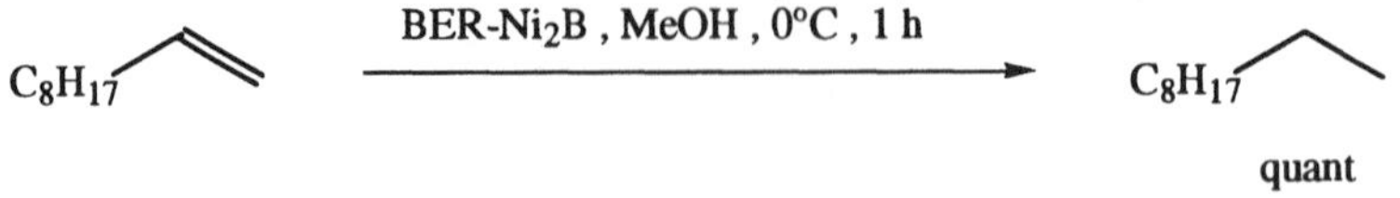

C8H17
BER-Ni2B , MeOH , 0°C , 1 h
C8H17
quant

Choi, J.; Yoon, N.M. *Synthesis*, ***1996***, 597.

REVIEW:
"Asymmetric Transfer Hydrogenation Catalyzed by Chiral Ruthenium Complexes."
Noyori, R.; Hasiguichi, S. *Accts. Chem. Res.*, ***1997***, *30*, 97.

SECTION 74B: Formation of Aryls

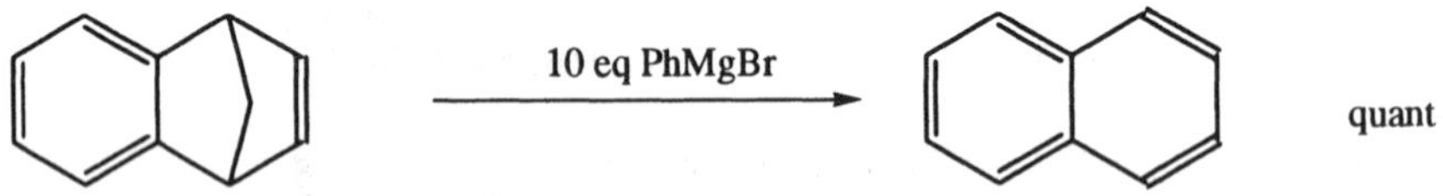

10 eq PhMgBr
quant

Blank, D.H.; Gribble, G.W. *Tetrahedron Lett.*, ***1997***, *38*, 4761.

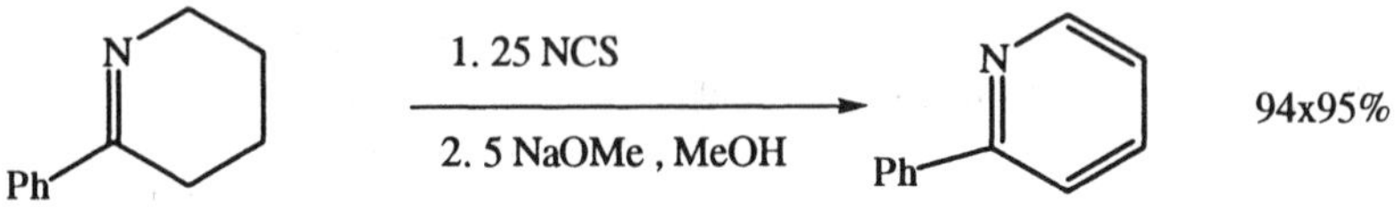

N
Ph
1. 25 NCS
2. 5 NaOMe , MeOH
N
Ph
94x95%

DeKimpe, N.; Keppens, M.; Fonck, G. *Chem. Commun.*, ***1996***, 635.

SECTION 74C: Alkylations and Arylations of Alkenes

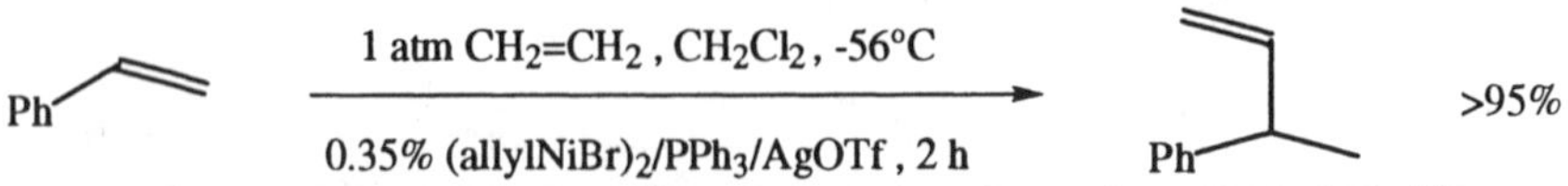

Ph
1 atm CH2=CH2 , CH2Cl2 , -56°C
0.35% (allylNiBr)2/PPh3/AgOTf , 2 h
Ph
>95%

Nomura, N.; Jin, J.; Park, H.; Rajanbabu, J.V. *J. Am. Chem. Soc.*, ***1998***, *120*, 459.

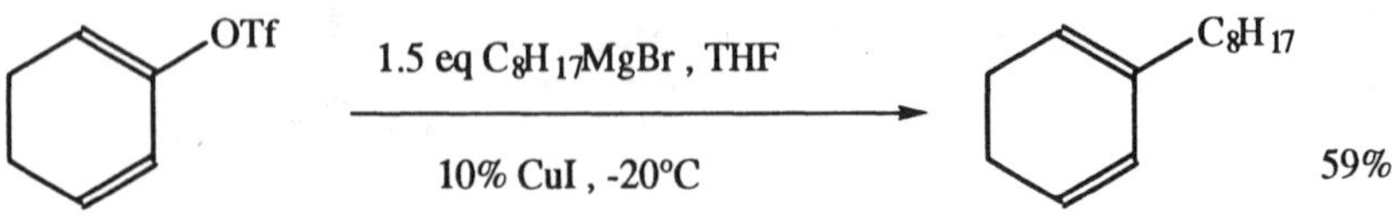

OTf
1.5 eq C8H17MgBr , THF
10% CuI , -20°C
C8H17
59%

Karlström, A.S.E.; Rönn, M.; Thorarensen, A.; Bäckvall, J.–E. *J. Org. Chem.*, ***1998***, *63*, 2517

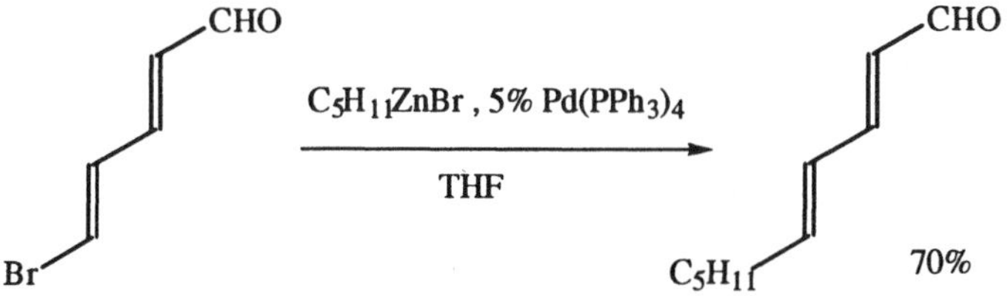
CHO
C5H11ZnBr , 5% Pd(PPh3)4
THF
Br
CHO
C5H11

70%

Vicart, N.; Castet-Caillabet, D.; Ramondency, Y.; Plé, G.; Cuhamel, L. *SynLett*, ***1998***, 411

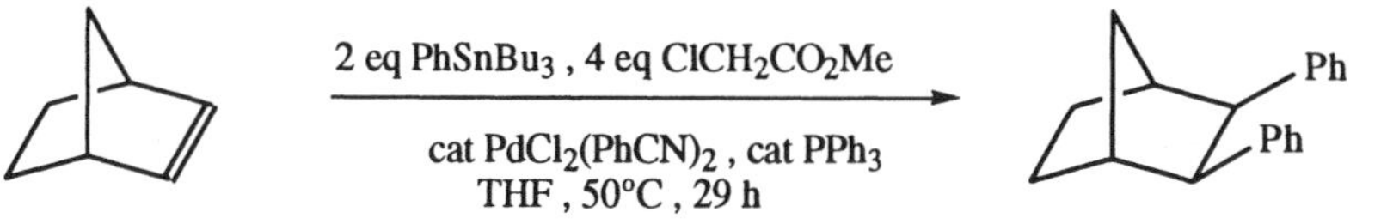
2 eq PhSnBu3 , 4 eq ClCH2CO2Me
cat PdCl2(PhCN)2 , cat PPh3
THF , 50°C , 29 h
Ph
Ph

58%

Fugami, K.; Hagiwara, S.; Oda, H.; Kosugi, M. *SynLett*, ***1998***, 477.

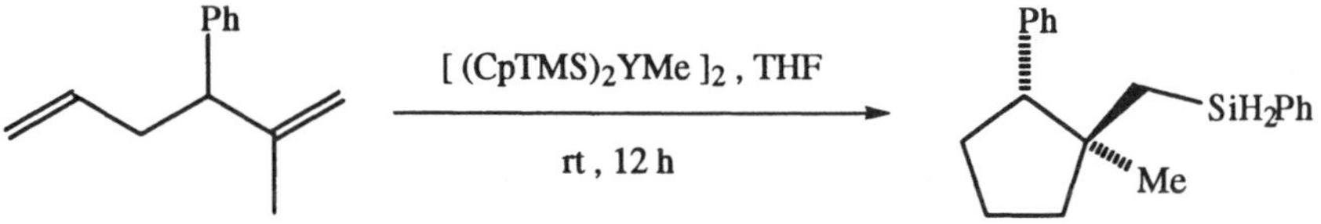
Ph
[(CpTMS)2YMe]2 , THF
rt , 12 h
Ph
SiH2Ph
Me

92%

Molander, G.A.; Dowdy, E.D.; Schmumann, H. *J. Org. Chem.*, ***1998***, *63*, 3386.

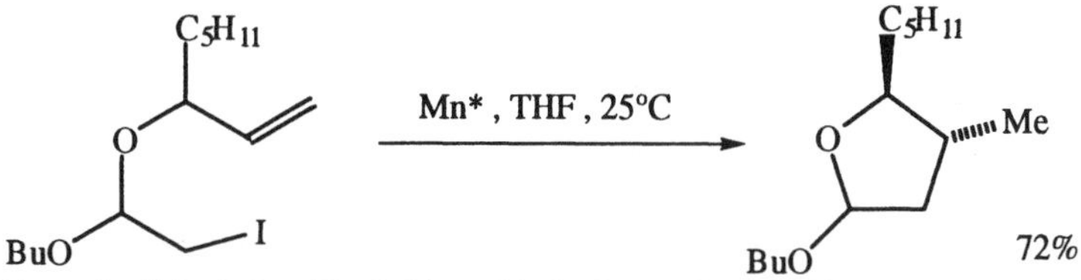
C5H11
O
BuO
I
Mn* , THF , 25°C
C5H11
O
Me
BuO

72% (1:1 *syn:anti*)

Tang, J.; Shinokubo, H.; Oshima, K. *SynLett*, ***1998***, 1075.

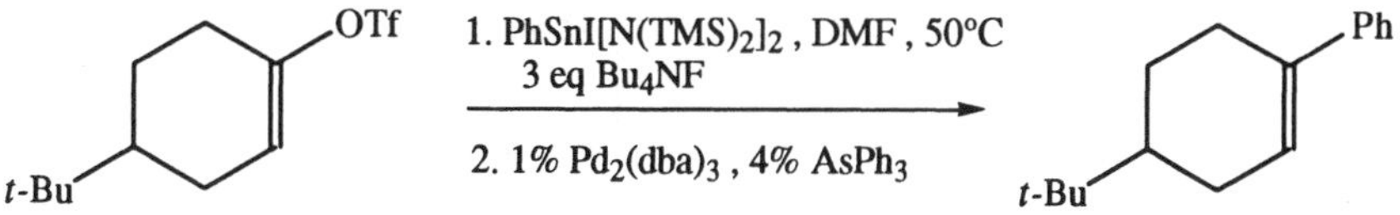
OTf
t-Bu
1. PhSnI[N(TMS)2]2 , DMF , 50°C
3 eq Bu4NF
2. 1% Pd2(dba)3 , 4% AsPh3
Ph
t-Bu

83%

Fouquet, E.; Rodriguez, A.L. *SynLett*, ***1998***, 1323.

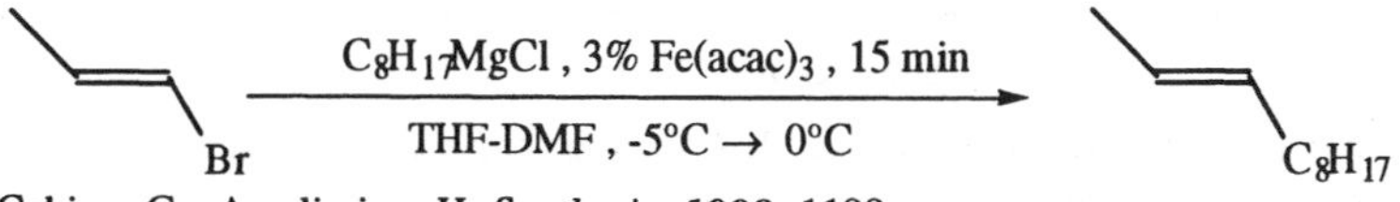
Br
C8H17MgCl , 3% Fe(acac)3 , 15 min
THF-DMF , -5°C → 0°C
C8H17

81%

Cahiez, G.; Avedissian, H. *Synthesis*, ***1998***, 1199.

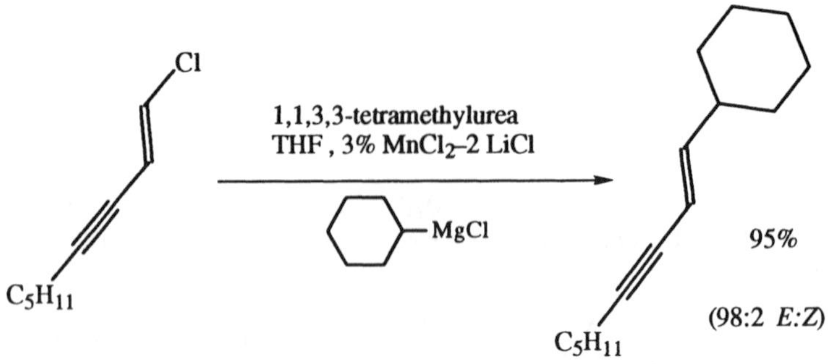
Cl
1,1,3,3-tetramethylurea
THF, 3% MnCl2-2 LiCl
MgCl
C5H11
95%
(98:2 E:Z)
C5H11

Alami, M.; Ramiandrasoa, P.; Cahiez, G. *SynLett*, ***1998***, 325.

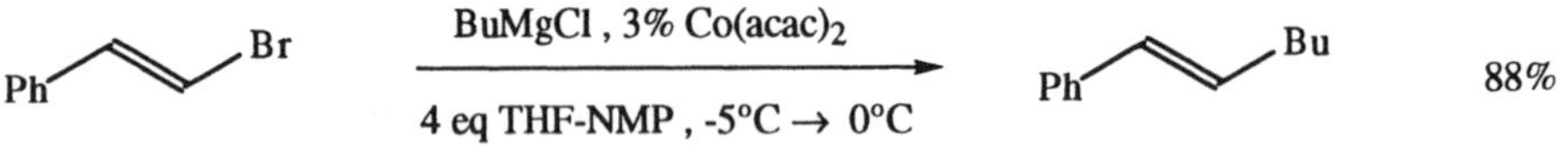
Ph
Br
BuMgCl, 3% Co(acac)2
4 eq THF-NMP, -5°C → 0°C
Ph
Bu
88%

Cahiez, G.; Avedissian, H. *Tetrahedron Lett.*, ***1998***, *39*, 6159.
Avedissian, H.; Bérillon, L.; Cahiez, G.; Knochel, P. *Tetrahedron Lett.*, ***1998***, *39*, 6163 (with R—Zn derivatives).

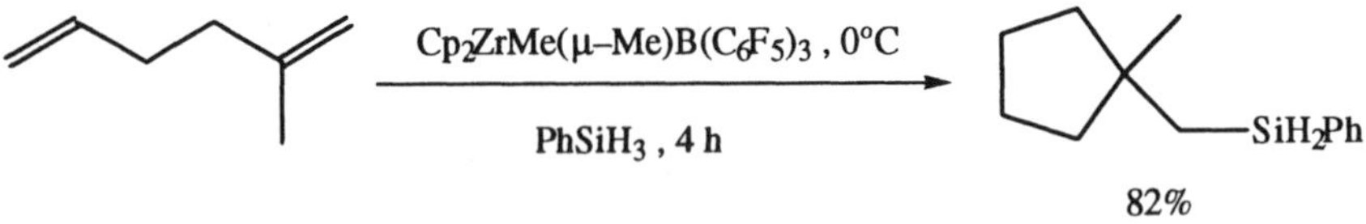
Cp2ZrMe(μ-Me)B(C6F5)3, 0°C
PhSiH3, 4 h
SiH2Ph
82%

Molander, G.A.; Corrette, C.P. *Tetrahedron Lett*, ***1998***, *39*, 5011.

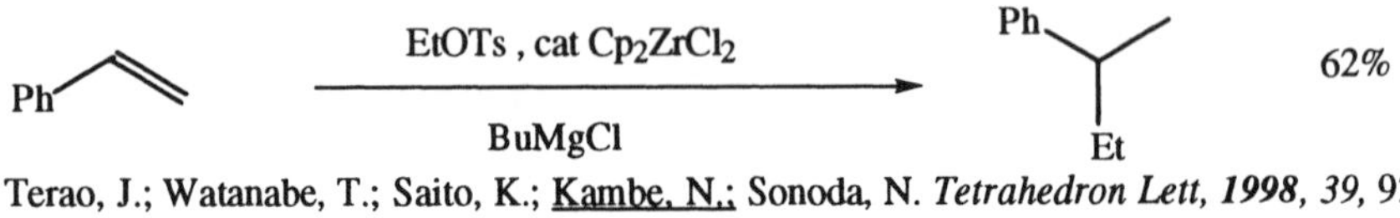
Ph
EtOTs, cat Cp2ZrCl2
BuMgCl
Ph
Et
62%

Terao, J.; Watanabe, T.; Saito, K.; Kambe, N.; Sonoda, N. *Tetrahedron Lett*, ***1998***, *39*, 9201.

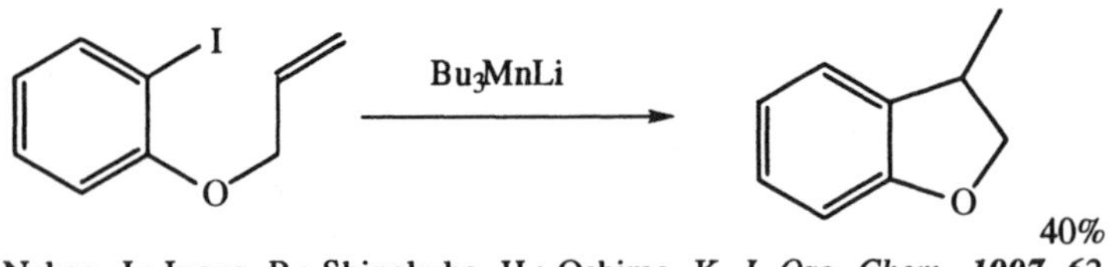
I
O
Bu3MnLi
O
40%

Nakao, J.; Inoue, R.; Shinokubo, H.; Oshima, K. *J. Org. Chem.*, ***1997***, *62*, 1910.

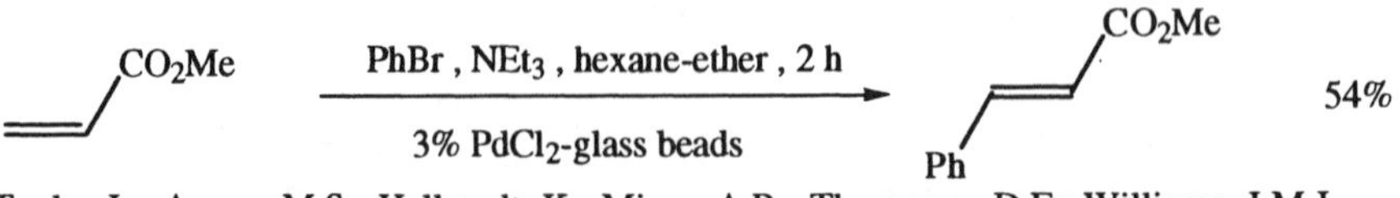
CO2Me
PhBr, NEt3, hexane-ether, 2 h
3% PdCl2-glass beads
CO2Me
Ph
54%

Tonks, L.; Anson, M.S.; Hellgardt, K.; Mirza, A.R.; Thompson, D.F.; Williams, J.M.J. *Tetrahedron Lett.*, ***1997***, *38*, 4319.

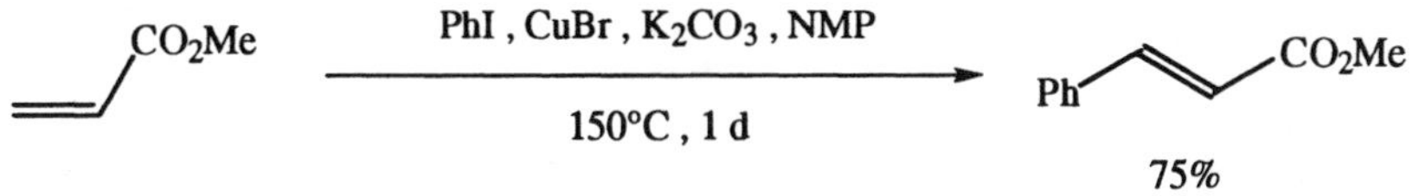
CO2Me
PhI , CuBr , K2CO3 , NMP
150°C , 1 d
Ph
CO2Me
75%

Iyer, S.; Ramesh, C.; Sarkar, A.; Wadgaonkar, P.P. *Tetrahedron Lett.*, ***1997***, *38*, 8113.

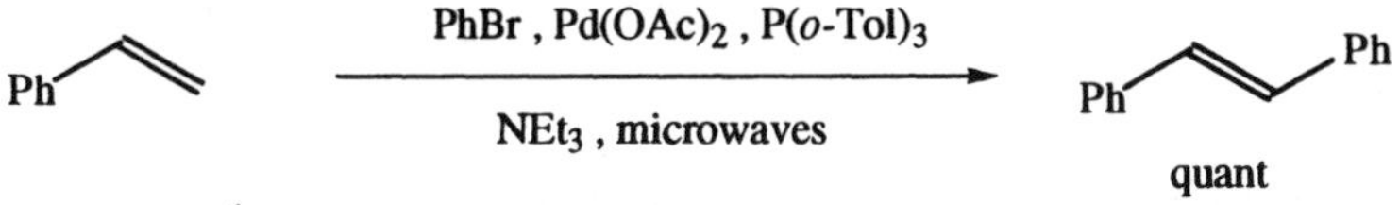
Ph
PhBr , Pd(OAc)2 , P(o-Tol)3
NEt3 , microwaves
Ph
Ph
quant

Díaz-Ortiz, Á.; Prieto, P.; Vázquez, E. *SynLett*, ***1997***, 269.

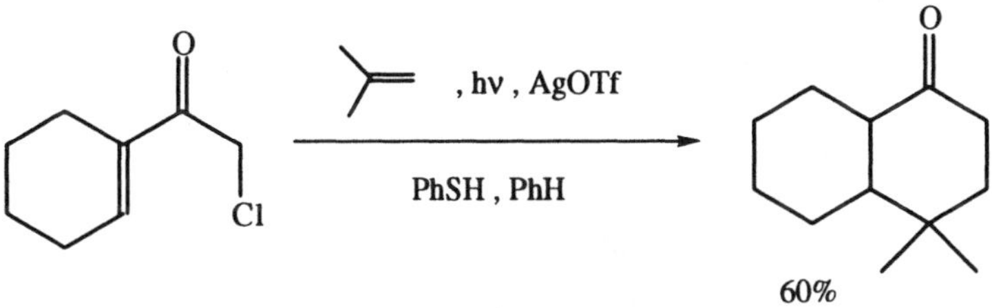
O
Cl
, hν , AgOTf
PhSH , PhH
O
60%

Sawayanagi, Y.; Sato, T.; Shimizu, I. *Chem. Lett.*, ***1997***, 843.

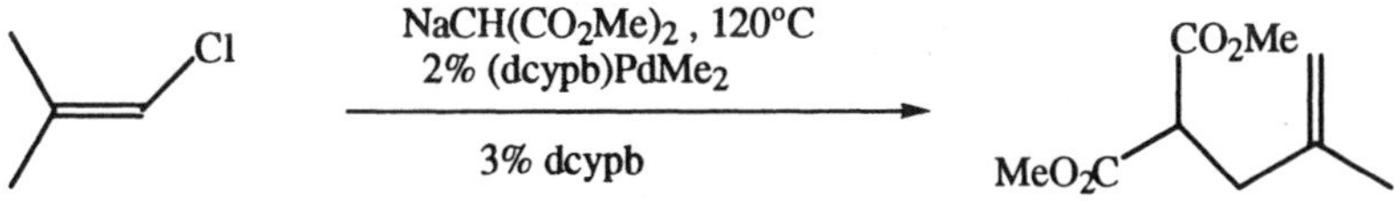
Cl
NaCH(CO2Me)2 , 120°C
2% (dcypb)PdMe2
3% dcypb
CO2Me
MeO2C

dcypb = 1,4-bis(dicyclohexylphosphino)butane

76%

Reetz, M.T.; Wanninger, K.; Hermes, M. *Chem. Commun.*, ***1997***, 535.

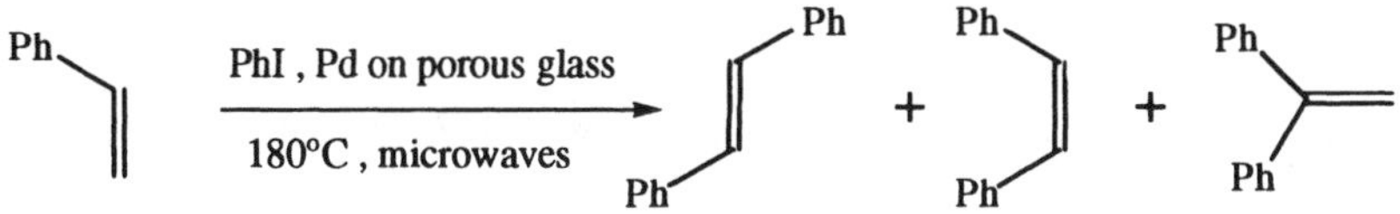
Ph
PhI , Pd on porous glass
180°C , microwaves
Ph
Ph
+
Ph
Ph
+
Ph
Ph

(83 : 1 : 16) quant

Li, J.; Mau, A.W.–H.; Strauss, C.R. *Chem. Commun.*, ***1997***, 1275.

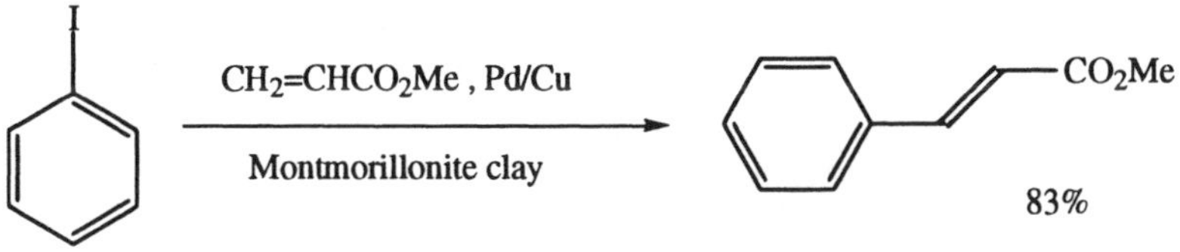
I
CH2=CHCO2Me , Pd/Cu
Montmorillonite clay
CO2Me
83%

Ramchandani, R.K.; Uphade, B.S.; Vinod, M.P.; Wakharkar, R.D.; Choudhary, V.R.; Sudalai, A. *Chem. Commun.*. ***1997***. 2071.

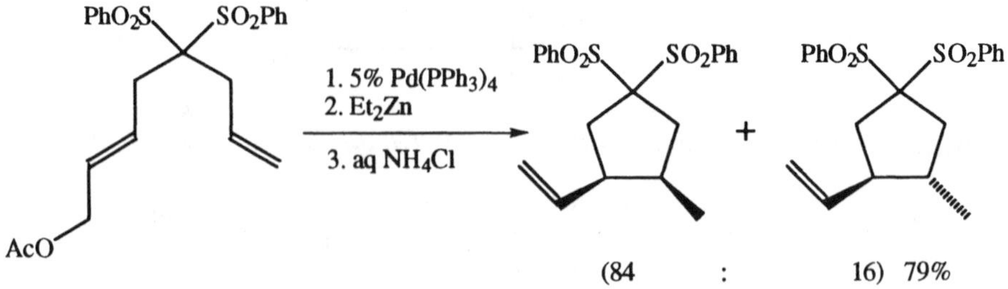
PhO2S
SO2Ph
AcO
1. 5% Pd(PPh3)4
2. Et2Zn
3. aq NH4Cl
PhO2S
SO2Ph
+
PhO2S
SO2Ph

(84 : 16) 79%

Oppolzer, W.; Schröder, F.; Kahl, S. *Helv. Chim. Acta*, ***1997***, *80*, 2047.

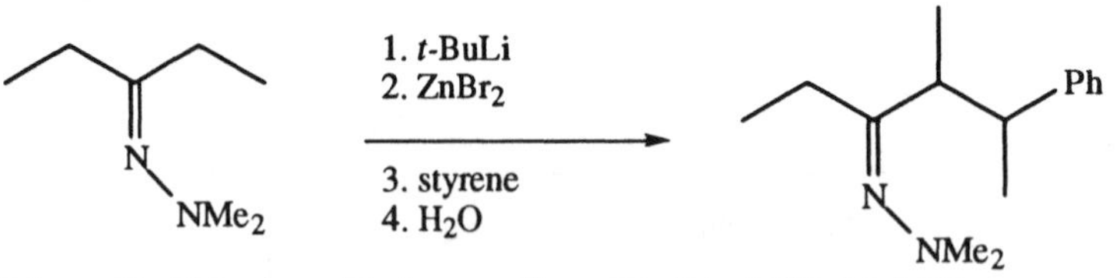
N
NMe2
1. t-BuLi
2. ZnBr2
3. styrene
4. H2O
Ph
N
NMe2

46%

Kubota, K.; Nakamura, E. *Angew. Chem. Int. Ed.*, ***1997***, *36*, 2491.

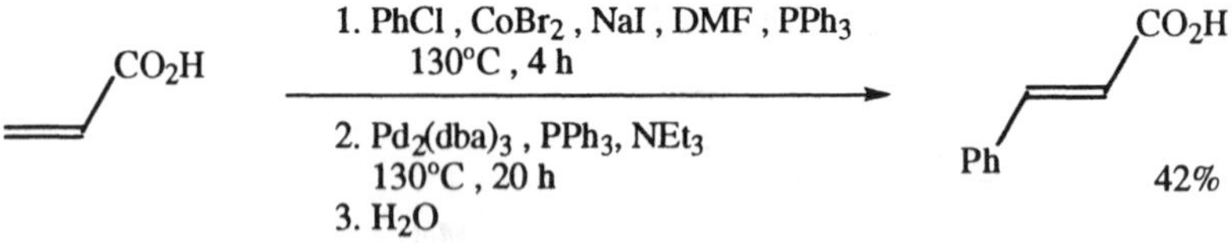
CO2H
1. PhCl , CoBr2 , NaI , DMF , PPh3
130°C , 4 h
2. Pd2(dba)3 , PPh3, NEt3
130°C , 20 h
3. H2O
CO2H
Ph

42%

Mitra, J.; Mitra, A.K. *J. Indian Chem. Soc.*, ***1997***, *74*, 146.

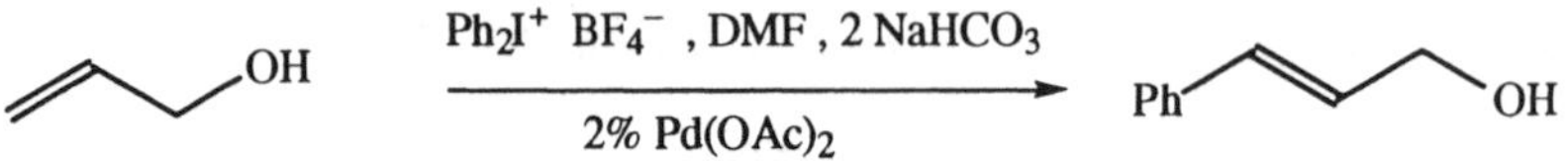
OH
Ph2I+ BF4− , DMF , 2 NaHCO3
2% Pd(OAc)2
Ph
OH

87%

Kang, S.-K.; Lee, H.-W.; Jang, S.B.; Kim, T.-H.; Pyun, S.-J. *J. Org. Chem.*, ***1996***, *61*, 2604

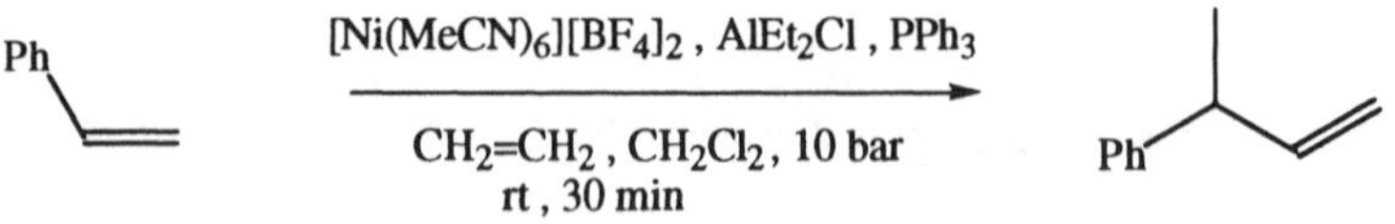
Ph
[Ni(MeCN)6][BF4]2 , AlEt2Cl , PPh3
CH2=CH2 , CH2Cl2 , 10 bar
rt , 30 min
Ph

97%

Monteiro, A.L.; Seferin, M.; Dupont, J.; de Souza, R.F. *Tetrahedron Lett.*, ***1996***, *37*, 1157.

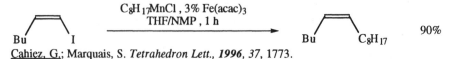

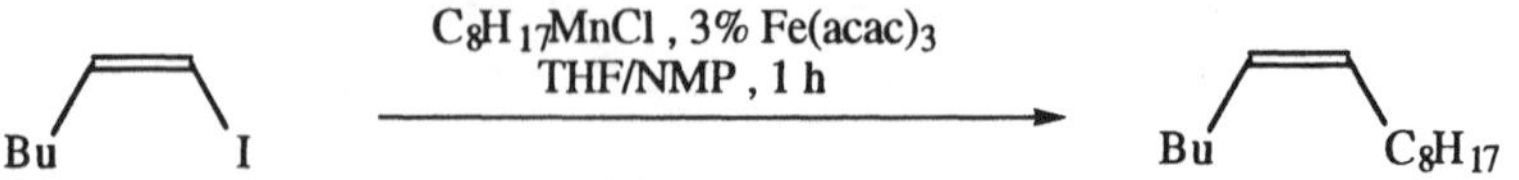
Bu
I
C8H17MnCl , 3% Fe(acac)3
THF/NMP , 1 h
Bu
C8H17

90%

Cahiez, G.; Marquais, S. *Tetrahedron Lett.*, ***1996***, *37*, 1773.

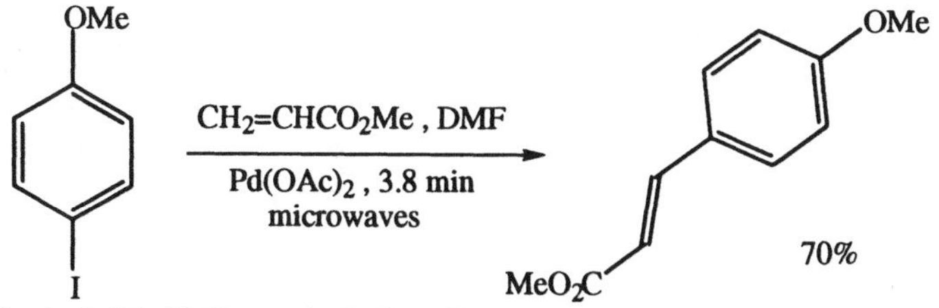
OMe
CH2=CHCO2Me , DMF
Pd(OAc)2 , 3.8 min
microwaves
I
OMe
MeO2C
70%

Larhed, M.; Hallberg, A. *J. Org. Chem.*, ***1996***, *61*, 9582.

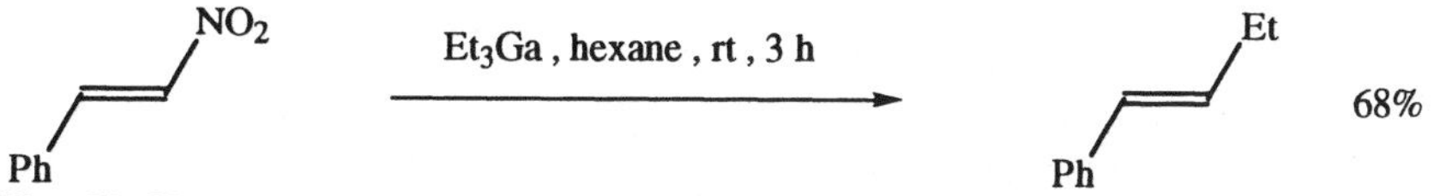
NO2
Ph
Et3Ga , hexane , rt , 3 h
Et
Ph
68%

Han, Y.; Huang, Y.-Z.; Zhou, C.-M. *Tetrahedron Lett.*, ***1996***, *37*, 3347.

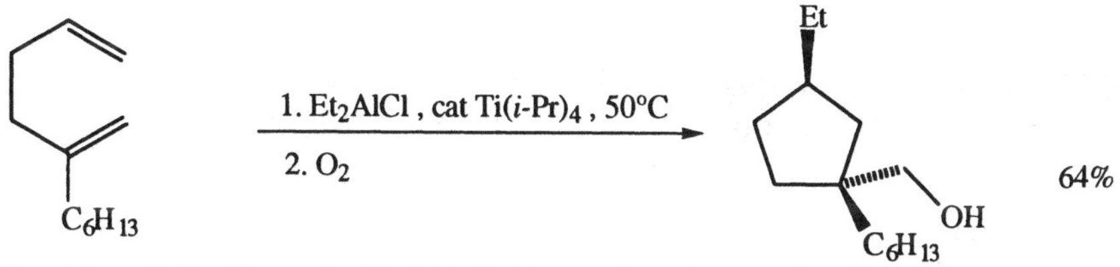
C6H13
1. Et2AlCl , cat Ti(i-Pr)4 , 50°C
2. O2
Et
OH
C6H13
64%

Kondakov, D.Y.; Wang, S.; Negishi, E. *Tetrahedron Lett.*, ***1996***, *37*, 3803.

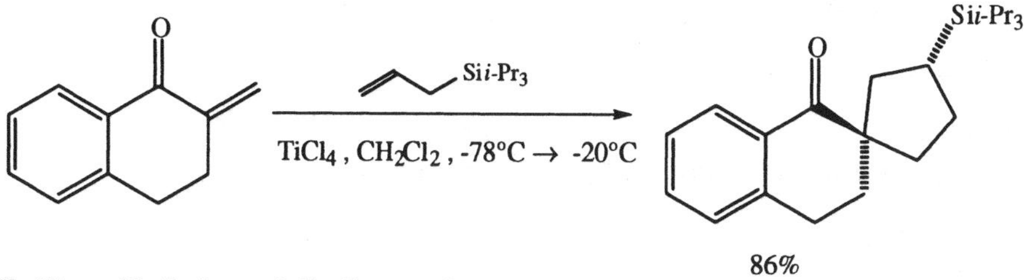
O
Sii-Pr3
TiCl4 , CH2Cl2 , -78°C → -20°C
O
Sii-Pr3
86%

Knölkner, H.-J.; Jones, P.G.; Graf, R. *SynLett*, ***1996***, 1155.

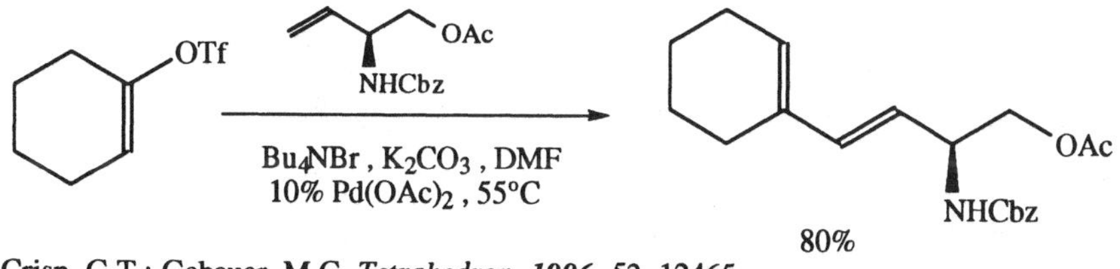
OTf
OAc
NHCbz
Bu4NBr , K2CO3 , DMF
10% Pd(OAc)2 , 55°C
OAc
NHCbz
80%

Crisp, G.T.; Gebauer, M.G. *Tetrahedron*, ***1996***, *52*, 12465.

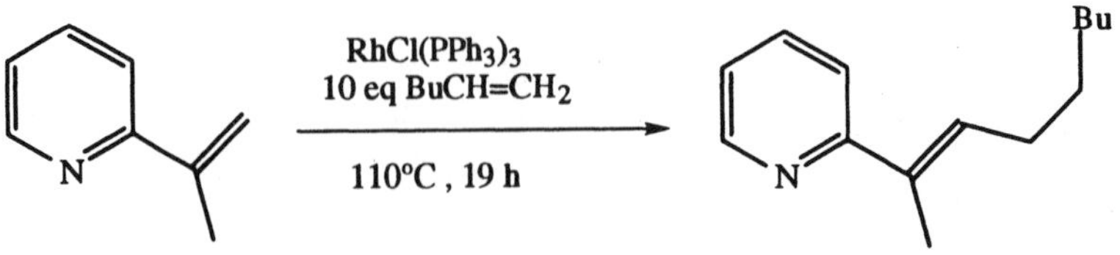
RhCl(PPh3)3
10 eq BuCH=CH2
110°C, 19 h
Bu
N
N

99% (93:7 *trans:cis*)

Lim, Y.–G.; Kang, J.–B.; Kim, Y.H. *Chem. Commun*, ***1996***, 585.

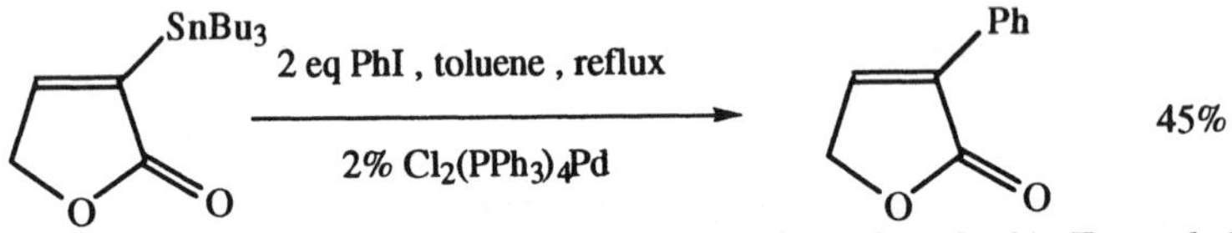
SnBu3
2 eq PhI, toluene, reflux
2% Cl2(PPh3)4Pd
Ph
O
O
O
O
45%

Holingworth, G.J.; Perkins, G.; Sweeney, J. *J. Chem. Soc., Perkin Trans. 1*, ***1996***, 1913.

SECTION 74D: Conjugate Reduction of α,β-Unsaturated Carbonyl Compounds and Nitriles

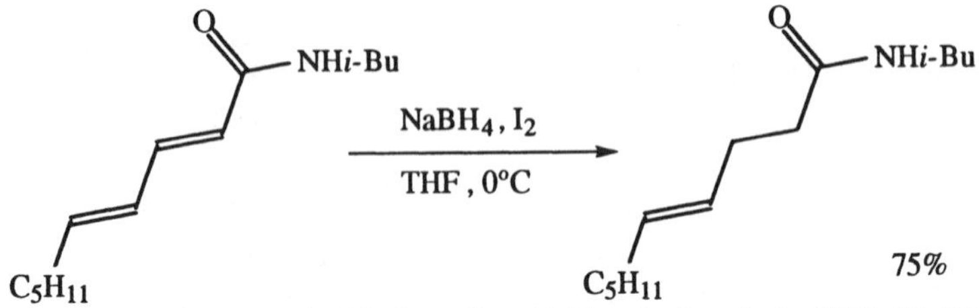
O
NHi-Bu
NaBH4, I2
THF, 0°C
O
NHi-Bu
C5H11
C5H11
75%

Das, B.; Kashinatham, A.; Madhusudhan, P. *Tetrahedron Lett.*, ***1998***, *39*, 677.

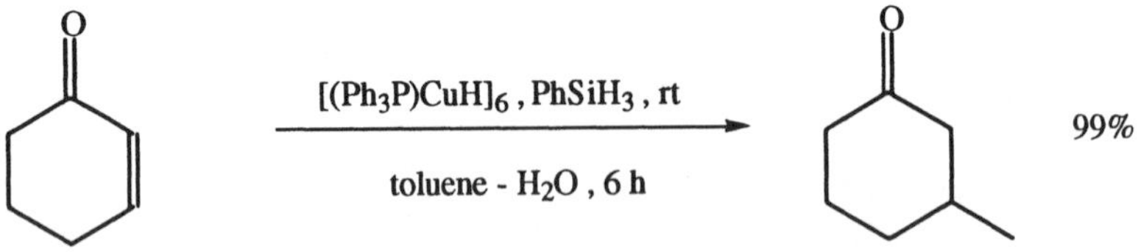
O
[(Ph3P)CuH]6, PhSiH3, rt
toluene - H2O, 6 h
O
99%

Lipshutz, B.H.; Keith, J.; Papa, P.; Vivian, R. *Tetrahedron Lett*, ***1998***, *39*, 4627.

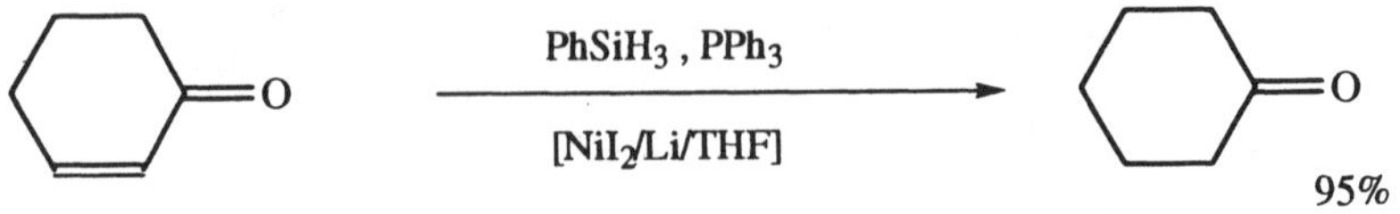
O
PhSiH3, PPh3
[NiI2/Li/THF]
O
95%

Boudjouk, P.; Choi, S.–B.; Hauck, B.J.; Rajkumar, A.B. *Tetrahedron Lett.*, ***1998***, *39*, 3951.

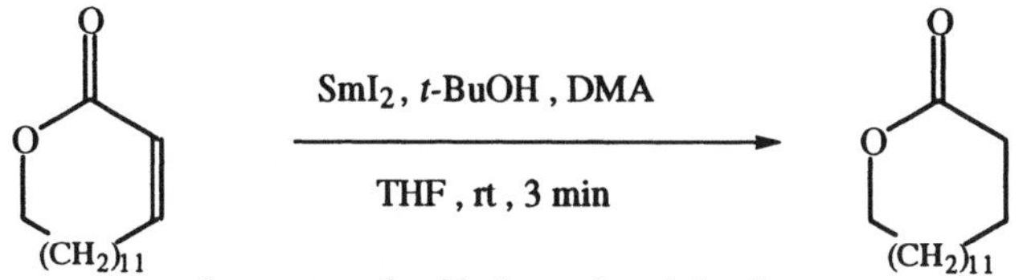
O
O
(CH2)11
SmI2 , t-BuOH , DMA
THF , rt , 3 min
O
O
(CH2)11

88%

does not work with 6-membered ring lactones

Fujita, Y.; Fukuzumi, S.; Otera, J. *Tetrahedron Lett.*, ***1997***, *38*, 2121.

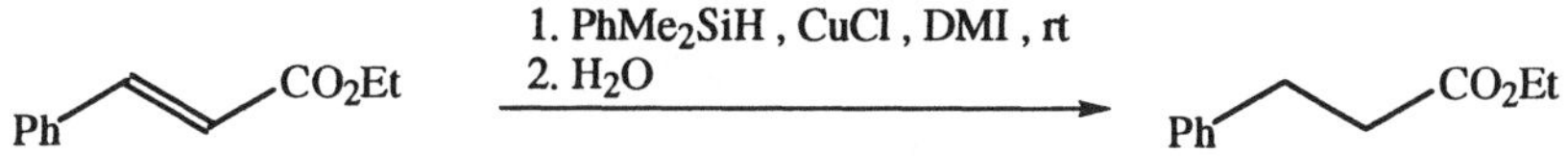
Ph
CO2Et
1. PhMe2SiH , CuCl , DMI , rt
2. H2O
Ph
CO2Et

DMI = 1,3-dimethylimidazolidinone 96%

Ito, H.; Ishizuka, T.; Arimoto, K.; Miura, K.; Hosomi, A. *Tetrahedron Lett.*, ***1997***, *38*, 8887

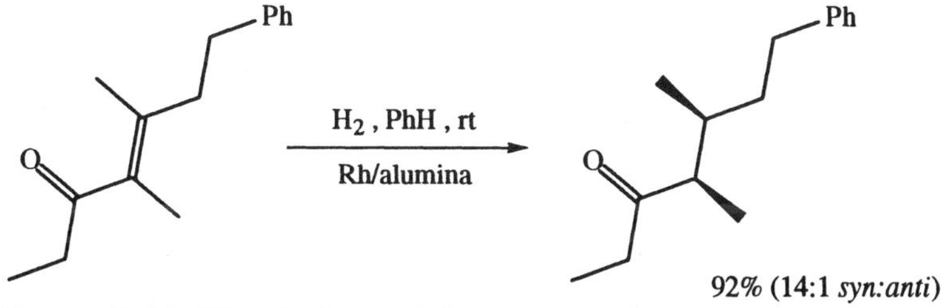
Ph
O
H2 , PhH , rt
Rh/alumina
Ph
O

92% (14:1 *syn:anti*)

Yamaguchi, M.; Nitta, A.; Reddy, R.S.; Hirama, M. *SynLett*, ***1997***, 117.

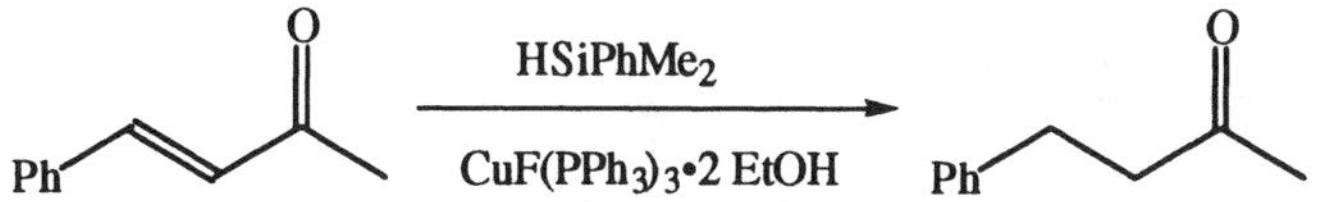
O
Ph
HSiPhMe2
CuF(PPh3)3•2 EtOH
O
Ph

> 99%

Mori, A.; Fujita, A.; Nishihara, Y.; Higama, T. *Chem. Commun.*, ***1997***, 2159.

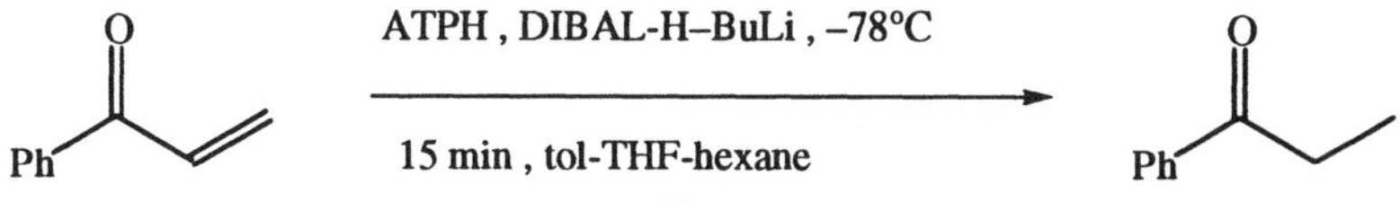
O
Ph
ATPH , DIBAL-H–BuLi , –78°C
15 min , tol-THF-hexane
O
Ph

ATPH =

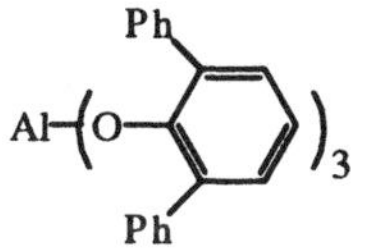
Ph
Al
O
3
Ph

99%

Saito, S.; Yamamoto, H. *J. Org. Chem.*, ***1996***, *61*, 2928.

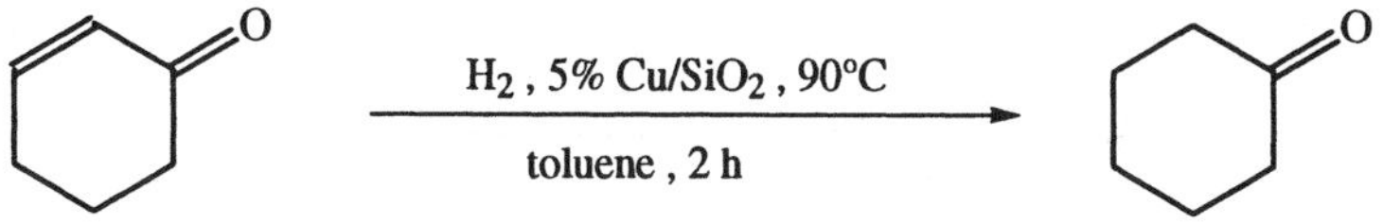
O
H2 , 5% Cu/SiO2 , 90°C
toluene , 2 h
O

quant

Ravasio, N.; Antenori, M.; Gargano, M.; Mastrorilli, P. *Tetrahedron Lett.*, ***1996***, *37*, 3529.

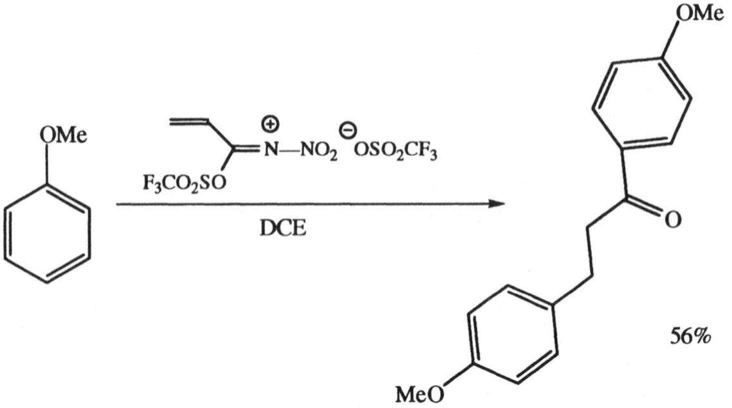
OMe
F3CO2SO
N—NO2
OSO2CF3
DCE
OMe
O
MeO
56%

Nenajdenko, V.G.; Baraznenok, I.L.; Balenkova, E.S. *Tetrahedron Lett.*, ***1996***, *37*, 4199.

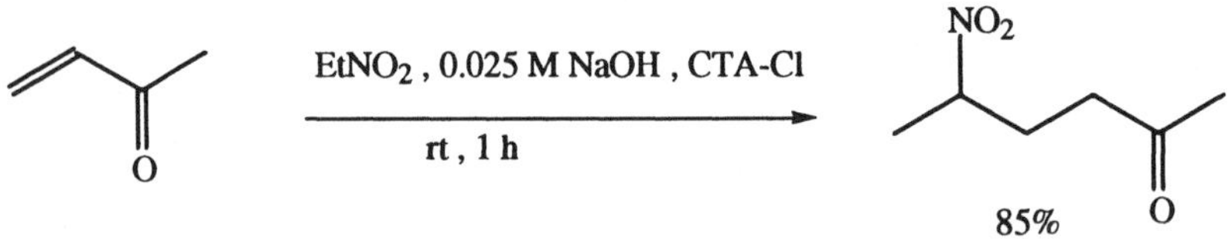
O
EtNO2 , 0.025 M NaOH , CTA-Cl
rt , 1 h
NO2
O
85%

Ballini, R.; Boscia, G. *Tetrahedron Lett.*, ***1996***, *37*, 8027.

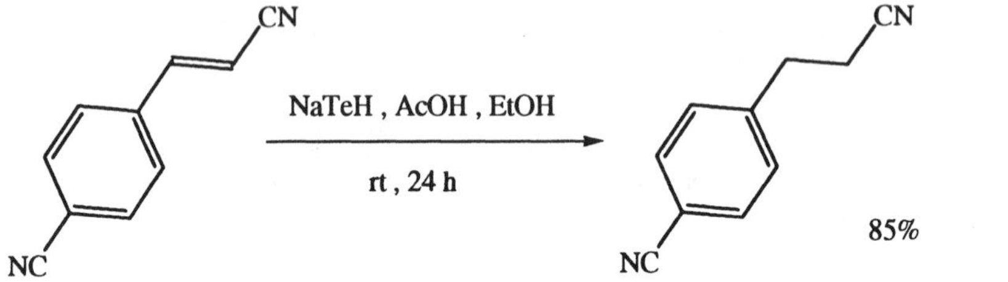
CN
NC
NaTeH , AcOH , EtOH
rt , 24 h
CN
NC
85%

Blay, G.; Cardona, L.; García, B.; Lahoz, L.; Pedro, J.R. *Tetrahedron*, ***1996***, *52*, 8611.

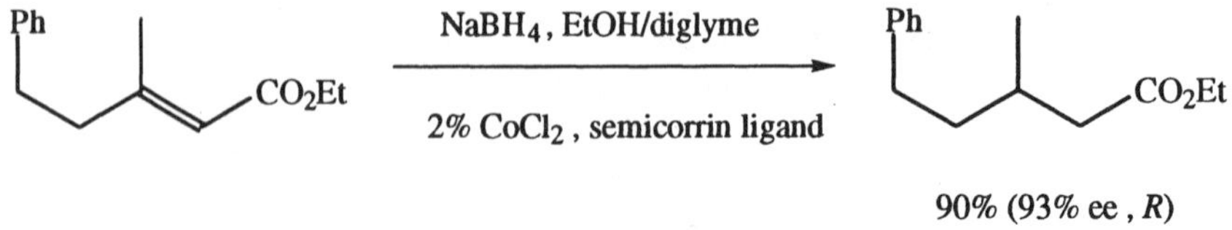
Ph
CO2Et
NaBH4, EtOH/diglyme
2% CoCl2 , semicorrin ligand
Ph
CO2Et
90% (93% ee , R)

Misun, M.; Pfaltz, A. *Helv. Chim. Acta*, ***1996***, *79*, 961.

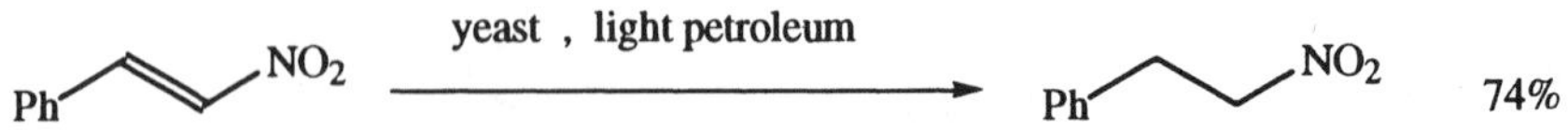
Ph
NO2
yeast , light petroleum
Ph
NO2
74%

Bak, R.R.; McAnda, A.F.; Smallridge, A.J.; Trewhella, M.A. *Aust. J. Chem.*, ***1996***, *49*, 1257

SECTION 74E: Conjugate Alkylations

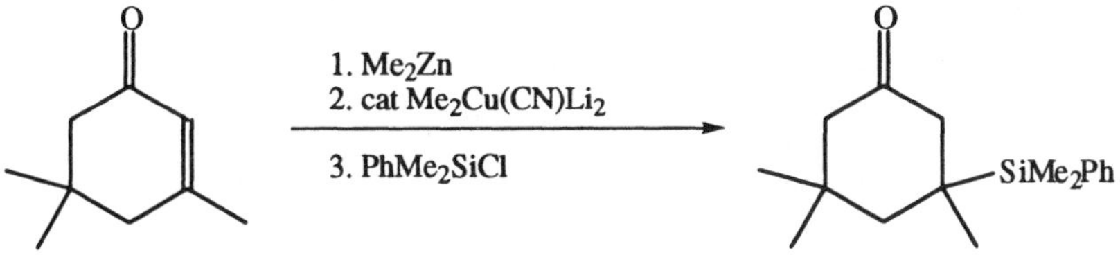
1. Me_2Zn
2. cat $Me_2Cu(CN)Li_2$
3. $PhMe_2SiCl$
$SiMe_2Ph$

90%

Lipshutz, B.H.; Sclafani, J.A.; Takanami, T. *J. Am. Chem. Soc.*, ***1998***, *120*, 4021.

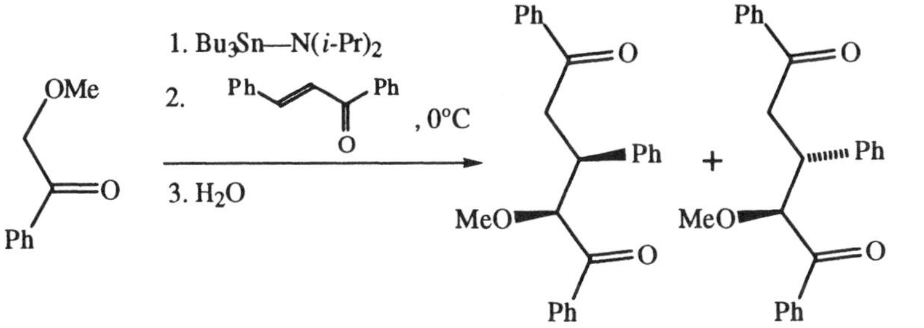
1. Bu_3Sn—N(*i*-Pr)$_2$
2.
, 0°C
3. H_2O

(22 : 78) 66%

with Cl_2BnSn—N(*i*-Pr)$_2$ (82 : 18) 99%

Shibata, I.; Yasuda, K.; Tanaka, Y.; Yasuda, M.; Baba, A. *J. Org. Chem.*, ***1998***, *63*, 1334.

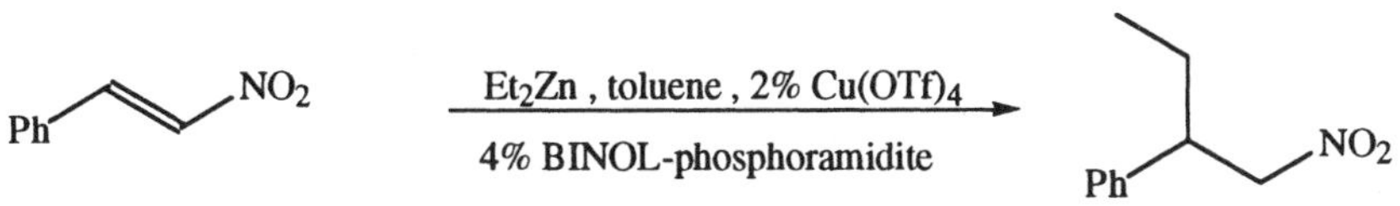
Et_2Zn , toluene , 2% $Cu(OTf)_4$
4% BINOL-phosphoramidite

90% (48% ee)

Sewald, N.; Wendisch, V. *Tetrahedron Asymmetry*, ***1998***, *9*, 1341.

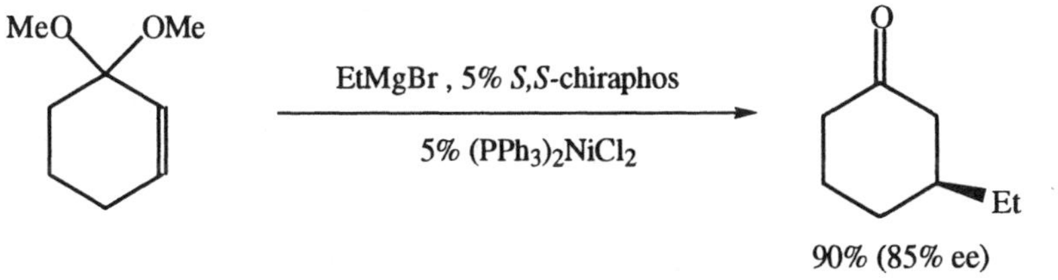
EtMgBr , 5% *S,S*-chiraphos
5% $(PPh_3)_2NiCl_2$

90% (85% ee)

Gomez-Bengoa, E.; Heron, N.M.; Didiuk, M.T.; Luchaco, C.A.; Hoveyda, A.H. *J. Am. Chem. Soc.*, ***1998***, *120*, 7649.

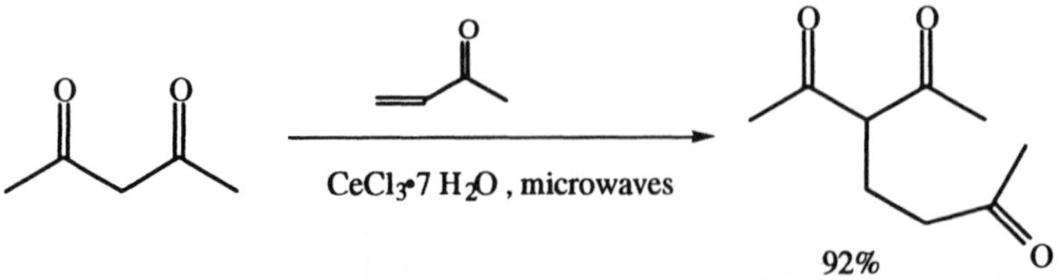
CeCl$_3$•7 H$_2$O , microwaves

92%

Boruah, A.; Baruah, M.; Prajapati, D.; Sandhu, J.S. *Synth. Commun.*, ***1998***, *28*, 653.

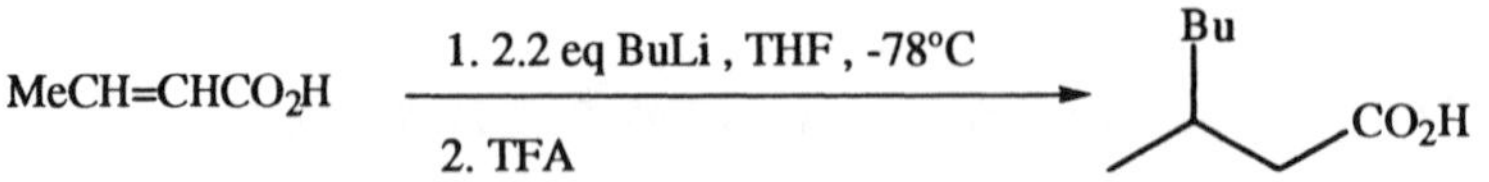
MeCH=CHCO$_2$H
1. 2.2 eq BuLi , THF , -78°C
2. TFA
Bu
CO$_2$H

80%

Plunian, B.; Vaultier, M.; Mortier, J. *Chem. Commun.*, ***1998***, 81.

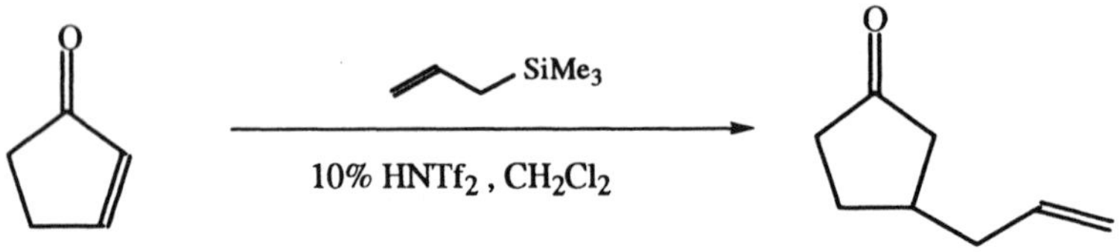
SiMe$_3$
10% HNTf$_2$, CH$_2$Cl$_2$

95%

Kuhnert, N.; Peverley, J.; Robertson, J. *Tetrahedron Lett.*, ***1998***, *39*, 3215.

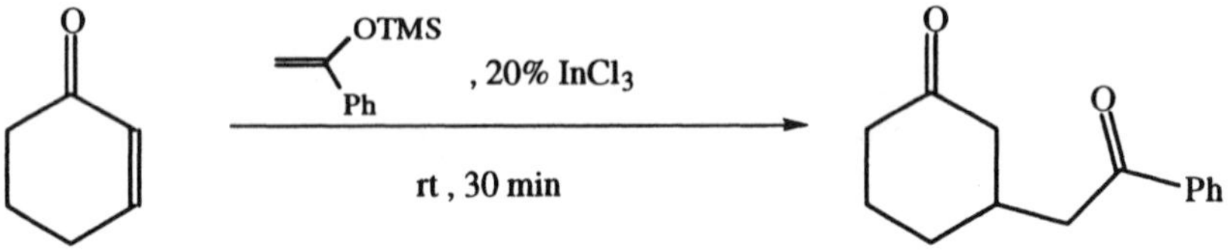
OTMS
Ph
, 20% InCl$_3$
rt , 30 min
Ph

67%

Loh, T.-P.; Wei, L.-L. *Tetrahedron*, ***1998***, *54*, 7615.

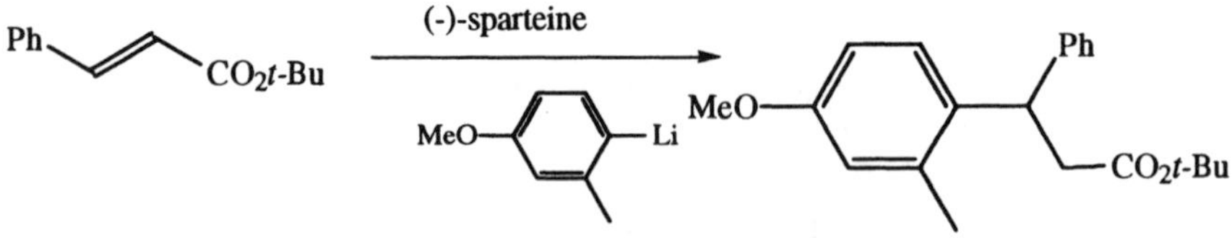
Ph
CO$_2$*t*-Bu
(-)-sparteine
MeO
Li
MeO
Ph
CO$_2$*t*-Bu

78% (25% ee)

Xu, F.; Tillyer, R.D.; Tschaen, D.M.; Grabowski, E.J.J.; Reider, P.J. *Tetrahedron Asymmetry*, ***1998***, *9*, 1651.

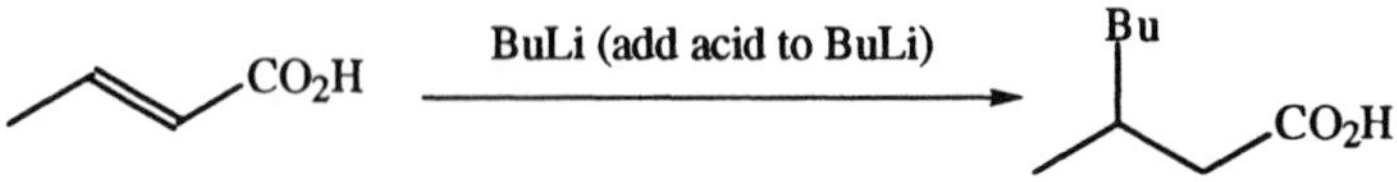
CO$_2$H
BuLi (add acid to BuLi)
Bu
CO$_2$H

65%

Aurell, M.J.; Mestres, R.; Muñoz, E. *Tetrahedron Lett.*, ***1998***, *39*, 6351.

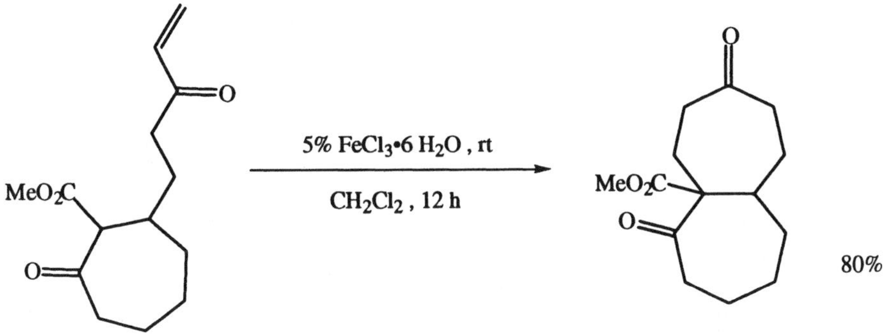
5% $FeCl_3$•6 H_2O , rt
CH_2Cl_2 , 12 h
MeO_2C
O
80%

Christoffers, J. *Tetrahedron Lett.*, ***1998***, *39*, 7083.

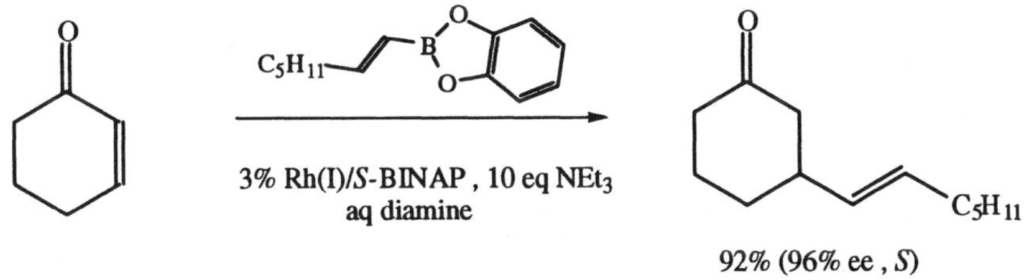
C_5H_{11}
3% Rh(I)/*S*-BINAP , 10 eq NEt_3
aq diamine
92% (96% ee , *S*)

Takaya, Y.; Ogasawara, M.; Hayashi, T. *Tetrahedron Lett*, ***1998***, *39*, 8479.

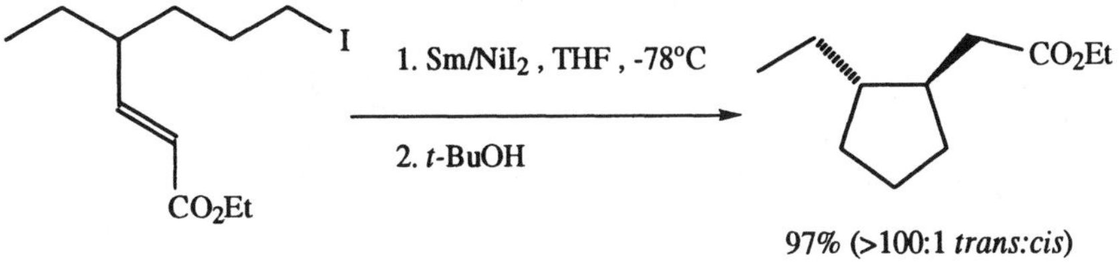
I
1. Sm/NiI_2 , THF , -78°C
2. *t*-BuOH
CO_2Et
97% (>100:1 *trans:cis*)

Molander, G.A.; Harris, C.R. *J. Org. Chem.*, ***1997***, *62*, 7418.

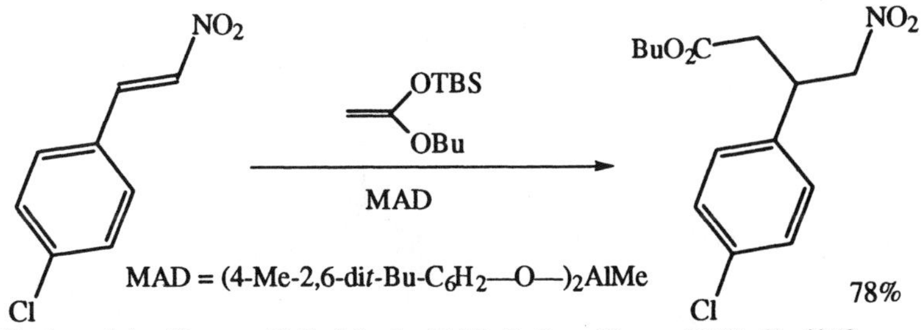
NO_2
OTBS
OBu
MAD
BuO_2C
Cl
MAD = (4-Me-2,6-di*t*-Bu-C_6H_2—O—$)_2$AlMe
78%

Tucker, J.A.; Clayton, T.C.; Morda, D.M. *J. Org. Chem.*, ***1997***, *62*, 4370.

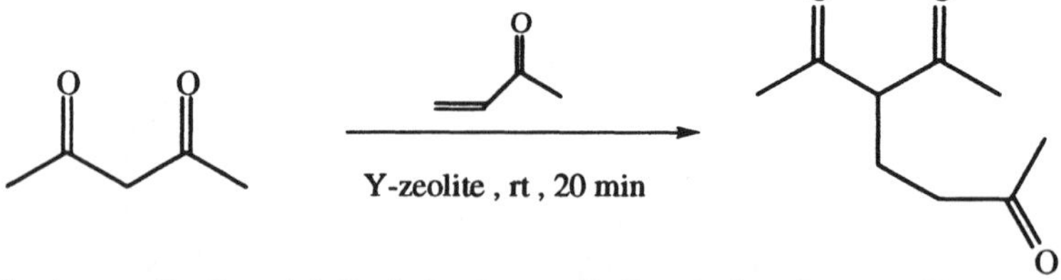
Y-zeolite , rt , 20 min

82%

Sreekumar, R.; Rugmini, P.; Padmakumar, R. *Tetrahedron Lett.*, ***1997***, *38*, 6557.

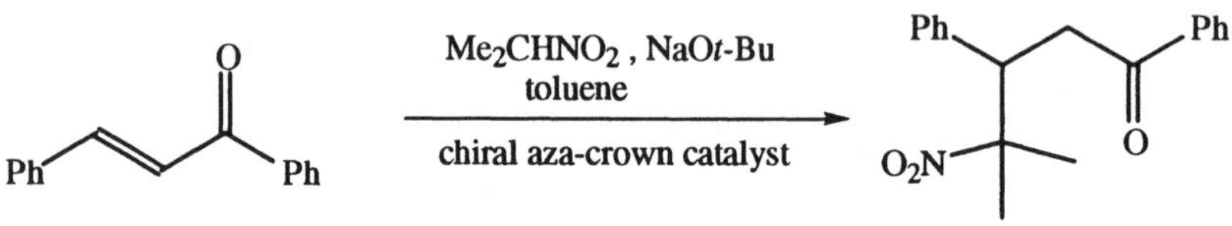
Me2CHNO2 , NaOt-Bu
toluene
chiral aza-crown catalyst
Ph
Ph
O2N
O

73% 961% ee , *S*)

Bakó, P.; Kiss, T.; Tõke, L. *Tetrahedron Lett.*, ***1997***, *38*, 7259.

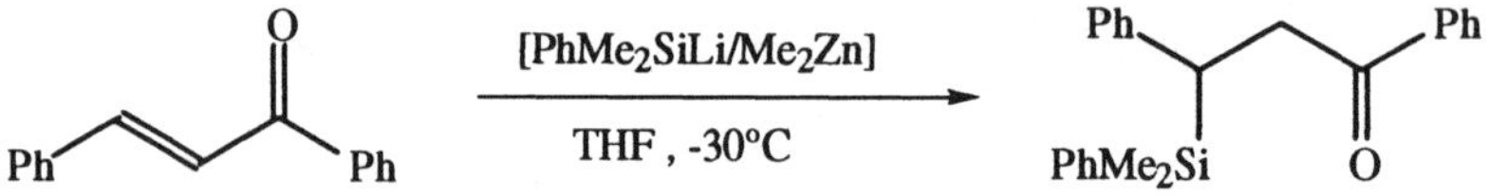
[PhMe2SiLi/Me2Zn]
THF , -30°C
PhMe2Si

91%

MacLean, B.L.; Henningar, K.A.; Kells, K.W.; Singer, R.D. *Tetrahedron Lett.*, ***1997***, *38*, 7313

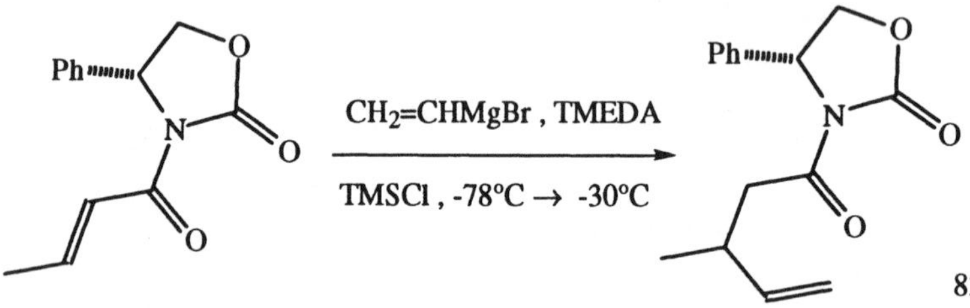
CH2=CHMgBr , TMEDA
TMSCl , -78°C → -30°C

83% (>90% de)

Han, Y.; Hruby, V.J. *Tetrahedron Lett.*, ***1997***, *38*, 7317.

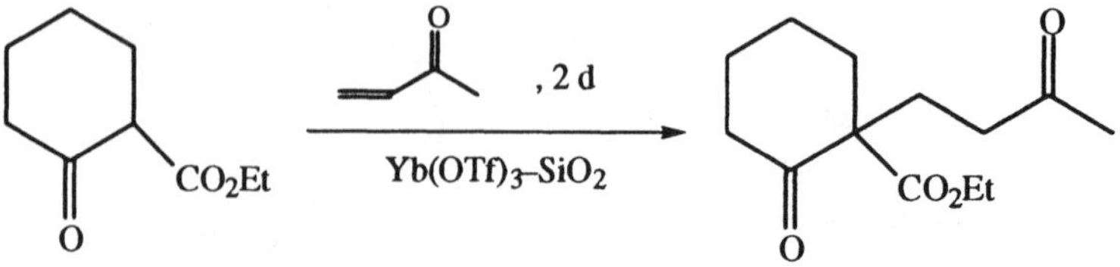
, 2 d
Yb(OTf)3-SiO2
CO2Et

95%

Kotsuki, H.; Arimura, K. *Tetrahedron Lett*, ***1997***, *38*, 7583.

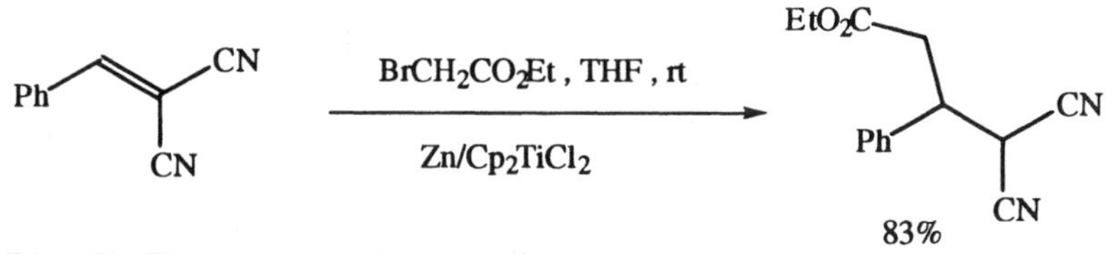
Ph
CN
CN
BrCH2CO2Et , THF , rt
Zn/Cp2TiCl2
EtO2C
Ph
CN
CN
83%

Ding, Y.; Zhao, Z.; Zhou, C. *Tetrahedron*, ***1997***, *53*, 2899.

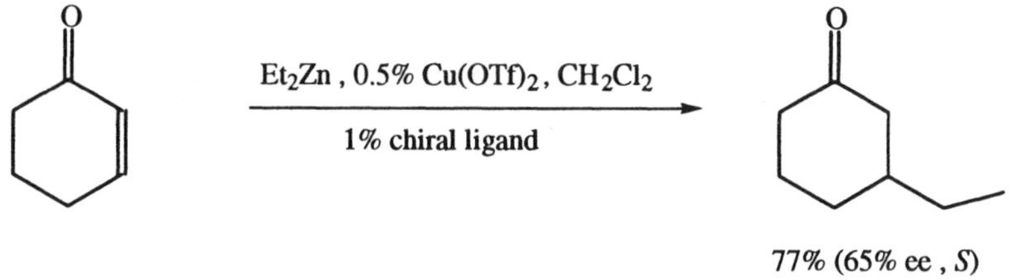
O
Et2Zn , 0.5% Cu(OTf)2 , CH2Cl2
1% chiral ligand
O
77% (65% ee , S)

Alexakis, A.; Vastra, J.; Burton, J.; Mangeney, P. *Tetrahedron Asymmetry*, ***1997***, *8*, 3193.

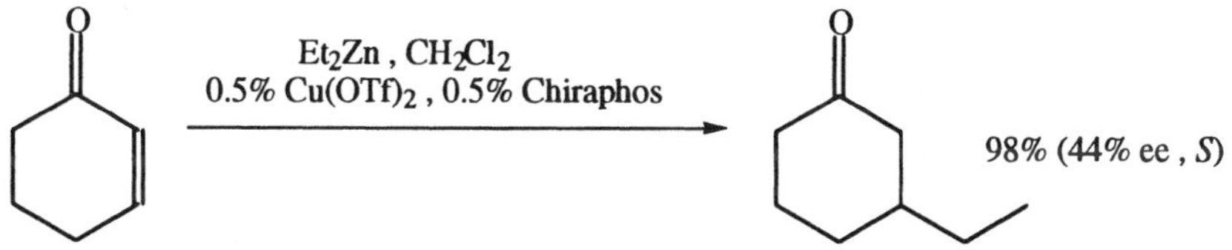
O
Et2Zn , CH2Cl2
0.5% Cu(OTf)2 , 0.5% Chiraphos
O
98% (44% ee , S)

Alexakis, A.; Burton, J.; Vastra, J.; Mangeney, P. *Tetrahedron Asymmetry*, ***1997***, *8*, 3987.

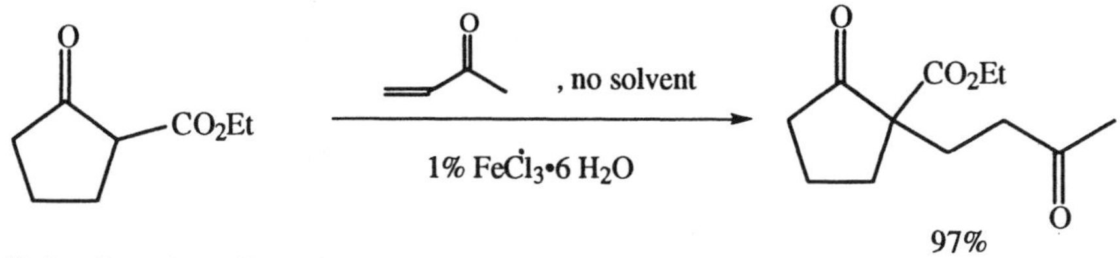
O
CO2Et
O
, no solvent
1% FeCl3•6 H2O
O
CO2Et
O
97%

Christoffers, J. *J. Chem. Soc., Perkin Trans. 1*, ***1997***, 3141.
Christoffers, J. *Chem. Commun.*, ***1997***, 943.

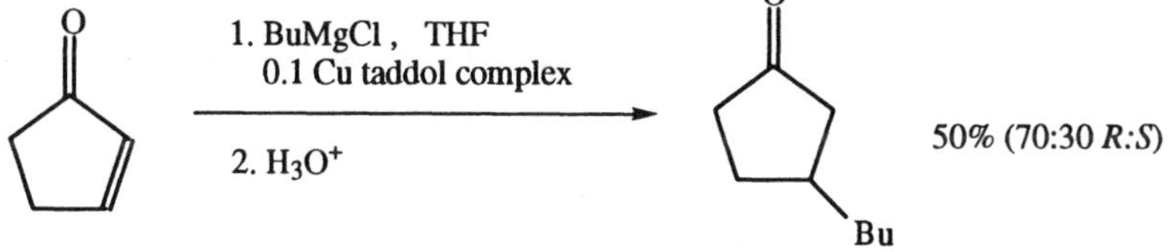
O
1. BuMgCl , THF
0.1 Cu taddol complex
2. H3O+
O
Bu
50% (70:30 R:S)

Seebach, D.; Jaeschke, G.; Pichota, A.; Audergon, L. *Helv. Chim. Acta*, ***1997***, *80*, 2515.

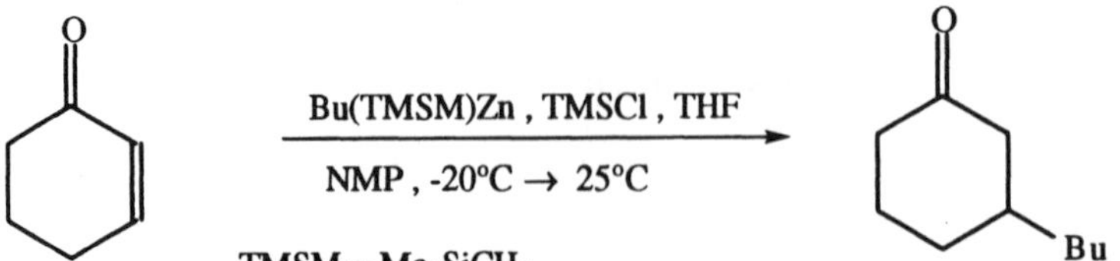
Bu(TMSM)Zn , TMSCl , THF
NMP , -20°C → 25°C
Bu

83%

TMSM = Me_3SiCH_2

Berger, S.; Langer, F.; Lutz, C.; Knochel, P.; Mobley, T.A.; Reddy, C.K. *Angew. Chem. Int. Ed.*, ***1997***, *36*, 1496.

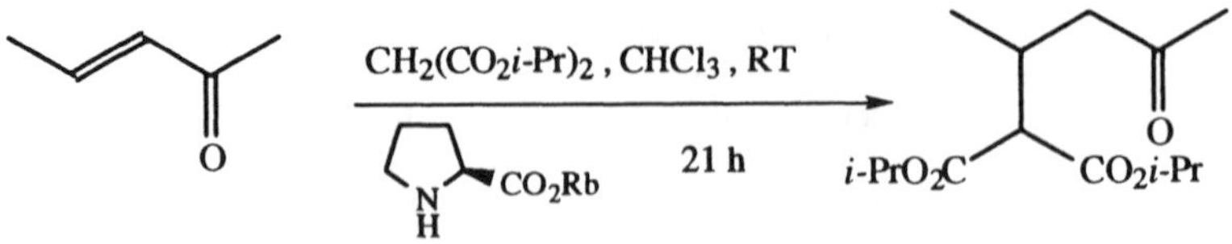
CH2(CO2i-Pr)2 , CHCl3 , RT
CO2Rb
21 h
i-PrO2C
CO2i-Pr

71%
(76% ee , (*S*))

Yamaguchi, M.; Shiraishi, T.; Hirama, M. *J. Org. Chem.*, ***1996***, *61*, 3520.

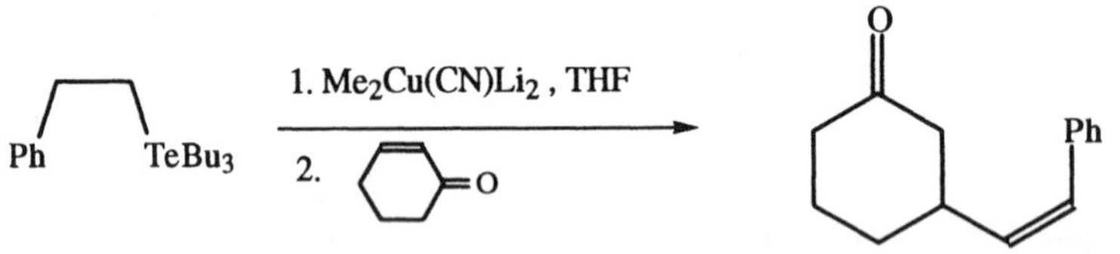
Ph
TeBu3
1. Me2Cu(CN)Li2 , THF
2.
Ph

90%

Tucci, F.C.; Chieffi, A.; Comasseto, J.V.; Marino, J.P. *J. Org. Chem.*, ***1996***, *61*, 4975.

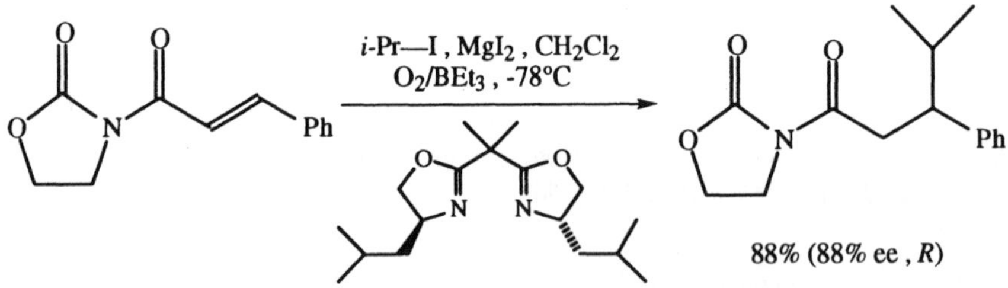
Ph
i-Pr—I , MgI2 , CH2Cl2
O2/BEt3 , -78°C
Ph

88% (88% ee , *R*)

Sibi, M.P.; Ji, J.; Wu, J.H.; Gürtler, S.; Porter, N.A. *J. Am. Chem. Soc.*, ***1996***, *118*, 9200.

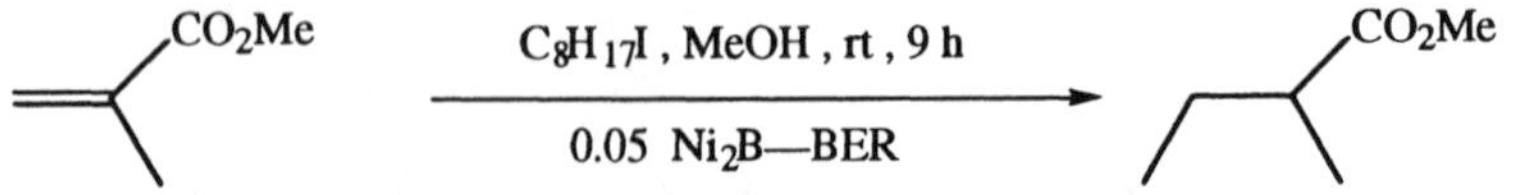
CO2Me
C8H17I , MeOH , rt , 9 h
0.05 Ni2B—BER
CO2Me

88%

BER = borohydride exchange resin

Sim, T.B.; Choi, J.; Yoon, N.M. *Tetrahedron Lett.*, ***1996***, *37*, 3137.

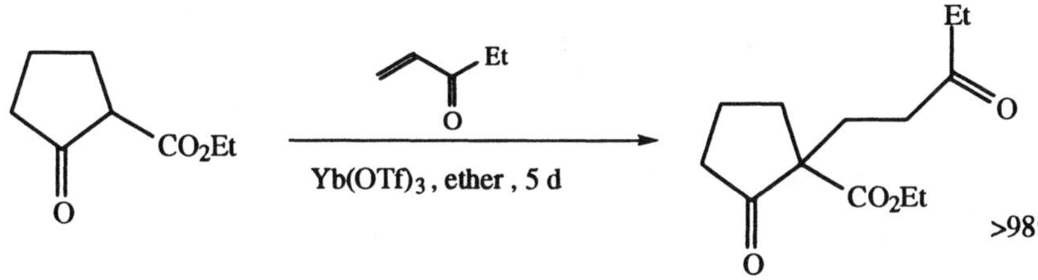
Et
O
CO2Et
Yb(OTf)3 , ether , 5 d
O
Et
O
CO2Et
O

>98%

Keller, E.; Feringa, B.L. *Tetrahedron Lett.*, ***1996***, *37*, 1879.

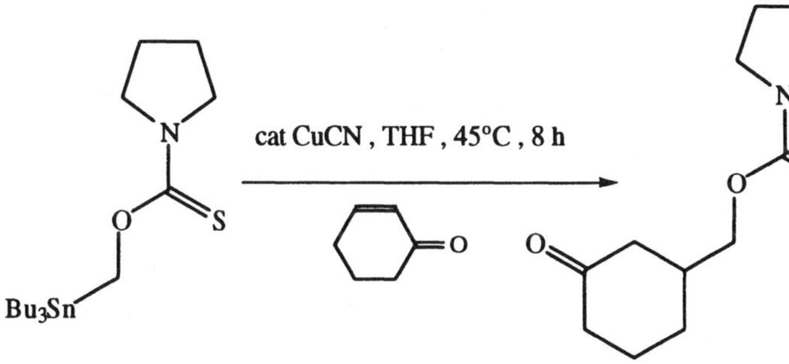
N
O
S
Bu3Sn
cat CuCN , THF , 45°C , 8 h
O
N
O
S
O

86%

Bhatt, R.K.; Ye, J.; Falck, J.R. *Tetrahedron Lett.* ***1996***, *37*, 3811.

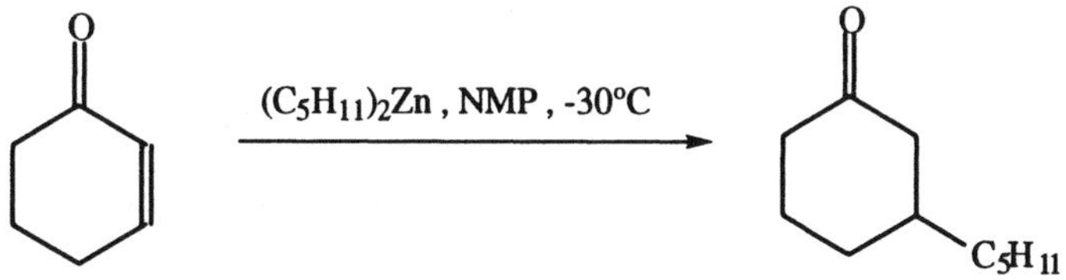
O
(C5H11)2Zn , NMP , -30°C
O
C5H11

83%

Reddy, Ch.K.; Devasagayara, A.; Knochel, P. *Tetrahedron Lett.*, ***1996***, *37*, 4495.

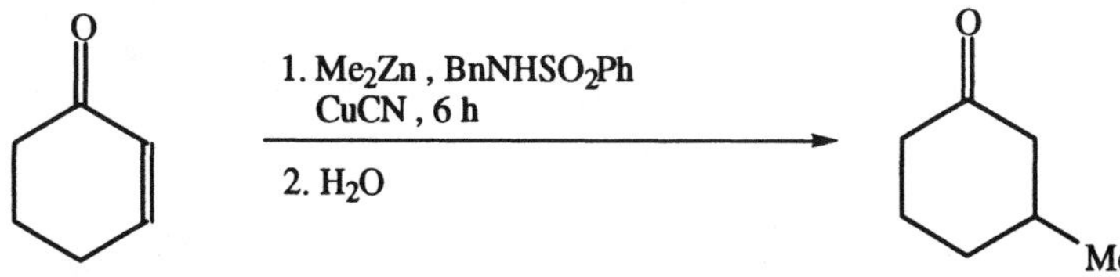
O
1. Me2Zn , BnNHSO2Ph
CuCN , 6 h
2. H2O
O
Me

90%

Kitamura, M.; Miki, T.; Nakano, K.; Noyori, R. *Tetrahedron Lett.*, ***1996***, *37*, 5141.

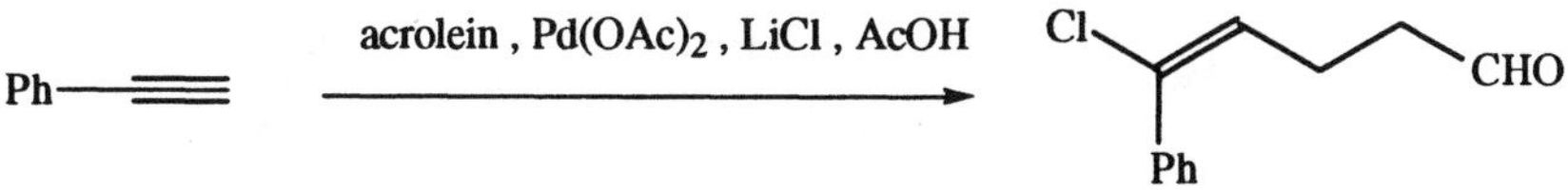
Ph
acrolein , Pd(OAc)2 , LiCl , AcOH
Cl
CHO
Ph

77% (95:5 *Z*:*E*)

Wang, Z.; Lu, X. *Chem. Commun.*, ***1996***, 535.

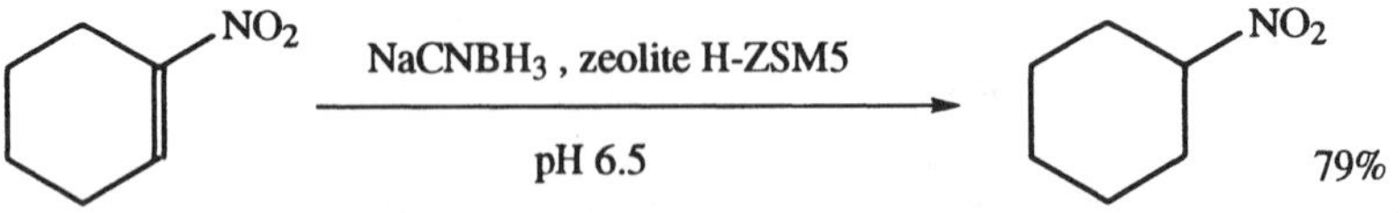
NO2
NaCNBH3 , zeolite H-ZSM5
pH 6.5
NO2
79%

Gupta, A.; Haque, A.; Vankar, Y.D. *Chem. Commun.*, ***1996***, 1653.

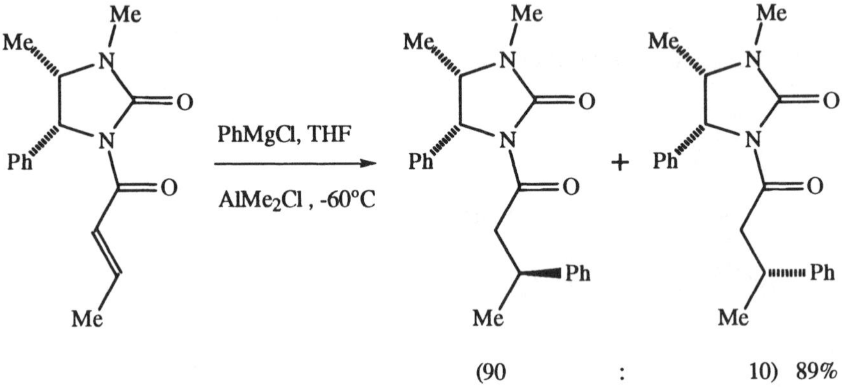
PhMgCl, THF
AlMe2Cl , -60°C
(90 : 10) 89%

Bongini, A.; Cardillo, G.; Mingardi, A.; Tomasini, C. *Tetrahedron Asymmetry, **1996**, 7,* 1457

SECTION 74F: Cyclopropanations, including Halocyclopropanations

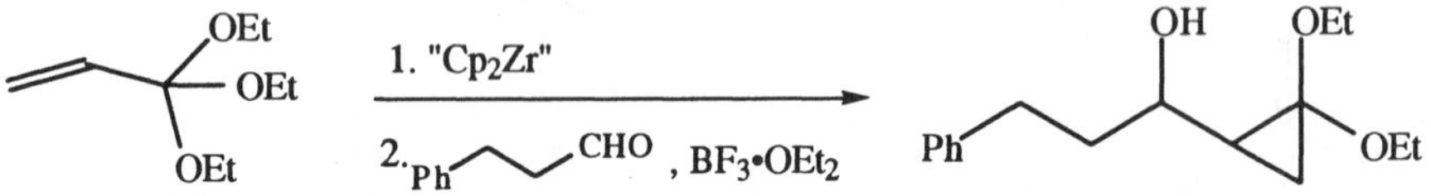
1. "Cp2Zr"
2. Ph CHO , BF3•OEt2

Ito, H.; Kuro, H.; Ding, H.; Taguchi, T. *J. Am. Chem. Soc.*, ***1998***, *120*, 6623.

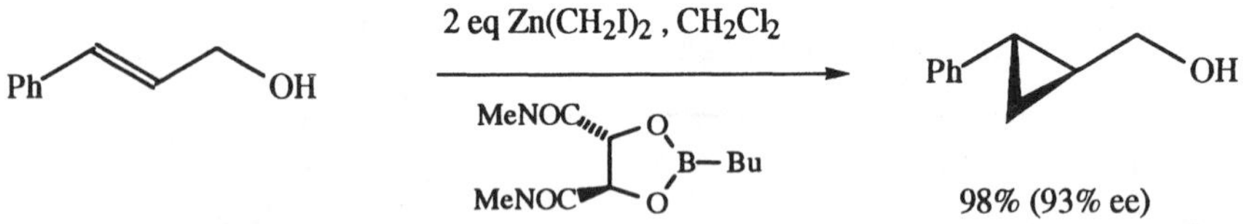
2 eq Zn(CH2I)2 , CH2Cl2
98% (93% ee)

Charette, A.B.; Juteau, H.; Lebel, H.; Molinaro, C. *J. Am. Chem. Soc.*, ***1998***, *120*, 11942.

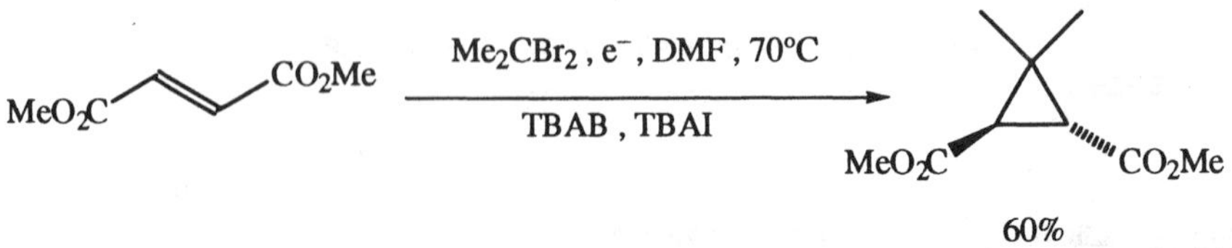
Me2CBr2 , e⁻, DMF , 70°C
TBAB , TBAI
60%

Léonel, E.; Paugam, J.P.; Condon-Gueugnot, S.; Nédélec, J.-J. *Tetrahedron*, ***1998***, *54*, 3207.

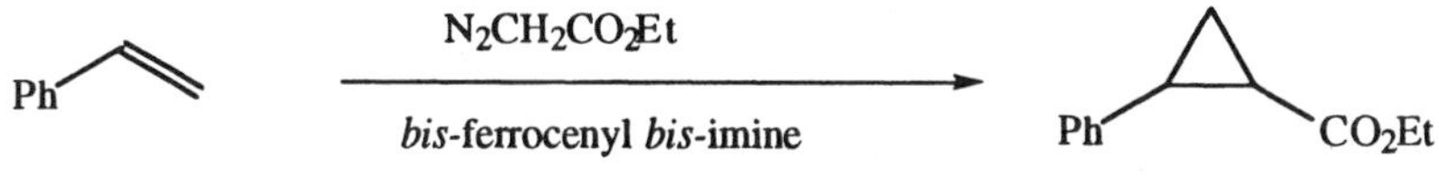
N2CH2CO2Et
bis-ferrocenyl bis-imine
Ph
Ph
CO2Et

84% (64:36 *trans:cis*)

Cho, D.–J.; Jeon, S.–J.; Kim, H.–S.; Kim, T.–J. *SynLett*, ***1998***, 617.

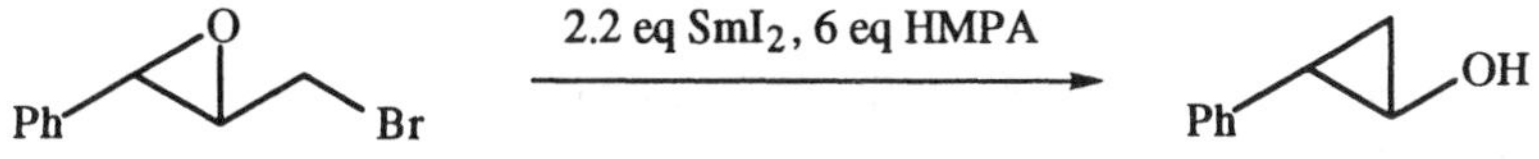
O
Ph
Br
2.2 eq SmI2, 6 eq HMPA
Ph
OH

40%

Park, H.S.; Chung, S.H.; Kim, Y.H. *SynLett*, ***1998***, 1073.

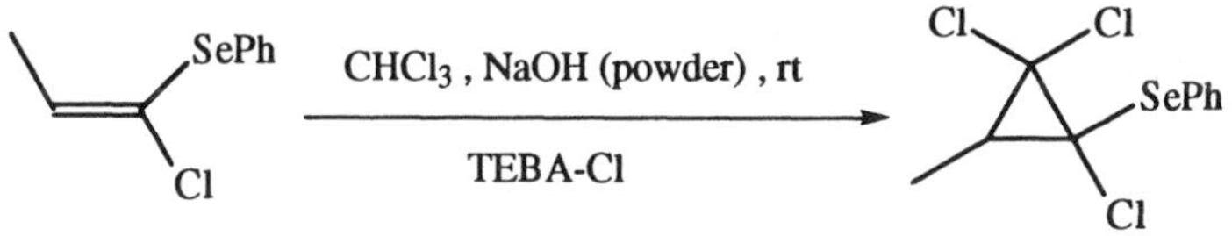
SePh
Cl
CHCl3, NaOH (powder), rt
TEBA-Cl
Cl
Cl
SePh
Cl

76%

Stefani, H.A.; Petragnani, N.; Comasseto, J.V.; Braga, A.; Menezes, P.H. *Synth. Commun.*, ***1998***, *28*, 1667.

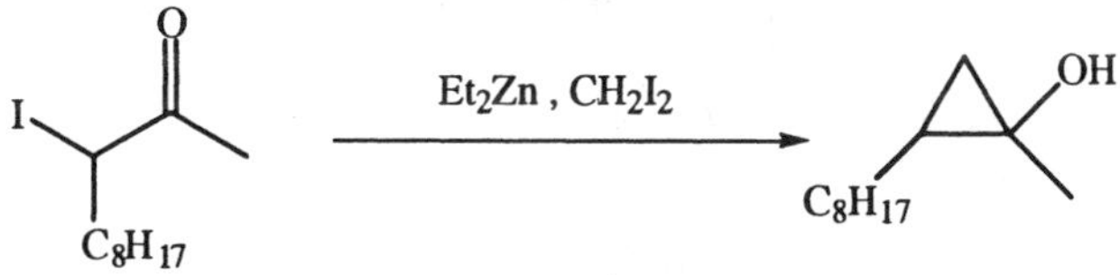
O
I
C8H17
Et2Zn, CH2I2
OH
C8H17

82% (93:7 *cis:trans*)

Ito, S.; Shinokubo, H.; Oshima, K. *Tetrahedron Lett.*, ***1998***, *39*, 5253.

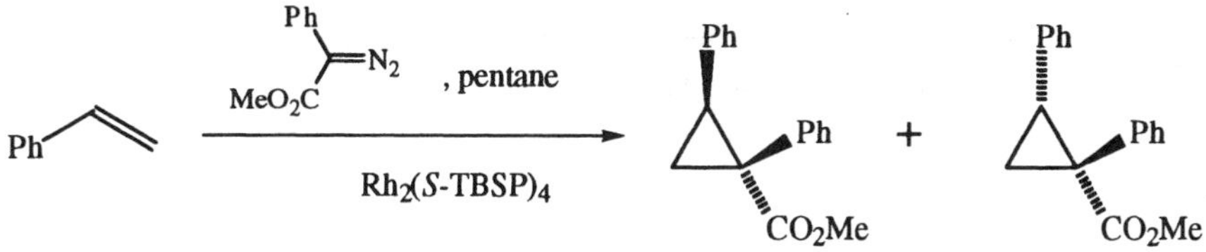
Ph
=N2
MeO2C
, pentane
Rh2(S-TBSP)4
Ph
Ph
Ph
CO2Me
+
Ph
Ph
CO2Me

(97 : 3) 72%

86% ee

Davies, H.M.L.; Rusiniak, L. *Tetrahedron Lett*, ***1998***, *39*, 8811.

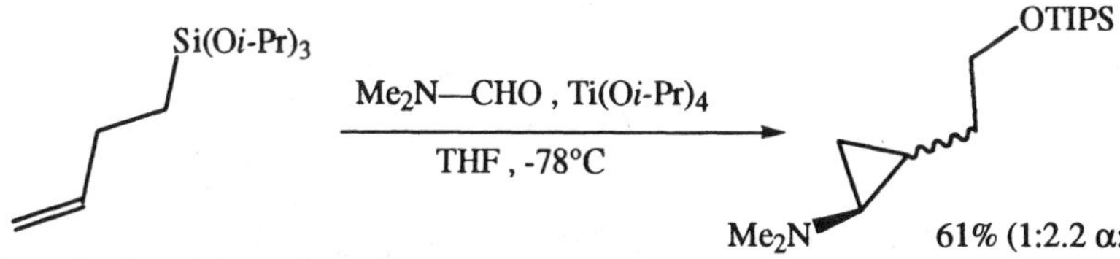
Si(Oi-Pr)3
Me2N—CHO, Ti(Oi-Pr)4
THF, -78°C
OTIPS
Me2N

61% (1:2.2 α:β)

Lee, J.; Cha, J.K. *J. Org. Chem.*, ***1997***, *62*, 1584.

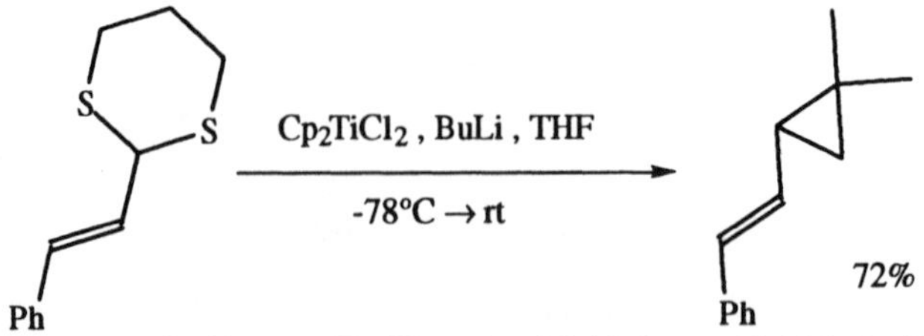
S
S
Cp_2TiCl_2 , BuLi , THF
-78°C → rt
72%
Ph
Ph

Horikawa, Y.; Nomura, T.; Watanabe, M.; Fujiwara, T.; Takeda, T.
J. Org. Chem., ***1997***, *62*, 3678.

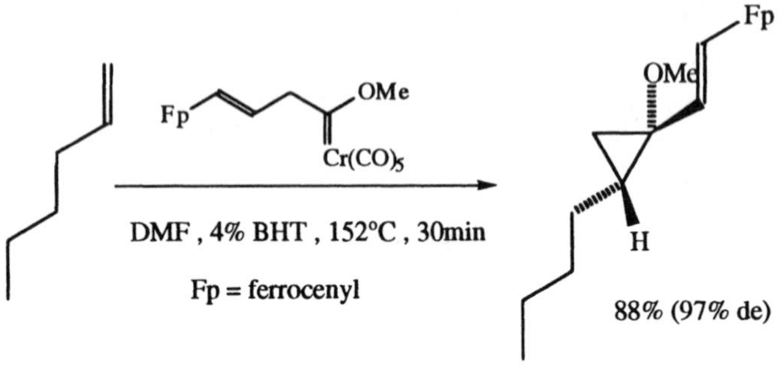
Fp
OMe
$Cr(CO)_5$
DMF , 4% BHT , 152°C , 30min
Fp = ferrocenyl
Fp
OMe
H
88% (97% de)

Barluenga, J.; Fernández–Acebes, A.; Trananco, A.A.; Flórez, J.
J. Am. Chem. Soc., ***1997***, *119*, 7591.

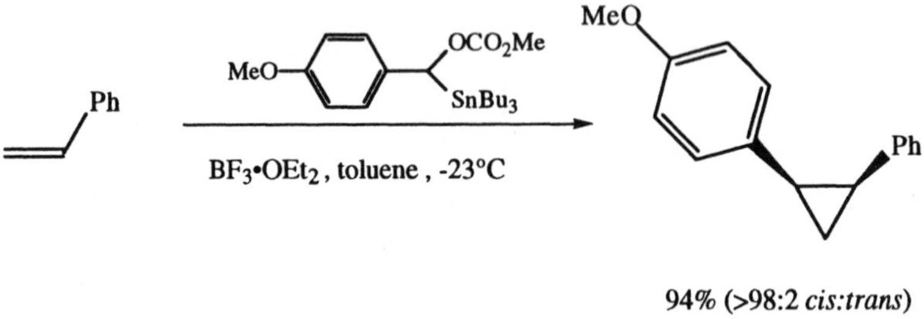
Ph
MeO
OCO_2Me
$SnBu_3$
$BF_3•OEt_2$, toluene , -23°C
MeO
Ph
94% (>98:2 *cis:trans*)

Sugawara, M.; Yosida, J. *J. Am. Chem. Soc.*, ***1997***, *119*, 11986.

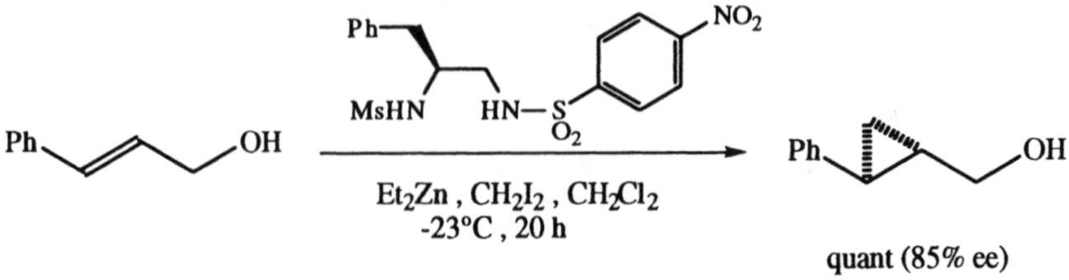
Ph
MsHN
HN—S
O_2
NO_2
Ph
OH
Et_2Zn , CH_2I_2 , CH_2Cl_2
-23°C , 20 h
Ph
OH
quant (85% ee)

Imai, N.; Sakamoto, K.; Maeda, M.; Kouge, K.; Yoshizane, K.; Nokami, J.
Tetrahedron Lett., ***1997***, *38*, 1423.

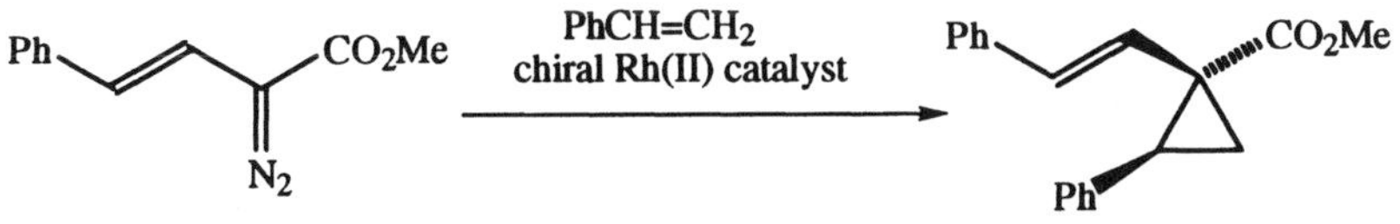

60% (74 ee , *1S,2S*)

Davies, H.M.L.; Kong, N. *Tetrahedron Lett.*, ***1997***, *38*, 4203.

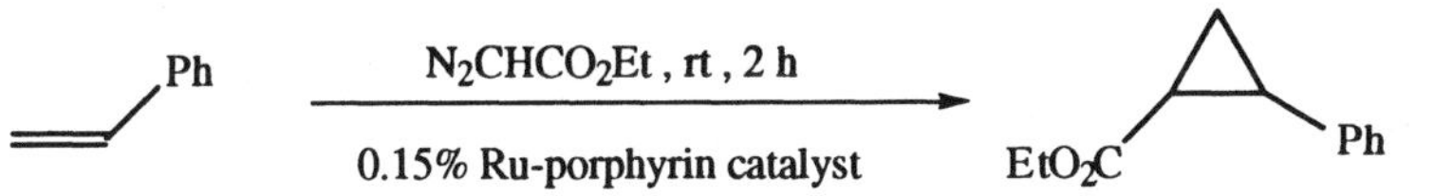

quant
(96:4 *trans:cis*)

Frauenkron, M.; Berkessel, A. *Tetrahedron Lett.*, ***1997***, *38*, 7175.

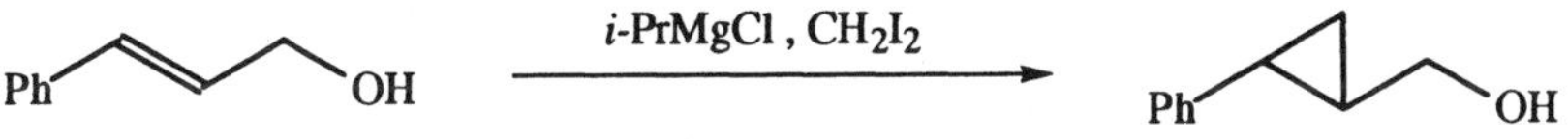

35%

Bolm, C.; Pupowicz, D. *Tetrahedron Lett.*, ***1997***, *38*, 7349.

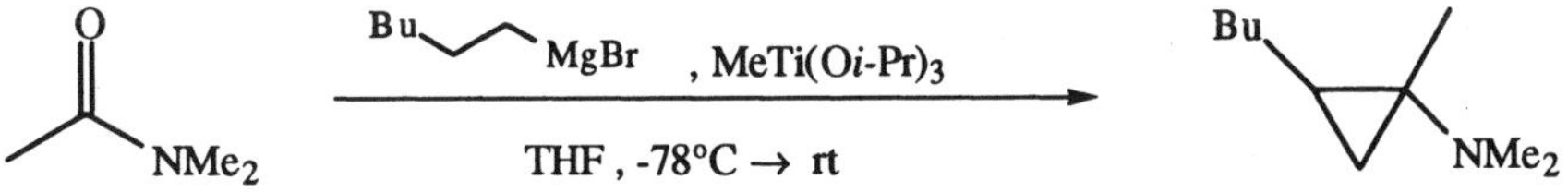

51%

Chaplinski, V.; Winsel, H.; Kordes, M.; de Meijere, A. *SynLett*, ***1997***, 111.

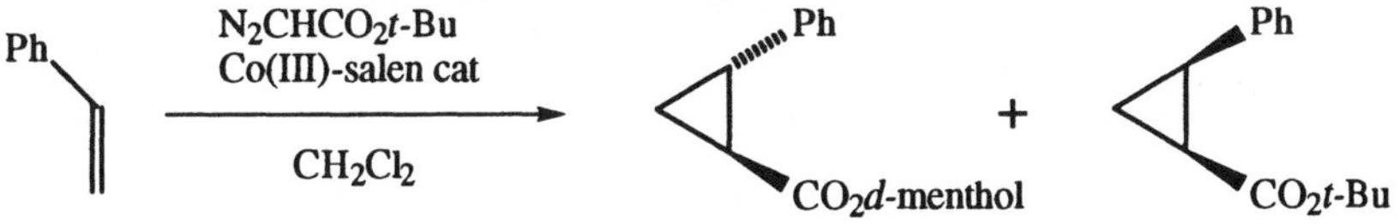

(82 : 5) 79%
60% ee 51% ee

Fukuda, T.; Katsuki, T. *Tetrahedron*, ***1997***, *53*, 7201.

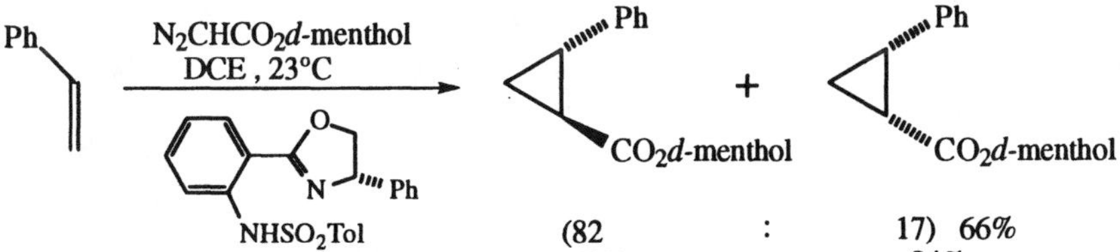

(82 : 17) 66%
60% ee 84% ee

Ichiyanagi, T.; Shimizu, M.; Fujisawa, T. *Tetrahedron*, ***1997***, *53*, 9599.

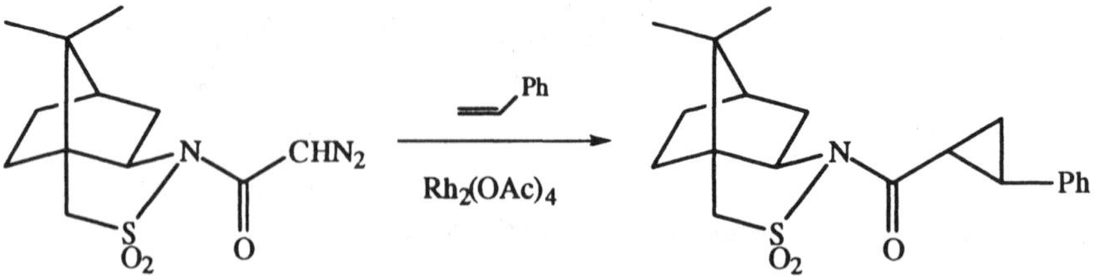

67% (68:32 *trans:cis*) 62% *de*:70%*de*

Haddad, N.; Galili, N. *Tetrahedron Asymmetry*, ***1997***, *8*, 3367.

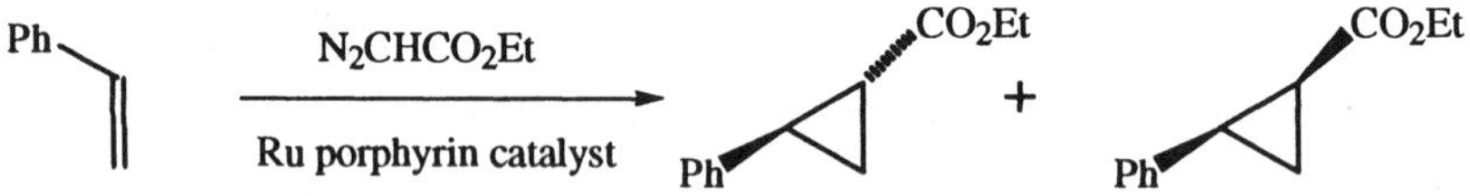

(14 : 1) no yield

Galardon, E.; LeMaux, P.; Simonneaux, G. *Chem. Commun.*, ***1997***, 927.

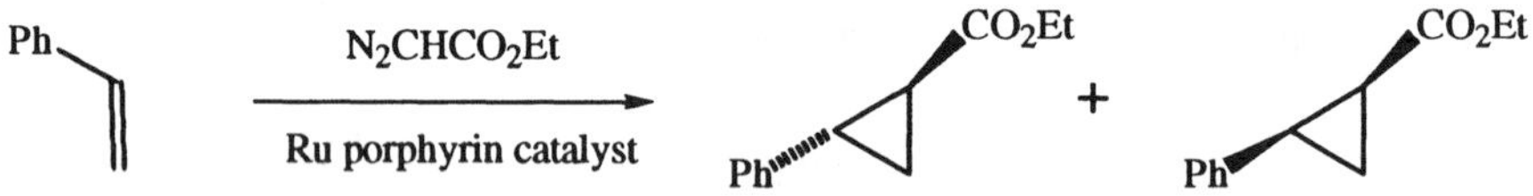

(7.1 : 1) 68%

Lo, W.C.; Che, C.M.; Cheng, K.-F.; Mak, T.C.W. *Chem. Commun.*, ***1997***, 1205.

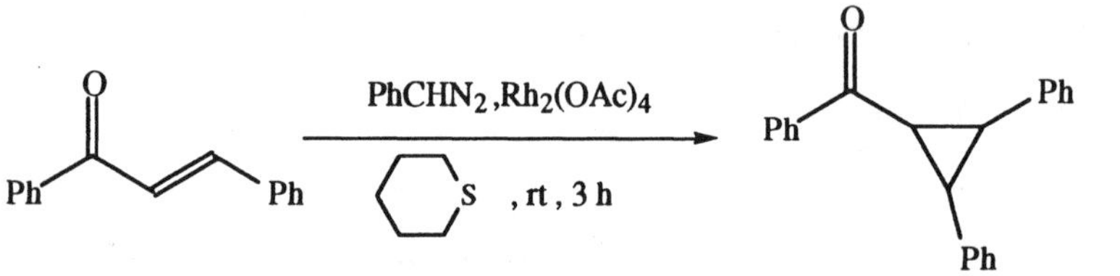

92% (4:1 *cis:trans*)

Aggarwal, V.K.; Smith, H.W.; Jones, R.V.H.; Fieldhouse, R. *Chem. Commun.*, ***1997***, 1785

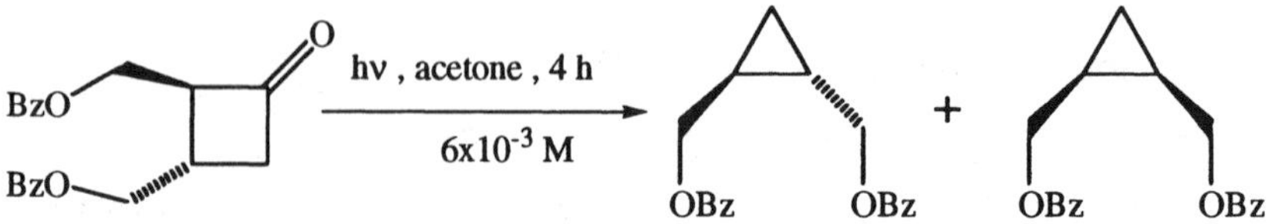

53% 17%

\+ 30% alkenes

Ramnauth, J.; Lee-Ruff, E. *Can. J. Chem.*, ***1997***, *75*, 518.

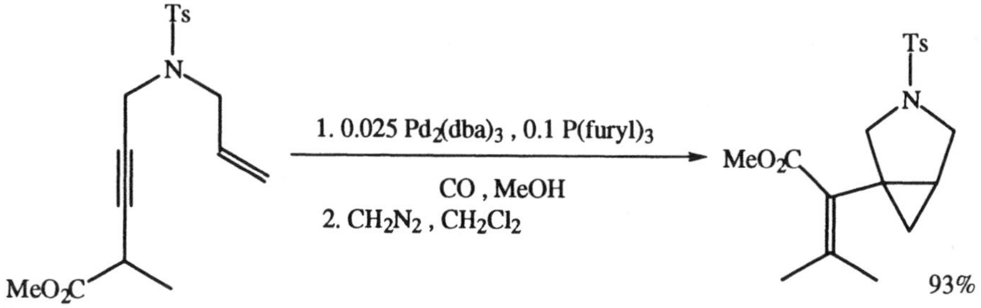
Ts
N
1. 0.025 Pd2(dba)3 , 0.1 P(furyl)3
CO , MeOH
2. CH2N2 , CH2Cl2
MeO2C
Ts
N
MeO2C
93%

Oppolzer, W.; Pimm, A.; Stammen, B.; Hume, W.E. *Helv. Chim. Acta,* ***1997****, 80,* 623.

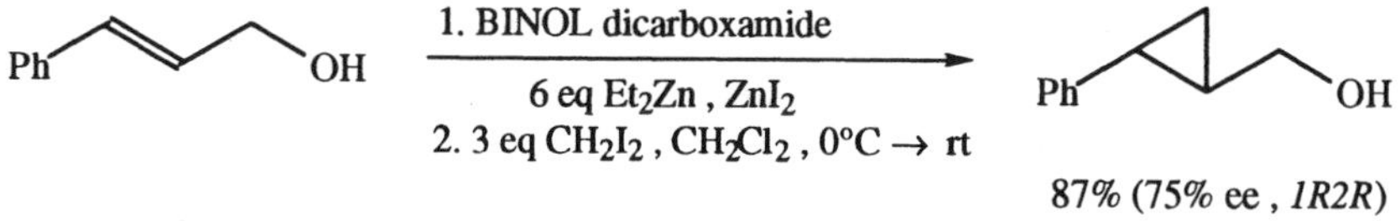
Ph
OH
1. BINOL dicarboxamide
6 eq Et2Zn , ZnI2
2. 3 eq CH2I2 , CH2Cl2 , 0°C → rt
Ph
OH
87% (75% ee , 1R2R)

Kitajima, H.; Ito, K.; Aoki, Y.; Katsuki, T. *Bull. Chem. Soc. Jpn.,* ***1997****, 70,* 207.

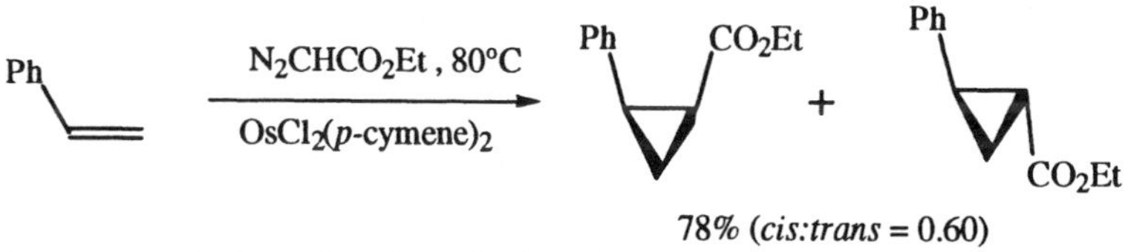
Ph
N2CHCO2Et , 80°C
OsCl2(p-cymene)2
Ph
CO2Et
+
Ph
CO2Et
78% (cis:trans = 0.60)

Demonceau, A.; Lemoine, C.A.; Noels, A.F. *Tetrahedron Lett.,* ***1996****, 37,* 1025.

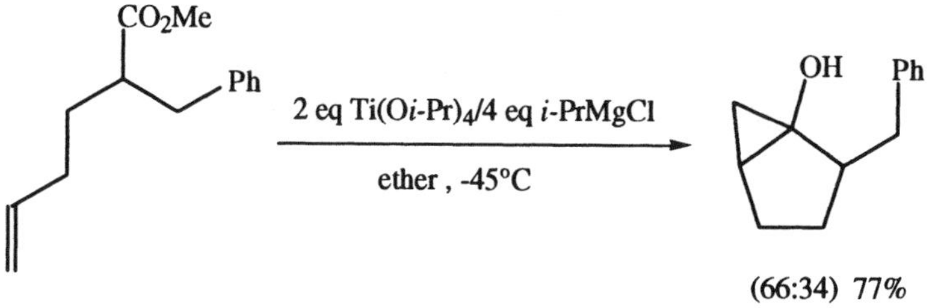
CO2Me
Ph
2 eq Ti(Oi-Pr)4/4 eq i-PrMgCl
ether , -45°C
OH
Ph
(66:34) 77%

Kasatkin, A.; Kobayashi, K.; Okamoto, S.; Sato, F. *Tetrahedron Lett.,* ***1996****, 37,* 1849.

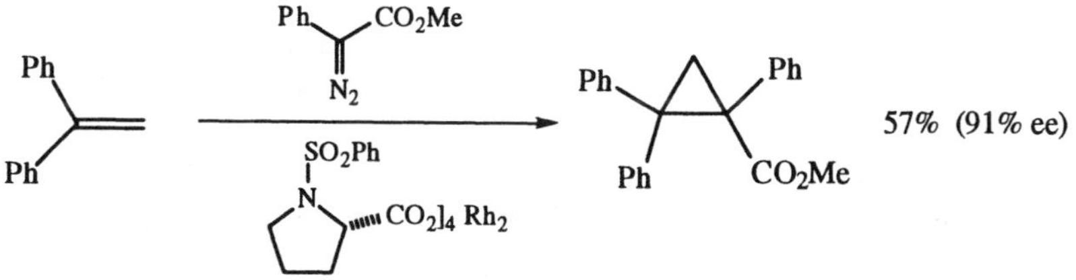
Ph
Ph
Ph
CO2Me
N2
SO2Ph
N
CO2]4 Rh2
Ph
Ph
Ph
CO2Me
57% (91% ee)

Doyle, M.P.; Zhou, Q.–L.; Charnsangavej, C.; Longoria, M.A.; McKervey, M.A.; Garcia, C.F *Tetrahedron Lett.,* ***1996****, 37,* 4129.

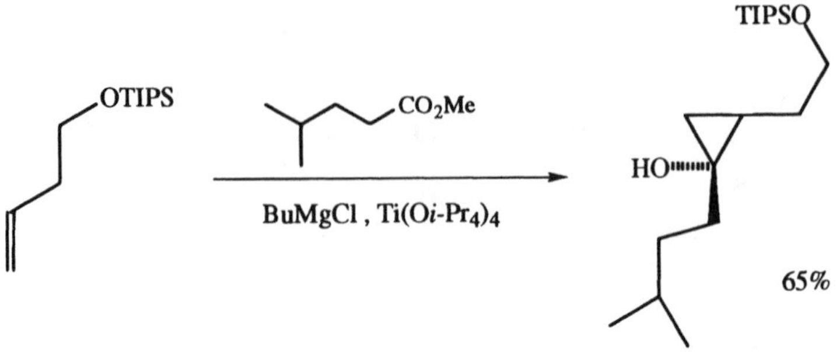
OTIPS
CO2Me
BuMgCl, Ti(Oi-Pr4)4
TIPSO
HO
65%

Lee, J.; Kim, H.; Cha, J.K. *J. Am. Chem. Soc.*, ***1996***, *118*, 4198.

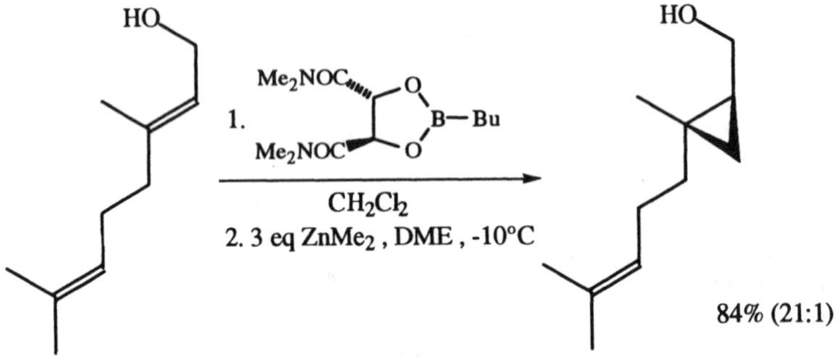
HO
1.
Me2NOC
Me2NOC
O
B—Bu
O
CH2Cl2
2. 3 eq ZnMe2, DME, -10°C
HO
84% (21:1)

Charette, A.B.; Juteau, H.; Lebel, H.; Deschênes, D. *Tetrahedron Lett.*, ***1996***, *37*, 7925.

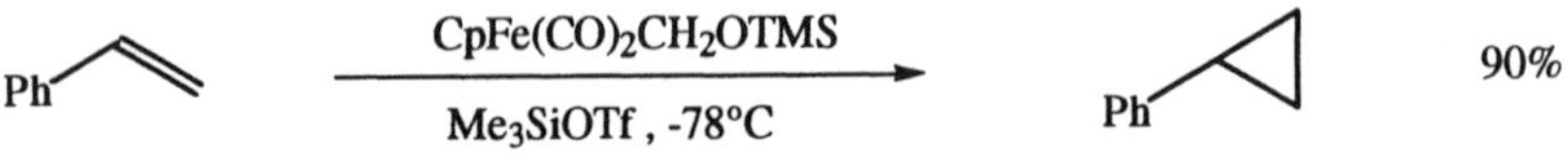
Ph
CpFe(CO)2CH2OTMS
Me3SiOTf, -78°C
Ph
90%

Du, H.; Yang, F.; Hossain, M.A. *Synth. Commun.*, ***1996***, *26*, 1371.

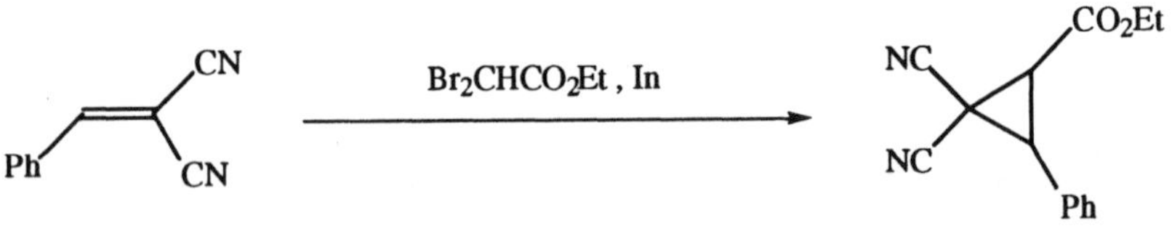
CN
Ph
CN
Br2CHCO2Et, In
CO2Et
NC
NC
Ph

69% (45:51 *cis:trans*)

Araki, S.; Hirashita, T.; Shimizu, K.; Ikeda, T.; Butsugan, Y. *Tetrahedron*, ***1996***, *52*, 2803.

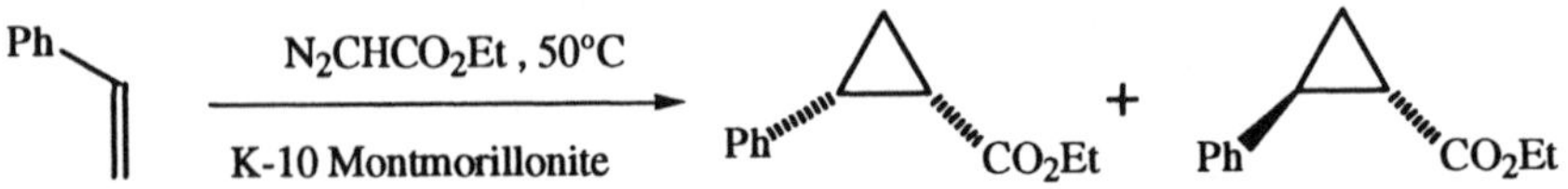
Ph
N2CHCO2Et, 50°C
K-10 Montmorillonite
Ph
CO2Et
+
Ph
CO2Et

61% conversion, *trans:cis* = 1:1

Fraile, J.M.; García, J.I.; Mayoral, J.A. *Chem. Commun.*, ***1996***, 1319.

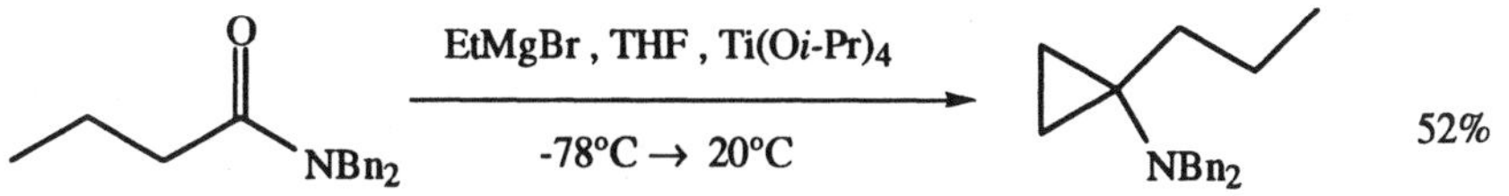
O
NBn2
EtMgBr , THF , Ti(Oi-Pr)4
-78°C → 20°C
NBn2
52%

Chaplinski, V.; de Meijere, A. *Angew. Chem. Int. Ed.*, ***1996***, *35*, 413.

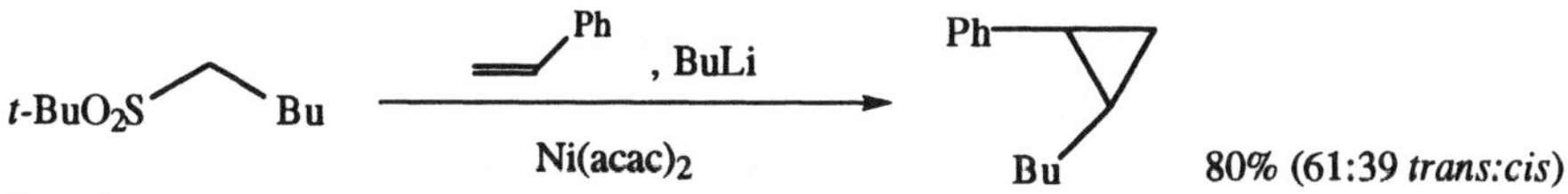
t-BuO2S
Bu
Ph
, BuLi
Ni(acac)2
Ph
Bu
80% (61:39 trans:cis)

Gai, Y.; Julia, M.; Verpeaux, J.–N. *Bull. Soc. Chim. Fr.*, ***1996***, *133*, 817.

REVIEW:
"Catalytic Enantioselective Cyclopropanation of Olefins Using Carbenoid Chemistry."
Singh, V.K.; Datta Gupta, A.; Sekar, G. *Synthesis*, ***1997***, 137.

SECTION 75: ALKYLS, METHYLENES AND ARYLS FROM MISCELLANEOUS COMPOUNDS

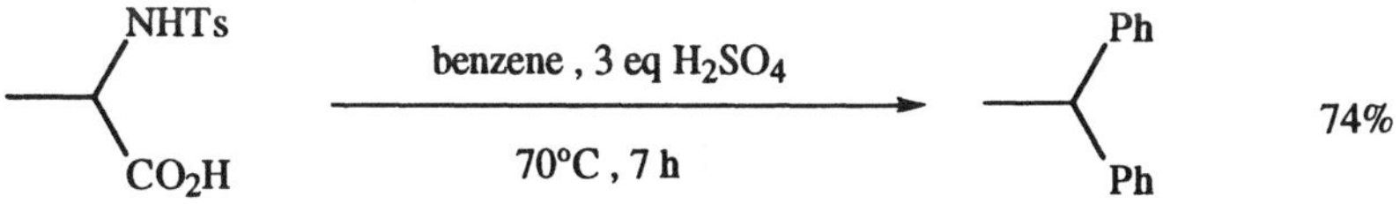
NHTs
CO2H
benzene , 3 eq H2SO4
70°C , 7 h
Ph
Ph
74%

Seong, M.R.; Lee, H.J.; Kim, J.N. *Tetrahedron Lett.*, ***1998***, *39*, 6219.

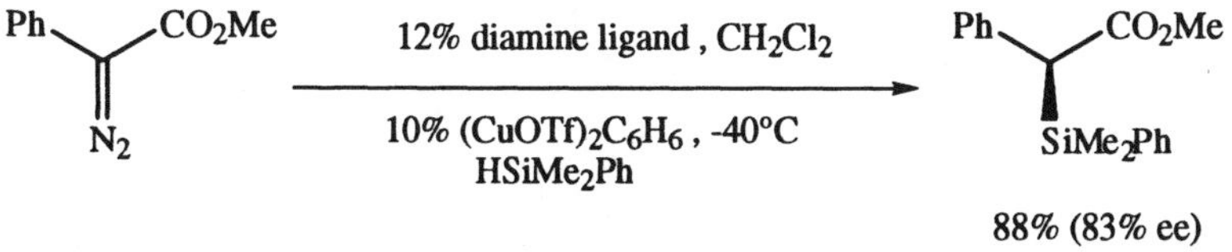
Ph
CO2Me
N2
12% diamine ligand , CH2Cl2
10% (CuOTf)2C6H6 , -40°C
HSiMe2Ph
Ph
CO2Me
SiMe2Ph
88% (83% ee)

Dakin, L.A.; Schaus, S.E.; Jacobsen, E.N.; Panek, J.S. *Tetrahedron Lett*, ***1998***, *39*, 8947.

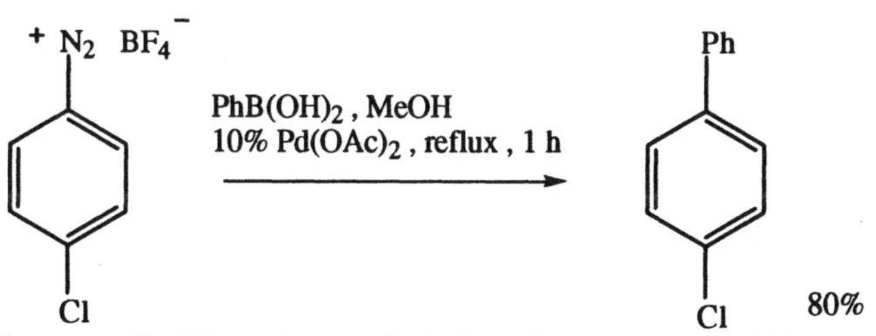
+ N2 BF4 −
Cl
PhB(OH)2 , MeOH
10% Pd(OAc)2 , reflux , 1 h
Ph
Cl
80%

Sengupta, S.; Bhattacharyya, S. *J. Org. Chem.*, ***1997***, *62*, 3405.

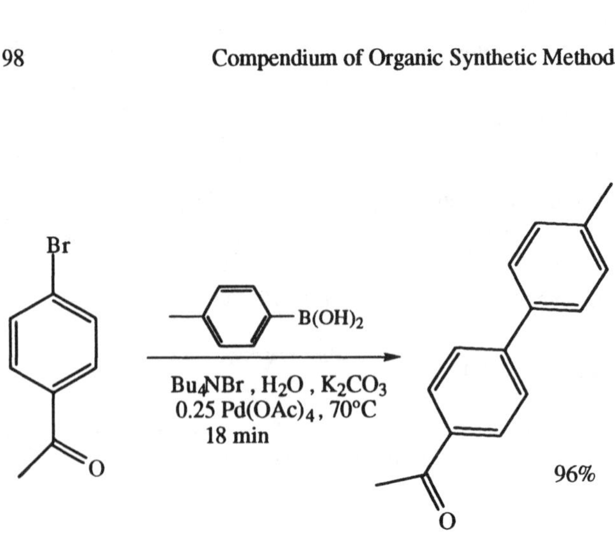
Br
B(OH)2
Bu4NBr , H2O , K2CO3
0.25 Pd(OAc)4 , 70°C
18 min
O
96%
O

Badone, D.; Baroni, M.; Cardamone, R.; Ielmini, A.; Guzzi, U. *J. Org. Chem.* ***1997***, *62*, 7170

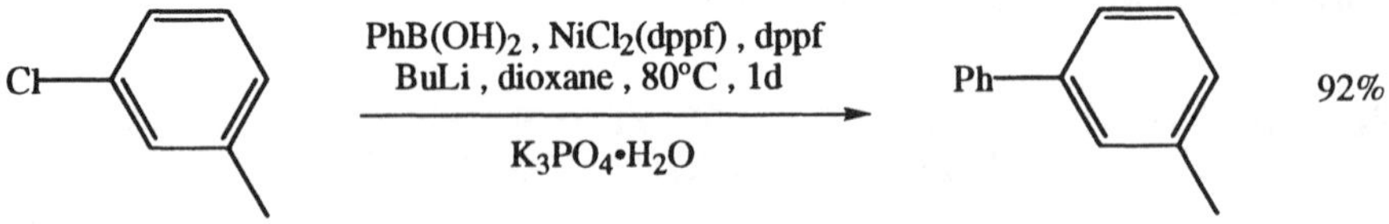
Cl
PhB(OH)2 , NiCl2(dppf) , dppf
BuLi , dioxane , 80°C , 1d
K3PO4•H2O
Ph
92%

Saito, S.; Oh-tani, S.; Miyaura, N. *J. Org. Chem.*, ***1997***, *62*, 8024.

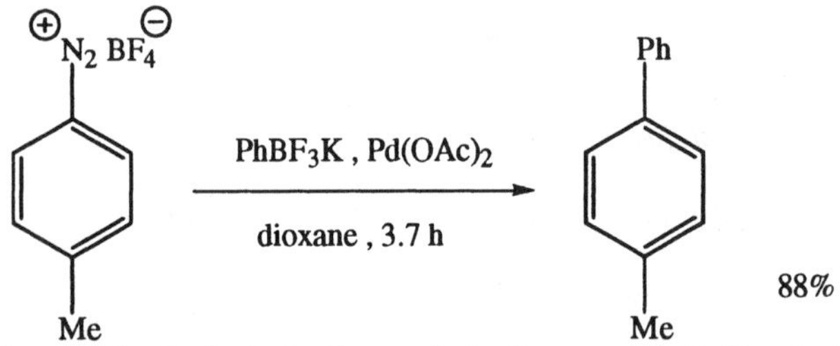
N2 BF4
PhBF3K , Pd(OAc)2
dioxane , 3.7 h
Ph
88%
Me
Me

Darses, S.; Genêt, J.-P.; Brayer, J.-L.; Demoute, J.-P. *Tetrahedron Lett.*, ***1997***, *38*, 4393.

$PhB(OH)_2$ —[$Ph_2I^+ BF_4^-$, 2% CuI, 35°C / Na_2CO_3, aq DME, 20 min]→ Ph—Ph 99%

Kang, S.-K.; Yamaguchi, T.; Kim, T.-H.; Ho, P.-S. *J. Org. Chem.*, ***1996***, *61*, 9082.

$PhPb(OAc)_3$ —[5% $Pd_2(dba)_3$•$CHCl_3$, $CHCl_3$ / rt]→ Ph—Ph 68%

Kang, S.-K.; Shivkumar, U.; Ahn, C.; Choi, S.-C.; Kim, J.S. *Synth. Commun.*, ***1997***, *27*, 1893.

$Ph_2I^+ BF_4^-$ —[$Pd(OAc)_2$, Et_2Zn]→ Ph—Ph 80%

Kang, S.-K.; Hong, R.-K.; Kim, T.-H.; Pyun, S.-J. *Synth. Commun.*, ***1997***, *27*, 2351.

$PhSiMe_2Cl$ —(5% CuI, TBAF, MeCN, rt, 5 min)→ Ph—Ph 73%

Kang, S.-K.; Kim, T.-H.; Pyun, S.-J. *J. Chem. Soc., Perkin Trans. 1*, ***1997***, 797.

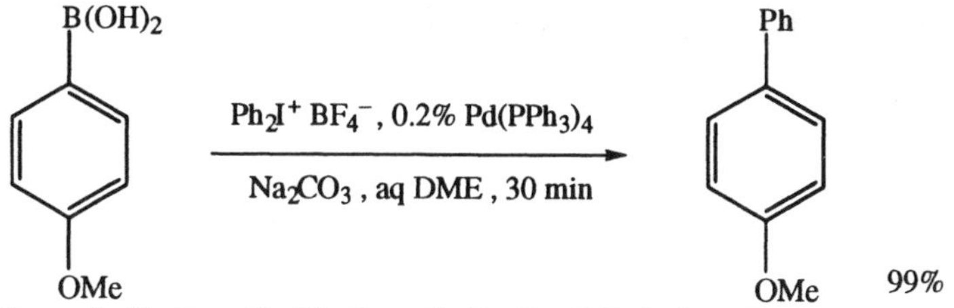
B(OH)2
Ph2I+ BF4−, 0.2% Pd(PPh3)4
Na2CO3, aq DME, 30 min
OMe
Ph
OMe
99%

Kang, S.-K.; Lee, H.-W.; Jang, S.-B.; Ho, P.S. *J. Org. Chem.*, ***1996***, *61*, 4720.

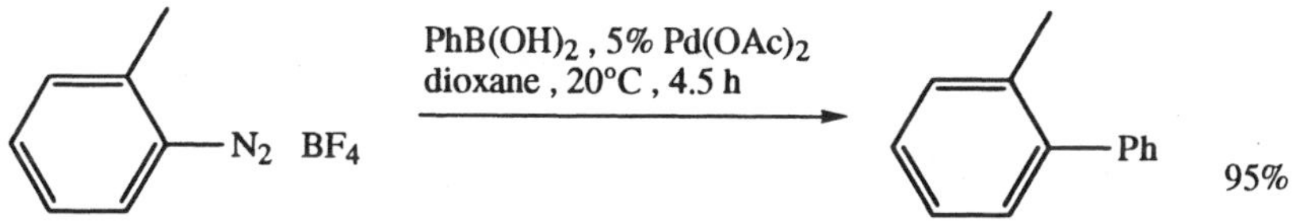
N2 BF4
PhB(OH)2, 5% Pd(OAc)2
dioxane, 20°C, 4.5 h
Ph
95%

Darses, S.; Jeffery, T.; Genet, J.-P.; Brayer, J.-l.; DeMoute, J.-P. *Tetrahedron Lett.*, ***1996***, *37*, 3857.

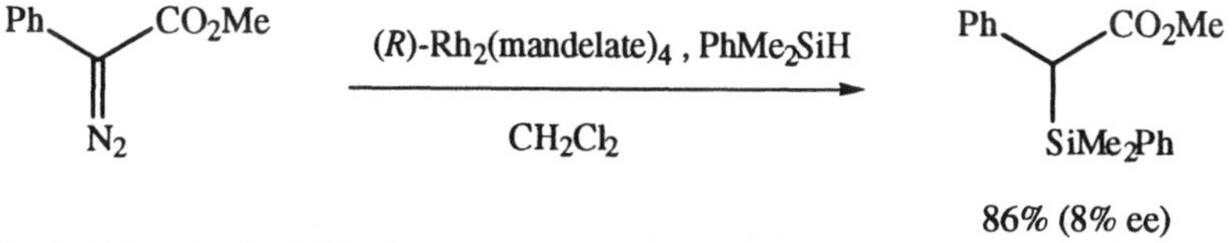
Ph
CO2Me
N2
(R)-Rh2(mandelate)4, PhMe2SiH
CH2Cl2
Ph
CO2Me
SiMe2Ph
86% (8% ee)

Buck, R.T.; Doyle, M.P.; Drysdale, M.J.; Ferris, L.; Forbes, D.C.; Haigh, D.; Moody, C.J.; Pearson, N.D.; Zhou, Q.-L. *Tetrahedron Lett.*, ***1996***, *37*, 7631.

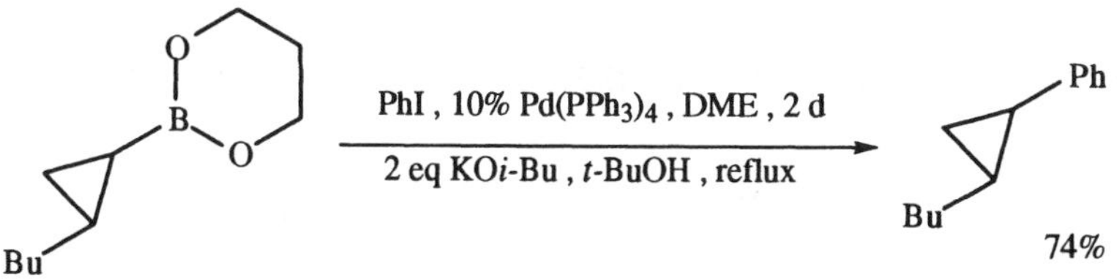
O
B
O
Bu
PhI, 10% Pd(PPh3)4, DME, 2 d
2 eq KOi-Bu, t-BuOH, reflux
Ph
Bu
74%

Hildebrand, J.P.; Marsden, S.P. *SynLett*, ***1996***, 893.

CHAPTER 6
PREPARATION OF AMIDES

SECTION 76: AMIDES FROM ALKYNES

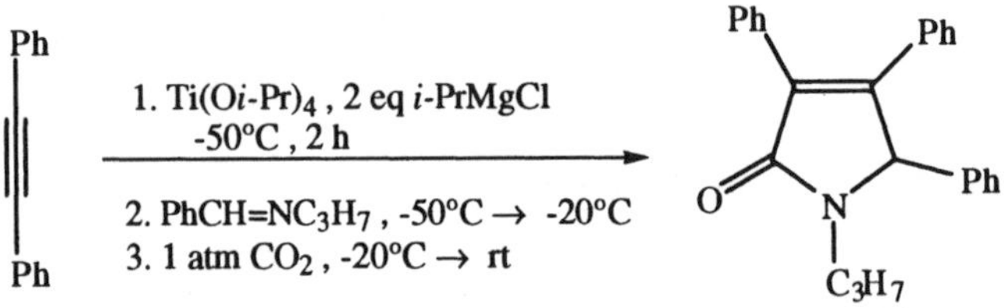
Ph
Ph
1. Ti(O*i*-Pr)$_4$, 2 eq *i*-PrMgCl
-50°C , 2 h
2. PhCH=NC$_3$H$_7$, -50°C → -20°C
3. 1 atm CO$_2$, -20°C → rt
Ph
Ph
Ph
O
N
C$_3$H$_7$

56%

Gao, Y.; Shirai, M.; Sato, F. *Tetrahedron Lett.*, ***1997***, *38*, 6849.

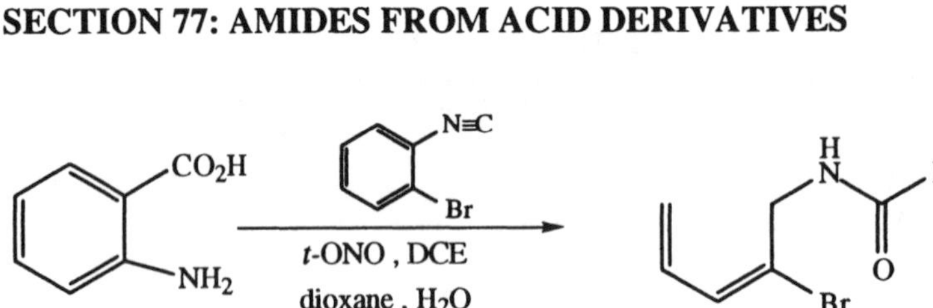
Me$_3$Si
1. BuLi , Se
2. AcOH
3. PhCHMeNH$_2$
Se
N
H
Ph

84%

Murai, T.; Ezaka, T.; Niwa, N.; Kanda, T.; Kato, S. *SynLett*, ***1996***, 865.

SECTION 77: AMIDES FROM ACID DERIVATIVES

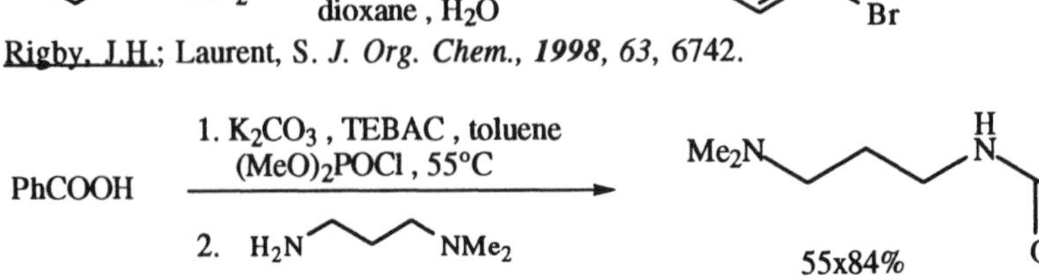
CO$_2$H
NH$_2$
N≡C
Br
t-ONO , DCE
dioxane , H$_2$O
H
N
Ph
O
Br

Rigby, J.H.; Laurent, S. *J. Org. Chem.*, ***1998***, *63*, 6742.

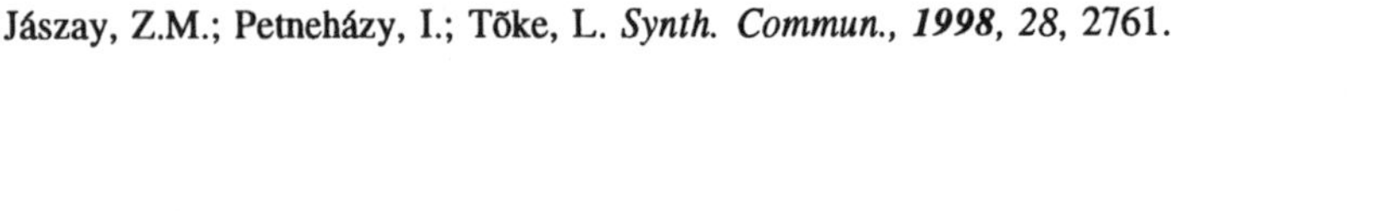
PhCOOH
1. K$_2$CO$_3$, TEBAC , toluene
(MeO)$_2$POCl , 55°C
2. H$_2$N NMe$_2$
Me$_2$N
H
N
Ph
O

55x84%

Jászay, Z.M.; Petneházy, I.; Tõke, L. *Synth. Commun.*, ***1998***, *28*, 2761.

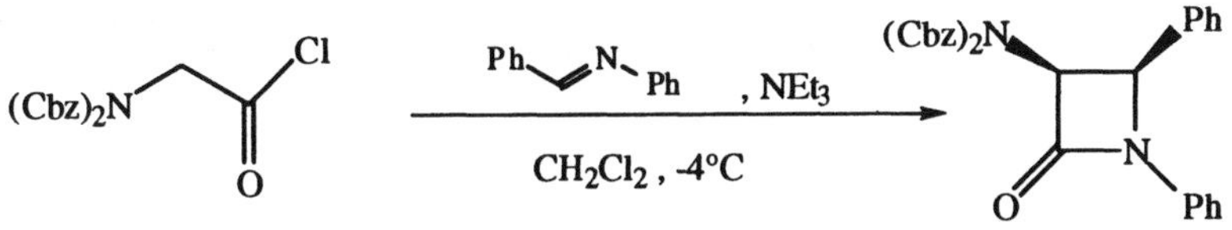

57% (90:10 *cis:trans*)

Cainelli, G.; Gallenti, P.; Giacomini, D. *SynLett*, ***1998***, 611.

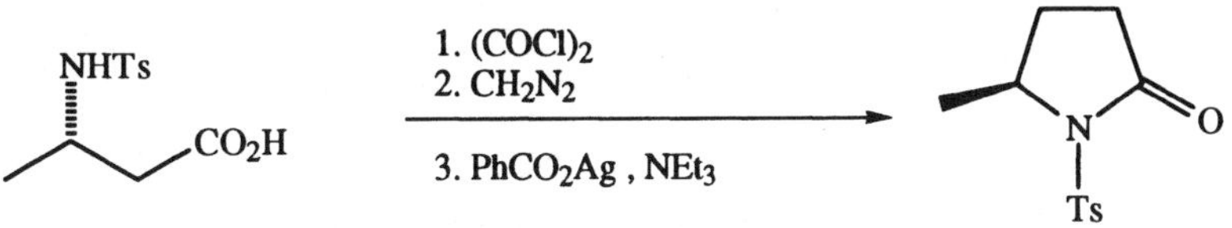

89%

Wang, J.; Hou, Y. *J. Chem. Soc., Perkin Trans. 1*, ***1998***, 1919.

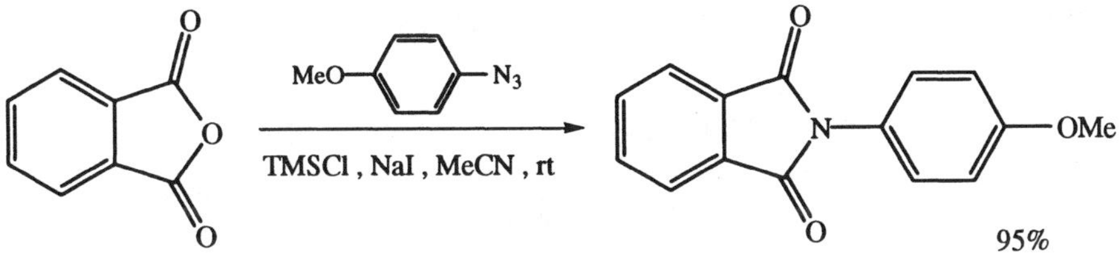

95%

Kamal, A.; Laxman, E.; Laxman, N.; Rao, N.V. *Tetrahedron Lett*, ***1998***, *39*, 8733.

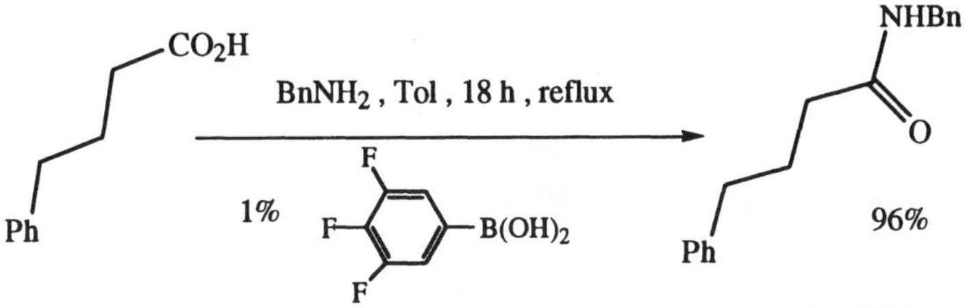

96%

Ishihara, K.; Ohara, S.; Yamamoto, H. *J. Org. Chem.*, ***1996***, *61*, 4196.

SECTION 78: AMIDES FROM ALCOHOLS AND THIOLS

BuOH —(TsNHMe, TMAD-TBP, PhH, rt)→ TsN(Me)Bu quant

TMAD = N,N,N',N'-tetramethylazodecarboxamide
TBP = Bu_3P

Tsunoda, T.; Kawamura, Y.; Uemoto, K.; Itô, S. *Heterocycles*, ***1998***, *47*, 177.

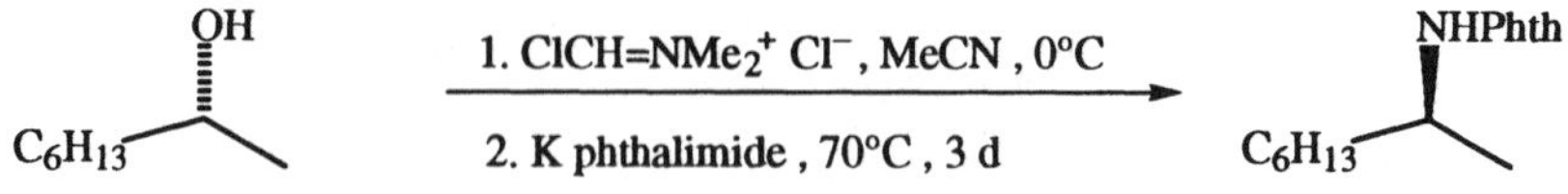
OH
C_6H_{13}
1. $ClCH=NMe_2^+ Cl^-$, MeCN, 0°C
2. K phthalimide, 70°C, 3 d
NHPhth
C_6H_{13}

95%

Barrett, A.G.M.; Braddock, D.C.; James, R.A.; Procopiou, P.A. *Chem. Commun.*, ***1997***, 433.

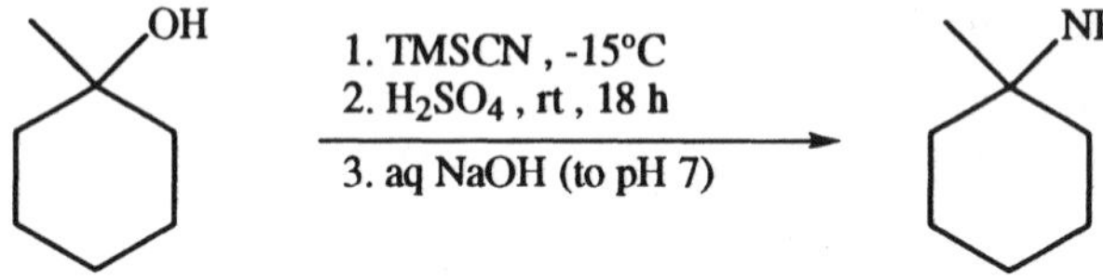
OH
1. TMSCN, -15°C
2. H_2SO_4, rt, 18 h
3. aq NaOH (to pH 7)
NHCHO

86%

Chen, H.G.; Goel, O.P.; Kesten, S.; Knobelsdorf, J. *Tetrahedron Lett.*, ***1996***, *37*, 8129.

SECTION 79: AMIDES FROM ALDEHYDES

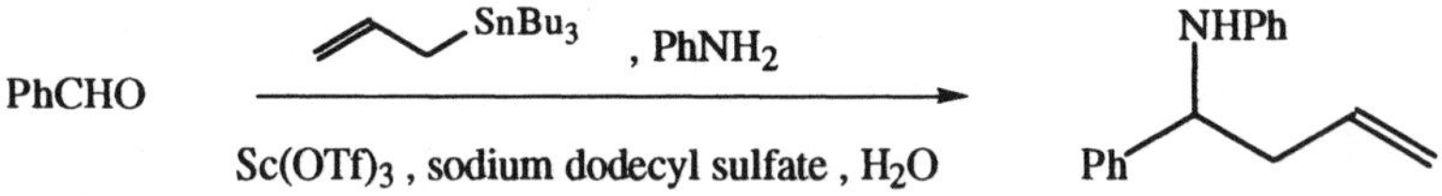
PhCHO
$SnBu_3$
, $PhNH_2$
$Sc(OTf)_3$, sodium dodecyl sulfate, H_2O
NHPh
Ph

80%

Kobayashi, S.; Busujima, T.; Nagayama, S. *Chem. Commun.*, ***1998***, 19.

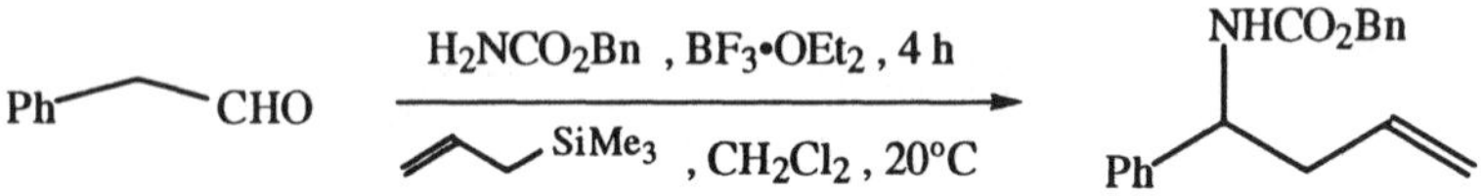
Ph
CHO
H_2NCO_2Bn, $BF_3•OEt_2$, 4 h
$SiMe_3$
, CH_2Cl_2, 20°C
$NHCO_2Bn$
Ph

80%

Veenstra, S.J.; Schmid, P. *Tetrahedron Lett.*, ***1997***, *38*, 997.

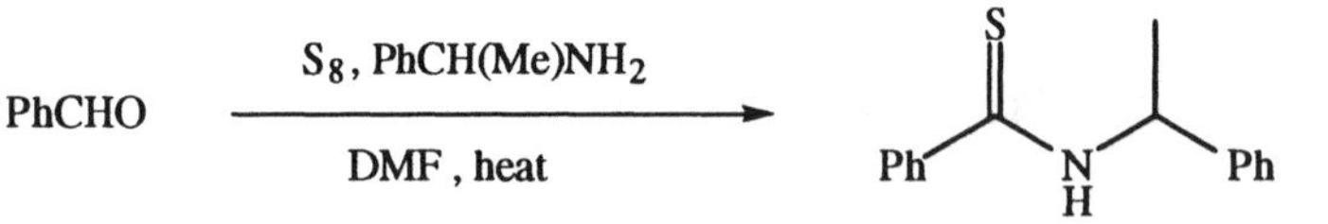
PhCHO
S_8, PhCH(Me)NH_2
DMF, heat
S
Ph
N
H
Ph

99%

Kanyonyo, M.R.; Gozzo, A.; Lambert, D.M.; Lesieur, D.; Poupaert, J.H. *Bull. Soc. Chim. Belg.*, ***1997***, *106*, 39.

SECTION 80: AMIDES FROM ALKYLS, METHYLENES AND ARYLS

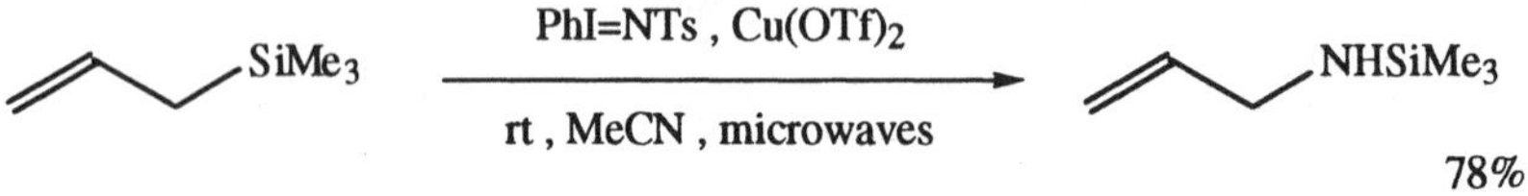
$SiMe_3$
PhI=NTs, $Cu(OTf)_2$
rt, MeCN, microwaves
$NHSiMe_3$

78%

Kim, D.Y.; Choi, J.S.; Rhie, D.Y.; Chang, S.K.; Kim, I.K. *Synth. Commun.*, ***1997***, 27, 2753

SECTION 81: AMIDES FROM AMIDES

Conjugate reductions of unsaturated amides are listed in Section 74D (Alkyls from Alkenes)

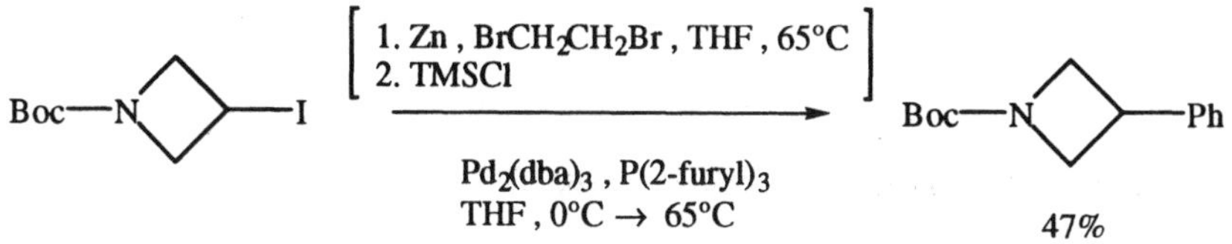
Boc
N
I
1. Zn , BrCH2CH2Br , THF , 65°C
2. TMSCl
Pd2(dba)3 , P(2-furyl)3
THF , 0°C → 65°C
Boc
N
Ph
47%

Billotte, S. *SynLett*, ***1998***, 379.

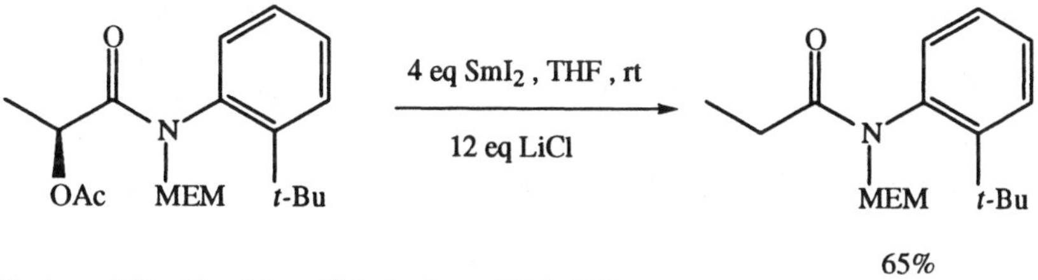
O
N
OAc
MEM
t-Bu
4 eq SmI2 , THF , rt
12 eq LiCl
O
N
MEM
t-Bu
65%

Hughes, A.D.; Simpkins, N.S. *SynLett*, ***1998***, 967.

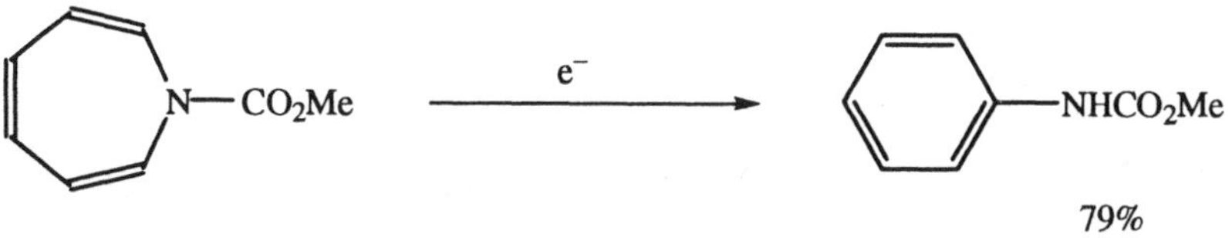
N— CO2Me
e−
NHCO2Me
79%

Kondo, S.; Suzuki, H.; Hattori, T.; Ido, T.; Saito, K. *Heterocycles*, ***1998***, *48*, 1151.

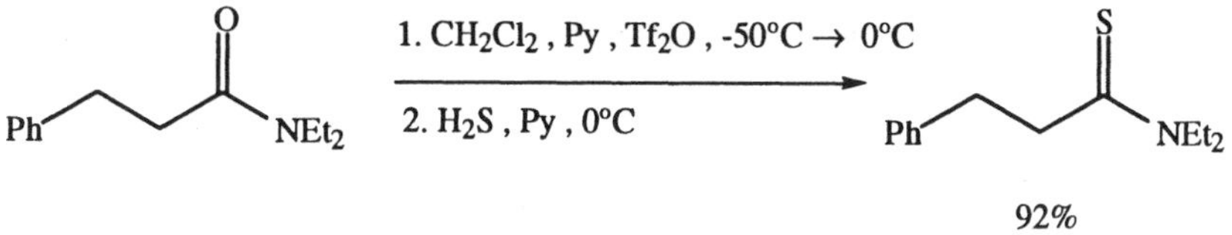
O
Ph
NEt2
1. CH2Cl2 , Py , Tf2O , -50°C → 0°C
2. H2S , Py , 0°C
S
Ph
NEt2
92%

Charette, A.B.; Chua, P. *Tetrahedron Lett*, ***1998***, *39*, 245.

O
Ph
t-Bu
N
Ph
PhCl , NBS , AIBN , reflux
O
H
t-Bu
N
Ph
96%

Baker, S.R.; Parsons, A.F.; Wilson, M. *Tetrahedron Lett.*, ***1998***, *39*, 331.

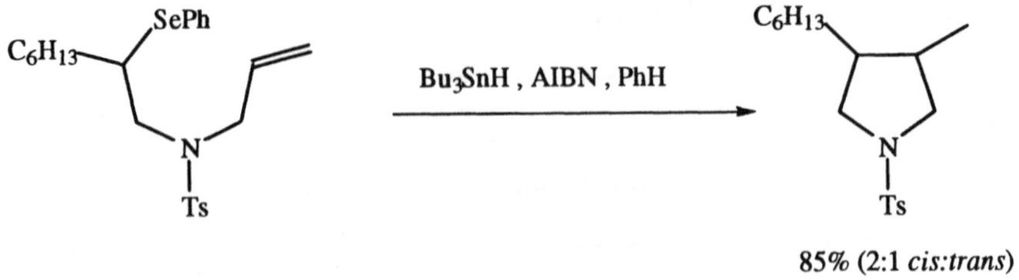
SePh
C6H13
N
Ts
Bu3SnH , AIBN , PhH
C6H13
N
Ts
85% (2:1 cis:trans)

Gupta, V.; Besev, M.; Engman, L. *Tetrahedron Lett*, ***1998***, *39*, 2429.

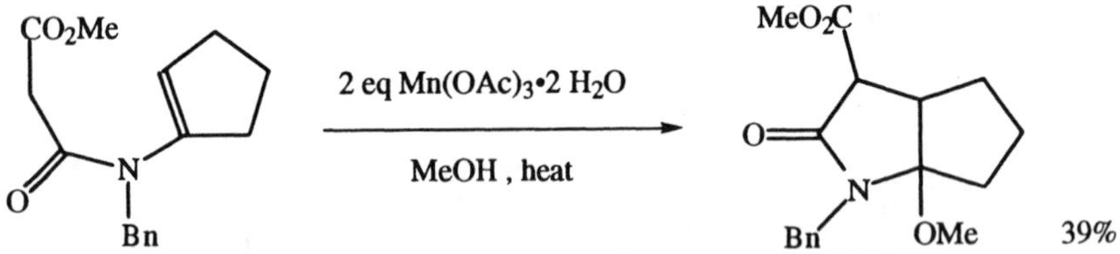
CO2Me
O
N
Bn
2 eq Mn(OAc)3•2 H2O
MeOH , heat
MeO2C
O
N
Bn
OMe
39%

Davies, D.T.; Kapur, N.; Parsons, A.F. *Tetrahedron Lett.*, ***1998***, *39*, 4397.

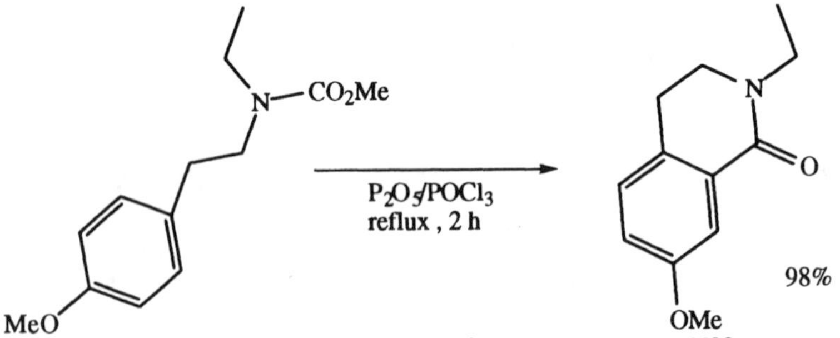
N
CO2Me
MeO
P2O5/POCl3
reflux , 2 h
N
O
OMe
98%

Wang, X.-j.; Tan, J.; Grozinger, K. *Tetrahedron Lett.*, ***1998***, *39*, 6609.

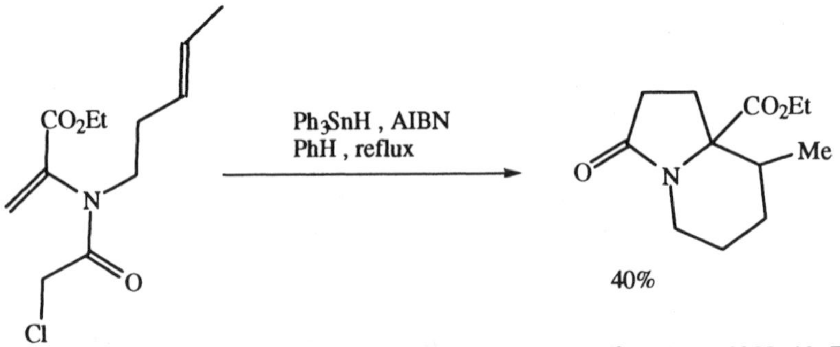
CO2Et
N
O
Cl
Ph3SnH , AIBN
PhH , reflux
O
N
CO2Et
Me
40%

Baker, S.R.; Parsons, A.F.; Pons, J.-F.; Wilson, M. *Tetrahedron Lett.*, ***1998***, *39*, 7197.

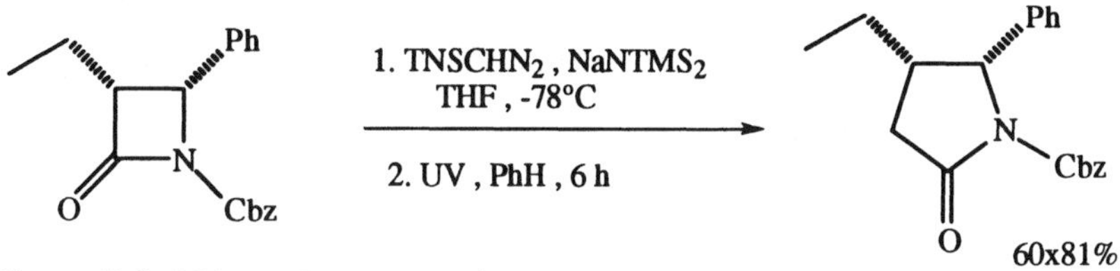
Ph
O
N
Cbz
1. TNSCHN2 , NaNTMS2
THF , -78°C
2. UV , PhH , 6 h
Ph
N
Cbz
O

60x81%

Varma, R.S.; Naicker, K.P. *Tetrahedron Lett*, ***1998***, *39*, 7463.

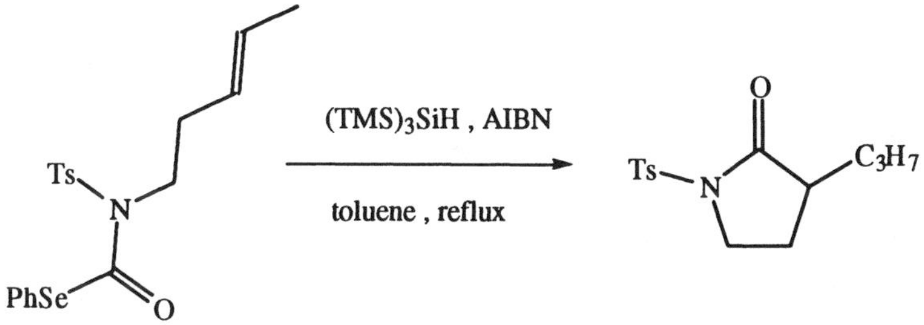
Ts
N
PhSe
O
(TMS)3SiH , AIBN
toluene , reflux
O
Ts
N
C3H7

68%

Rigby, J.H.; Danca, D.M.; Horner, J.H. *Tetrahedron Lett.*, ***1998***, *39*, 8413.

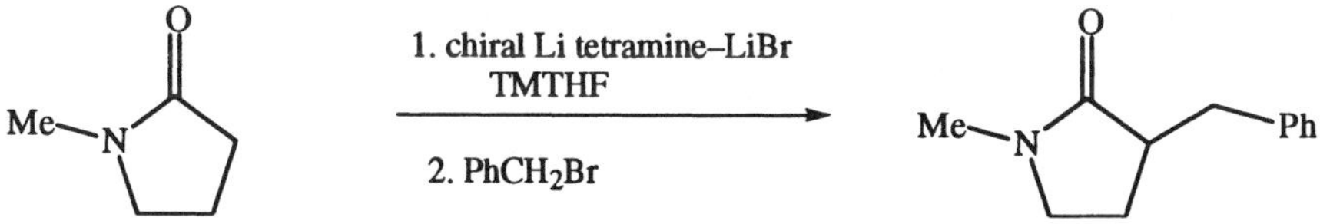
O
Me
N
1. chiral Li tetramine–LiBr
TMTHF
2. PhCH2Br
O
Me
N
Ph

64% (98% ee , *R*)

Matsuo, J.–i.; Kobayashi, S.; Koga, K. *Tetrahedron Lett*, ***1998***, *39*, 9723.

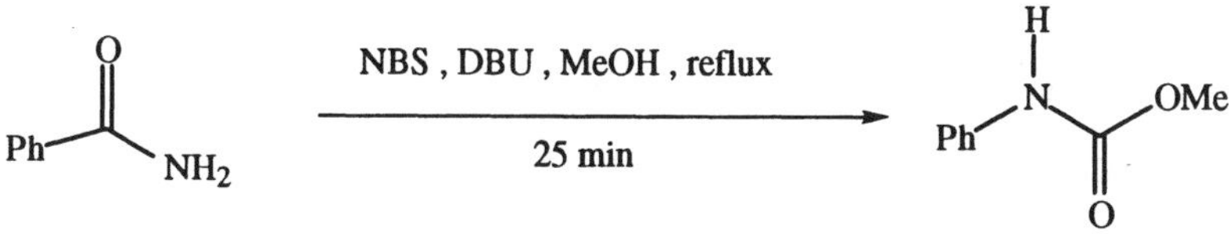
O
Ph
NH2
NBS , DBU , MeOH , reflux
25 min
H
N
Ph
OMe
O

95%

Huang, X.; Seid, M.; Keillor, J.W. *J. Org. Chem.*, ***1997***, *62*, 7495.

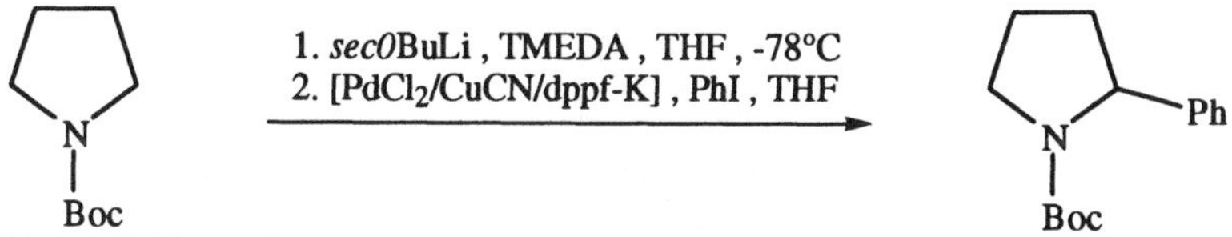
N
Boc
1. secOBuLi , TMEDA , THF , -78°C
2. [PdCl2/CuCN/dppf-K] , PhI , THF
N
Ph
Boc

79%

Dieter, R.K.; Li, S.J. *J. Org. Chem.*, ***1997***, *62*, 7726.

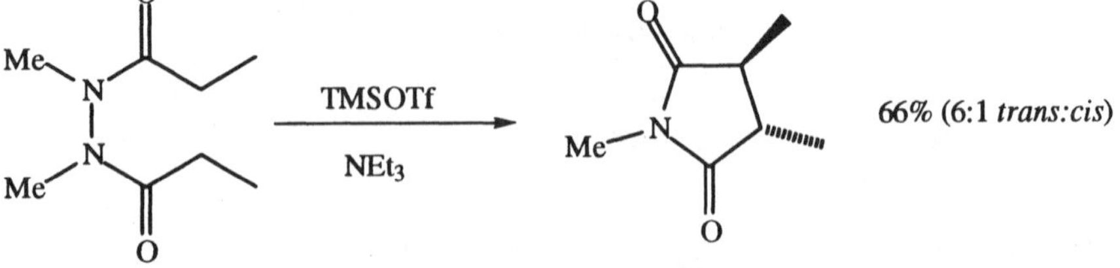

TMSOTf
NEt3
66% (6:1 trans:cis)

<u>Miller, S.J.</u>; Bayne, C.D. *J. Org. Chem.*, ***1997***, *62*, 5680.

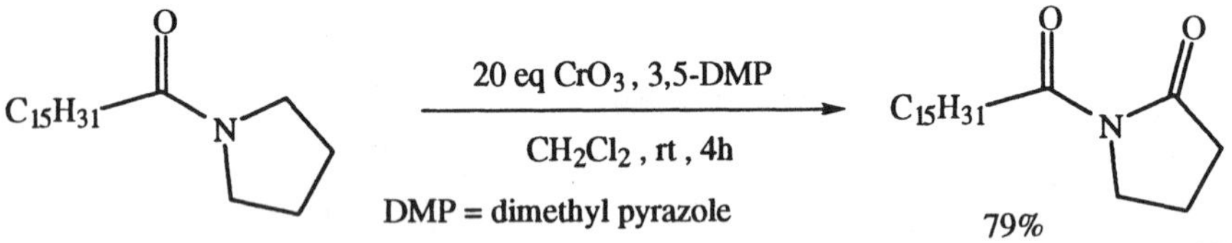

20 eq CrO3, 3,5-DMP
CH2Cl2, rt, 4h
DMP = dimethyl pyrazole
79%

Blay, G.; Cardona, L.; García, B.; García, C.L.; <u>Pedro, J.R.</u> *Tetrahedron Lett*, ***1997***, *38*, 8257

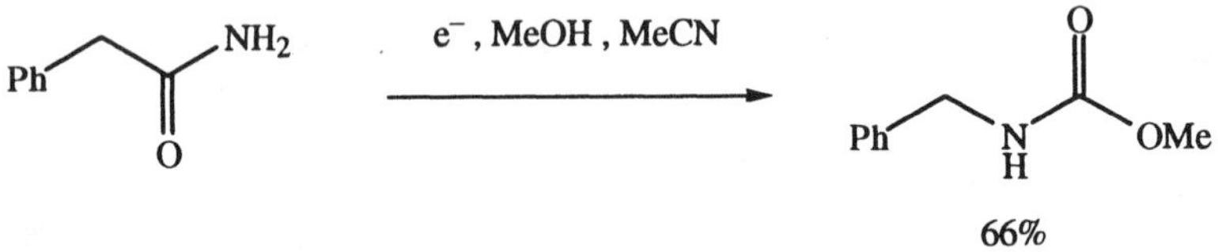

e⁻, MeOH, MeCN
66%

<u>Matsumura, Y.</u>; Maki, T.; Satoh, Y. *Tetrahedron Lett*, ***1997***, *38*, 8879.

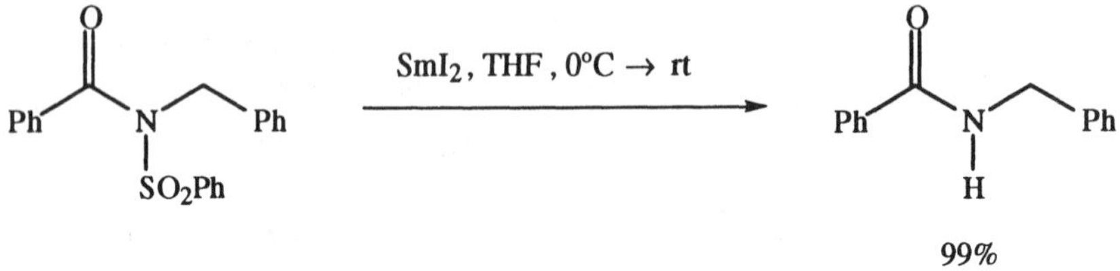

SmI2, THF, 0°C → rt
99%

Knowles, H.; <u>Parsons, A.F.</u>; Pettifer, R.M. *SynLett*, ***1997***, 271.

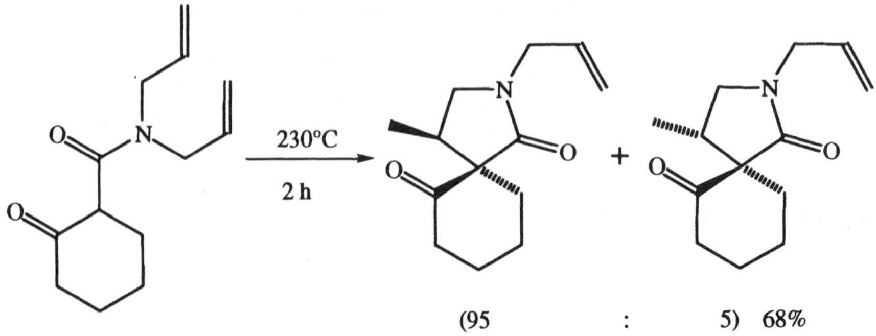

230°C
2 h
(95 : 5) 68%

<u>Cossy, J.</u>; Bouzide, A. *Tetrahedron*, ***1997***, *53*, 5775.

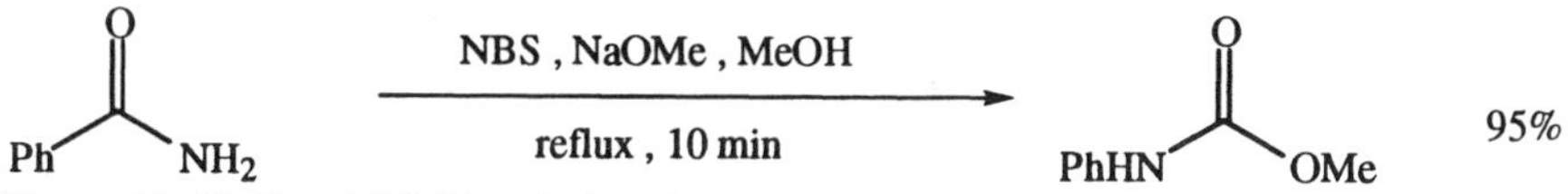
O
Ph
NH2
NBS , NaOMe , MeOH
reflux , 10 min
O
PhHN
OMe
95%

Huang, X.; Keillor, J.W. *Tetrahedron Lett.*, ***1997***, *38*, 313.

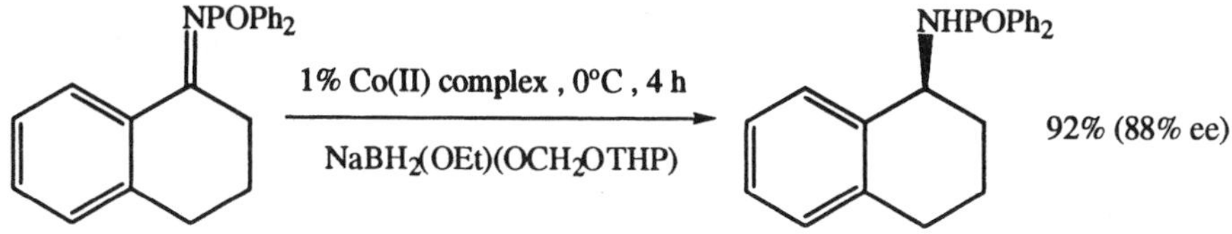
NPOPh2
1% Co(II) complex , 0°C , 4 h
NaBH2(OEt)(OCH2OTHP)
NHPOPh2
92% (88% ee)

Sugi, K.D.; Nagata, T.; Yamada, T.; Mukaiyama, T. *Chem. Lett.*, ***1997***, 493.

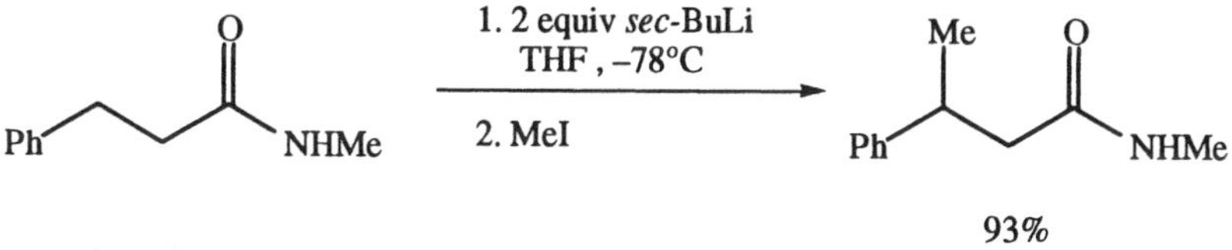
O
Ph
NHMe
1. 2 equiv sec-BuLi
THF , –78°C
2. MeI
Me
O
Ph
NHMe
93%

Lutz, G.P.; Du, H.; Gallagher, D.J.; Beak, P. *J. Org. Chem.*, ***1996***, *61*, 4542.

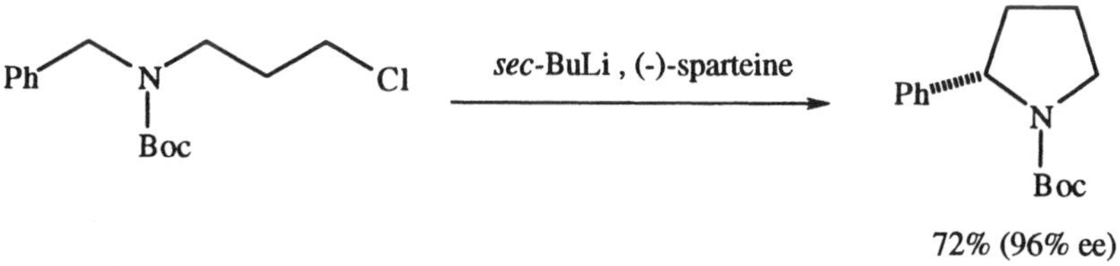
Ph
N
Cl
Boc
sec-BuLi , (-)-sparteine
Ph
N
Boc
72% (96% ee)

Wu, S.; Lee, S.; Beak, P. *J. Am. Chem. Soc.*, ***1996***, *118*, 715.

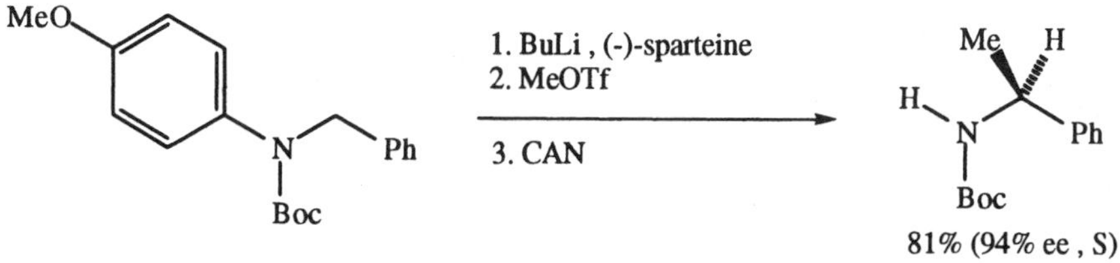
MeO
N
Ph
Boc
1. BuLi , (-)-sparteine
2. MeOTf
3. CAN
Me
H
H
N
Ph
Boc
81% (94% ee , S)

Park, Y.S.; Boys, M.L.; Beak, P. *J. Am. Chem. Soc.*, ***1996***, *118*, 3757.

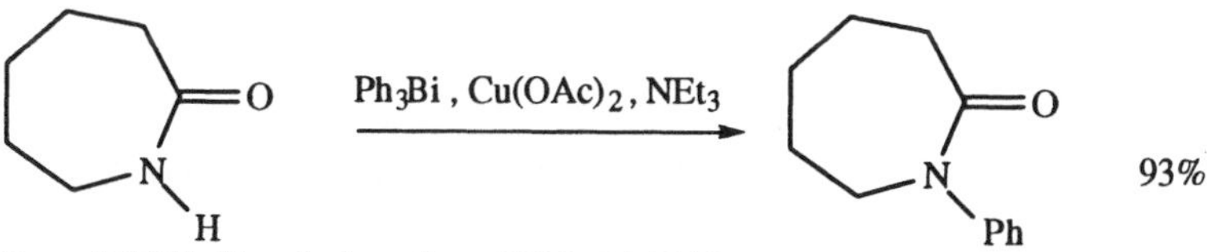
O
N
H
Ph3Bi , Cu(OAc)2 , NEt3
O
N
Ph
93%

Chan, D.M.T. *Tetrahedron Lett*, ***1996***, *37*, 9013.

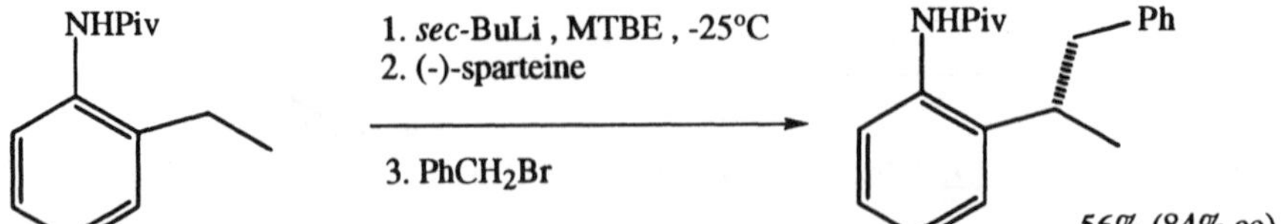
NHPiv
1. sec-BuLi , MTBE , -25°C
2. (-)-sparteine
3. PhCH2Br
NHPiv
Ph

56% (84% ee)

DuBois, J.; Hong, J.; Carreira, E.M.; Day, M.W. *J. Am. Chem. Soc.*, ***1996***, *118*, 915.

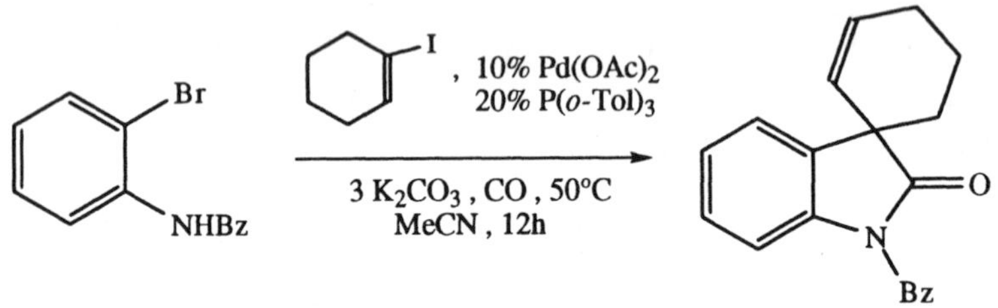
I
, 10% Pd(OAc)2
20% P(o-Tol)3
Br
NHBz
3 K2CO3 , CO , 50°C
MeCN , 12h
O
N
Bz

61%

Grigg, R.; Putnikovic, B.; Urch, C.J. *Tetrahedron Lett.* ***1996***, *37*, 695.

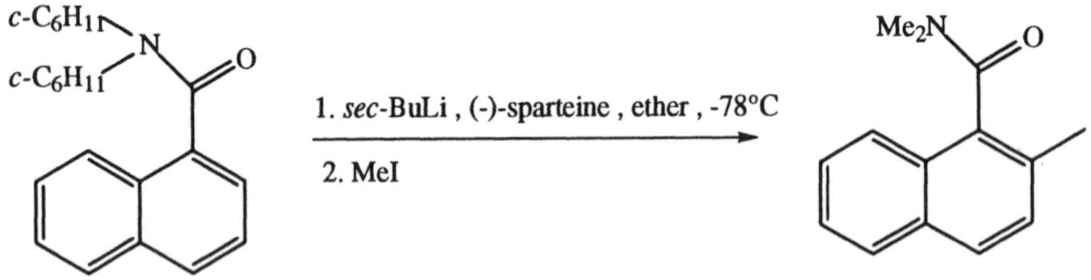
c-C6H11
c-C6H11
N
O
1. sec-BuLi , (-)-sparteine , ether , -78°C
2. MeI
Me2N
O

35% (30% ee)

Thayumanavan, S.; Beak, P.; Curran, D.P. *Tetrahedron Lett.*, ***1996***, *37*, 2899.

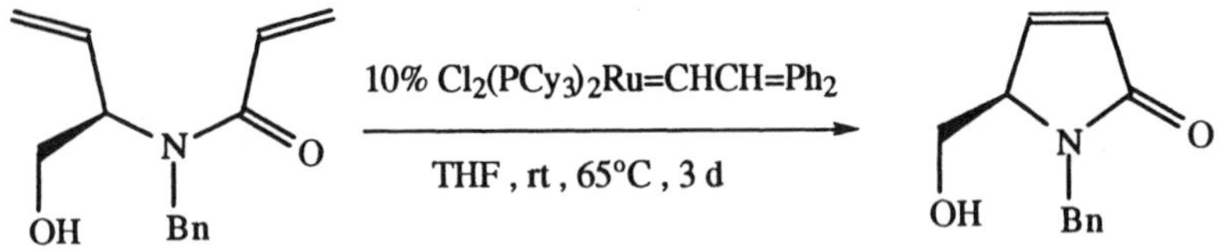
N
O
OH
Bn
10% Cl2(PCy3)2Ru=CHCH=Ph2
THF , rt , 65°C , 3 d
N
O
OH
Bn

41%

Huwe, C.M.; Kiehl, O.C.; Blechert, S. *SynLett*, ***1996***, 65.

SECTION 82: AMIDES FROM AMINES

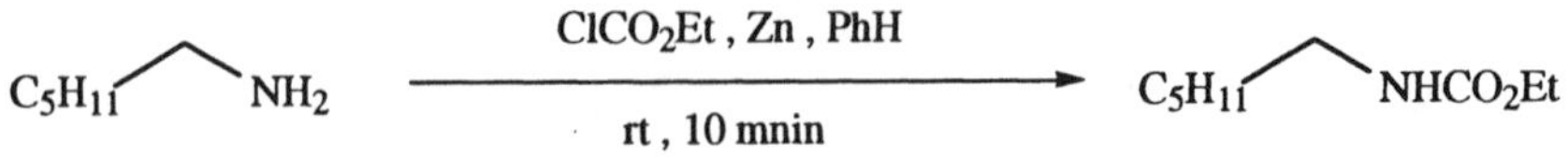
C5H11
NH2
ClCO2Et , Zn , PhH
rt , 10 mnin
C5H11
NHCO2Et

96%

Yadav, J.S.; Reddy, G.S.; Reddy, M.M.; Meshram, H.M. *Tetrahedron Lett.*, ***1998***, *39*, 3259.

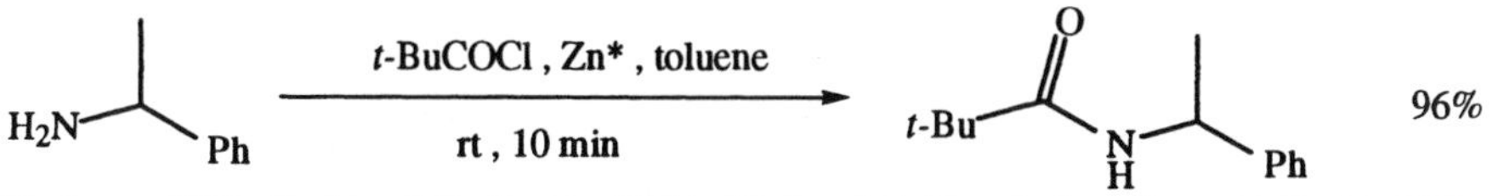
H2N
Ph
t-BuCOCl , Zn* , toluene
rt , 10 min
O
t-Bu
N
H
Ph
96%

Meshram, H.M.; Reddy, G.S.; Reddy, M.M.; Yadav, J.S. *Tetrahedron Lett.*, ***1998***, *39*, 4103.

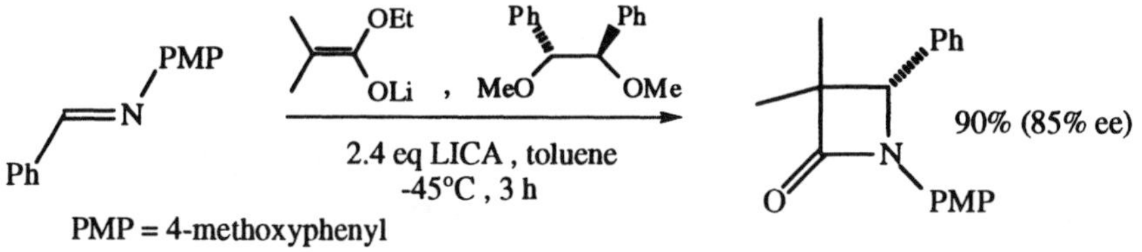
PMP
N
Ph
OEt
OLi
Ph
Ph
MeO
OMe
2.4 eq LICA , toluene
-45°C , 3 h
Ph
N
O
PMP
90% (85% ee)
PMP = 4-methoxyphenyl

Kambara, T.; Hussein, M.A.; Fujieda, H.; Iida, A.; Tomioka, K.
Tetrahedron Lett, ***1998***, *39*, 9055.

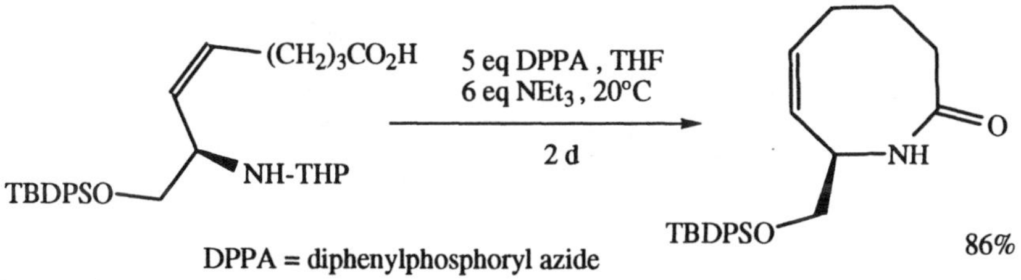
(CH2)3CO2H
NH-THP
TBDPSO
5 eq DPPA , THF
6 eq NEt3 , 20°C
2 d
O
NH
TBDPSO
86%
DPPA = diphenylphosphoryl azide

Derrer, S.; Feeder, N.; Teat, S.J.; Davies, J.E.; Holmes, A.B. *Tetrahedron Lett*, ***1998***, *39*, 9309

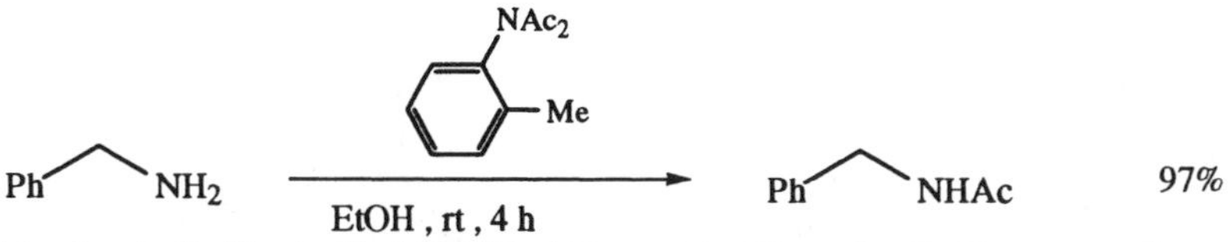
NAc2
Me
Ph
NH2
EtOH , rt , 4 h
Ph
NHAc
97%

Murakami, Y.; Kondo, K.; Miki, K.; Akiyama, Y.; Watanabe, T.; Yokeyama, Y.
Tetrahedron Lett., ***1997***, *38*, 3751.

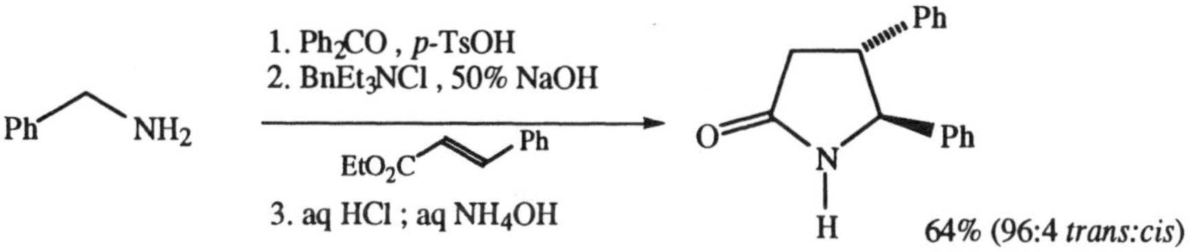
Ph
NH2
1. Ph2CO , p-TsOH
2. BnEt3NCl , 50% NaOH
EtO2C
Ph
3. aq HCl ; aq NH4OH
Ph
O
Ph
N
H
64% (96:4 trans:cis)

Yee, N.K. *Tetrahedron Lett.*, ***1997***, *38*, 5091.

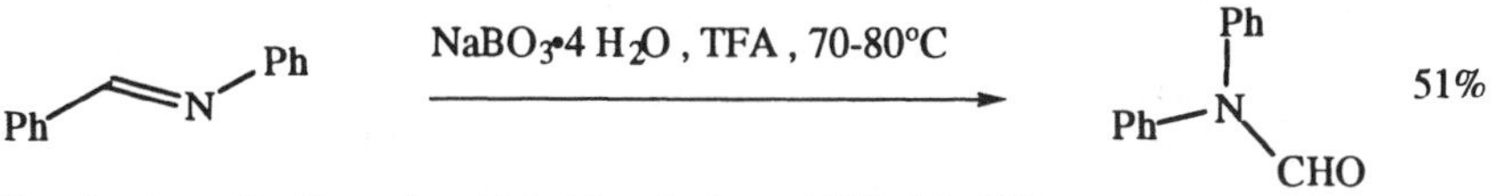
Ph
N
Ph
NaBO3•4 H2O , TFA , 70-80°C
Ph
Ph
N
CHO
51%

Nongkunsarn, P.; Ramsden, C.A. *Tetrahedron*, ***1997***, *53*, 3805.

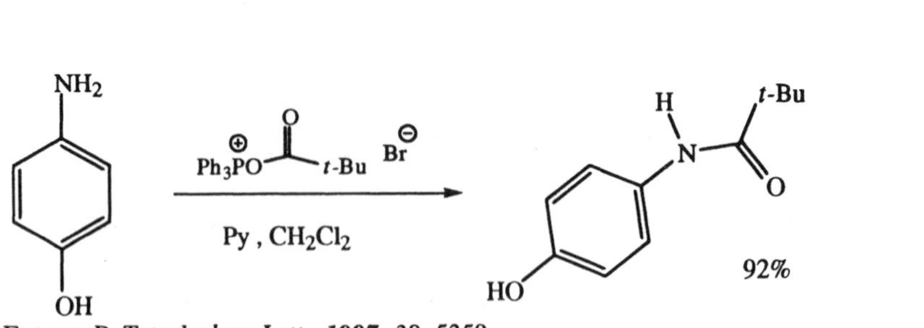
NH2
OH
Ph3PO
O
t-Bu
Br
Py , CH2Cl2
H
t-Bu
N
O
HO
92%

Frøyen, P. *Tetrahedron Lett.*, ***1997***, *38*, 5359.

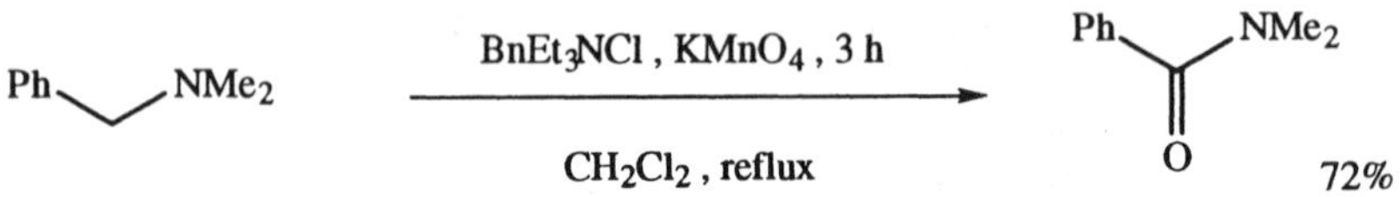
Ph
NMe2
BnEt3NCl , KMnO4 , 3 h
CH2Cl2 , reflux
Ph
NMe2
O
72%

Markgraf, J.H.; Sangani, P.K.; Finkelstein, M. *Synth. Commun.*, ***1997***, *27*, 1285.

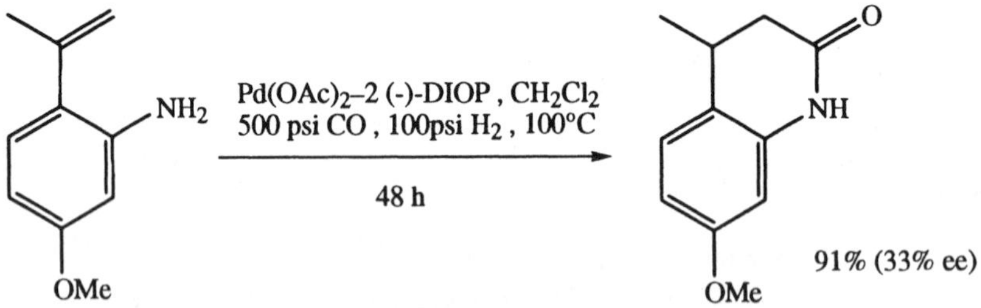
NH2
OMe
Pd(OAc)2–2 (-)-DIOP , CH2Cl2
500 psi CO , 100psi H2 , 100°C
48 h
O
NH
OMe
91% (33% ee)

Okuro, K.; Kai, H.; Alper, H. *Tetrahedron Asymmetry*, ***1997***, *8*, 2307.

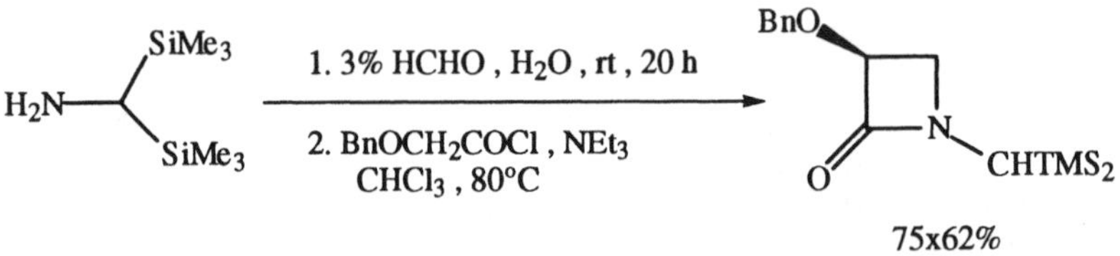
SiMe3
H2N
SiMe3
1. 3% HCHO , H2O , rt , 20 h
2. BnOCH2COCl , NEt3
CHCl3 , 80°C
BnO
N
CHTMS2
O
75x62%

Palomo, C.; Aizpurua, J.M.; Legido, M.; Galarza, R. *Chem. Commun.*, ***1997***, 233.

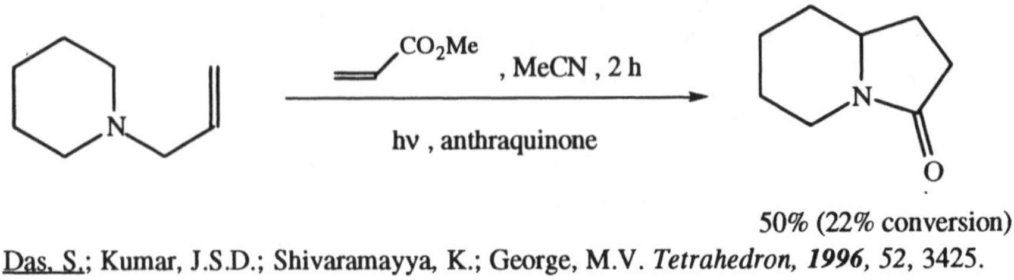
N
CO2Me
, MeCN , 2 h
hv , anthraquinone
N
O
50% (22% conversion)

Das, S.; Kumar, J.S.D.; Shivaramayya, K.; George, M.V. *Tetrahedron*, ***1996***, *52*, 3425.

$PhNH_2 \xrightarrow{e^-,\ CO_2,\ EtI,\ MeCN,\ Et_4NClO_4} PhNHCO_2Et$ 89%

Casadei, M.A.; Inesi, A.; Moracci, F.M.; Rossi, L. *Chem. Commun.*, ***1996***, 2575.

SECTION 83: AMIDES FROM ESTERS

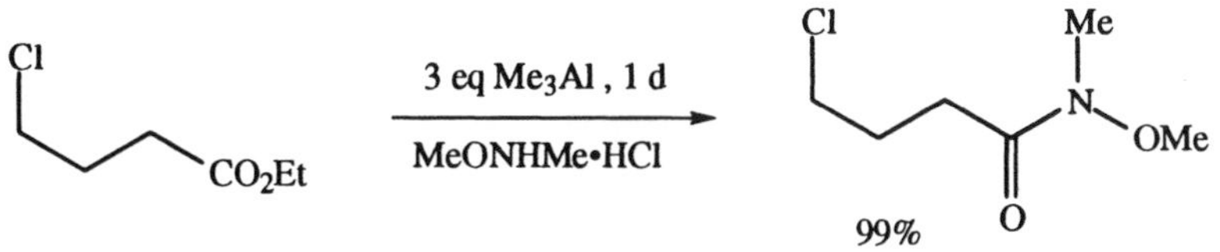

Shimizu, T.; Osako, K.; Nakata, T. *Tetrahedron Lett.*, ***1997***, *38*, 2685.

SECTION 84: AMIDES FROM ETHERS, EPOXIDES AND THIOETHERS

NO ADDITIONAL EXAMPLES

SECTION 85: AMIDES FROM HALIDES AND SULFONATES

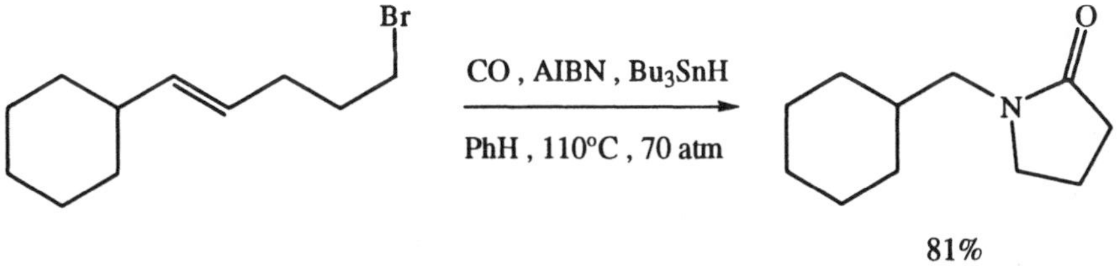

Ryu, I.; Matsu, K.; Minakata, S.; Komatsu, M. *J. Am. Chem. Soc.*, ***1998***, *120*, 5838.

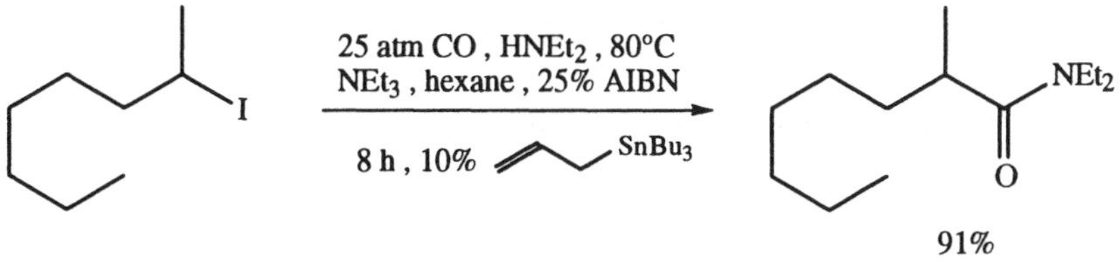

Ryu, I.; Nagahara, K.; Kambe, N.; Sonoda, N.; Kreimerman, S.; Komatsu, M. *Chem. Commun.*, ***1998***, 1953.

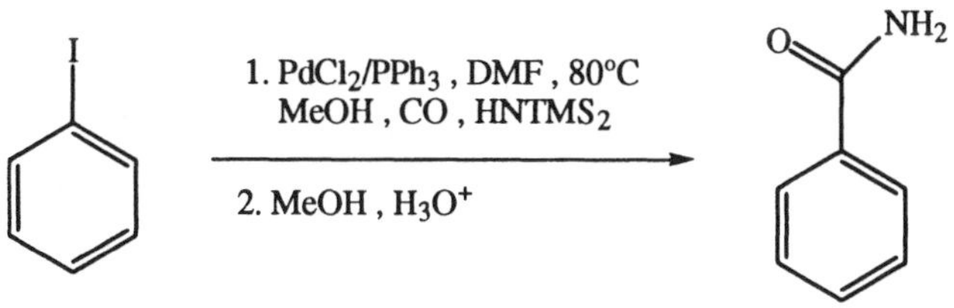

Morera, E.; Ortar, G. *Tetrahedron Lett.*, ***1998***, *39*, 2835.

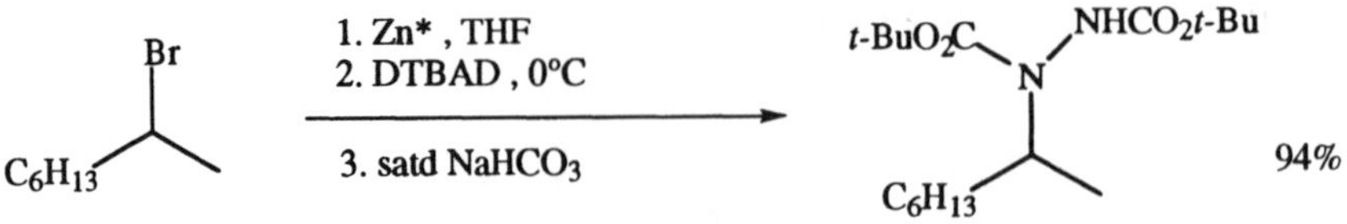
Br
C_6H_{13}
1. Zn* , THF
2. DTBAD , 0°C
3. satd $NaHCO_3$
t-BuO_2C
$NHCO_2t$-Bu
N
C_6H_{13}
94%

DTAB = di-*t*-butylazodicarboxylate

Velarde–Ortiz, R.; Guijarro, A.; Rieke, R.D. *Tetrahedron Lett.*, ***1998***, *39*, 9157.

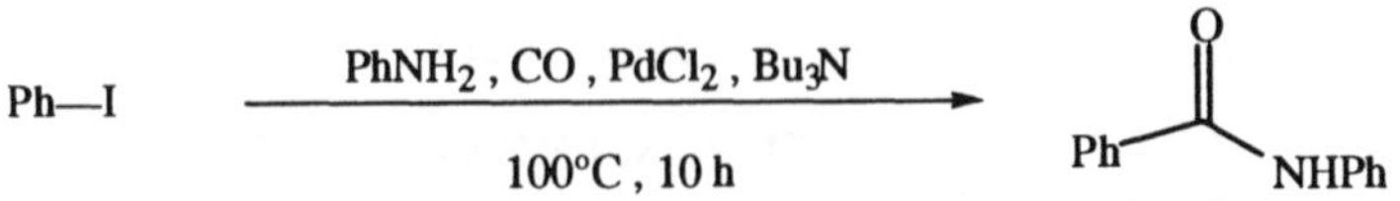
Ph—I
$PhNH_2$, CO , $PdCl_2$, Bu_3N
100°C , 10 h
O
Ph
NHPh

Cai, M.–Z.; Song, C.–S.; Huang, X. *Synth. Commun.*, ***1997***, *27*, 361.

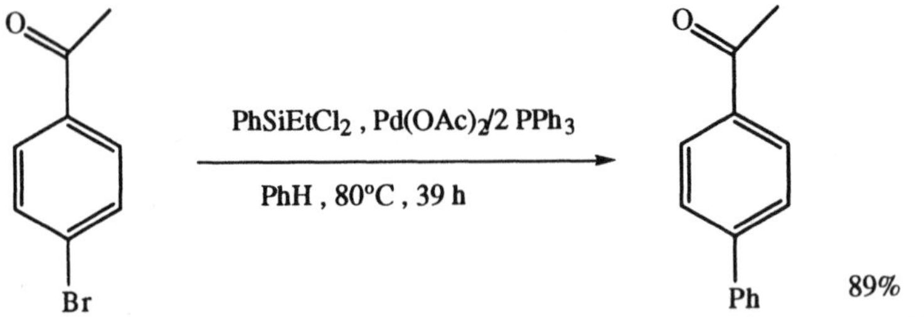
O
Br
$PhSiEtCl_2$, $Pd(OAc)_2$/2 PPh_3
PhH , 80°C , 39 h
O
Ph
89%

Hagiwara, E.; Gouda, K.–i.; Hatanaka, Y.; Hiyama, T. *Tetrahedron Lett.*, ***1997***, *38*, 439.

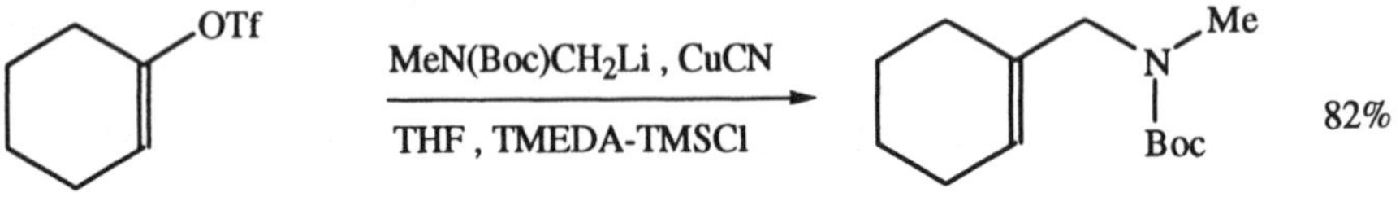
OTf
$MeN(Boc)CH_2Li$, CuCN
THF , TMEDA-TMSCl
Me
N
Boc
82%

Dieter, R.K.; Dieter, J.W.; Alexander, C.W.; Bhinderwala, N.S. *J. Org. Chem.*, ***1996***, *61*, 2930

SECTION 86: AMIDES FROM HYDRIDES

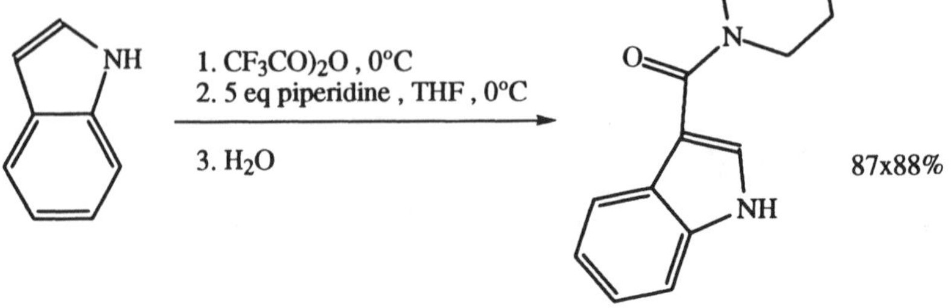
NH
1. $CF_3CO)_2O$, 0°C
2. 5 eq piperidine , THF , 0°C
3. H_2O
O
N
NH
87x88%

Hassinger, H.L.; Soll, R.M.; Gribble, G.W. *Tetrahedron Lett.*, ***1998***, *39*, 3095.

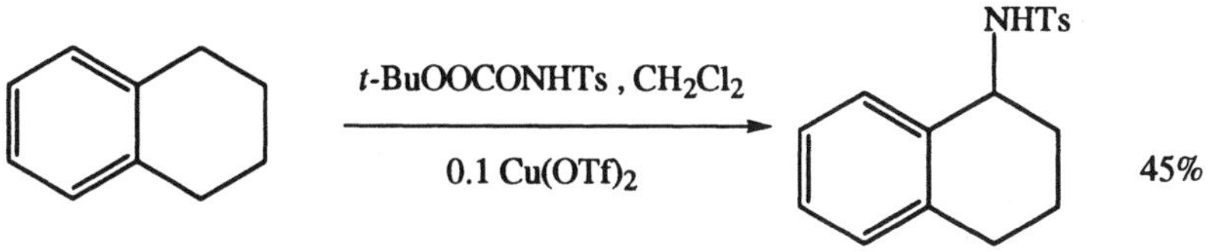
NHTs
t-BuOOCONHTs , CH2Cl2
0.1 Cu(OTf)2
45%

Kohmura, Y.; Kawasaki, K.; Katsuki, T. *SynLett*, ***1997***, 1456.

SECTION 87: AMIDES FROM KETONES

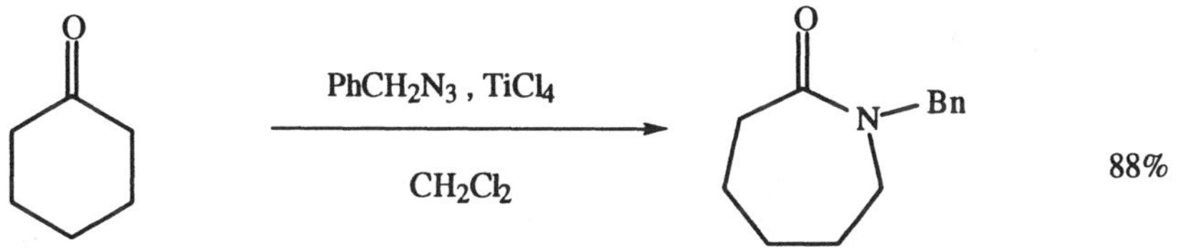
O
PhCH2N3 , TiCl4
CH2Cl2
O
N
Bn
88%

Schildknegt, K.; Agrios, K.A.; Aubé, J. *Tetrahedron Lett.*, ***1998***, *39*, 7687.

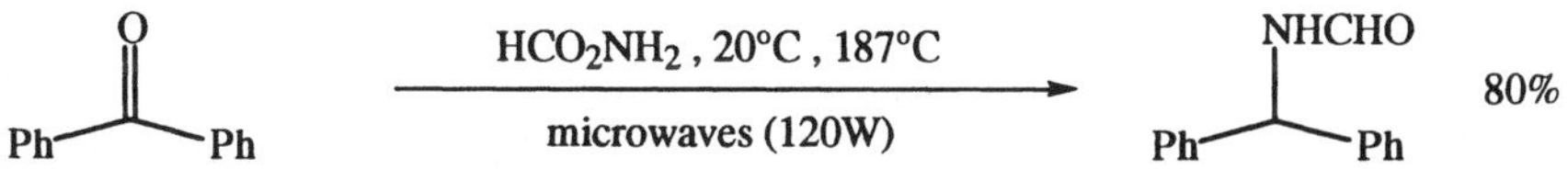
O
Ph
Ph
HCO2NH2 , 20°C , 187°C
microwaves (120W)
NHCHO
Ph
Ph
80%

Loupy, A.; Monteux, D.; Petit, A.; Zirpurua, J.M.; Domínguez, E.; Palomo, C. *Tetrahedron Lett.*, ***1996***, *37*, 8177.

SECTION 88: AMIDES FROM NITRILES

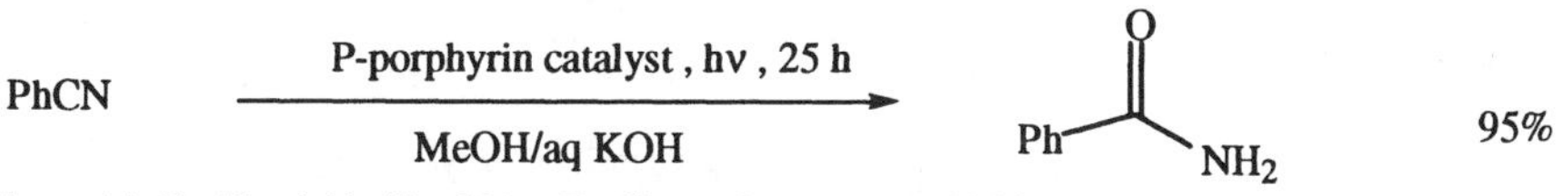
PhCN
P-porphyrin catalyst , hv , 25 h
MeOH/aq KOH
O
Ph
NH2
95%

Igarashi, T.; Konishi, K.; Aida, T. *Chem. Lett.*, ***1998***, 1039.

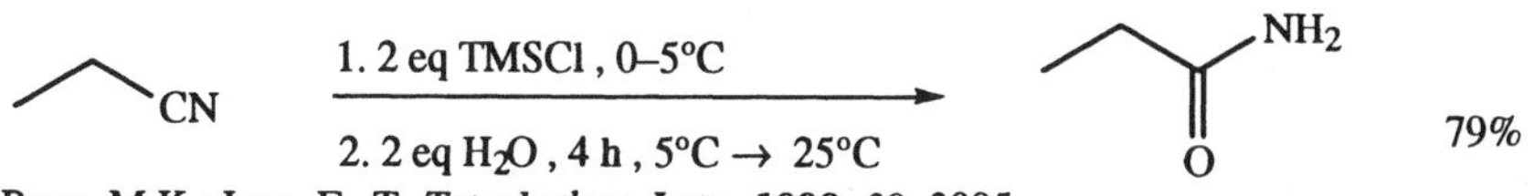
CN
1. 2 eq TMSCl , 0–5°C
2. 2 eq H2O , 4 h , 5°C → 25°C
NH2
O
79%

Basu, M.K.; Luo, F.-T. *Tetrahedron Lett.*, ***1998***, *39*, 3005.

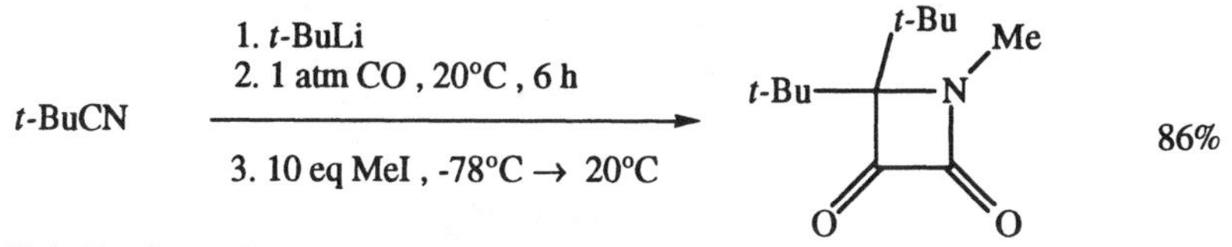
t-BuCN
1. t-BuLi
2. 1 atm CO , 20°C , 6 h
3. 10 eq MeI , -78°C → 20°C
t-Bu
Me
t-Bu
N
O
O
86%

Kai, H.; Orita, A.; Murai, S. *Synth. Commun.*, ***1998***, *28*, 1947.

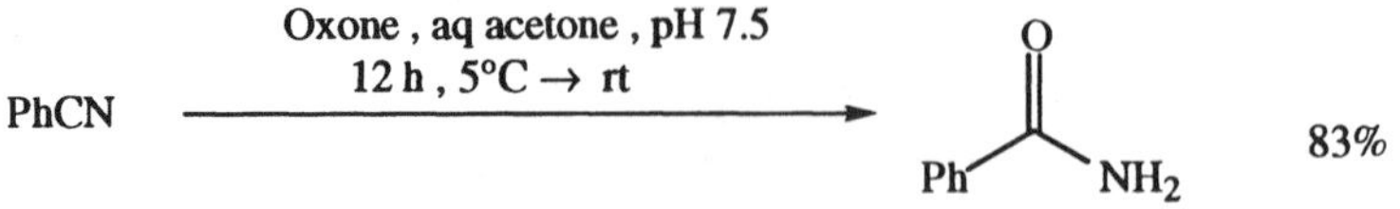
Oxone , aq acetone , pH 7.5
12 h , 5°C → rt
PhCN
O
Ph
NH2
83%

Bose, D.S.; Baquer, S.M. *Synth. Commun.*, ***1997***, *27*, 3119.

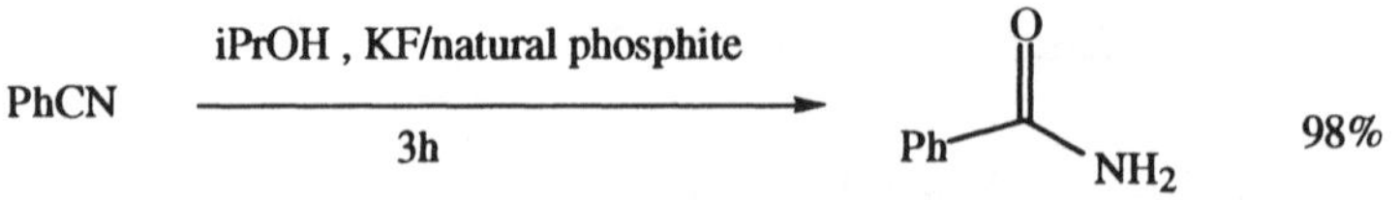
iPrOH , KF/natural phosphite
3h
PhCN
O
Ph
NH2
98%

Sebti, S.; Rhihil, A.; Saber, A.; Hanafi, N. *Tetrahedron Lett.*, ***1996***, *37*, 6555.

SECTION 89: AMIDES FROM ALKENES

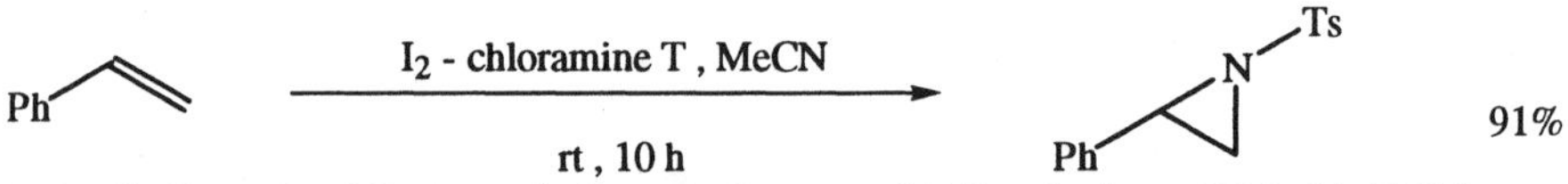
I2 - chloramine T , MeCN
rt , 10 h
Ph
Ts
N
Ph
91%

Abdo, T.; Kano, D.; Minakata, S.; Ryu, I.; Komatsu, M. *Tetrahedron*, ***1998***, *54*, 13485.

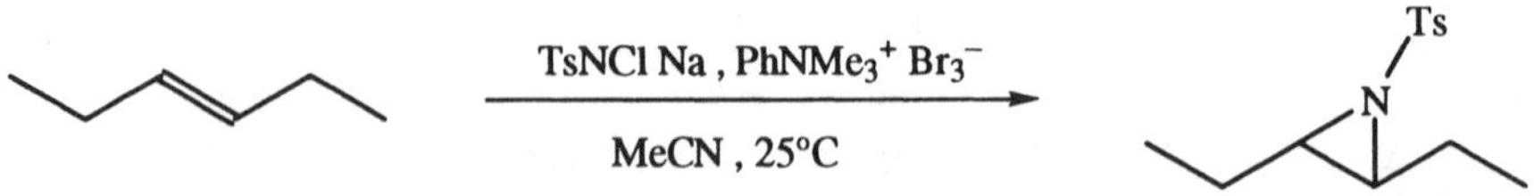
TsNCl Na , PhNMe3+ Br3−
MeCN , 25°C
Ts
N

Jeong, J.U.; Tao, B.; Sagasser, I.; Henniges, H.; Sharpless, K.B.
J. Am. Chem. Soc., ***1998***, *120*, 6844.

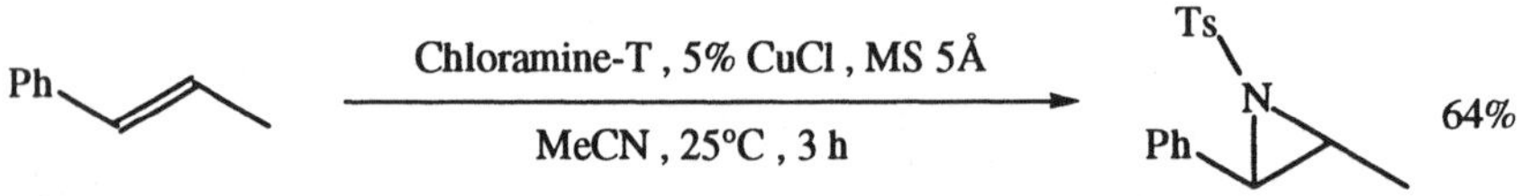
Chloramine-T , 5% CuCl , MS 5Å
MeCN , 25°C , 3 h
Ph
Ts
N
Ph
64%

Ando, T.; Minakata, S.; Ryu, I.; Komatsu, M. *Tetrahedron Lett*, ***1998***, *39*, 309.

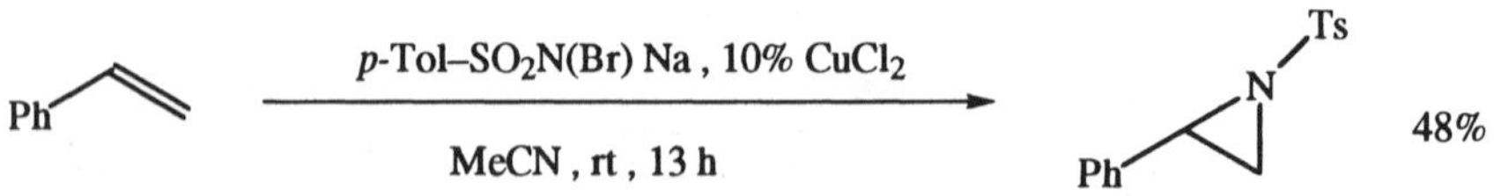
p-Tol–SO2N(Br) Na , 10% CuCl2
MeCN , rt , 13 h
Ph
Ts
N
Ph
48%

Vyas, R.; Chanda, B.M.; Bedekar, A.V. *Tetrahedron Lett.*, ***1998***, *39*, 4715.

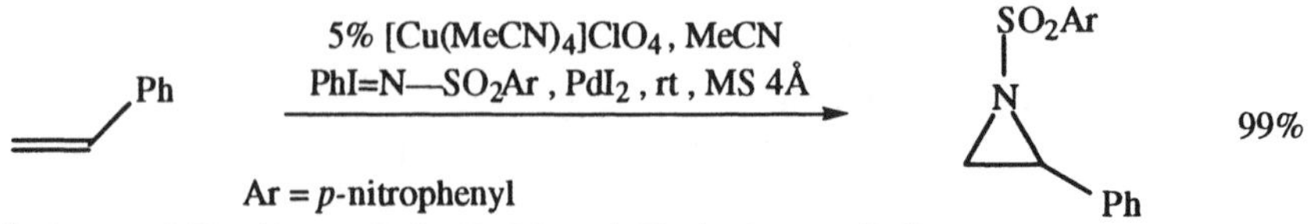
5% [Cu(MeCN)4]ClO4 , MeCN
PhI=N—SO2Ar , PdI2 , rt , MS 4Å
Ph
Ar = p-nitrophenyl
SO2Ar
N
Ph
99%

Södergren, M.J.; Alonso, D.A.; Bedekar, A.V.; Andersson, P.G.
Tetrahedron Lett.. ***1997***. *38*. 6897.

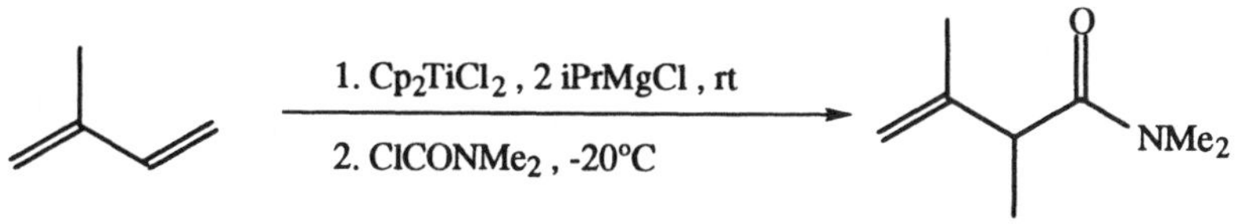
1. Cp2TiCl2 , 2 iPrMgCl , rt
2. ClCONMe2 , -20°C
O
NMe2

75%

Szymoniak, K.; Felix, D.; Moïse, C. *Tetrahedron Lett.*, ***1996***, *37*, 6603.

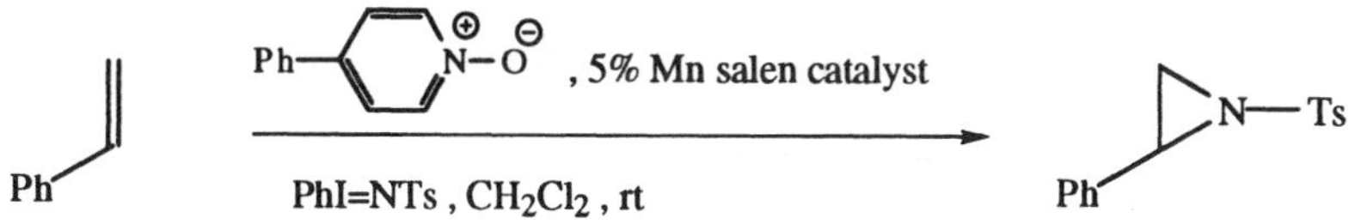
Ph
N—O
, 5% Mn salen catalyst
PhI=NTs , CH2Cl2 , rt
Ph
N—Ts

76% (94% ee, *S*)

Nishikori, H.; Katsuki, T. *Tetrahedron Lett.*, ***1996***, *37*, 9245.

SECTION 90: AMIDES FROM MISCELLANEOUS COMPOUNDS

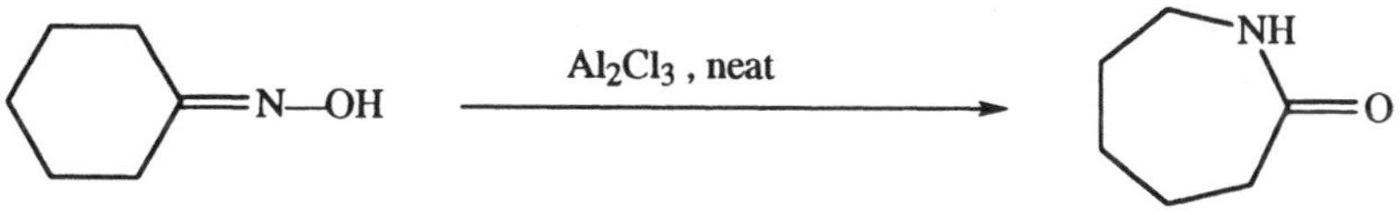
=N—OH
Al2Cl3 , neat
NH
=O

quant

Ghiaci, M.; Imanzadeh, G.H. *Synth. Commun.*, ***1998***, *28*, 2275.

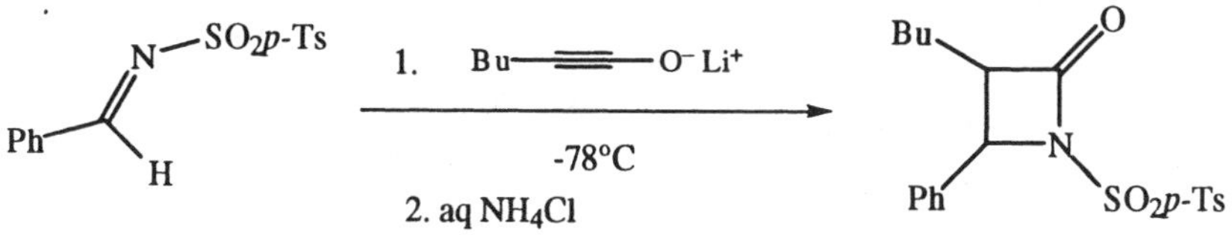
N—SO2p-Ts
Ph
H
1. Bu—≡—O⁻ Li⁺
-78°C
2. aq NH4Cl
Bu
O
N
Ph
SO2p-Ts

58%

Shindo, M.; Oya, S.; Sato, Y.; Shishido, K. *Heterocycles*, ***1998***, *49*, 113.

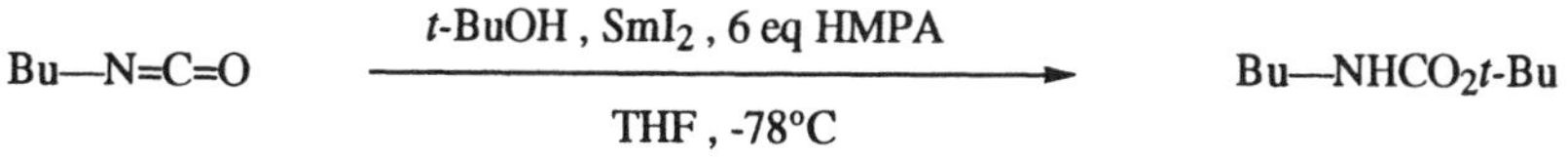
Bu—N=C=O
t-BuOH , SmI2 , 6 eq HMPA
THF , -78°C
Bu—NHCO2t-Bu

98%

Kim, Y.H.; Park, H.S. *SynLett*, ***1998***, 261.

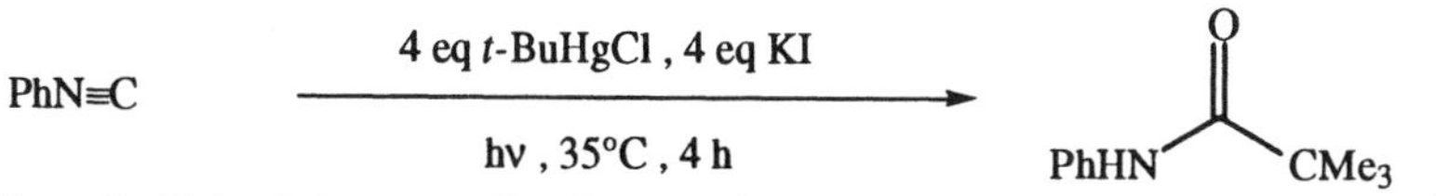
PhN≡C
4 eq t-BuHgCl , 4 eq KI
hν , 35°C , 4 h
O
PhHN
CMe3

84%

Russell, G.A.; Rajaratnam, R.; Chen, P. *Acta Chem. Scand. B.*, ***1998***, *52*, 528.

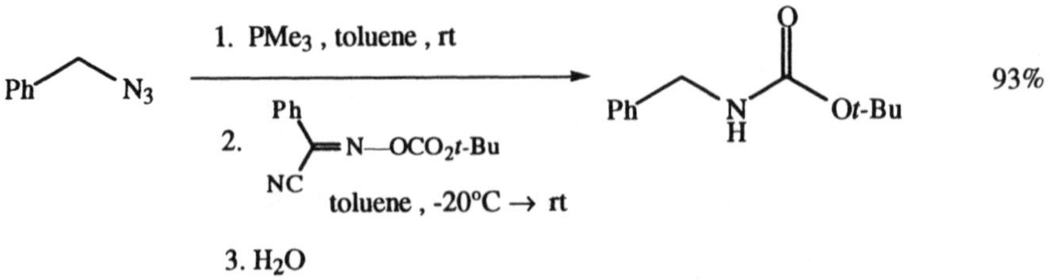
1. PMe_3 , toluene , rt
2.
Ph
NC
N—OCO_2t-Bu
toluene , -20°C → rt
3. H_2O
Ph
N_3
Ph
N
H
O
Ot-Bu
93%

Arizo, X.; Urpí, F.; Viladomat, C.; Vilarrasa, J. *Tetrahedron Lett.*, ***1998***, *39*, 9101.

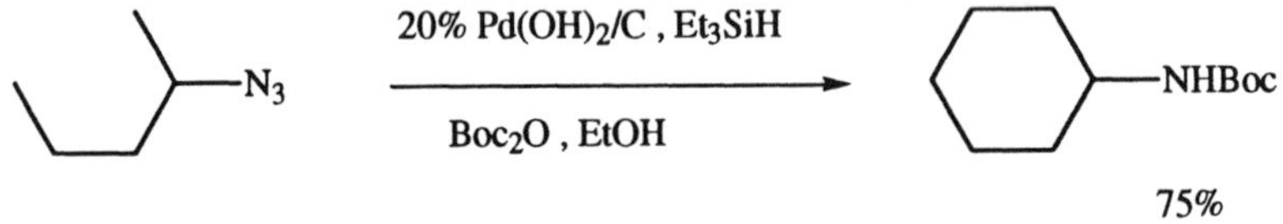
N_3
20% $Pd(OH)_2/C$, Et_3SiH
Boc_2O , EtOH
NHBoc
75%

Kotsuki, H.; Ohishi, T.; Araki, T. *Tetrahedron Lett.*, ***1997***, *38*, 2129.

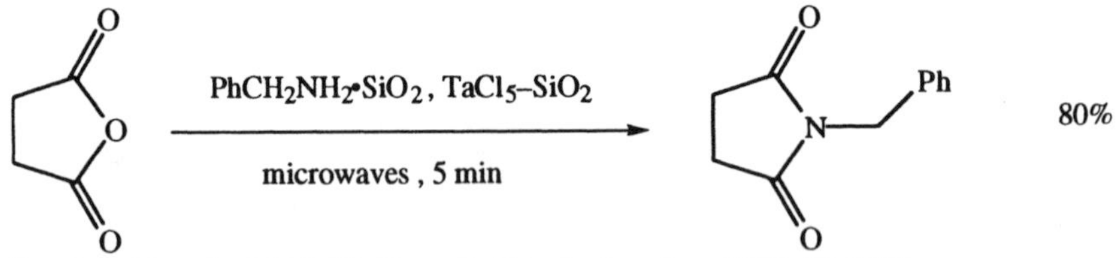
O
O
O
$PhCH_2NH_2$•SiO_2 , $TaCl_5$–SiO_2
microwaves , 5 min
O
N
Ph
O
80%

Chandrasekhar, S.; Jakhi, M.; Uma, G. *Tetrahedron Lett*, ***1997***, *38*, 8089.

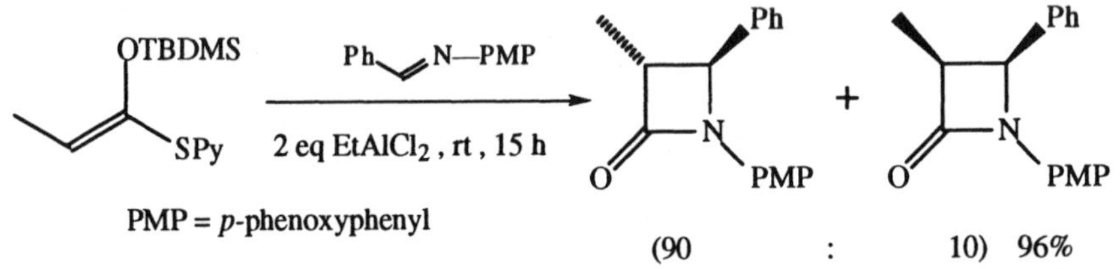
OTBDMS
SPy
Ph
N—PMP
2 eq $EtAlCl_2$, rt , 15 h
PMP = *p*-phenoxyphenyl
Ph
N
PMP
O
+
Ph
N
PMP
O
(90 : 10) 96%

Annunziata, R.; Cinquini, M.; Cozzi, F.; Molteni, V.; Schupp, O. *Tetrahedron*, ***1996***, *52*, 2573
Annunziata, R.; Benaglia, M.; Cinquini, M.; Cozzi, F.; Martini, O.; Molteni, V. *Tetrahedron*, ***1996***, *52*, 2583.

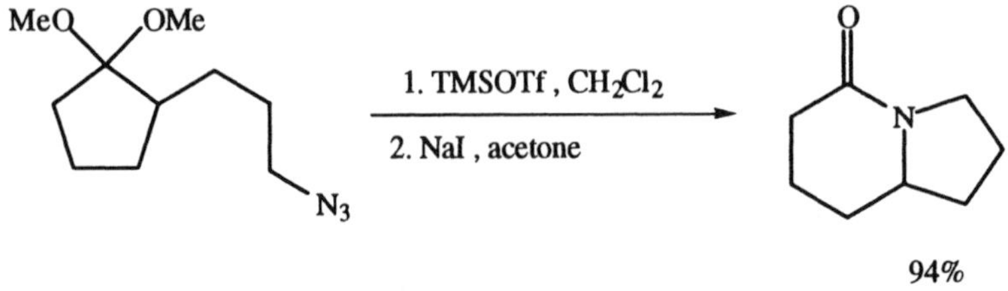
MeO
OMe
N_3
1. TMSOTf , CH_2Cl_2
2. NaI , acetone
O
N
94%

Mossman, C.J.; Aubé, J. *Tetrahedron*, ***1996***, *52*, 3403.

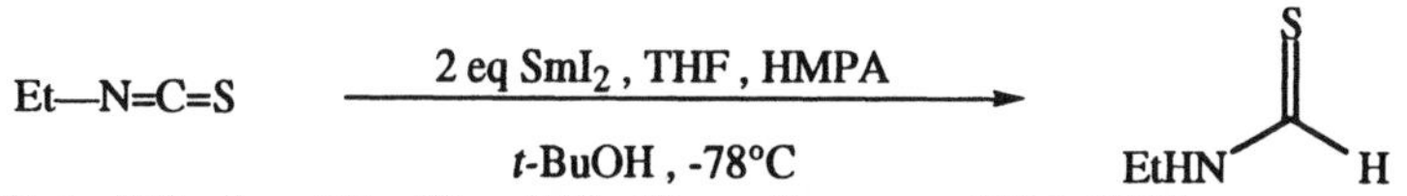
Et—N=C=S
2 eq SmI_2 , THF , HMPA
t-BuOH , -78°C
S
EtHN
H

92%

Park, H.S.; Lee, I.S.; Kim, Y.H. *Chem. Commun.*, ***1996***, 1805.

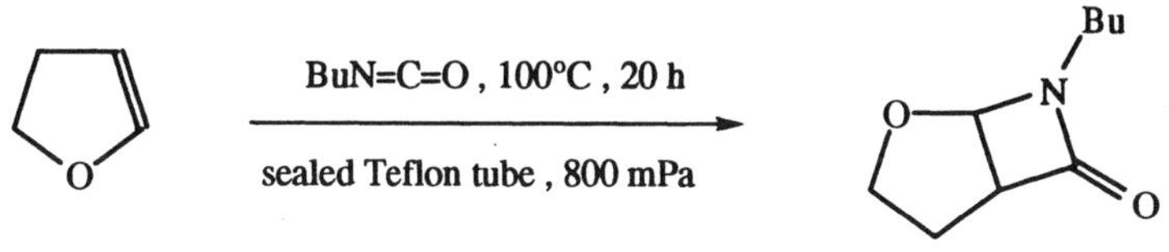
O
BuN=C=O , 100°C , 20 h
sealed Teflon tube , 800 mPa
Bu
N
O
O

77%

Taguchi, Y.; Tsuchiya, T.; Oishi, A.; Shibuya, I. *Bull. Chem. Soc. Jpn.*, ***1996***, *69*, 1667.

SECTION 90A: PROTECTION OF AMIDES

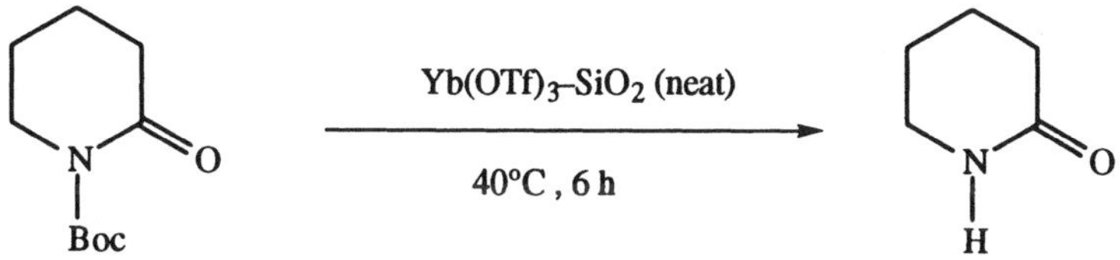
N
O
Boc
$Yb(OTf)_3$–SiO_2 (neat)
40°C , 6 h
N
O
H

96%

Kotsuki, H.; Ohishi, T.; Araki, T.; Arimura, K. *Tetrahedron Lett.*, ***1998***, *39*, 4869.

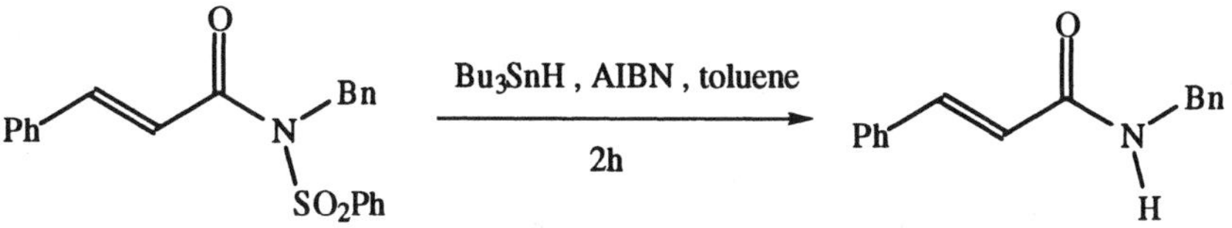
O
Ph
N
Bn
SO_2Ph
Bu_3SnH , AIBN , toluene
2h
O
Ph
N
Bn
H

94%

Parsons, A.F.; Pettifer, R.M. *Tetrahedron Lett.*, ***1996*** *37*, 1667.

CHAPTER 7
PREPARATION OF AMINES

SECTION 91: AMINES FROM ALKYNES

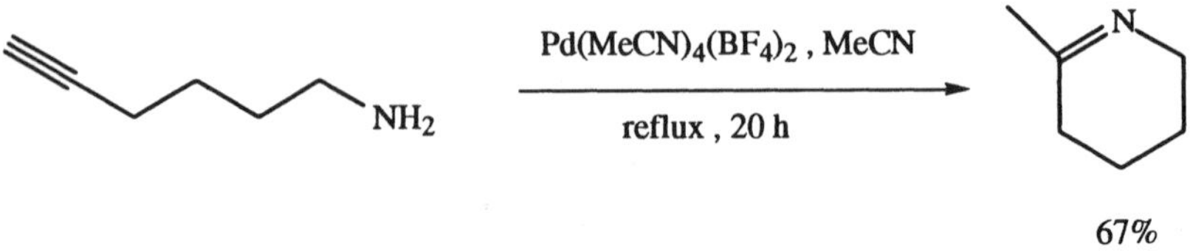

NH2
Pd(MeCN)4(BF4)2 , MeCN
reflux , 20 h
N
67%

Müller, T.E. *Tetrahedron Lett.*, ***1998***, *39*, 5961.

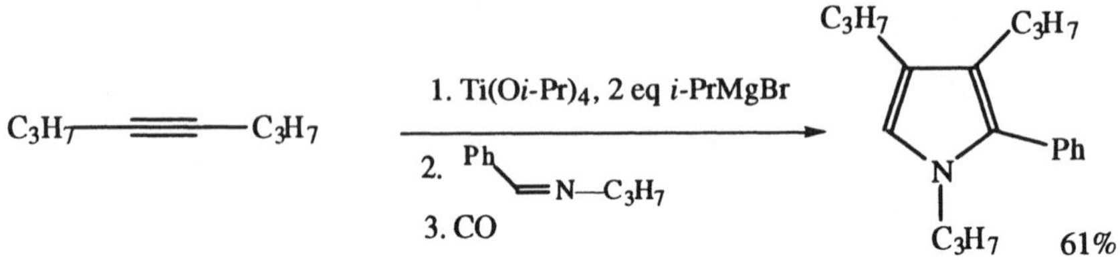

C3H7
C3H7
1. Ti(Oi-Pr)4, 2 eq i-PrMgBr
2. Ph
=N—C3H7
3. CO
C3H7
C3H7
Ph
N
C3H7
61%

Gao, Y.; Shirai, M.; Sato, F. *Tetrahedron Lett.*, ***1996***, *37*, 7787.

SECTION 92: AMINES FROM ACID DERIVATIVES

NO ADDITIONAL EXAMPLES

SECTION 93: AMINES FROM ALCOHOLS AND THIOLS

NO ADDITIONAL EXAMPLES

SECTION 94: AMINES FROM ALDEHYDES

Related Methods: Section 102 (Amines from Ketones)

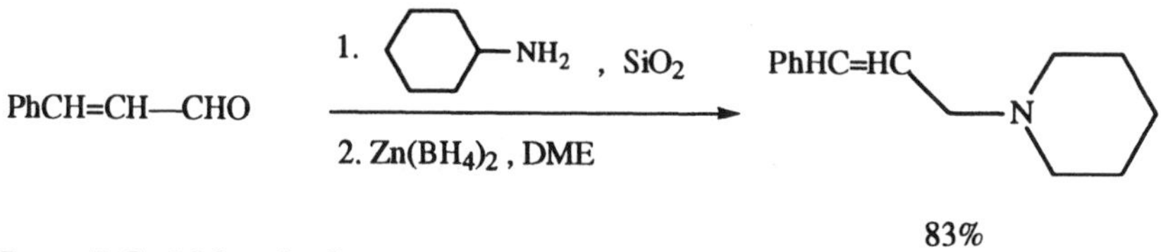
PhCH=CH—CHO
1. NH2 , SiO2
2. Zn(BH4)2 , DME
PhHC=HC
N
83%

Ranu, B.C.; Majee, A.; Sarkar, A. *J. Org. Chem.*, ***1998***, *63*, 370.

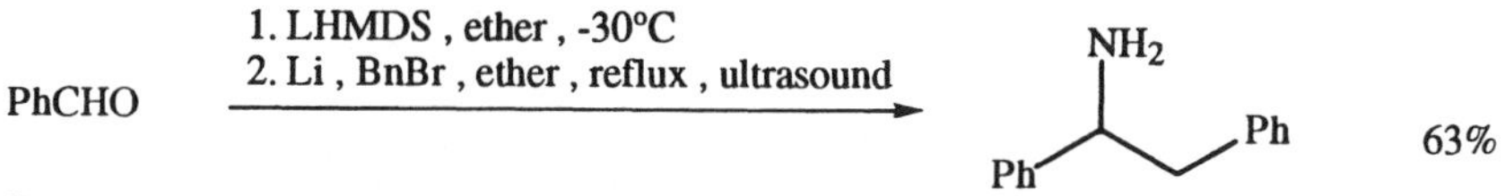
PhCHO
1. LHMDS , ether , -30°C
2. Li , BnBr , ether , reflux , ultrasound
NH2
Ph
Ph
63%

Gyenes, F.; Bergmann, K.E.; Welch, J.T. *J. Org. Chem.*, ***1998***, *63*, 2824.

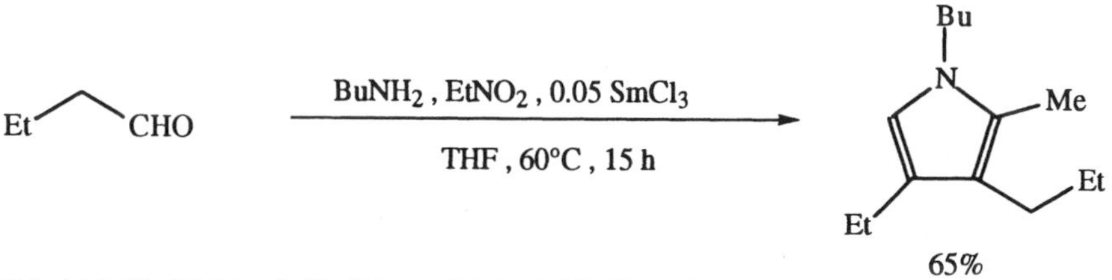
Et
CHO
BuNH2 , EtNO2 , 0.05 SmCl3
THF , 60°C , 15 h
Bu
N
Me
Et
Et
65%

Shiraishi, H.; Nishitani, T.; Sakaguchi, S.; Ishii, Y. *J. Org. Chem.*, ***1998***, *63*, 6234.

PhCHO $\xrightarrow{PhNH_2\text{ , }Bu_2SnClH\text{–}HMPA}$ $PhCH_2NHPh$ 81%

Shibata, I.; Suwa, T.; Sugiyama, E.; Baba, A. *SynLett*, ***1998***, 1081.

PhCHO $\xrightarrow[\text{5 min}]{PhNH_2\text{ , K10- clay , microwaves}}$ $PhCH_2NHPh$ 97%

Varma, R.S.; Dahiya, R. *Tetrahedron*, ***1998***, *54*, 6293.

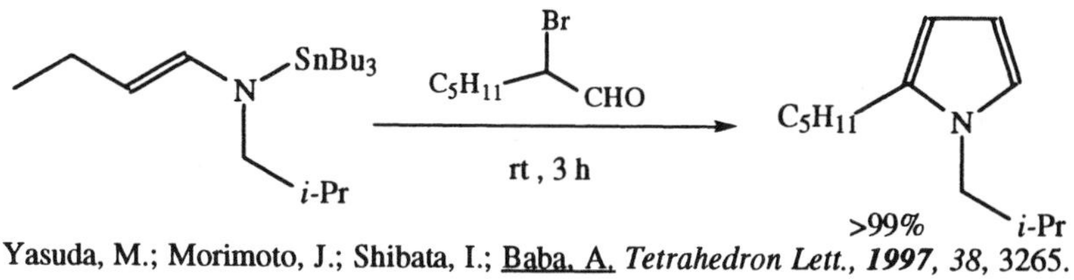
SnBu3
N
i-Pr
Br
C5H11
CHO
rt , 3 h
C5H11
N
i-Pr
>99%

Yasuda, M.; Morimoto, J.; Shibata, I.; Baba, A. *Tetrahedron Lett.*, ***1997***, *38*, 3265.

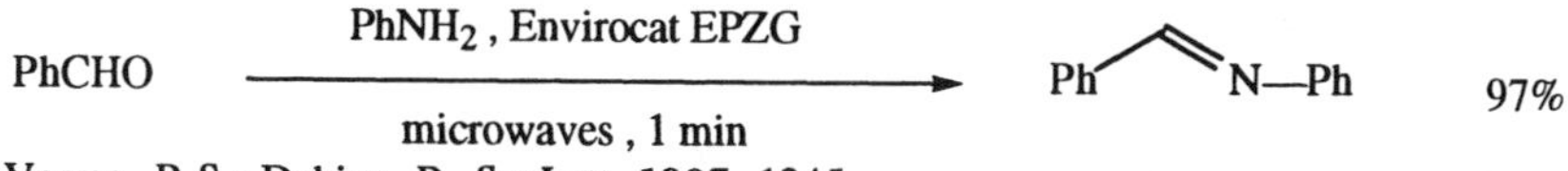
PhCHO
PhNH2 , Envirocat EPZG
microwaves , 1 min
Ph
N—Ph
97%

Varma, R.S.; Dahiya, R. *SynLett*, ***1997***, 1245.

SECTION 95: AMINES FROM ALKYLS, METHYLENES AND ARYLS

NO ADDITIONAL EXAMPLES

SECTION 96: AMINES FROM AMIDES

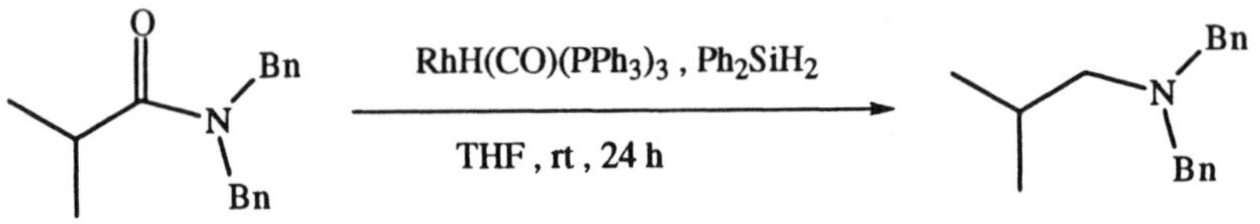
Bn
Bn
RhH(CO)(PPh3)3 , Ph2SiH2
THF , rt , 24 h
Bn
Bn

90%

Kuwano, R.; Takahashi, M.; Ito, Y. *Tetrahedron Lett.*, ***1998***, *39*, 1017.

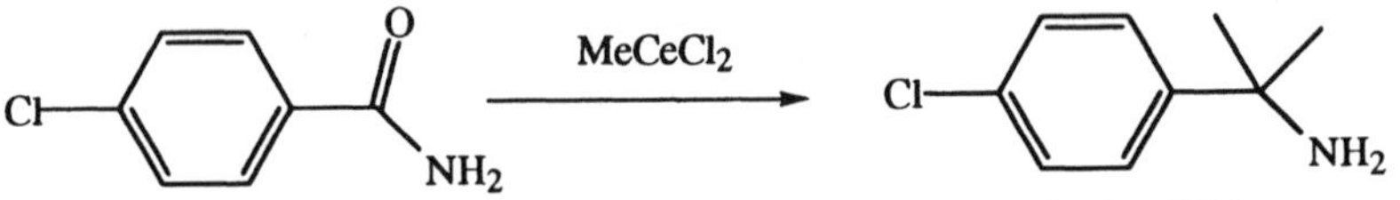
Cl
NH2
MeCeCl2
Cl
NH2

66%

Calderwood, D.J.; Davies, R.V.; Rafferty, P.; Twigger, H.L.; Whelan, H.M. *Tetrahedron Lett.*, ***1997***, *38*, 1241.

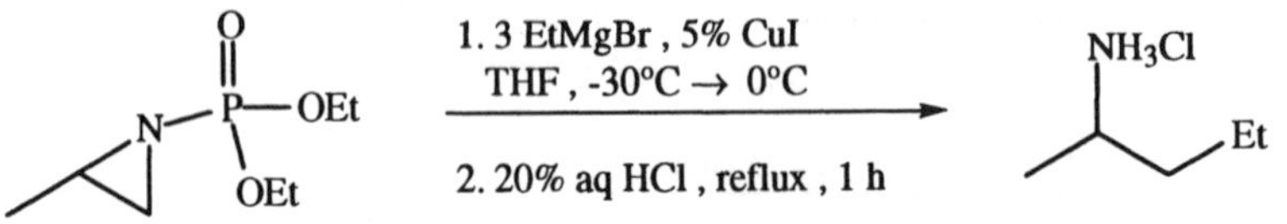
OEt
OEt
1. 3 EtMgBr , 5% CuI
THF , -30°C → 0°C
2. 20% aq HCl , reflux , 1 h
NH3Cl
Et

66%

Gajda, T.; Napieraj, A.; Osowska–Pacewicka, K.; Zawadzki, S.; Zwierzak, A. *Tetrahedron*, ***1997***, *53*, 4935.

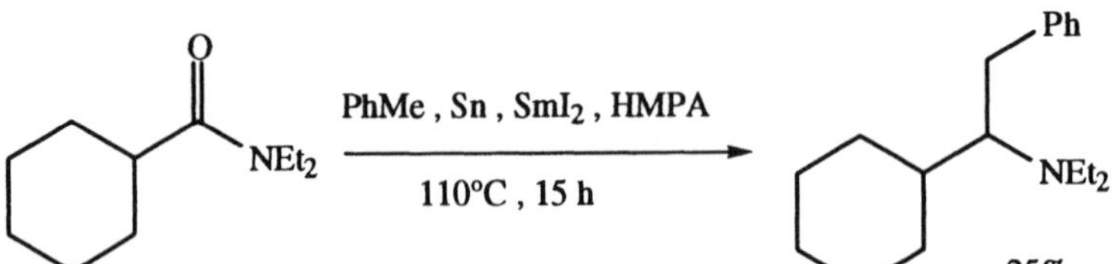
NEt2
PhMe , Sn , SmI2 , HMPA
110°C , 15 h
Ph
NEt2

25%

Ogawa, A.; Takami, N.; Nanke, T.; Ohya, S.; Hirao, T.; Sonoda, N. *Tetrahedron*, ***1997***, *53*, 12895.

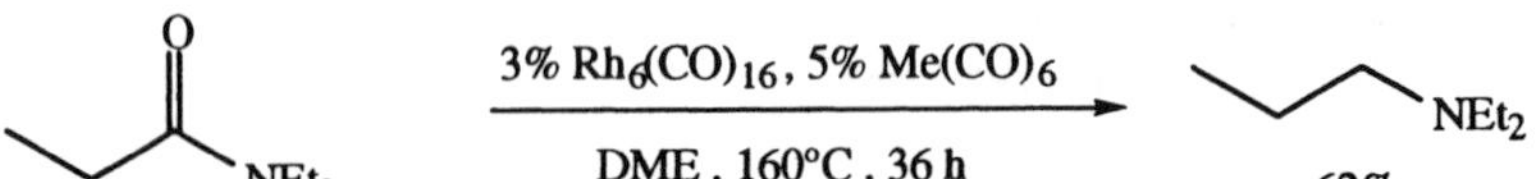
NEt2
3% Rh6(CO)16, 5% Me(CO)6
DME , 160°C , 36 h
NEt2

62%

Hirosawa, C.; Wakasa, N.; Fuchikami, T. *Tetrahedron Lett.*, ***1996***, *37*, 6749.

Related Methods: Section 105A (Protection of Amines)

SECTION 97: AMINES FROM AMINES

N—Ph
Ph
Ph
O
Bu2SnClH•HMPA, rt, 20 h
NHPh
Ph
Ph
O
73%

Shibata, I.; Moriuchi-Kawakami, T.; Tankzawa, D.; Suwa, T.; Sugiyama, E.; Matsuda, H.; Baba, A. *J. Org. Chem.*, ***1998***, *63*, 383.

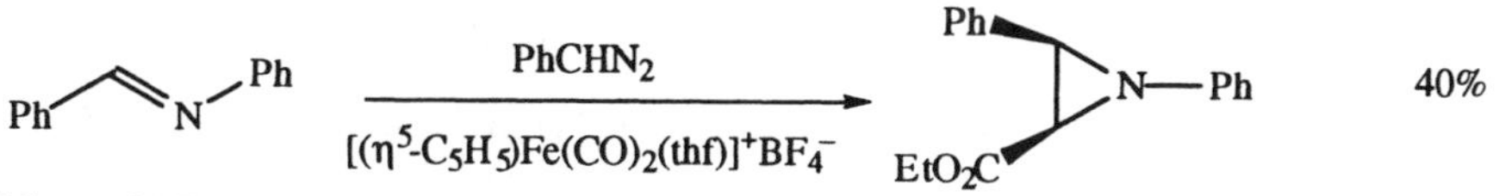
Ph
Ph
N
PhCHN2
[(η5-C5H5)Fe(CO)2(thf)]+BF4−
Ph
N—Ph
EtO2C
40%

Mayer, M.F.; Hossain, M.M. *J. Org. Chem.*, ***1998***, *63*, 6839.

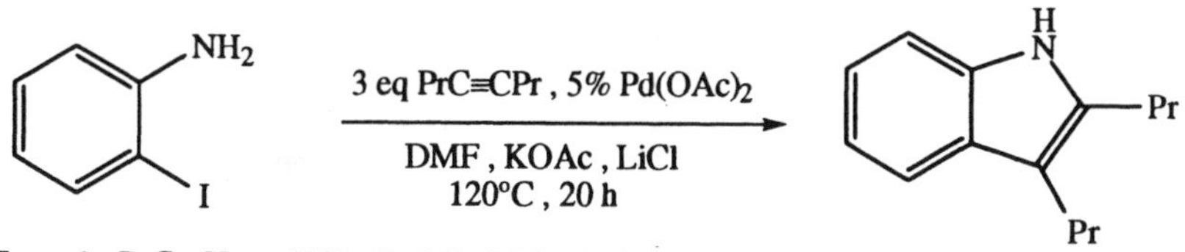
NH2
I
3 eq PrC≡CPr, 5% Pd(OAc)2
DMF, KOAc, LiCl
120°C, 20 h
H
N
Pr
Pr

Larock, R.C.; Yum, E.K.; Refvik, M.D. *J. Org. Chem.*, ***1998***, *63*, 7652.

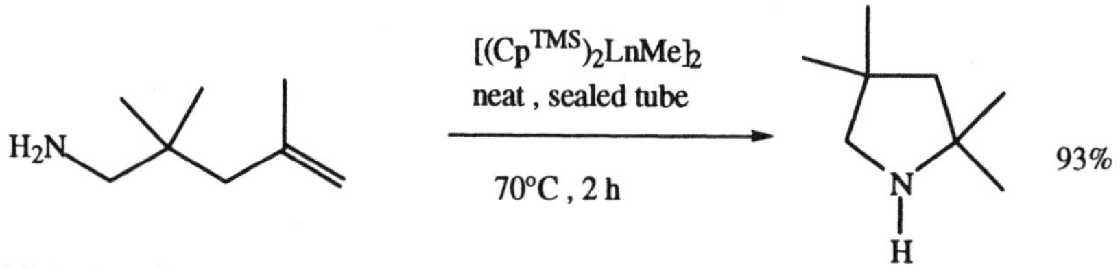
H2N
[(CpTMS)2LnMe]2
neat, sealed tube
70°C, 2 h
N
H
93%

Molander, G.A.; Dowdy, E.D. *J. Org. Chem.*, ***1998***, *63*, 8983.

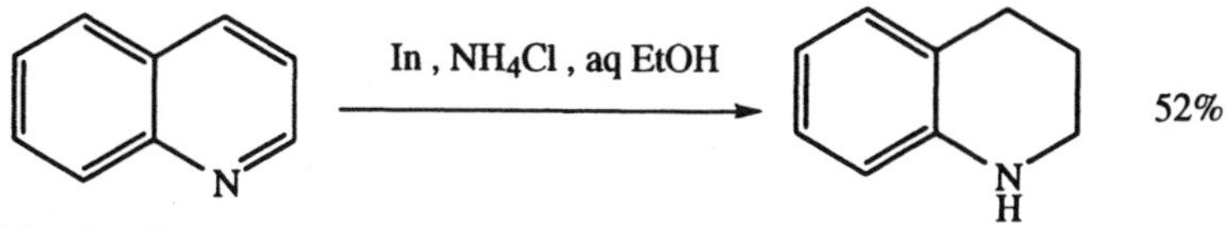
N
In, NH4Cl, aq EtOH
N
H
52%

Moody, C.J.; Pitts, M.R. *SynLett*, ***1998***, 1029.

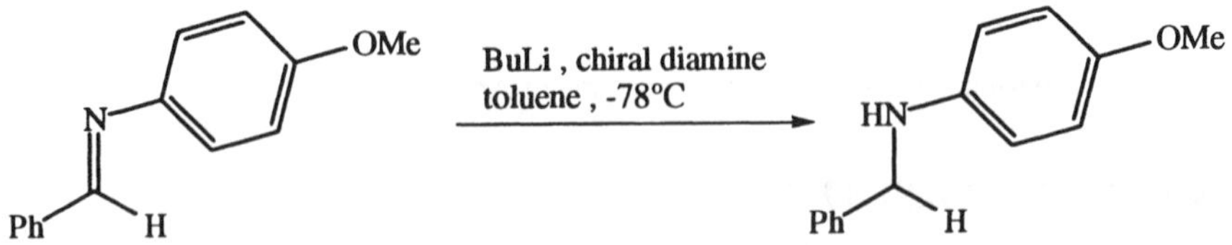

44% (68% ee , *R*)

Andersson, P.G.; Johansson, F.; Tanner, D. *Tetrahedron*, ***1998***, *54*, 11549.

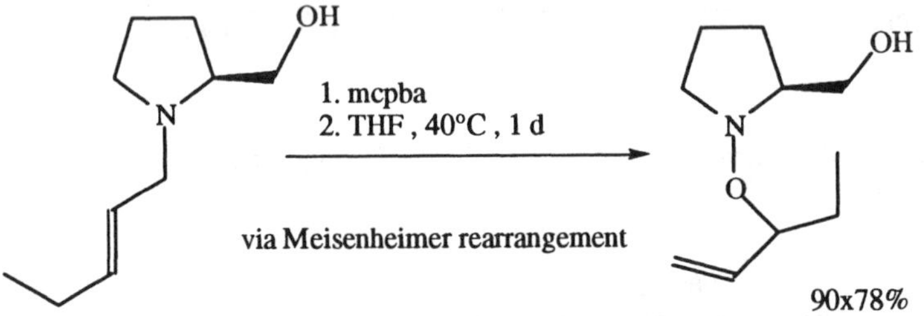

90x78%

Buston, J.E.H.; Coldham, I.; Mulholland, K.R. *Tetrahedron Asymmetry*, ***1998***, *9*, 1995.

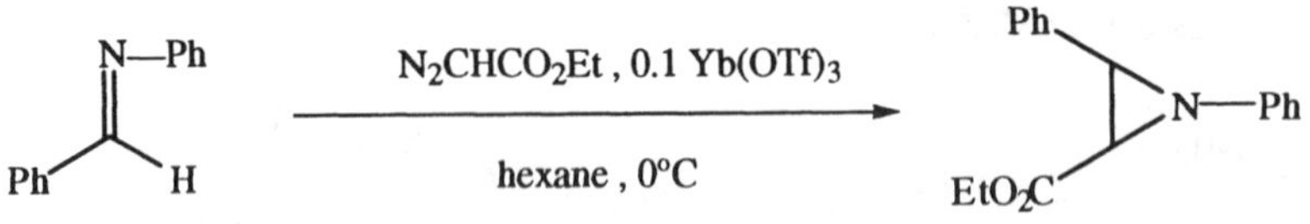

77% (20% ee)

Nagayama, S.; Kobayashi, S. *Chem. Lett.*, ***1998***, 685.

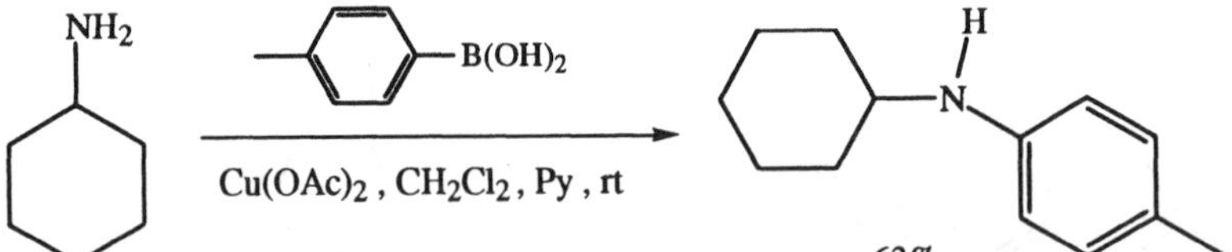

63%

Chan, D.M.T.; Monaco, K.L.; Wang, R.–P.; Winters, M.P. *Tetrahedron Lett*, ***1998***, *39*, 2937

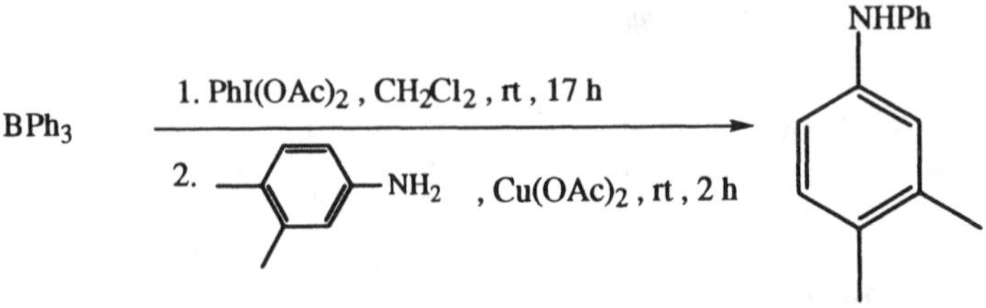

86%

Combes, S.; Finet, J.-P. *Tetrahedron*, ***1998***, *54*, 4313.

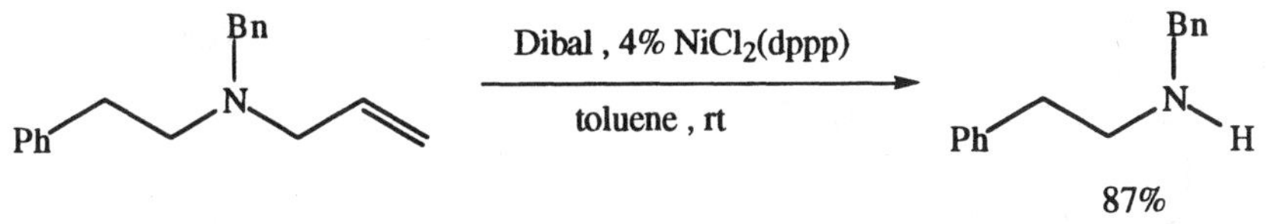
Bn
Ph
N
Dibal , 4% NiCl2(dppp)
toluene , rt
Bn
Ph
N
H
87%

Taniguchi, T.; Ogasawara, K. *Tetrahedron Lett.*, ***1998***, *39*, 4679.

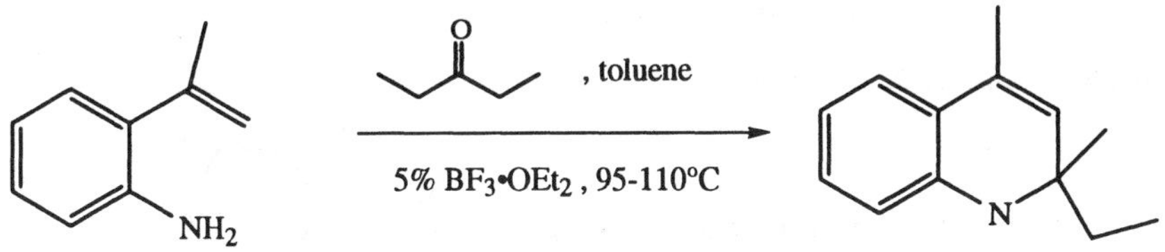
NH2
O
, toluene
5% BF3•OEt2 , 95-110°C
N

Edwards, J.P.; Ringgenberg, J.D.; Jones, T.K. *Tetrahedron Lett.*, ***1998***, *39*, 5139.

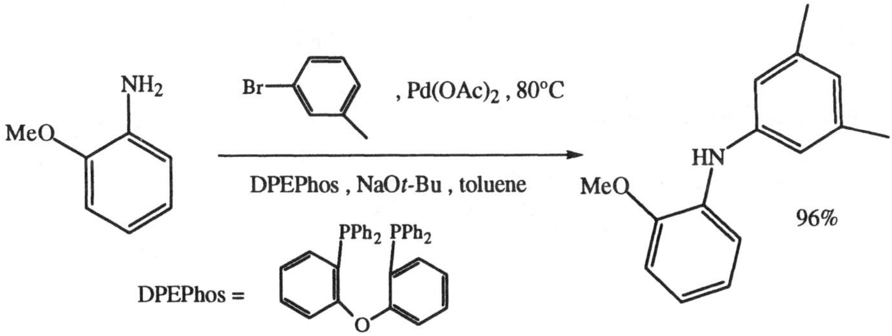
NH2
MeO
Br
, Pd(OAc)2 , 80°C
DPEPhos , NaOt-Bu , toluene
HN
MeO
96%
DPEPhos =
PPh2 PPh2
O

Sadighi, J.P.; Harris, M.C.; Buchwald, S.L. *Tetrahedron Lett.*, ***1998***, *39*, 5327.

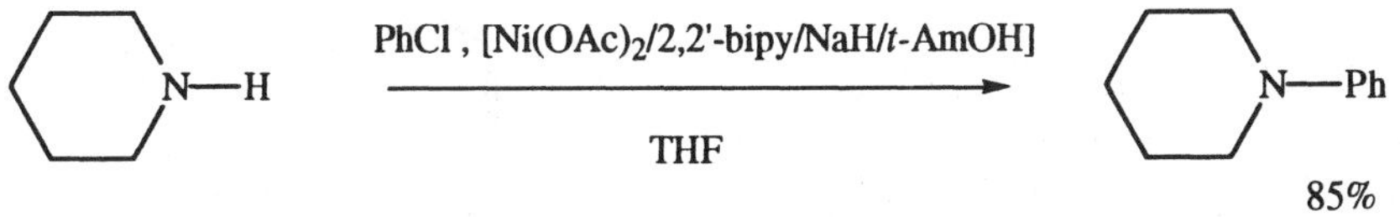
N—H
PhCl , [Ni(OAc)2/2,2'-bipy/NaH/t-AmOH]
THF
N—Ph
85%

Brenner, E.; Fort, Y. *Tetrahedron Lett.*, ***1998***, *39*, 5359.

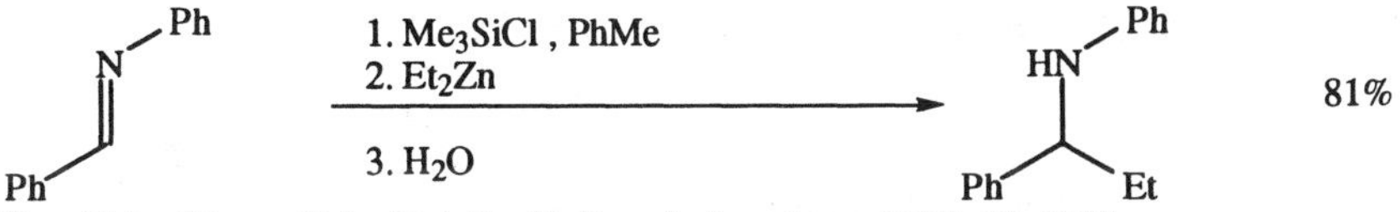
Ph
N
Ph
1. Me3SiCl , PhMe
2. Et2Zn
3. H2O
HN
Ph
Ph
Et
81%

Hou, X.L.; Zheng, X.L.; Dai, L.-X. *Tetrahedron Lett.*, ***1998***, *39*, 6949.

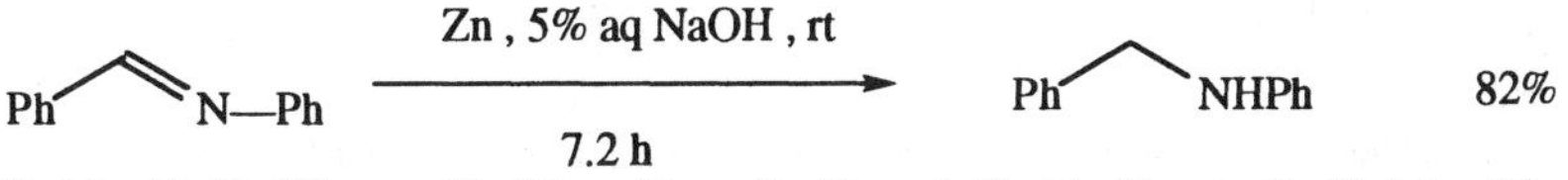
Ph
N—Ph
Zn , 5% aq NaOH , rt
7.2 h
Ph
NHPh
82%

Tsukinoki, T.; Mitoma, Y.; Nagashima, S.; Kawaji, T.; Hashimoto, I.; Tashiro, M. *Tetrahedron Lett.*, ***1998***, *39*, 8873.

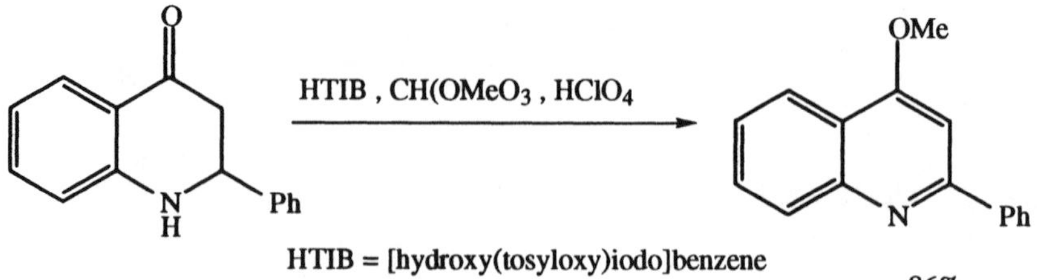
O
HTIB , CH(OMeO3 , HClO4
OMe
N
H
Ph
N
Ph
HTIB = [hydroxy(tosyloxy)iodo]benzene

86%

Varma, R.S.; Kumar, D. *Tetrahedron Lett.*, ***1998***, *39*, 9113.

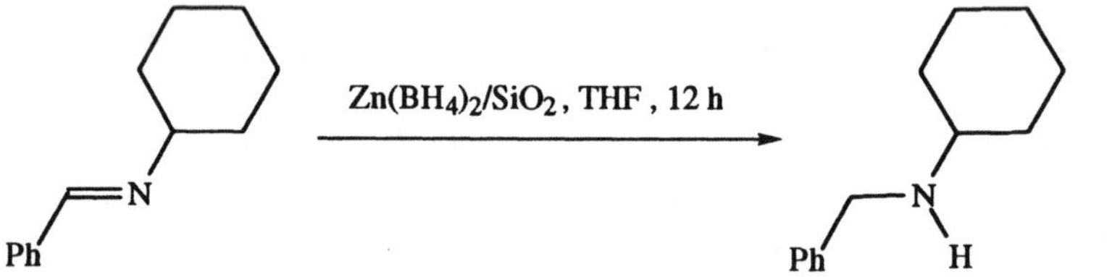
Zn(BH4)2/SiO2 , THF , 12 h
N
Ph
N
H
Ph

92%

Ranu, B.C.; Sarkar, A.; Majee, A. *J. Org. Chem.*, ***1997***, *62*, 1841.

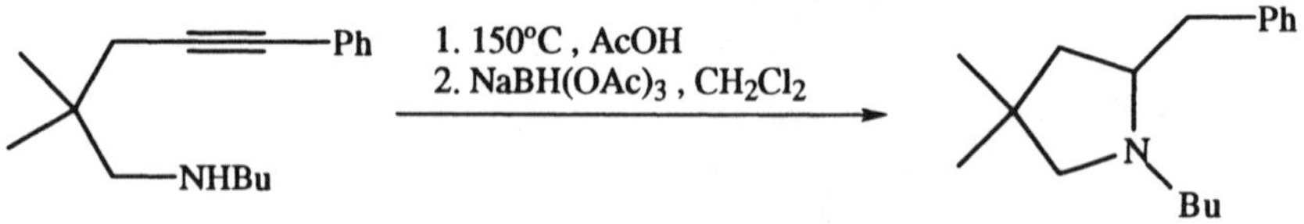
Ph
NHBu
1. 150°C , AcOH
2. NaBH(OAc)3 , CH2Cl2
Ph
N
Bu

95%

Cossy, J.; Belotti, D.; Bellosta, V.; Boggio, C. *Tetrahedron Lett.*, ***1997***, *38*, 2677.

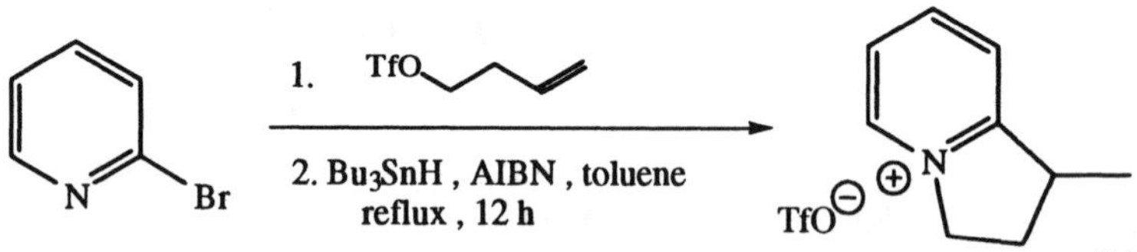
N
Br
1. TfO
2. Bu3SnH , AIBN , toluene
reflux , 12 h
TfO
N

74%

Dobbs, A.P.; Jones, K.; Veal, K.T. *Tetrahedron Lett.*, ***1997***,.*38*, 5383.

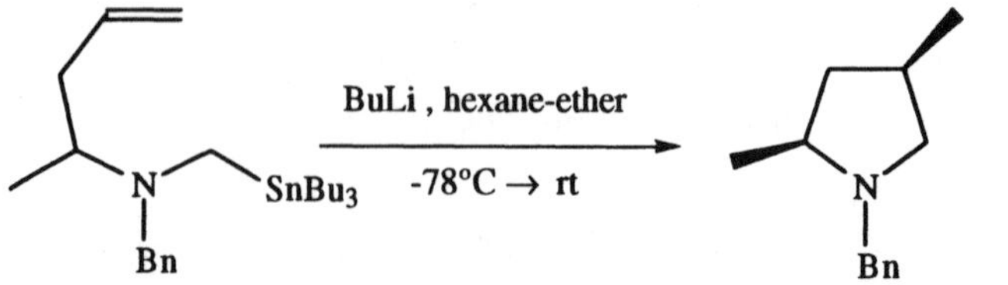
N
Bn
SnBu3
BuLi , hexane-ether
-78°C → rt
N
Bn

78% (7:1 *cis:trans*)

Coldham, I.; Hufton, R.; Rathmell, R.E. *Tetrahedron Lett.*, ***1997***, *38*, 7617.

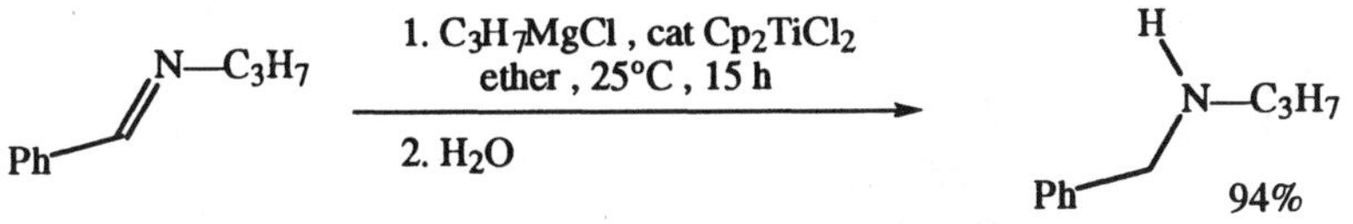

Amin, SK.R.; Crowe, W.E. *Tetrahedron Lett*, ***1997***, *38*, 7487.

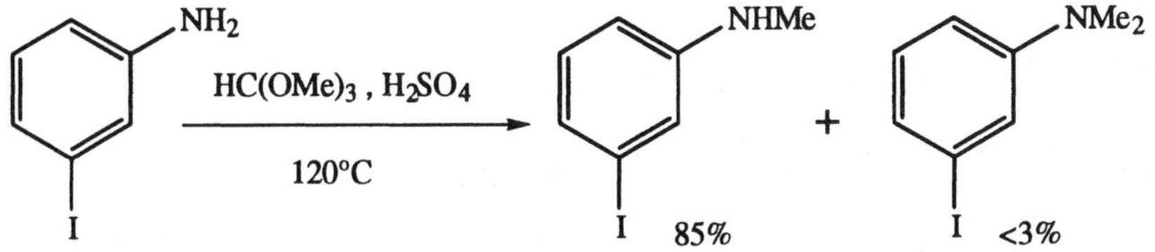

Padmanabhan, S.; Reddy, N.L.; Durant, G.J. *Synth. Commun.*, ***1997***, *27*, 695.

PhNHLi —— $BuCu(CN)Li$, O_2, THF, –78°C ——→ PhNHBu 62%

Alberti, A.; Canè, F.; Dembech, P.; Lazzari, D.; Ricci, A.; Seconi, G. *J. Org. Chem.*, ***1996***, *61*, 1677.

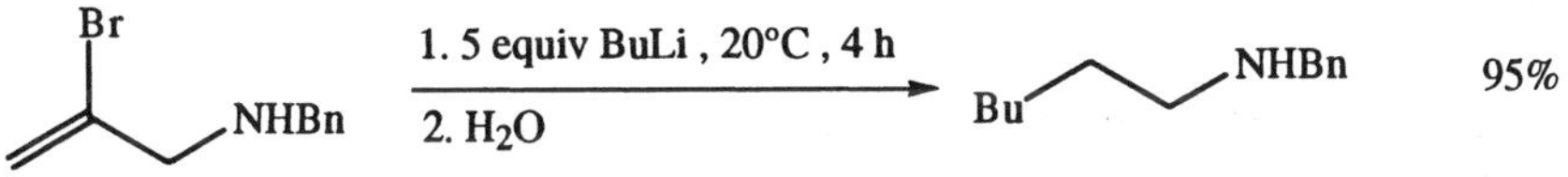

Barluenga, J.; Cantel, R.–M.; Flórez, J. *J. Org. Chem.*, ***1996***, *61*, 3646.

$PhNH_2$ —— $RuCl_2(PPh_3)_3$, EtOH, 180°C, 5 h ——→ PhNHEt (13%) + $PhNEt_2$ (74%)

Watanabe, Y.; Morisaki, Y.; Kondo, T.; Mitsudo, T. *J. Org. Chem.*, ***1996***, *61*, 4214.

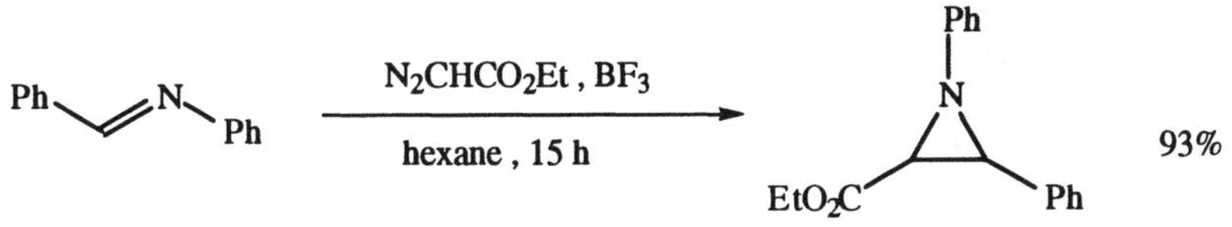

(93:7 *cis:trans*)

Casarrubios, L.; Pérez, J.A.; Brookhart, M.; Templeton, J.L. *J. Org. Chem.*, ***1996***, *61*, 8358

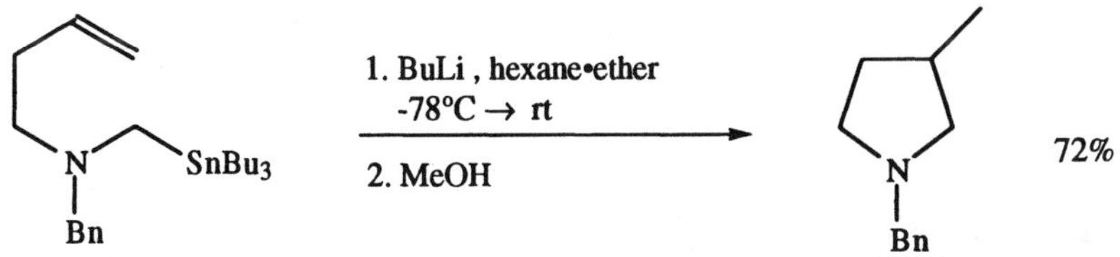

Coldham, I.; Hufton, R.; Snowden, D.J. *J. Am. Chem. Soc.*, ***1996***, *118*, 5322.

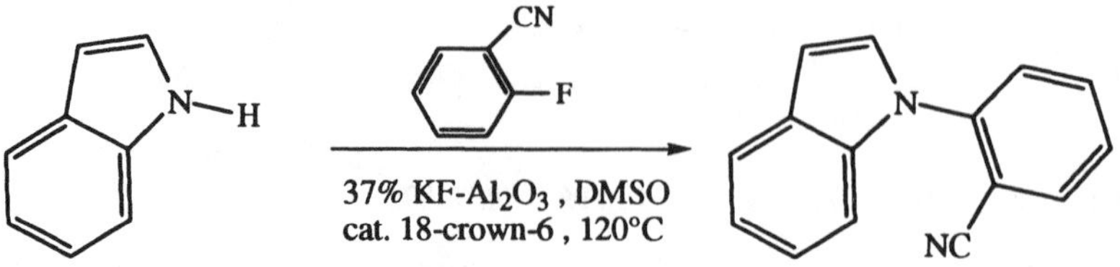

93%

Smith III, W.J.; Sawyer, J.S. *Tetrahedron Lett.*, ***1996***, *37*, 299.

$PhNH_2$ $\xrightarrow{Cu(II)(OPiv)_2,\ AlEt_2Cl,\ PhH}$ PhNHEt + $PhNEt_2$

(80 : 20) quant

Barton, D.H.R.; Doris, E. *Tetrahedron Lett.*, ***1996***, *37*, 3295.

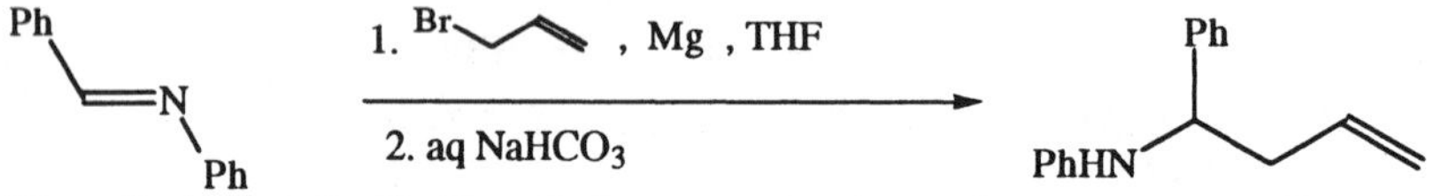

99%

Wang, D.–k.; Dai, L.–X.; Hou, X.–L.; Zhang, Y. *Tetrahedron Lett.*, ***1996***, *37*, 4187.

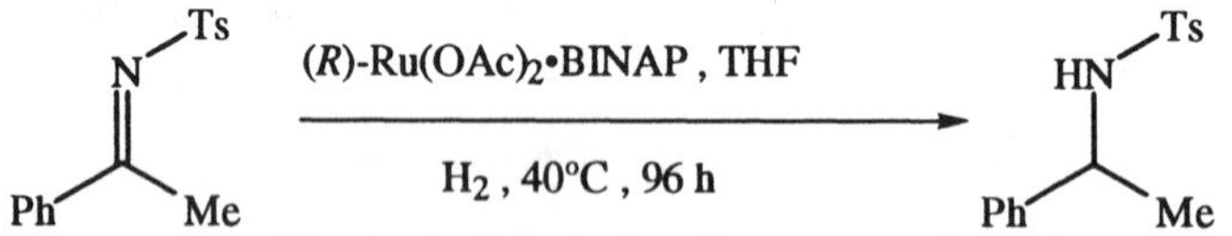

82% (62% ee , R)

Charette, A.B.; Giroux, A. *Tetrahedron Lett.*, ***1996***, *37*, 6669.

$PhNH_2$ $\xrightarrow{EtOH,\ sealed\ tube,\ microwaves}$ PhNHEt 77%

Jiang,. Y.–L.; Hu, Y.–Q.; Feng, S.–Q.; Wu, J.–S.; Wu, Z.W.; Yuan, Y.–C.; Liu, J.–M.; Hao, Q.–S.; Li, D.–P.; Yuan, Y.–C.; Liu, J.–M.; Hao, Q.–S.; Li, D.–P.
Synth. Commun., ***1996***, *26*, 161.

Bu_2NH $\xrightarrow[3.\ KOH]{1.\ (CH_2O)_n,\ microwaves,\ HCOOH;\ 2.\ 6N\ HCl}$ Bu_2N—Me 67%

Barbry, D.; Torchy, S. *Synth. Commun.*, ***1996***, *26*, 3919.

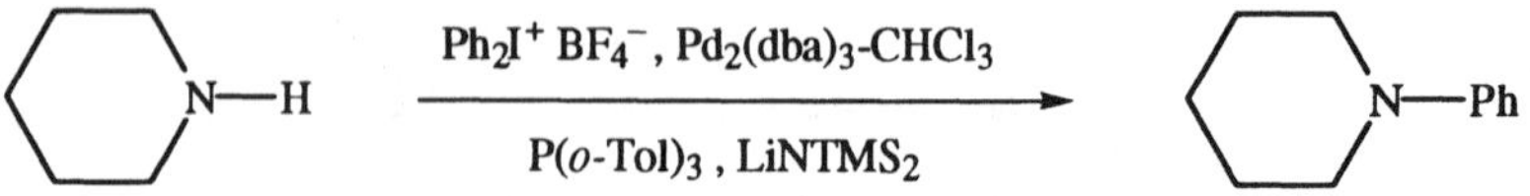

62%

Kang, S.–K.; Lee, H.–W.; Choi, W.–K.; Hong, R.–K.; Kin, J.–S.
Synth. Commun., ***1996***, *26*, 4219.

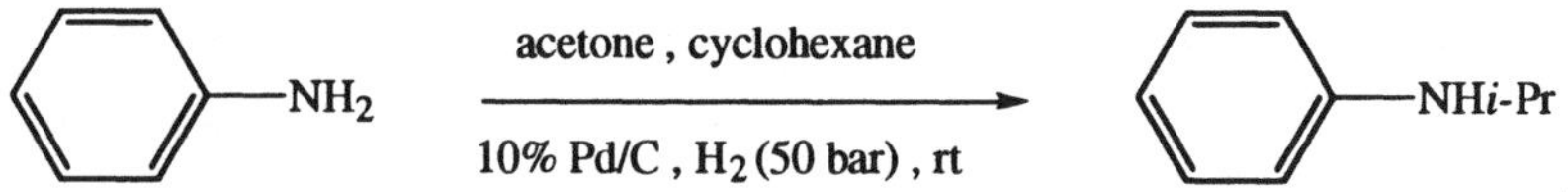
NH2
acetone , cyclohexane
10% Pd/C , H2 (50 bar) , rt
NHi-Pr

95%

Fache, F.; Valot, F.; Milenkovic, A.; LeMaire, M. *Tetrahedron*, ***1996***, *52*, 9777.

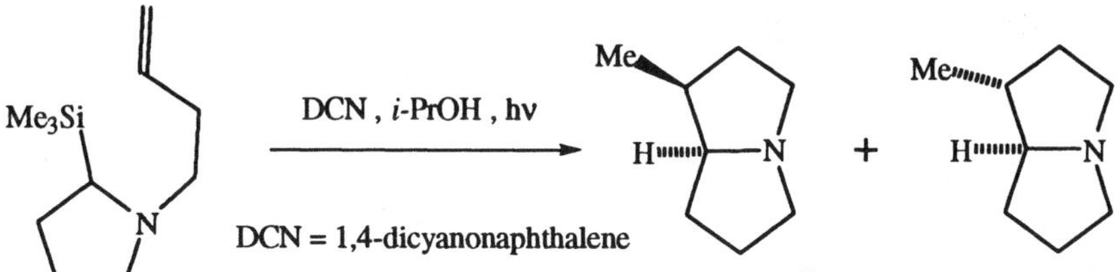
Me3Si
N
DCN , i-PrOH , hν
DCN = 1,4-dicyanonaphthalene
Me
H
N
+
Me
H
N

(97 : 3) 90%

Pandey, G.; Reddy, G.D.; Chakrabarti, D. *J. Chem. Soc., Perkin Trans. 1*, ***1996***, 219.

$PhNH_2$ —(1. BuLi; 2. BuBr)→ PhNHBu 98%

Vitale, A.A.; Chiocconi, A.A. *J. Chem. Res. (S)*, ***1996***, 336.

SECTION 98: AMINES FROM ESTERS

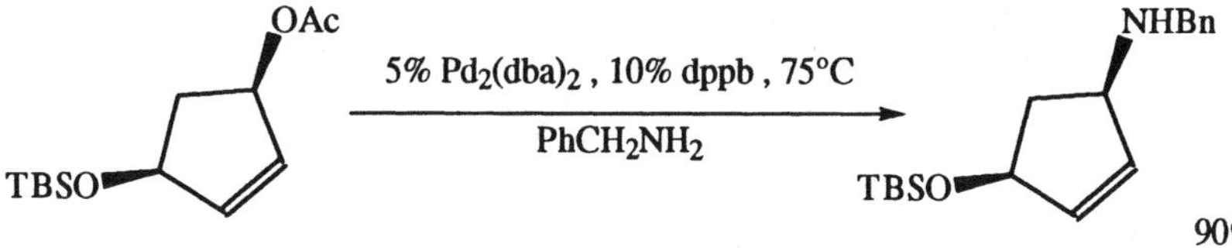
OAc
TBSO
5% Pd2(dba)2 , 10% dppb , 75°C
PhCH2NH2
NHBn
TBSO

90%

Sirisoma, N.S.; Woster, P.M. *Tetrahedron Lett.*, ***1998***, *39*, 1489.

SECTION 99: AMINES FROM ETHERS, EPOXIDES AND THIOETHERS

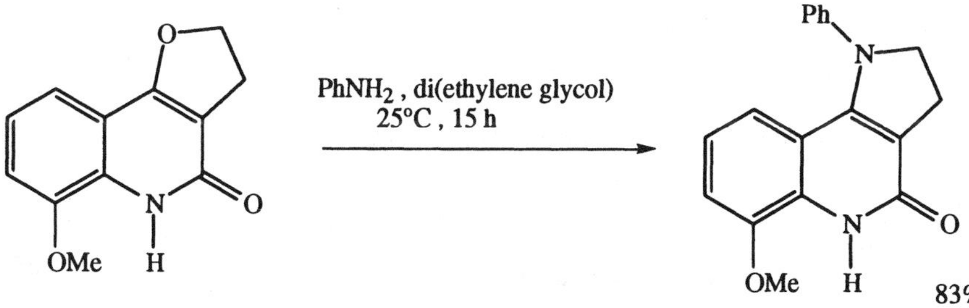
O
N
O
OMe
H
PhNH2 , di(ethylene glycol)
25°C , 15 h
Ph
N
N
O
OMe
H

83%

Kim, S.S.; Cheon, H.G.; Kang, S.K.; Yum, E.K.; Choi, J.-K. *Heterocycles*, ***1998***, *48*, 221.

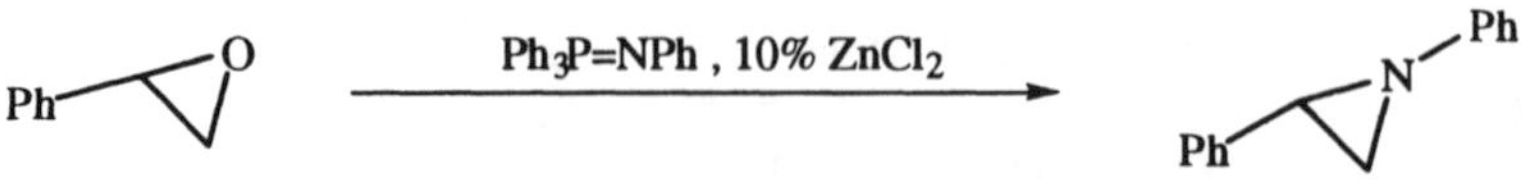
O
Ph
Ph_3P=NPh , 10% $ZnCl_2$
Ph
N
Ph

72%

Kühnau, D.; Thomsen, I.; Jørgensen, K.A. *J. Chem. Soc., Perkin Trans. 1*, ***1996***, 1167.

SECTION 100: AMINES FROM HALIDES AND SULFONATES

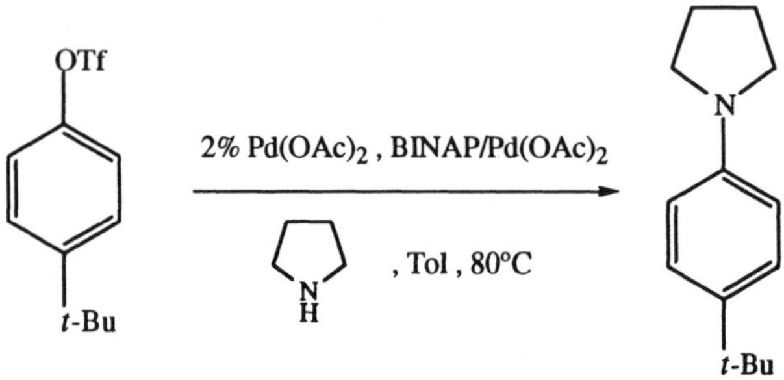
OTf
t-Bu
2% $Pd(OAc)_2$, BINAP/$Pd(OAc)_2$
N
H
, Tol , 80°C
N
t-Bu

67%

Wolfe, J.P.; Buchwald, S.L. *J. Org. Chem.*, ***1997***, *62*, 1264.

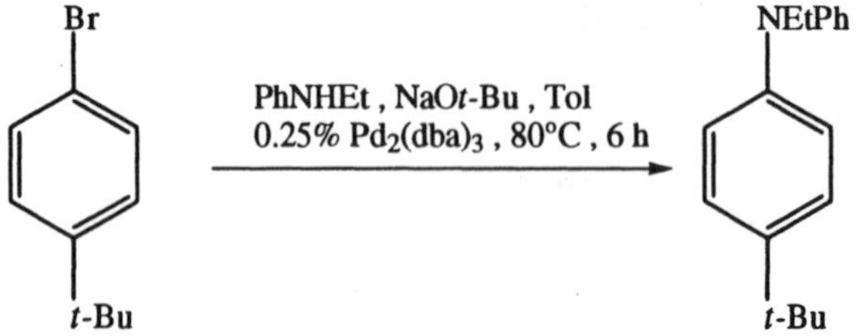
Br
t-Bu
PhNHEt , NaOt-Bu , Tol
0.25% $Pd_2(dba)_3$, 80°C , 6 h
NEtPh
t-Bu

91%

Marcoux, J.–F.; Wagaw, S.; Buchwald, S.L. *J. Org. Chem.*, ***1997***, *62*, 1568.

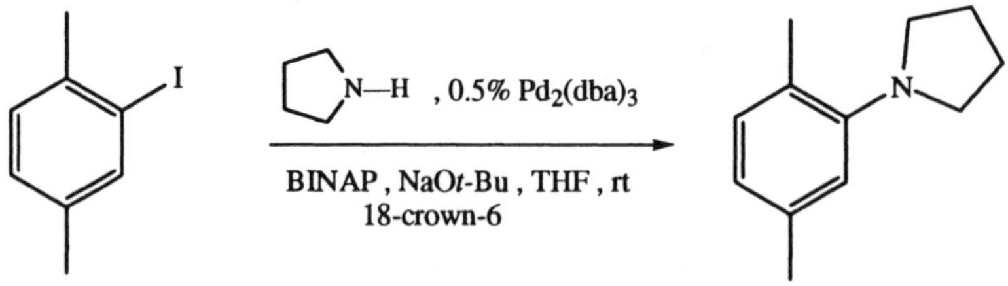
I
N—H , 0.5% $Pd_2(dba)_3$
BINAP , NaOt-Bu , THF , rt
18-crown-6
N

82%

Wolfe, J.P.; Buchwald, S.L. *J. Org. Chem.*, ***1997***, *62*, 6066.

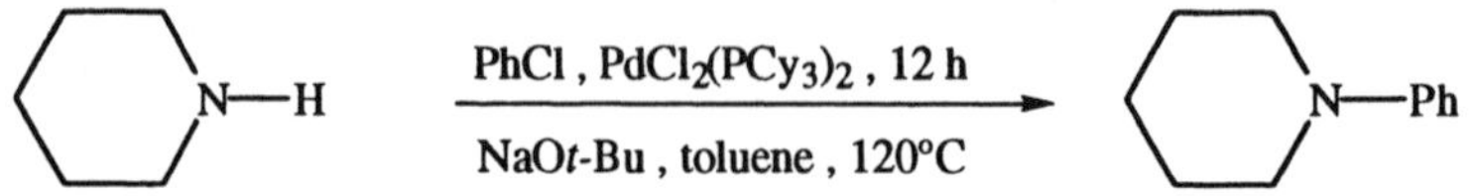
N—H
PhCl , $PdCl_2(PCy_3)_2$, 12 h
NaOt-Bu , toluene , 120°C
N—Ph

65%

Reddy, N.P.; Tanaka, M. *Tetrahedron Lett.*, ***1997***, *38*, 4807.

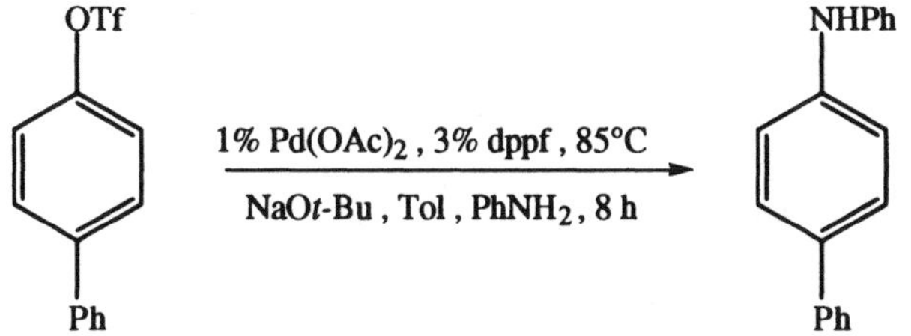

96%

Louie, J.; Driver, M.S.; Hamann, B.C.; Hartwig, J.F. *J. Org. Chem.*, ***1997***, *62*, 1268.

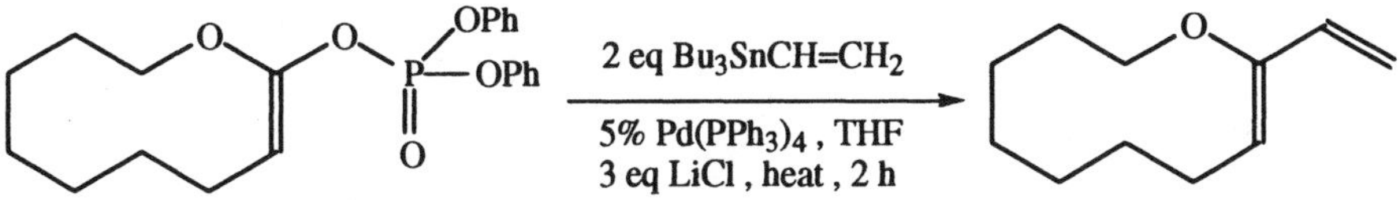

80%

Nicolaou, K.C.; Shi, G.–Q.; Gunzner, J.L.; Gärtner, P.; Yang, Z.
J. Am. Chem. Soc., ***1997***, *119*, 5467.

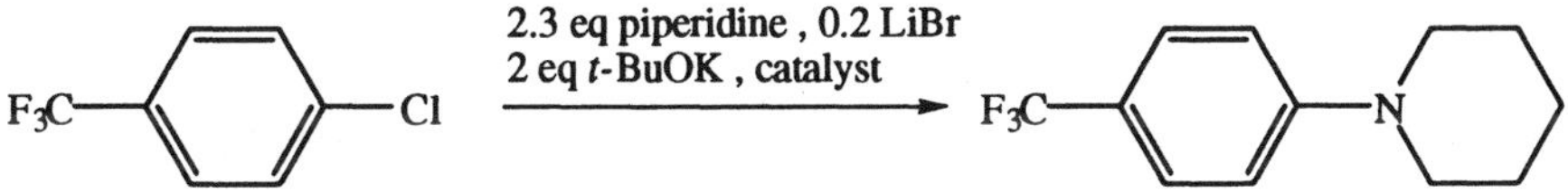

catalyst = *trans*=di(m-acetalo)-*bis*-
[*o*-(di-*o*-tolylphophino)benzyl] palladium (II)

74% (7:1 *p*:*m*)

Beller, M.; Riermeier, T.H.; Reisinger, C.–P.; Herrmann, W.A.
Tetrahedron Lett., ***1997***, *38*, 2073.

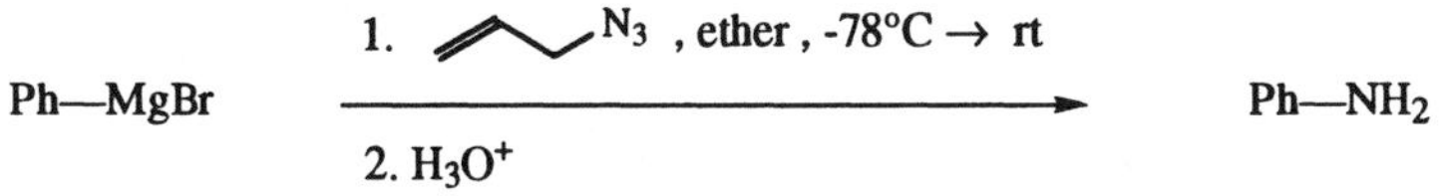

76%

Kabalka, G.W.; Li, G. *Tetrahedron Lett.*, ***1997***, *38*, 5777.

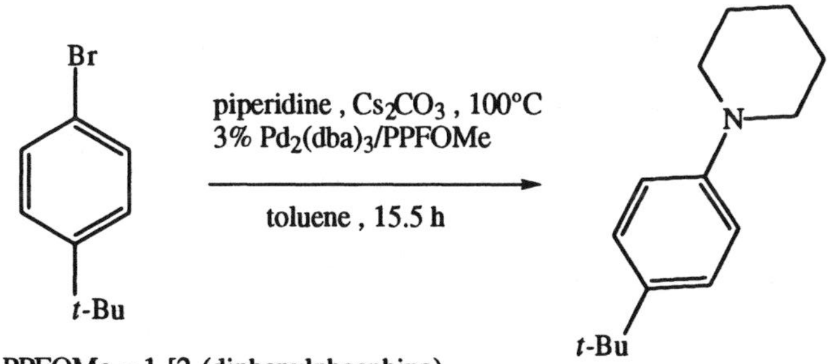

92%

PPFOMe = 1-[2-(diphenylphosphino)
ferrocenyl]ethyl methyl ether

Wolfe, J.P.; Buchwald, S.L. *Tetrahedron Lett.*, ***1997***, *38*, 6359.

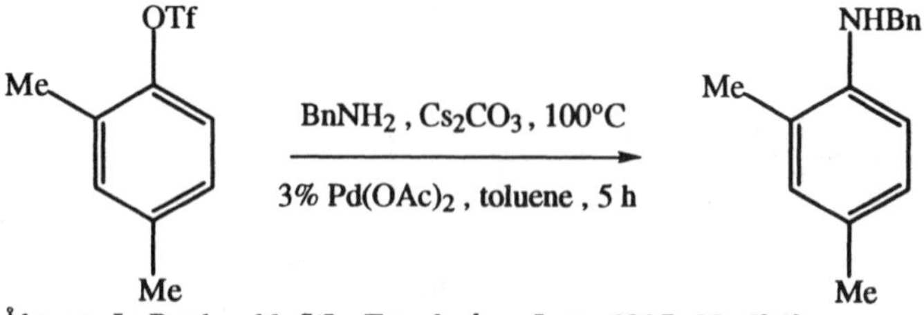
OTf
Me
Me
$BnNH_2$, Cs_2CO_3 , 100°C
3% $Pd(OAc)_2$, toluene , 5 h
NHBn
Me
Me

90%

Åhman, J.; Buchwald, S.L. *Tetrahedron Lett.*, ***1997***, *38*, 6363.

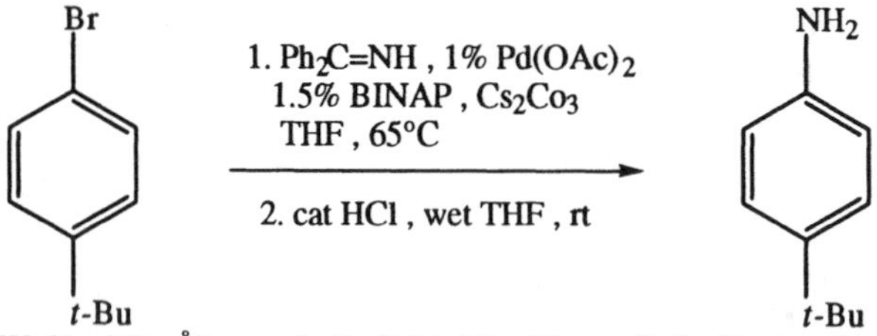
Br
t-Bu
1. Ph_2C=NH , 1% $Pd(OAc)_2$
1.5% BINAP , Cs_2Co_3
THF , 65°C
2. cat HCl , wet THF , rt
NH_2
t-Bu

90x84%

Wolfe, J.P.; Åhman, J.; Sadighi, J.P.; Singer, R.A.; Buchwald, S.L. *Tetrahedron Lett.*, ***1997***, *38*, 6367.

$PhCH_2Br$ —— PPh_2Cl , $NiCl_2$(dppe) , Zn / 110°C , DMF ——→ $PhCH_2PPh_2$ 90%

Ager, D.J.; East, M.B.; Eisenstadt, A.; Laneman, S.A. *Chem. Commun.*, ***1997***, 2359.

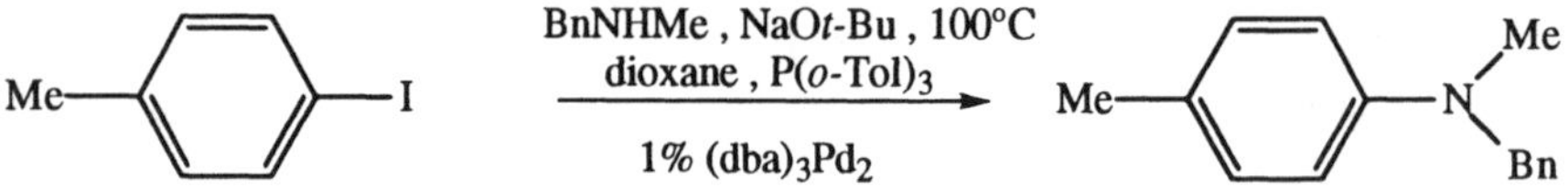
Me
I
BnNHMe , NaO*t*-Bu , 100°C
dioxane , P(*o*-Tol)$_3$
1% $(dba)_3Pd_2$
Me
N
Me
Bn

78%

Wolfe, J.P.; Buchwald, S.L. *J. Org. Chem.*, ***1996***, *61*, 1133.

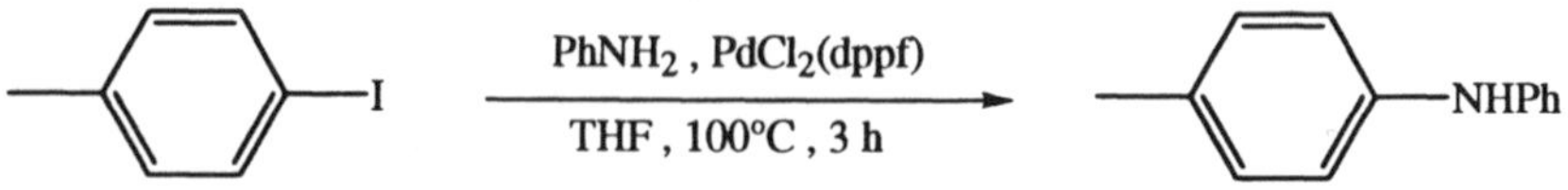
I
$PhNH_2$, $PdCl_2$(dppf)
THF , 100°C , 3 h
NHPh

92%

Driver, M.S.; Hartwig, J.F. *J. Am. Chem. Soc.*, ***1996***, *118*, 7217.

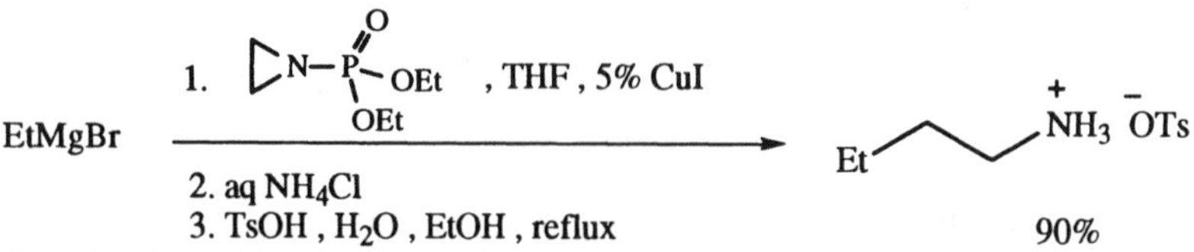
EtMgBr
1.
N
P
O
OEt
OEt
, THF , 5% CuI
2. aq NH_4Cl
3. TsOH , H_2O , EtOH , reflux
Et
$\overset{+}{N}H_3$ $\overset{-}{O}Ts$
90%

Osowska-Pacewicka, K.; Zwierzak, A. *Synthesis*, ***1996***, 333.

SECTION 101: AMINES FROM HYDRIDES

NO ADDITIONAL EXAMPLES

SECTION 102: AMINES FROM KETONES

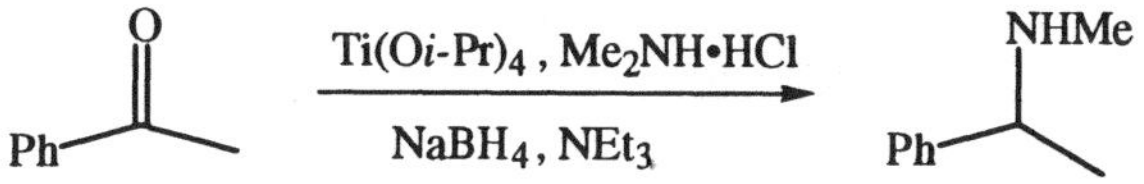
O
Ph
Ti(O*i*-Pr)$_4$, Me$_2$NH•HCl
NaBH$_4$, NEt$_3$
NHMe
Ph

86%

Neidigh, K.A.; Avery, M.A.; Williamson, J.S.; Bhattacharyya, S. *J. Chem. Soc., Perkin Trans. 1*, ***1998***, 2527.

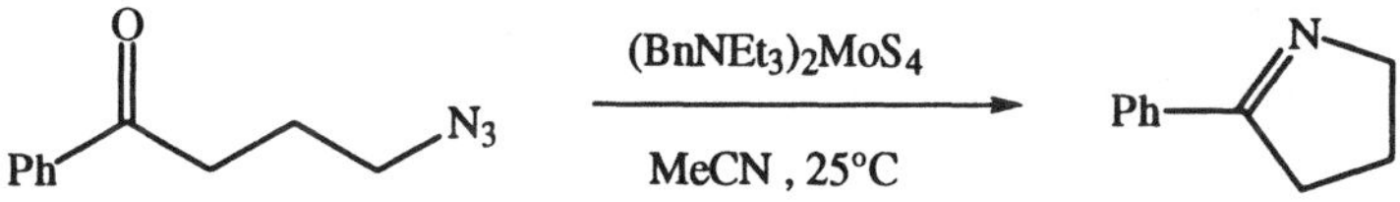
O
Ph
N$_3$
(BnNEt$_3$)$_2$MoS$_4$
MeCN , 25°C
N
Ph

81%

Prabhu, K.P.; Sivanand, P.S.; Chandrasekaran, S. *SynLett*, ***1998***, 47.

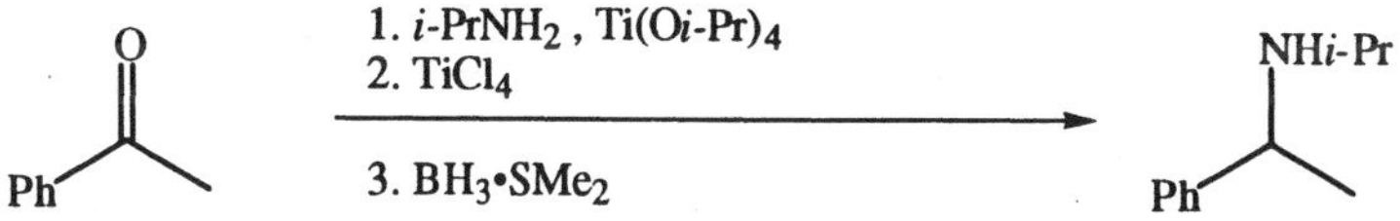
O
Ph
1. *i*-PrNH$_2$, Ti(O*i*-Pr)$_4$
2. TiCl$_4$
3. BH$_3$•SMe$_2$
NH*i*-Pr
Ph

79%

Johansson, A.; Lindstedt, E.–L.; Olsson, T. *Acta Chem. Scand. B.*, ***1997***, *51*, 351.

Related Methods: Section 94 (Amines from Aldehydes)

SECTION 103: AMINES FROM NITRILES

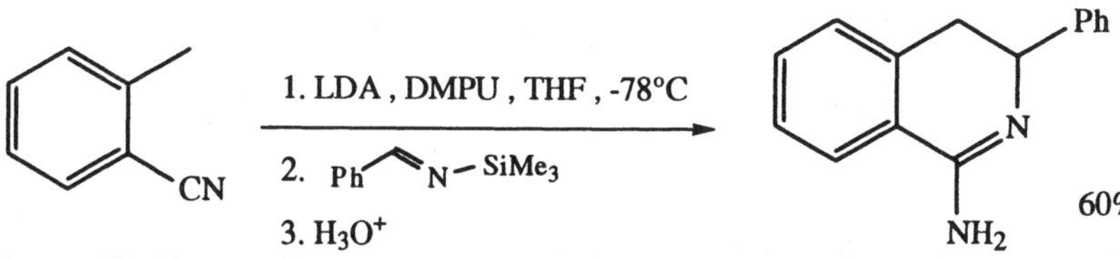
CN
1. LDA , DMPU , THF , -78°C
2. Ph N SiMe$_3$
3. H$_3$O$^+$
Ph
N
NH$_2$

60%

Baeton, H.; Hamley, P.; Tinker, A.C. *Tetrahedron Lett.*, ***1998***, *39*, 1227.

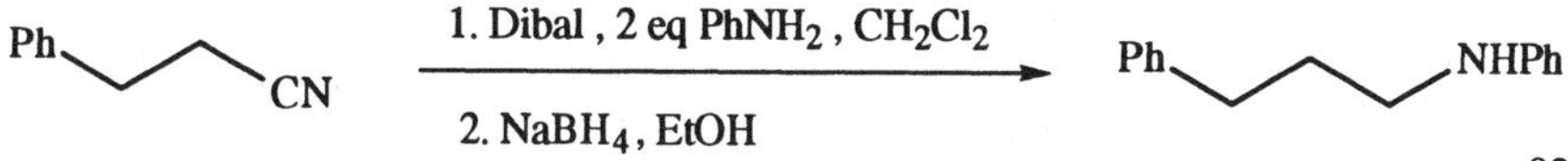
Ph
CN
1. Dibal , 2 eq PhNH$_2$, CH$_2$Cl$_2$
2. NaBH$_4$, EtOH
Ph
NHPh

92%

Barmore, R.M.; Logan, S.R.; Van Wagenen, B.C. *Tetrahedron Lett.*, ***1998***, *34*, 3451.

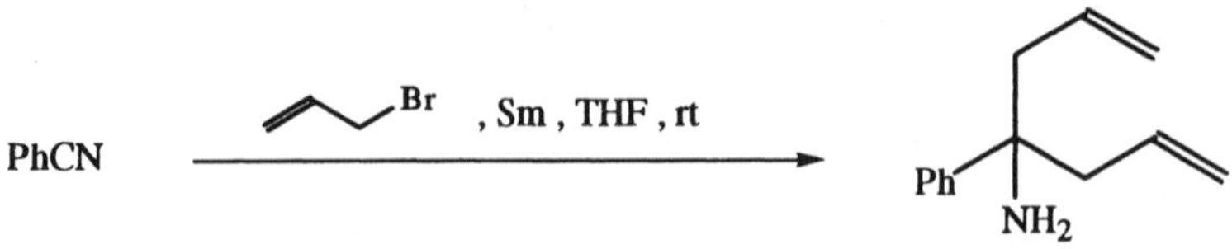
PhCN
Br
, Sm , THF , rt
Ph
NH2

78%

Yu, M.; Zhang, Y.; Guo, H. *Synth. Commun.*, ***1997***, *27*, 1495.

SECTION 104: AMINES FROM ALKENES

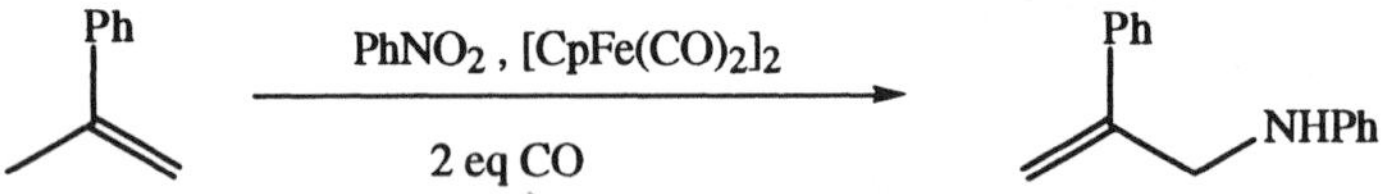
Ph
PhNO2 , [CpFe(CO)2]2
2 eq CO
Ph
NHPh

92%

Srivastava, R.S.; Nicholas, K.M. *Chem. Commun.*, ***1998***, 2705.

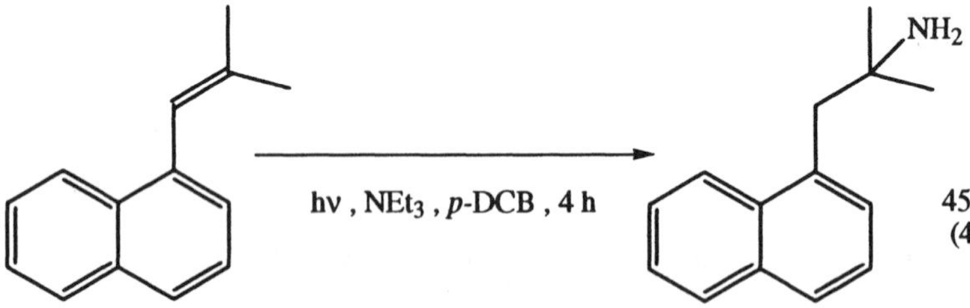
hν , NEt3 , p-DCB , 4 h
NH2

45%
(46% conversion)

Yasuda, M.; Kojima, R.; Ohira, R.; Shiragami, T.; Shima, K.
Bull. Chem. Soc. Jpn., ***1998***, *71*, 1655.

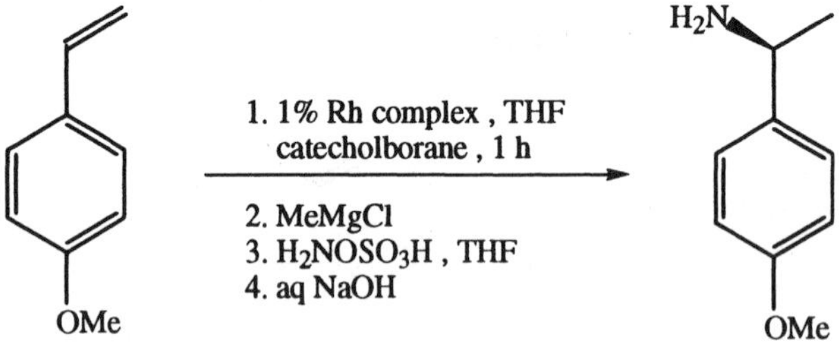
1. 1% Rh complex , THF
catecholborane , 1 h
2. MeMgCl
3. H2NOSO3H , THF
4. aq NaOH
OMe
H2N
OMe

56% (98% ee)
>98% regioselective

Fernandez, E.; Hooper, M.W.; Knight, F.I.; Brown, J.M. *Chem. Commun.*, ***1997***, 173.

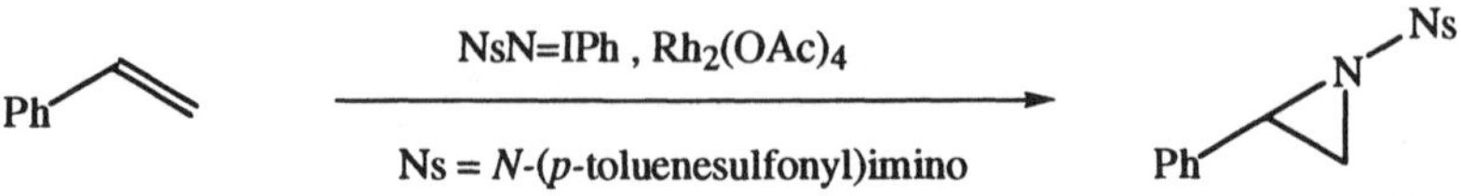
Ph
NsN=IPh , Rh2(OAc)4
Ns = N-(p-toluenesulfonyl)imino
Ns
N
Ph

85%

Müller, P.; Baud, C.; Jacquier, Y. *Tetrahedron*, ***1996***, *52*, 1543.

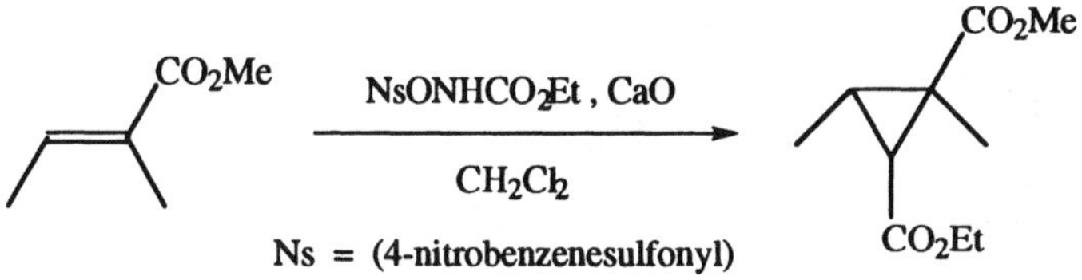
CO_2Me
$NsONHCO_2Et$, CaO
CH_2Cl_2
CO_2Me
CO_2Et

70%

Ns = (4-nitrobenzenesulfonyl)

Carducci, M.; Fioravanti, S.; Loreto, M.A.; Pellacani, L.; Tardella, P.A. *Tetrahedron Lett.*, ***1996***, *37*, 3777.

SECTION 105: AMINES FROM MISCELLANEOUS COMPOUNDS

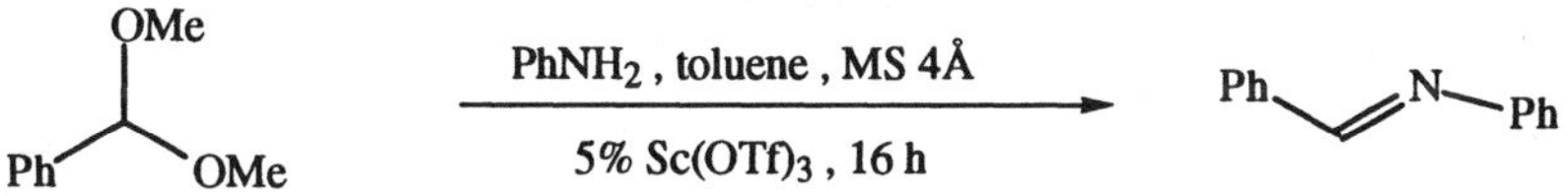
OMe
Ph
OMe
$PhNH_2$, toluene, MS 4Å
5% $Sc(OTf)_3$, 16 h
Ph
N
Ph

92%

Heaney, H.; Simcox, M.T.; Slawin, A.M.Z.; Giles, R.G. *SynLett*, ***1998***, 640.

$$PhN_3 \xrightarrow[\text{rt, 30 min}]{NH_2\text{-}NMe_2,\ FeCl_3\bullet 6\ H_2O} PhNH_2 \quad 82\%$$

Kamal, A.; Reddy, B.S.N. *Chem. Lett.*, ***1998***, 593.

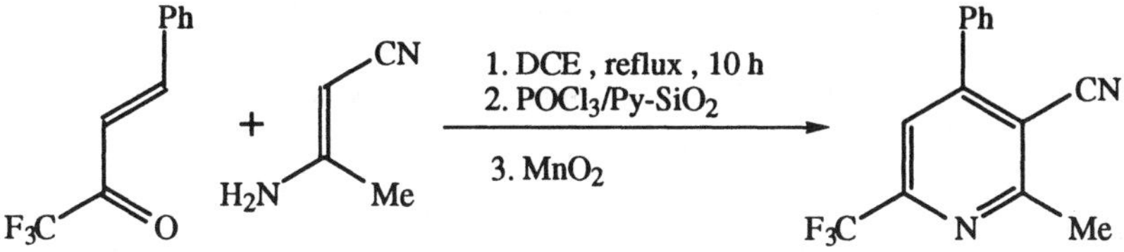
Ph
F_3C
O
+
CN
H_2N
Me
1. DCE, reflux, 10 h
2. $POCl_3$/Py-SiO_2
3. MnO_2
Ph
CN
F_3C
N
Me

83%

Katsuyama, I.; Ogawa, S.; Nakamura, H.; Yamaguchi, Y.; Funabiki, K.; Matsui, M.; Muramatsu, H.; Shibata, K. *Heterocycles*, ***1998***, *48*, 779.

O
P
Ph
Ph
AlH_3•THF
P
Ph
Ph

95%

Griffin, S.; Heath, L.; Wyatt, P. *Tetrahedron Lett.*, ***1998***, *39*, 4405.

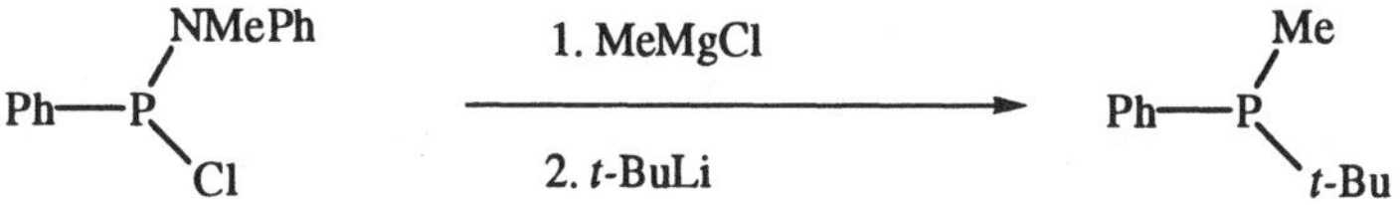
NMePh
Ph—P
Cl
1. MeMgCl
2. *t*-BuLi
Me
Ph—P
t-Bu

44%

Singh, S.; Nicholas, K.M. *Chem. Commun.*, ***1998***, 149.

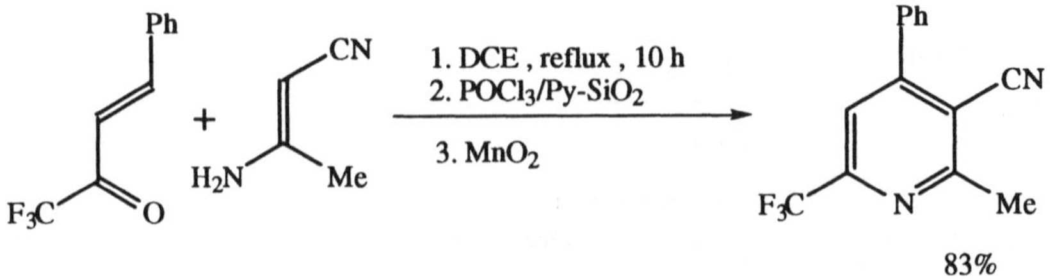
Ph
CN
+
H_2N
Me
F_3C
O
1. DCE , reflux , 10 h
2. $POCl_3$/Py-SiO_2
3. MnO_2
Ph
CN
F_3C
N
Me
83%

Katsuyama, I.; Ogawa, S.; Nakamura, H.; Yamaguchi, Y.; Funabiki, K.; Matsui, M.; Muramatsu, H.; Shibata, K. *Heterocycles, **1998**, 48*, 779.

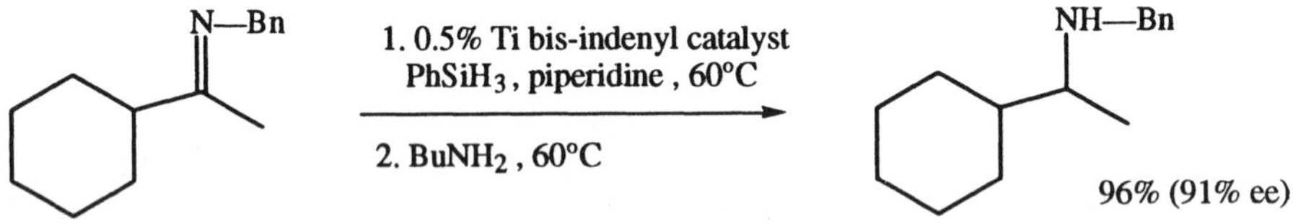
N—Bn
1. 0.5% Ti bis-indenyl catalyst
$PhSiH_3$, piperidine , 60°C
2. $BuNH_2$, 60°C
NH—Bn
96% (91% ee)

Verdaguer, X.; Lange, U.E.W.; Buchwald, S.L. *Angew. Chem. Int. Ed., **1998**, 37*, 1103.

PhN_3 —— Zn—$CoCl_2$•6 H_2O , THF , 45 min ——→ $PhNH_2$ 95%

Baruah, M.; Hussain, A.; Prajapati, D.; Sandhu, J.S. *Chem. Lett., **1997***, 789.

PhN_3 —— bakers yeast , phosphate buffer / pH 7.2 , aq EtOH , 8 h ——→ $PhNH_2$ 83-92%

Kamal, A.; Damayanthi, Y.; Reddy, B.S.N.; Lakminarayana, B.; Reddy, B.S.P. *Chem. Commun., **1997***, 1015.

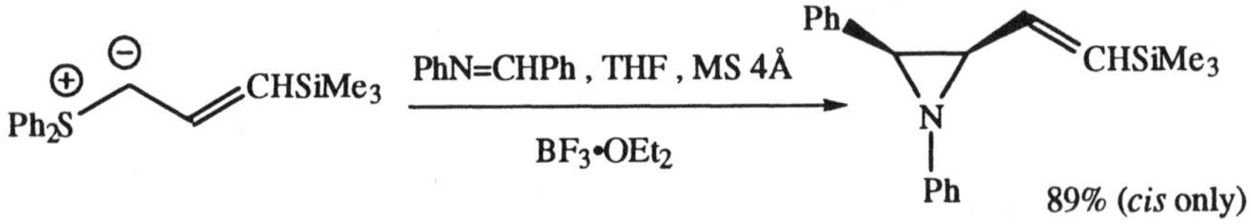
Ph_2S
$CHSiMe_3$
PhN=CHPh , THF , MS 4Å
BF_3•OEt_2
Ph
$CHSiMe_3$
N
Ph
89% (*cis* only)

Wang, D.-K.; Dai, L.-X.; Hou, X.-L. *Chem. Commun., **1997***, 1231.

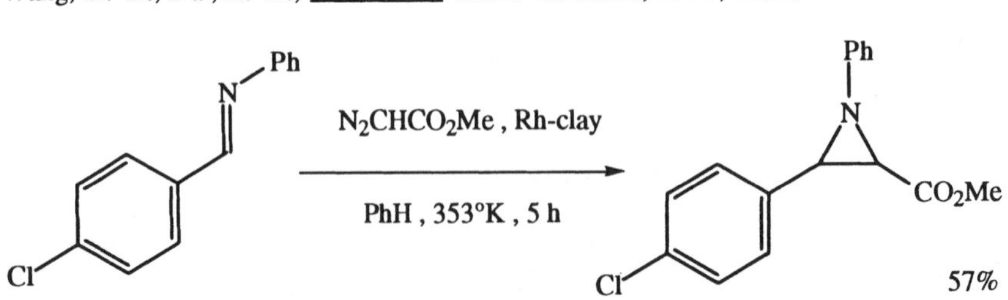
Ph
N
Cl
N_2CHCO_2Me , Rh-clay
PhH , 353°K , 5 h
Ph
N
CO_2Me
Cl
57%

Mohan, J.M.; Uphade, T.S.S.; Choudhary, V.R.; Ravindranathan, T.; Sudalai, A. *Chem. Commun.. **1997***. 1429.

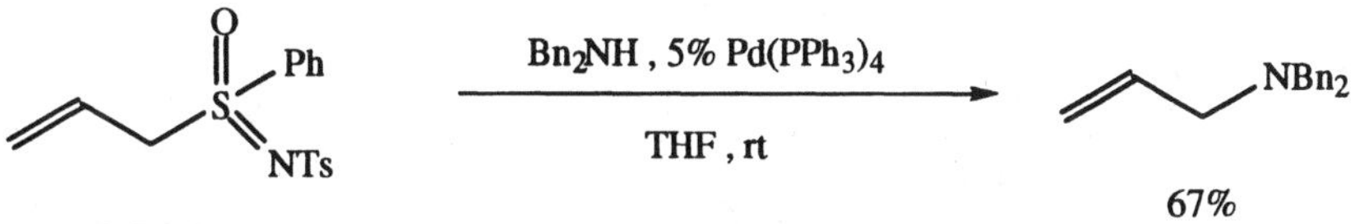
Ph
NTs
Bn2NH , 5% Pd(PPh3)4
THF , rt
NBn2
67%

Pyne, S.G.; O'Meara, G.; David, D.M. *Tetrahedron Lett.*, ***1997***, *38*, 3623.

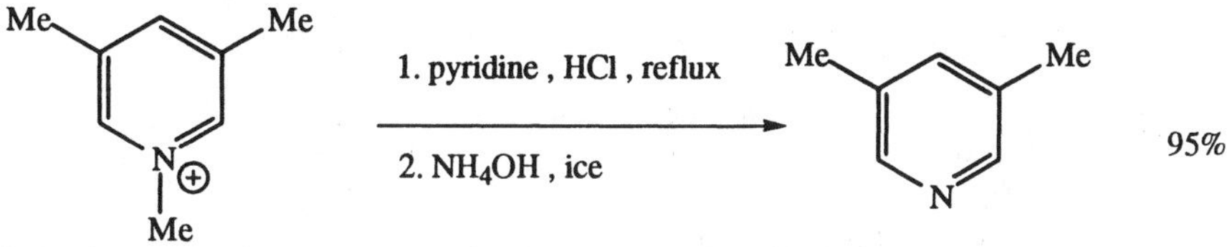
Me
Me
1. pyridine , HCl , reflux
2. NH4OH , ice
N
Me
95%

Ruiz, A.; Rocca, P.; Marsais, F.; Godard, A.; Quéguiner, G. *Tetrahedron Lett.*, ***1997***, *38*, 6205

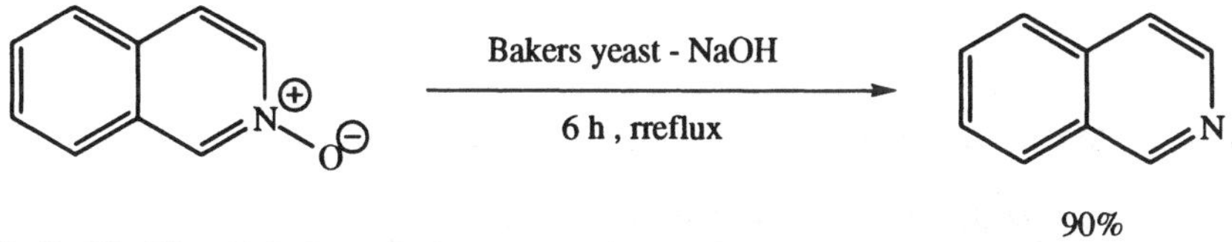
Bakers yeast - NaOH
6 h , rreflux
N
90%

Baik, W.; Kim, D.I.; Koo, S.; Rhee, J.U.; Shin, S.H.; Kim, B.H. *Tetrahedron Lett.*, ***1997***, *38*, 845.

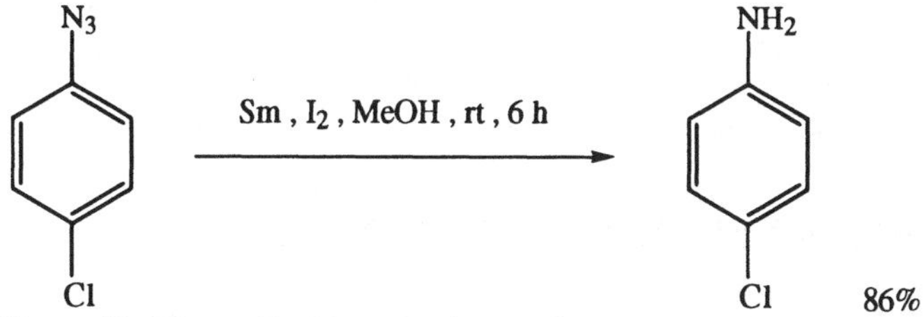
N3
Sm , I2 , MeOH , rt , 6 h
NH2
Cl
Cl
86%

Huang, Y.; Zhang, Y.; Wang, Y. *Tetrahedron Lett.*, ***1997***, *38*, 1065.

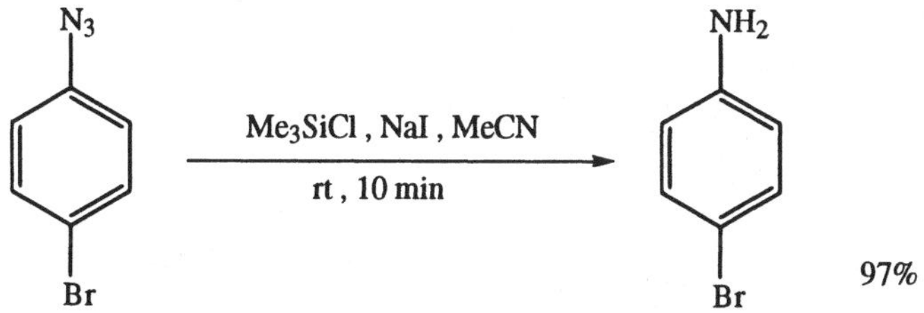
N3
Me3SiCl , NaI , MeCN
rt , 10 min
NH2
Br
Br
97%

Kamal, A.; Rao, N.V.; Laxman, E. *Tetrahedron Lett.*, ***1997***, *38*, 6945.

PhN_3 —— Zn–$NiCl_2$•6 H_2O, THF , 2 h ——→ $PhNH_2$ 90%

Boruah, A.; Baruah, M.; Prajapati, D.; Sandhu, J.S. *SynLett*, ***1997***, 1253.

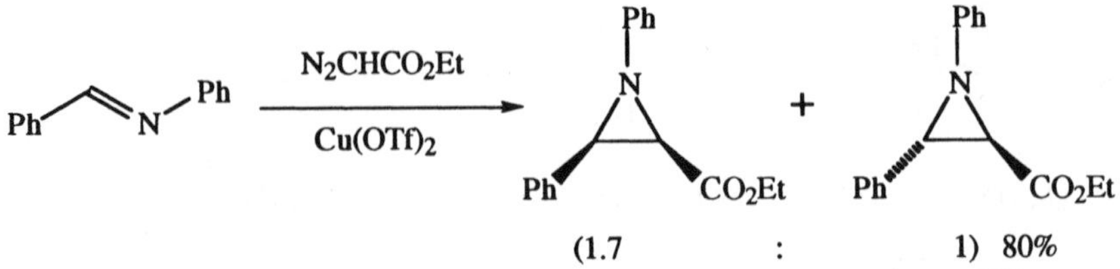

Ph
N
Ph
N2CHCO2Et
Cu(OTf)2
Ph
N
Ph
CO2Et
+
Ph
N
Ph
CO2Et
(1.7 : 1) 80%

Rasmussen, K.G.; Jørgensen, K.A. *J. Chem. Soc., Perkin Trans. 1*, ***1997***, 1287.

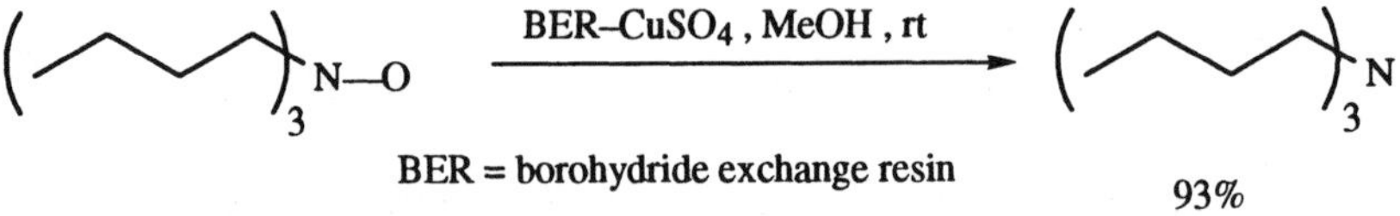

N—O
3
BER–CuSO4 , MeOH , rt
BER = borohydride exchange resin
N
3
93%

Sim, T.B.; Ahn, J.H.; Yoon, N.M. *Synthesis*, ***1996***, 324.

PhN_3 —(Bakers yeast)→ $PhNH_3$ 90%

Baruah, M.; Boruah, A.; Prajapati, D.; Sandhu, J.S. *SynLett*, ***1996***, 1193.

Ph—N_3 —($Fe/NiCl_2 \bullet 6\ H_2O$, THF , 30 min)→ Ph—NH_2 85%

Baruah, M.; Boruah, A.; Prajapati, D.; Sandhu, J.S.; Ghosh, A.C. *Tetrahedron Lett*, ***1996***, *37*, 4559.

PhN_3 —(Cp_2TiCl_2–Sm , *t*-BuOH , rt , 5 min)→ $PhNH_2$ 86%

Huang, Y.; Zhang, Y.; Wang, Y. *Synth. Commun.*, ***1996***, *26*, 2911.

AMINES FROM NITRO COMPOUNDS

C_6H_{13}—NO_2 —(1. 3 $ClP(OEt)_2$, *i*-Pr_2NEt , $CHCl_3$, 20°C; 2. $HCl_{(g)}$, MeOH; 3. basic workup)→ C_6H_{13}—NH_2 >95%

Fischer, B.; Sheihet, L. *J. Org. Chem.*, ***1998***, *63*, 393.

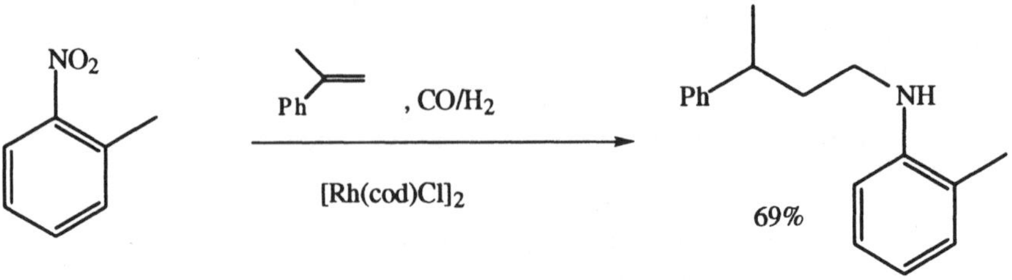

NO2
Ph
, CO/H2
[Rh(cod)Cl]2
Ph
NH
69%

Rische, T.; Eilbracht, P. *Tetrahedron*, ***1998***, *54*, 8441.

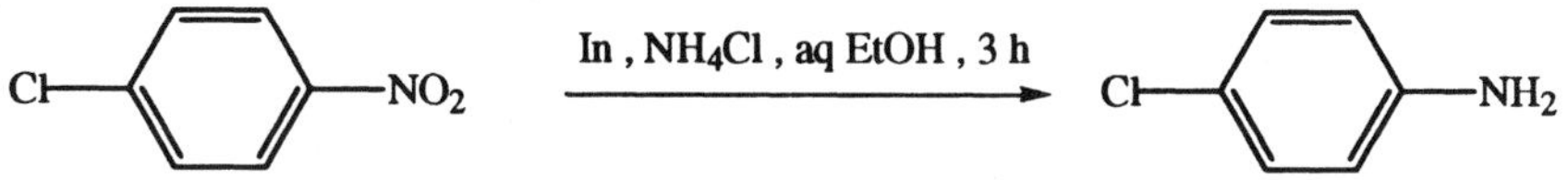
Cl
NO2
In , NH4Cl , aq EtOH , 3 h
Cl
NH2

95%

Moody, C.J.; Pitts, M.R. *SynLett*, ***1998***, 1028.

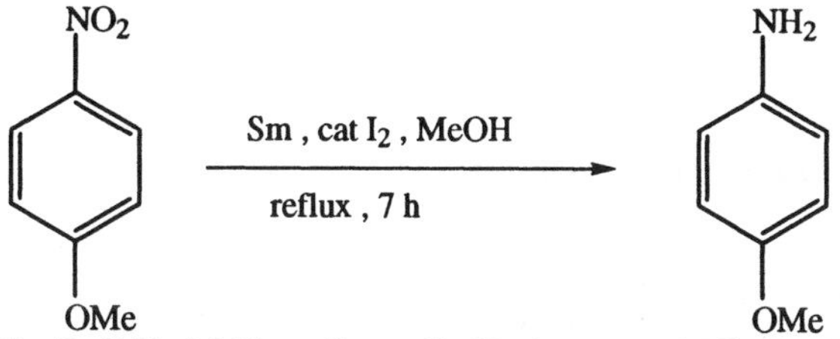
NO2
Sm , cat I2 , MeOH
reflux , 7 h
NH2
OMe
OMe

88%

Banik, B.K.; Mukhopadhyay, C.; Venkatraman, M.S.; Becker, F.F. *Tetrahedron Lett*, ***1998***, *39*, 7243.

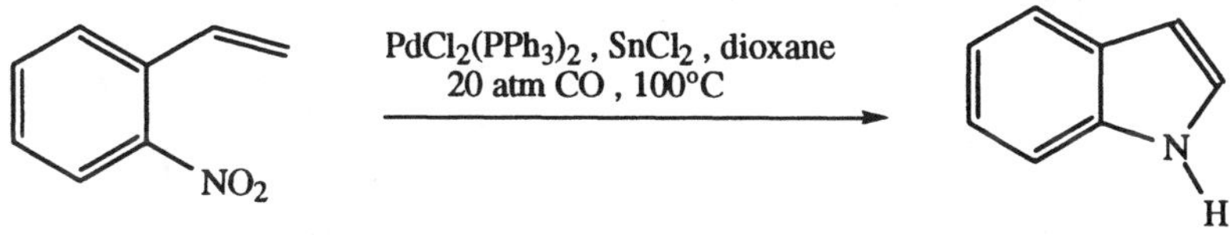
NO2
PdCl2(PPh3)2 , SnCl2 , dioxane
20 atm CO , 100°C
N
H

60%

Söderbereg, B.C.; Shriver, J.A. *J. Org. Chem.*, ***1997***, *62*, 5838.

$PhNO_2$ —— Cp_2TiCl_2/Sm/THF; MeOH , rt , 10min ——→ $PhNH_2$ 87%

Huang, Y.; Liao, P.; Zhang, Y.; Wang, Y. *Synth. Commun.*, ***1997***, *27*, 1059.

$PhNO_2$ —— Ni* , THF , ultrasound , rt; 50% $N_2H_4 \cdot H_2O$, 5 min ——→ $PhNH_2$ quant

Li, H.; Zhang, R.; Wang, H.; Pan, Y.; Shi, Y. *Synth. Commun.*, ***1997***, *27*, 3047.

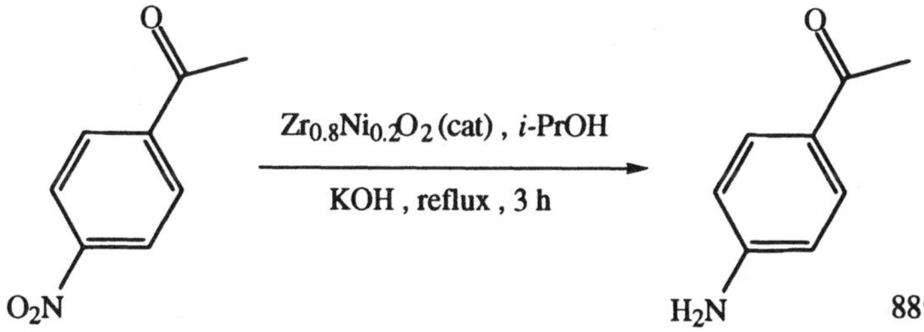
O
Zr0.8Ni0.2O2 (cat) , i-PrOH
KOH , reflux , 3 h
O
O2N
H2N

88%

Upadhya, T.T.; Katdare, S.P.; Sadbe, D.P.; Ramaswamy, V.; Sudalai, A. *Chem. Commun.*. ***1997***. 1119.

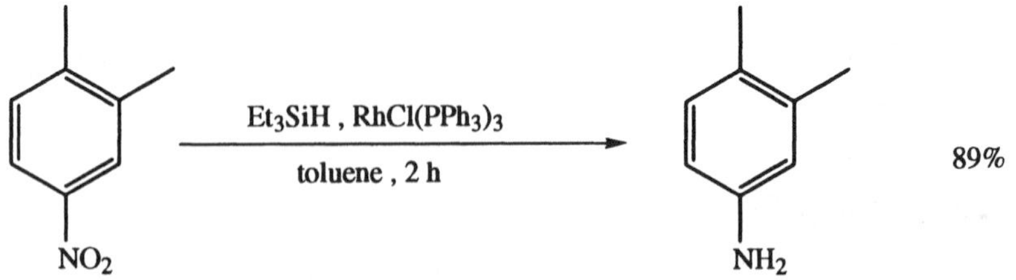
Et3SiH , RhCl(PPh3)3
toluene , 2 h
NO2
NH2
89%

Brinkman, H.R. *Synth. Commun.*, ***1996***, *26*, 973.

SECTION 105A: PROTECTION OF AMINES

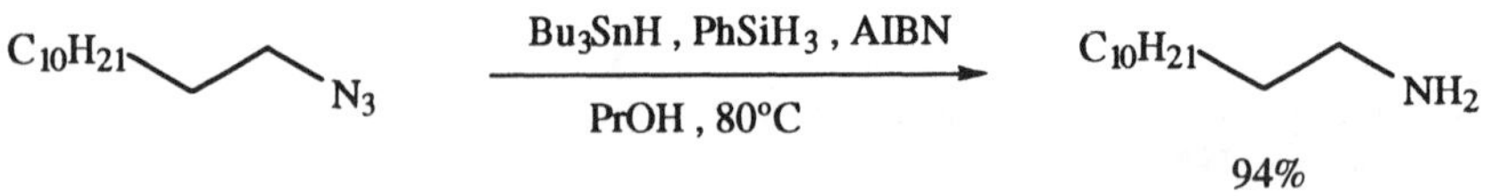
C10H21
N3
Bu3SnH , PhSiH3 , AIBN
PrOH , 80°C
C10H21
NH2
94%

Hays, D.S.; Fu, G.C. *J. Org. Chem.*, ***1998***, *63*, 2796.

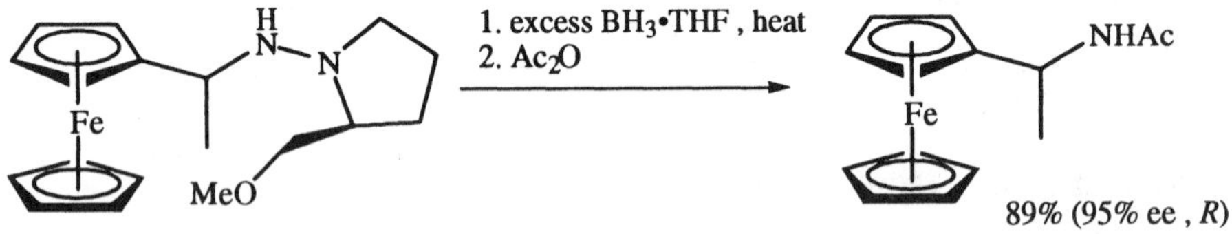
H
N
N
Fe
MeO
1. excess BH3•THF , heat
2. Ac2O
NHAc
Fe
89% (95% ee , R)

Enders, D.; Lochtman, R.; Meiers, M.; Müller, S.; Lazny, R. *SynLett*, ***1998***, 1182.

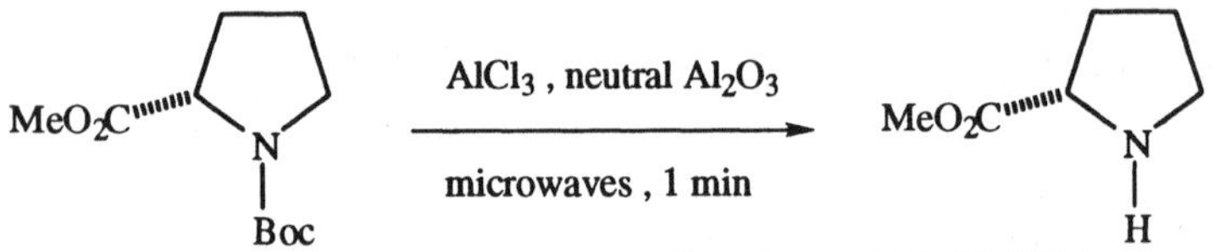
MeO2C
N
Boc
AlCl3 , neutral Al2O3
microwaves , 1 min
MeO2C
N
H

Bose, D.S.; Lakshminarayana, V. *Tetrahedron Lett.*, ***1998***, *39*, 5631.

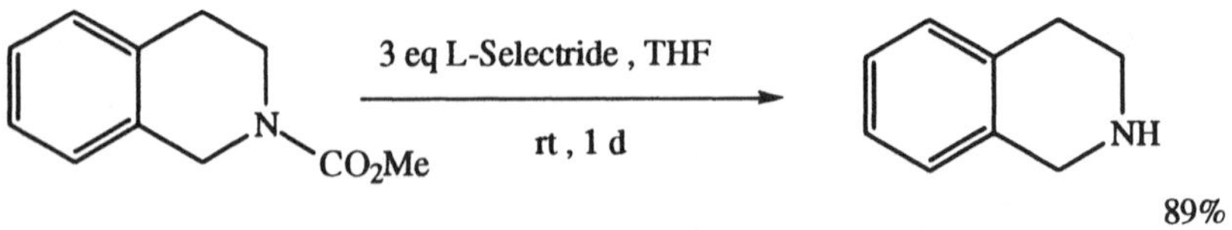
N
CO2Me
3 eq L-Selectride , THF
rt , 1 d
NH
89%

Coop, A.; Rice, K.C. *Tetrahedron Lett*, ***1998***, *39*, 8933.

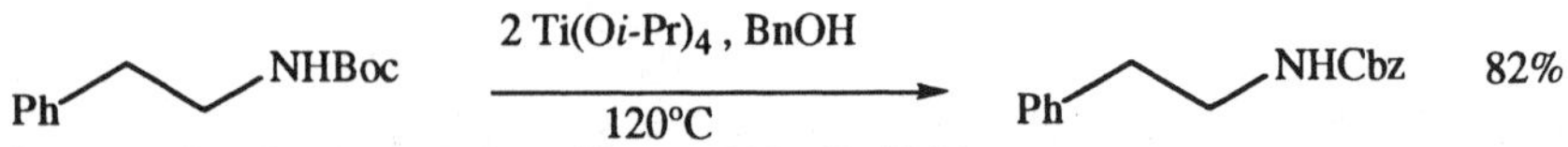
Ph
NHBoc
2 Ti(Oi-Pr)4 , BnOH
120°C
Ph
NHCbz
82%

Shapiro, G.; Marzi, M. *J. Org. Chem.* ***1997***, *62*, 7096.

CHAPTER 8
PREPARATION OF ESTERS

SECTION 106: ESTERS FROM ALKYNES

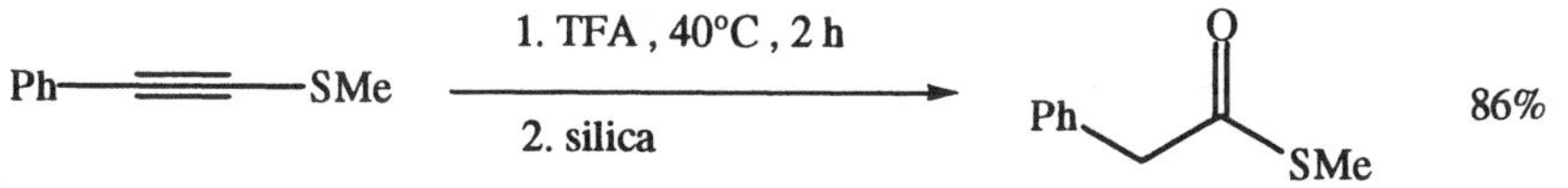

Ph
SMe
1. TFA , 40°C , 2 h
2. silica
O
Ph
SMe
86%

Braga, A.L.; Rodregues, O.E.D.; de Avila, E.; Silveira, C.C. *Tetrahedron Lett.*, ***1998***, *39*, 3395

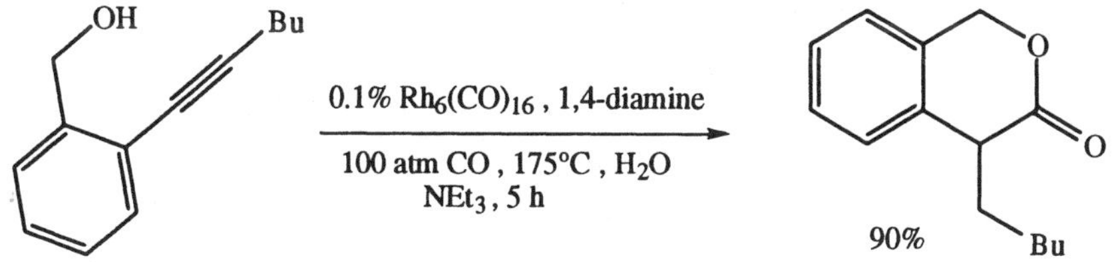

OH
Bu
0.1% Rh6(CO)16 , 1,4-diamine
100 atm CO , 175°C , H2O
NEt3 , 5 h
O
O
90%
Bu

Yoneda, E.; Kaneko, T.; Zhang, S.–W.; Takahashi, S. *Tetrahedron Lett.*, ***1998***, *39*, 5061.

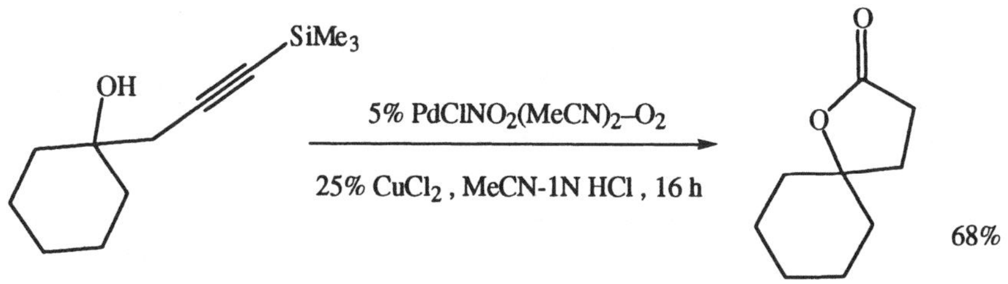

SiMe3
OH
5% PdClNO2(MeCN)2–O2
25% CuCl2 , MeCN-1N HCl , 16 h
O
O
68%

Compain, P.; Goré, J.; Vatèle, J.–M. *Tetrahedron*, ***1996***, *52*, 10405.

SECTION 107: ESTERS FROM ACID DERIVATIVES

The following types of reactions are found in this section:

1. Esters from the reaction of alcohols with carboxylic acids, acid halides and anhydrides.
2. Lactones from hydroxy acids
3. Esters from carboxylic acids and halides, sulfoxides and miscellaneous compounds

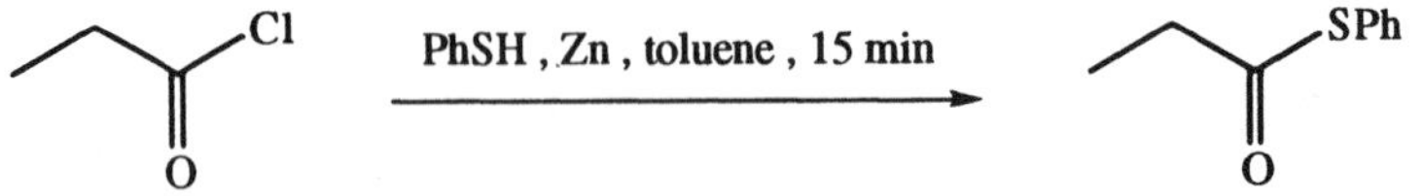
Cl
PhSH , Zn , toluene , 15 min
O
SPh
O

91%

Meshram, H.M.; Reddy, G.S.; Bindu, K.H.; Yadav, J.S. *SynLett*, ***1998***, 877.

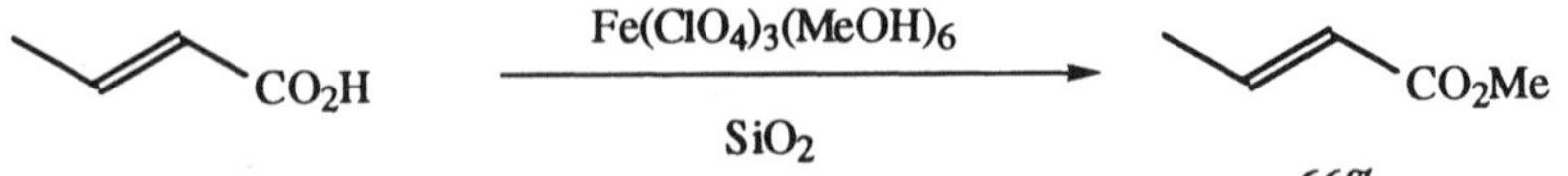
CO2H
Fe(ClO4)3(MeOH)6
SiO2
CO2Me

66%

Parmar, A.; Kaur, J.; Goyal, R.; Kumar, B.; Kumar, H. *Synth. Commun.*, ***1998***, *28*, 2821.

PhCOOH —(EtI , CsF-Celite , MeCN , reflux)→ PhCOOEt 96%

Lee, J.C.; Choi, Y. *Synth. Commun.*, ***1998***, *28*, 2021.

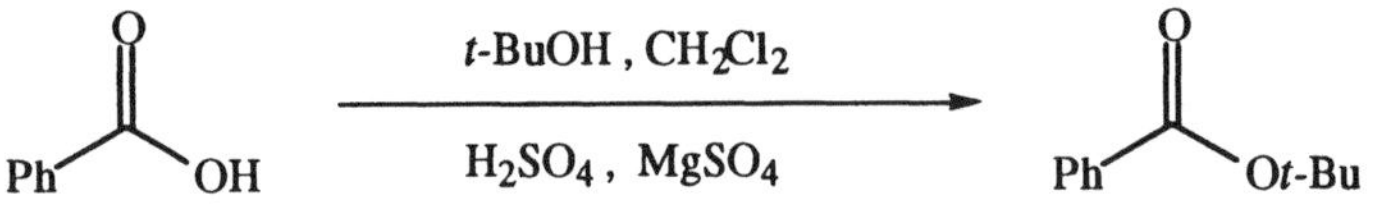
O
Ph
OH
t-BuOH , CH2Cl2
H2SO4 , MgSO4
O
Ph
Ot-Bu

93%

Can form ethers from alcohols

Wright, S.W.; Hageman, D.L.; Wright, A.S.; McClure, L.D. *Tetrahedron Lett.*, ***1997***, *38*, 7345

$C_{11}H_{23}COOH$ —(20% TCNE , MeOH , rt)→ $C_{11}H_{23}COOMe$ 99%

Masaki, Y.; Tanaka, N.; Miura, Y. *Chem. Lett.*, ***1997***, 55.

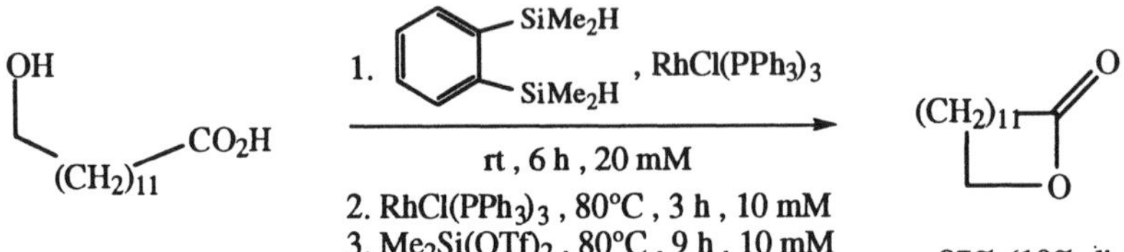
OH
CO2H
(CH2)11
SiMe2H
1.
, RhCl(PPh3)3
SiMe2H
rt , 6 h , 20 mM
2. RhCl(PPh3)3 , 80°C , 3 h , 10 mM
3. Me2Si(OTf)2 , 80°C , 9 h , 10 mM
(CH2)11
O
O

87% (10% diolide)

Mukaiyama, T.; Izumi, J.; Shiina, I. *Chem. Lett*, ***1997***, 187.

Ph—CO_2H —(EtI , Cs_2CO_3 , MeCN , reflux , 1 h)→ Ph—CO_2Et 96%

Lee, J.C.; Oh, Y.S.; Cho, S.H.; Lee, J.I. *Org. Prep. Proceed. Int.*, ***1996***, *28*, 480.

Ph—CO_2H —(1. PCl_3; 2. CH_2N_2; 3. $PhCO_2Ag$, NEt_3, MeOH , ultrasound)→ Ph—CH_2CO_2Me Arndt-Eistert

Winum, Y.–Y.; Kamal, M.; Leydet, A.; Rogue, J.–P.; Montero, J.–L. *Tetrahedron Lett.*. ***1996***. *37*. 1781.

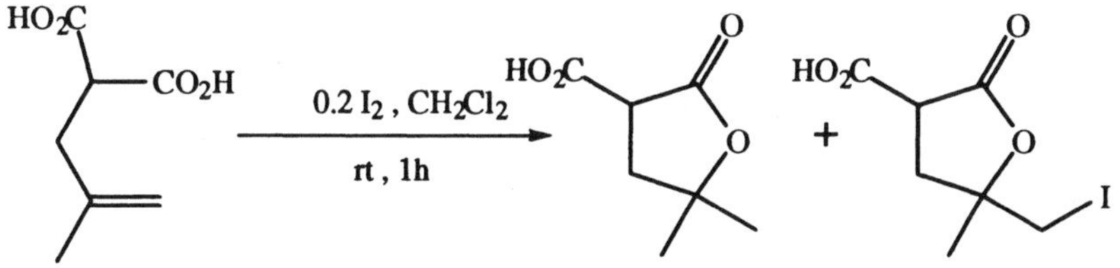
HO2C
CO2H
0.2 I2 , CH2Cl2
rt , 1h
HO2C
O
O
+
HO2C
O
O
I

(98 : 2) 85%

Kim, K.M.; Ryu, E.K. *Tetrahedron Lett.*, ***1996***, *37*, 14411.

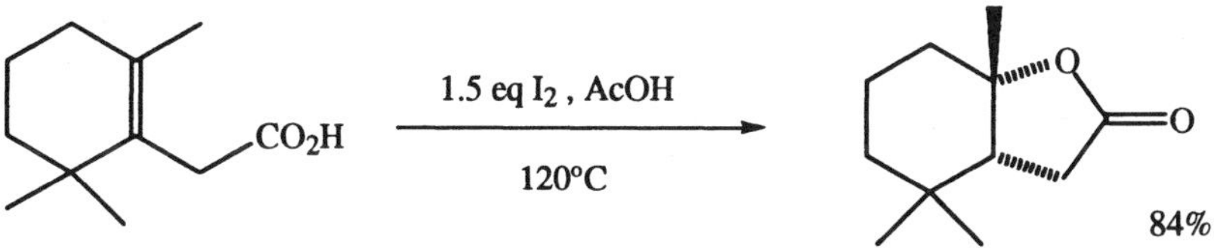
CO2H
1.5 eq I2 , AcOH
120°C
O
O

84%

Yaguchi, Y.; Akiba, M.; Harada, M.; Kato, T. *Heterocycles*, ***1996***, *43*, 601.

Further examples of the reaction RCO_2H + R'OH → RCO_2R' are included in Section 108 (Esters from Alcohols and Phenols) and in Section 30A (Protection of Carboxylic Acids).

SECTION 108: ESTERS FROM ALCOHOLS AND THIOLS

Further examples of the reaction ROH → RCO_2R' are included in Section 107 (Esters from Acid Derivatives) and in Section 45A (Protection of Alcohols and Phenols).

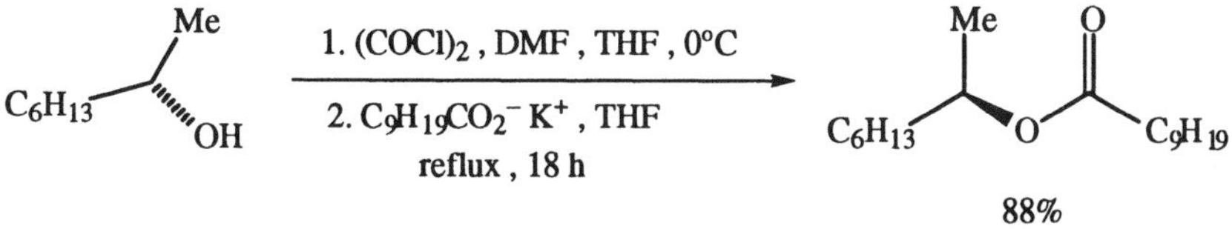
Me
C6H13
OH
1. (COCl)2 , DMF , THF , 0°C
2. C9H19CO2⁻ K⁺ , THF
reflux , 18 h
Me
O
C6H13
O
C9H19

88%

Barrett, A.G.M.; Braddock, D.C.; James, R.A.; Koike, N.; Procopiou, P.A. *J. Org. Chem.*, ***1998***, *63*, 6273.

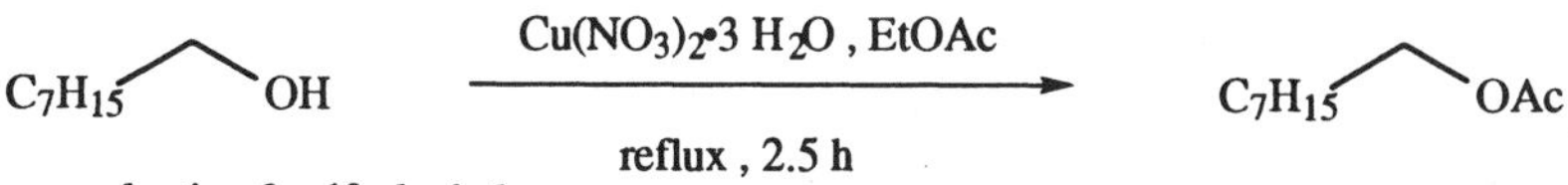
C7H15
OH
Cu(NO3)2•3 H2O , EtOAc
reflux , 2.5 h
C7H15
OAc

selective for 1° alcohols

91%

Iranpoor, N.; Firouzabadi, H.; Zolfigol, M.A. *Synth. Commun.*, ***1998***, *28*, 1923.

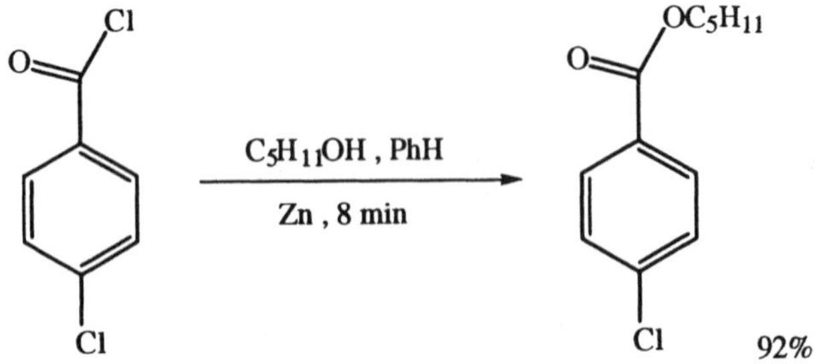
Cl
O
$C_5H_{11}OH$, PhH
Zn , 8 min
OC_5H_{11}
O
Cl
Cl
92%

Yadav, J.S.; Reddy, G.S.; Srinivas, D.; Himabindu, K. *Synth. Commun.*, ***1998***, *28*, 2337.

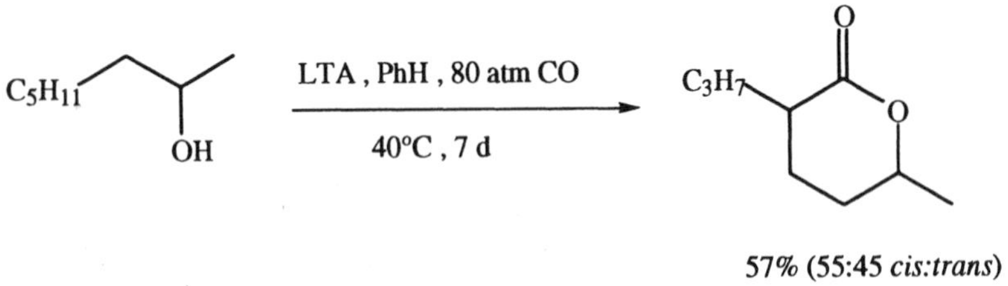
C_5H_{11}
OH
LTA , PhH , 80 atm CO
40°C , 7 d
O
C_3H_7
O
57% (55:45 *cis:trans*)

Tsunoi, S.; Ryu, I; Okuda, T.; Tanaka, M.; Komatsu, M.; Sonoda, N.
J. Am. Chem. Soc., ***1998***, *120*, 8692.

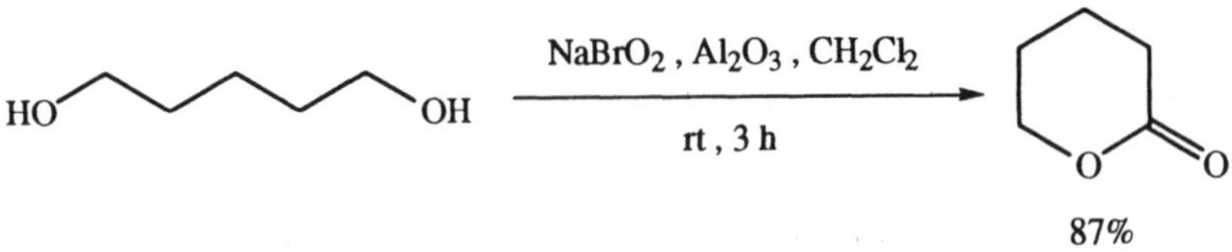
HO
OH
$NaBrO_2$, Al_2O_3 , CH_2Cl_2
rt , 3 h
O
O
87%

Hirano, M.; Yakabe, S.; Morimoto, T. *Synth. Commun.*, ***1998***, *28*, 127.

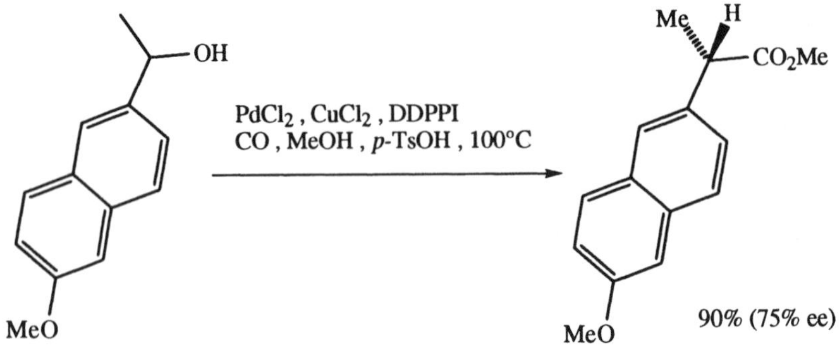
OH
MeO
$PdCl_2$, $CuCl_2$, DDPPI
CO , MeOH , *p*-TsOH , 100°C
Me
H
CO_2Me
MeO
90% (75% ee)

Xie, B.–H.; Xia, C.–G.; Lu, S.–J.; Chen, K.–J.; Kou, Y.; Yin, Y.–Q.
Tetrahedron Lett., ***1998***, *39*, 7365.

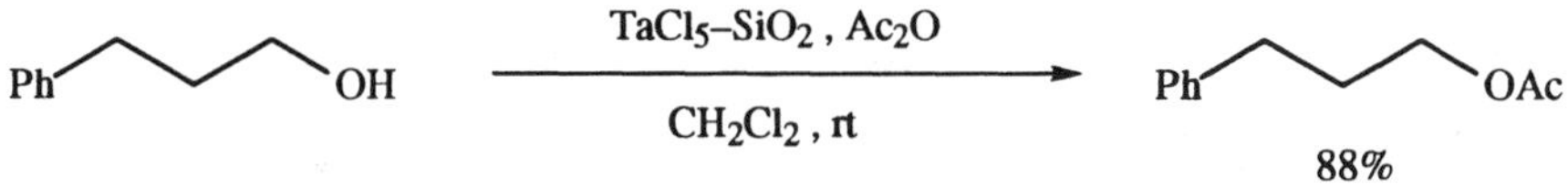
Ph
OH
TaCl5–SiO2 , Ac2O
CH2Cl2 , rt
Ph
OAc
88%

Chandrasekhar, S.; Ramachander, T.; Takhi, M. *Tetrahedron Lett.*, ***1998***, *39*, 3263.

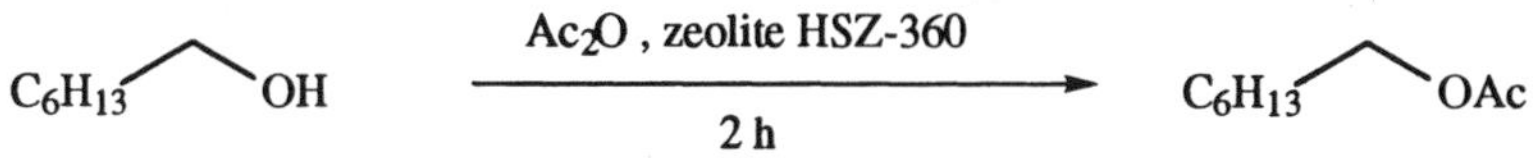
C6H13
OH
Ac2O , zeolite HSZ-360
2 h
C6H13
OAc

Ballini, R.; Bosica, G.; Carloni, S.; Ciaralli, L.; Maggi, R.; Sartori, G. *Tetrahedron Lett.*, ***1998***, *39*, 6049.

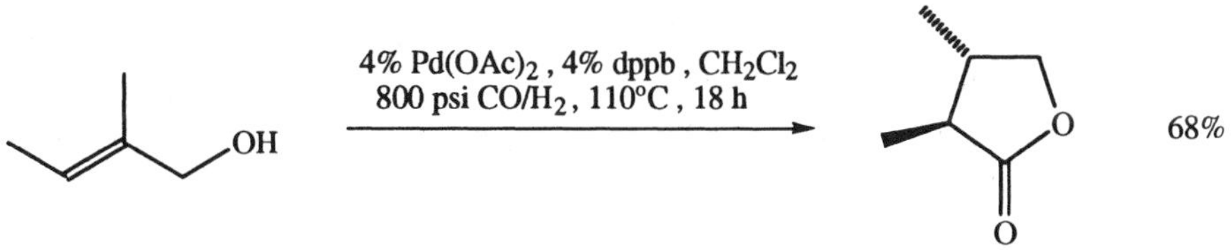
OH
4% Pd(OAc)2 , 4% dppb , CH2Cl2
800 psi CO/H2 , 110°C , 18 h
O
O
68%

Brunner, M.; Alper, H. *J. Org. Chem.*, ***1997***, *62*, 7565.

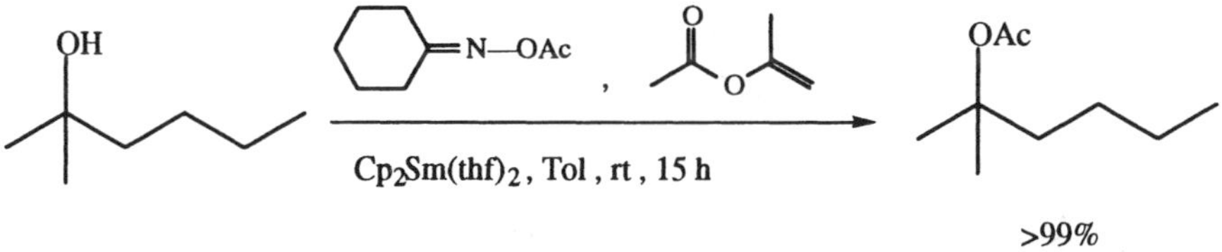
OH
N—OAc
O
O
Cp2Sm(thf)2 , Tol , rt , 15 h
OAc
>99%

Tashiro, D.; Kawasaki, Y.; Sakaguchi, S.; Ishii, Y. *J. Org. Chem.*, ***1997***, *62*, 8141.

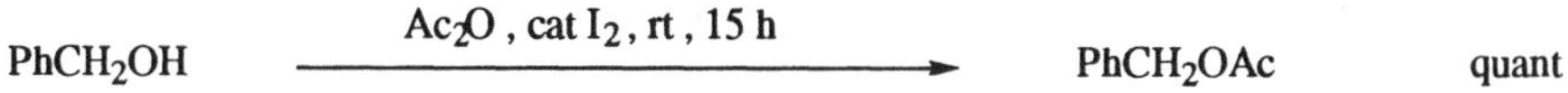
PhCH2OH
Ac2O , cat I2 , rt , 15 h
PhCH2OAc
quant

Borah, R.; Deka, N.; Sarma, J.C. *J. Chem. Soc. (S)*, ***1997***, 110.

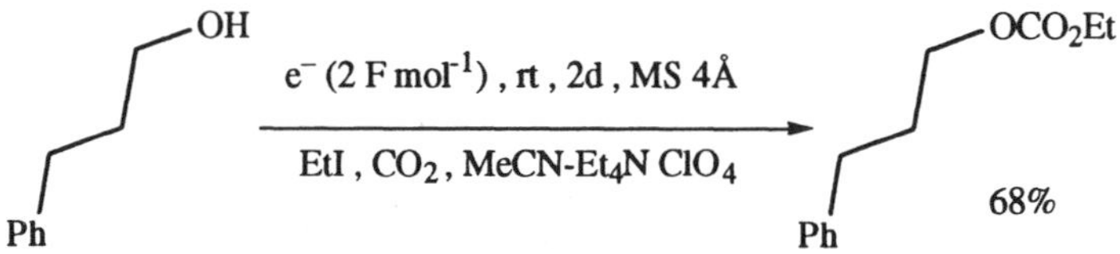
OH
e⁻ (2 F mol⁻¹) , rt , 2d , MS 4Å
EtI , CO2 , MeCN-Et4N ClO4
Ph
OCO2Et
68%
Ph

Casadei, M.A.; Inesi, A.; Rossi, L. *Tetrahedron Lett.*, ***1997***, *38*, 3565.

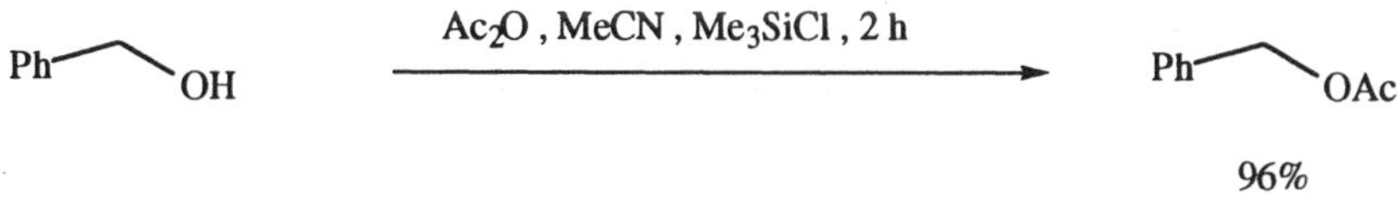
Ph
OH
Ac2O , MeCN , Me3SiCl , 2 h
Ph
OAc
96%

Kumareswaran, R.; Gupta, A.; Vankar, Y.D. *Synth. Commun.*, ***1997***, *27*, 277.

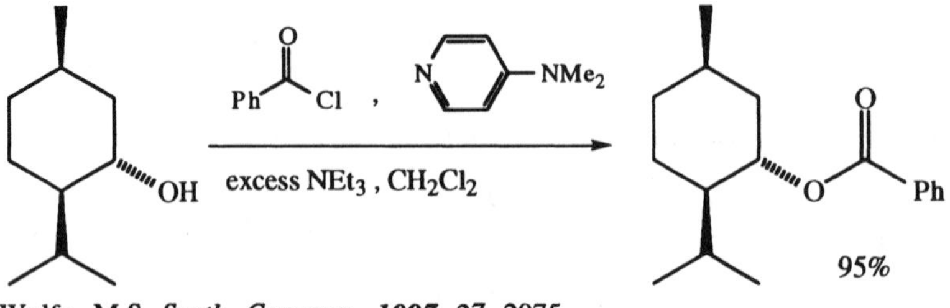
O
Ph Cl
N
NMe2
OH
excess NEt3 , CH2Cl2
O
O
Ph
95%

Wolfe, M.S. *Synth. Commun.*, ***1997***, *27*, 2975.

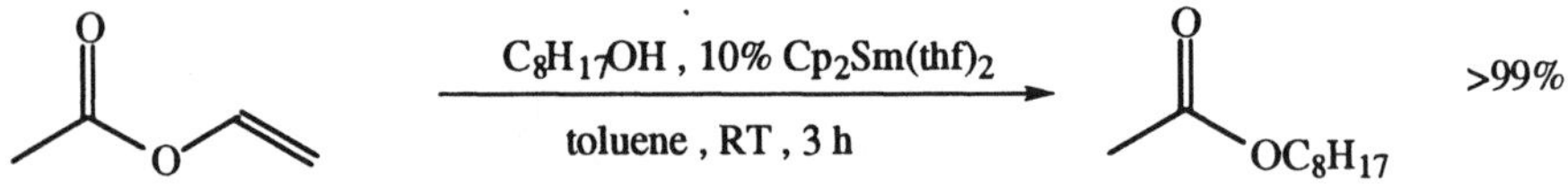
O
O
C8H17OH , 10% Cp2Sm(thf)2
toluene , RT , 3 h
O
OC8H17
>99%

Ishii, Y.; Takeno, M.; Kawasaki, Y.; Muromachi, A.; Nishiyama, Y.; Sakaguchi, S. *J. Org. Chem.*, ***1996***, *61*, 3088.

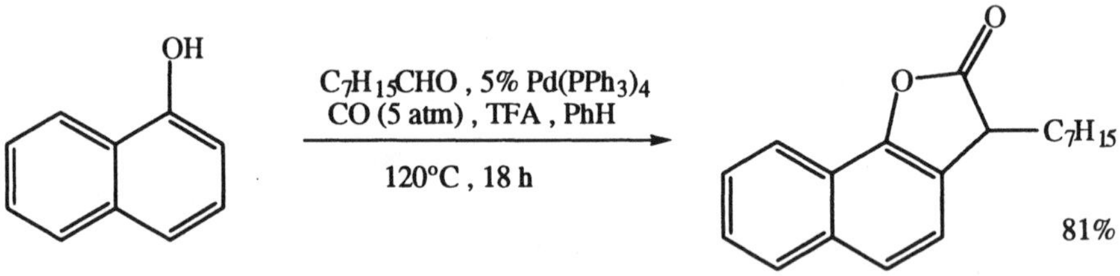
OH
C7H15CHO , 5% Pd(PPh3)4
CO (5 atm) , TFA , PhH
120°C , 18 h
O
O
C7H15
81%

Satoh, T.; Tsuda, T.; Kushino, Y.; Miura, M.; Nomura, M. *J. Org. Chem.*, ***1996***, *61*, 6476.

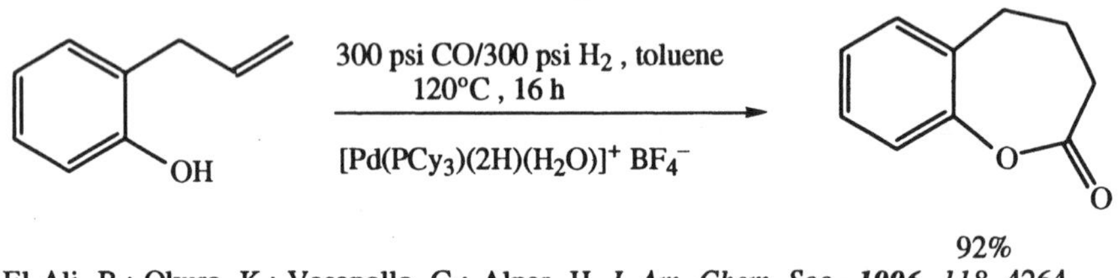
OH
300 psi CO/300 psi H2 , toluene
120°C , 16 h
[Pd(PCy3)(2H)(H2O)]+ BF4−
O
O
92%

El Ali, B.; Okuro, K.; Vasapollo, G.; Alper, H. *J. Am. Chem. Soc.*, ***1996***, *118*, 4264.

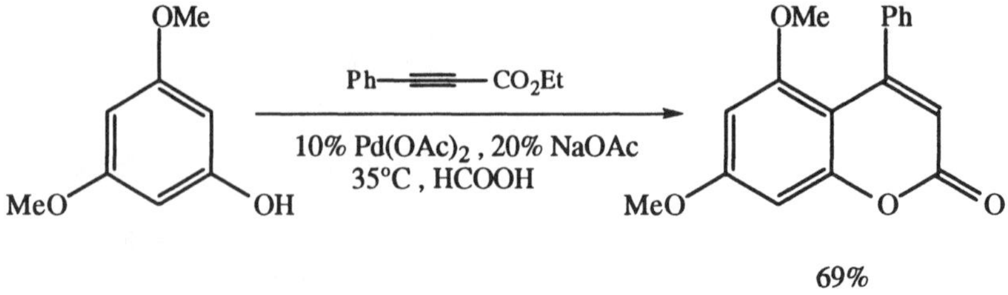
OMe
MeO
OH
Ph CO2Et
10% Pd(OAc)2 , 20% NaOAc
35°C , HCOOH
OMe
Ph
MeO
O
O
69%

Trost, B.M.; Toste, F.D. *J. Am. Chem. Soc.*, ***1996***, *118*, 6305.

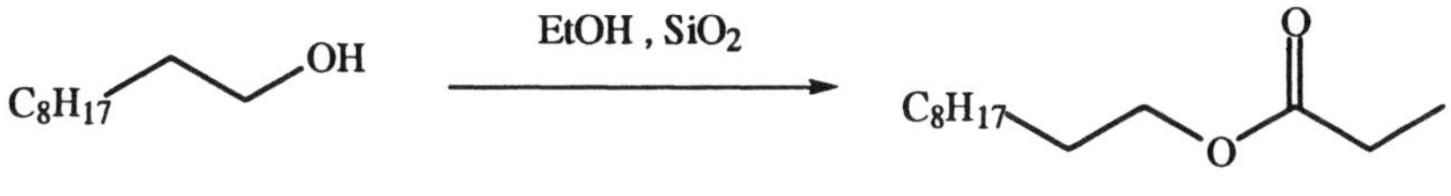
C_8H_{17}
OH
EtOH , SiO_2
C_8H_{17}
O
O

83%

Nascimento, M. de G.; Zanotto, S.P.; Scremin, M.; Rezende, M.C.
Synth. Commun., ***1996***, *26*, 2715.

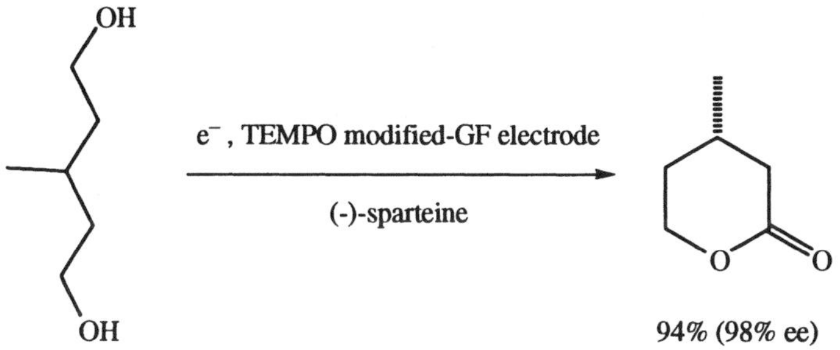
OH
e^- , TEMPO modified-GF electrode
(-)-sparteine
O
O
OH

94% (98% ee)

Yanagisawa, Y.; Kashiwagi, Y.; Kurashima, F.; Anazai, J.; Osa, T.; Bobbitt, J.M.
Chem. Lett, ***1996***, 1043.

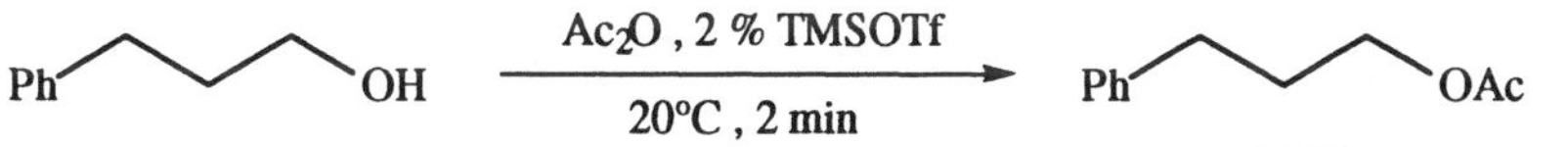
Ph
OH
Ac_2O , 2 % TMSOTf
20°C , 2 min
Ph
OAc

quant

Procopiou, P.A.; Baugh, S.P.D.; Flack, S.S.; Inglis, G.G.A. *Chem. Commun.*, ***1996***, 2625

SECTION 109: ESTERS FROM ALDEHYDES

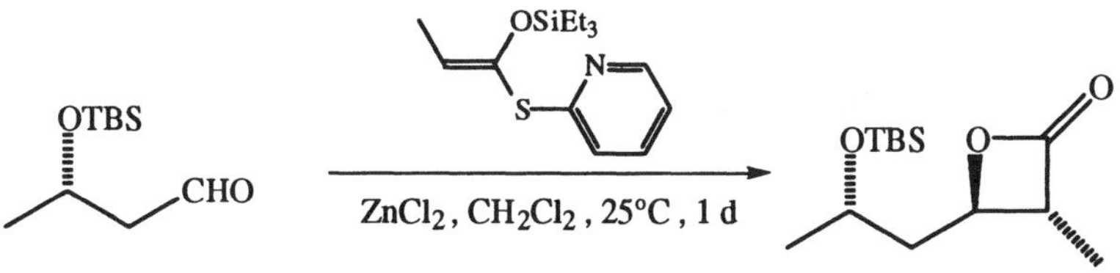
OSiEt3
N
S
OTBS
CHO
ZnCl2 , CH2Cl2 , 25°C , 1 d
OTBS
O
O

69% (>19:1 *trans:cis*)
1:5.3 (*syn:anti*) 94% ee

Yang, H.W.; Romo, D. *J. Org. Chem.*, ***1998***, *63*, 1344.

PhCHO → $PhCO_2Me$ (95%)

1. $HC(OMe)_3$, Amberlyst-15 , $CHCl_3$
2. cat $BF_3 \cdot OEt_2$, mcpba
3. DBU

Rhee, H.; Kim, J.Y. *Tetrahedron Lett*, ***1998***, *39*, 1365.

$PhCHO \xrightarrow{MnO_2\text{ , NaCN , MeOH , 14 h}} PhCO_2Me$ 92%

Lai, G.; Anderson, W.K. *Synth. Commun.*, ***1997***, *27*, 1281.

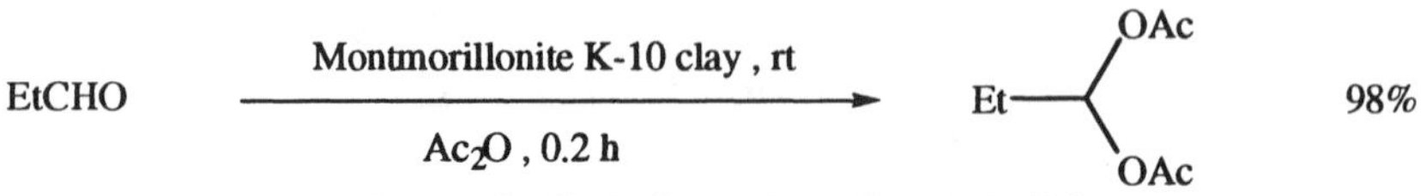
EtCHO
Montmorillonite K-10 clay , rt
Ac2O , 0.2 h
Et
OAc
OAc
98%

Zhang, Z.–H.; Li, T.–S.; Fu, C.–G. *J. Chem. Res. (S)*, ***1997***, 174.

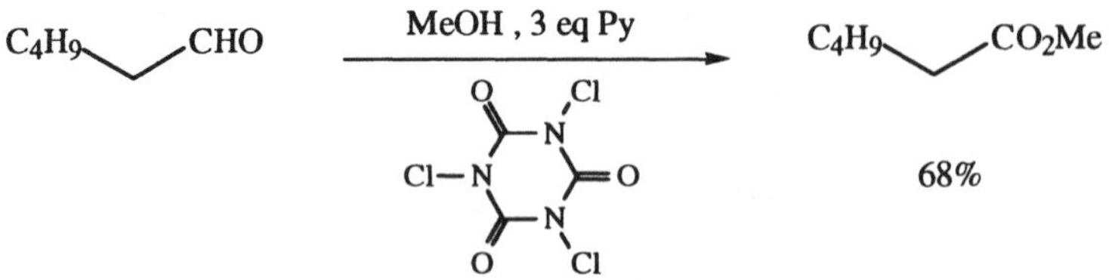
C4H9
CHO
MeOH , 3 eq Py
C4H9
CO2Me
O
Cl
N
Cl—N
O
N
O
Cl
68%

Hiegel, G.A.; Bayne, C.D.; Donde, Y.; Tamashiro, G.S.; Hilberath, L.A.
Synth. Commun., ***1996***, *26*, 2633.

Related Methods: Section 117 (Esters from Ketones)

SECTION 110: ESTERS FROM ALKYLS, METHYLENES AND ARYLS

No examples of the reaction R-R → RCO_2R' or $R'CO_2R$ (R,R' = alkyl, aryl, etc.) occur in the literature. For the reaction R-H → RCO_2R' or $R'CO_2R$, see Section 116 (Esters from Hydrides).

NO ADDITIONAL EXAMPLES

SECTION 111: ESTERS FROM AMIDES

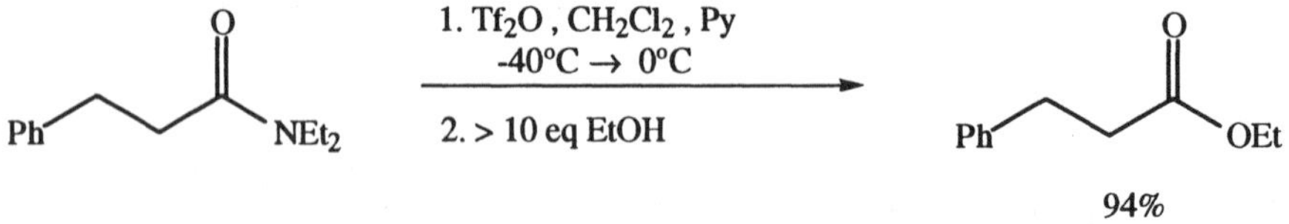
O
Ph
NEt2
1. Tf2O , CH2Cl2 , Py
-40°C → 0°C
2. > 10 eq EtOH
O
Ph
OEt
94%

Charette, A.B.; Chua, P. *SynLett*, ***1998***, 163.

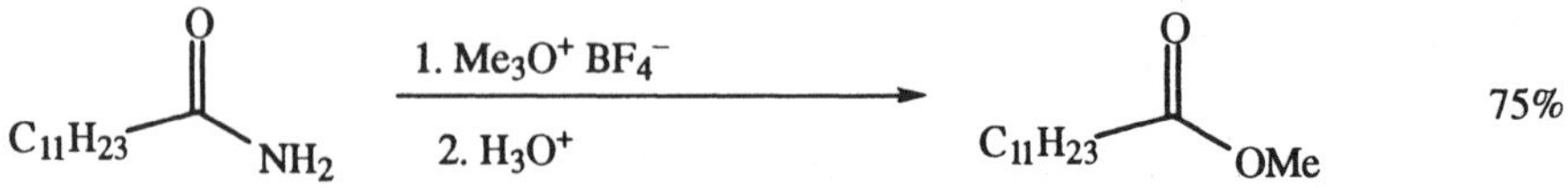
O
C11H23
NH2
1. Me3O+ BF4−
2. H3O+
O
C11H23
OMe
75%

Kiessling, A.J.; McClure, C.K. *Synth. Commun.*, ***1997***, *27*, 923.

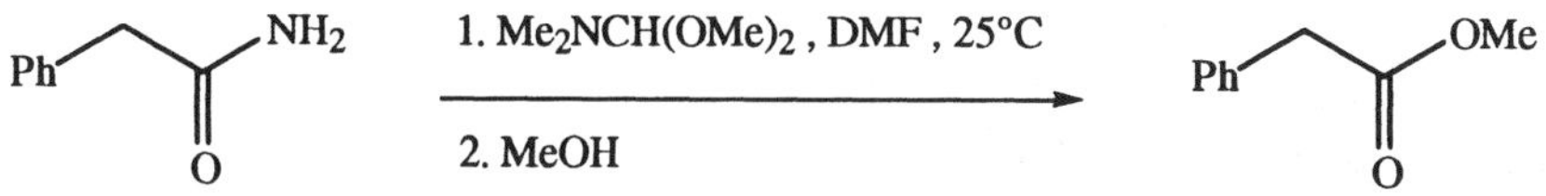
NH2
Ph
O
1. Me2NCH(OMe)2 , DMF , 25°C
2. MeOH
OMe
Ph
O

94%

Anelli, P.L.; Brocchetta, M.; Palano, D.; Visigalli, M. *Tetrahedron Lett.*, ***1997***, *38*, 2367.

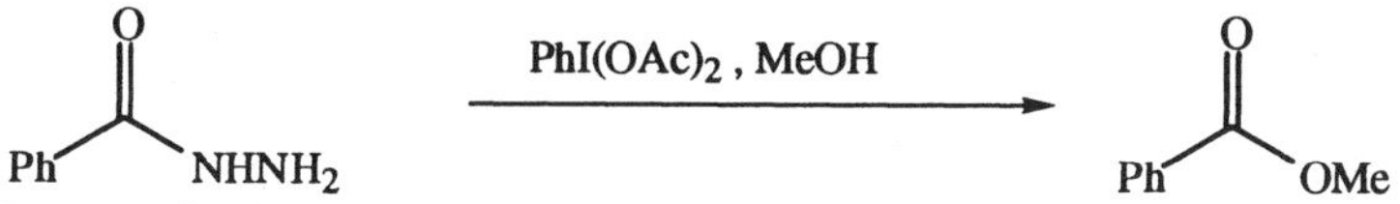
O
Ph
NHNH2
PhI(OAc)2 , MeOH
O
Ph
OMe

40%

Prakash, O.; Sharma, V.; Sadana, A. *J. Chem. Soc., (S),* ***1996***, 100.

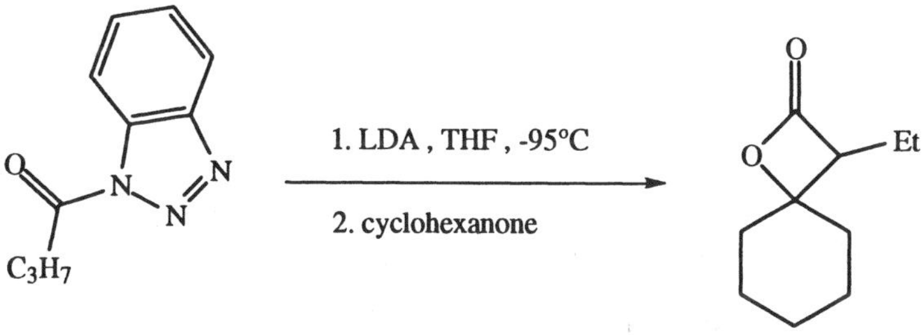
O
N
N
N
C3H7
1. LDA , THF , -95°C
2. cyclohexanone
O
O
Et

78%

Wedler, C.; Kleiner, K.; Kunath, A.; Schick, H. *Liebigs Ann. Chem.*, ***1996***, 881.

SECTION 112: ESTERS FROM AMINES

NO ADDITIONAL EXAMPLES

SECTION 113: ESTERS FROM ESTERS

Conjugate reductions and conjugate alkylations of unsaturated esters are found in Section 74 (Alkyls from Alkenes).

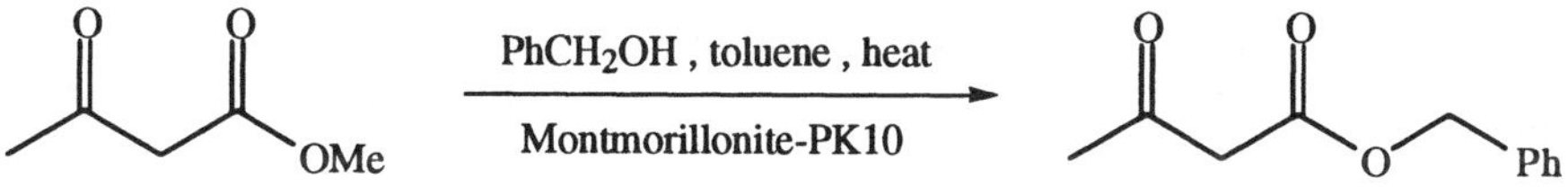
O
O
OMe
PhCH2OH , toluene , heat
Montmorillonite-PK10
O
O
O
Ph

86%

Ponde, D.E.; Deshpande, V.H.; Bulbule, V.J.; Sudalai, A.; Gajare, A.S. *J. Org. Chem.*, ***1998***, *63*, 1058.

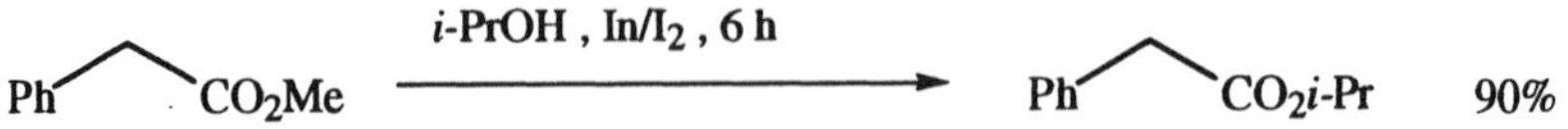
i-PrOH , In/I2 , 6 h
Ph
CO2Me
Ph
CO2i-Pr
90%

Ranu, B.C.; Dutta, P.; Sarkar, A. *J. Org. Chem.*, ***1998***, *63*, 6027.

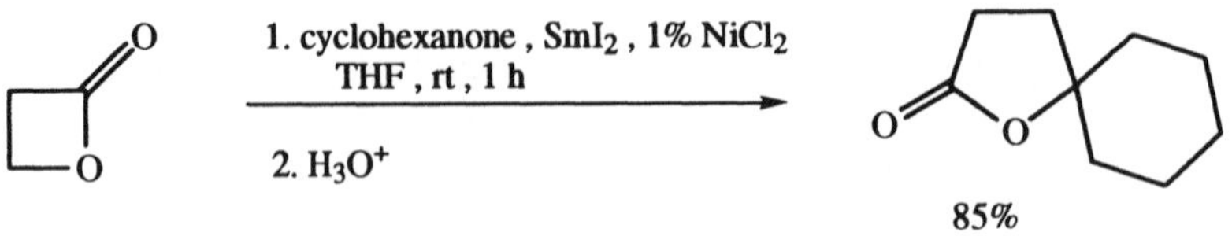
O
O
1. cyclohexanone , SmI2 , 1% NiCl2
THF , rt , 1 h
2. H3O+
O
O
85%

Machrouhi, F.; Namy, J.-L. *Tetrahedron*, ***1998***, *54*, 11111.

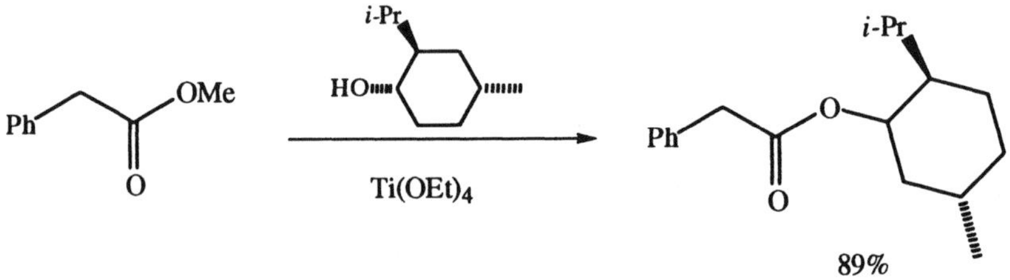
Ph
OMe
O
i-Pr
HO
Ti(OEt)4
Ph
O
O
i-Pr
89%

Krasik, P. *Tetrahedron Lett.*, ***1998***, *39*, 4223.

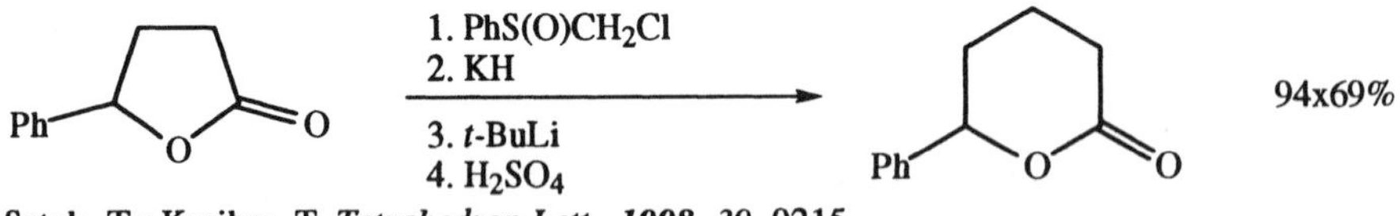
Ph
O
O
1. PhS(O)CH2Cl
2. KH
3. t-BuLi
4. H2SO4
Ph
O
O
94x69%

Satoh, T.; Kurihra, T. *Tetrahedron Lett.*, ***1998***, *39*, 9215.

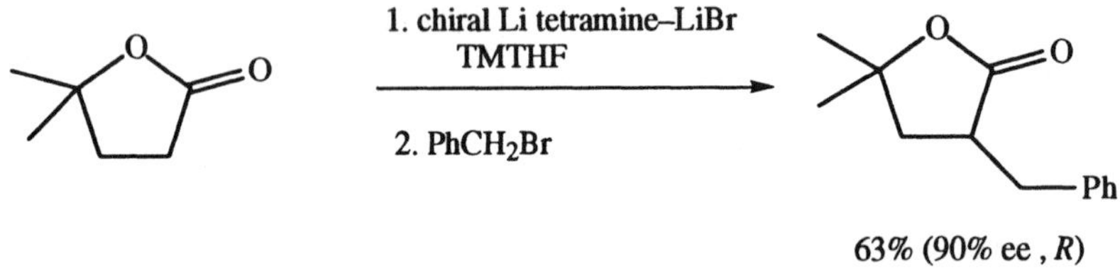
O
O
1. chiral Li tetramine–LiBr
TMTHF
2. PhCH2Br
O
O
Ph
63% (90% ee , R)

Matsuo, J.-i.; Kobayashi, S.; Koga, K. *Tetrahedron Lett*, ***1998***, *39*, 9723.

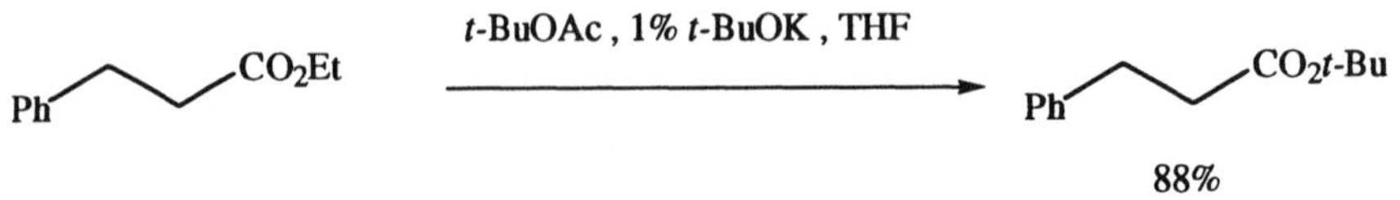
Ph
CO2Et
t-BuOAc , 1% t-BuOK , THF
Ph
CO2t-Bu
88%

Stanton, M.G.; Gagné, M.R. *J. Org. Chem.*, ***1997***, *62*, 8240.

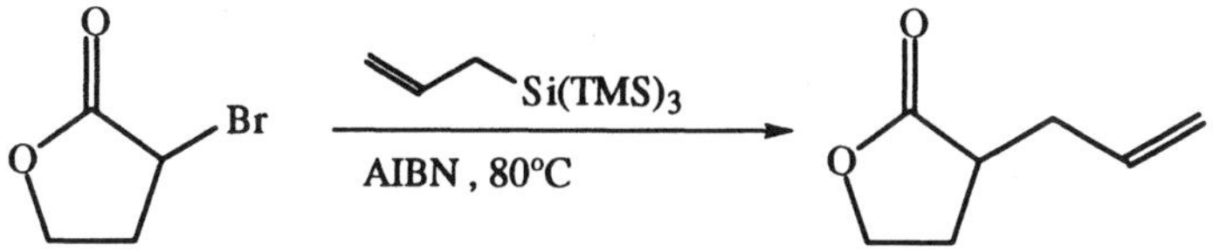
O
Br
O
Si(TMS)3
AIBN, 80°C
O
O

87%

Chatgilialoglu, C.; Ferreri, C.; Ballestri, M.; Curran, D.P. *Tetrahedron Lett.*, ***1996***, *37*, 6387.
Chatgilialoglu, C.; Alberti, A.; Ballestri, M.; MacCiatelli, D.; Curran, D.P. *Tetrahedron Lett.*, ***1996***, *37*, 6391.

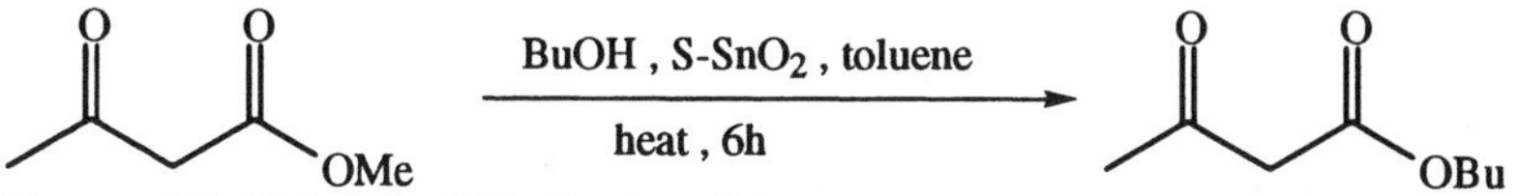
O
O
OMe
BuOH, S-SnO2, toluene
heat, 6h
O
O
OBu

97%

Chavan, S.P.; Zubaidha, P.K.; Dantale, S.W.; Keshavaraja, A.; Ramaswamy, A.V.; Ravindranathan, T. *Tetrahedron Lett.*, ***1996***, *37*, 233.

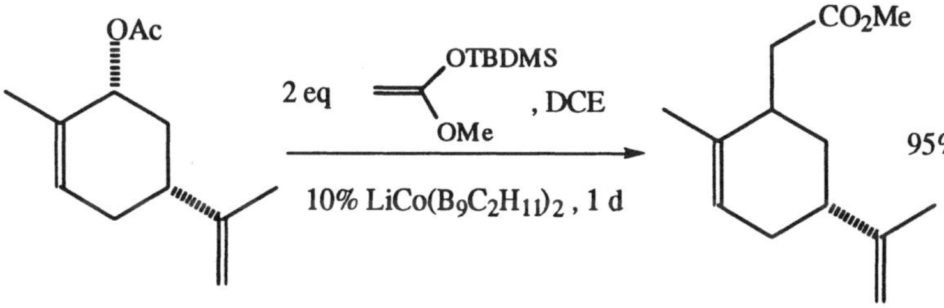
OAc
2 eq
OTBDMS
OMe
, DCE
10% LiCo(B9C2H11)2, 1 d
CO2Me

95% (5:7 *cis:trans*)

Grieco, P.A.; DuBay, W.J.; Todd, L.J. *Tetrahedron Lett.*, ***1996***, *37*, 8707.

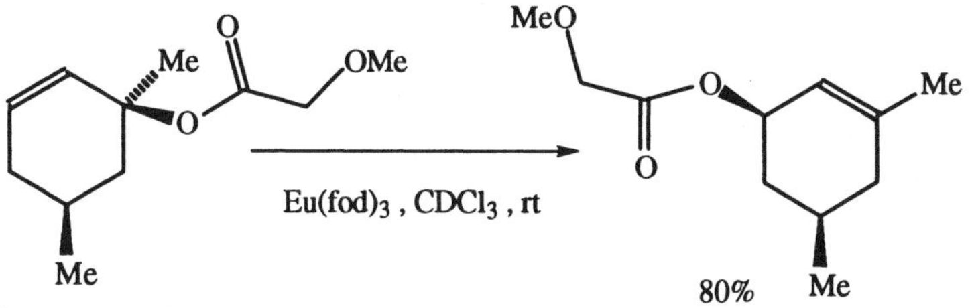
Me
O
OMe
O
Me
Eu(fod)3, CDCl3, rt
MeO
O
O
Me
Me

80%

Shull, B.K.; Sakai, T.; Koreeda, M. *J. Am. Chem. Soc.*, ***1996***, *118*, 11690.

SECTION 114: ESTERS FROM ETHERS, EPOXIDES AND THIOETHERS

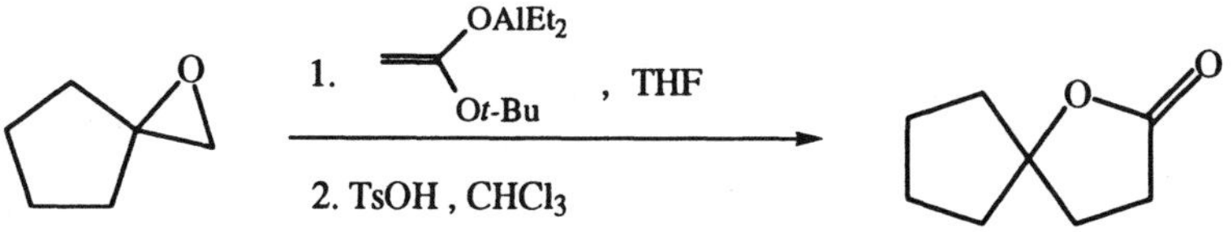
O
1.
OAlEt2
Ot-Bu
, THF
2. TsOH, CHCl3
O
O

76x81%

Taylor, S.K.; Chmiel, N.H.; Mann, E.E.; Silver, M.E.; Vyvyan, J.R. *Synthesis*, ***1998***, 1009.

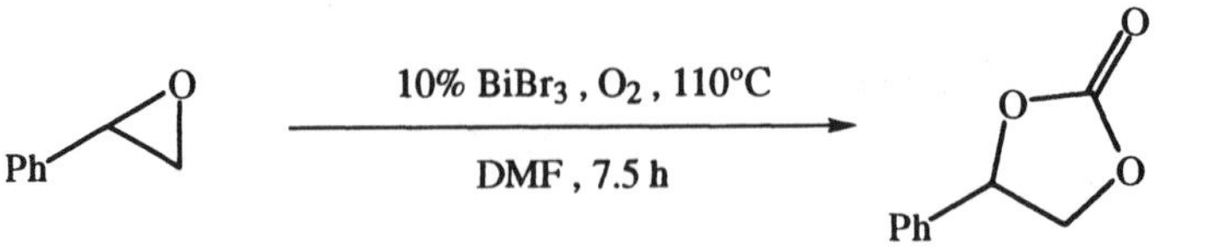
10% $BiBr_3$, O_2 , 110°C
DMF , 7.5 h

56%

LeBoisselier, V.; Postel, M.; Duñach, E. *Chem. Commun.*, ***1997***, 95.

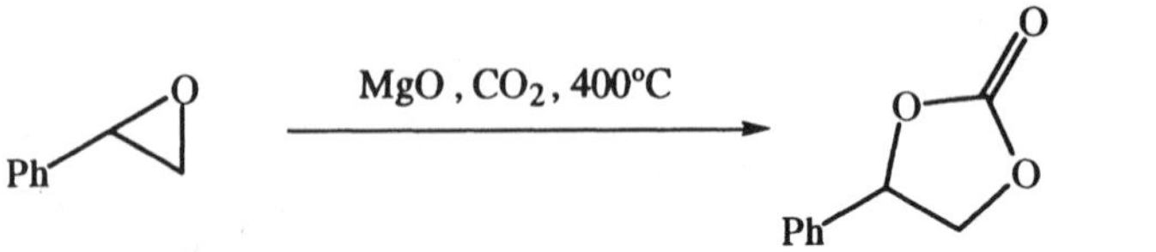
MgO , CO_2 , 400°C

60%

Yano, T.; Matsui, H.; Koike, T.; Ishiguro, H.; Fujihara, H.; Yoshihara, M.; Maeshima, T. *Chem. Commun.*, ***1997***, 1129.

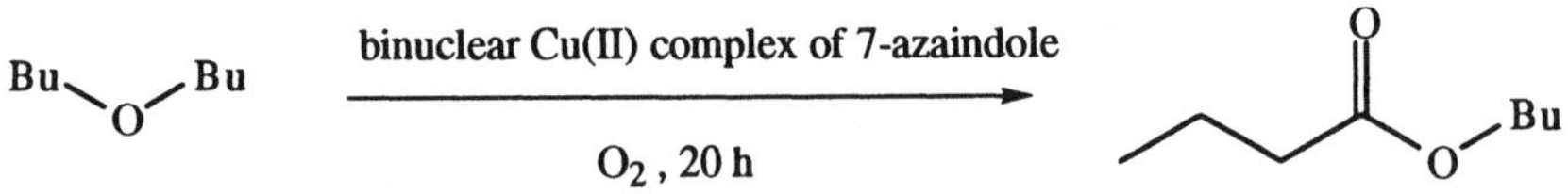
binuclear Cu(II) complex of 7-azaindole
O_2 , 20 h

19 turnovers

Minakata, S.; Imai, E.; Ohshima, Y.; Inaki, K.; Ryu, I.; Komatsu, M.; Ohshiro, Y. *Chem. Lett.*, ***1996***, 19.

SECTION 115: ESTERS FROM HALIDES AND SULFONATES

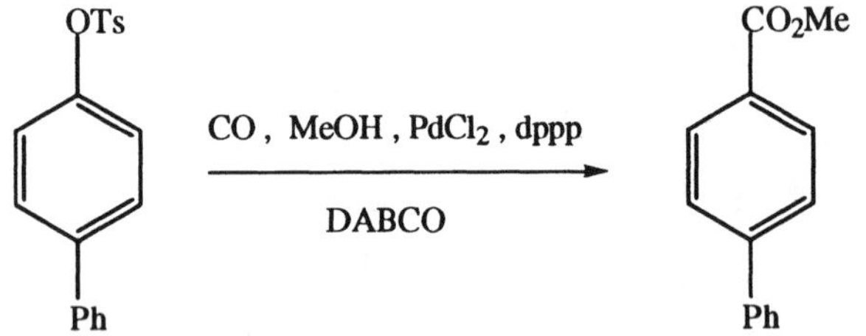
CO , MeOH , $PdCl_2$, dppp
DABCO

81%

Kubota, Y.; Nakada, S.; Sugi, Y. *SynLett*, ***1998***, 183.

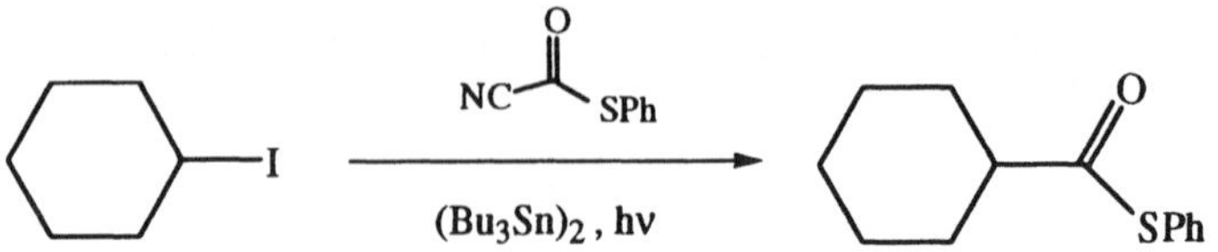
$(Bu_3Sn)_2$, hv

56%

Kim, S.; Jon, S.Y. *Chem. Commun.*, ***1998***, 815.

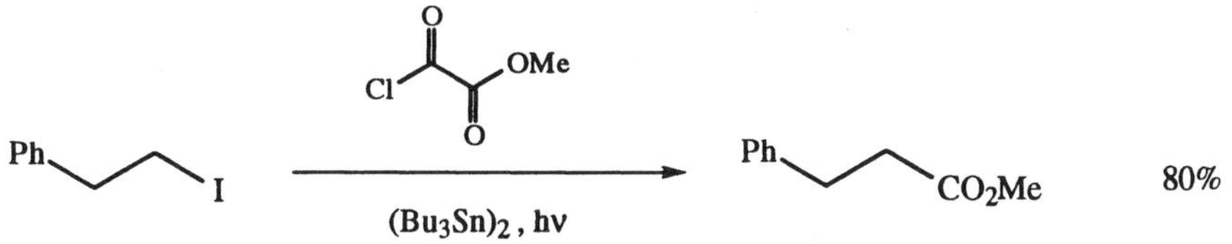
OMe
Cl
O
O
Ph
I
(Bu3Sn)2 , hv
Ph
CO2Me
80%

Kim, S.; Jon, S.Y. *Tetrahedron Lett.*, ***1998***, *39*, 7317.

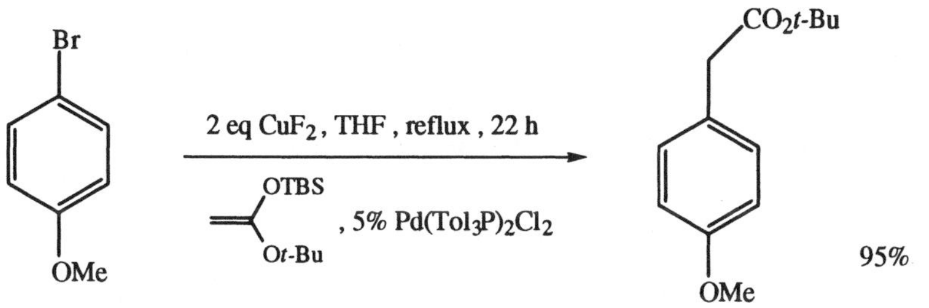
Br
OMe
2 eq CuF2 , THF , reflux , 22 h
OTBS
Ot-Bu
, 5% Pd(Tol3P)2Cl2
CO2t-Bu
OMe
95%

Agnelli, F.; Sulikowski, G.A. *Tetrahedron Lett*, ***1998***, *39*, 8807.

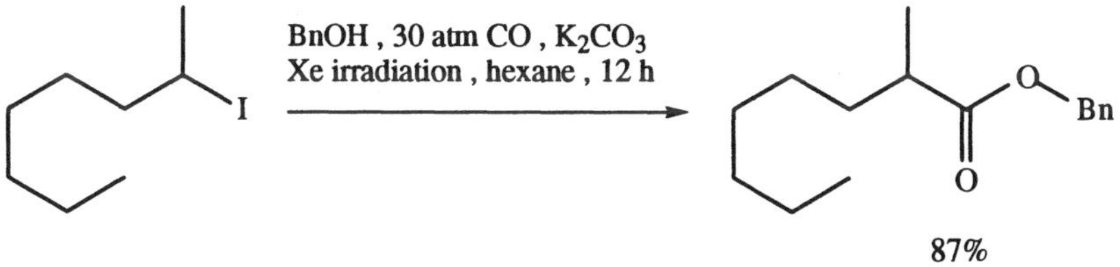
I
BnOH , 30 atm CO , K2CO3
Xe irradiation , hexane , 12 h
O
Bn
O
87%

Nagahara, K.; Ryu, I.; Komatsu, M.; Sonoda, N. *J. Am. Chem. Soc.*, ***1997***, *118*, 5465.

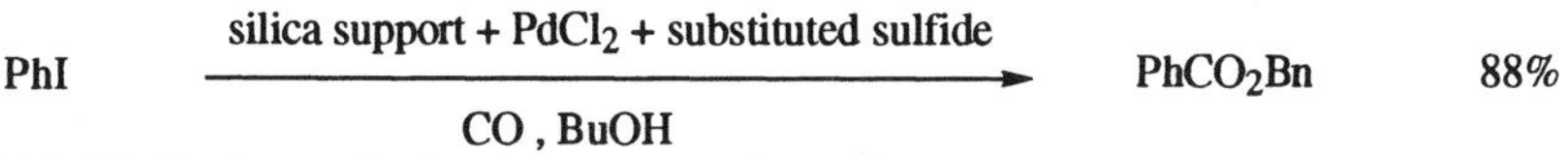
PhI
silica support + PdCl2 + substituted sulfide
CO , BuOH
PhCO2Bn
88%

Cai, M.–Z.; Song, C.–S.; Huang, X. *J. Chem. Soc., Perkin Trans. 1*, ***1997***, 2273.

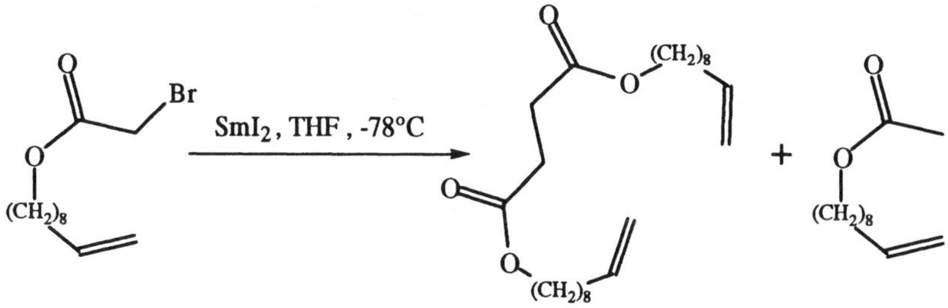
O
Br
O
(CH2)8
SmI2 , THF , -78°C
O
O
(CH2)8
O
O
(CH2)8
+
O
O
(CH2)8

add ester to SmI_2	(18	:	58)
add SmI_2 to ester	(59	:	3)

Balaux, É; Ruel, R. *Tetrahedron Lett.*, ***1996***, *37*, 801.

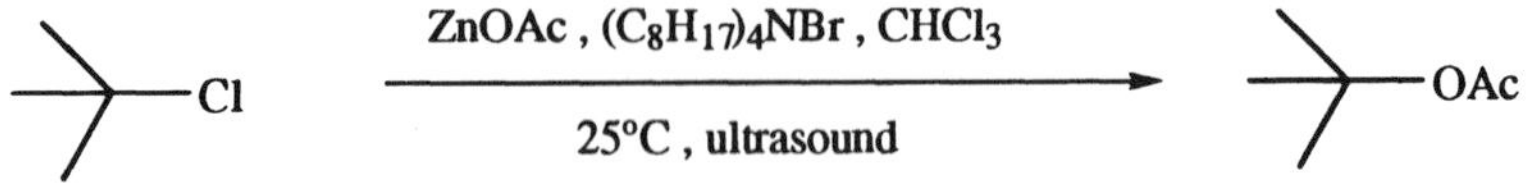
Cl
ZnOAc , $(C_8H_{17})_4NBr$, $CHCl_3$
25°C , ultrasound
OAc

75%

Jayasree, J.; Rao, J.M. *Synth. Commun*, ***1996***, *26*, 1103.

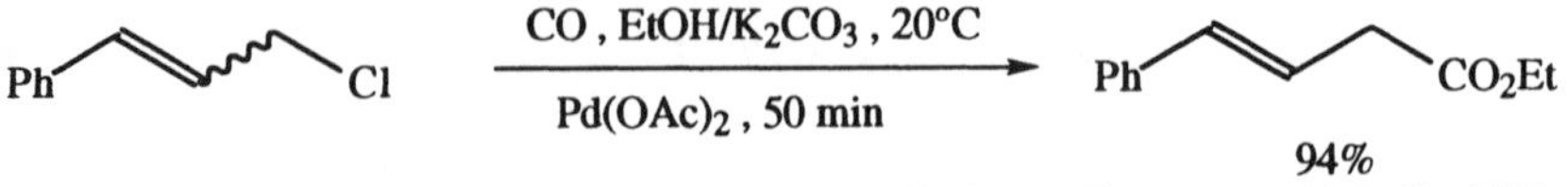
Ph
Cl
CO , $EtOH/K_2CO_3$, 20°C
$Pd(OAc)_2$, 50 min
Ph
CO_2Et

94%

Kiji, J.; Okano, T.; Higashimae, Y.; Fukui, Y. *Bull. Chem. Soc. Jpn.*, ***1996***, *69*, 1029.

Related Methods: Section 25 (Acid Derivatives from Halides).

SECTION 116: ESTERS FROM HYDRIDES

This section contains examples of the reaction R-H → RCO_2R' or $R'CO_2R$ (R = alkyl, aryl, etc.).

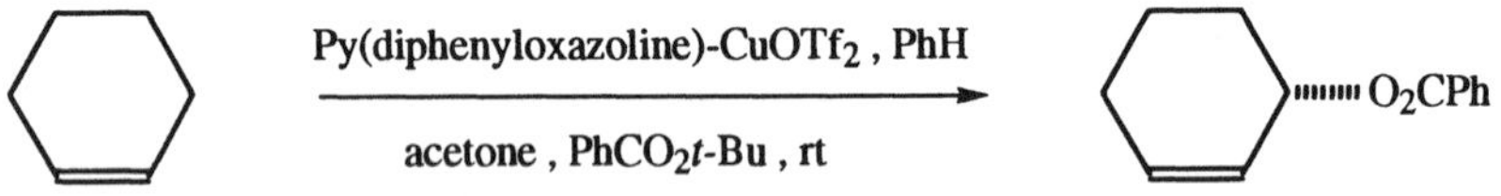
Py(diphenyloxazoline)-$CuOTf_2$, PhH
acetone , $PhCO_2t$-Bu , rt
O_2CPh

87% (73% ee)

Sekar, G.; Datta Gupta, A.; Singh, V.K. *J. Org. Chem.*, ***1998***, *63*, 2961.

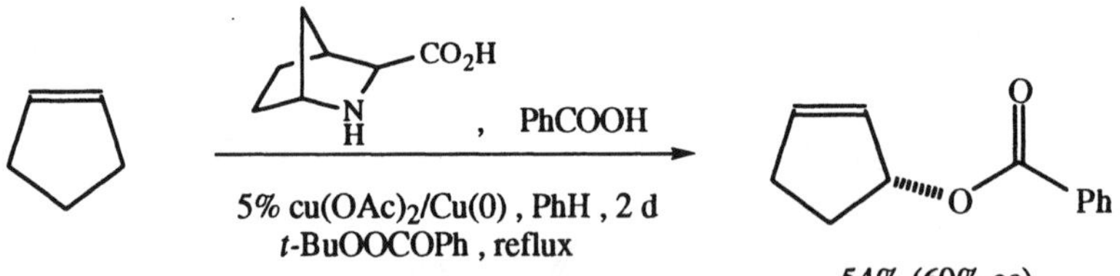
CO_2H
N
H
, PhCOOH
5% $cu(OAc)_2/Cu(0)$, PhH , 2 d
t-BuOOCOPh , reflux
O
O
Ph

54% (60% ee)

Södergren, M.J.; Andersson, P.G. *Tetrahedron Lett.*, ***1996***, *37*, 7577.

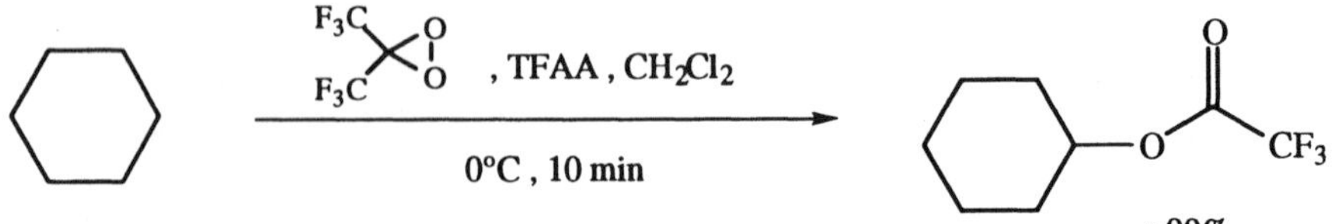
F_3C
F_3C
O
O
, TFAA , CH_2Cl_2
0°C , 10 min
O
O
CF_3

>99%

Asensio, G.; Mello, R.; González-Nuñez, M.E.; Castellano, G.; Corral, J. *Angew. Chem. Int. Ed.*, ***1996***, *35*, 217.

Also via: Section 26 (Acid Derivatives) and Section 41 (Alcohols).

SECTION 117: ESTERS FROM KETONES

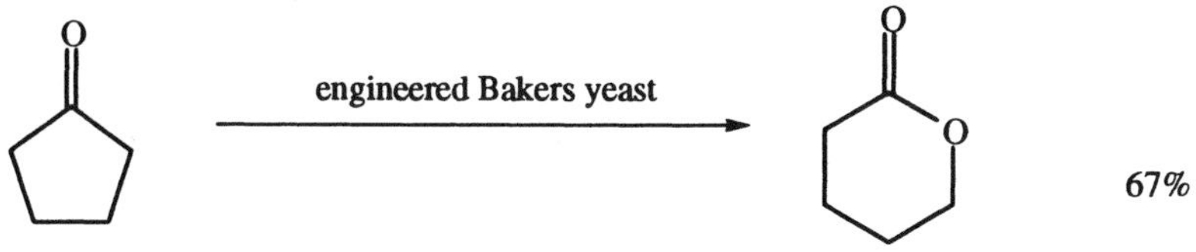
engineered Bakers yeast
67%

Stewart, J.D.; Reed, K.W.; Martinez, C.A.; Zhu, J.; Chen, G.; Kayser, M.M.
J. Am. Chem. Soc., ***1998***, *120*, 3541.

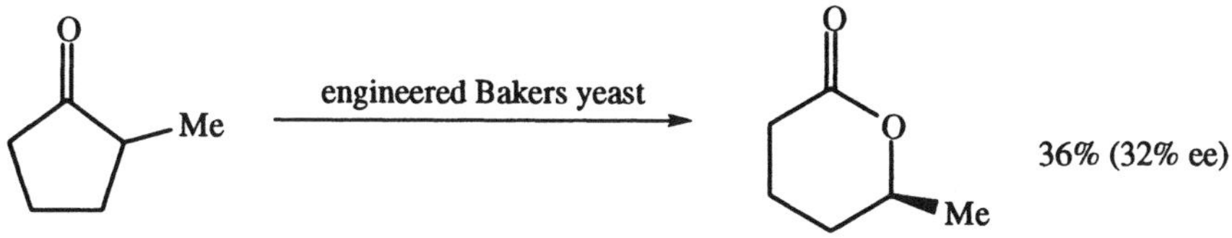
Me
engineered Bakers yeast
Me
36% (32% ee)

Yayser, M.M.; Chen, G.; Stewart, J.D. *J. Org. Chem.*, ***1998***, *63*, 7103.

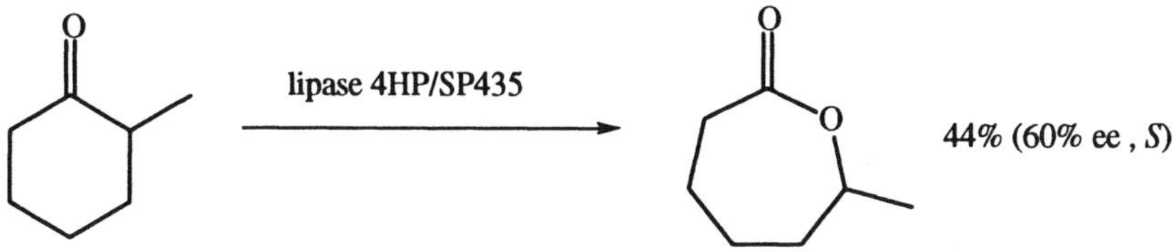
lipase 4HP/SP435
44% (60% ee , *S*)

Pchelka, B.K.; Gelo–Pujic, M.; Guibé-Jampel, E. *J. Chem. Soc., Perkin Trans. 1*, ***1998***, 2625

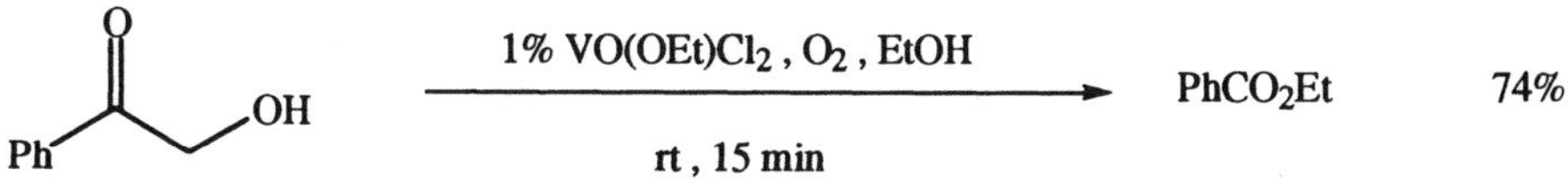
OH
Ph
1% VO(OEt)Cl2 , O2 , EtOH
rt , 15 min
PhCO2Et
74%

Dang, H.–S.; Roberts, B.P. *J. Chem. Soc., Perkin Trans. 1*, ***1998***, 67.

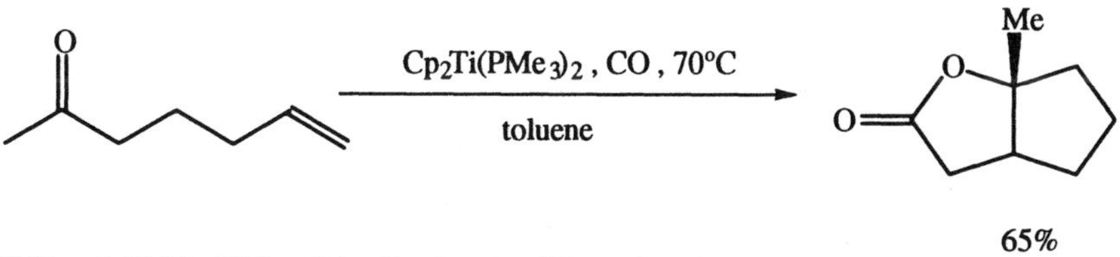
Cp2Ti(PMe3)2 , CO , 70°C
toluene
Me
65%

Kablaoui, N.M.; Hicks, F.A.; Buchwald, S.L. *J. Am. Chem. Soc.*, ***1997***, *119*, 4424.
Kablaoui, N.M.; Hicks, F.A.; Buchwald, S.L. *J. Am. Chem. Soc.*, ***1996***, *118*, 5818.

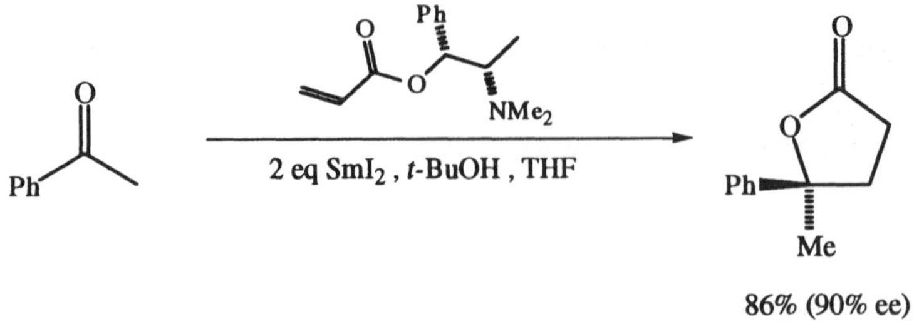
Ph
O
NMe2
2 eq SmI2 , t-BuOH , THF
Ph
Me
86% (90% ee)

Fukuzawa, S.; Seki, K.; Tatsuzawa, M.; Mutoh, K. *J. Am. Chem. Soc.*, ***1997***, *119*, 1482.

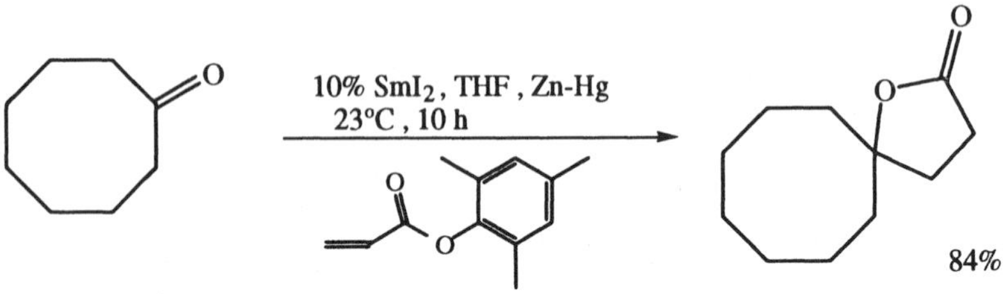
10% SmI2 , THF , Zn-Hg
23°C , 10 h
O
84%

Corey, E.J.; Zheng, G.Z. *Tetrahedron Lett.*, ***1997***, *38*, 2045.

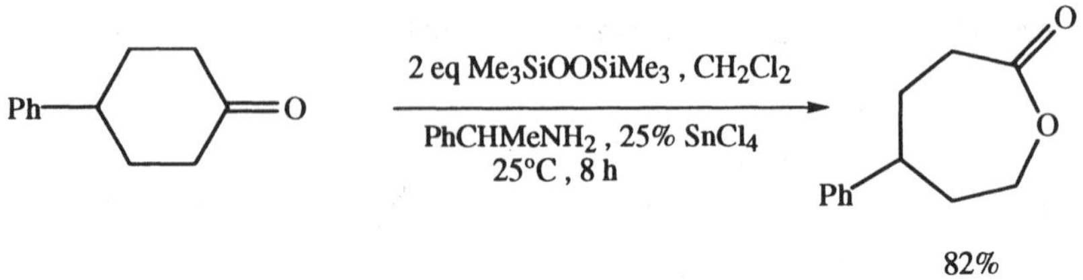
Ph
O
2 eq Me3SiOOSiMe3 , CH2Cl2
PhCHMeNH2 , 25% SnCl4
25°C , 8 h
Ph
82%

Göttlich, R.; Yamakoshi, K.; Sasai, H.; Shibasaki, M. *SynLett*, ***1997***, 971.

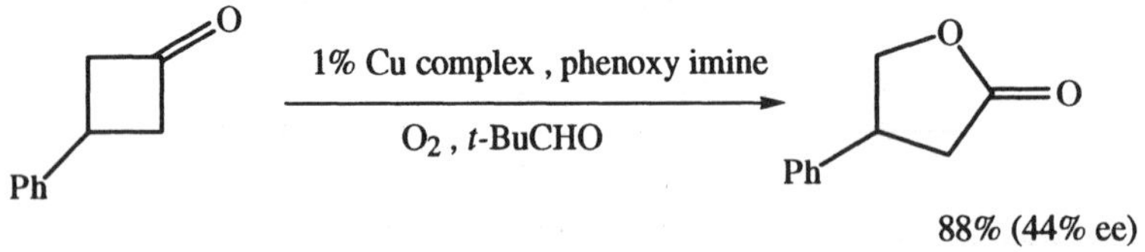
O
Ph
1% Cu complex , phenoxy imine
O2 , t-BuCHO
Ph
88% (44% ee)

Bolm, C.; Luong, T.K.K.; Schlingloff, G. *SynLett*, ***1997***, 1151.

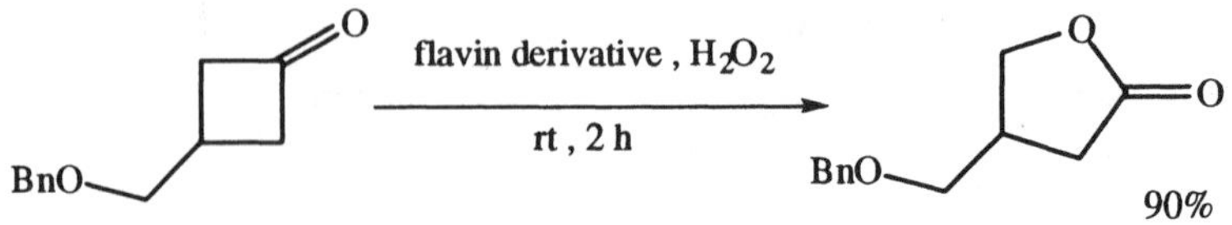
O
BnO
flavin derivative , H2O2
rt , 2 h
BnO
90%

Mazzini, C.; Lebreton, J.; Furstoss, R. *Heterocycles*, ***1997***, *45*, 1161.

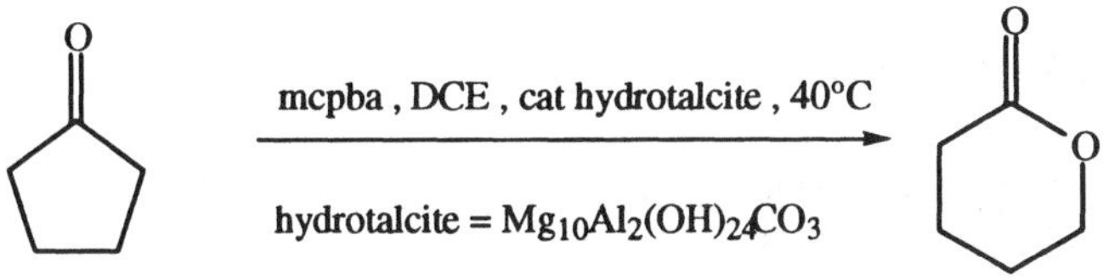
mcpba , DCE , cat hydrotalcite , 40°C
hydrotalcite = Mg10Al2(OH)24CO3

heterogenous catalyst for
Baeyer-Villiger oxidation

76% (30% without catalyst)

Kaneda, K.; Yamashita, T. *Tetrahedron Lett*, ***1996***, *37*. 4555.

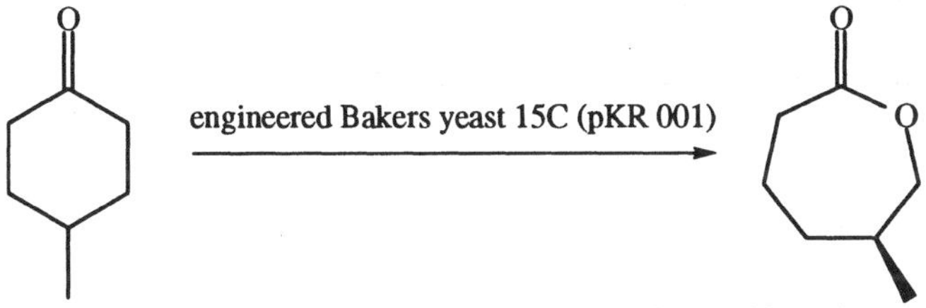
engineered Bakers yeast 15C (pKR 001)

83% (>98% ee)

Stewart, J.D.; Reed, K.W.; Kayser, M.M. *J. Chem. Soc., Perkin Trans. 1*, ***1996***, 755.

Also via: Section 27 (Acid Derivatives).

SECTION 118: ESTERS FROM NITRILES

NO ADDITIONAL EXAMPLES

SECTION 119: ESTERS FROM ALKENES

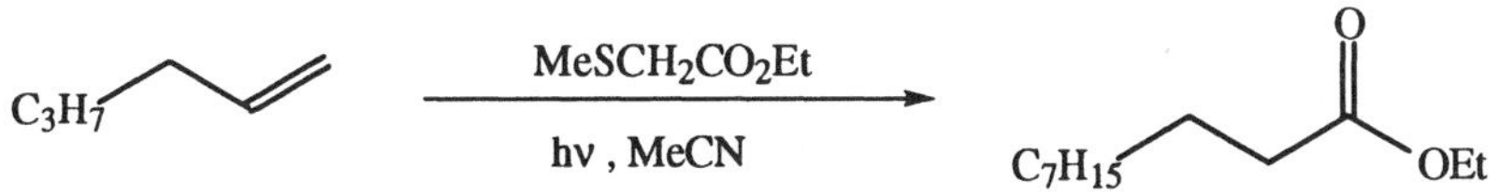
C3H7
MeSCH2CO2Et
hv , MeCN
C7H15
OEt

70%

Deng, L.X.; Kutateladze, A.G. *Tetrahedron Lett*, ***1997***, *38*, 7829.

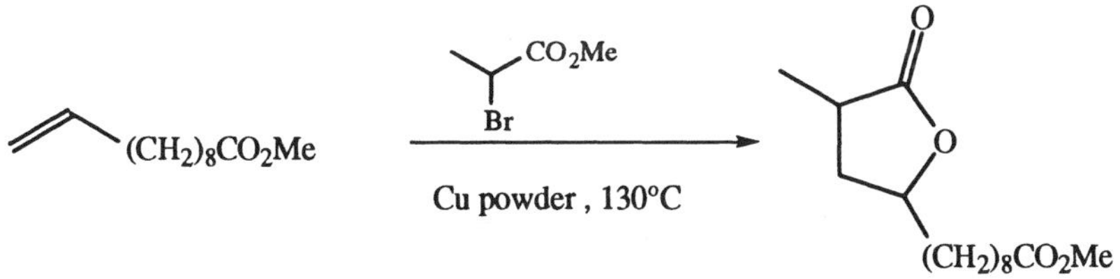
(CH2)8CO2Me
CO2Me
Br
Cu powder , 130°C
(CH2)8CO2Me

78%

Metzger, J.O.; Mahler, R.; Francke, G. *Liebigs Ann. Chem.*, ***1997***, 2303.

Also via: Section 44 (Alcohols).

SECTION 120: ESTERS FROM MISCELLANEOUS COMPOUNDS

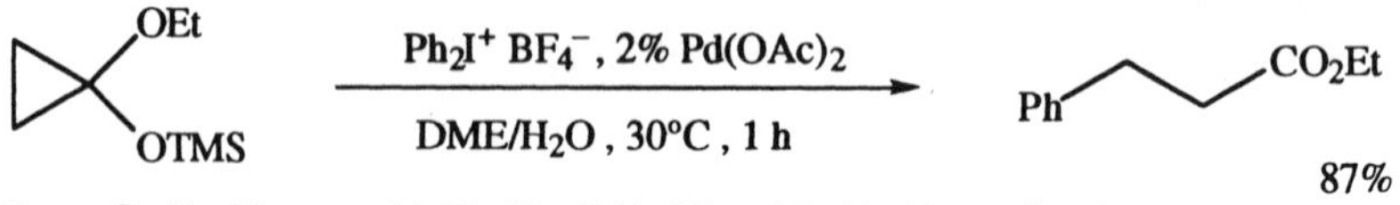
OEt
OTMS
Ph2I+ BF4−, 2% Pd(OAc)2
DME/H2O, 30°C, 1 h
Ph
CO2Et
87%

Kang, S.-K.; Yamaguchi, T.; Ho, P.S.; Kim, W.-Y.; Yoon, S.-K. *Tetrahedron Lett.*, *1997*, *38*, 1947.

CHAPTER 9

PREPARATION OF ETHERS, EPOXIDES AND THIOETHERS

SECTION 121: ETHERS, EPOXIDES AND THIOETHERS FROM ALKYNES

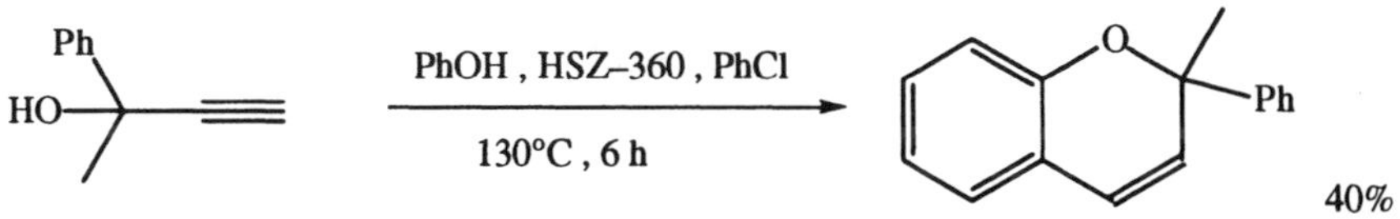
Ph
HO
PhOH , HSZ–360 , PhCl
130°C , 6 h
O
Ph
40%

HSZ = a zeolite

Bigi, F.; Carloni, S.; Maggi, R.; Muchetti, C.; Sartori, G. *J. Org. Chem.*, ***1997***, *62*, 7024.

SECTION 122: ETHERS, EPOXIDES AND THIOETHERS FROM ACID DERIVATIVES

NO ADDITIONAL EXAMPLES

SECTION 123: ETHERS, EPOXIDES AND THIOETHERS FROM ALCOHOLS AND THIOLS

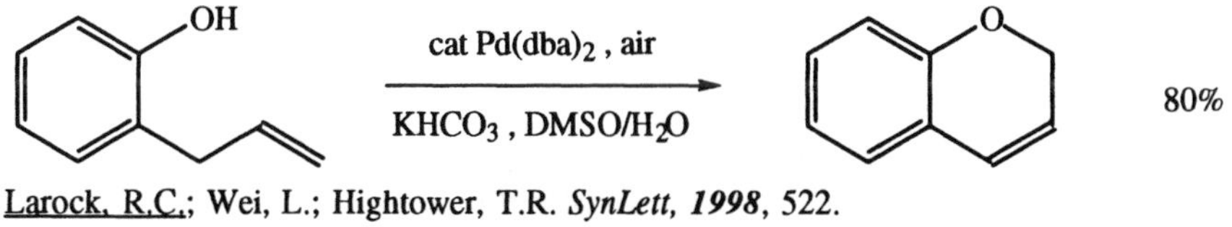
OH
cat Pd(dba)2 , air
KHCO3 , DMSO/H2O
O
80%

Larock, R.C.; Wei, L.; Hightower, T.R. *SynLett*, ***1998***, 522.

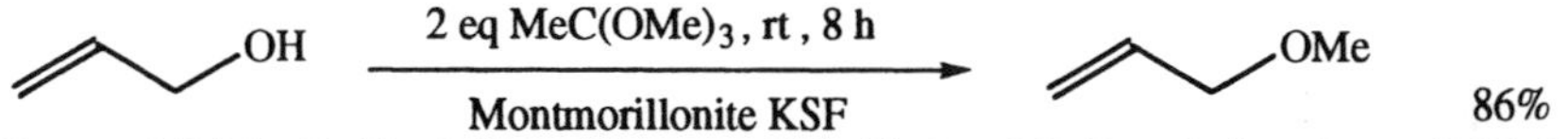
OH
2 eq MeC(OMe)3 , rt , 8 h
Montmorillonite KSF
OMe
86%

Kumar, H.M.S.; Reddy, B.V.S.; Mohanty, P.K.; Yadav, J.S. *Tetrahedron Lett.*, ***1997***, *38*, 3619

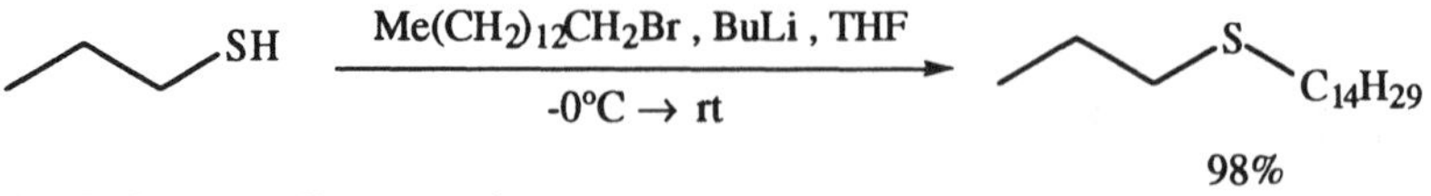
SH
Me(CH2)12CH2Br , BuLi , THF
-0°C → rt
S
C14H29
98%

Yin, J.; Pidgeon, C. *Tetrahedron Lett*, ***1997***, *38*, 5953.

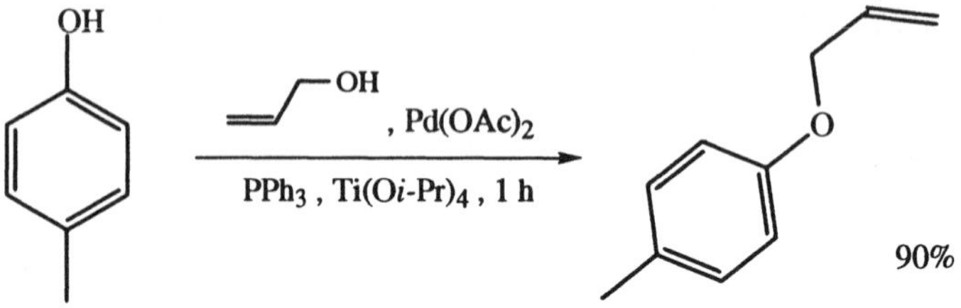
OH
OH
, Pd(OAc)2
PPh3 , Ti(Oi-Pr)4 , 1 h
O
90%

Satoh, T.; Ikeda, M.; Miura, M.; Nomura, M. *J. Org. Chem.*, ***1997***, *62*, 4877.

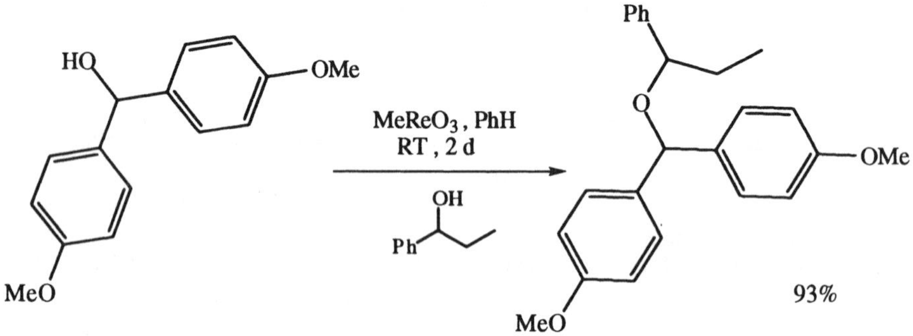
HO
OMe
MeO
MeReO3 , PhH
RT , 2 d
OH
Ph
Ph
O
OMe
MeO
93%

Zhu, Z.; Espenson, J.H. *J. Org. Chem.*, ***1996***, *61*, 324.

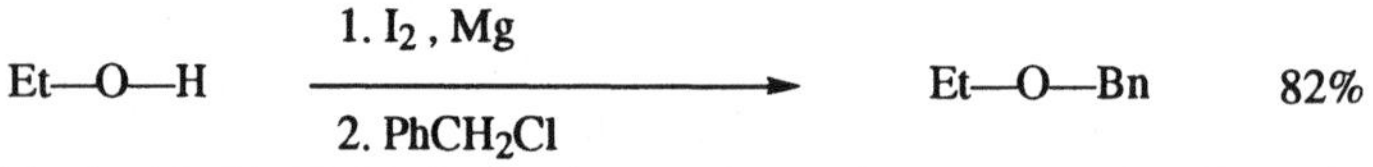
1. I2 , Mg
Et—O—H
2. PhCH2Cl
Et—O—Bn
82%

Lin, J.-M.; Li, H.-H.; Zhou, A.-M. *Tetrahedron Lett.*, ***1996***, *37*, 5159.

SECTION 124: ETHERS, EPOXIDES AND THIOETHERS FROM ALDEHYDES

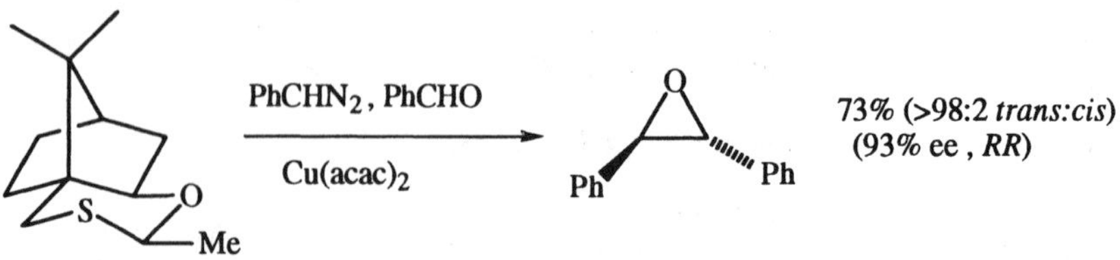
PhCHN2 , PhCHO
Cu(acac)2
O
S
Me
O
Ph
Ph
73% (>98:2 trans:cis)
(93% ee , RR)

Aggarwal, V.K. *SynLett*, ***1998***, 329.

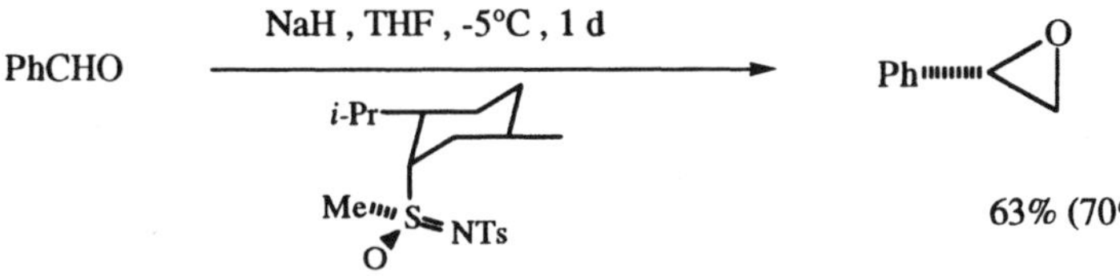
PhCHO
NaH , THF , -5°C , 1 d
i-Pr
Me
S
NTs
O
Ph
O

63% (70% ee)

Baird, C.P.; Taylor, P.C. *J. Chem. Soc., Perkin Trans. 1*, ***1998***, 3399.

Ph2S
Ph
PhCHO , THF
O
Ph
Ph

>98:2 *trans:cis*

Aggarwal, V.K.; Calamai, S.; Ford, J.G. *J. Chem. Soc., Perkin Trans. 1*, ***1997***, 593.

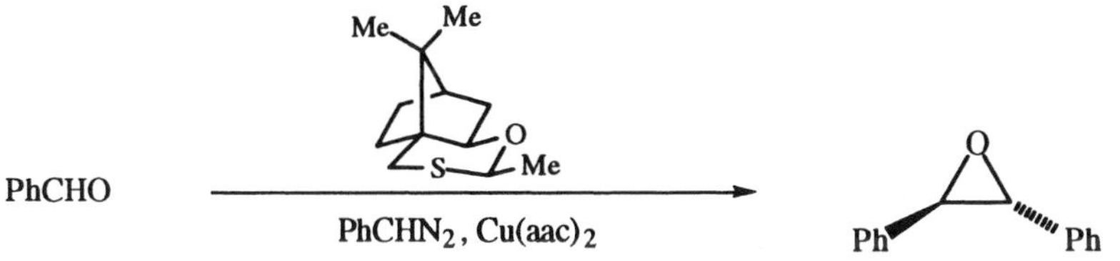
Me
Me
O
S
Me
PhCHO
PhCHN2 , Cu(aac)2
O
Ph
Ph

73% (93% ee , *RR*)
>98:2 *trans:cis*

Aggarwal, V.K.; Ford, J.G.; Thompson, A.; Jones, R.V.H.; Standen, M.C.H. *J. Am. Chem. Soc.*, ***1996***, *118*, 7004.

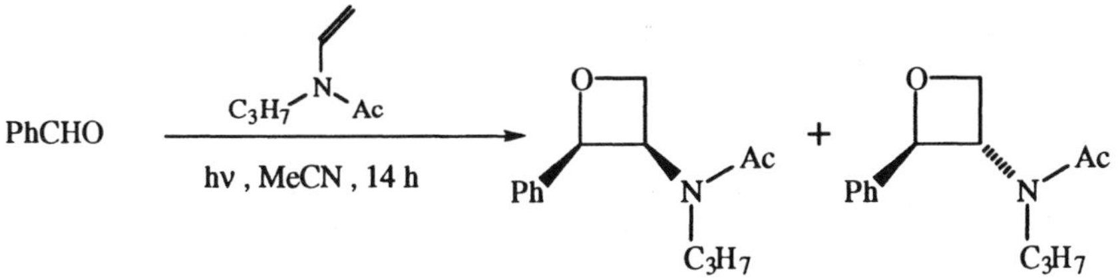
N
C3H7
Ac
PhCHO
hν , MeCN , 14 h
O
Ph
N
Ac
C3H7
+
O
Ph
N
Ac
C3H7

(>90 : 10) 70%

Bach, T. *Angew. Chem. Int. Ed.*, ***1996***, *35*, 884.

SECTION 125: ETHERS, EPOXIDES AND THIOETHERS FROM ALKYLS, METHYLENES AND ARYLS

NO ADDITIONAL EXAMPLES

SECTION 126: ETHERS, EPOXIDES AND THIOETHERS FROM AMIDES

NO ADDITIONAL EXAMPLES

SECTION 127: ETHERS, EPOXIDES AND THIOETHERS FROM AMINES

NO ADDITIONAL EXAMPLES

SECTION 128: ETHERS, EPOXIDES AND THIOETHERS FROM ESTERS

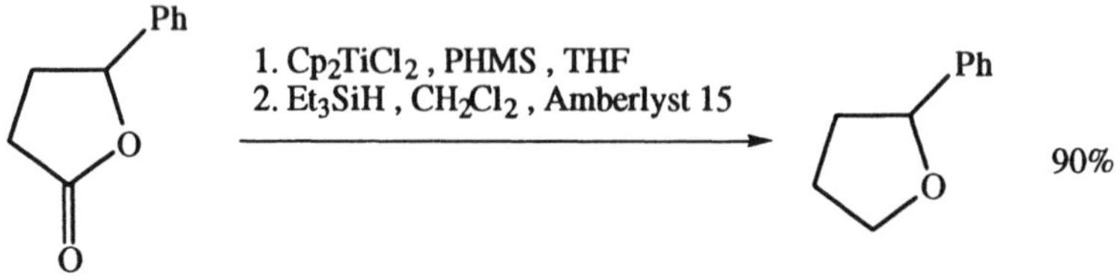

Ph
O
O
1. Cp_2TiCl_2 , PHMS , THF
2. Et_3SiH , CH_2Cl_2 , Amberlyst 15
Ph
O
90%

Hansen, M.C.; Verdaguer, X.; Buchwald, S.L. *J. Org. Chem.*, ***1998***, *63*, 2360.

SECTION 129: ETHERS, EPOXIDES AND THIOETHERS FROM ETHERS, EPOXIDES AND THIOETHERS

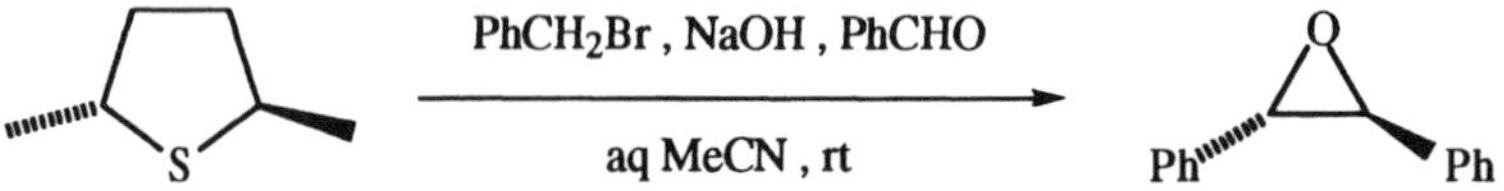

S
$PhCH_2Br$, NaOH , PhCHO
aq MeCN , rt
O
Ph
Ph

92% (88% *trans*/84% ee , *SS*)

Julienne, K.; Metzner, P. *J. Org. Chem.*, ***1998***, *63*, 4532.

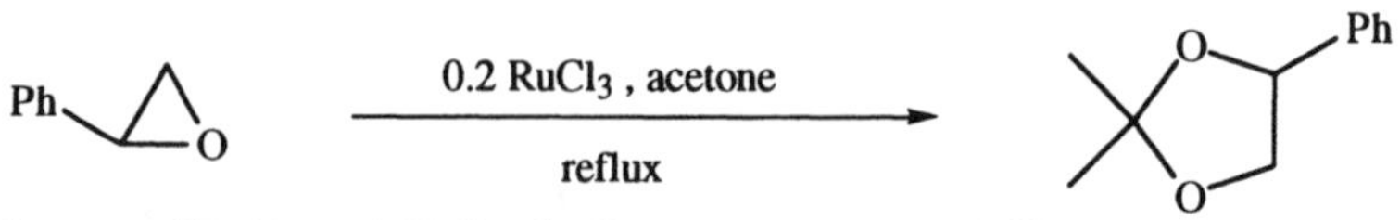

Ph
O
0.2 $RuCl_3$, acetone
reflux
O
O
Ph

90%

Iranpoor, N.; Kazemi, F. *Synth. Commun.*, ***1998***, *28*, 3189.

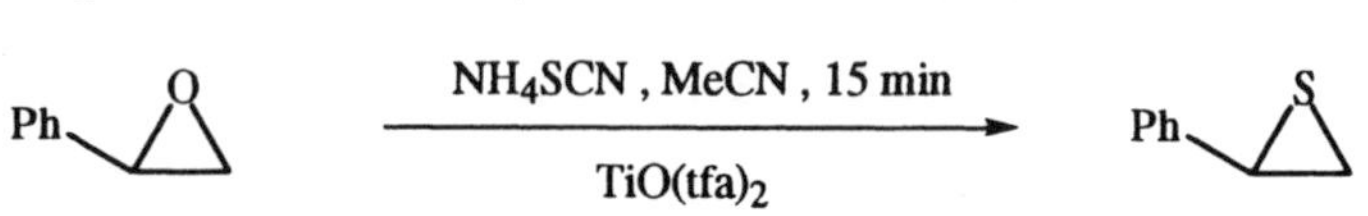

Ph
O
NH_4SCN , MeCN , 15 min
$TiO(tfa)_2$
Ph
S

97%

Iranpoor, N.; Zeynizadeh, B. *Synth. Commun.*, ***1998***, *28*, 3913.

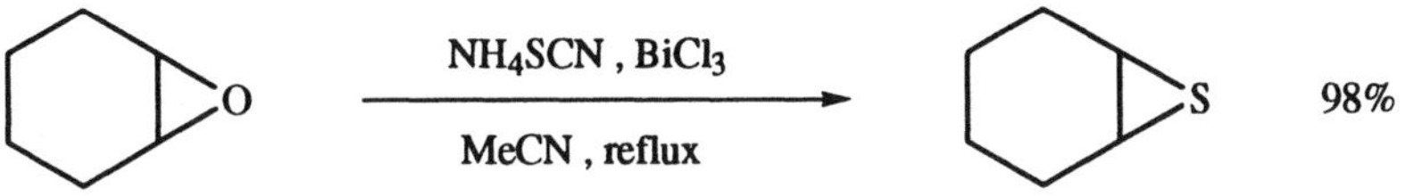
NH_4SCN, $BiCl_3$
MeCN, reflux
O
S
98%

Mohammadpoor–Baltork, I.; Aliyan, H. *Synth. Commun.*, ***1998***, *28*, 3943.

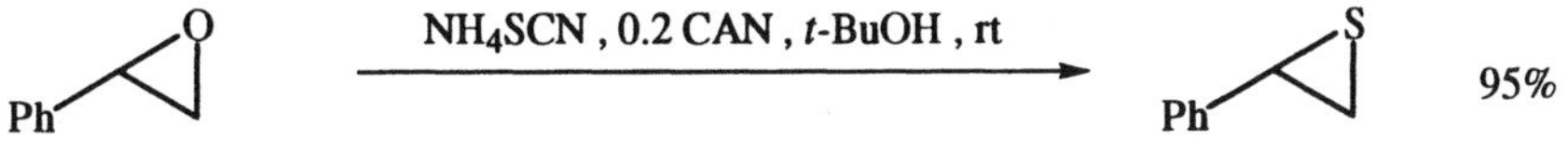
NH_4SCN, 0.2 CAN, *t*-BuOH, rt
Ph
O
S
95%

Iranpoor, N.; Kazemi, F. *Synthesis*, ***1996***, 821.

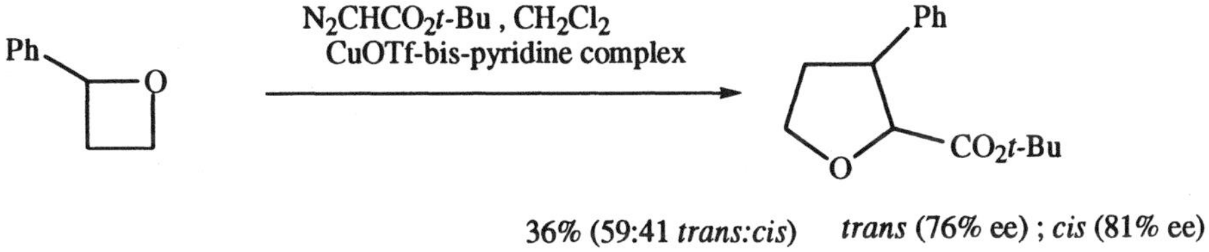
N_2CHCO_2*t*-Bu, CH_2Cl_2
CuOTf-bis-pyridine complex
Ph
O
CO_2*t*-Bu
36% (59:41 *trans:cis*)
trans (76% ee) ; *cis* (81% ee)

Ito, K.; Yoshitake, M.; Katsuki, T. *Heterocycles*, ***1996***, *42*, 305.

SECTION 130: ETHERS, EPOXIDES AND THIOETHERS FROM HALIDES AND SULFONATES

Se
1. BER, MeOH, rt
2. 2 eq BnBr
Bn
Se
Bn
96%

Yanada, K.; Fujita, T.; Yanada, R. *SynLett*, ***1998***, 971.

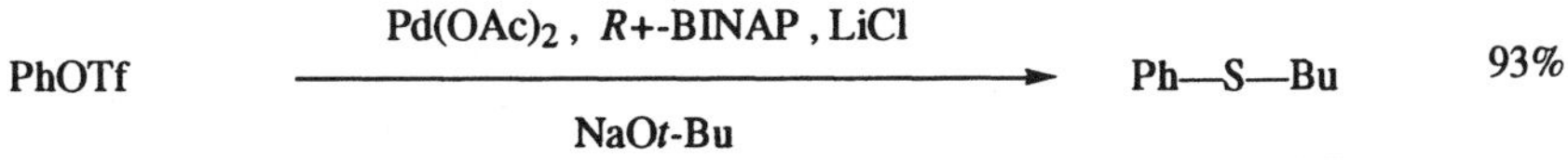
PhOTf
$Pd(OAc)_2$, *R*+-BINAP, LiCl
NaO*t*-Bu
Ph—S—Bu
93%

Zheng, N.; McWilliams, J.C.; Fleitz, F.J.; Armstrong III, J.D.; Volante, R.P. *J. Org. Chem.*, ***1998***, *63*, 9606.

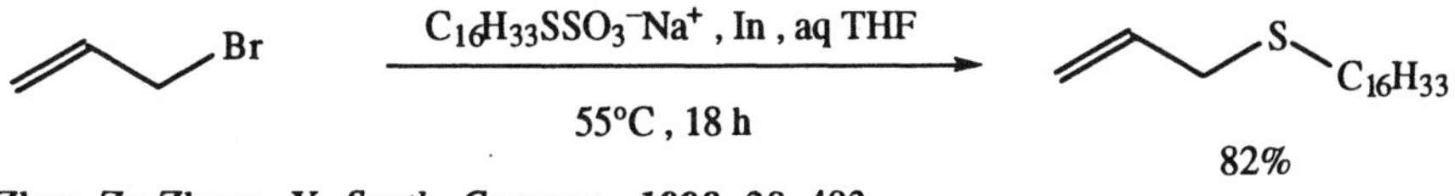
Br
$C_{16}H_{33}SSO_3^-Na^+$, In, aq THF
55°C, 18 h
S
$C_{16}H_{33}$
82%

Zhan, Z.; Zhang, Y. *Synth. Commun.*, ***1998***, *28*, 493.

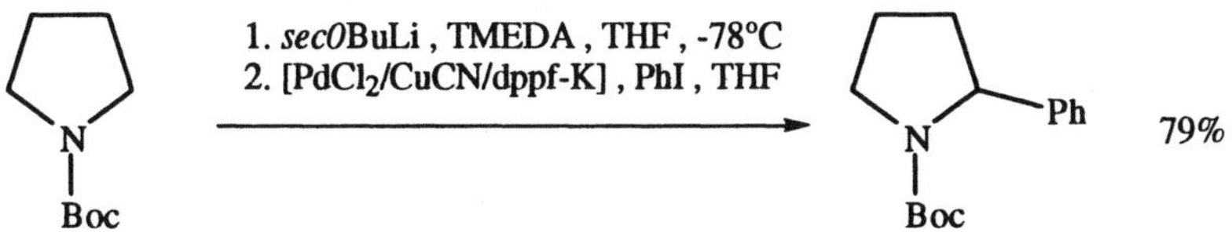
1. *sec0*BuLi, TMEDA, THF, -78°C
2. [$PdCl_2$/CuCN/dppf-K], PhI, THF
N
Boc
Ph
79%

Larhed, M.; Hallberg, A. *J. Org. Chem.*, ***1997***, *62*, 7858.

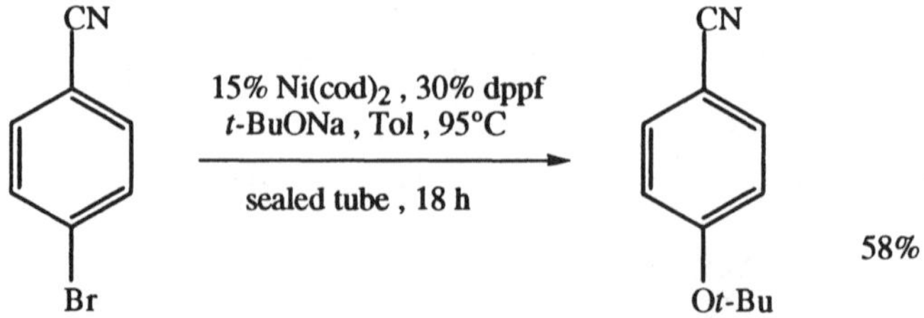
CN
15% Ni(cod)2 , 30% dppf
t-BuONa , Tol , 95°C
sealed tube , 18 h
CN
58%
Br
Ot-Bu

Mann, G.; Hartwig, J.F. *J. Org. Chem.*, ***1997***, *62*, 5413.

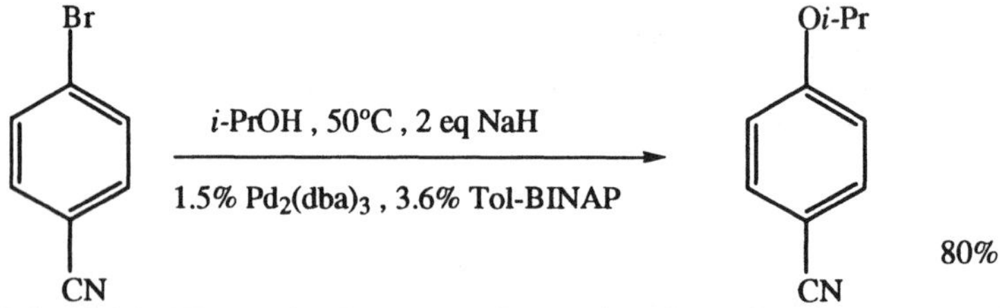
Br
Oi-Pr
i-PrOH , 50°C , 2 eq NaH
1.5% Pd2(dba)3 , 3.6% Tol-BINAP
80%
CN
CN

Palucki, M.; Wolfe, J.P.; Buchwald, S.L. *J. Am. Chem. Soc.*, ***1997***, *119*, 3395.

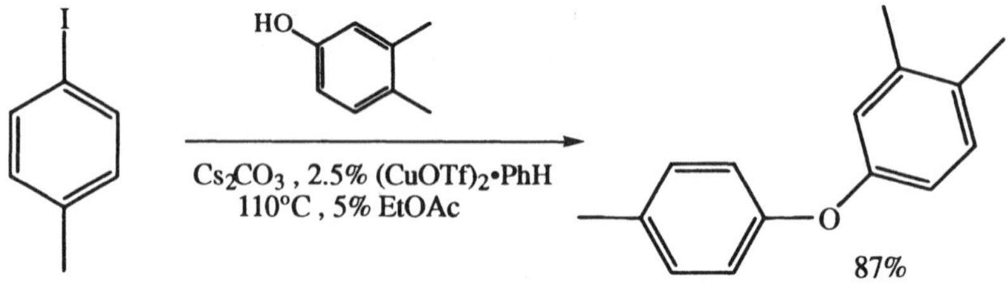
I
HO
Cs2CO3 , 2.5% (CuOTf)2•PhH
110°C , 5% EtOAc
O
87%

Marcoux, J.–F.; Doye, S.; Buchwald, S.L. *J. Am. Chem. Soc.*, ***1997***, *119*, 10539.

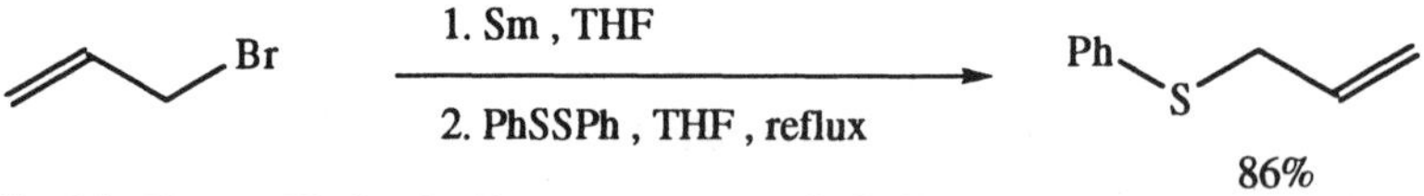
Br
1. Sm , THF
2. PhSSPh , THF , reflux
Ph
S
86%

Yu, M.; Zhang, Y. *Synth. Commun.*, ***1997***, *27*, 2743.

Related Methods: Section 123 (Ethers from Alcohols).

SECTION 131: ETHERS, EPOXIDES AND THIOETHERS FROM HYDRIDES

NO ADDITIONAL EXAMPLES

SECTION 132: ETHERS, EPOXIDES AND THIOETHERS FROM KETONES

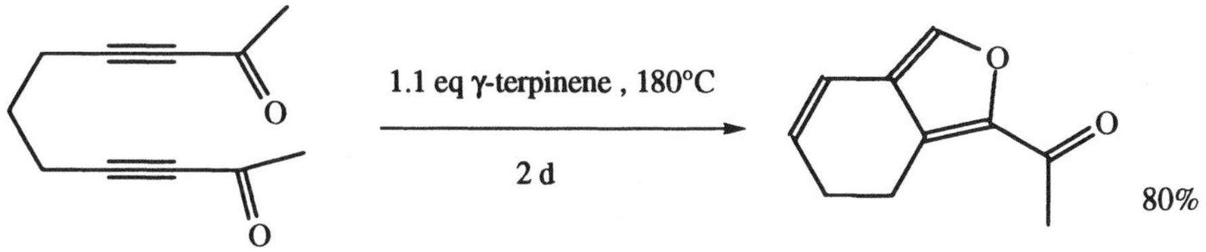

Wills, M.S.B.; Danheiser, R.L. *J. Am. Chem. Soc.*, ***1998***, *120*, 9378.

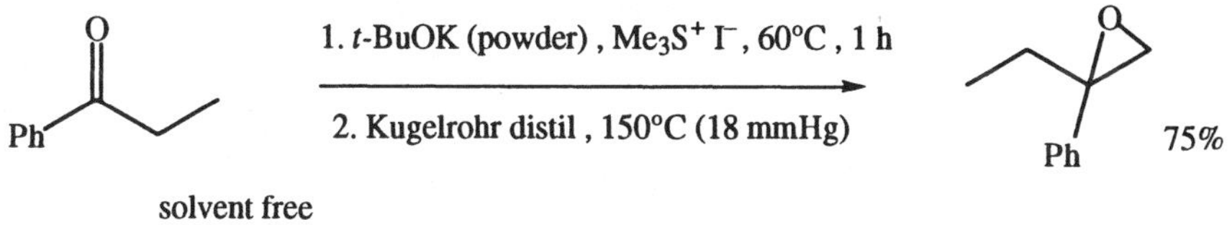

Toda, F.; Kanemoto, K. *Heterocycles*, ***1997***, *46*, 185.

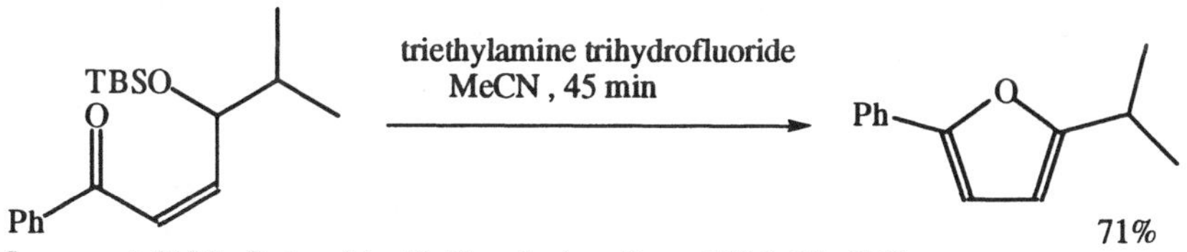

Sammond, D.M.; Sammakia, T. *Tetrahedron Lett.*, ***1996***, *37*, 6065.

Related Methods: Section 124 (Epoxides from Aldehydes).

SECTION 133: ETHERS, EPOXIDES AND THIOETHERS FROM NITRILES

NO ADDITIONAL EXAMPLES

SECTION 134: ETHERS, EPOXIDES AND THIOETHERS FROM ALKENES

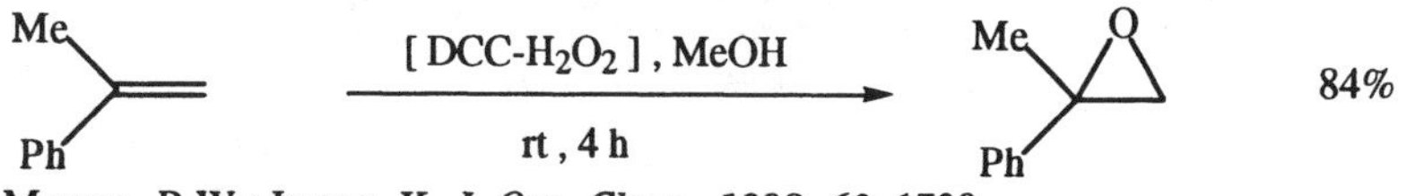

Murray, R.W.; Iyanar, K. *J. Org. Chem.*, ***1998***, *63*, 1730.

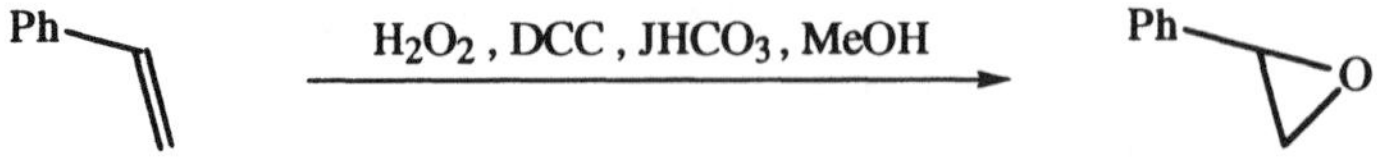

Ph
H_2O_2 , DCC , $JHCO_3$, MeOH
Ph
O

75%

Majetich, G.; Hicks, R.; Sun, G.–R.; McGill, P. *J. Org. Chem.*, ***1998***, *63*, 2564.

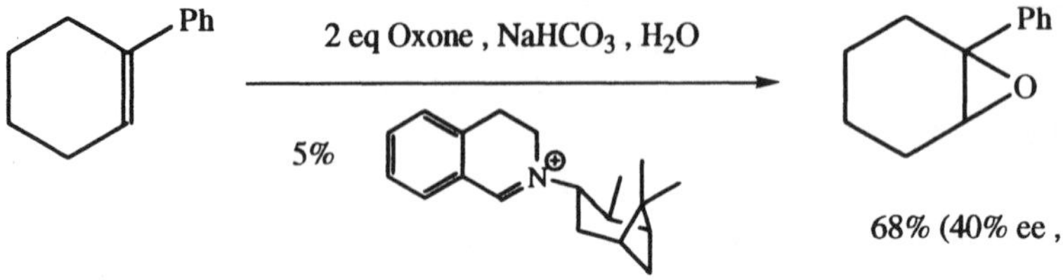

Ph
2 eq Oxone , $NaHCO_3$, H_2O
5%
N⊕
Ph
O

68% (40% ee , *RR*)

Page, P.C.B.; Rassias, G.A.; Bethell, D.; Schilling, M.B. *J. Org. Chem.*, ***1998***, *63*, 2774.

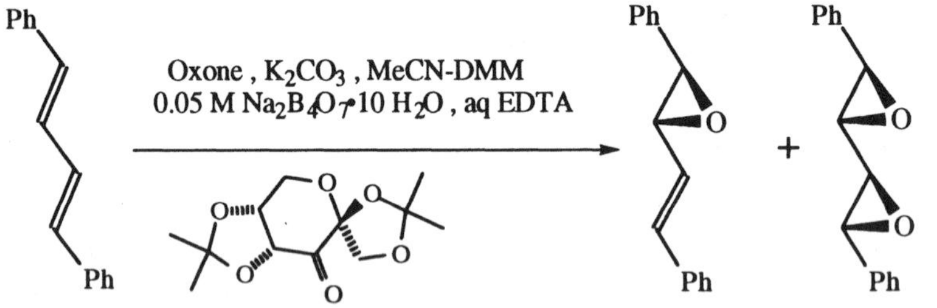

Ph
Ph
Oxone , K_2CO_3 , MeCN-DMM
0.05 M $Na_2B_4O_7$•10 H_2O , aq EDTA
O
O
O
O
O
O
Ph
O
Ph
Ph
O
O
Ph
+

(22 : 1) 77%
97% ee

94% conversion

Frohn, M.; Dalkiewicz, M.; Tu, Y.; Wang, Z.–X.; Shi, Y. *J. Org. Chem.*, ***1998***, *63*, 2948.

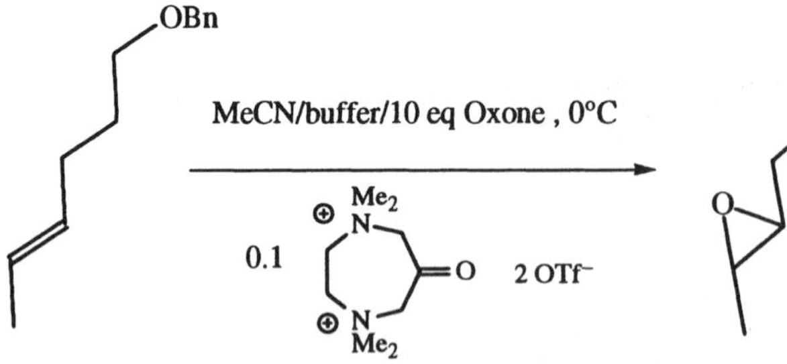

OBn
MeCN/buffer/10 eq Oxone , 0°C
0.1
⊕ Me2 N
O
⊕ N Me2
2 OTf⁻
OBn
O

quant

Denmark, S.E.; Wu, Z. *J. Org. Chem.*, ***1998***, *63*, 2810.

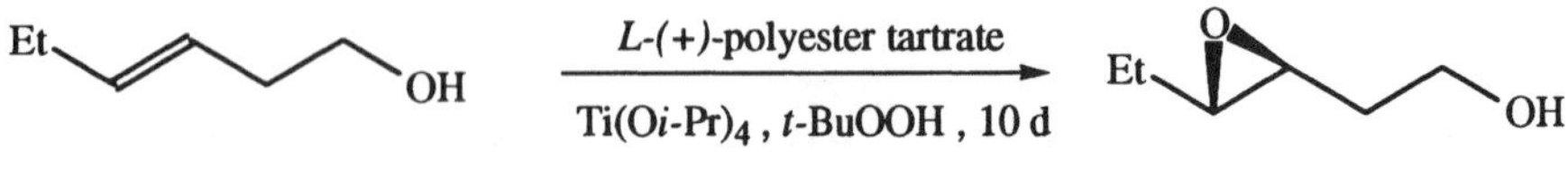

Et
OH
L-(+)-polyester tartrate
Ti(O*i*-Pr)4 , *t*-BuOOH , 10 d
Et
O
OH

58% (29% ee)

Karjalainen, J.K.; Hormi, O.E.D.; Sherrington, D.C. *Tetrahedron Asymmetry*, ***1998***, *9*, 3895

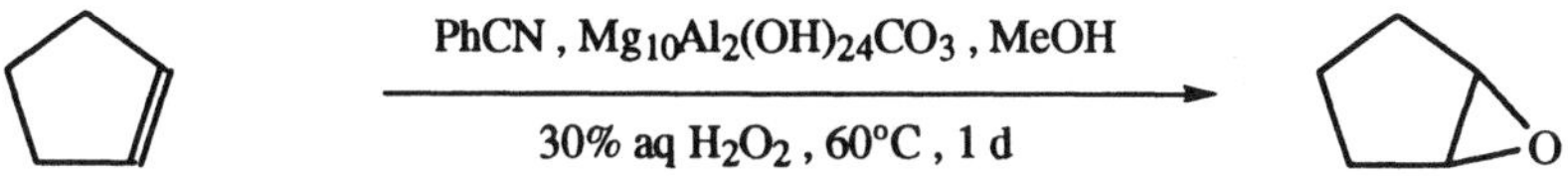
PhCN , $Mg_{10}Al_2(OH)_{24}CO_3$, MeOH
30% aq H_2O_2 , 60°C , 1 d
O

Ueno, S.; Yamaguchi, K.; Yoshida, K.; Ebitani, K.; Kaneda, K. *Chem. Commun.*, ***1998***, 295

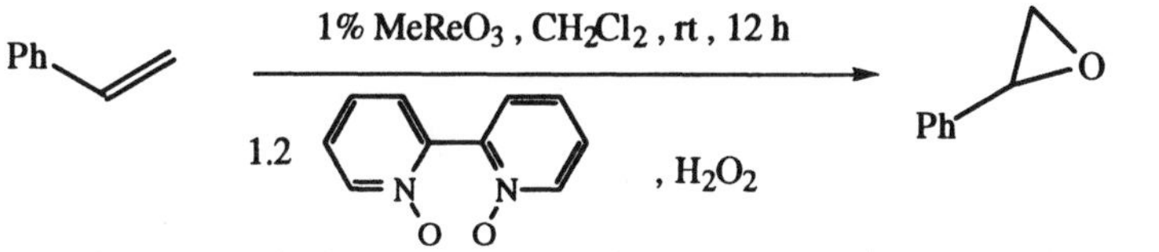
Ph
1% $MeReO_3$, CH_2Cl_2 , rt , 12 h
1.2
N
N
O O
, H_2O_2
O
Ph

95%

Nakajima, M.; Sasaki, Y.; Iwamoto, H.; Hashimoto, S. *Tetrahedron Lett.*, ***1998***, *39*, 87.

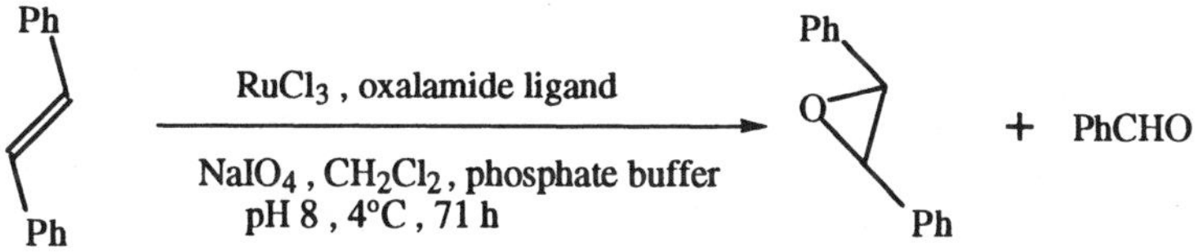
Ph
Ph
$RuCl_3$, oxalamide ligand
$NaIO_4$, CH_2Cl_2 , phosphate buffer
pH 8 , 4°C , 71 h
Ph
O
Ph
+ PhCHO

42% 21%

End, N.; Pfaltz, A. *Chem. Commun.*, ***1998***, 589.

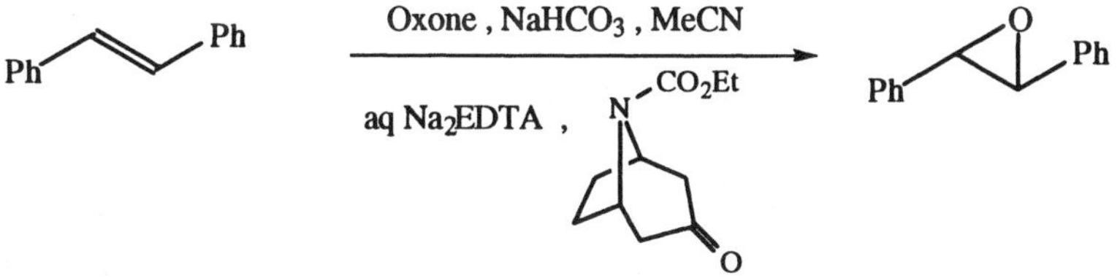
Ph
Ph
Oxone , $NaHCO_3$, MeCN
aq Na_2EDTA ,
N
CO_2Et
O
O
Ph
Ph

quant

Armstrong, A.; Hayter, B.R. *Chem. Commun.*, ***1998***, 621.

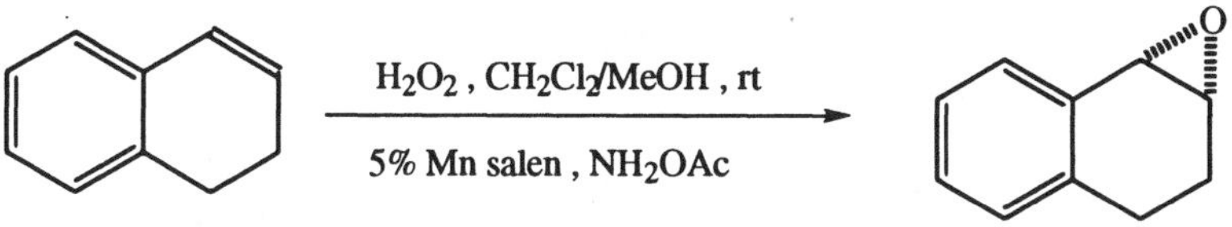
H_2O_2 , CH_2Cl_2/MeOH , rt
5% Mn salen , NH_2OAc
O

73% (61% ee)

Pietikäinen, P. *Tetrahedron*, ***1998***, *54*, 4319.

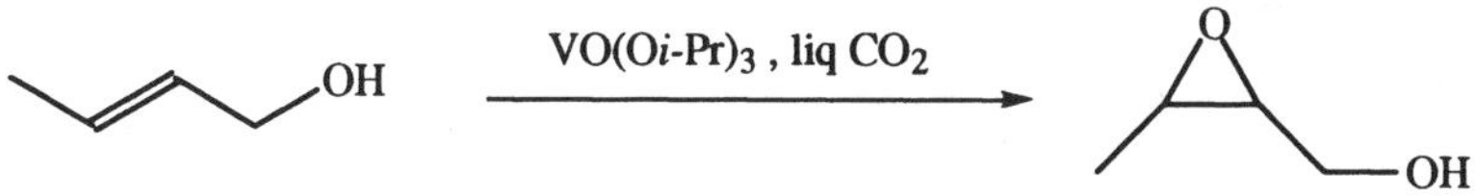
OH
VO(O*i*-Pr)$_3$, liq CO_2
O
OH

85%

Pesiri, D.R.; Morita, D.K.; Glaze, W.; Tumas, W. *Chem. Commun.*, ***1998***, 1015.

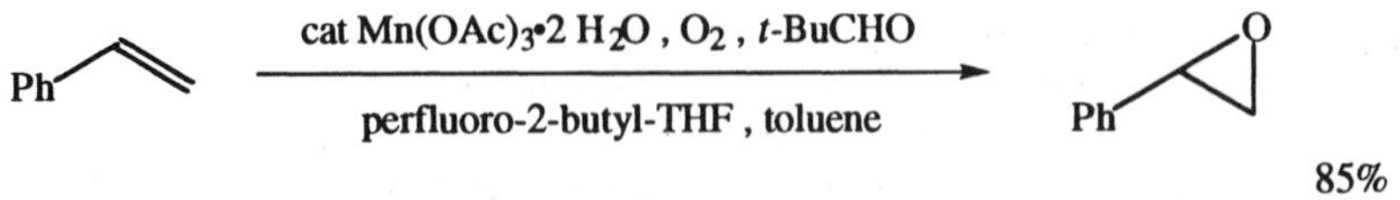
cat Mn(OAc)3•2 H2O , O2 , t-BuCHO
perfluoro-2-butyl-THF , toluene
Ph
Ph
O
85%

Ravikumar, K.S.; Barbier, F.; Bégué, J.–P.; Bonnet–Delpon, D. *Tetrahedron*, ***1998***, *54*, 7457.

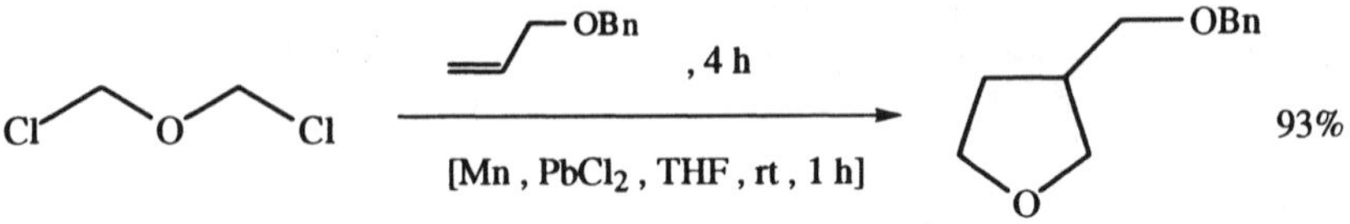
OBn
, 4 h
Cl
O
Cl
[Mn , PbCl2 , THF , rt , 1 h]
OBn
O
93%

Hojo, M.; Aihara, H.; Suginohara, Y.; Sakata, K.; Nakamura, S.; Murakami, C.; Hosomi, A. *J. Org. Chem.*, ***1997***, *62*, 8610.

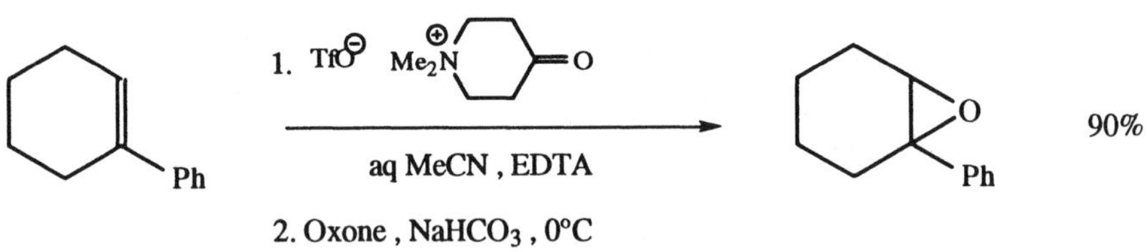
1. TfO Me2N O
aq MeCN , EDTA
2. Oxone , NaHCO3 , 0°C
Ph
O
Ph
90%

Denmark, S.E.; Wu, Z. *J. Org. Chem.*, ***1997***, *62*, 8767.

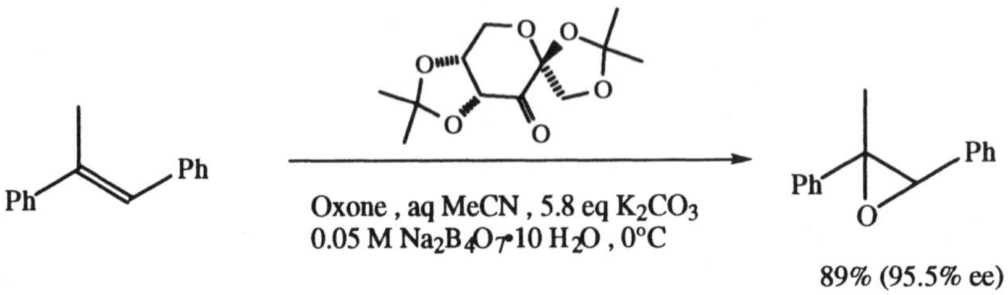
Ph
Ph
Oxone , aq MeCN , 5.8 eq K2CO3
0.05 M Na2B4O7•10 H2O , 0°C
Ph
Ph
O
89% (95.5% ee)

Wang, Z.–X.; Tu, Y.; Frohn, M.; Zhang, J.–R.; Shi, Y. *J. Am. Chem. Soc.*, ***1997***, *119*, 11224.

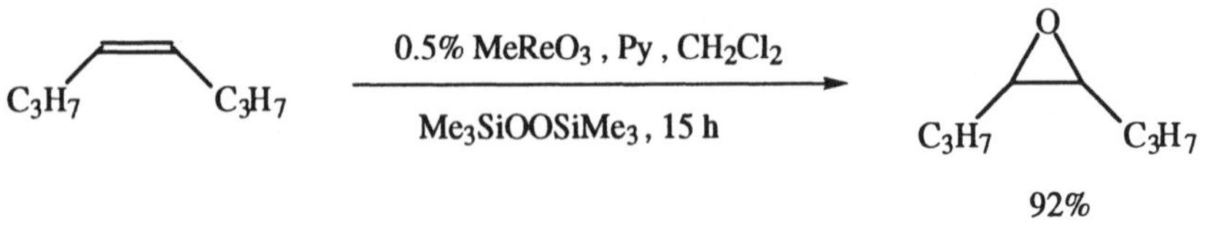
C3H7
C3H7
0.5% MeReO3 , Py , CH2Cl2
Me3SiOOSiMe3 , 15 h
O
C3H7
C3H7
92%

Yudin, A.K.; Sharpless, K.B. *J. Am. Chem. Soc.*, ***1997***, *119*, 11536.

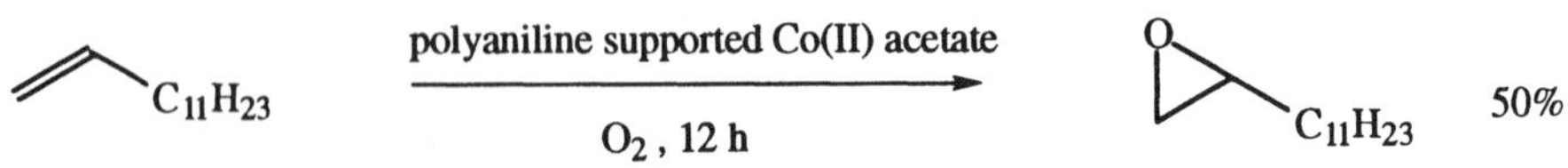
C11H23
polyaniline supported Co(II) acetate
O2 , 12 h
O
C11H23
50%

Das, B.C.; Iqbal, J. *Tetrahedron Lett.*, ***1997***, *38*, 1235.

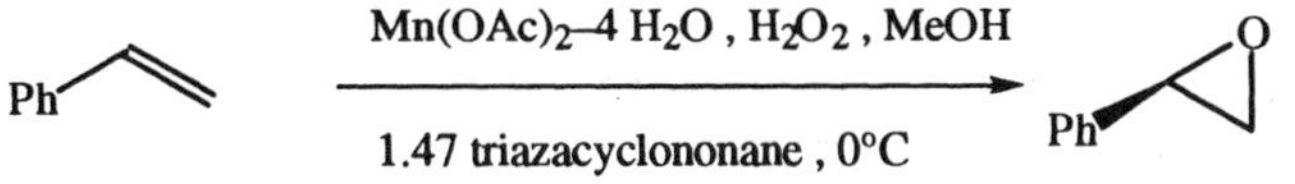
Mn(OAc)2–4 H2O , H2O2 , MeOH
1.47 triazacyclononane , 0°C
Ph
Ph
O

43% ee

Bolm, C.; Kadereit, D.; Valacchi, M. *SynLett*, ***1997***, 687.

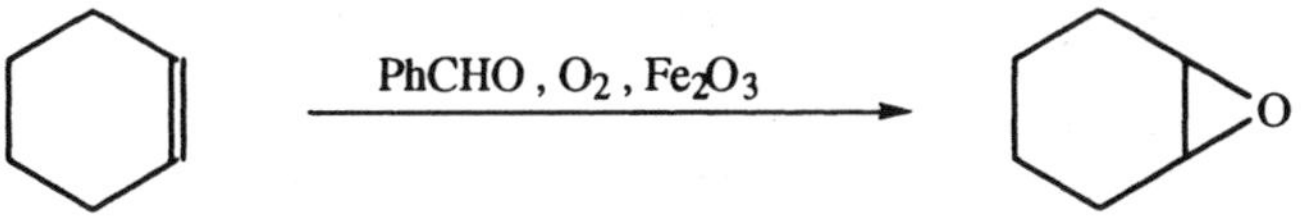
PhCHO , O2 , Fe2O3
O

97%

Li, X.; Wang, F.; Lu, X.; Song, G.; Zhang, H. *Synth. Commun.*, ***1997***, *27*, 2075.

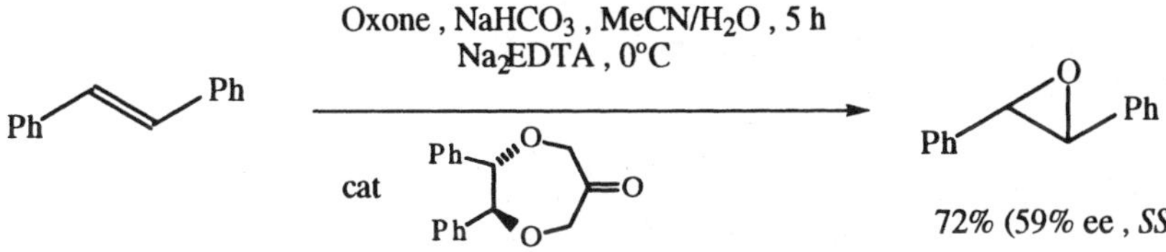
Oxone , NaHCO3 , MeCN/H2O , 5 h
Na2EDTA , 0°C
cat
Ph
Ph
Ph
Ph
O
O
O
Ph
Ph
O

72% (59% ee , *SS*)

Song, C.E.; Kim, Y.H.; Lee, K.C.; Lee, S.–g.; Jin, B.W.
Tetrahedron Asymmetry,, ***1997***, *8*, 2921.

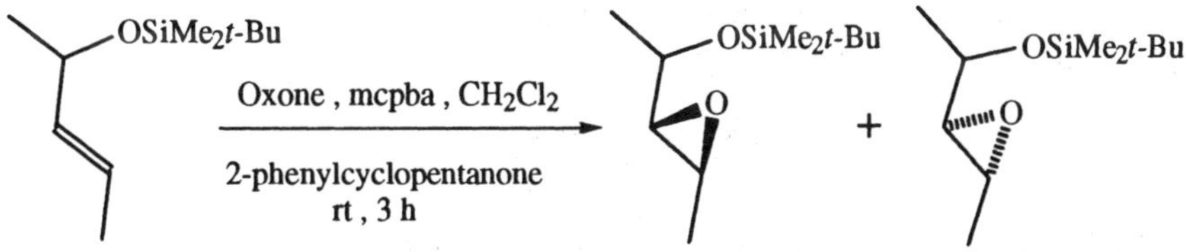
OSiMe2t-Bu
Oxone , mcpba , CH2Cl2
2-phenylcyclopentanone
rt , 3 h
OSiMe2t-Bu
O
+
OSiMe2t-Bu
O

	(22	:	78) 84%
without ketone	66	:	34) 90%

Kurihara, M.; Ishii, K.; Kasahara, Y.; Kameda, M.; Pathak, A.K.; Miyata, N. *Chem. Lett.*, ***1997***, 1015.

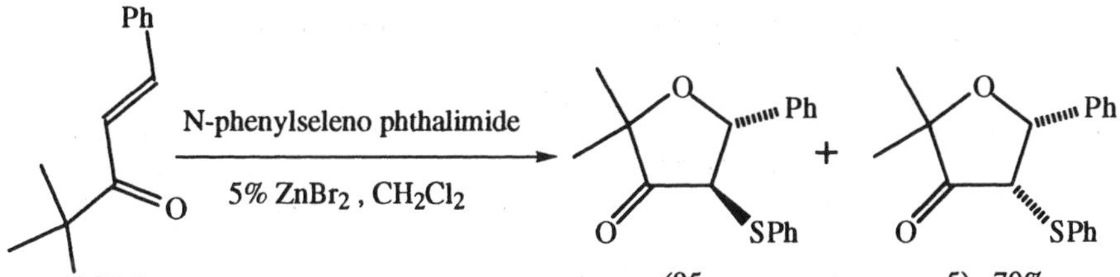
Ph
N-phenylseleno phthalimide
5% ZnBr2 , CH2Cl2
O
OSiMe3
O
Ph
O
SPh
+
O
Ph
O
SPh

(95 : 5) 70%

Colins, C.C.; Cronin, M.F.; Moynihan, H.A.; McCarthy, D.G.
J. Chem. Soc.. Perkin Trans. 1. ***1997***. 1267.

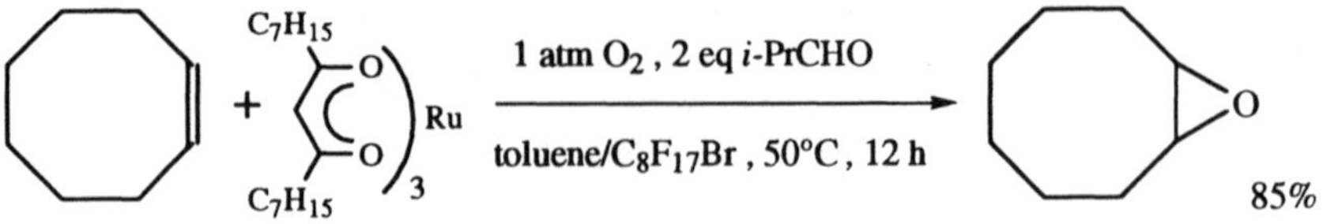
C7H15
Ru
3
1 atm O2 , 2 eq i-PrCHO
toluene/C8F17Br , 50°C , 12 h
O
85%

Klement, I.; Lütjens, H.; Knochel, P. *Angew. Chem. Int. Ed.*, ***1997***, *36*, 1454.

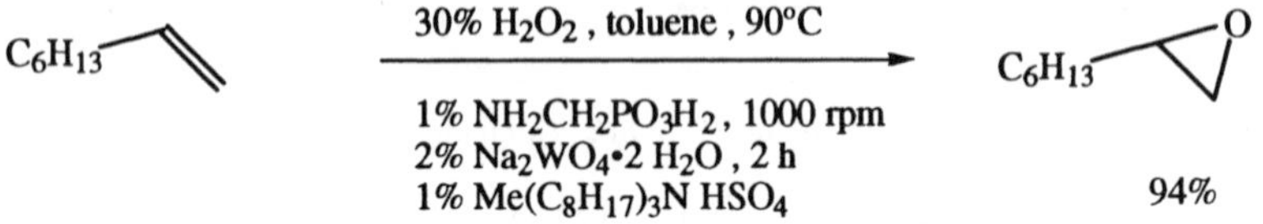
C6H13
30% H2O2 , toluene , 90°C
1% NH2CH2PO3H2 , 1000 rpm
2% Na2WO4•2 H2O , 2 h
1% Me(C8H17)3N HSO4
94%

Sato, K.; Aoki, M.; Ogawa, M.; Hashimoto, T.; Panyella, D.; Noyori, R. *Bull. Chem. Soc. Jpn.*, ***1997***, *70*, 905.

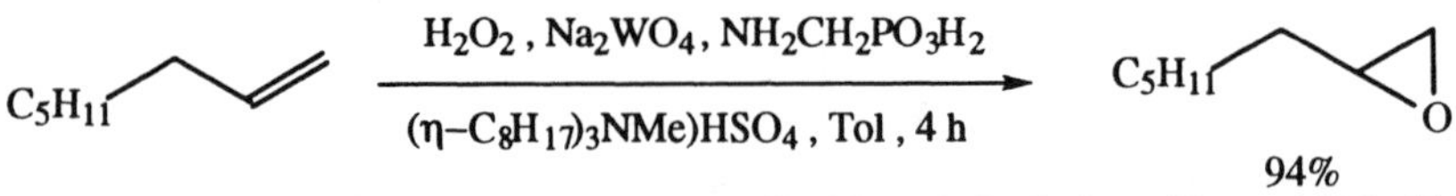
C5H11
H2O2 , Na2WO4 , NH2CH2PO3H2
(η–C8H17)3NMe)HSO4 , Tol , 4 h
94%

Sato, K.; Aoki, M.; Ogawa, M.; Hashimoto, T.; Noyori, R. *J. Org. Chem.*, ***1996***, *61*, 8310.

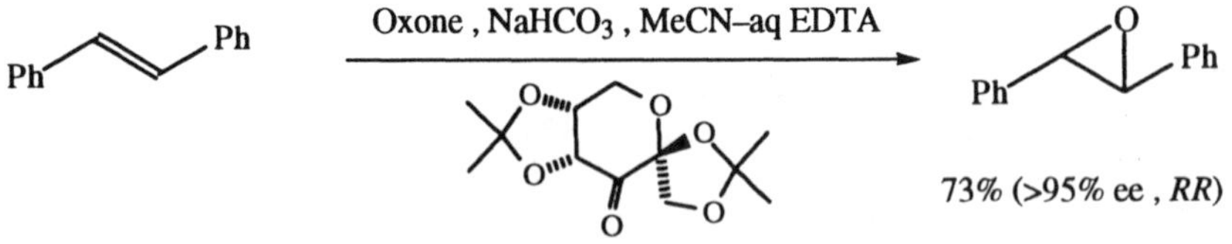
Ph
Oxone , NaHCO3 , MeCN–aq EDTA
73% (>95% ee , RR)

Tu, Y.; Wang, Z.-X.; Shi, Y. *J. Am. Chem. Soc.*, ***1996***, *118*, 9806.

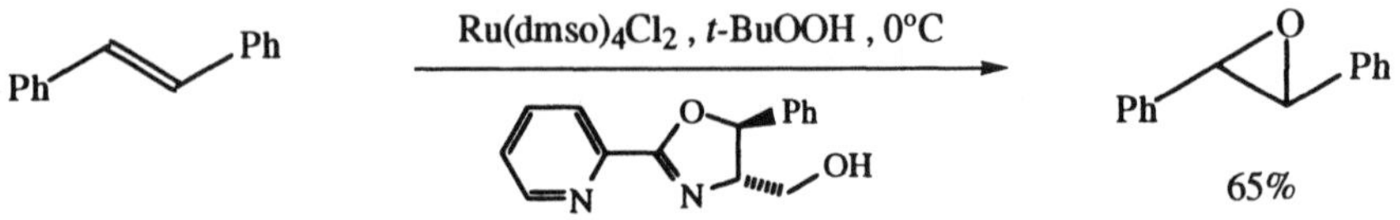
Ph
Ru(dmso)4Cl2 , t-BuOOH , 0°C
OH
65%

Barf, G.A.; van den Hoek, D.; Sheldon, R.A. *Tetrahedron*, ***1996***, *52*, 12971.

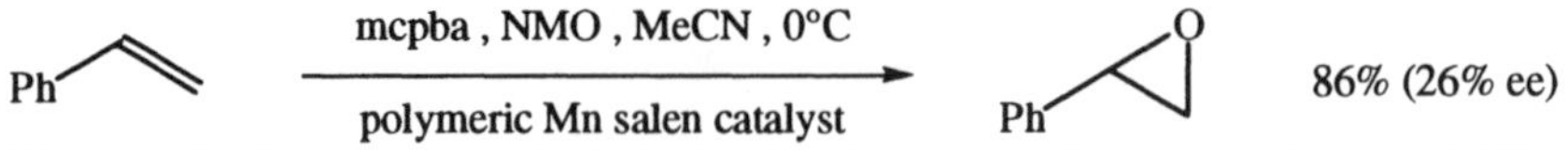
Ph
mcpba , NMO , MeCN , 0°C
polymeric Mn salen catalyst
86% (26% ee)

Minutolo, F.; Pini, D.; Petri, A.; Salvadori, P. *Tetrahedron Asymmetry*, ***1996***, *7*, 2293.

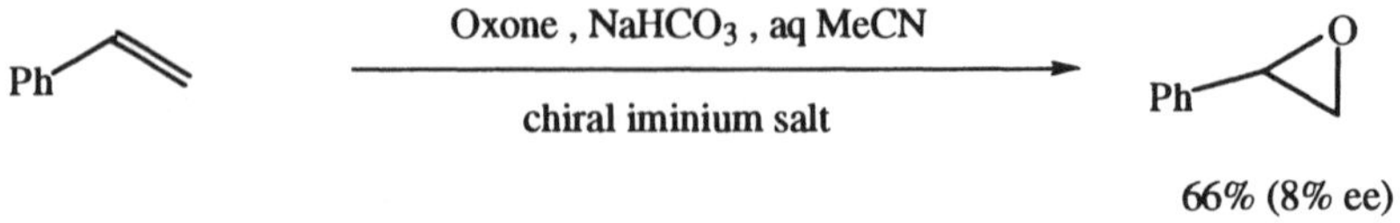
Ph
Oxone , NaHCO3 , aq MeCN
chiral iminium salt
66% (8% ee)

Aggarwal, V.K.; Wang, M.F. *Chem. Commun.*, ***1996***, 191.

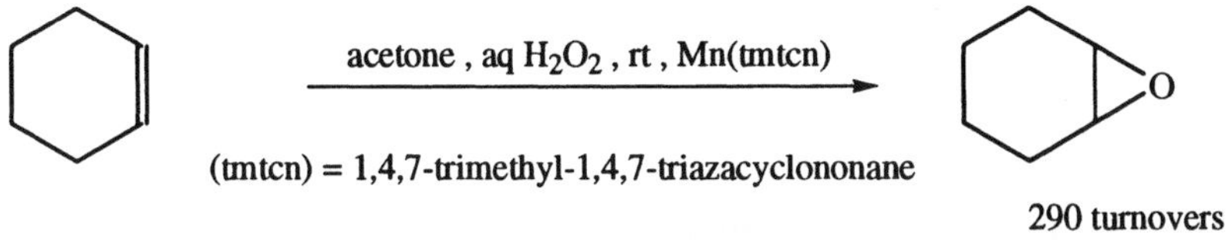
acetone , aq H2O2 , rt , Mn(tmtcn)
(tmtcn) = 1,4,7-trimethyl-1,4,7-triazacyclononane
O
290 turnovers

DeVos, D.; Bein, T. *Chem. Commun.*, ***1996***, 917.

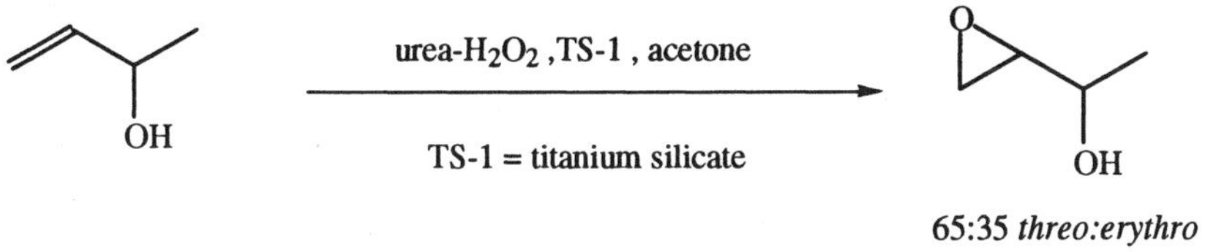
OH
urea-H2O2 ,TS-1 , acetone
TS-1 = titanium silicate
O
OH
65:35 threo:erythro

Adam, W.; Kumar, R.; Reddy, T.T.; Renz, M. *Angew. Chem. Int. Ed.*, ***1996***, *35*, 880.

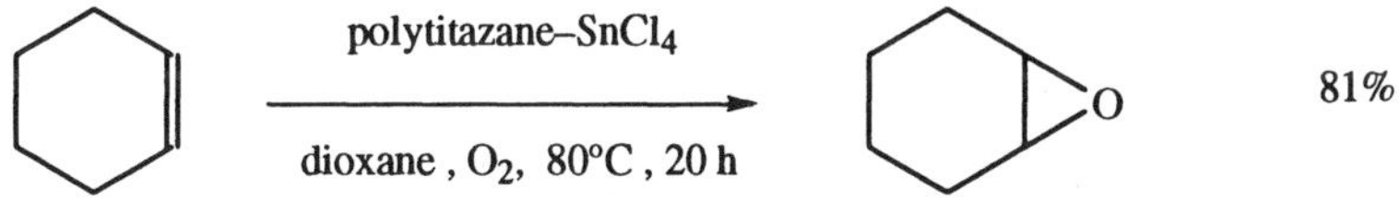
polytitazane–SnCl4
dioxane , O2, 80°C , 20 h
O
81%

Wang, T.–J.; Ma, Z.–H.; Yan, Y.–Y.; Huang, M.–Y.; Jiang, Y.–Y.
Chem. Commun., ***1996***, 1335.

SECTION 135: ETHERS, EPOXIDES AND THIOETHERS FROM MISCELLANEOUS COMPOUNDS

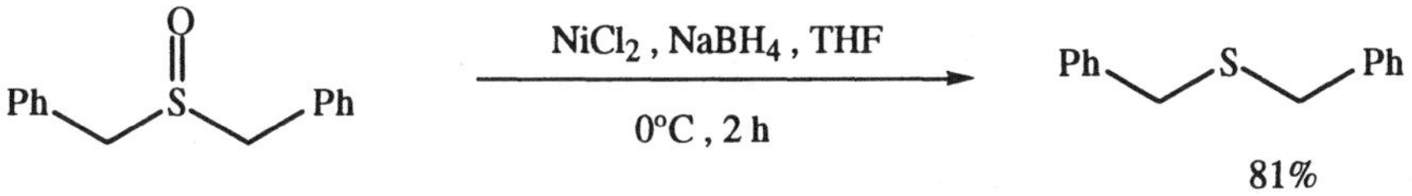
O
Ph
S
Ph
NiCl2 , NaBH4 , THF
0°C , 2 h
Ph
S
Ph
81%

Khurana, J.M.; Ray, A.; Singh, S. *Tetrahedron Lett.*, ***1998***, *39*, 3829.

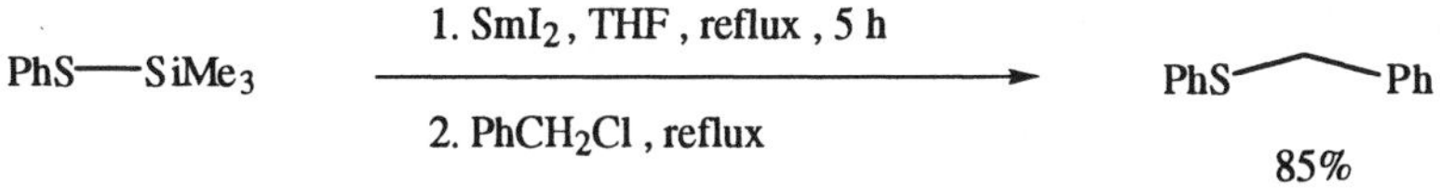
PhS—SiMe3
1. SmI2 , THF , reflux , 5 h
2. PhCH2Cl , reflux
PhS
Ph
85%

Zhang, S.; Zhang, Y.M. *J. Chem. Res. (S)*, ***1998***, 48.

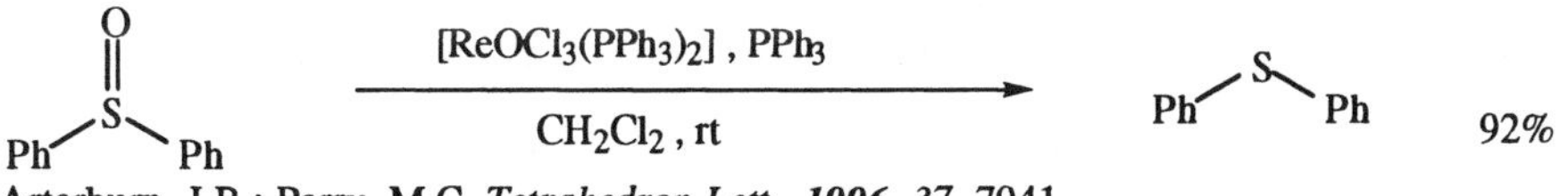
O
S
Ph
Ph
[ReOCl3(PPh3)2] , PPh3
CH2Cl2 , rt
S
Ph
Ph
92%

Arterburn, J.B.; Perry, M.C. *Tetrahedron Lett.*, ***1996***, *37*, 7941.

CHAPTER 10

PREPARATION OF HALIDES AND SULFONATES

SECTION 136: HALIDES AND SULFONATES FROM ALKYNES

NO ADDITIONAL EXAMPLES

SECTION 137: HALIDES AND SULFONATES FROM ACID DERIVATIVES

NO ADDITIONAL EXAMPLES

SECTION 138: HALIDES AND SULFONATES FROM ALCOHOLS AND THIOLS

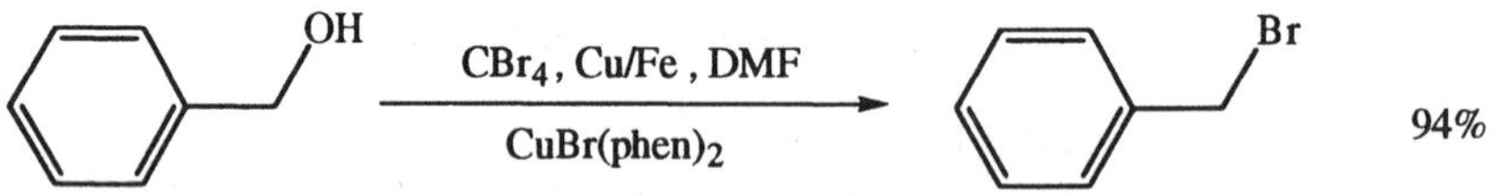
OH
CBr4, Cu/Fe, DMF
CuBr(phen)2
Br
94%

Léonel, E.; Paugam, J.P.; Nédélec, J.Y. *J. Org. Chem.*, ***1997***, *62*, 7061.

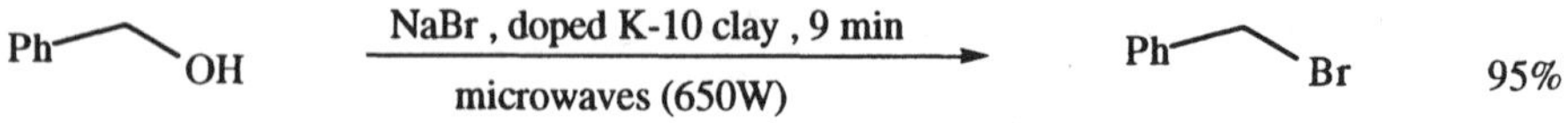
Ph
OH
NaBr, doped K-10 clay, 9 min
microwaves (650W)
Ph
Br
95%

Kad, G.L.; Singh, V.; Kaur, K.P.; Singh, J. *Tetrahedron Lett.*, ***1997***, *38*, 1079.

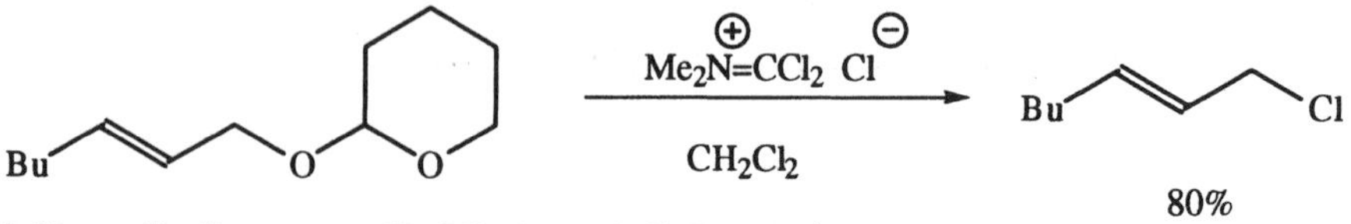
Bu
O
O
Me2N=CCl2 Cl
CH2Cl2
Bu
Cl
80%

Schlama, T.; Gouverneur, V.; Misokowski, C. *Tetrahedron Lett.*, ***1997***, *38*, 3517.

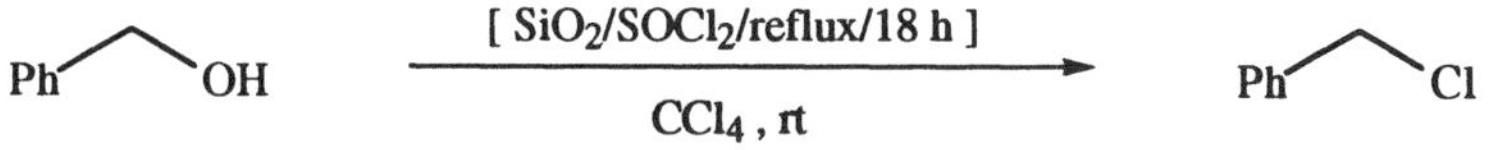
Ph
OH
[$SiO_2/SOCl_2$/reflux/18 h]
CCl_4 , rt
Ph
Cl

90%

Mohanazadeh, F.; Momeni, A.R. *Org. Prep. Proceed. Int.*, ***1996***, *28*, 492.

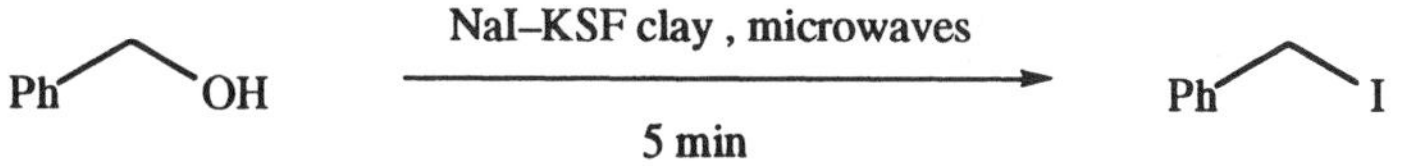
Ph
OH
NaI–KSF clay , microwaves
5 min
Ph
I

55%

Kad, G.L.; Kaur, J.; Bansal, P.; Singh, J. *J. Chem. Res. (S)*, ***1996***, 188.

SECTION 139: HALIDES AND SULFONATES FROM ALDEHYDES

NO ADDITIONAL EXAMPLES

SECTION 140: HALIDES AND SULFONATES FROM ALKYLS, METHYLENES AND ARYLS

For the conversion R-H → R-Halogen, see Section 146 (Halides from Hydrides).

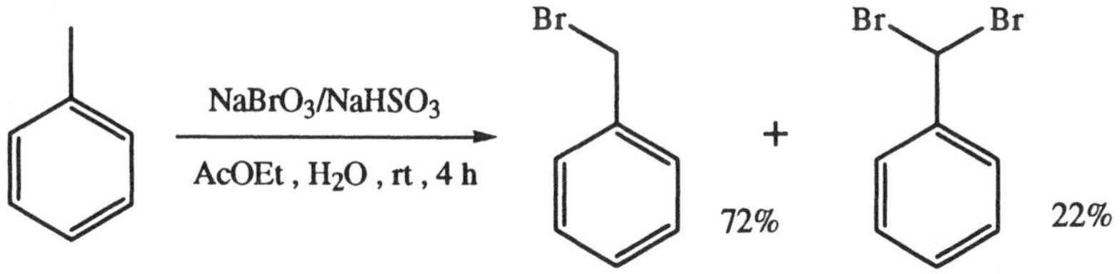
$NaBrO_3/NaHSO_3$
AcOEt , H_2O , rt , 4 h
Br
Br
Br
+
72%
22%

Kikuchi, D.; Sakaguchi, S.; Ishii, Y. *J. Org. Chem.*, ***1998***, *63*, 6023.

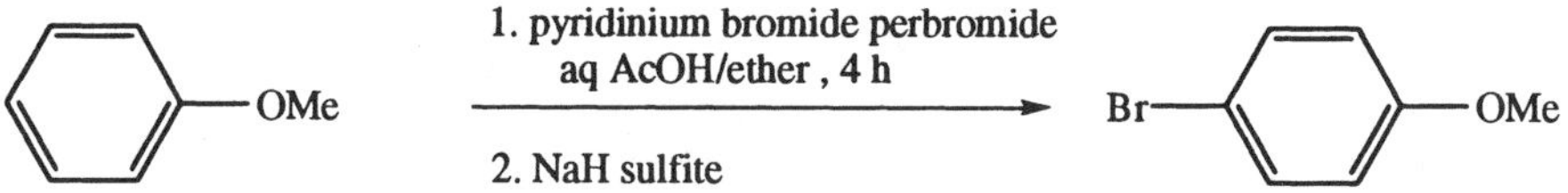
OMe
1. pyridinium bromide perbromide
aq AcOH/ether , 4 h
2. NaH sulfite
Br
OMe

95%

Reeves, W.P.; Lu, C.V.; Schlmeier, B.; Jonas, L.; Hatlevik, O. *Synth. Commun.*, ***1998***, *28*, 499.

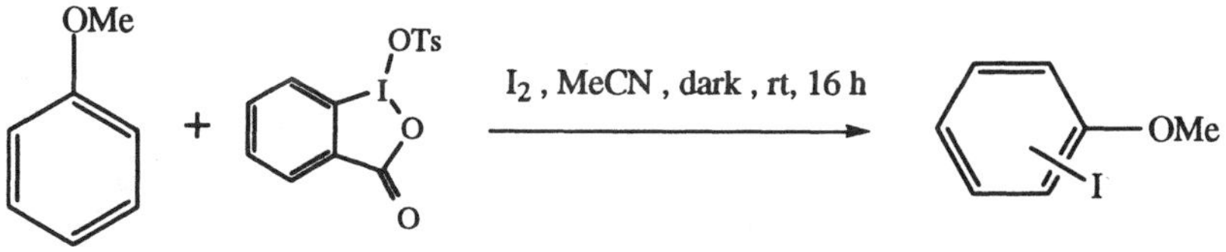
OMe
+
OTs
I
O
O
I_2 , MeCN , dark , rt, 16 h
OMe
I

o + *p* 80%

Muraki, T.; Togo, H.; Yokoyama, M. *SynLett*, ***1998***, 286.

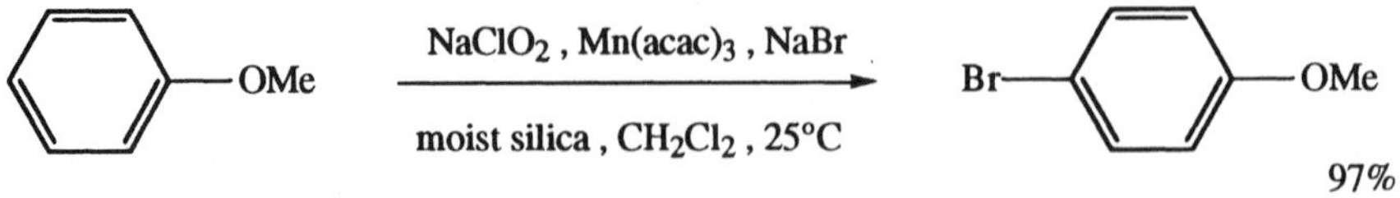

OMe
NaClO2 , Mn(acac)3 , NaBr
moist silica , CH2Cl2 , 25°C
Br
OMe
97%

Hirano, M.; Monobe, H.; Takabe, S.; Morimoto, T. *Synth. Commun.*, ***1998***, *28*, 1463.

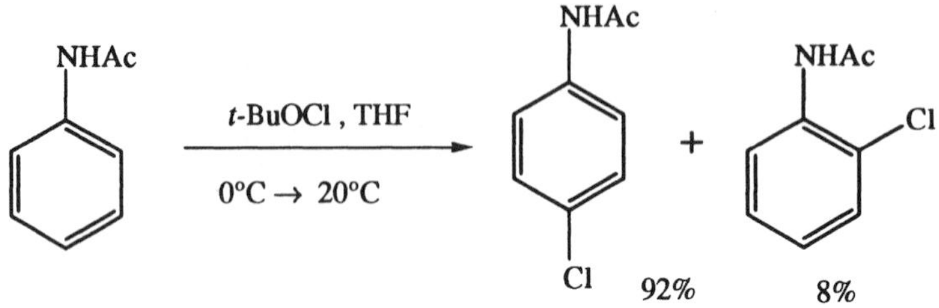

NHAc
t-BuOCl , THF
0°C → 20°C
NHAc
Cl
92%
+
NHAc
Cl
8%

Lengyel, I.; Cesare, V.; Stephani, R. *Synth. Commun.*, ***1998***, *28*, 1891.

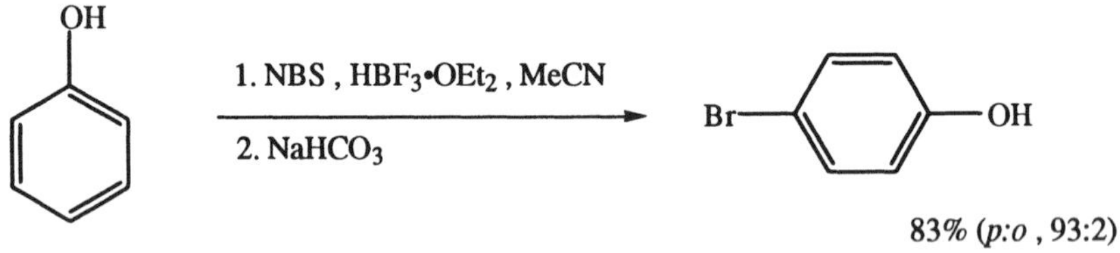

OH
1. NBS , HBF3•OEt2 , MeCN
2. NaHCO3
Br
OH
83% (p:o , 93:2)

Oberhauser, T. *J. Org. Chem.*, ***1997***, *62*, 4504.

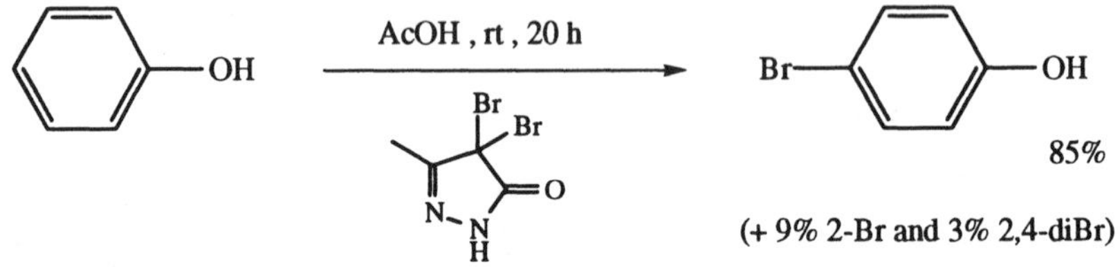

OH
AcOH , rt , 20 h
Br
Br
O
N
N
H
Br
OH
85%
(+ 9% 2-Br and 3% 2,4-diBr)

Mashraqui, S.H.; Mudaliar, C.D.; Hariharasubrahmanian, H. *Tetrahedron Lett.*, ***1997***, *38*, 4865

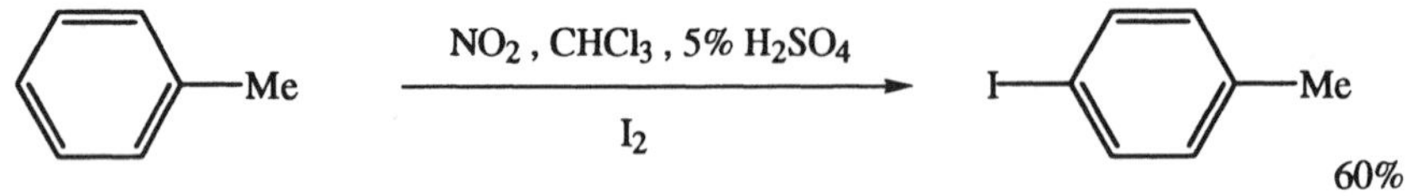

Me
NO2 , CHCl3 , 5% H2SO4
I2
I
Me
60%

Noda, Y.; Kashima, M. *Tetrahedron Lett.*, ***1997***, *38*, 6225.

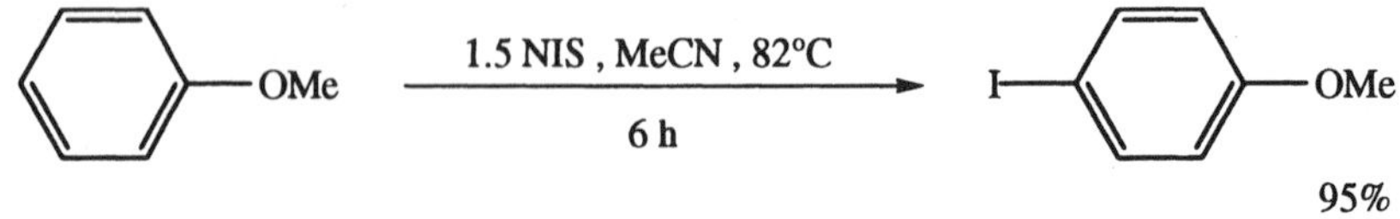

OMe
1.5 NIS , MeCN , 82°C
6 h
I
OMe
95%

Carreño, M.C.; García Ruano, J.L.; Sanz, G.; Toledo, M.A.; Urbano, A. *Tetrahedron Lett.*, ***1996***, *37*, 4081.

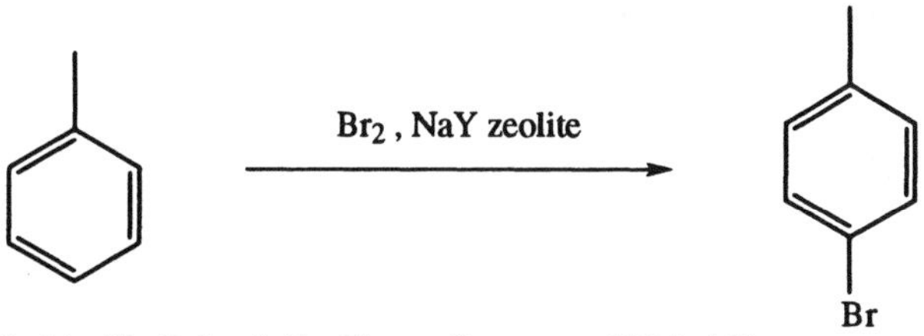
Br2 , NaY zeolite
Br

93% *para*

Smith, K.; Bahzad, D. *Chem. Commun.*, ***1996***, 467.

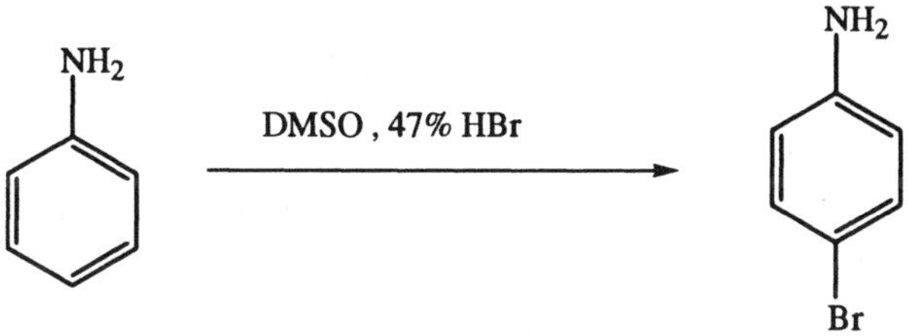
NH2
DMSO , 47% HBr
NH2
Br

91%

Srivastava, S.K.; Chauhan, P.M.S.; Bhaduri, A.P. *Chem. Commun.*, ***1996***, 2679.

SECTION 141: HALIDES AND SULFONATES FROM AMIDES

NO ADDITIONAL EXAMPLES

SECTION 142: HALIDES AND SULFONATES FROM AMINES

$$PhNH_2 \xrightarrow[\text{2. heat}]{\text{1. 1.5 eq } SiF_4\text{, } t\text{-BuONO, } CH_2Cl_2} PhF \quad 69\%$$

Tamura, M.; Shibakami, M.; Sekiya, A. *Eur. J. Org. Chem.*, ***1998***, 725.

SECTION 143: HALIDES AND SULFONATES FROM ESTERS

NO ADDITIONAL EXAMPLES

SECTION 144: HALIDES AND SULFONATES FROM ETHERS, EPOXIDES AND THIOETHERS

NO ADDITIONAL EXAMPLES

SECTION 145: HALIDES AND SULFONATES FROM HALIDES AND SULFONATES

$C_8H_{17}Br \xrightarrow{Bu_nNHF_2\text{ , Py , dioxane , }80^\circ C\text{ , 1 d}} C_8H_{17}F$ 86%

Moughamir, K.; Atmani, A.; Mestdagh, H.; Rolando, C.; Francesch, C. *Tetrahedron Lett.*, ***1998***, *39*, 7305.

$Me(CH_2)_6CH_2Br \xrightarrow[90^\circ C\text{ , 1 h}]{TMSCl\text{ , 2 eq imidazole , DMF}} Me(CH_2)_6CH_2Cl$ 94%

Peyrat, J.–F.; Figadère, B.; Cavé, A. *Synth. Commun.*, ***1996***, *26*, 4563.

SECTION 146: HALIDES AND SULFONATES FROM HYDRIDES

α-Halogenations of aldehydes, ketones and acids are found in Sections 338 (Halide-Aldehyde), 369 (Halide-Ketone), 359 (Halide-Esters) and 319 (Halide-Acids).

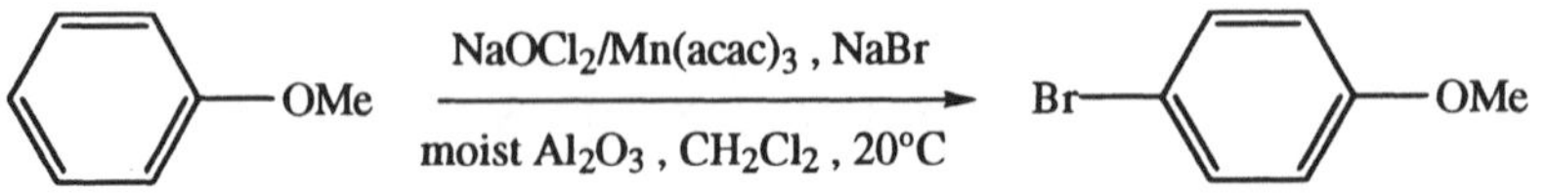
OMe
NaOCl2/Mn(acac)3 , NaBr
moist Al2O3 , CH2Cl2 , 20°C
Br
OMe

95%

Yakabe, S.; Hirano, M.; Morimoto, T. *Org. Prep. Proceed. Int.*, ***1998***, *30*, 218.

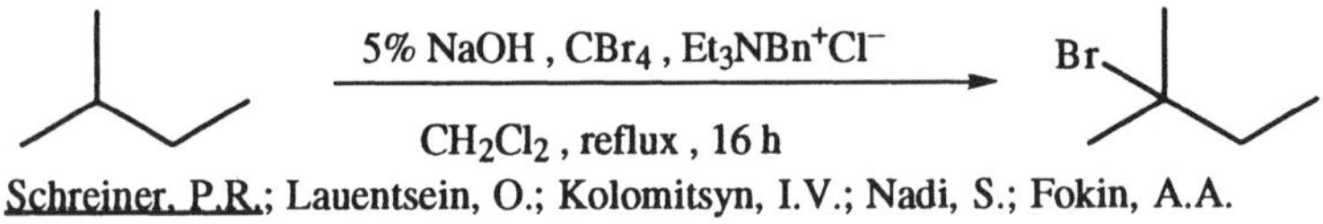
5% NaOH , CBr4 , Et3NBn+Cl−
CH2Cl2 , reflux , 16 h
Br

37%

Schreiner, P.R.; Lauentsein, O.; Kolomitsyn, I.V.; Nadi, S.; Fokin, A.A. *Angew. Chem. Int. Ed.*, ***1998***, *37*, 1895.

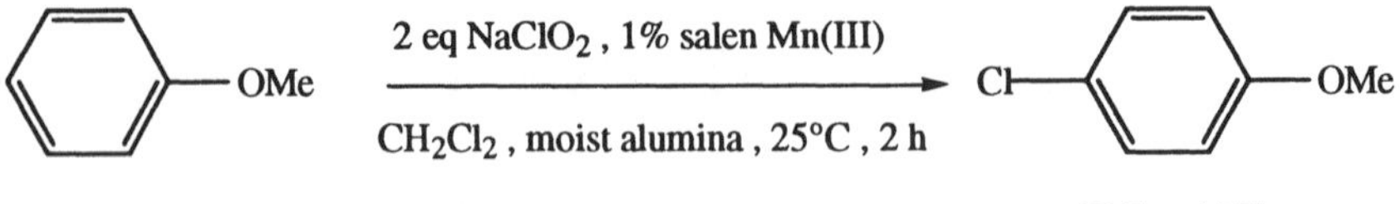
OMe
2 eq NaClO2 , 1% salen Mn(III)
CH2Cl2 , moist alumina , 25°C , 2 h
Cl
OMe

93% + 4.5% *o*

Hirano, M; Yakabe, S.; Monobe, H.; Morimoto, T. *Can. J. Chem.* ***1997***, *75*, 1905.

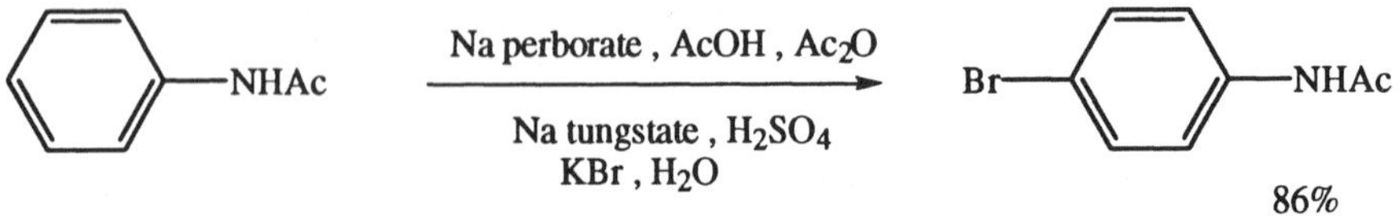
NHAc
Na perborate , AcOH , Ac2O
Na tungstate , H2SO4
KBr , H2O
Br
NHAc

86%

Hanson, J.R.; Harpel, S.; Medina, I.C.R.; Rose, D. *J. Chem. Res. (S)*, ***1997***, 432.

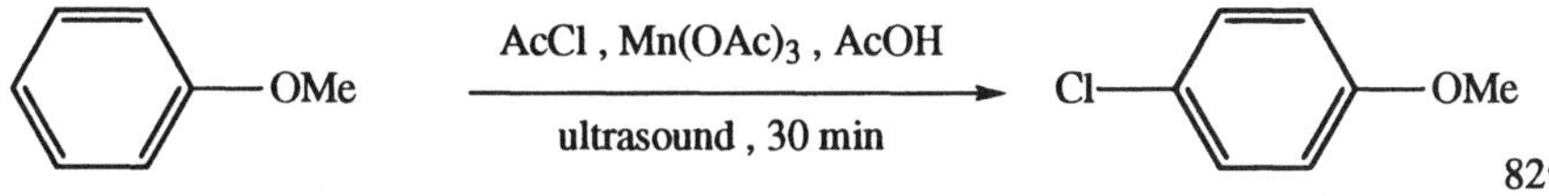

82%

Prokes, I.; Tomna, S.; Luche, J.-L. *J. Chem. Res. (S)*, ***1996***, 164.

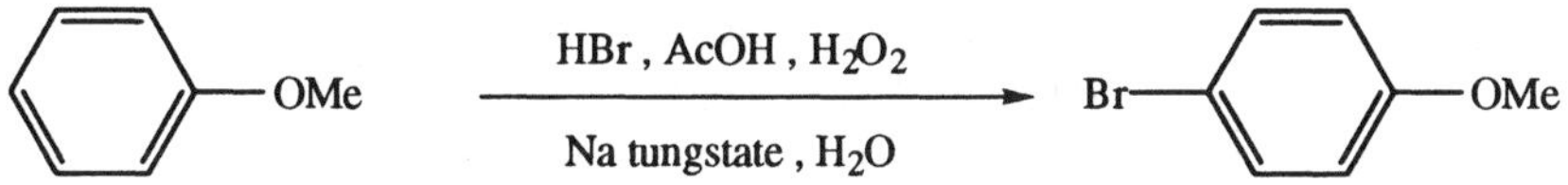

80%

Bezodis, P.; Hanson, J.R.; Petit, P. *J. Chem. Res. (S)*, ***1996***, 334.

SECTION 147: HALIDES AND SULFONATES FROM KETONES

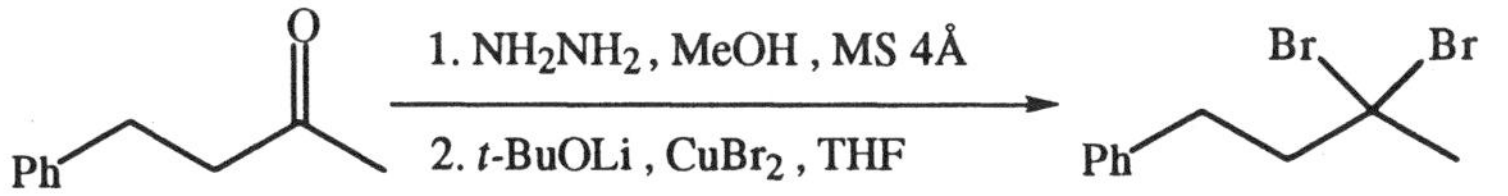

82%

Takeda, T.; Sasaki, R.; Nakamura, A.; Yamauchi, S.; Fujiwara, T. *SynLett*, ***1996***, 273.

SECTION 148: HALIDES AND SULFONATES FROM NITRILES

NO ADDITIONAL EXAMPLES

SECTION 149: HALIDES AND SULFONATES FROM ALKENES

For halocyclopropanations, see Section 74E (Alkyls from Alkenes).

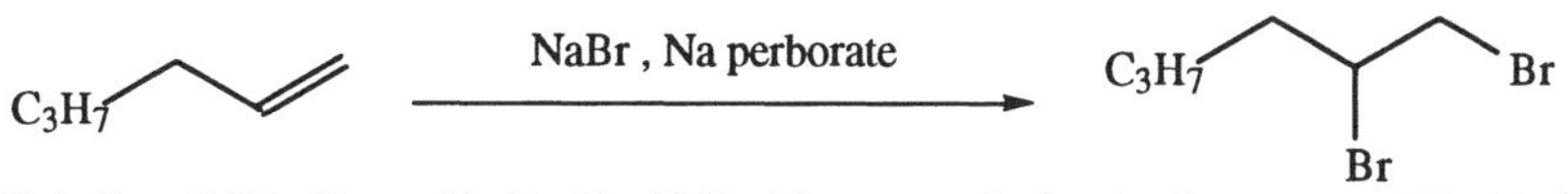

81%

Kabalka, G.W.; Yang, K.; Reddy, N.K.; Narayana, C. *Synth. Commun.*, ***1998***, *28*, 925.

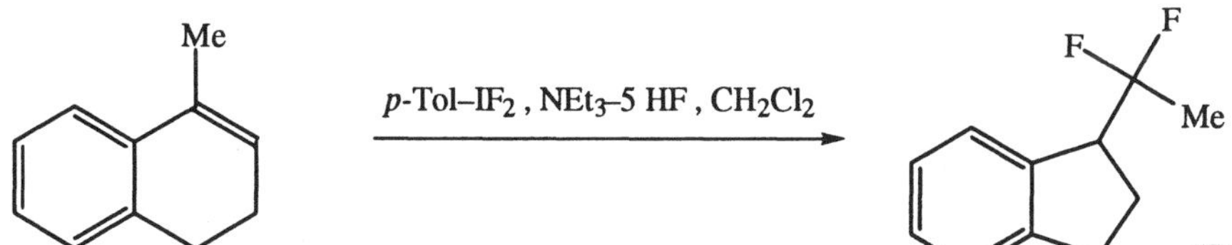

67%

Hara, S.; Nakahigashi, J.; Ishi-i, K.; Fukuhara, T.; Yoneda, N.
Tetrahedron Lett.. ***1998***. *39*. 2589.

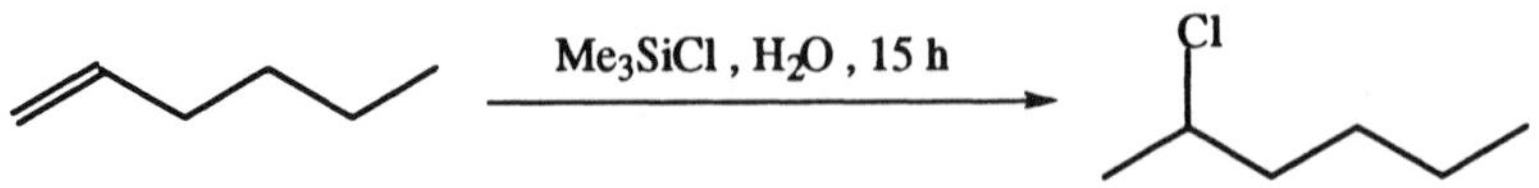

Me3SiCl , H2O , 15 h
Cl

78%

Boudjouk, P.; Kim, B.–K.; Han, B.–H. *Synth. Commun.*, ***1996***, *26*, 3479.

SECTION 150: HALIDES AND SULFONATES FROM MISCELLANEOUS COMPOUNDS

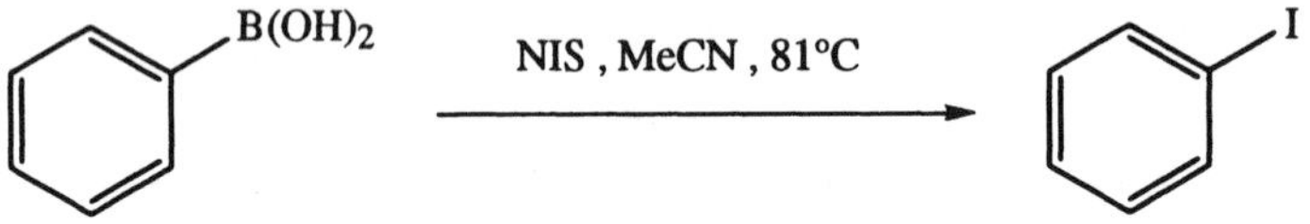

B(OH)2
NIS , MeCN , 81°C
I

61%

Thiebes, C.; Prakash, G.K.S.; Petasis, N.A.; Olah, G.A. *SynLett*, ***1998***, 141.

CHAPTER 11

PREPARATION OF HYDRIDES

This chapter lists hydrogenolysis and related reactions by which functional groups are replaced by hydrogen: e.g. $RCH_2X \rightarrow RCH_2\text{-}H$ or R-H.

SECTION 151: HYDRIDES FROM ALKYNES

NO ADDITIONAL EXAMPLES

SECTION 152: HYDRIDES FROM ACID DERIVATIVES

This section lists examples of decarboxylations ($RCO_2H \rightarrow$ R-H) and related reactions

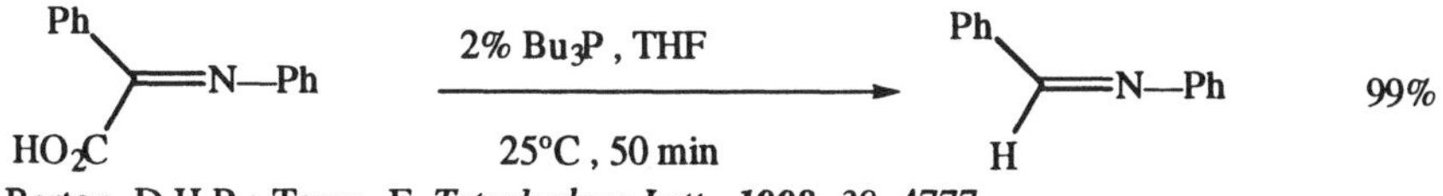

Ph
N—Ph
HO2C
2% Bu3P , THF
25°C , 50 min
Ph
N—Ph
H
99%

Barton, D.H.R.; Taran, F. *Tetrahedron Lett.*, ***1998***, *39*, 4777.

SECTION 153: HYDRIDES FROM ALCOHOLS AND THIOLS

This section lists examples of the hydrogenolysis of alcohols and phenols (ROH $\rightarrow$ R-H).

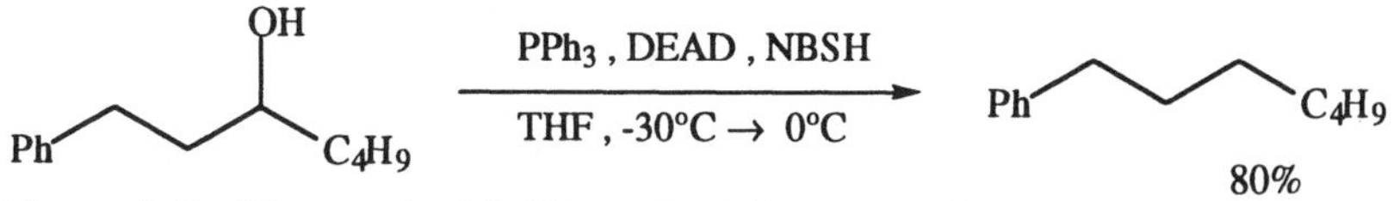

OH
Ph
C4H9
PPh3 , DEAD , NBSH
THF , -30°C → 0°C
Ph
C4H9
80%

Myers, A.G.; Movassaghe, M.; Zheng, B. *J. Am. Chem. Soc.*, ***1997***, *119*, 8572.

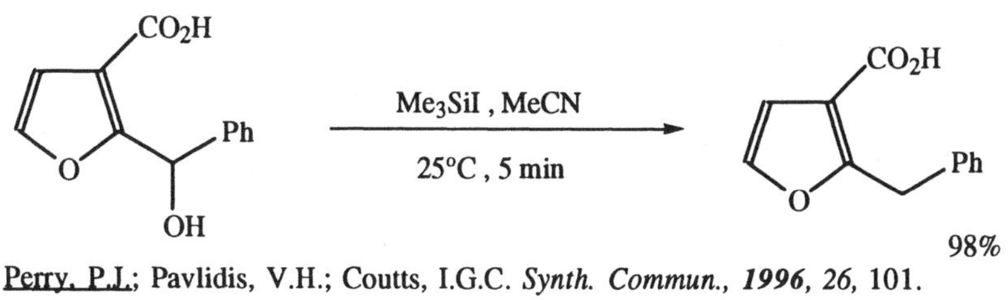

CO2H
Ph
O
OH
Me3SiI , MeCN
25°C , 5 min
CO2H
Ph
O
98%

Perry, P.J.; Pavlidis, V.H.; Coutts, I.G.C. *Synth. Commun.*, ***1996***, *26*, 101.

Also via: Section 160 (Halides and Sulfonates).

SECTION 154: HYDRIDES FROM ALDEHYDES

For the conversion RCHO → R-Me, etc., see Section 64 (Alkyls from Aldehydes).

NO ADDITIONAL EXAMPLES

SECTION 155: HYDRIDES FROM ALKYLS, METHYLENES AND ARYLS

NO ADDITIONAL EXAMPLES

SECTION 156: HYDRIDES FROM AMIDES

NO ADDITIONAL EXAMPLES

SECTION 157: HYDRIDES FROM AMINES

This section lists examples of the conversion RNH_2 (or R_2NH) → R-H.

$PhNH_2$ —(20 eq NO , THF)→ Ph—H 85%

Itoh, T.; Matsuya, Y.; Nagata, K.; Ohsawa, A. *Tetrahedron Lett.*, ***1996***, *37*, 4165.

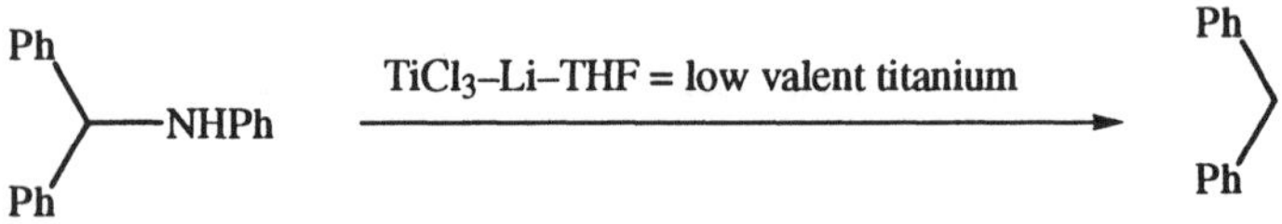
Ph
NHPh
Ph
TiCl3–Li–THF = low valent titanium
Ph
Ph

48%

Talukdar, S.; Banerji, A. *Synth. Commun.*, ***1996***, *26*, 1051.

SECTION 158: HYDRIDES FROM ESTERS

This section lists examples of the reactions RCO_2R' → R-H and RCO_2R' → R'H.

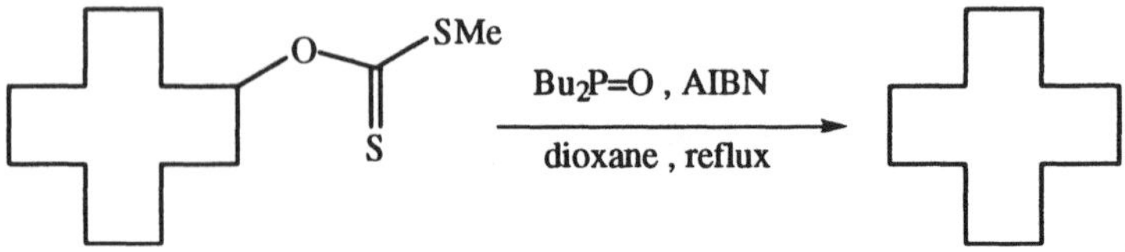
O
SMe
S
Bu2P=O , AIBN
dioxane , reflux

94%

Jang, D.O.; Cho, D.H.; Barton, D.H.R. *SynLett*, ***1998***, 39.

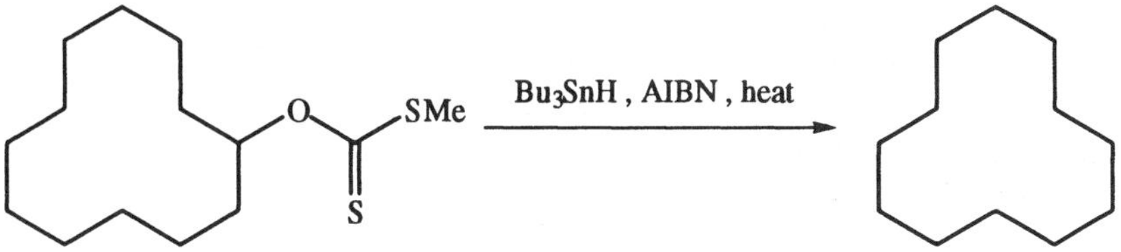
SMe
Bu3SnH , AIBN , heat

76%

Jang, D.O.; Cho, D.H.; Kim, J. *Synth. Commun.*, ***1998***, *28*, 3559.

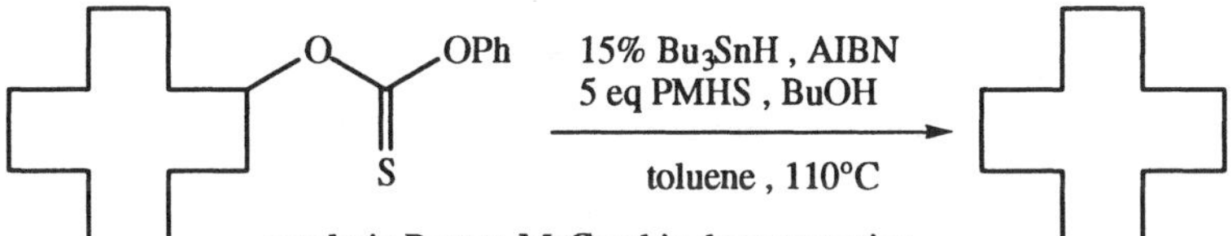
OPh
15% Bu3SnH , AIBN
5 eq PMHS , BuOH
toluene , 110°C

catalytic Barton-McCombie deoxygenation

66%

Lopez, R.M.; Hays, D.S.; Fu, G.C. *J. Am. Chem. Soc.*, ***1997***, *119*, 6949.

SECTION 159: HYDRIDES FROM ETHERS, EPOXIDES AND THIOETHERS

This section lists examples of the reaction R-O-R' → R-H.

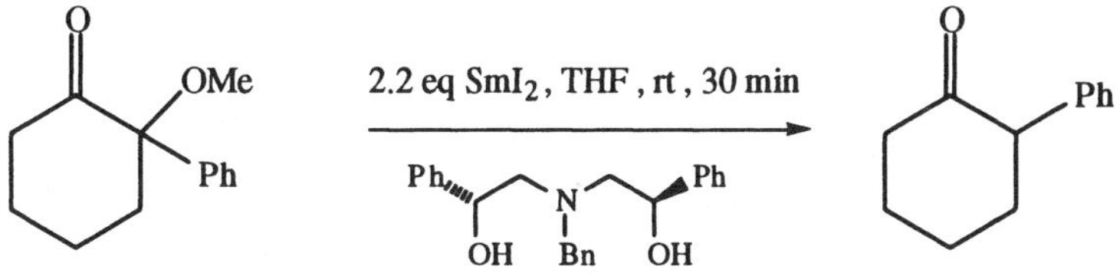
OMe
Ph
2.2 eq SmI2 , THF , rt , 30 min
Ph
N
Bn
OH
OH
Ph
Ph

78% (67% ee , *S*)

Mikami, K.; Yamaoka, M.; Yoshida, A. *SynLett*, ***1998***, 607.

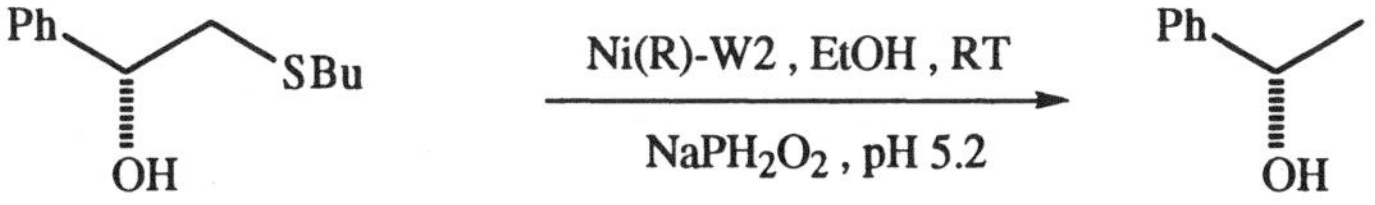
Ph
SBu
OH
Ni(R)-W2 , EtOH , RT
NaPH2O2 , pH 5.2
Ph
OH

84%

no loss of optical purity

Nishide, K.; Shigeta, Y.; Obata, K.; Inoue, T.; Node, M. *Tetrahedron Lett.*, ***1998***, *37*, 2271.

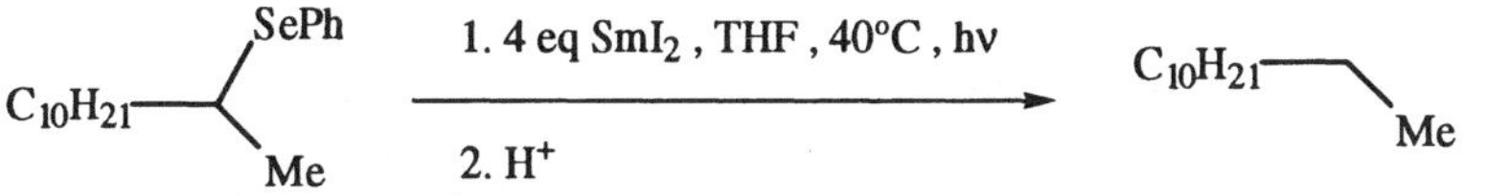
SePh
C10H21
Me
1. 4 eq SmI2 , THF , 40°C , hν
2. H+
C10H21
Me

83%

Ogawa, A.; Ohya, S.; Doi, M.; Sumino, Y.; Sonoda, N.; Hirao, T. *Tetrahedron Lett.* ***1998***. *39*. 6341.

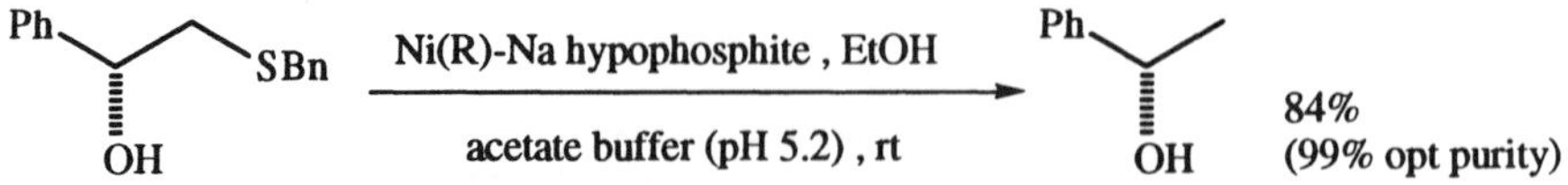
Ph
SBn
OH
Ni(R)-Na hypophosphite , EtOH
acetate buffer (pH 5.2) , rt
Ph
OH
84%
(99% opt purity)

Node, M.; Nishide, K.; Shigeta, Y.; Obata, K.; Shiraki, H.; Kunishige, H. *Tetrahedron*, ***1997***, *53*, 12883.

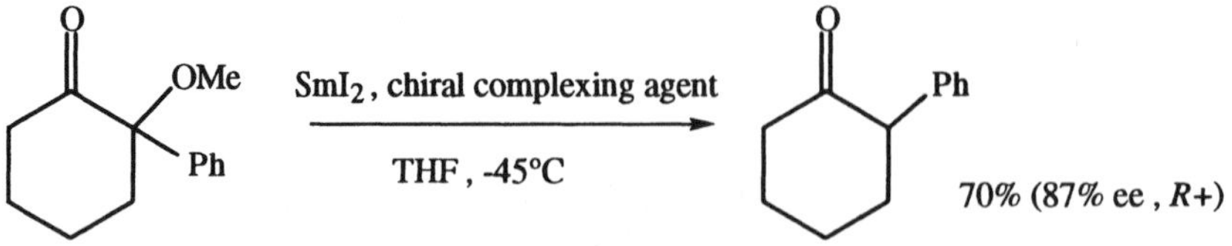
O
OMe
Ph
SmI2 , chiral complexing agent
THF , -45°C
O
Ph
70% (87% ee , R+)

Nakamura, Y.; Takeuchi, S.; Ohgo, Y.; Yamaoka, M.; Yoshida, A.; Mikami, K. *Tetrahedron Lett.*, ***1997***, *38*, 2709.

SECTION 160: HYDRIDES FROM HALIDES AND SULFONATES

This section lists the reductions of halides and sulfonates, R-X → R-H.

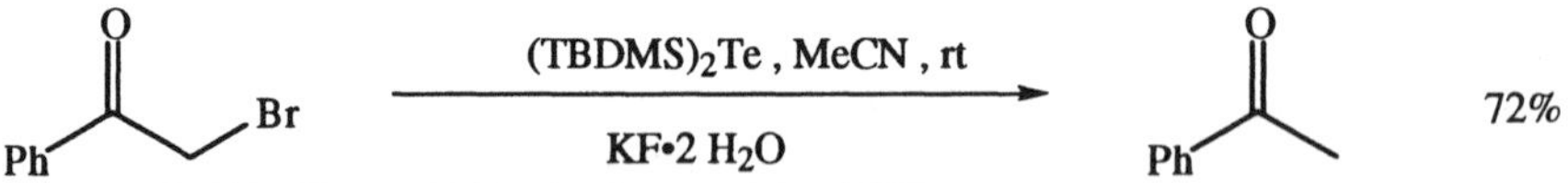
O
Br
Ph
(TBDMS)2Te , MeCN , rt
KF•2 H2O
O
Ph
72%

Schultz, E.K.V.; Harpp, D.N. *Synthesis*, ***1998***, 1137.

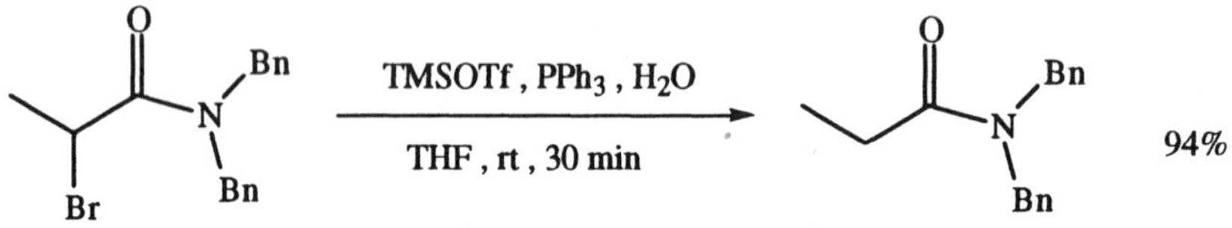
O
Bn
N
Bn
Br
TMSOTf , PPh3 , H2O
THF , rt , 30 min
O
Bn
N
Bn
94%

Kagoshima, H.; Hashimoto, Y.; Oguro, D.; Kutsuna, T.; Saigo, K. *Tetrahedron Lett.*, ***1998***, *39*, 1203.

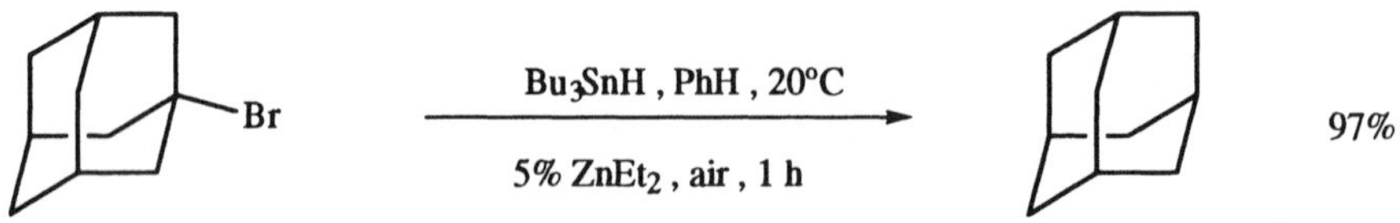
Br
Bu3SnH , PhH , 20°C
5% ZnEt2 , air , 1 h
97%

Ryu, I.; Araki, F.; Minakata, S.; Komatsu, M. *Tetrahedron Lett*, ***1998***, *39*, 6335.

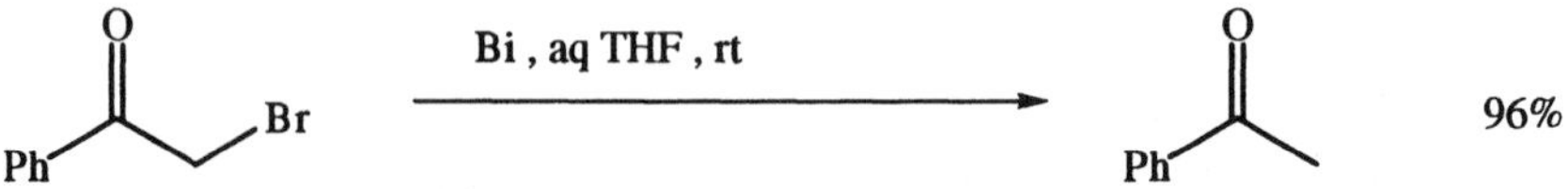
O
Br
Ph
Bi , aq THF , rt
O
Ph
96%

Ren, P.–D.; Jin, Q.–H.; Yao, Z.–P. *Synth. Commun.*, ***1997***, *27*, 2577.

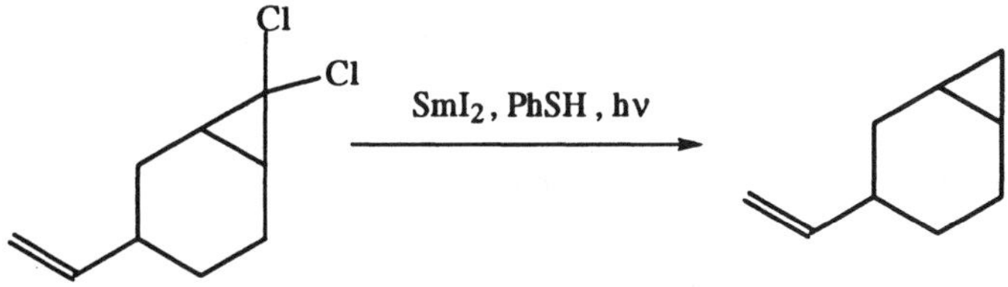
Cl
Cl
SmI2, PhSH, hν

65%

Ogawa, A.; Ohya, S.; Hirao, T. *Chem. Lett.*, ***1997***, 275.

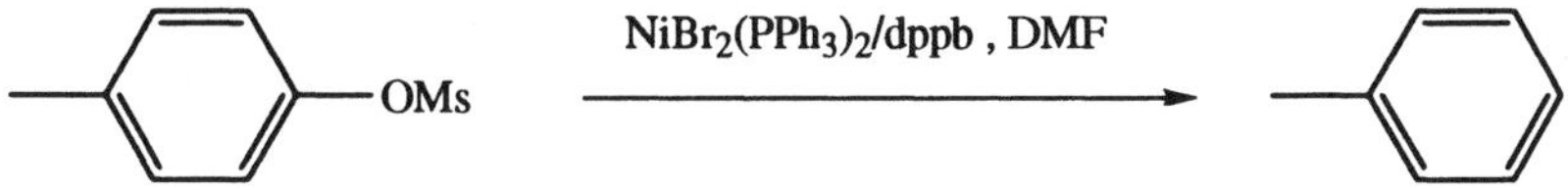
OMs
NiBr2(PPh3)2/dppb, DMF

82%

Sasaki, K.; Kubo, T.; Sakai, M.; Kuroda, Y. *Chem. Lett.*, ***1997***, 617.

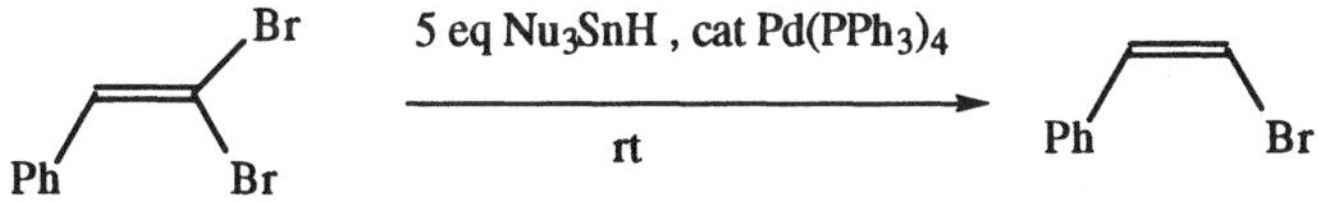
Br
Ph
Br
5 eq Nu3SnH, cat Pd(PPh3)4
rt
Ph
Br

76%

Uenishi, J.; Kawahama, R.; Shiga, Y.; Yonemitsu, O.; Tsuji, J. *Tetrahedron Lett.*, ***1996***, *37*, 6759.

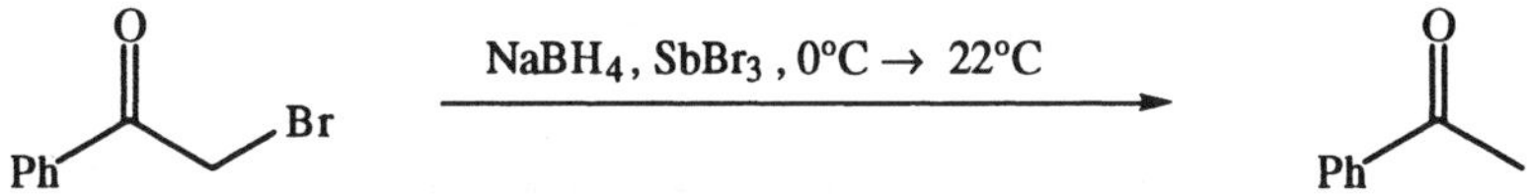
O
Ph
Br
NaBH4, SbBr3, 0°C → 22°C
O
Ph

88%

Sayama, S.; Inamura, Y. *Chem. Lett.*, ***1996***, 633.

Ph—I $\xrightarrow[\text{2. } H_3O^+]{\text{1. } Fe(CO)_5\text{, aq NaOH/PhH; TBAB, CO, 65°C, 65 h}}$ Ph—H 80%

Brunet, J.-J.; El Zaizi, A. *Bull. Soc. Chim. Fr.*, ***1996***, *133*, 75.

SECTION 161: HYDRIDES FROM HYDRIDES

NO ADDITIONAL EXAMPLES

SECTION 162: HYDRIDES FROM KETONES

This section lists examples of the reaction $R_2C\text{-}(C{=}O)R \rightarrow R_2C\text{-}H$.

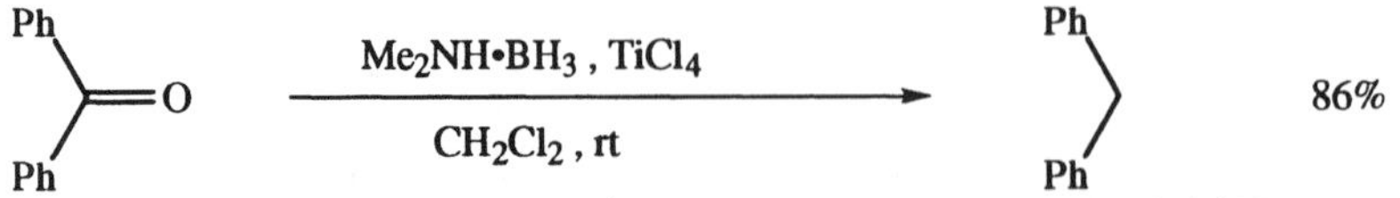
Ph
Ph
O
Me2NH•BH3 , TiCl4
CH2Cl2 , rt
Ph
Ph
86%

Dehmlow, E.V.; Niemann, T.; Kraft, A. *Synth. Commun.*, ***1996***, *26*, 1467.

SECTION 163: HYDRIDES FROM NITRILES

This section lists examples of the reaction, R-C≡N → R-H (includes reactions of isonitriles (R-N≡C).

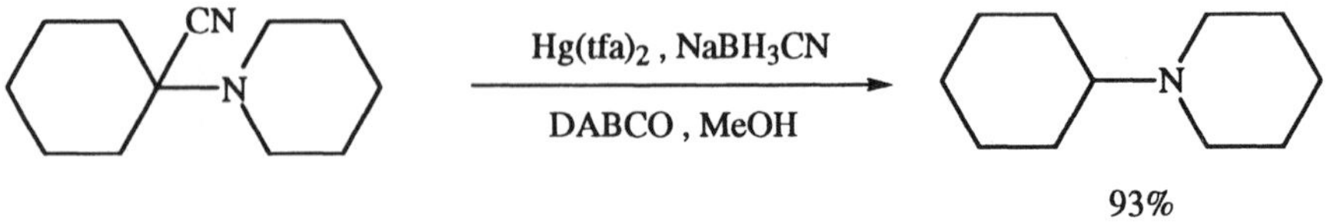
CN
N
Hg(tfa)2 , NaBH3CN
DABCO , MeOH
N
93%

Sassman, M.B. *Tetrahedron*, ***1996***, *52*, 10835.

SECTION 164: HYDRIDES FROM ALKENES

NO ADDITIONAL EXAMPLES

SECTION 165: HYDRIDES FROM MISCELLANEOUS COMPOUNDS

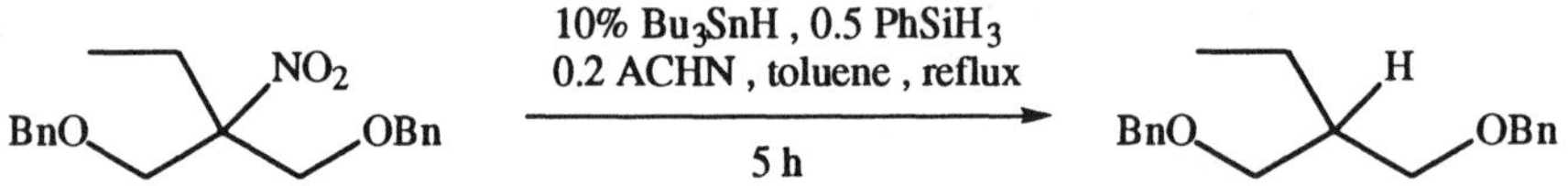
NO2
BnO
OBn
10% Bu3SnH , 0.5 PhSiH3
0.2 ACHN , toluene , reflux
5 h
H
BnO
OBn

Tormo, J.; Hays, D.S.; Fu, G.C. *J. Org. Chem.*, ***1998***, *63*, 5296.

CHAPTER 12
PREPARATION OF KETONES

SECTION 166: KETONES FROM ALKYNES

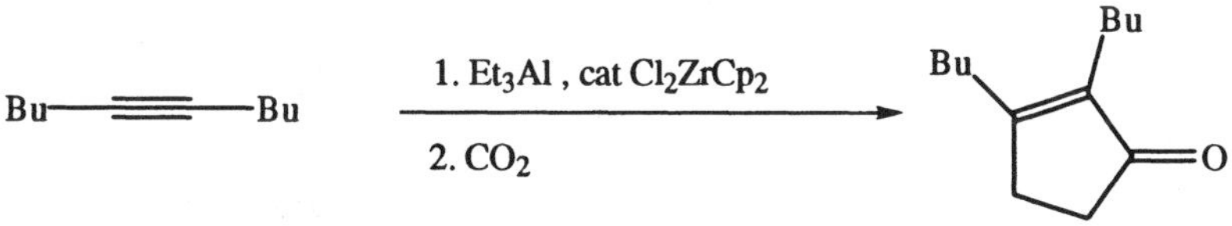

Bu
Bu
1. Et_3Al , cat Cl_2ZrCp_2
Bu
Bu
2. CO_2
O

76%

Negishi, E.; Montchamp, J.-L.; Anastasia, L.; Elizarov, A.; Choueiry, D. *Tetrahedron Lett.*, ***1998***, *39*, 2503.

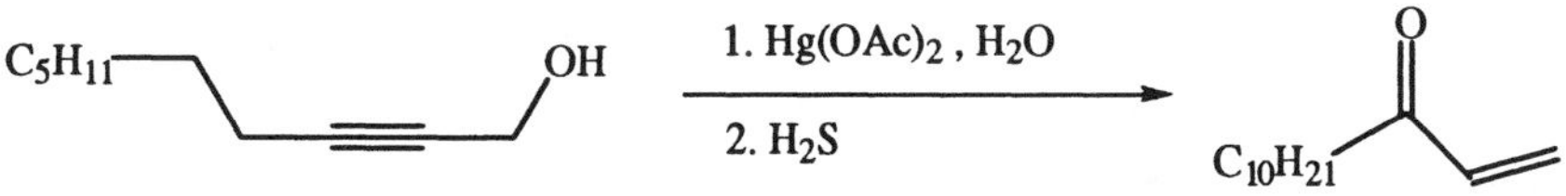

C_5H_{11}
OH
1. $Hg(OAc)_2$, H_2O
2. H_2S
O
$C_{10}H_{21}$

80%

Yadav, J.S.; Prahlad, V.; Muralidhar, B. *Synth. Commun.*, ***1997***, *27*, 3415.

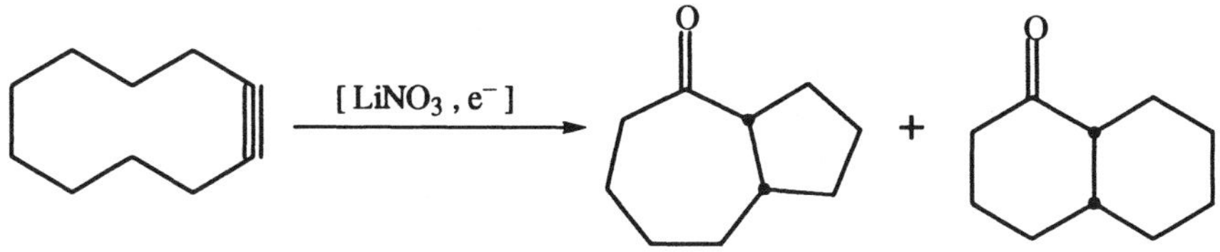

[$LiNO_3$, e^-]
O
O
+

(3 : 1) 70%

Wille, U.; Plath, C. *Liebigs Ann. Chem.*, ***1997***, 111.

SECTION 167: KETONES FROM ACID DERIVATIVES

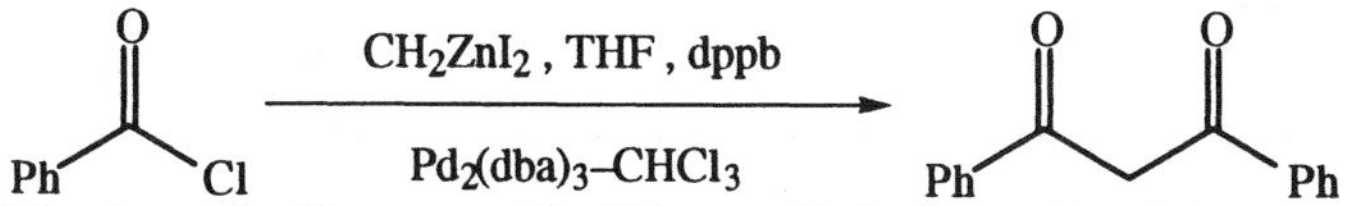

O
Ph
Cl
CH_2ZnI_2 , THF , dppb
$Pd_2(dba)_3$–$CHCl_3$
O
O
Ph
Ph

64%

Matsubara, S.; Kawamoto, K.; Utimoto, K. *SynLett*, ***1998***, 267.

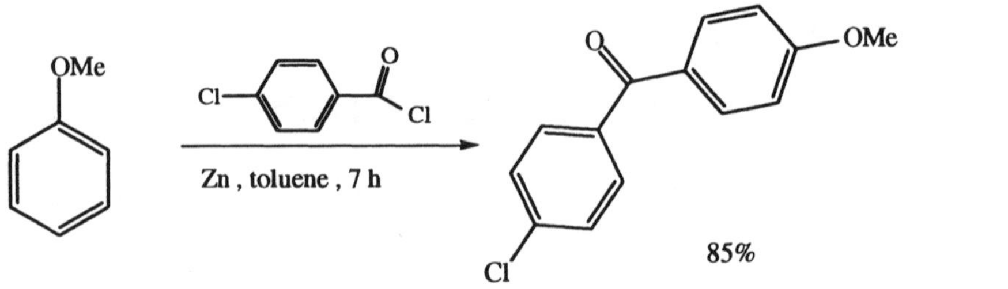
OMe
Cl
O
Cl
Zn , toluene , 7 h
O
OMe
Cl
85%

Meshram, H.M.; Reddy, G.S.; Reddy, M.M.; Yadav, J.S. *Synth. Commun.*, ***1998***, *28*, 2203.

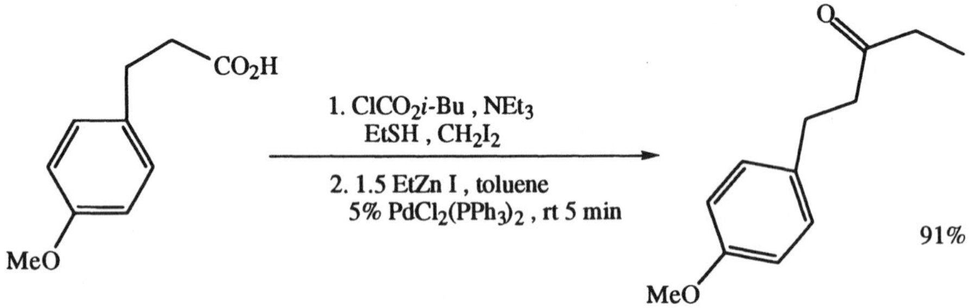
CO2H
MeO
1. ClCO2i-Bu , NEt3
EtSH , CH2I2
2. 1.5 EtZn I , toluene
5% PdCl2(PPh3)2 , rt 5 min
O
MeO
91%

Tokuyama, H.; Yokoshima, S.; Yamashita, T.; Fukuyama, T.
Tetrahedron Lett., ***1998***, *39*, 3189.

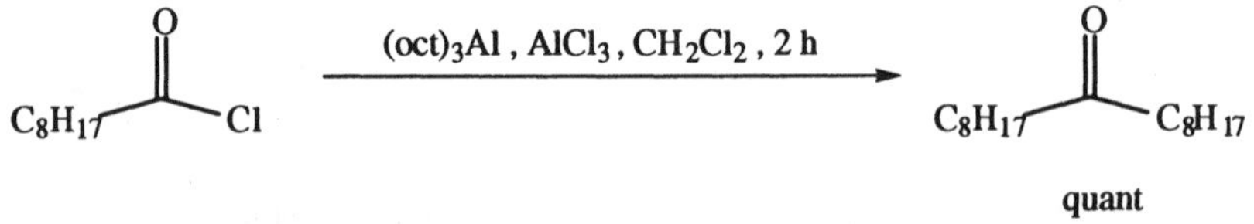
O
C8H17
Cl
(oct)3Al , AlCl3 , CH2Cl2 , 2 h
O
C8H17
C8H17
quant

Arisawa, M.; Torisawa, Y.; Kawahara, M.; Yamanaka, M.; Nishida, A.; Nakagawa, M.
J. Org. Chem., ***1997***, *62*, 4327.

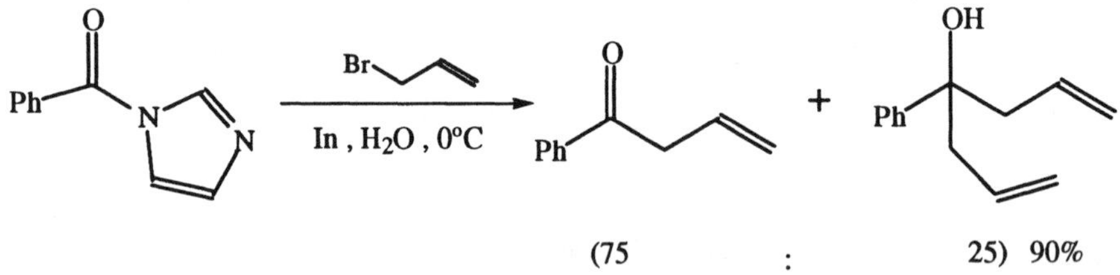
O
Ph
N
N
Br
In , H2O , 0°C
O
Ph
+
OH
Ph
(75 : 25) 90%

Bryann, V.J.; Chan, T.-H. *Tetrahedron Lett*, ***1997***, *38*, 6493.

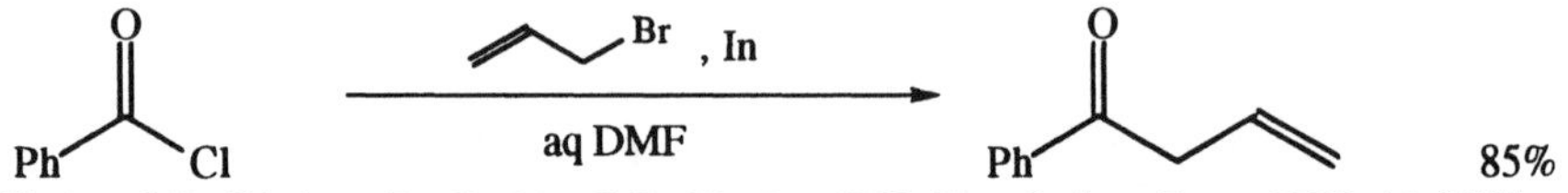
O
Ph
Cl
Br , In
aq DMF
O
Ph
85%

Yadav, J.S.; Srinivas, D.; Reddy, G.S.; Bindhu, K.H. *Tetrahedron Lett.*, ***1997***, *38*, 8745.

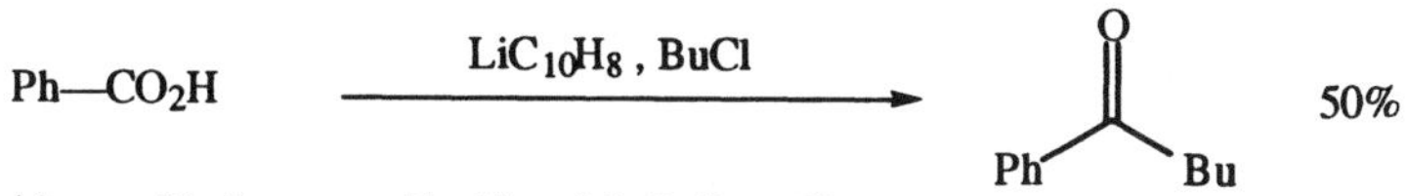
Ph—CO2H
LiC10H8 , BuCl
O
Ph
Bu
50%

Alonso, F.; Lorenzo, E.; Yus, M. *J. Org. Chem.*, ***1996***, *61*, 6058.

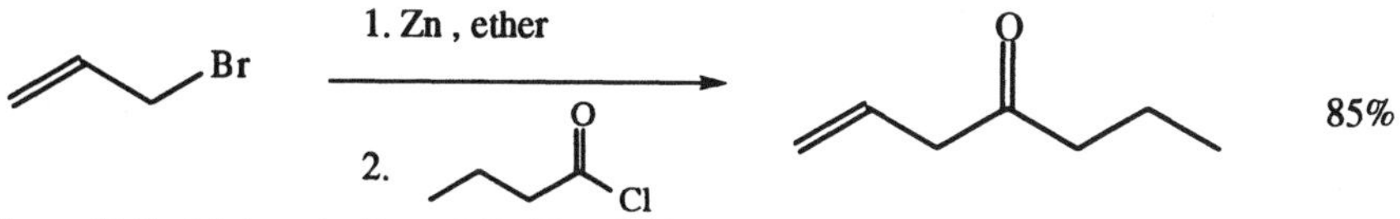
Br
1. Zn , ether
2.
O
Cl
O
85%

Ranu, B.C.; Majee, A.; Das, A.R. *Tetrahedron Lett.*, ***1996***, *37*, 1109.

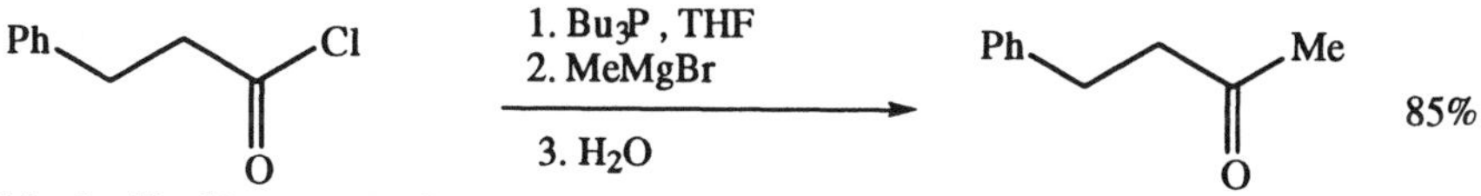
Ph
Cl
O
1. Bu3P , THF
2. MeMgBr
3. H2O
Ph
Me
O
85%

Maeda, H.; Okamoto, J.; Ohmori, H. *Tetrahedron Lett.*, ***1996***, *37*, 5381.

SECTION 168: KETONES FROM ALCOHOLS AND THIOLS

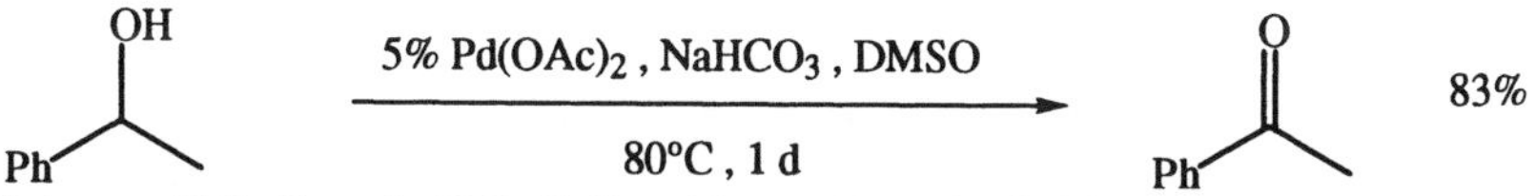
OH
Ph
5% Pd(OAc)2 , NaHCO3 , DMSO
80°C , 1 d
O
Ph
83%

Peterson, K.P.; Larock, R.C. *J. Org. Chem.*, ***1998***, *63*, 3185.

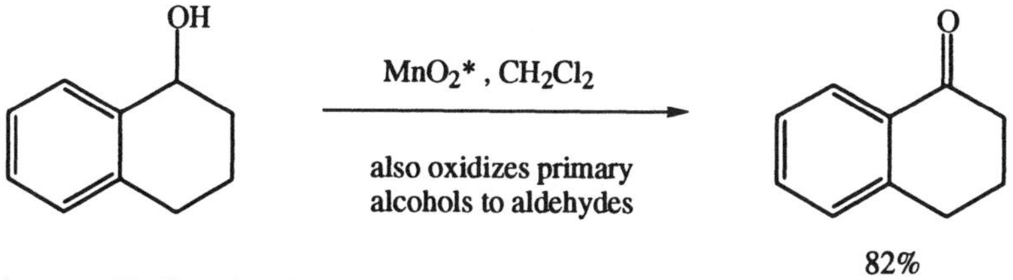
OH
MnO2* , CH2Cl2
also oxidizes primary
alcohols to aldehydes
O
82%

Aoyama, T.; Sonoda, N.; Yamauchi, M.; Toriyama, K.; Anzai, M.; Ando, A.; Shioiri, T. *SynLett*, ***1998***, 35.

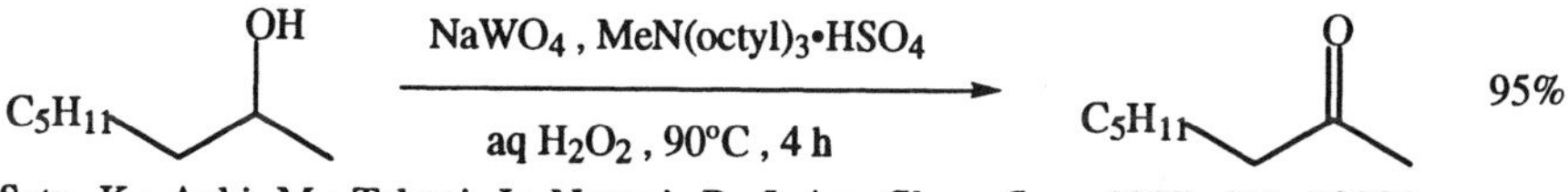
OH
C5H11
NaWO4 , MeN(octyl)3•HSO4
aq H2O2 , 90°C , 4 h
O
C5H11
95%

Sato, K.; Aoki, M.; Takagi, J.; Noyori, R. *J. Am. Chem. Soc.*, ***1997***, *119*, 12386.

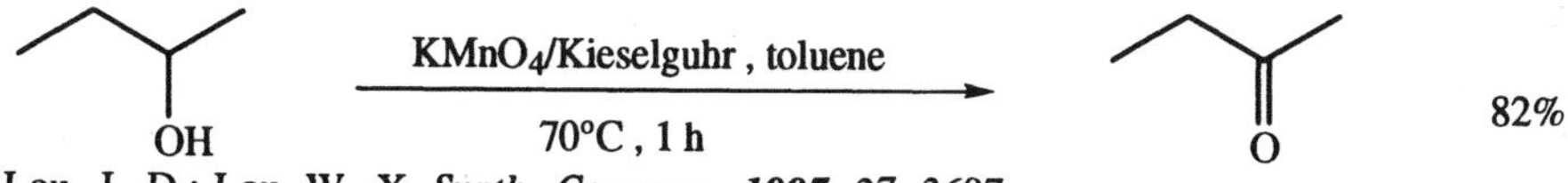
OH
KMnO4/Kieselguhr , toluene
70°C , 1 h
O
82%

Lou, J.-D.; Lou, W.-X. *Synth. Commun.*, ***1997***, *27*, 3697.

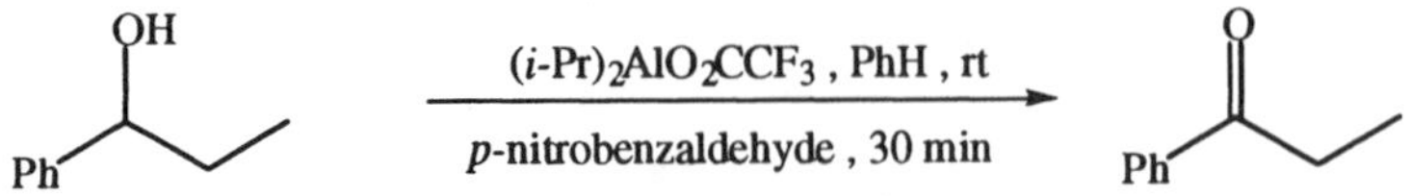
(*i*-Pr)$_2$AlO$_2$CCF$_3$, PhH , rt
p-nitrobenzaldehyde , 30 min

new Oppenauer oxidation catalyst

Akamanchi, K.G.; Chaudhari, B.A. *Tetrahedron Lett.*, ***1997***, *38*, 6925.

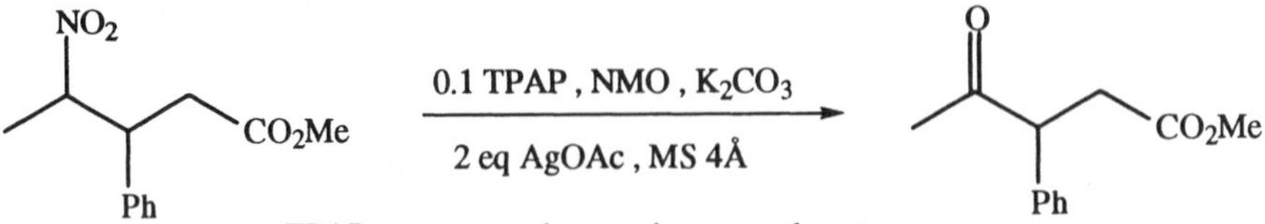
0.1 TPAP , NMO , K$_2$CO$_3$
2 eq AgOAc , MS 4Å

TPAP = tetrapropylammonium perruthenate

89%

Tokunaga, Y.; Thara, M.; Fukumoto, K. *J. Chem. Soc., Perkin Trans. 1*, ***1997***, 207.

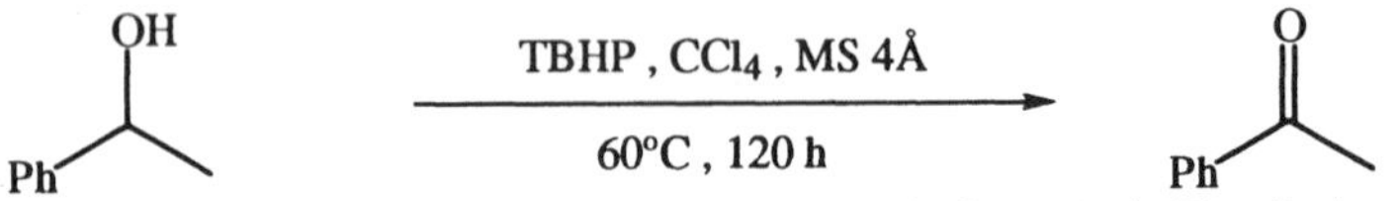
TBHP , CCl$_4$, MS 4Å
60°C , 120 h

55%

Palombi, L.; Arista, L.; Lattanzi, A.; Bonadies, F.; Screttri, A. *Tetrahedron Lett.*, ***1996***, *37*, 7849.

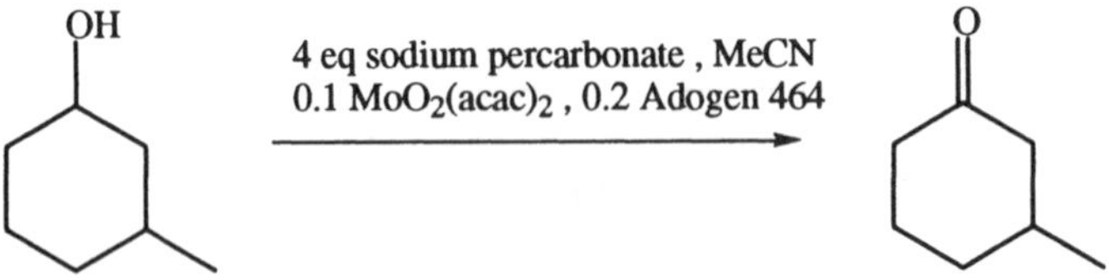
4 eq sodium percarbonate , MeCN
0.1 MoO$_2$(acac)$_2$, 0.2 Adogen 464

91%

Maignien, S.; Aït–Mohand, S.; Muzart, J. *SynLett*, ***1996***, 439.

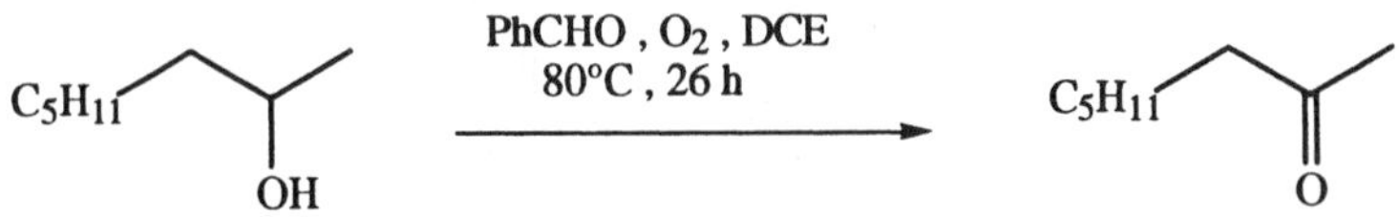
PhCHO , O$_2$, DCE
80°C , 26 h

80%

Choudary, B.M.; Sudha, Y. *Synth. Commun.*, ***1996***, *26*, 1651.

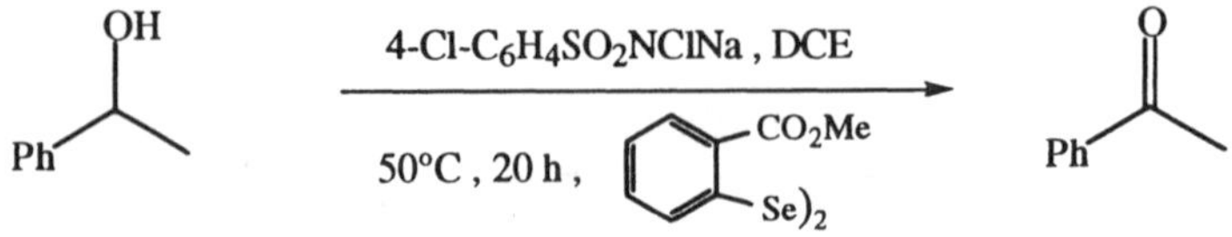
4-Cl-C$_6$H$_4$SO$_2$NClNa , DCE
50°C , 20 h ,

94%

Onami, T.; Ikeda, M.; Woodward, S.S. *Bull. Chem. Soc. Jpn.*, ***1996***, *69*, 3601.

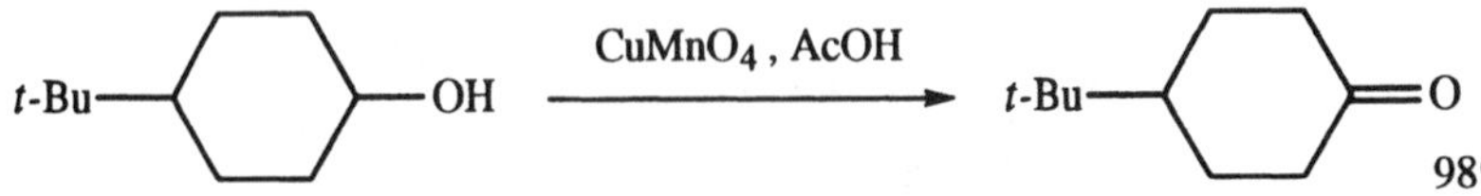
CuMnO$_4$, AcOH

98%

Ansari, M.A.; Craig, J.C. *Synth. Commun.*, ***1996***, *26*, 1789.

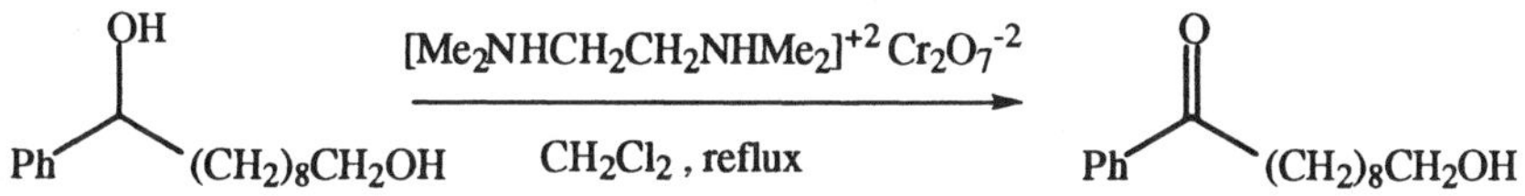

70%

Chandrasekhar, S.; Takhi, M.; Mohapatra, S. *Synth. Commun.*, ***1996***, *26*, 3947.

Related Methods: Section 48 (Aldehydes from Alcohols and Phenols).

SECTION 169: KETONES FROM ALDEHYDES

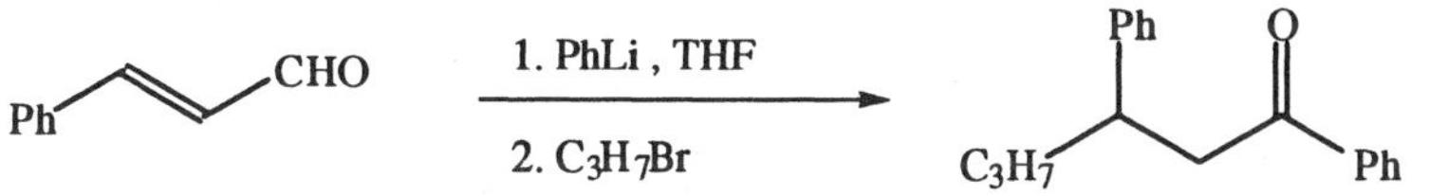

91%

Nudelman, N.S.; García, G.V.; Schulz, H.G. *J. Org. Chem.*, ***1998***, *63*, 5730.

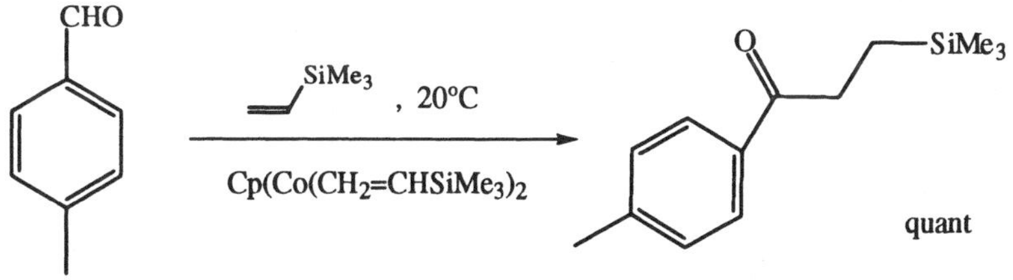

quant

Lenges, C.P.; Brookhart, M. *J. Am. Chem.. Soc.*, ***1997***, *119*, 3165.

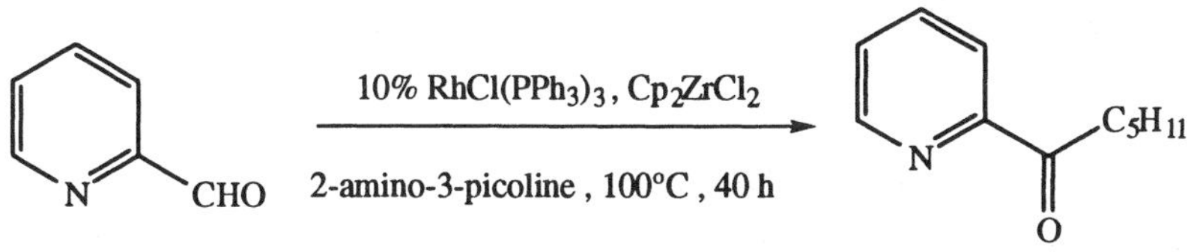

50%

Jun, C.-H.; Lee, D.-Y.; Hong, J.-B. *Tetrahedron Lett.*, ***1997***, *38*, 6673.

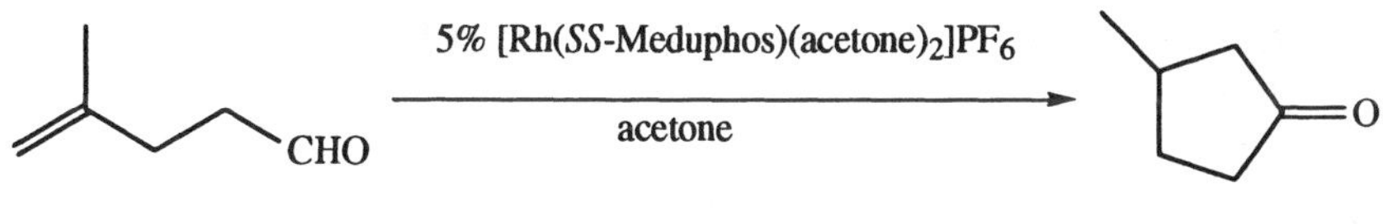

95% (94% ee)

Barnhart, R.W.; McMorran, D.A.; Bosnich, B. *Chem. Commun.*, ***1997***, 589.

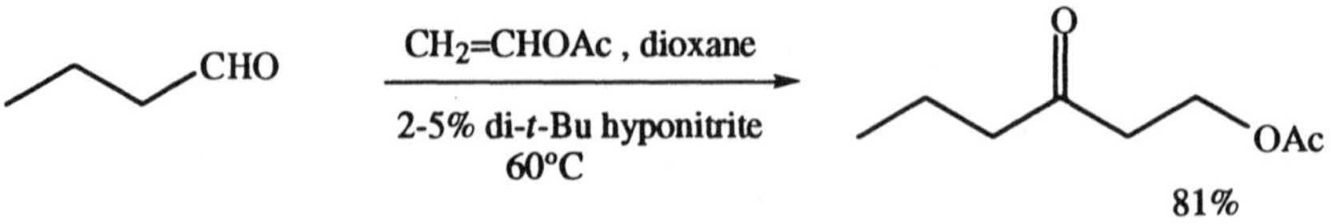
CHO
CH_2=CHOAc , dioxane
2-5% di-*t*-Bu hyponitrite
60°C
O
OAc
81%

Dang, H.–S.; Roberts, B.P. *Chem. Commun.*, ***1996***, 2201.

SECTION 170: KETONES FROM ALKYLS, METHYLENES AND ARYLS

This section lists examples of the reaction, R-CH_2-R' → R(C=O)-R'.

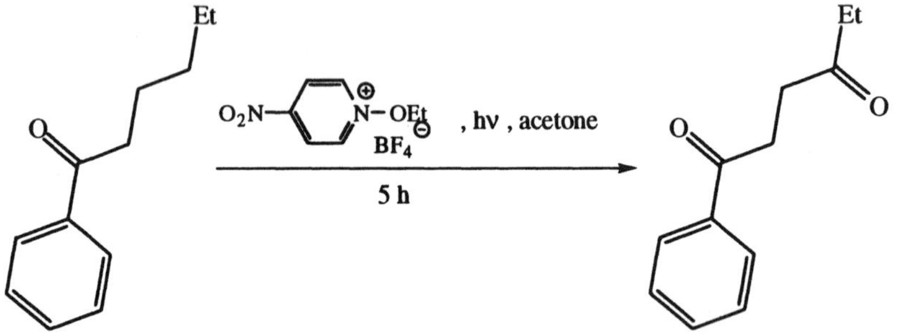
Et
O
O_2N
N−OEt
BF_4
, hν , acetone
5 h
Et
O
O

Negele, S.; Wieser, K.; Severin, T. *J. Org. Chem.*, ***1998***, *63*, 1138.

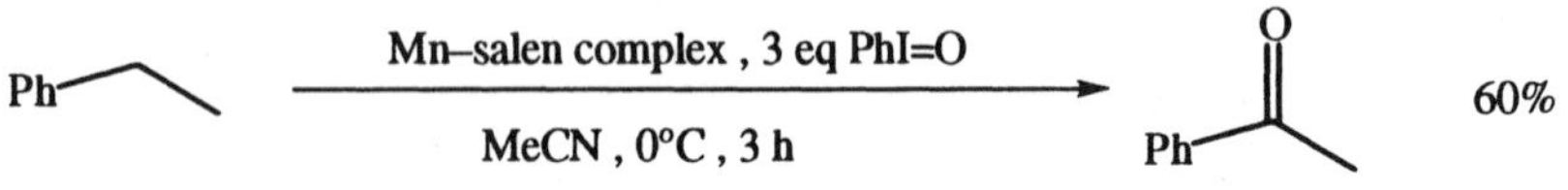
Ph
Mn–salen complex , 3 eq PhI=O
MeCN , 0°C , 3 h
O
Ph
60%

Lee, N.H.; Lee, C.–S.; Jung, D.–S. *Tetrahedron Lett*, ***1998***, *39*, 1385.

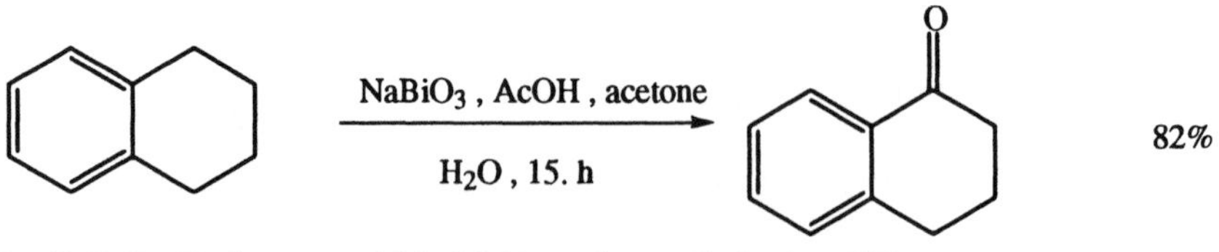
$NaBiO_3$, AcOH , acetone
H_2O , 15. h
O
82%

Banik, B.K.; Venkatraman, M.S.; Mukhopadhyay, C.; Becker, F.F. *Tetrahedron Lett.*, ***1998***, *39*, 7247.

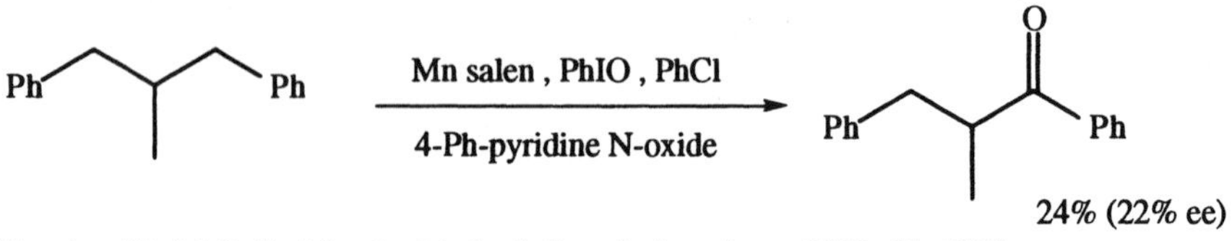
Ph
Ph
Mn salen , PhIO , PhCl
4-Ph-pyridine N-oxide
O
Ph
Ph
24% (22% ee)

Komiya, N.; Noji, S.; Murahashi, S.–I. *Tetrahedron Lett.*, ***1998***, *39*, 7921.

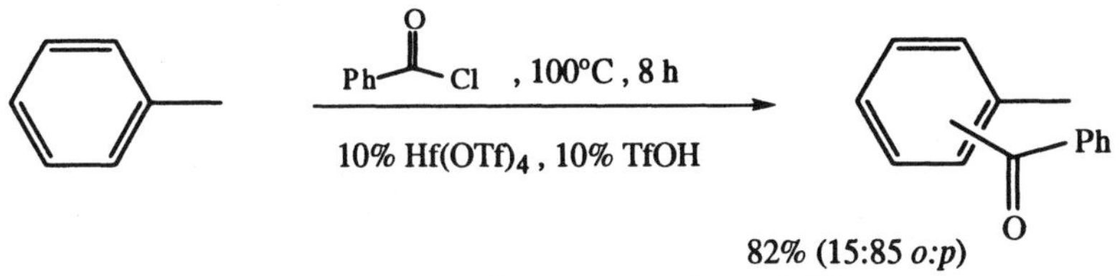
Ph Cl , 100°C , 8 h
10% Hf(OTf)4 , 10% TfOH
Ph
82% (15:85 o:p)

Kobayashi, S.; Iwamoto, S. *Tetrahedron Lett.*, ***1998***, *39*, 4697.

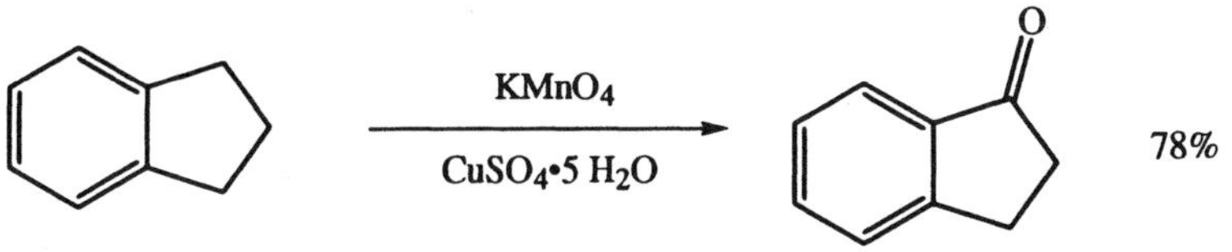
KMnO4
CuSO4•5 H2O
78%

Noureldin, N.A.; Zhao, D.; Lee, D.G. *J. Org. Chem.*, ***1997***, *62*, 8767.

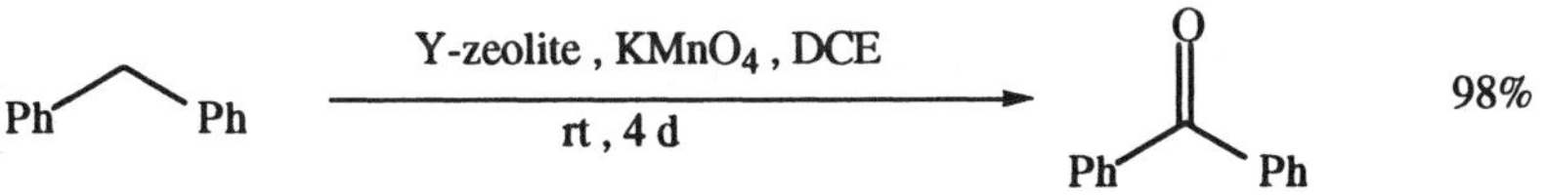
Ph Ph
Y-zeolite , KMnO4 , DCE
rt , 4 d
Ph Ph
98%

Sreekumar, R.; Padmakumar, R. *Tetrahedron Lett*, ***1997***, *38*, 5143.

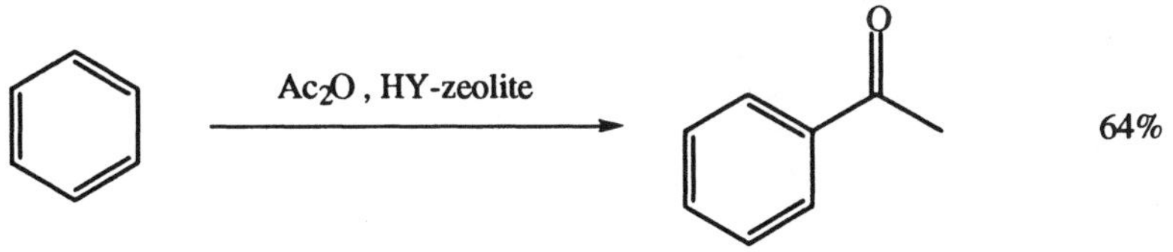
Ac2O , HY-zeolite
64%

Sreekumar, R.; Padmakumar, R. *Synth. Commun.*, ***1997***, *27*, 777.

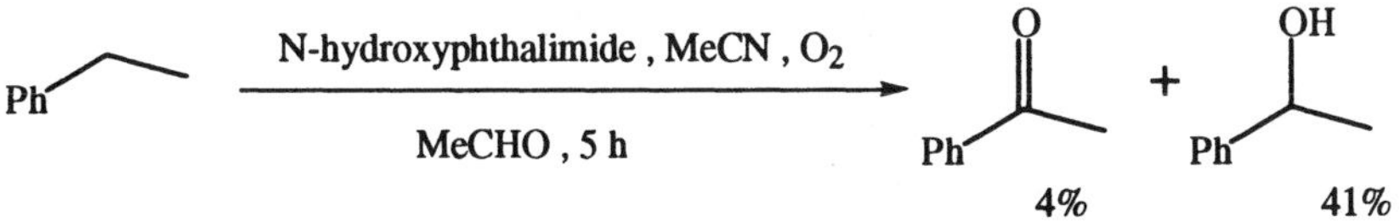
Ph
N-hydroxyphthalimide , MeCN , O2
MeCHO , 5 h
Ph
OH
Ph
4%
41%

Einhorn, C.; Einhorn, J.; Marcadal, C.; Pierre, J.–L. *Chem. Commun.*, ***1997***, 447.

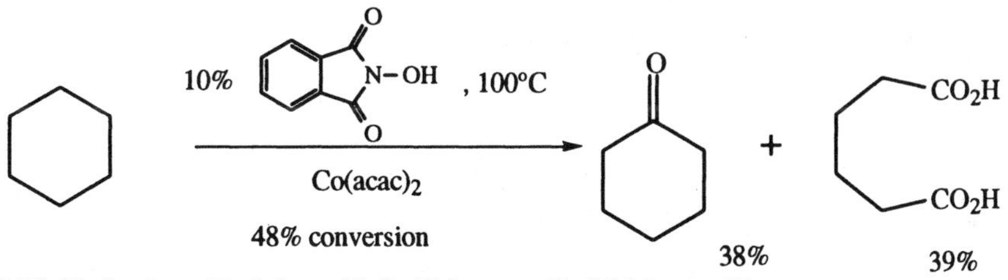
10%
N-OH
, 100°C
Co(acac)2
48% conversion
CO2H
CO2H
38%
39%

Ishii, Y.; Iwahara, T.; Sakaguchi, S.; Nakayama, K.; Nishiyama, Y.
J. Org. Chem., ***1996***, *61*, 4520.

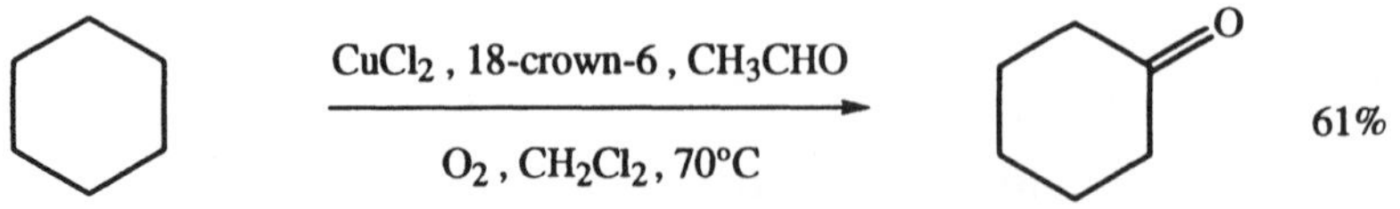
CuCl2 , 18-crown-6 , CH3CHO
O2 , CH2Cl2 , 70°C
O
61%

Koniya, N.; Naota, T.; Murahashi, S.-I. *Tetrahedron Lett.*, ***1996***, *37*, 1633.

SECTION 171: KETONES FROM AMIDES

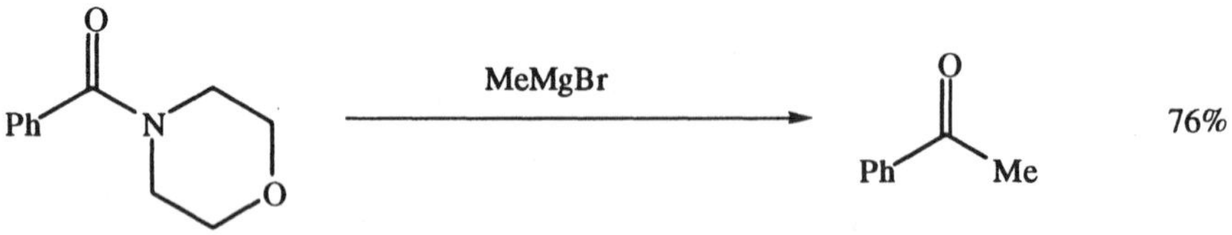
O
Ph
N
O
MeMgBr
O
Ph
Me
76%

Martín, R.; Romea, P.; Tey, C.; Urpí, F.; Vilarrasa, J. *SynLett*, ***1997***, 1414.

SECTION 172: KETONES FROM AMINES

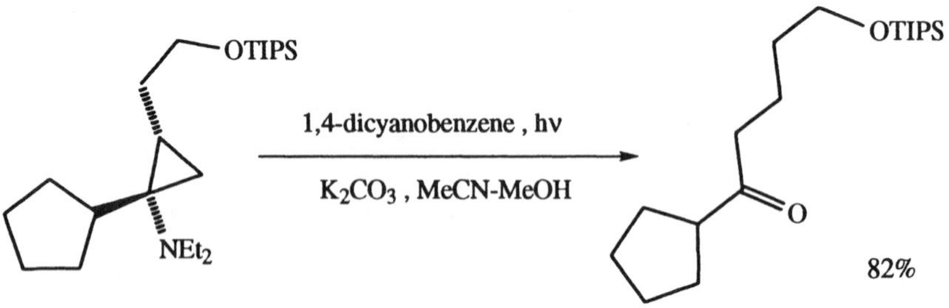
OTIPS
NEt2
1,4-dicyanobenzene , hν
K2CO3 , MeCN-MeOH
OTIPS
O
82%

Lee, J.; Sun, U.J.; Blackstock, S.C.; Cha, J.K. *J. Am. Chem. Soc.*, ***1997***, *119*, 10241.

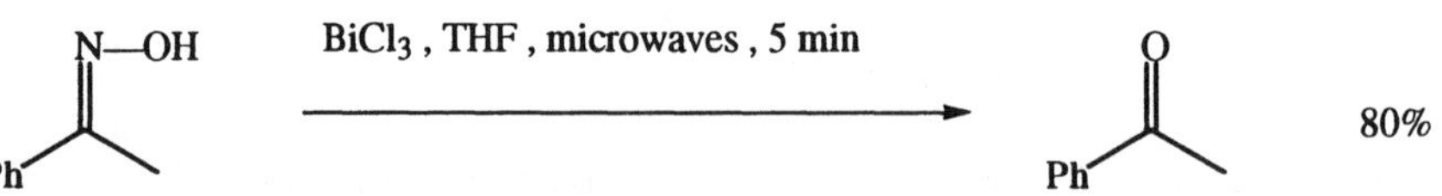
N—OH
Ph
BiCl3 , THF , microwaves , 5 min
O
Ph
80%

Boruah, A.; Baruah, B.; Prajapati, D.; Sandhu, J.S. *Tetrahedron Lett.*, ***1997***, *38*, 4267.

SECTION 173: KETONES FROM ESTERS

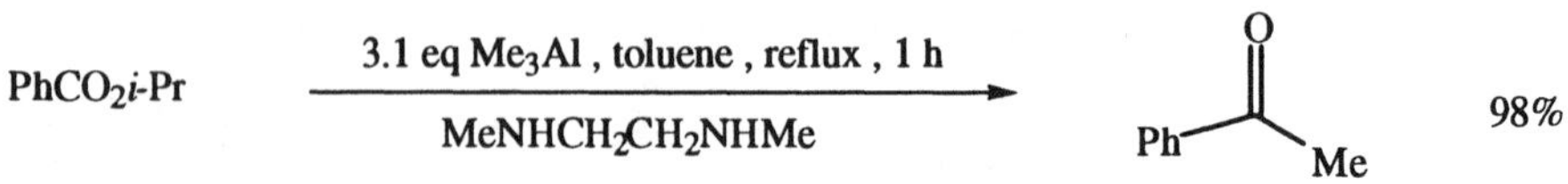
PhCO2i-Pr
3.1 eq Me3Al , toluene , reflux , 1 h
MeNHCH2CH2NHMe
O
Ph
Me
98%

Chung, E.-A.; Cho, C.-W.; Ahn, K.H. *J. Org. Chem.*, ***1998***, *63*, 7590.

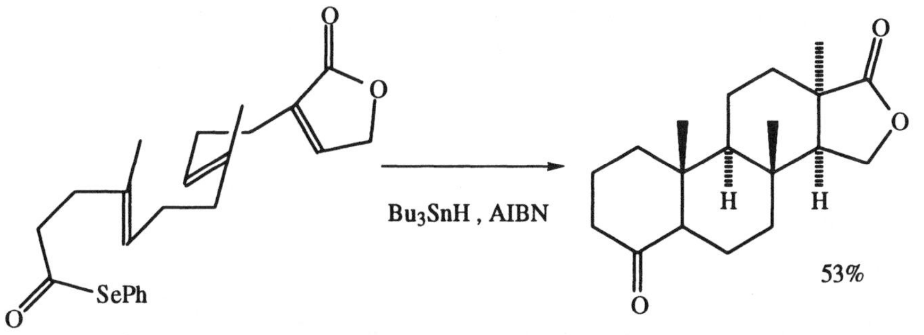

SePh
Bu3SnH , AIBN
H
H
53%

Pattenden, G.; Roberts, L.; Blake, A.J. *J. Chem. Soc., Perkin Trans. 1*, ***1998***, 8630.

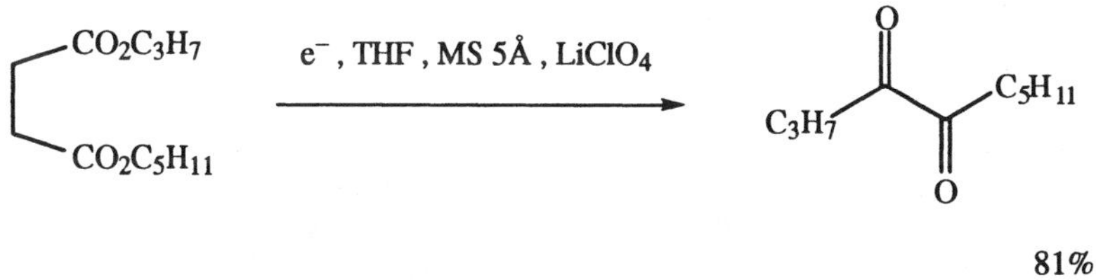

CO2C3H7
CO2C5H11
e⁻ , THF , MS 5Å , LiClO4
C3H7
C5H11
81%

Kashimura, S.; Murai, Y.; Washika, C.; Yoshihara, D.; Kataoka, Y.; Murase, H.; Shono, T. *Tetrahedron Lett*, ***1997***, *38*, 6717.

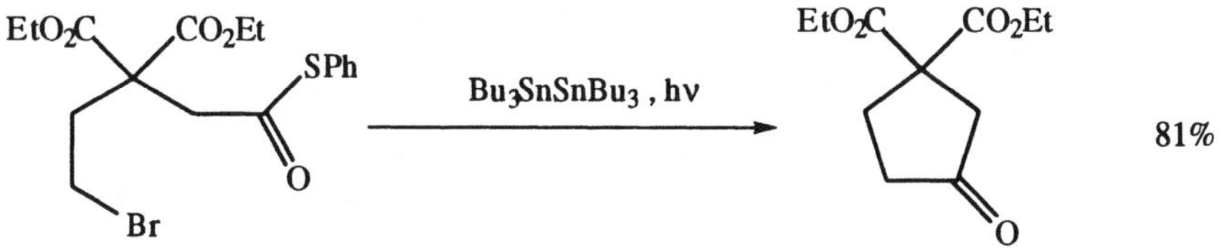

EtO2C
CO2Et
SPh
Br
Bu3SnSnBu3 , hν
81%

Kim, S.; Jon, S.Y. *Chem. Commun.*, ***1996***, 1335.

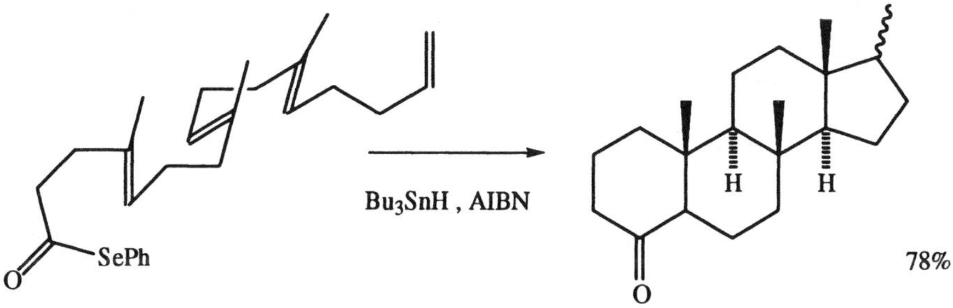

SePh
Bu3SnH , AIBN
H
H
78%

Batsanov, A.; Chen, L.; Gill, G.B.; Pattenden, G. *J. Chem. Soc., Perkin Trans. 1*, ***1996***, 45.

SECTION 174: KETONES FROM ETHERS, EPOXIDES AND THIOETHERS

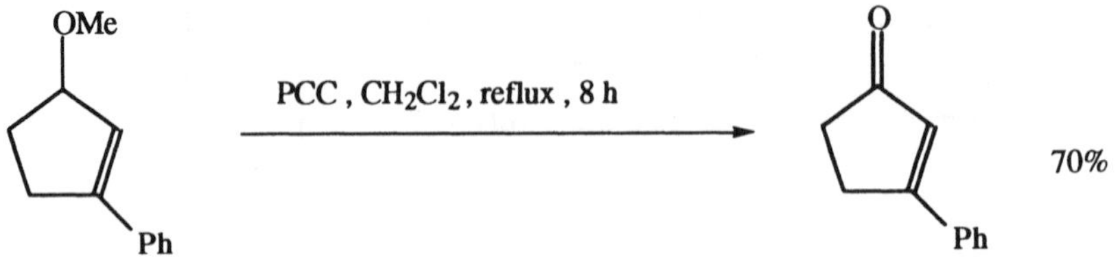
OMe
PCC , CH_2Cl_2 , reflux , 8 h
O
70%
Ph
Ph

Cossy, J.; Bouzbouz, S.; Lachgar, M.; Hakiki, A.; Tabyaoui, B. *Tetrahedron Lett.*, ***1998***, *39*, 2561.

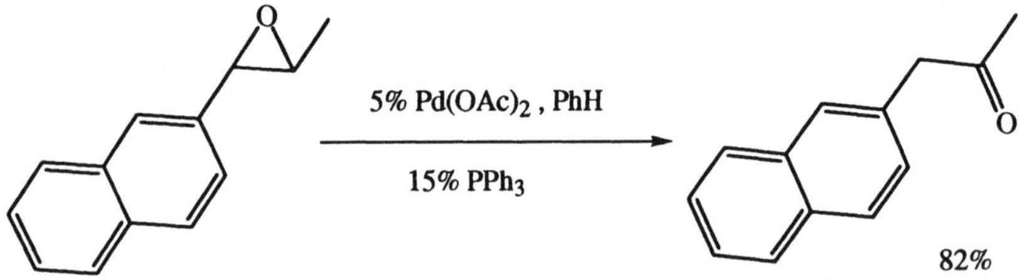
O
5% $Pd(OAc)_2$, PhH
15% PPh_3
O
82%

Kulasegaram, S.; Kulawiec, R.J. *J. Org. Chem.*, ***1997***, *62*, 6547.

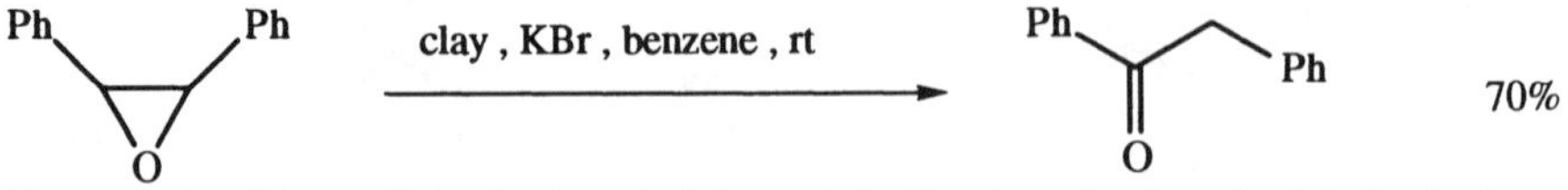
Ph
Ph
O
clay , KBr , benzene , rt
Ph
Ph
O
70%

Martínez, R.; Velasco, M.; Martínez, I.; Menconi, I.; Ramírez, A.; Angeles, E.; Regla, I.; López, R. *J. Heterocyclic Chem.*, ***1997***, *34*, 1865.

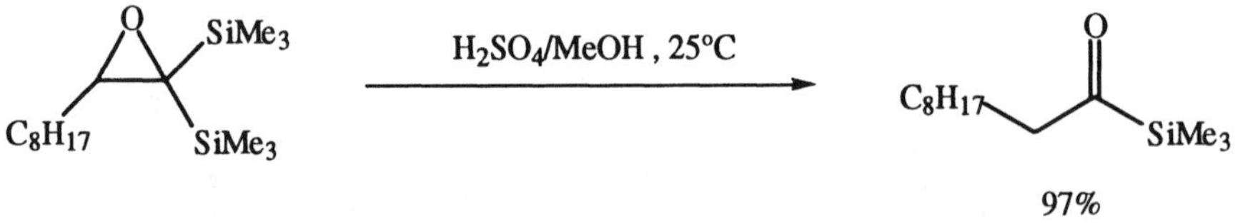
O
$SiMe_3$
C_8H_{17}
$SiMe_3$
H_2SO_4/MeOH , 25°C
O
C_8H_{17}
$SiMe_3$
97%

Hodgson, D.M.; Comina, P.J. *Chem. Commun.*, ***1996***, 755.

SECTION 175: KETONES FROM HALIDES AND SULFONATES

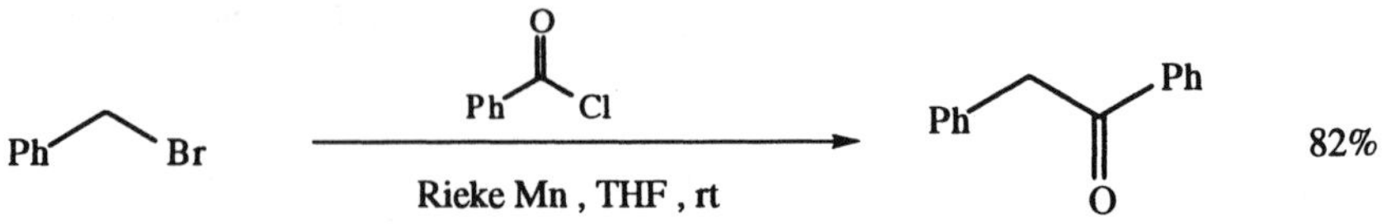
O
Ph
Cl
Ph
Br
Rieke Mn , THF , rt
Ph
Ph
O
82%

Kim, S.-H.; Rieke, R.D. *J. Org. Chem.*, ***1998***, *63*, 6766.

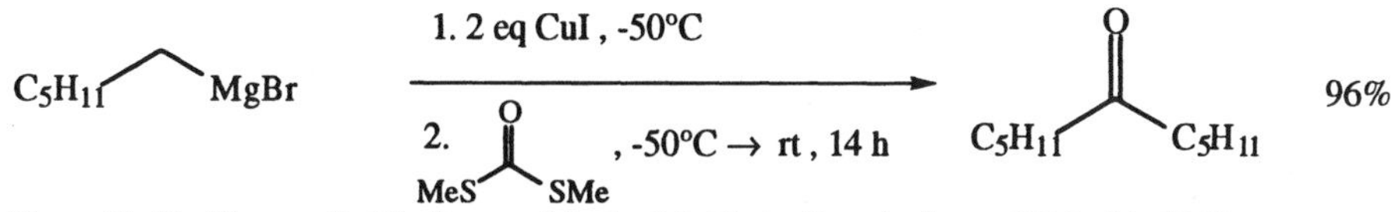
1. 2 eq CuI , -50°C
2. , -50°C → rt , 14 h
MeS SMe
96%

Chen, C.–D.; Huang, J.–W.; Leung, M.–k.; Li, H.–h. *Tetrahedron, **1998**, 54*, 4067.

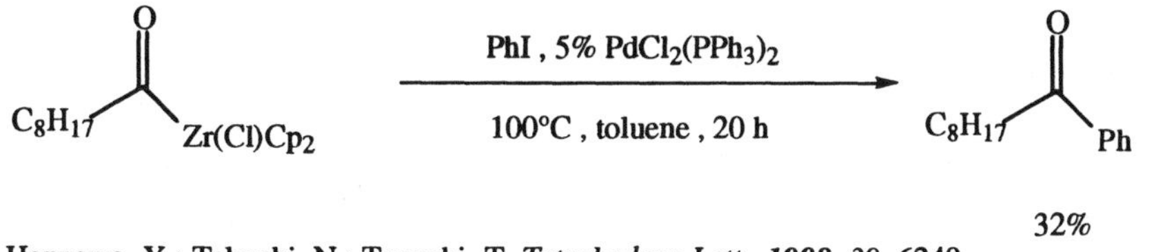
PhI , 5% PdCl2(PPh3)2
100°C , toluene , 20 h
32%

Hanzawa, Y.; Tabuchi, N.; Taguchi, T. *Tetrahedron Lett., **1998**, 39*, 6249.

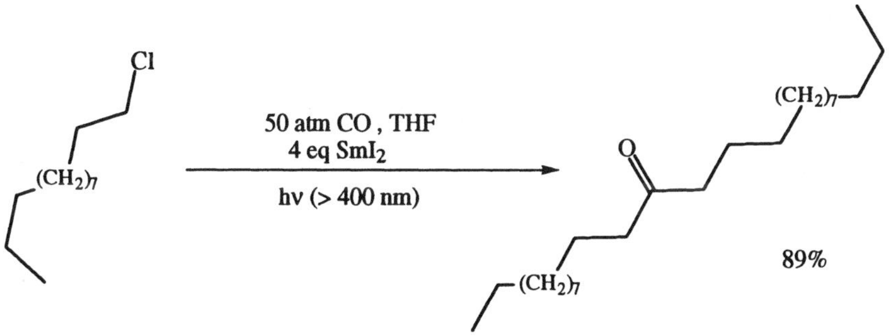
50 atm CO , THF
4 eq SmI2
hν (> 400 nm)
89%

Ogawa, A.; Sumino, Y.; Nanke, T.; Ohya, S.; Sonoda, N.; Hirao, T.
*J. Am. Chem. Soc., **1997**, 119*, 2745.

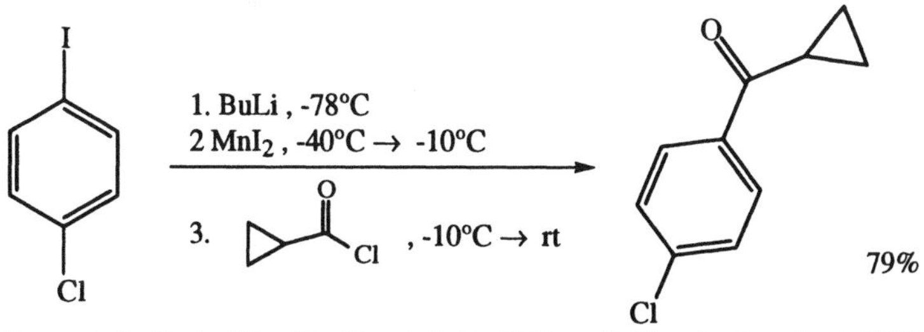
1. BuLi , -78°C
2 MnI2 , -40°C → -10°C
3. , -10°C → rt
79%

Klement, I.; Stadmüller, H.; Knochel, P.; Cahiez, G. *Tetrahedron Lett., **1997**, 38*, 1927.

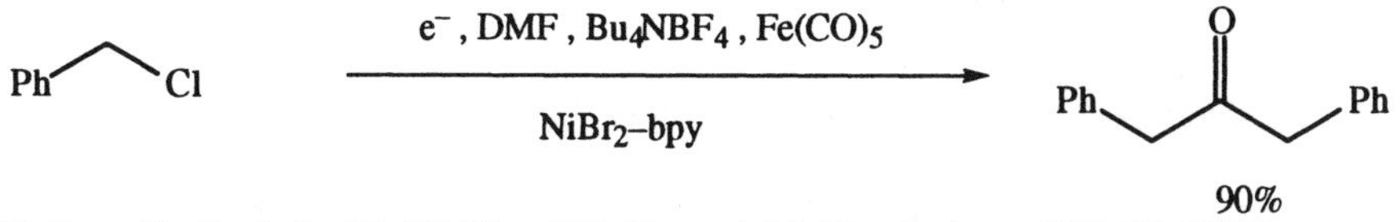
e⁻ , DMF , Bu4NBF4 , Fe(CO)5
NiBr2–bpy
90%

Dolhem, E.; Ocafrain, M.; Nédélec, J.Y.; Troupel, M. *Tetrahedron, **1997**, 53*, 17089.

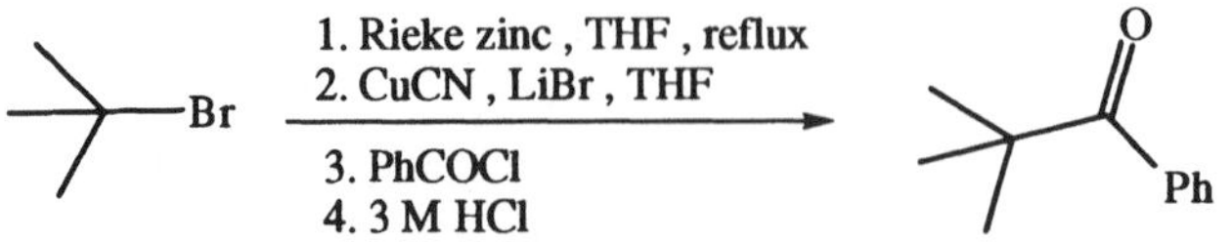
Br
1. Rieke zinc , THF , reflux
2. CuCN , LiBr , THF
3. PhCOCl
4. 3 M HCl
O
Ph

75%

Kang, S.-K.; Lee, H.-W.; Jang, S.-B.; Kim, T.-H.; Pyun, S.-J. *J. Org. Chem.*, ***1996***, *61*, 2604.

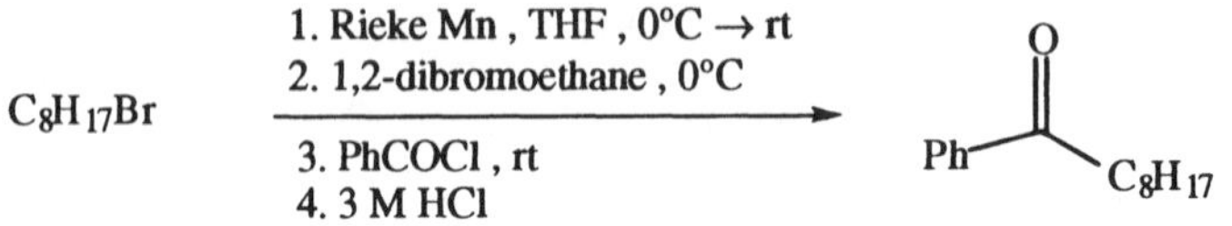
C8H17Br
1. Rieke Mn , THF , 0°C → rt
2. 1,2-dibromoethane , 0°C
3. PhCOCl , rt
4. 3 M HCl
O
Ph
C8H17

49%

Kim, S.-H.; Hanson, M.V.; Rieke, R.D. *Tetrahedron Lett.*, ***1996***, *37*, 2197.

Related Methods: Section 177 (Ketones from Ketones).
Section 55 (Aldehydes from Halides).

SECTION 176: KETONES FROM HYDRIDES

This section lists examples of the replacement of hydrogen by ketonic groups, R-H → R(C=O)-R'. For the oxidation of methylenes, R_2CH_2 → $R_2C=O$, see section 170 (Ketones from Alkyls).

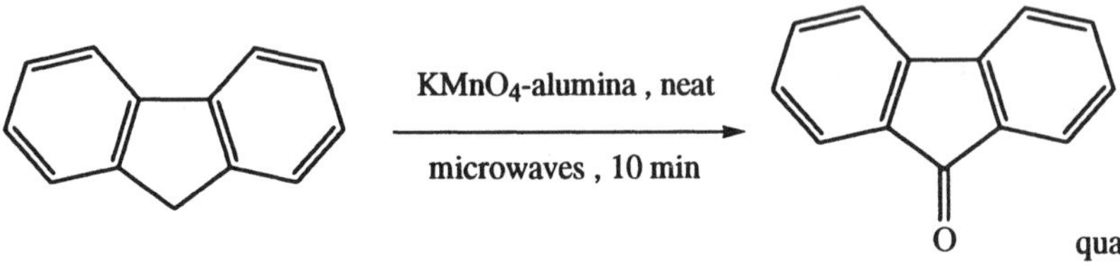
KMnO4-alumina , neat
microwaves , 10 min
O

quant

Oussaid, A.; Loupy, A. *J. Chem. Res. (S)*, ***1997***, 342.

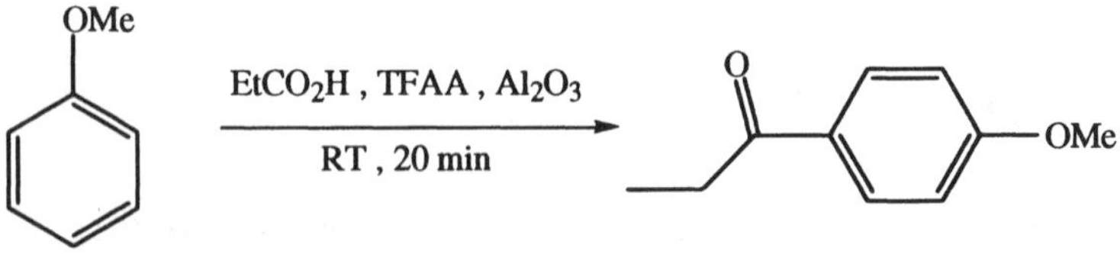
OMe
EtCO2H , TFAA , Al2O3
RT , 20 min
O
OMe

95%

Ranu, B.C.; Ghosh, K.; Jana, U. *J. Org. Chem.*, ***1996***, *61*, 9546.

SECTION 177: KETONES FROM KETONES

This section contains alkylations of ketones and protected ketones, ketone transpositions and annulations, ring expansions and ring openings and dimerizations. Conjugate reductions and Michael alkylations of enone are listed in Section 74 (Alkyls from Alkenes).

For the preparation of enamines or imines from ketones, see Section 356 (Amine-Alkene)

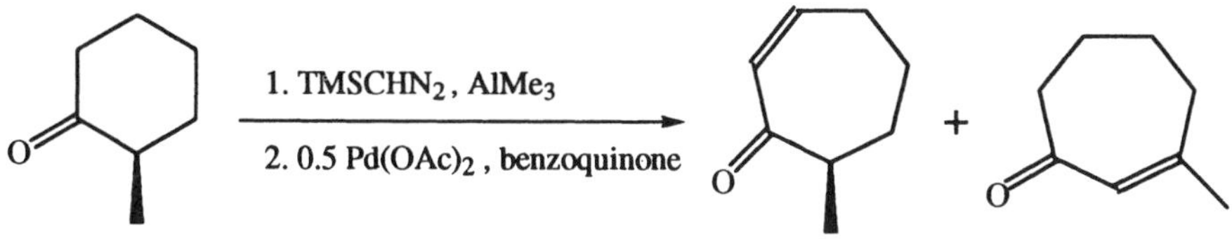
1. TMSCHN2 , AlMe3
2. 0.5 Pd(OAc)2 , benzoquinone
O
+

(2.1 : 1) 98x90%

Yang, S.; Hungerhoff, B.; Metz, P. *Tetrahedron Lett.*, ***1998***, *39*, 2097.

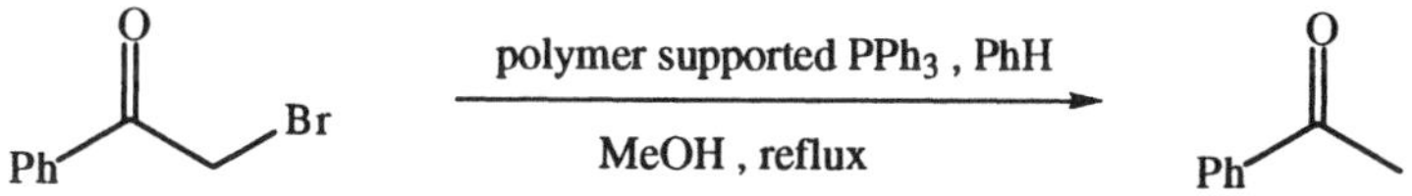
polymer supported PPh3 , PhH
MeOH , reflux
Ph
Br
O

45%

Dhuru, S.P.; Padiya, K.J.; Salunkhe, M.M. *J. Chem. Res. (S)*, ***1998***, 56.

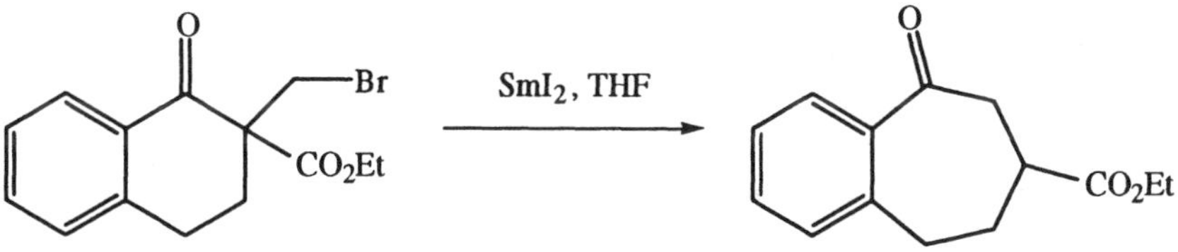
SmI2 , THF
Br
CO2Et
O

65%

Hasegawa, E.; Kitazume, T.; Suzuki, K.; Tosaka, E. *Tetrahedron Lett*, ***1998***, *39*, 4059.

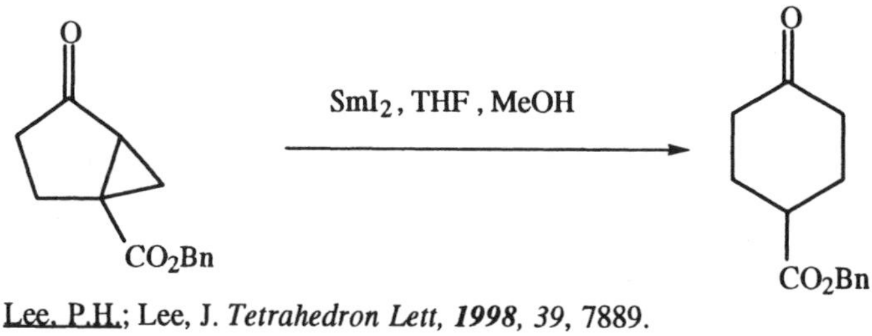
SmI2 , THF , MeOH
CO2Bn
O

83%

Lee, P.H.; Lee, J. *Tetrahedron Lett*, ***1998***, *39*, 7889.

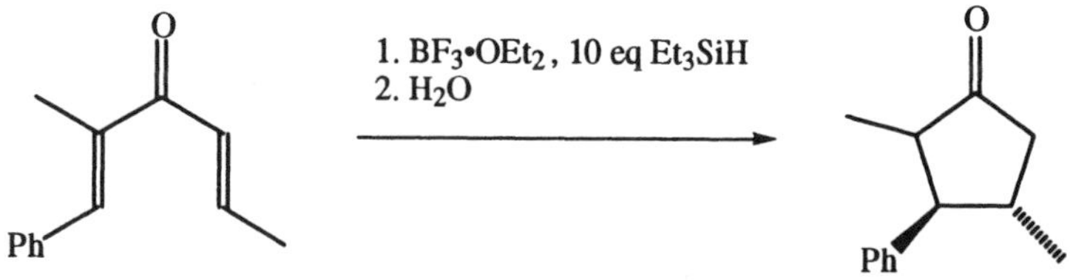
1. BF3•OEt2 , 10 eq Et3SiH
2. H2O
Ph
O

77% (3.3:1 β-Me:α-Me)

Giese, S.; West, F.G. *Tetrahedron Lett*, ***1998***, *39*, 8393.

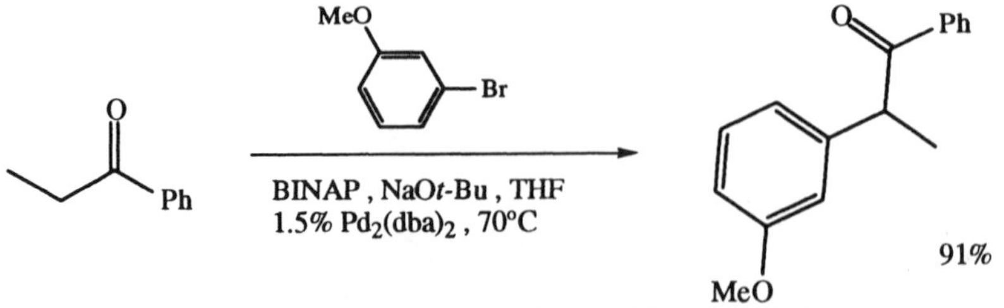
MeO
Br
O
Ph
BINAP , NaO*t*-Bu , THF
1.5% $Pd_2(dba)_2$, 70°C
O
Ph
MeO
91%

Palucki, M.; Buchwald, S.L. *J. Am. Chem. Soc.*, ***1997***, *119*, 11108.

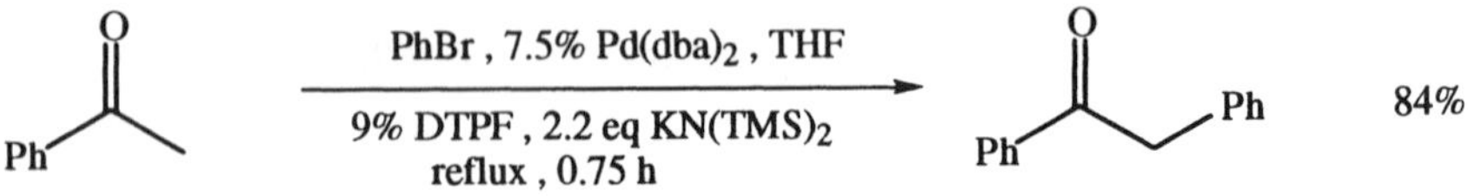
O
Ph
PhBr , 7.5% $Pd(dba)_2$, THF
9% DTPF , 2.2 eq $KN(TMS)_2$
reflux , 0.75 h
O
Ph
Ph
84%

DTPF = 1,1'-*bis*(di-*o*-tolylphosphino)ferrocene

Hamann, B.C.; Hartwig, J.F. *J. Am. Chem. Soc.*, ***1997***, *119*, 12382.

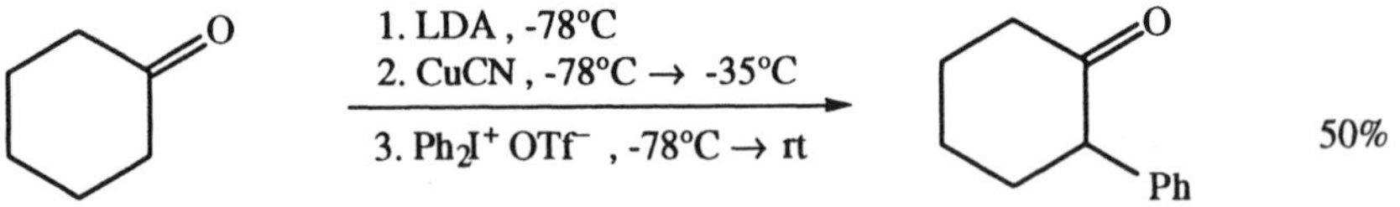
O
1. LDA , -78°C
2. CuCN , -78°C → -35°C
3. Ph_2I^+ OTf^- , -78°C → rt
O
Ph
50%

Ryan, J.H.; Stang, P.J. *Tetrahedron Lett.*, ***1997***, *38*, 5061.

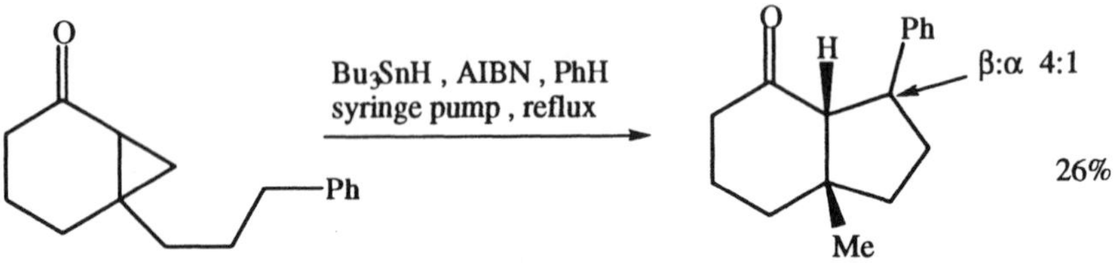
O
Ph
Bu_3SnH , AIBN , PhH
syringe pump , reflux
O
H
Ph
β:α 4:1
26%
Me

Chambournier, G.; Krishnamurthy, V.; Rawal, V.H. *Tetrahedron Lett.*, ***1997***, *38*, 6313.

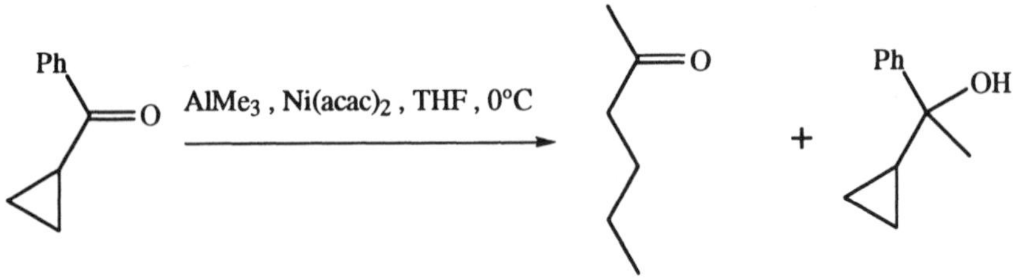
Ph
O
$AlMe_3$, $Ni(acac)_2$, THF , 0°C
O
+
Ph
OH

	76%	0%
with MeLi rather than $AlMe_3$	0%	99%

Ichiyanagi, T.; Kuniyama, S.; Shimizu, M.; Fujisawa, T. *Chem. Lett.*, ***1997***, 1149.

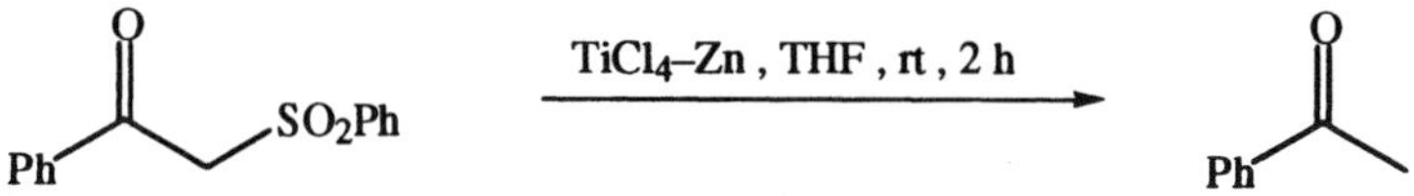
O
SO2Ph
Ph
TiCl4–Zn , THF , rt , 2 h
O
Ph

83%

Guo, H.; Ye, S.; Wang, J.; Zhang, Y. *J. Chem. Res. (S)*, ***1997***, 114.

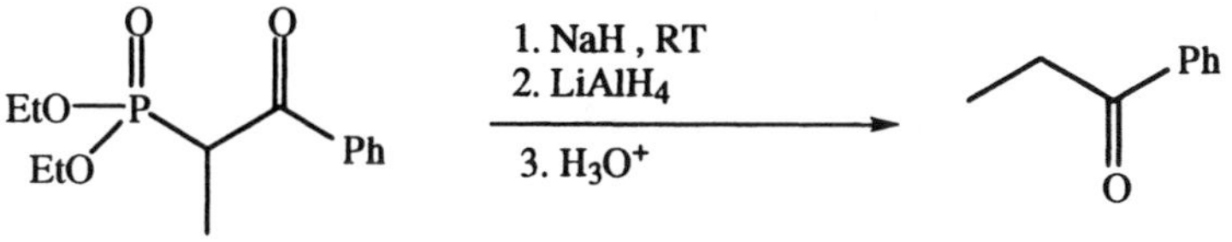
O
O
EtO
P
EtO
Ph
1. NaH , RT
2. LiAlH4
3. H3O+
Ph
O

96%

Hong, J.E.; Shin, W.S.; Jang, W.P.; Oh, D.Y. *J. Org. Chem.*, ***1996***, *61*, 2199.

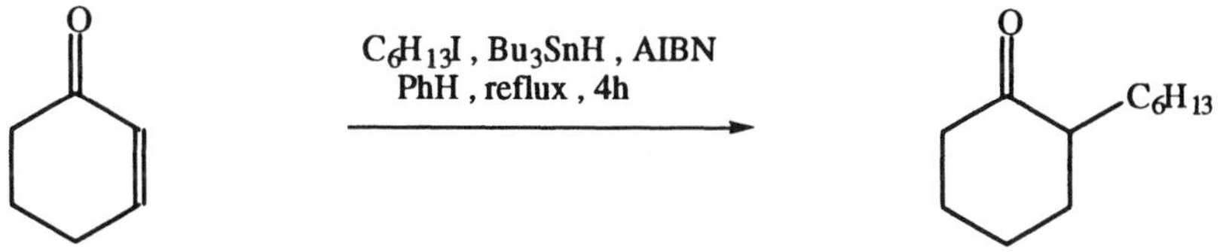
O
C6H13I , Bu3SnH , AIBN
PhH , reflux , 4h
O
C6H13

61%

Enholm, E.J.; Whitley, P.E. *Tetrahedron Lett*, ***1996***, *37*, 559.

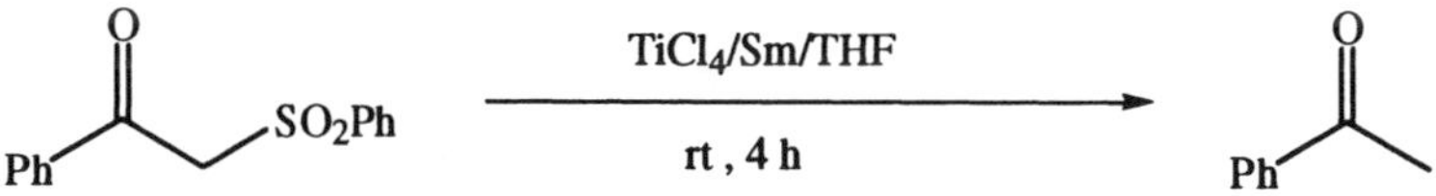
O
SO2Ph
Ph
TiCl4/Sm/THF
rt , 4 h
O
Ph

65%

Wang, J.; Zhang, Y. *Synth. Commun.*, ***1996***, *26*, 1931.

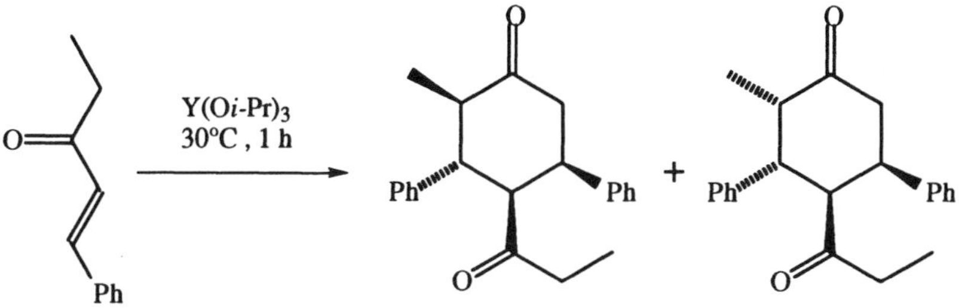
O
Ph
Y(O*i*-Pr)3
30°C , 1 h
O
Ph
Ph
O
+
O
Ph
Ph
O

(69 : 31) 95%

Okano, T.; Ohno, K.; Kiji, J. *Chem. Lett.*, ***1996***, 1041.

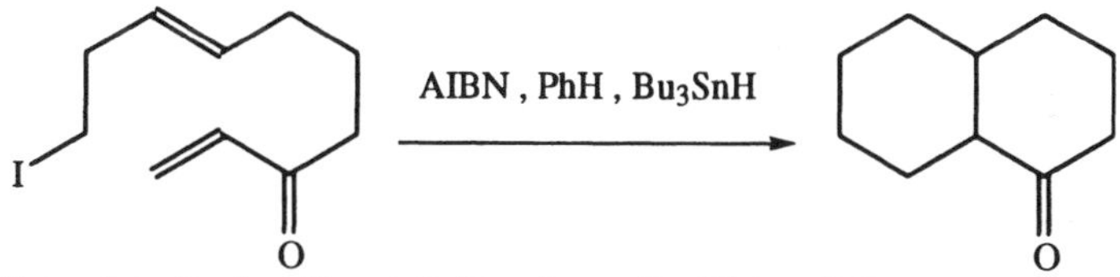
I
O
AIBN , PhH , Bu3SnH
O

72%
(2:1 *trans:cis*)

Pattenden, G.; Smithies, A.J.; Tapolczay, D.; Walter, D.S.
J. Chem. Soc.. Perkin Trans. 1. ***1996***. 7.

REVIEW:

"Conversion of the Thiocarbonyl Group Into the Carbonyl Group."
Corsaro, A.; Pistarà, V. *Tetrahedron*, ***1998***, *54*, 15027.

Related Methods: Section 49 (Aldehydes from Aldehydes).

SECTION 178: KETONES FROM NITRILES

NO ADDITIONAL EXAMPLES

SECTION 179: KETONES FROM ALKENES

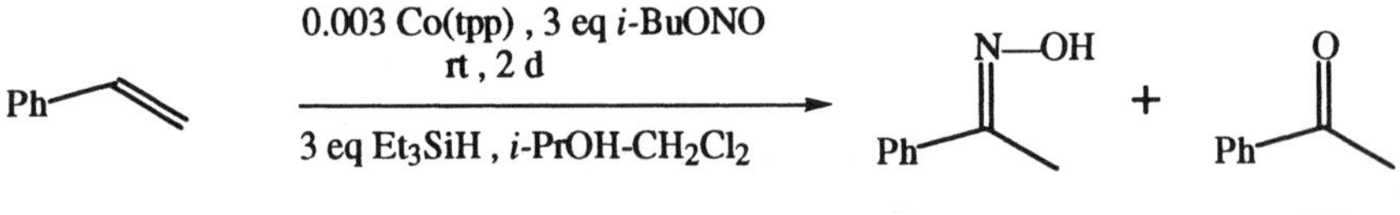
Ph
0.003 Co(tpp) , 3 eq i-BuONO
rt , 2 d
3 eq Et3SiH , i-PrOH-CH2Cl2
N—OH
Ph
+
O
Ph

67% 7%

Sugamoto, K.; Hamasuna, Y.; Matsushita, Y.–i. *SynLett*, ***1998***, 1270.

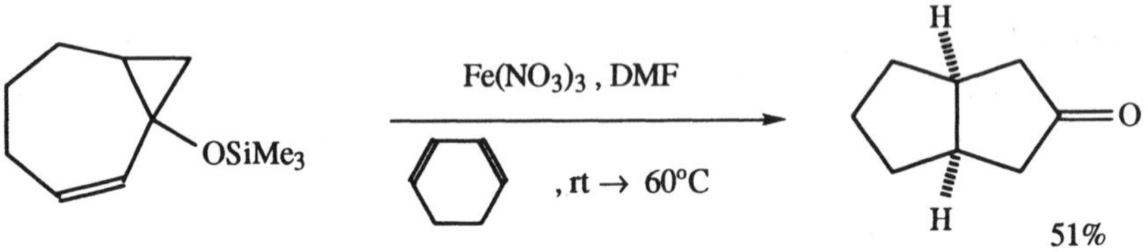
OSiMe3
Fe(NO3)3 , DMF
, rt → 60°C
H
O
H

51%

Booker-Milburn, K.I.; Dainty, R.F. *Tetrahedron Lett.*, ***1998***, *39*, 5097.

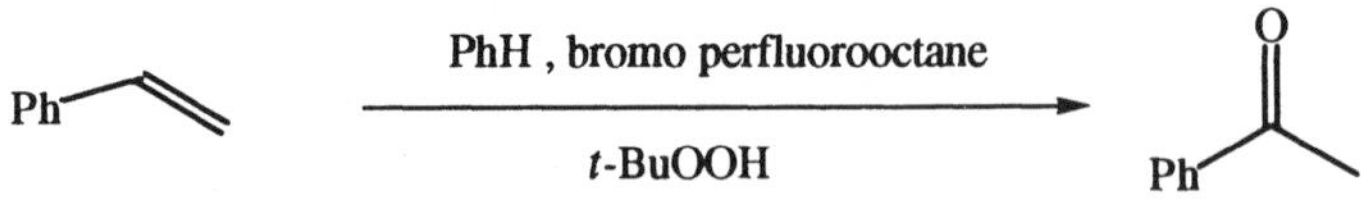
Ph
PhH , bromo perfluorooctane
t-BuOOH
O
Ph

95%

Betzemeier, B.; Lhermitte, F.; Knochel, P. *Tetrahedron Lett.*, ***1998***, *39*, 6667.

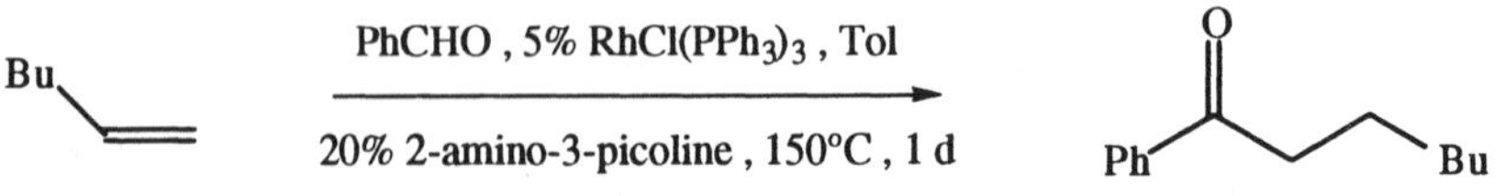
Bu
PhCHO , 5% RhCl(PPh3)3 , Tol
20% 2-amino-3-picoline , 150°C , 1 d
O
Ph
Bu

72%

Jun, C.-H.; Lee, H.; Hong, J.–B. *J. Org. Chem.*, ***1997***, *62*, 1200.

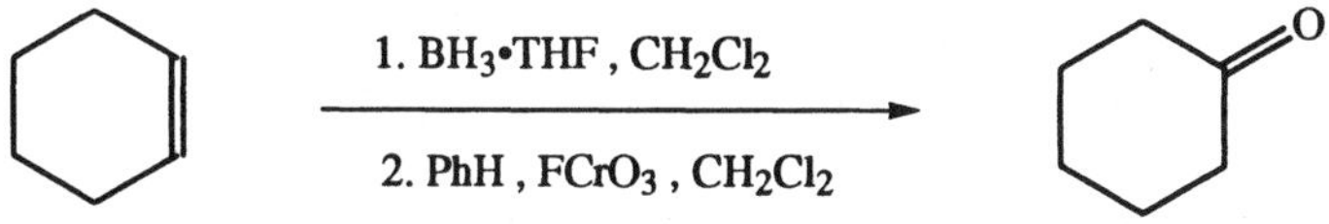
1. BH_3•THF , CH_2Cl_2
2. PhH , $FCrO_3$, CH_2Cl_2
O

78%

Parish, E.J.; Kizito, S.A.; Sun, H. *J. Chem. Res. (S)*, ***1997***, 64.

See also: Section 134 (Ethers from Alkenes).
Section 174 (Ketones from Ethers).

SECTION 180: KETONES FROM MISCELLANEOUS COMPOUNDS

Conjugate reductions and reductive alkylations of enones are listed in Section 74 (Alkyls from Alkenes).

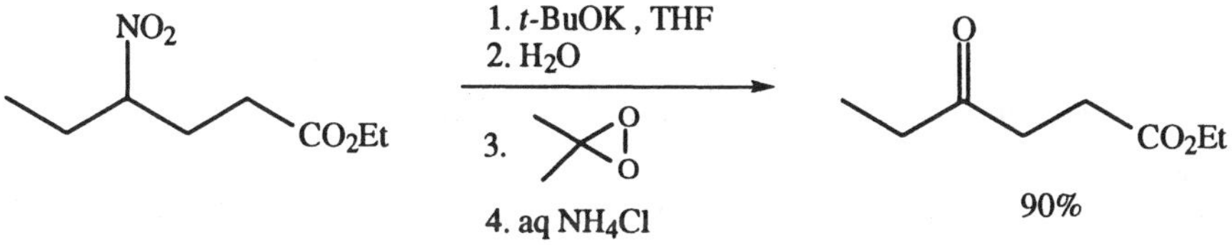
NO_2
CO_2Et
1. *t*-BuOK , THF
2. H_2O
3.
4. aq NH_4Cl
O
CO_2Et
90%

Adam, W.; Makosza, M.; Saha-Möller, C.R.; Zhao, C.-G. *SynLett*, ***1998***, 1335.

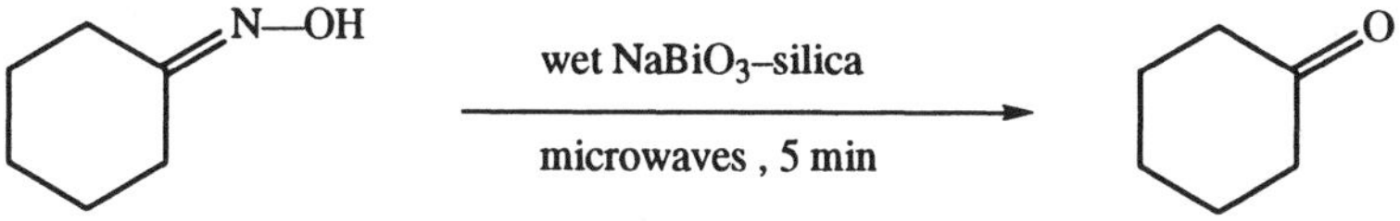
N—OH
wet $NaBiO_3$–silica
microwaves , 5 min
O

73%

Mitra, A.K.; De, A.; Karchaudhuri, N. *SynLett*, ***1998***, 1345.

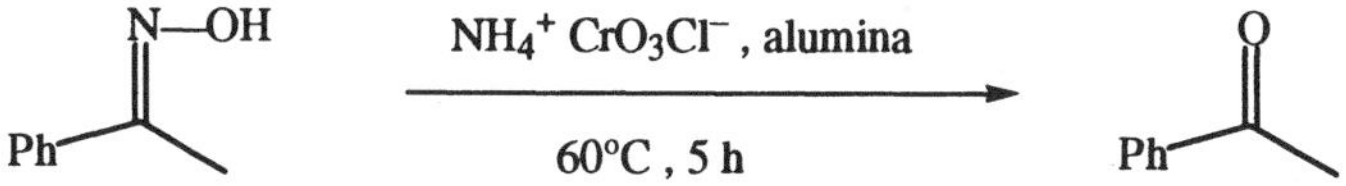
N—OH
Ph
NH_4^+ CrO_3Cl^- , alumina
60°C , 5 h
O
Ph

85%

Zhang, G.-S.; Yang, D.-H.; Chen, M.-F.; Cai, K. *Synth. Commun.*, ***1998***, *28*, 2221.

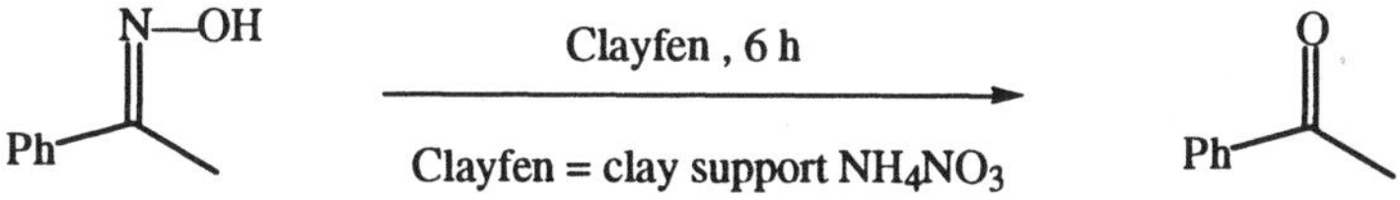
N—OH
Ph
Clayfen , 6 h
Clayfen = clay support NH_4NO_3
O
Ph

95%

Meshram, H.M.; Reddy, G.S.; Srinivas, D.; Yadav, J.S. *Synth. Commun.*, ***1998***, *28*, 2593.

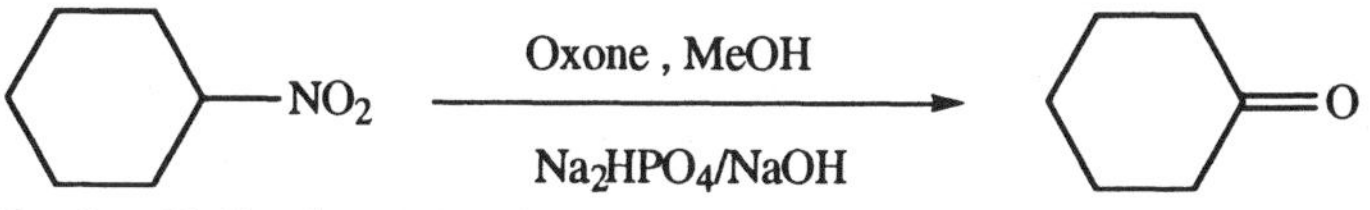
NO_2
Oxone , MeOH
Na_2HPO_4/NaOH
O

81%

Ceccherelli, P.; Curini, M.; Marcotullio, M.C.; Epifano, F.; Rosati, O. *Synth. Commun.*, ***1998***, *28*, 3057.

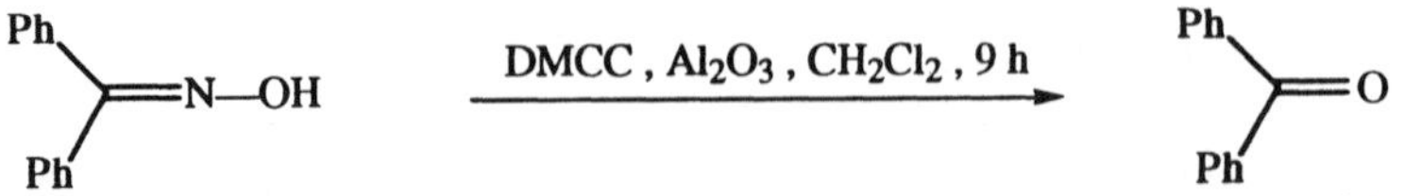

74%

Zhang, G.-S.; Yang, D.-H.; Chen, M.-F. *Synth. Commun.*, ***1998***, *28*, 3721.

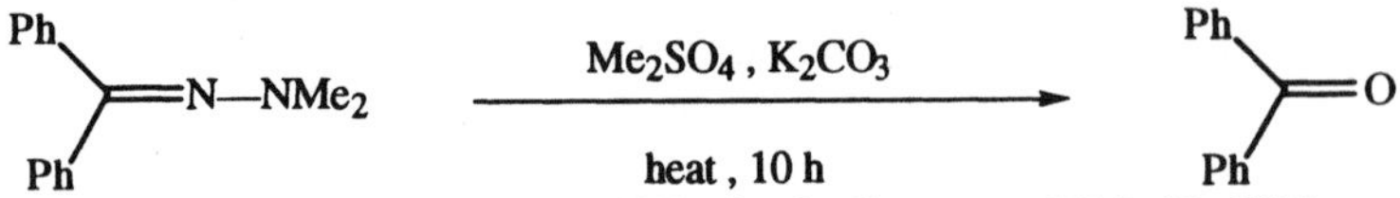

97%

Kamaol, A.; Arifuddin, M.; Rao, N.V. *Synth. Commun.*, ***1998***, *28*, 3927.

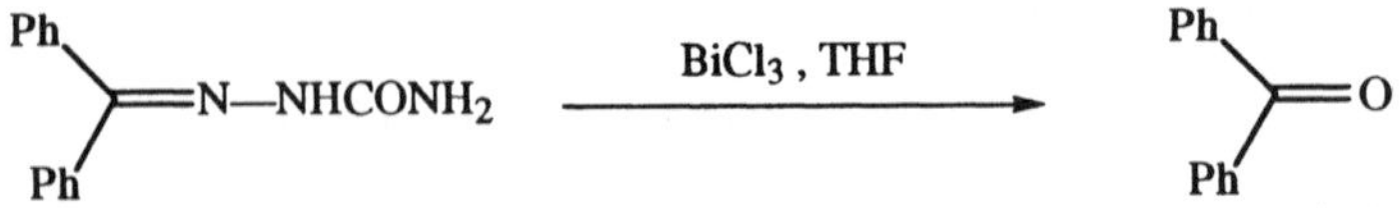

90%

Baruah, M.; Prajapati, D.; Sandhu, J.S. *Synth. Commun.*, ***1998***, *28*, 4157.

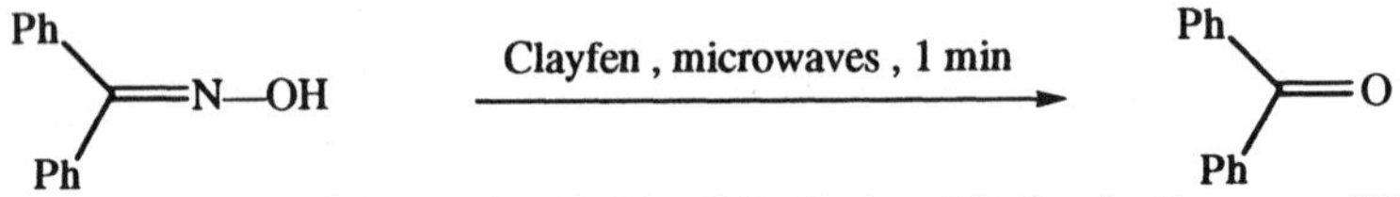

78%

Meshram, H.M.; Srinivas, D.; Reddy, G.S.; Yadav, J.S. *Synth. Commun.*, ***1998***, *38*, 4401.

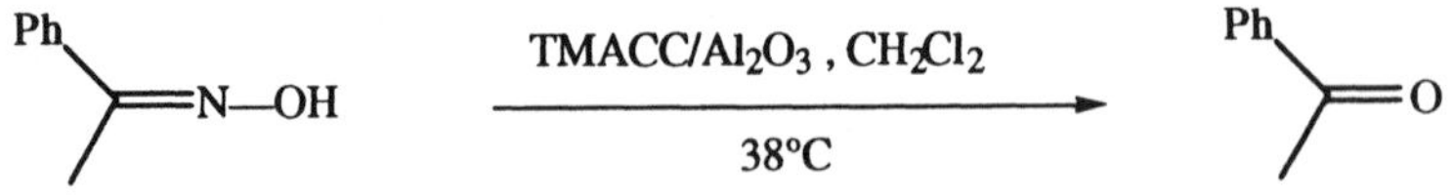

83%

TMACC = $Me_3NH^+ \ CrO_3Cl$

Zhang, G.-S.; Yang, D.-H.; Chen, M.-F. *Org. Prep. Proceed. Int.*, ***1998***, *30I*, 713.

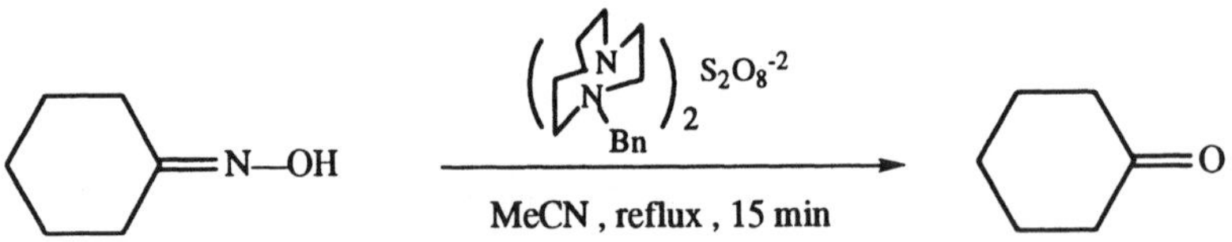

95%

Hajipour, A.R.; Mohammadpoor-Baltork, I.; Kianfar, G.
Bull. Chem. Soc. Jpn., ***1998***, *71*, 2655.

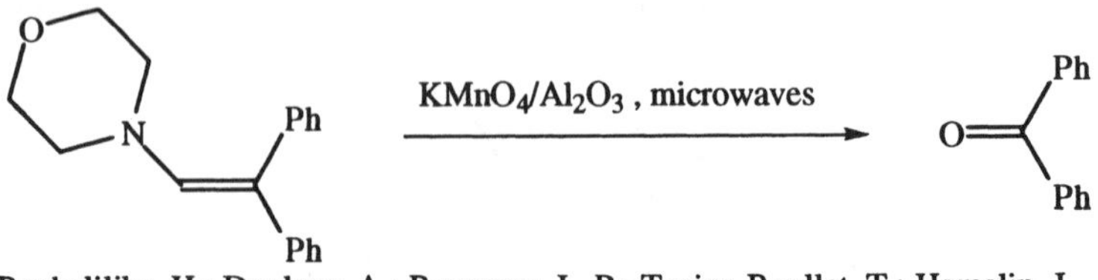

83%

Benhaliliba, H.; Derdour, A.; Bazureau, J.-P.; Texier-Boullet, T.; Hamelin, J.
Tetrahedron Lett., ***1998***, *39*, 541.

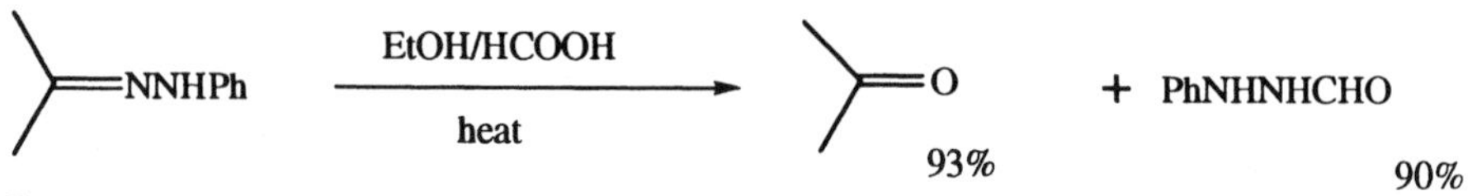
NNHPh
EtOH/HCOOH
heat
O
+ PhNHNHCHO
93%
90%

Chakrabarty, M.; Khasnobis, S. *Synth. Commun.*, ***1998***, *28*, 1361.

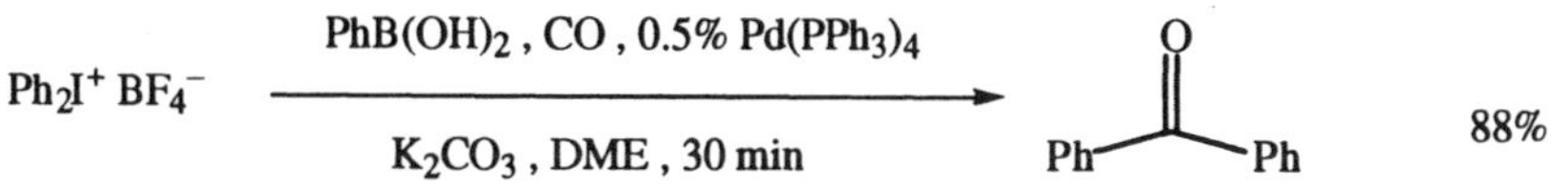
$Ph_2I^+ BF_4^-$
$PhB(OH)_2$, CO , 0.5% $Pd(PPh_3)_4$
K_2CO_3 , DME , 30 min
O
Ph
Ph
88%

Kang, S.–K.; Lim, K.–H.; Ho, P.–S.; Yoon, S.–K.; Son, H.–J.
Synth. Commun., ***1998***, *28*, 1481.

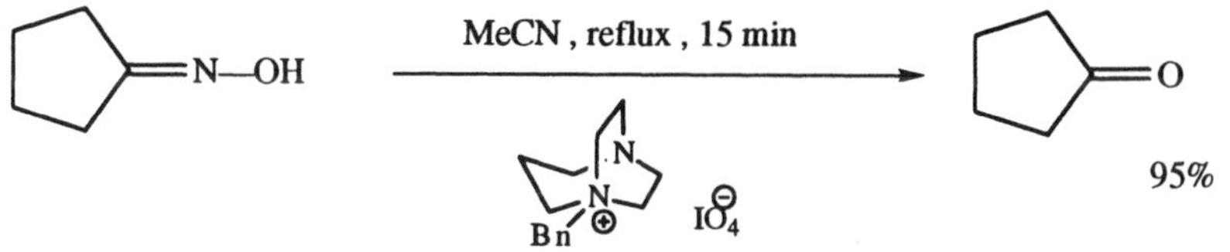
N—OH
MeCN , reflux , 15 min
N
N
Bn
IO_4
O
95%

Hajipour, A.R.; Mahboubghah, N. *J. Chem. Res. (S)*, ***1998***, 122.

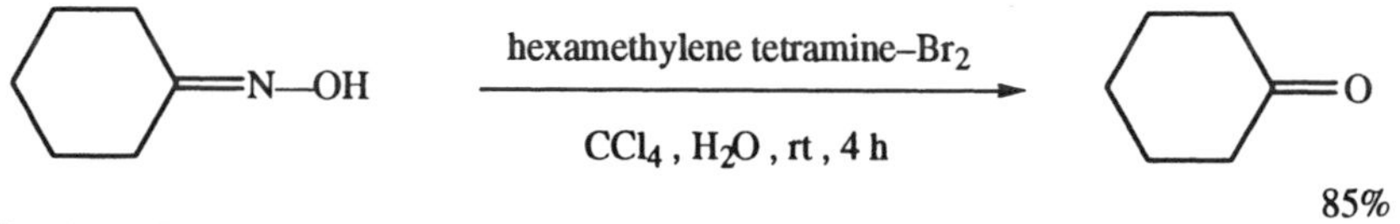
N—OH
hexamethylene tetramine–Br_2
CCl_4 , H_2O , rt , 4 h
O
85%

Bandgar, B.P.; Admane, S.B.; Jare, S.S. *J. Chem. Soc. (S)*, ***1998***, 154.

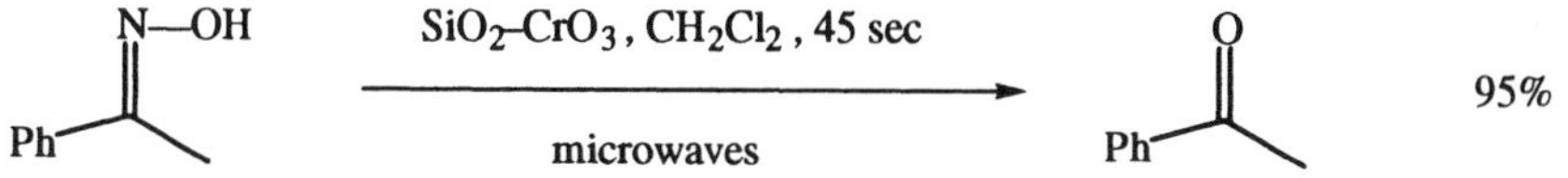
N—OH
Ph
SiO_2–CrO_3 , CH_2Cl_2 , 45 sec
microwaves
O
Ph
95%

Bendale, P.M.; Khadilkar, B.M. *Tetrahedron Lett.*, ***1998***, *39*, 5867.

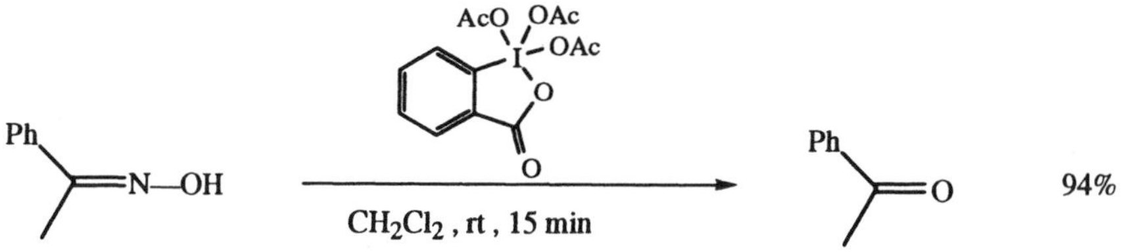
AcO
OAc
I
OAc
O
O
Ph
N—OH
CH_2Cl_2 , rt , 15 min
Ph
O
94%

Chaudhari, S.S.; Akamanchi, K.G. *Tetrahedron Lett.*, ***1998***, *39*, 3209.

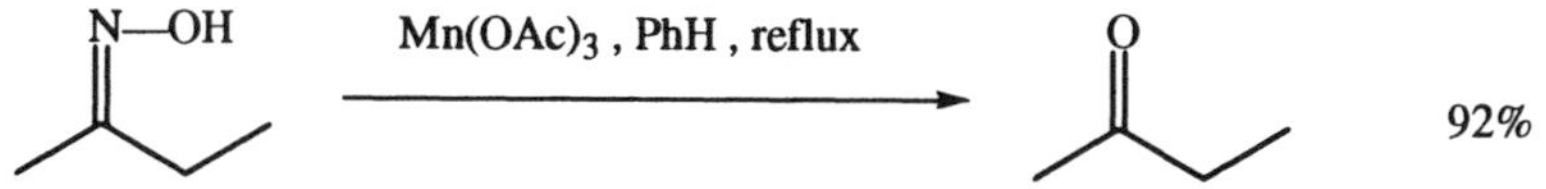
N—OH
$Mn(OAc)_3$, PhH , reflux
O
92%

Demir, A.S.; Tanyeli, C.; Altinel, E. *Tetrahedron Lett.*, ***1997***, *38*, 7267.

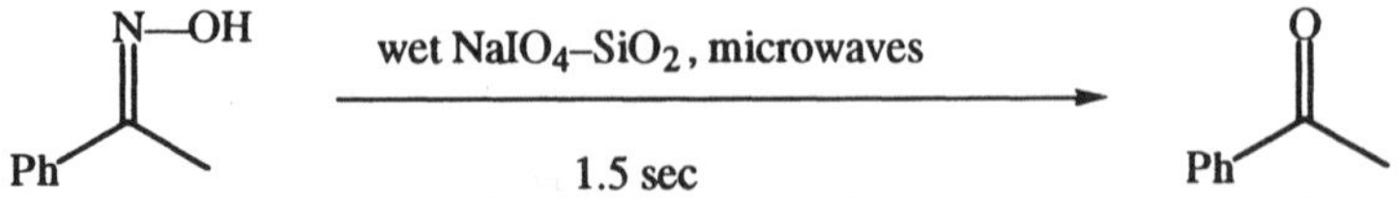
N—OH
wet $NaIO_4$–SiO_2, microwaves
1.5 sec
Ph
O
Ph

80%

Varma, R.S.; Dahiya, R.; Saini, P.K. *Tetrahedron Lett.*, ***1997***, *38*, 8819.

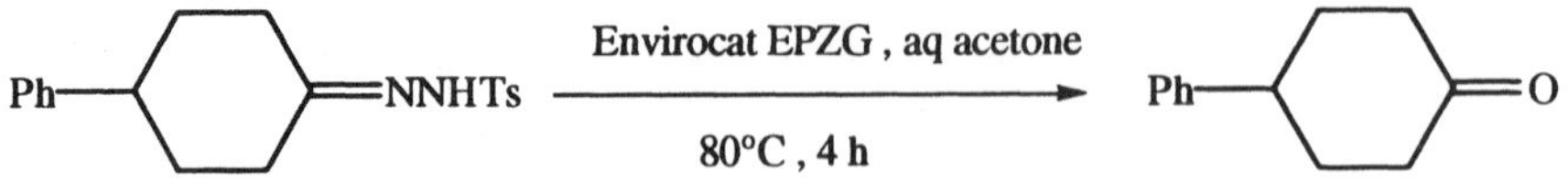
Ph
NNHTs
Envirocat EPZG , aq acetone
80°C , 4 h
Ph
O

90%

Ballini, R.; Bosica, G.; Maggi, R.; Sartori, G. *SynLett*, ***1997***, 795.

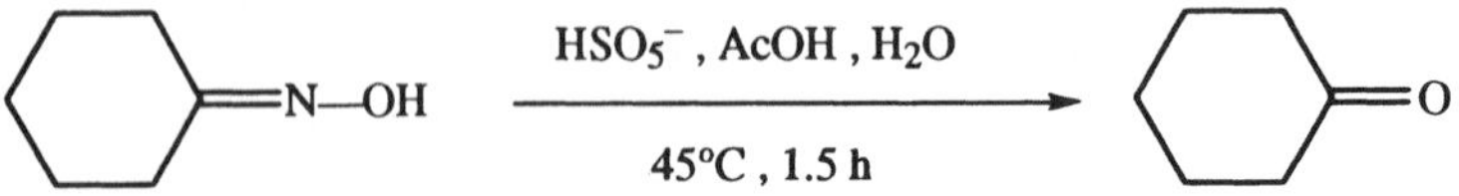
N—OH
HSO_5^- , AcOH , H_2O
45°C , 1.5 h
O

83%

Bose, D.S.; Srinivas, P. *Synth. Commun.*, ***1997***, *27*, 3835.

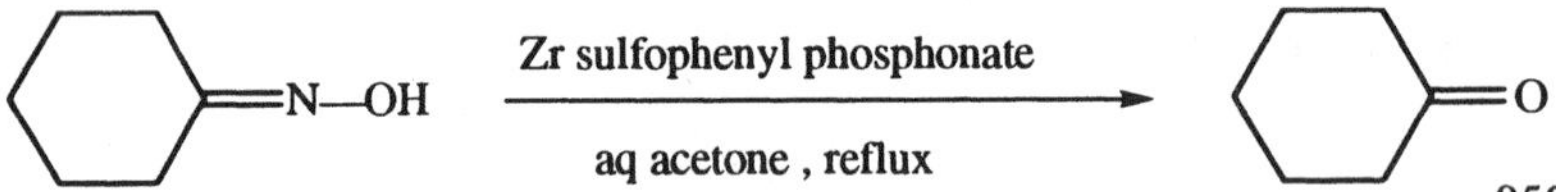
N—OH
Zr sulfophenyl phosphonate
aq acetone , reflux
O

95%

Curini, M.; Rosati, O.; Pisani, E.; Costantino, U. *SynLett*, ***1996***, 333.

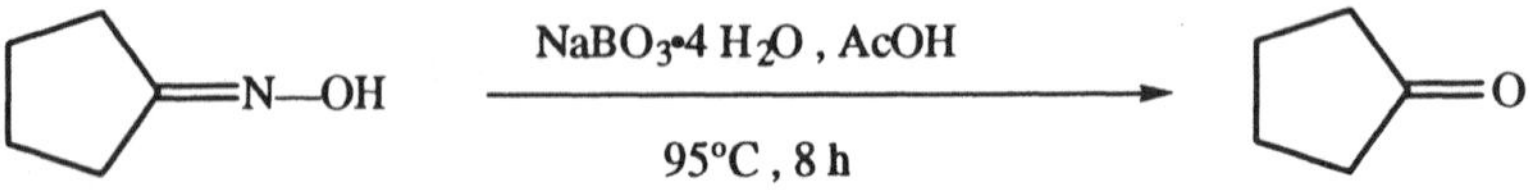
N—OH
$NaBO_3$•4 H_2O , AcOH
95°C , 8 h
O

89%

Bandgar, B.P.; Shaikh, S.I.; Iyer, S. *Synth. Commun.*, ***1996***, *26*, 1163.

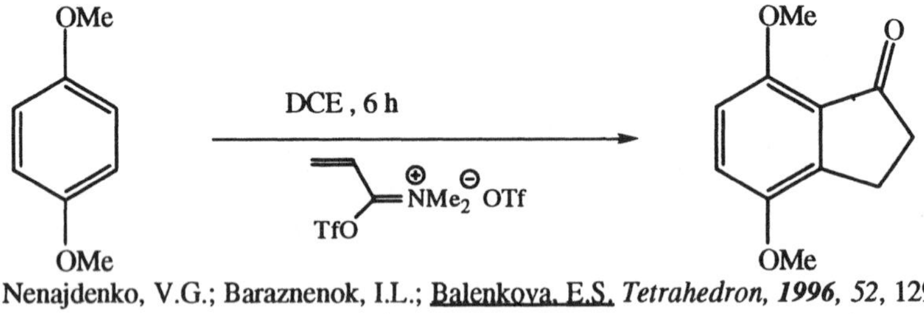
OMe
OMe
DCE , 6 h
NMe_2 OTf
TfO
OMe
O
OMe

60%

Nenajdenko, V.G.; Baraznenok, I.L.; Balenkova, E.S. *Tetrahedron*, ***1996***, *52*, 12993.

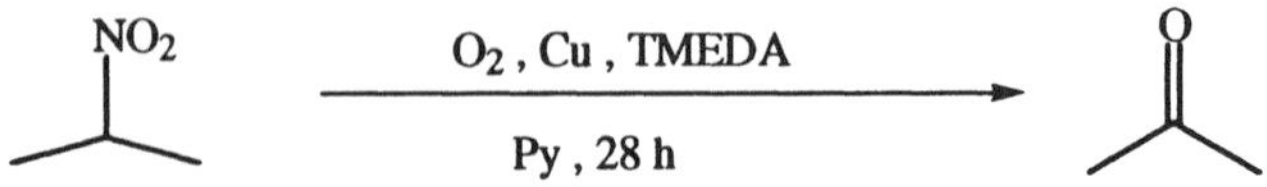
NO_2
O_2 , Cu , TMEDA
Py , 28 h
O

71%

Balogh–Hergovich, É.; Kaizer, J.; Speier, G. *Chem. Lett.*, ***1996***, 573.

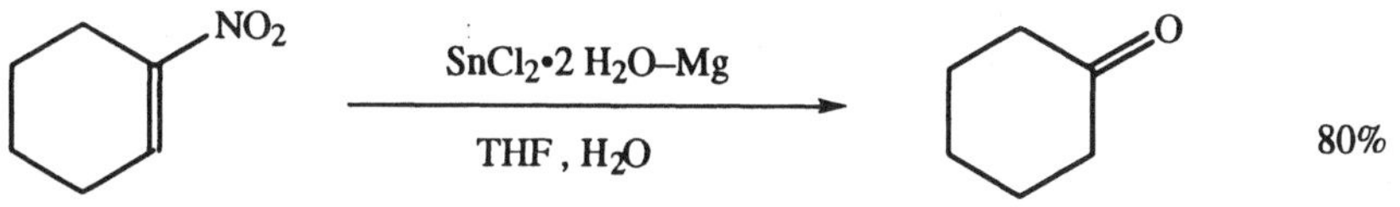
NO2
SnCl2•2 H2O–Mg
THF, H2O
O
80%

Das, N.B.; Sarangi, C.; Nanda, B.; Nayak, A.; Sharma, R.P. *J. Chem. Res (S)*, ***1996***, 28.

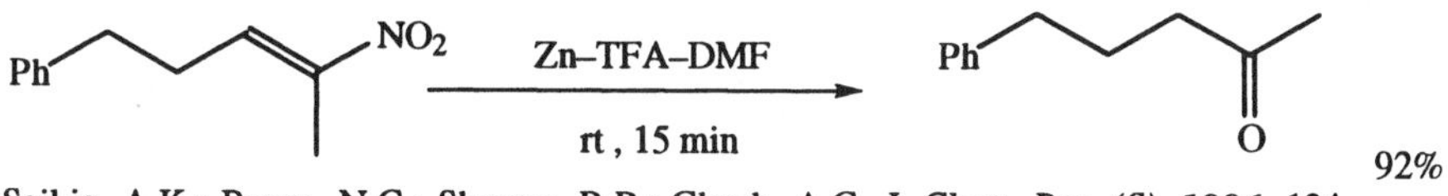
Ph
NO2
Zn–TFA–DMF
rt, 15 min
Ph
O
92%

Saikia, A.K.; Barua, N.C.; Sharma, R.P.; Ghosh, A.C. *J. Chem. Res. (S)*, ***1996***, 124.

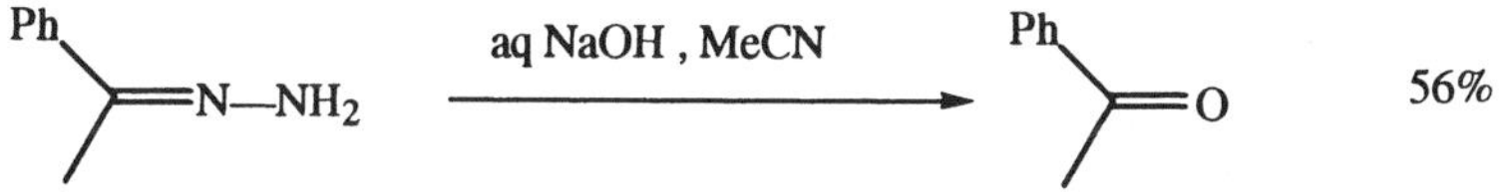
Ph
N—NH2
aq NaOH, MeCN
Ph
O
56%

Khurana, J.M.; Singh, S.; Panda, A.K. *J. Chem. Res. (S)*, ***1996***, 532.

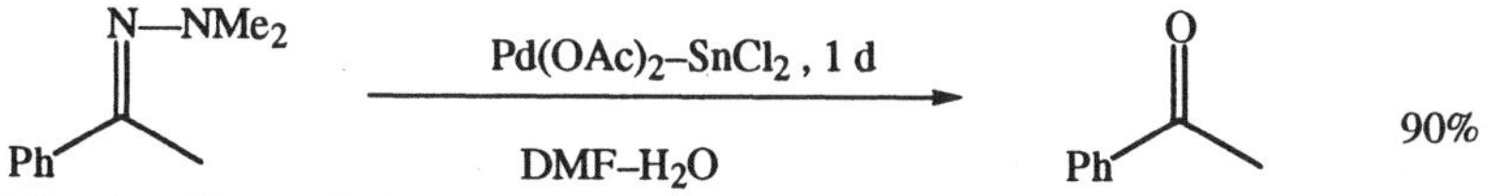
N—NMe2
Ph
Pd(OAc)2–SnCl2, 1 d
DMF–H2O
O
Ph
90%

Mino, T.; Hirota, T.; Yamashita, M. *SynLett*, ***1996***, 999.

REVIEW:
"Aldehydes and Ketones."
Lawrence, N.J. *J. Chem. Soc., Perkin Trans. 1*, ***1998***, 1739.

SECTION 180A: PROTECTION OF KETONES

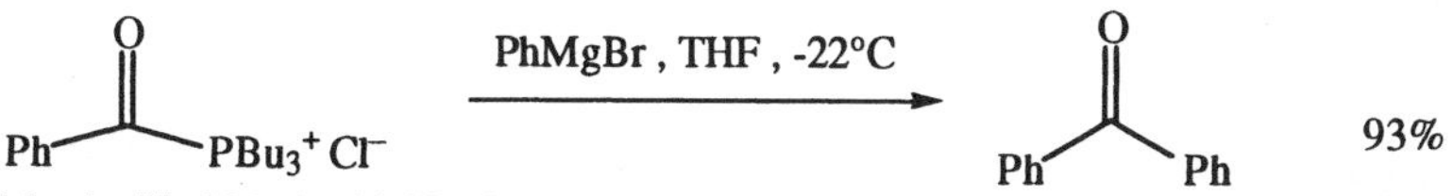
O
Ph
PBu3+ Cl−
PhMgBr, THF, -22°C
O
Ph
Ph
93%

Maeda, H.; Takahashi, K.; Ohmori, H. *Tetrahedron*, ***1998***, *54*, 12233.

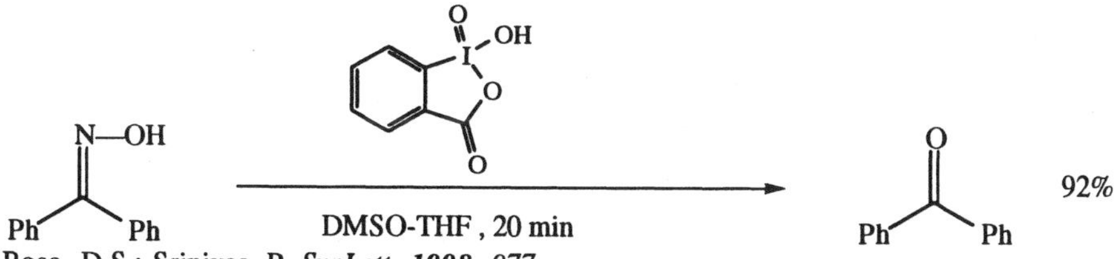
N—OH
Ph
Ph
O
OH
I
O
O
DMSO-THF, 20 min
O
Ph
Ph
92%

Bose, D.S.; Srinivas, P. *SynLett*, ***1998***, *977*.

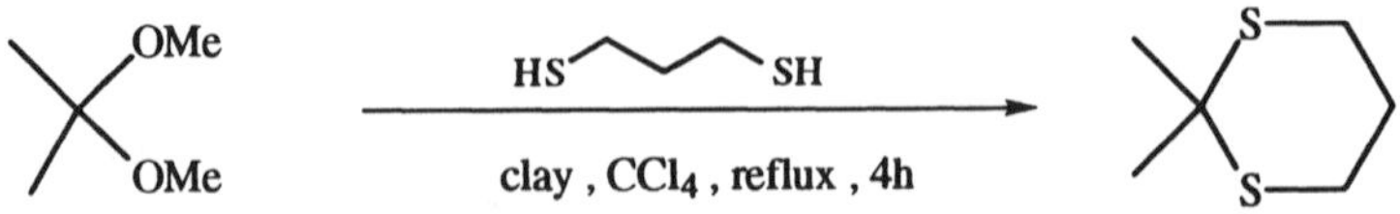
OMe
OMe
HS SH
clay , CCl_4 , reflux , 4h
S
S

86%

Jnaneshwara, G.K.; Barhate, N.B.; Sudalai, A.; Deshpande, V.H.; Wakharkar, R.D.; Gajare, A.S Shingare, M.S.; Sukumar, R. *J. Chem. Soc., Perkin Trans. 1*, ***1998***, 965.

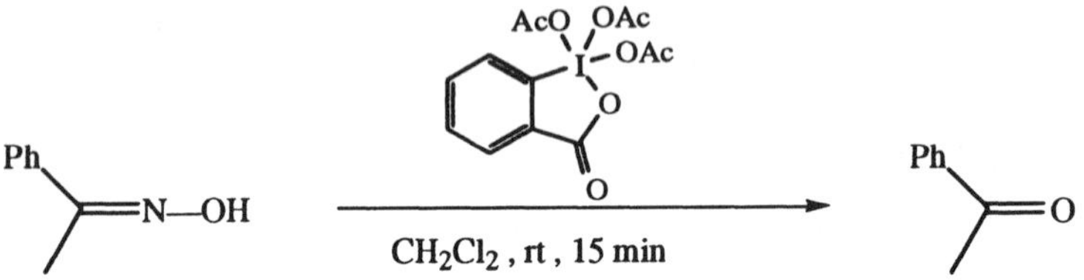
AcO OAc
I
OAc
O
O
Ph
N—OH
CH_2Cl_2 , rt , 15 min
Ph
O

94%

DeVos, D.E.; Sels, B.F.; Reynaers, M.; Subba Rao, Y.V.; Jacobs, P.A. *Tetrahedron Lett.*, ***1998***, *39*, 3221.

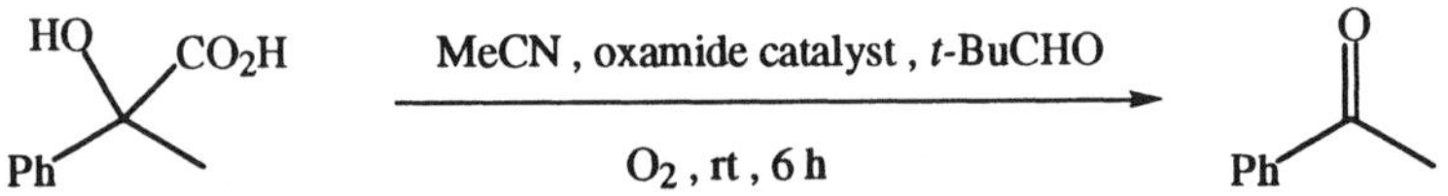
HO
CO_2H
Ph
MeCN , oxamide catalyst , t-BuCHO
O_2 , rt , 6 h
O
Ph

87%

Blay, G.; Fernández, I.; Formentin, P.; Pedro, J.R.; Roselló, A.L.; Ruiz, R.; Journaux, Y. *Tetrahedron Lett.*, ***1998***, *39*, 3327.

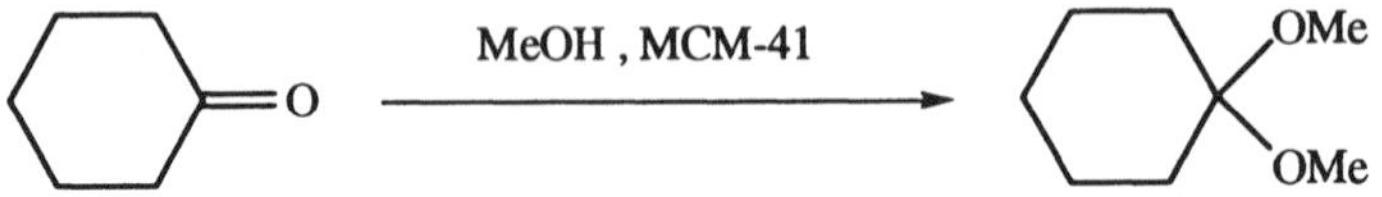
O
MeOH , MCM-41
OMe
OMe

89%

Tanaka, Y.; Sawamura, N.; Iwamoto, M. *Tetrahedron Lett.*, ***1998***, *39*, 9457.

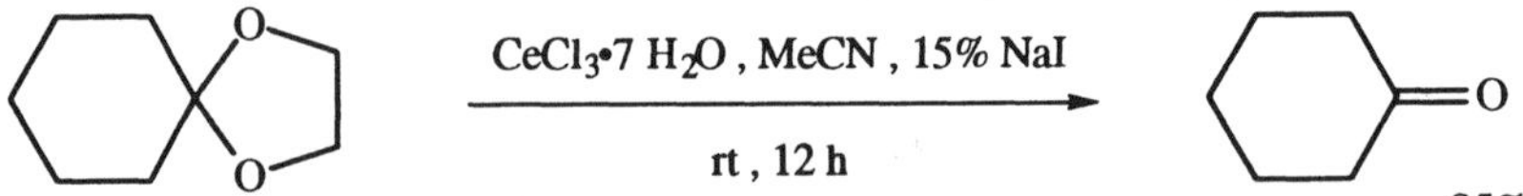
O
O
$CeCl_3$•7 H_2O , MeCN , 15% NaI
rt , 12 h
O

85%

Marcantoni, E.; Nobili, F.; Bartoli, G.; Bosco, M.; Sambri, L. *J. Org. Chem.*, ***1997***, *62*, 4183.

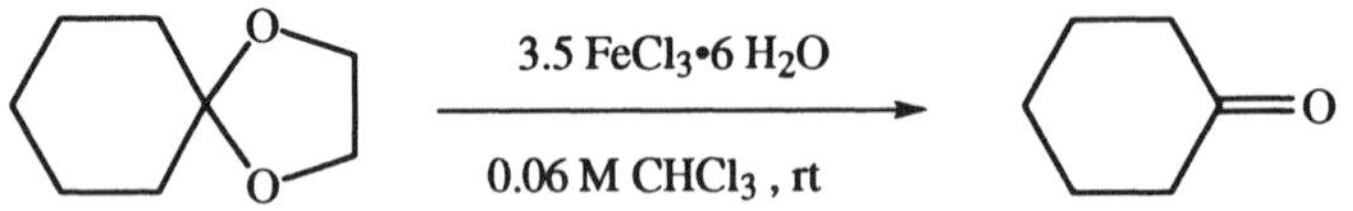
O
O
3.5 $FeCl_3$•6 H_2O
0.06 M $CHCl_3$, rt
O

96%

Sen, S.E.; Roach, S.L.; Boggs, J.K.; Ewing, G.J.; Magrath, J. *J. Org. Chem.*, ***1997***, *62*, 6684

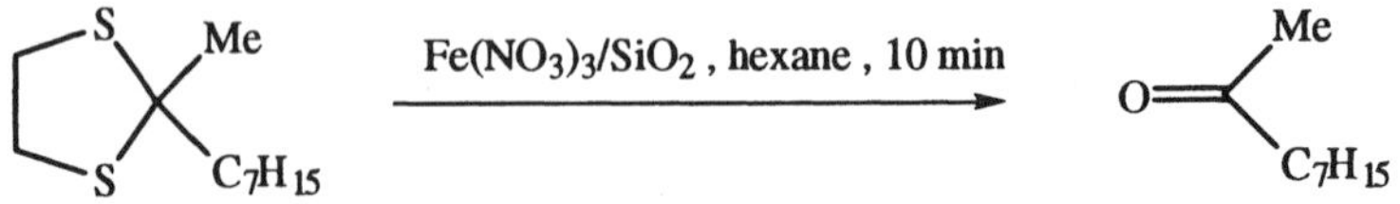
S
S
Me
C_7H_{15}
$Fe(NO_3)_3/SiO_2$, hexane , 10 min
Me
O
C_7H_{15}

99%

Hirano, M.; Ukawa, K.; Yakabe, S.; Morimoto, T. *Org. Prep. Proceed. Int.*, ***1997***, *29*, 480.

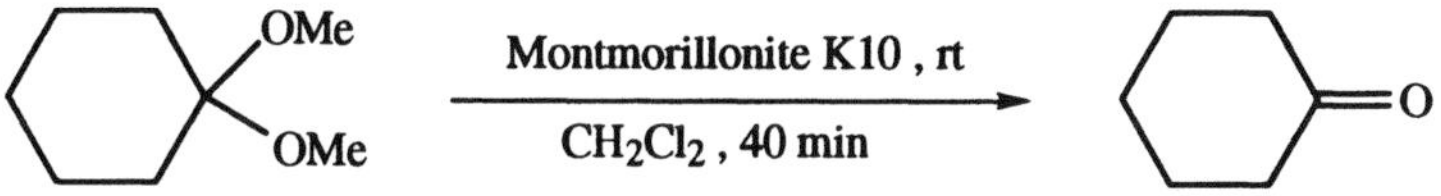
OMe
OMe
Montmorillonite K10 , rt
CH_2Cl_2 , 40 min
O

95%

Gautier, E.C.L.; Graham, A.E.; McKillop, A.; Standen, S.P.; Taylor, R.J.K. *Tetrahedron Lett.*, ***1997***, *38*, 1881.

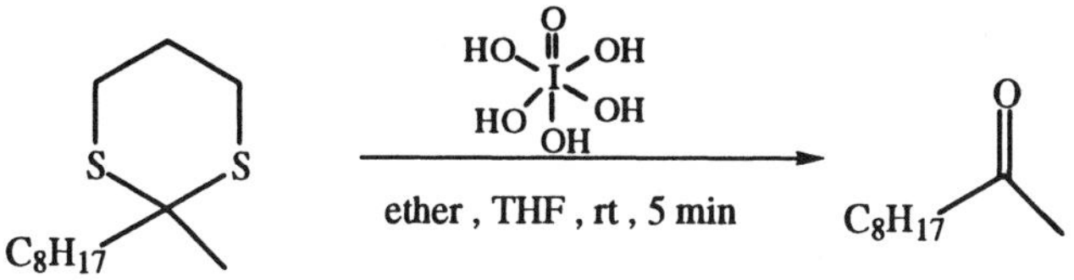
O
HO
OH
I
HO
OH
OH
S
S
C_8H_{17}
ether , THF , rt , 5 min
O
C_8H_{17}

94%

Shi, X.–X.; Khanapure, S.P.; Rokach, J. *Tetrahedron Lett.*, ***1996***, *37*, 4331.

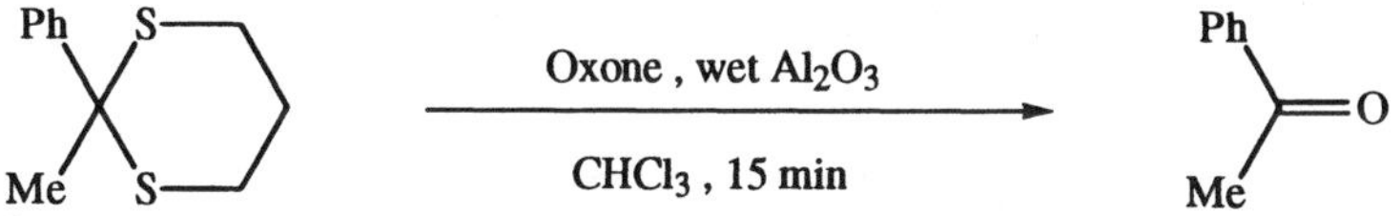
Ph
S
Me
S
Oxone , wet Al_2O_3
$CHCl_3$, 15 min
Ph
O
Me

96%

Ceccherelli, P.; Curini, M.; Marcotullio, M.C.; Epifano, F.; Rosati, O. *SynLett*, ***1996***, 767.

See Section 362 (Ester-Alkene) for the formation of enol esters and Section 367 (Ether-Alkenes) for the formation of enol ethers. Many of the methods in Section 60A (Protection of Aldehydes) are also applicable to ketones.

CHAPTER 13
PREPARATION OF NITRILES

SECTION 181: NITRILES FROM ALKYNES

NO ADDITIONAL EXAMPLES

SECTION 182: NITRILES FROM ACID DERIVATIVES

$$PhCOOH \xrightarrow[\text{microwaves , 10 min}]{CO(NH_2)_2\text{ , }H_2NSO_3H\text{ , }Al_2O_3} PhCN \quad 62\%$$

Juncai, F.; Bin, L.; Yang, L.; Chanchuan, L. *Synth. Commun.*, ***1996***, *26*, 4545.

SECTION 183: NITRILES FROM ALCOHOLS AND THIOLS

NO ADDITIONAL EXAMPLES

SECTION 184: NITRILES FROM ALDEHYDES

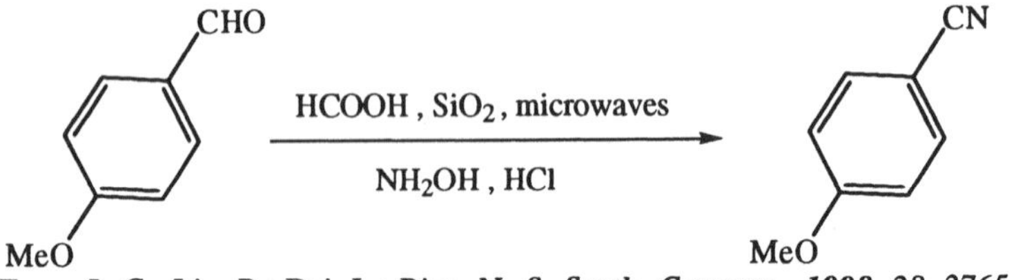

CHO
HCOOH , SiO2 , microwaves
NH2OH , HCl
MeO
CN
MeO

88%

Feng, J.-C.; Liu, B.; Dai, L.; Bian, N.-S. *Synth. Commun.*, ***1998***, *28*, 3765.

$$PhCHO \xrightarrow[\text{2. }Me_2SO_4\text{ , }K_2CO_3]{\text{1. }H_2NNMe_2} PhCN \quad 97\%$$

Kamal, A.; Arifuddin, M.; Rao, N.V. *Synth. Commun.*, ***1998***, *28*, 4507.

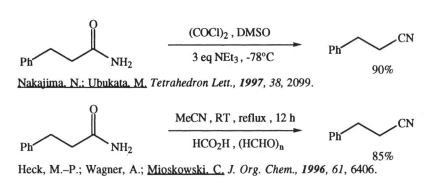

$C_8H_{17}CHO \xrightarrow[\text{2. phthalic anhydride}]{\text{1. } NH_2OH\text{ , }HCl/NEt_3\text{ , 8 h}} C_8H_{17}CN$ 80%

Wang, E.-C.; Lin, G.-J. *Tetrahedron Lett*, ***1998***, *39*, 4047.

$PhCHO \xrightarrow[Al_2O_3\text{ , microwaves , 7.5 min}]{NH_2OH\bullet HCl\text{ , peroxy monosulfate}} PhCN$ 77%

Bose, D.S.; Narsaiah, A.V. *Tetrahedron Lett.*, ***1998***, *39*, 6533.

SECTION 185: NITRILES FROM ALKYLS, METHYLENES AND ARYLS

NO ADDITIONAL EXAMPLES

SECTION 186: NITRILES FROM AMIDES

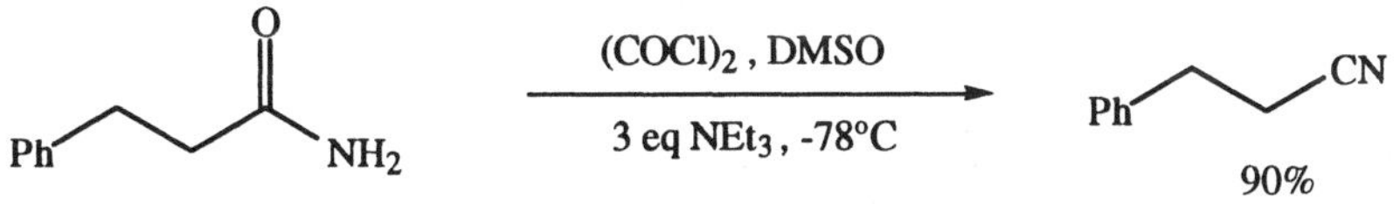
O
Ph
NH2
(COCl)2 , DMSO
3 eq NEt3 , -78°C
Ph
CN
90%

Nakajima, N.; Ubukata, M. *Tetrahedron Lett.*, ***1997***, *38*, 2099.

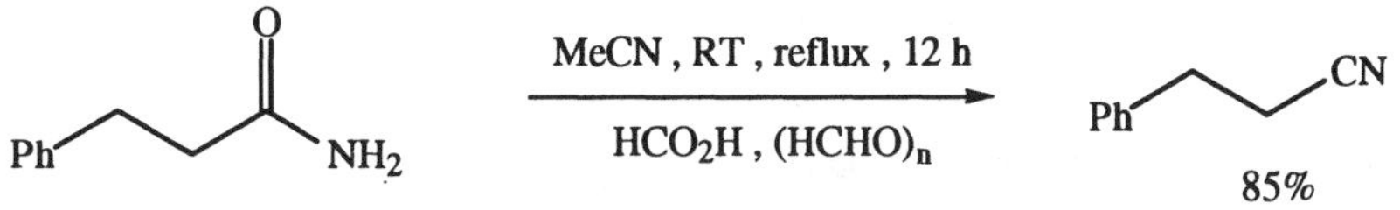
O
Ph
NH2
MeCN , RT , reflux , 12 h
HCO2H , (HCHO)n
Ph
CN
85%

Heck, M.-P.; Wagner, A.; Mioskowski, C. *J. Org. Chem.*, ***1996***, *61*, 6406.

SECTION 187: NITRILES FROM AMINES

$PhCH_2NH_2 \xrightarrow{NaOCl\text{ , EtOH , rt , 15 min}} PhCN$ 92%

Yamazaki, S. *Synth. Commun.*, ***1997***, *27*, 3559.

SECTION 188: NITRILES FROM ESTERS

NO ADDITIONAL EXAMPLES

SECTION 189: NITRILES FROM ETHERS, EPOXIDES AND THIOETHERS

NO ADDITIONAL EXAMPLES

SECTION 190: NITRILES FROM HALIDES AND SULFONATES

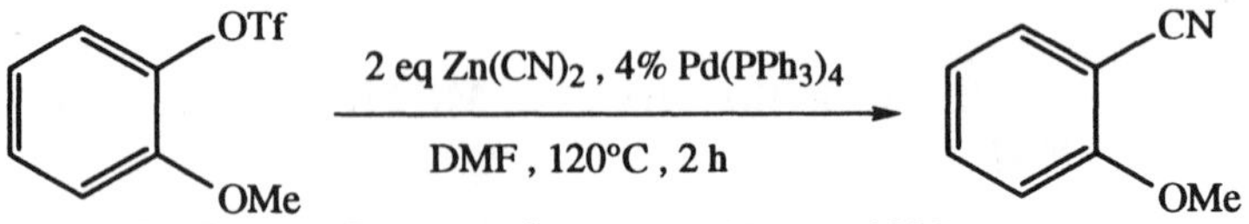
OTf
OMe
2 eq $Zn(CN)_2$, 4% $Pd(PPh_3)_4$
DMF , 120°C , 2 h
CN
OMe

81%

Kubota, H.; Rice, K.C. *Tetrahedron Lett.*, ***1998***, *39*, 2907.

$C_{11}H_{23}CH_2Br$ $\xrightarrow{\text{Li—CH}_2\text{CN , THF , -78°C , 1 h}}$ $C_{11}H_{23}CH_2CH_2CN$

79%

Taber, D.F.; Kong, S. *J. Org. Chem.*, ***1997***, *62*, 8575.

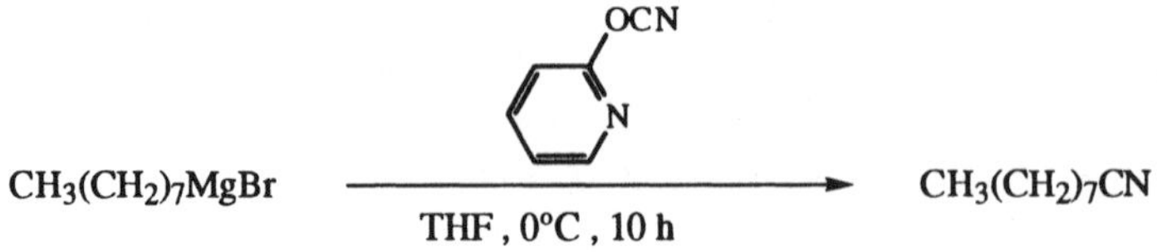
OCN
N
$CH_3(CH_2)_7MgBr$
THF , 0°C , 10 h
$CH_3(CH_2)_7CN$

97%

Koo, J.S.; Lee, J.I. *Synth. Commun.*, ***1996***, *26*, 3601.

SECTION 191: NITRILES FROM HYDRIDES

NO ADDITIONAL EXAMPLES

SECTION 192: NITRILES FROM KETONES

NO ADDITIONAL EXAMPLES

SECTION 193: NITRILES FROM NITRILES

Conjugate reductions and Michael alkylations of alkene nitriles are found in Section 74D (Alkyls from Alkenes).

NO ADDITIONAL EXAMPLES

SECTION 194: NITRILES FROM ALKENES

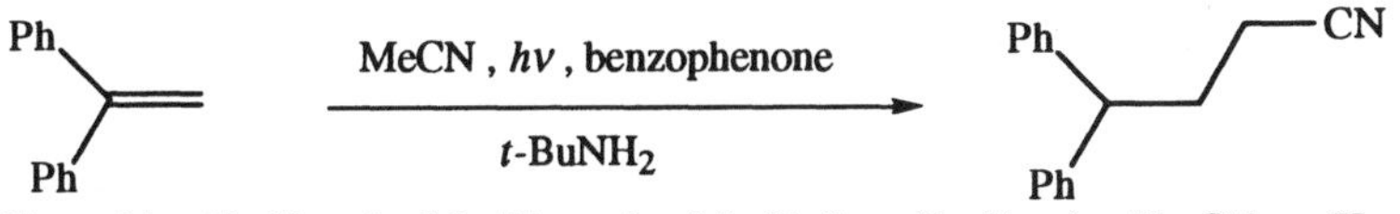
Ph
Ph
MeCN , hv , benzophenone
t-BuNH2
Ph
Ph
CN

77%

Yamashita, T.; Yasuda, M.; Watanabe, M.; Kojima, R.; Tanabe, K.; Shima, K.
J. Org. Chem., ***1996***, *61*, 6438.

SECTION 195: NITRILES FROM MISCELLANEOUS COMPOUNDS

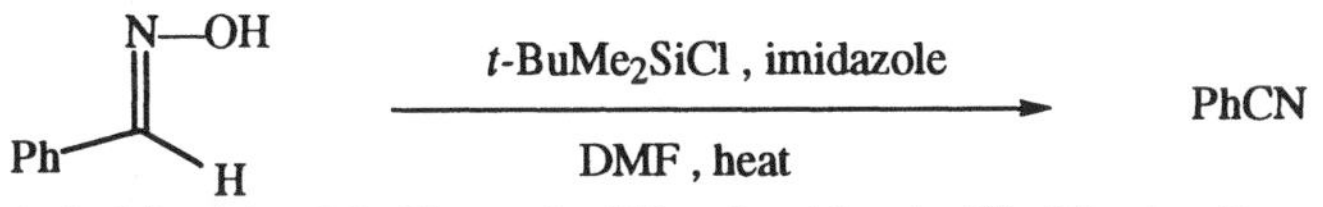
N—OH
Ph
H
t-BuMe2SiCl , imidazole
DMF , heat
PhCN

81%

Ortiz-Marciales, M.; Piñero, L.; Ufret, L.; Algarín, W.; Morales, J.
Synth. Commun., ***1998***, *28*, 2807.

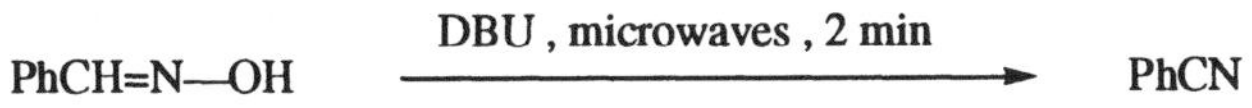
DBU , microwaves , 2 min
PhCH=N—OH
PhCN

92%

Sabitha, G.; Syamala, M. *Synth. Commun.*, ***1998***, *28*, 4577.

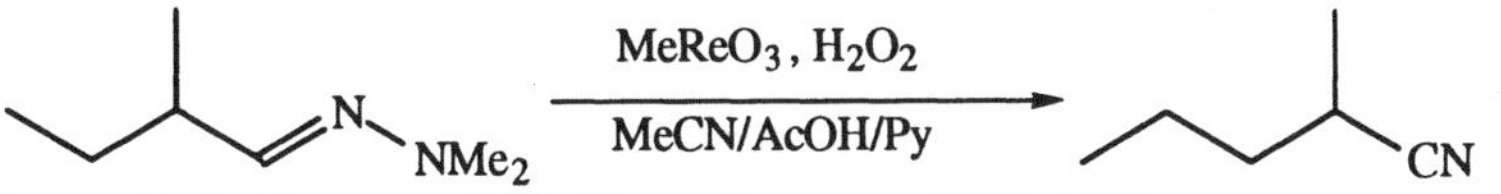
N
NMe2
MeReO3 , H2O2
MeCN/AcOH/Py
CN

90%

Stanković, S.; Espenson, J.H. *Chem. Commun.*, ***1998***, 1579.

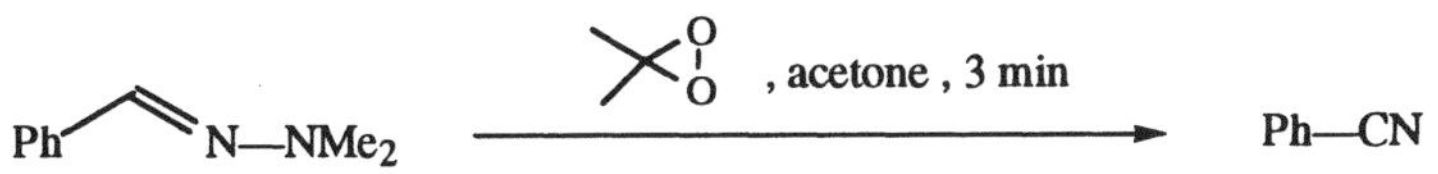
O
O
, acetone , 3 min
Ph
N—NMe2
Ph—CN

97%

Altamura, A.; D'Accolti, L.; Detomaso, A.; Dinoi, A.; Fiorentino, M.; Fusco, C.; Curci, R.
Synth. Commun., ***1998***, *39*, 2009.

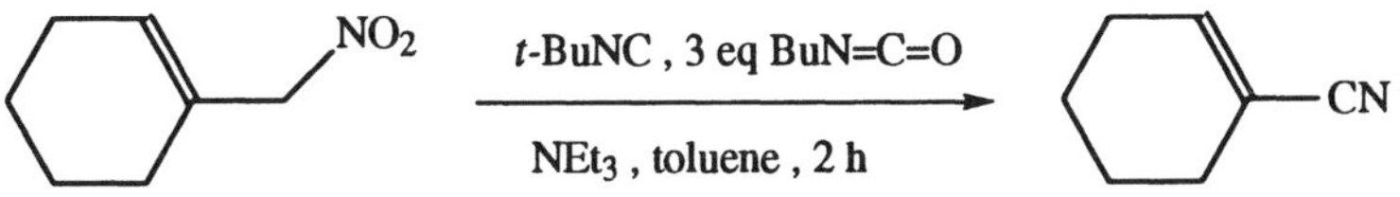
NO2
t-BuNC , 3 eq BuN=C=O
NEt3 , toluene , 2 h
CN

58%

El Kaim, L.; Gacon, A. *Tetrahedron Lett.*, ***1997***, *38*, 3391.

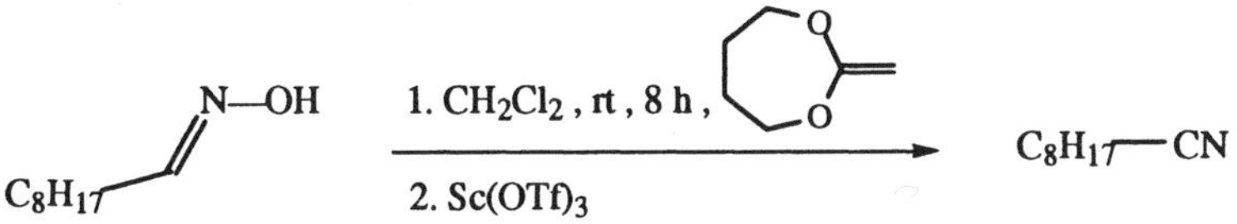
N—OH
C8H17
1. CH2Cl2 , rt , 8 h ,
O
O
2. Sc(OTf)3
C8H17—CN

85%

Fukuzama, S.; Yamaishi, Y.; Furuya, H.; Terao, K.; Iwasaki, F.
Tetrahedron Lett.. ***1997***. *38*. 7203.

PhCH=N—OH —(H_2SO_4/SiO_2, microwaves, 4 min)→ PhCN 76%

Kumar, H.M.S.; Mohanty, P.K.; Kumar, M.S.; Yadav, J.S. *Synth. Commun.*, ***1997***, *27*, 1327

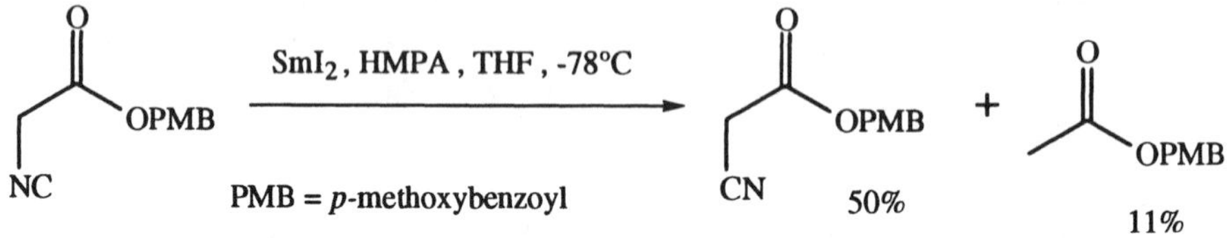

O
OPMB
NC
SmI_2, HMPA, THF, -78°C
PMB = *p*-methoxybenzoyl
O
OPMB
CN
50%
+
O
OPMB
11%

Kang, H.-Y.; Pae, A.N.; Cho, Y.S.; Koh, H.Y.; Chung, B.Y. *Chem. Commun.*, ***1997***, 821.

CHAPTER 14
PREPARATION OF ALKENES

SECTION 196: ALKENES FROM ALKYNES

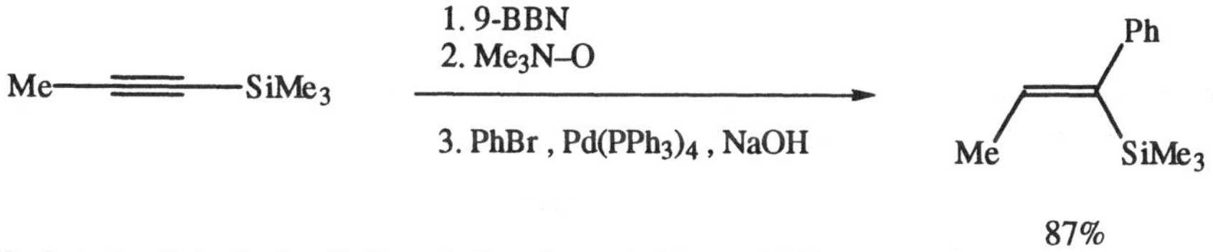

Me
SiMe3
1. 9-BBN
2. Me3N–O
3. PhBr , Pd(PPh3)4 , NaOH
Ph
Me
SiMe3
87%

Soderquist, J.A.; León, G. *Tetrahedron Lett.*, ***1998***, *39*, 3989.

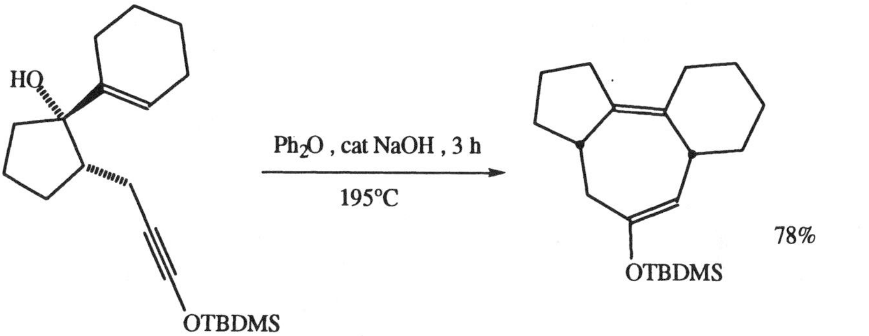

HO
OTBDMS
Ph2O , cat NaOH , 3 h
195°C
OTBDMS
78%

Ovaska, T.V.; Roark, J.L.; Shoemaker, C.M.; Bordner, J. *Tetrahedron Lett*, ***1998***, *39*, 5705

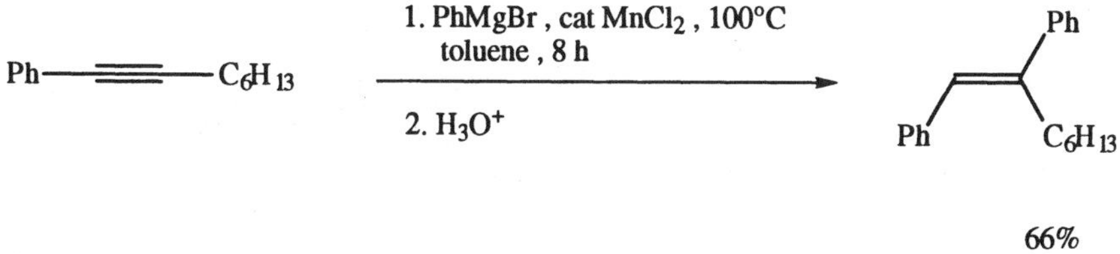

Ph
C6H13
1. PhMgBr , cat MnCl2 , 100°C
toluene , 8 h
2. H3O+
Ph
Ph
C6H13
66%

Yorimitsu, H.; Tang, J.; Okada, K.; Shinokubo, H.; Oshima, K. *Chem. Lett.*, ***1998***, 11.

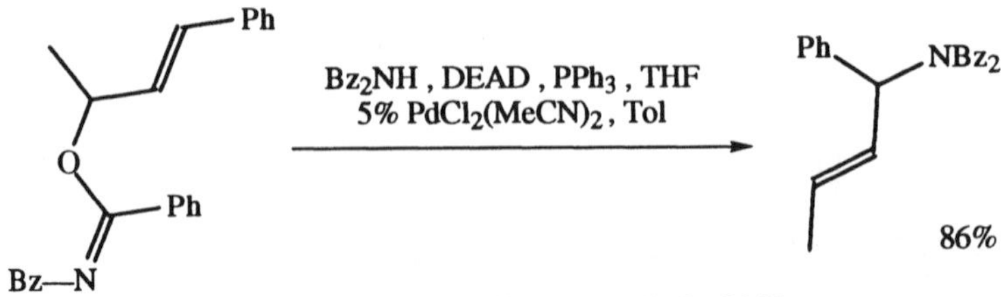
Ph
Ph
Bz—N
Bz_2NH , DEAD , PPh_3 , THF
5% $PdCl_2(MeCN)_2$, Tol
Ph
NBz_2
86%

Fujiwara, N.; Yamamoto, Y. *J. Org. Chem.*, ***1997***, *62*, 2318.

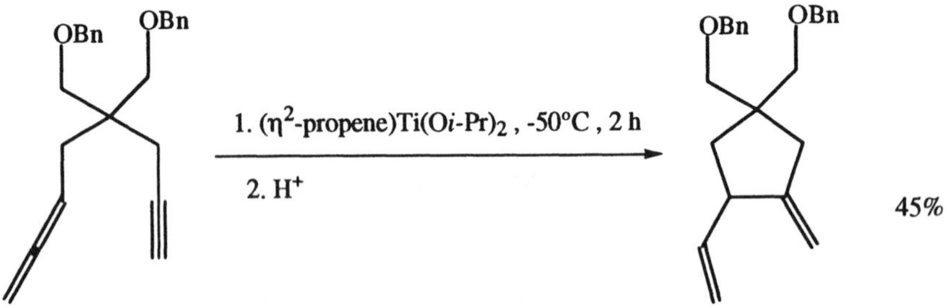
OBn
OBn
1. (η^2-propene)Ti(O*i*-Pr)$_2$, -50°C , 2 h
2. H^+
OBn
OBn
45%

Urabe, H.; Takeda, T.; Hideura, D.; Sato, F. *J. Am. Chem. Soc.*, ***1997***, *119*, 11295.

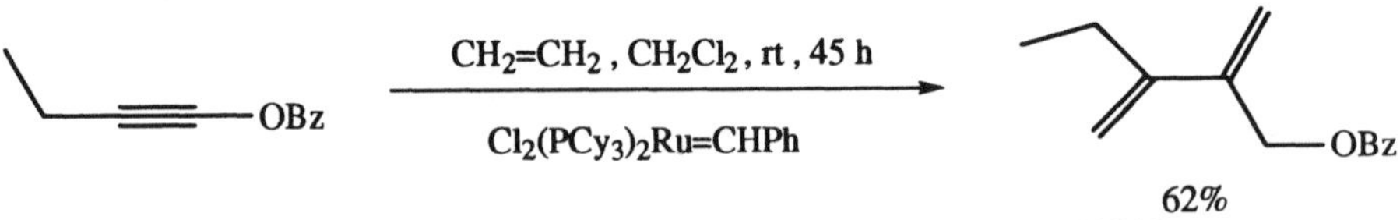
OBz
CH_2=CH_2 , CH_2Cl_2 , rt , 45 h
$Cl_2(PCy_3)_2Ru$=CHPh
OBz
62%

Kinoshita, A.; Sakakibara, N.; Mori, M. *J. Am. Chem. Soc.*, ***1997***, *119*, 12388.

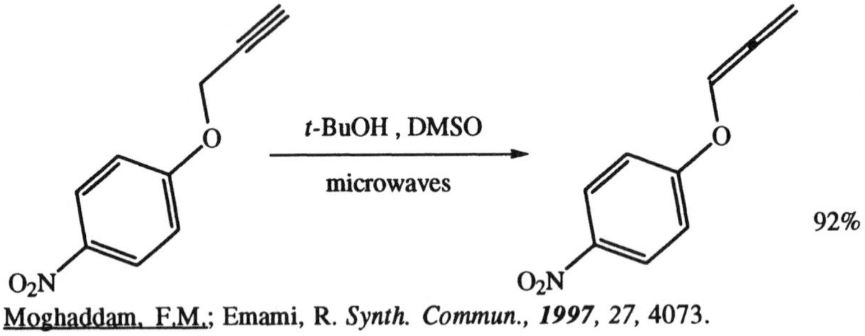
O
t-BuOH , DMSO
microwaves
O
O_2N
O_2N
92%

Moghaddam, F.M.; Emami, R. *Synth. Commun.*, ***1997***, *27*, 4073.

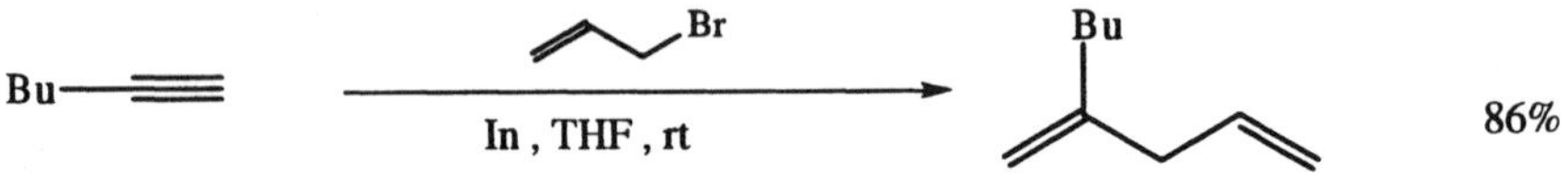
Bu
Br
In , THF , rt
Bu
86%

Ranu, B.C.; Majee, A. *Chem. Commun.*, ***1997***, 1225.

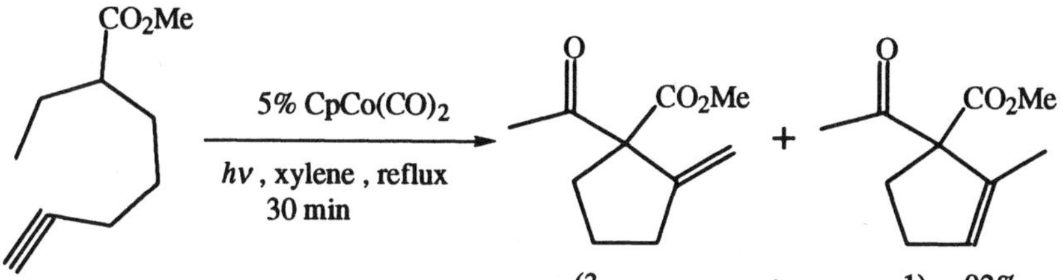
CO2Me
5% CpCo(CO)2
hv , xylene , reflux
30 min
O
CO2Me
+
O
CO2Me

(3 : 1) 92%

in benzene 93% 0%

Cruciani, P.; Stammler, R.; Aubert, C.; Malacria, M. *J. Org. Chem.*, ***1996***, *61*, 2699.

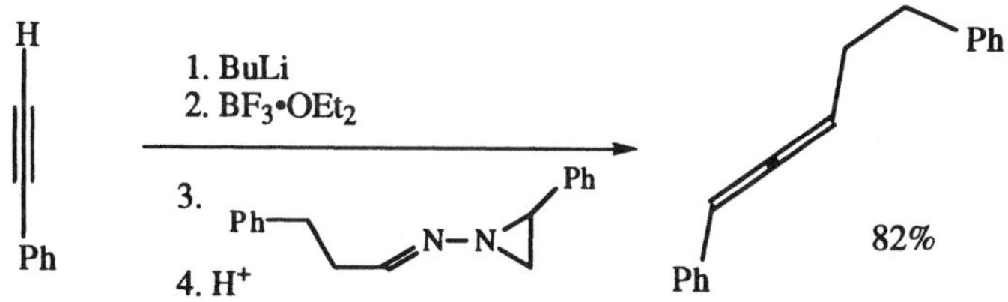
H
Ph
1. BuLi
2. BF3•OEt2
3.
Ph
Ph
N–N
4. H+
Ph
Ph
82%

Kim, S.; Cho, C.M.; Yoon, Y.–Y. *J. Org. Chem.*, ***1996***, *61*, 6018.

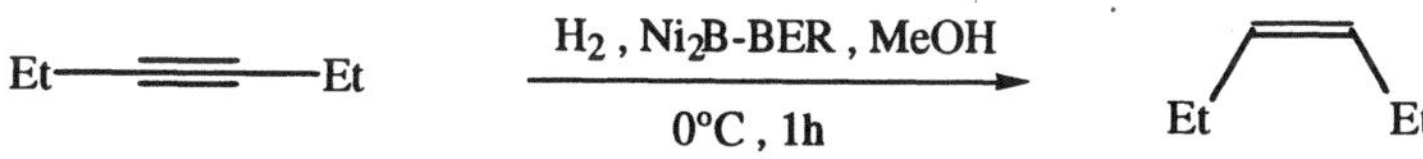
Et
Et
H2 , Ni2B-BER , MeOH
0°C , 1h
Et
Et

quant.

BER = borohydride exchange resin

Choi, J.; Yoon, N.M. *Tetrahedron Lett.*, ***1996***, *37*, 1057.

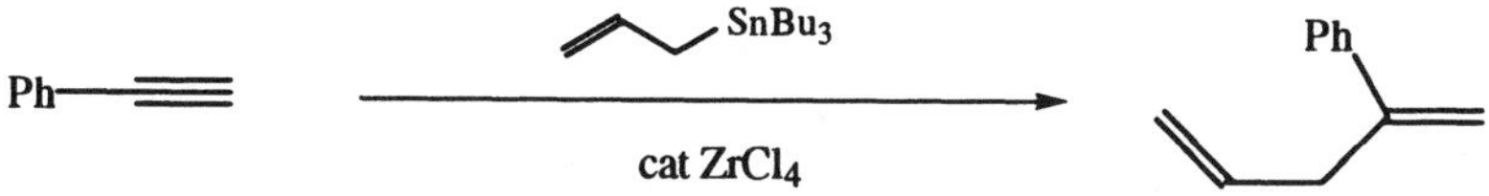
Ph
SnBu3
cat ZrCl4
Ph

87%

Asao, N.; Matsukawa, Y.; Yamamoto, Y. *Chem. Commun.*, ***1996***, 1513.

SECTION 197: ALKENES FROM ACID DERIVATIVES

NO ADDITIONAL EXAMPLES

SECTION 198: ALKENES FROM ALCOHOLS AND THIOLS

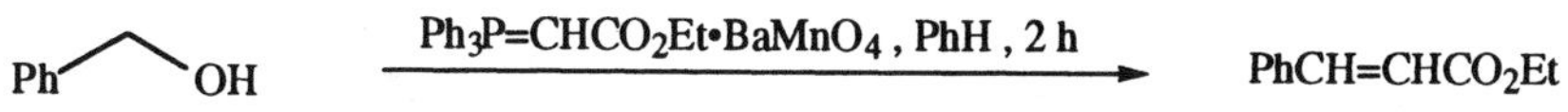
Ph
OH
Ph3P=CHCO2Et•BaMnO4 , PhH , 2 h
PhCH=CHCO2Et

99% ($E/Z = 23$)

Shuto, S.; Niizuma, S.; Matsuda, A. *J. Org. Chem.*, ***1998***, *63*, 4489.

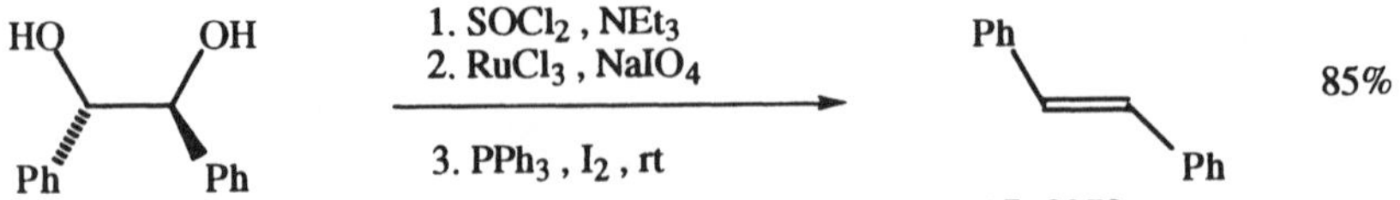
HO
OH
Ph
Ph
1. $SOCl_2$, NEt_3
2. $RuCl_3$, $NaIO_4$
3. PPh_3, I_2, rt
Ph
Ph
85%

Jang, D.O.; Joo, Y.H.; Cho, D.H. *Synth. Commun.*, ***1998***, *27*, 2379.

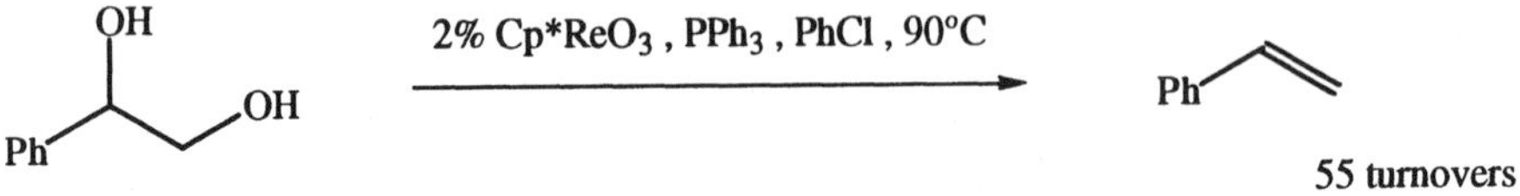
OH
OH
Ph
2% Cp^*ReO_3, PPh_3, PhCl, 90°C
Ph
55 turnovers

Cook, G.K.; Andrews, M.A. *J. Am. Chem. Soc.*, ***1996***, *118*, 9448.

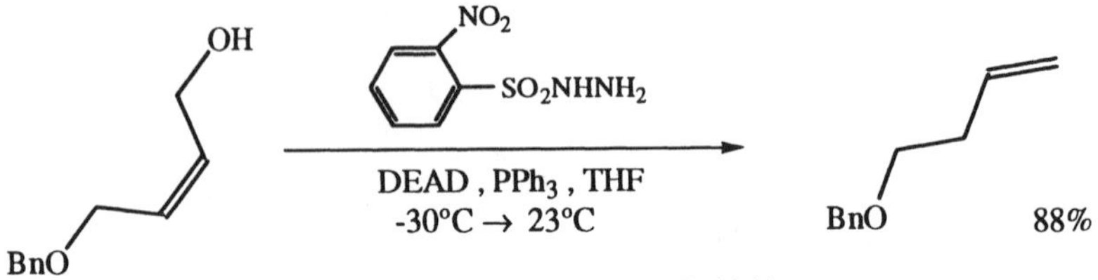
OH
BnO
NO_2
SO_2NHNH_2
DEAD, PPh_3, THF
-30°C → 23°C
BnO
88%

Myers, A.G.; Zheng, B. *Tetrahedron Lett.*, ***1996***, *37*, 4841.

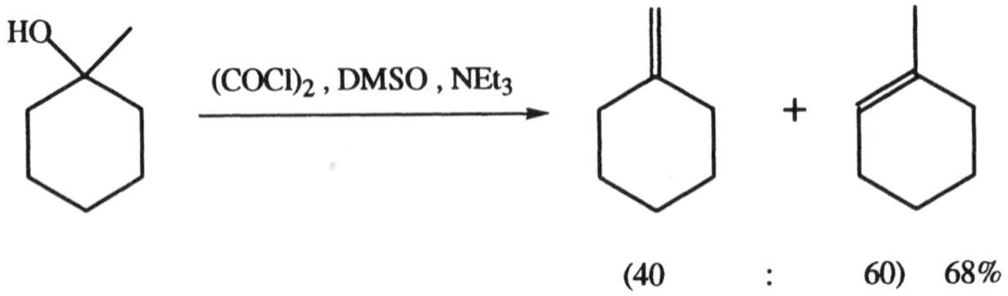
HO
$(COCl)_2$, DMSO, NEt_3
+
(40 : 60) 68%

Gleiter, R.; Herb, T.; Hofmann, J. *SynLett*, ***1996***, 987.

SECTION 199: ALKENES FROM ALDEHYDES

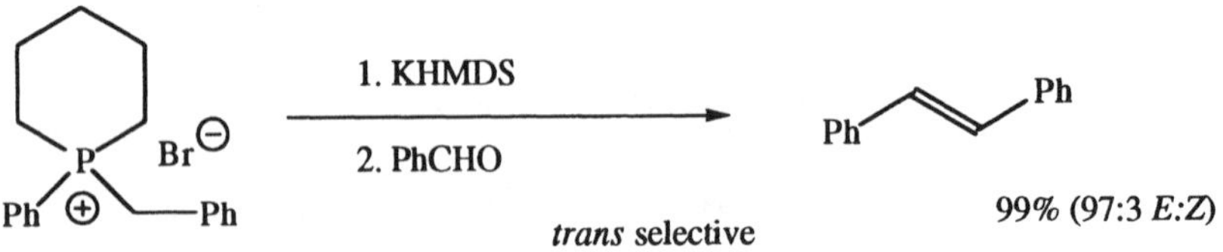
P
Br⊖
Ph ⊕ Ph
1. KHMDS
2. PhCHO
trans selective
Ph
Ph
99% (97:3 *E*:*Z*)

Lawrence, N.J.; Beynek, H. *SynLett*, ***1998***, 497.

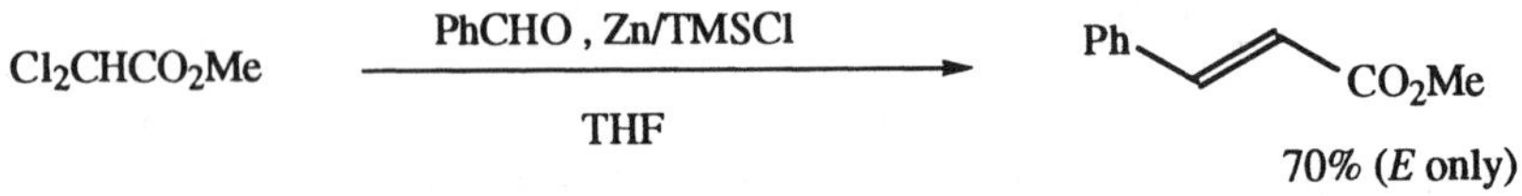
Cl_2CHCO_2Me
PhCHO, Zn/TMSCl
THF
Ph
CO_2Me
70% (*E* only)

Ishino, Y.; Mihara, M.; Nishihama, S.; Nishiguchi, I. *Bull. Chem. Soc. Jpn.*, ***1998***, *71*, 2669

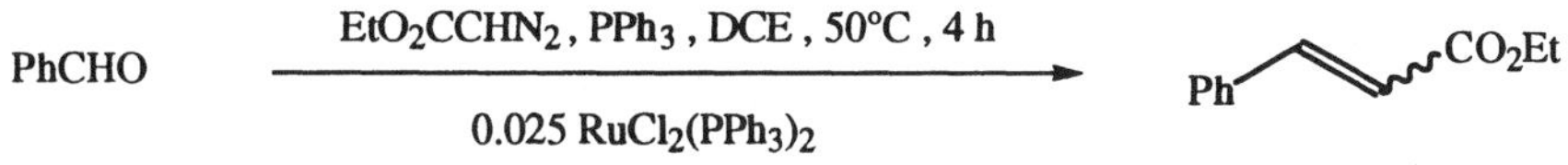
PhCHO
EtO2CCHN2, PPh3, DCE, 50°C, 4 h
0.025 RuCl2(PPh3)2
Ph
CO2Et

90% (97:3 *E:Z*)

Fujimura, O.; Honma, T. *Tetrahedron Lett*, ***1998***, *39*, 625.

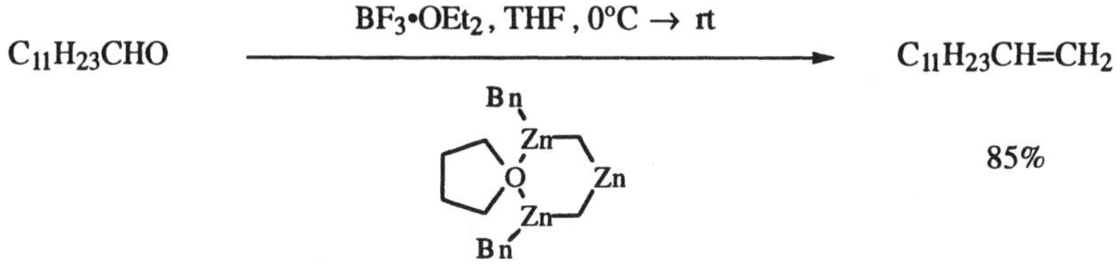
C11H23CHO
BF3•OEt2, THF, 0°C → rt
C11H23CH=CH2
Bn
Zn
O
Zn
Zn
Bn
85%

Matsubara, S.; Sugihara, M.; Utimoto, K. *SynLett*, ***1998***, 313.

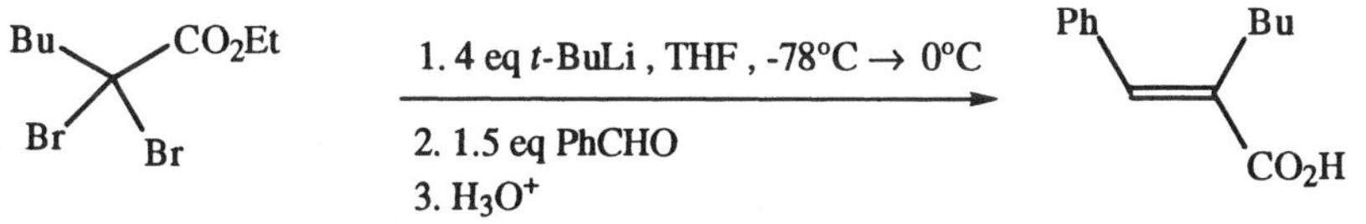
Bu
CO2Et
Br
Br
1. 4 eq t-BuLi, THF, -78°C → 0°C
2. 1.5 eq PhCHO
3. H3O+
Ph
Bu
CO2H

73% (>99:1 *E:Z*)

Shindo, M.; Sato, Y.; Shishido, K. *Tetrahedron Lett.*, ***1998***, *39*, 4857.

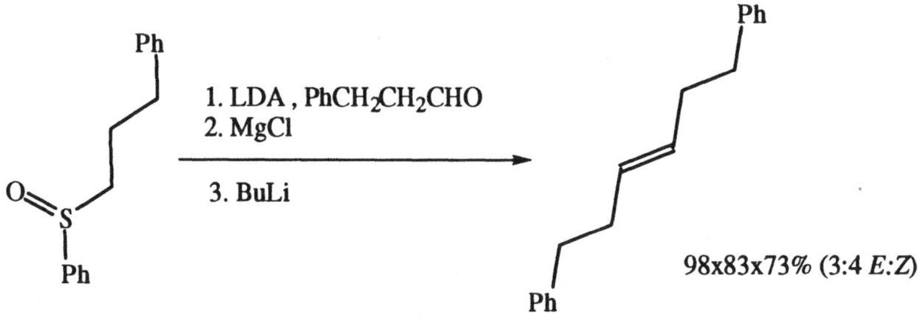
Ph
O
S
Ph
1. LDA, PhCH2CH2CHO
2. MgCl
3. BuLi
Ph
Ph
98x83x73% (3:4 E:Z)

Satoh, T.; Yamada, N.; Asano, T. *Tetrahedron Lett*, ***1998***, *39*, 6935.

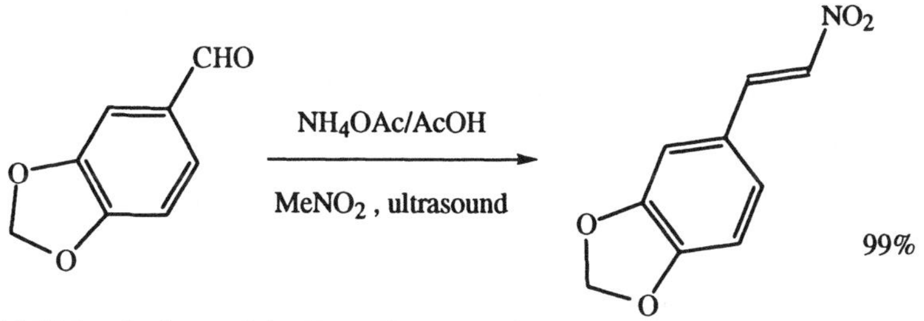
CHO
O
O
NH4OAc/AcOH
MeNO2, ultrasound
NO2
O
O
99%

McNulty, J.; Steere, J.A.; Wolf, S. *Tetrahedron Lett.*, ***1998***, *39*, 8013.

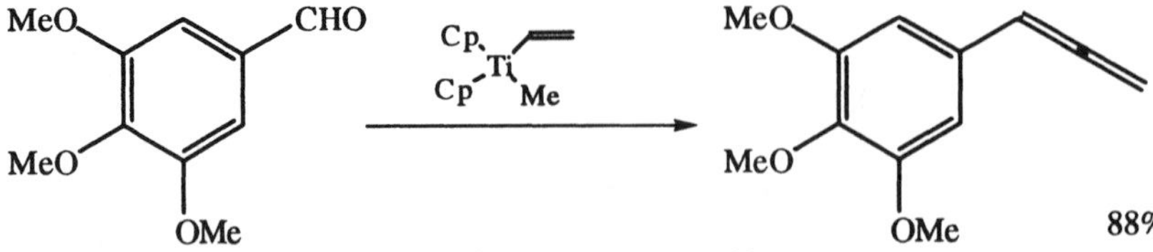
MeO
CHO
MeO
OMe
Cp
Ti
Cp
Me
MeO
MeO
OMe

88%

Petasis, N.A.; Hu, Y.–H. *J. Org. Chem.*, ***1997***, *62*, 782.

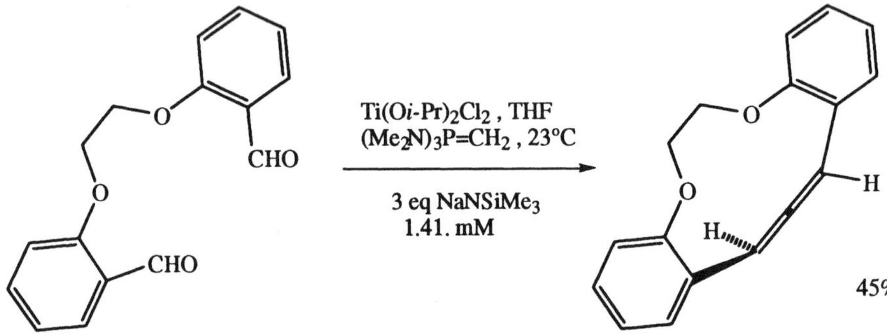
O
O
CHO
CHO
Ti(O*i*-Pr)2Cl2 , THF
(Me2N)3P=CH2 , 23°C
3 eq NaNSiMe3
1.41. mM
O
O
H
H

45%

Brody, M.S.; Williams, R.M.; Finn, M.G. *J. Am. Chem. Soc.*, ***1997***, *119*, 3429.

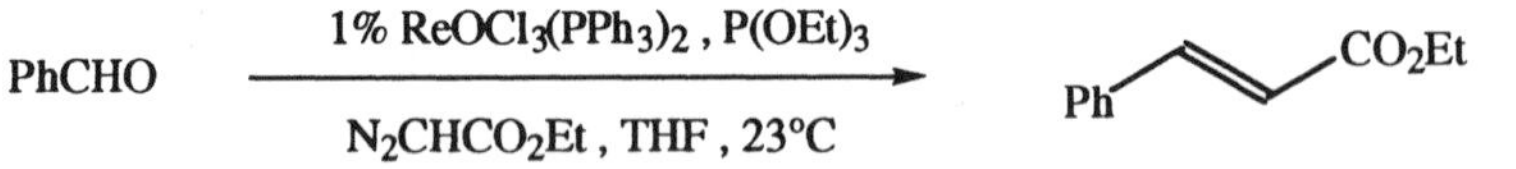
PhCHO
1% ReOCl3(PPh3)2 , P(OEt)3
N2CHCO2Et , THF , 23°C
Ph
CO2Et

90% (>20:1 *E:Z*)

Ledford, B.E.; Carreira, E.M. *Tetrahedron Lett.*, ***1997***, *38*, 8125.

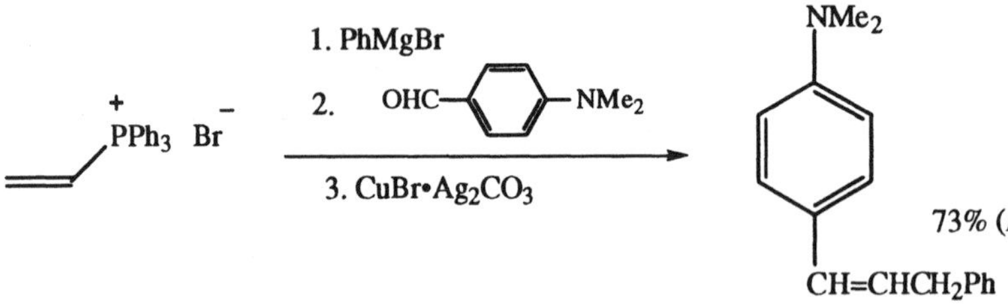
PPh3 Br
1. PhMgBr
2. OHC NMe2
3. CuBr•Ag2CO3
NMe2
CH=CHCH2Ph

73% (*E* only)

Shen, Y.; Yao, J. *J. Org. Chem.*, ***1996***, *61*, 8659.

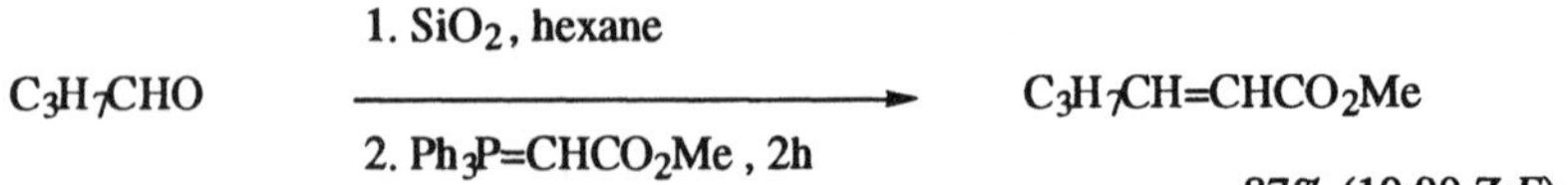
C3H7CHO
1. SiO2 , hexane
2. Ph3P=CHCO2Me , 2h
C3H7CH=CHCO2Me

87% (10:90 *Z:E*)

reaction time was 42 h without silica gel

Patil, V.J.; Mävers, U. *Tetrahedron Lett.*, ***1996***, *37*, 1281.

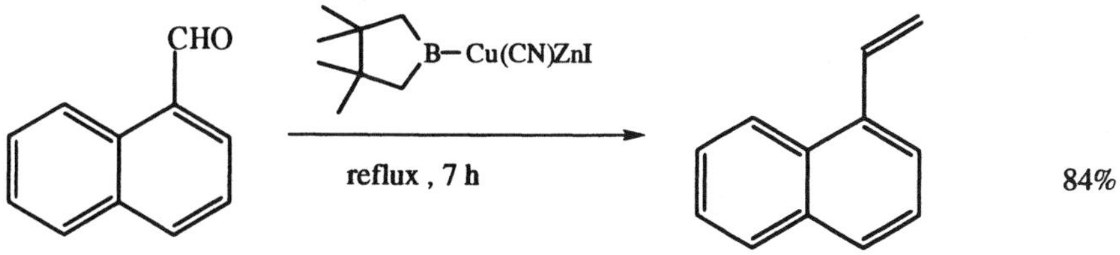
CHO
B—Cu(CN)ZnI
reflux , 7 h
84%

Sakai, M.; Saito, S.; Kanai, G.; Suzuki, A.; Miyaura, N. *Tetrahedron*, ***1996***, *52*, 915.

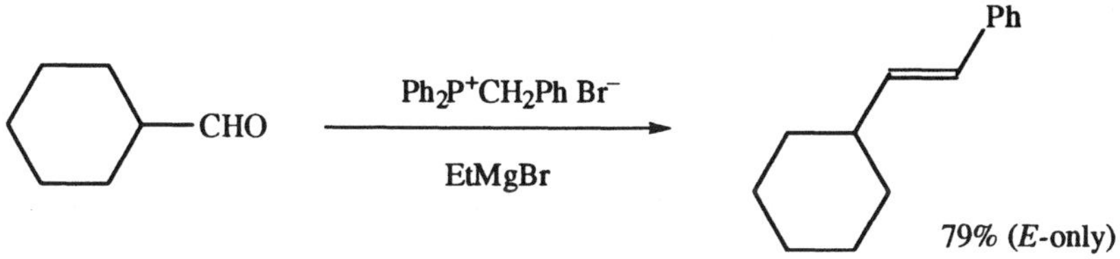
CHO
$Ph_2P^+CH_2Ph\ Br^-$
EtMgBr
Ph
79% (*E*-only)

Shen, Y.; Yao, J. *J. Chem. Res. (S)*, ***1996***, 394.

Related Methods: Section 207 (Alkenes from Ketones).

SECTION 200: ALKENES FROM ALKYLS, METHYLENES AND ARYLS

This section contains dehydrogenations to form alkenes and unsaturated ketones, esters and amides. It also includes the conversion of aromatic rings to alkenes. Reduction of aryls to dienes is found in Section 377 (Alkene-Alkene). Hydrogenation of aryls to alkanes and dehydrogenations to form aryls are included in Section 74 (Alkyls from Alkenes).

NO ADDITIONAL EXAMPLES

SECTION 201: ALKENES FROM AMIDES

Related Methods: Section 65 (Alkyls from Alkyls).
Section 74 (Alkyls from Alkenes).

NO ADDITIONAL EXAMPLES

SECTION 202: ALKENES FROM AMINES

NO ADDITIONAL EXAMPLES

SECTION 203: ALKENES FROM ESTERS

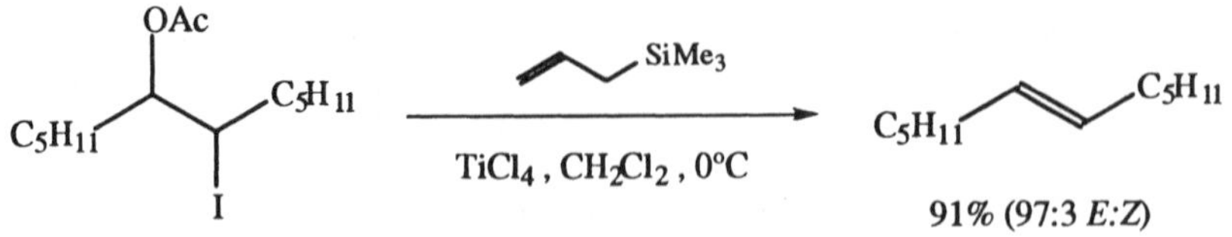
OAc
C_5H_{11}
C_5H_{11}
I
$SiMe_3$
$TiCl_4$, CH_2Cl_2, 0°C
C_5H_{11}
C_5H_{11}
91% (97:3 *E:Z*)

Yachi, K.; Maeda, K.; Shinokubo, H.; Oshima, K. *Tetrahedron Lett*, ***1997***, *38*, 5161.

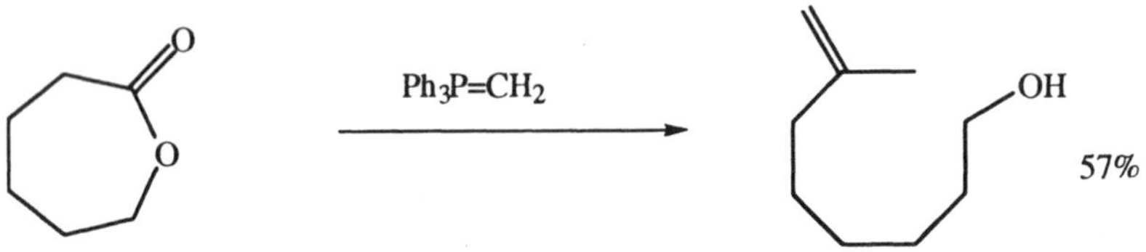
O
O
$Ph_3P=CH_2$
OH
57%

Brunel, Y.; Rousseau, G. *Tetrahedron Lett.*, ***1996***, *37*, 3853.

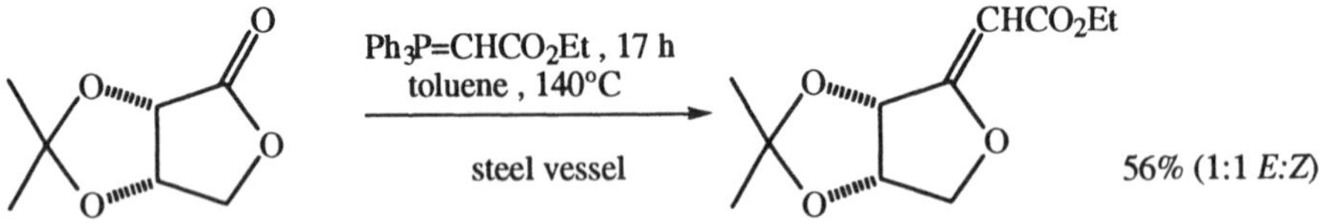
O
O
O
O
$Ph_3P=CHCO_2Et$, 17 h
toluene, 140°C
steel vessel
$CHCO_2Et$
O
O
O
56% (1:1 *E:Z*)

Lakhrissi, M.; Chapleur, Y. *Angew. Chem. Int. Ed.*, ***1996***, *35*, 750.

SECTION 204: ALKENES FROM ETHERS, EPOXIDES AND THIOETHERS

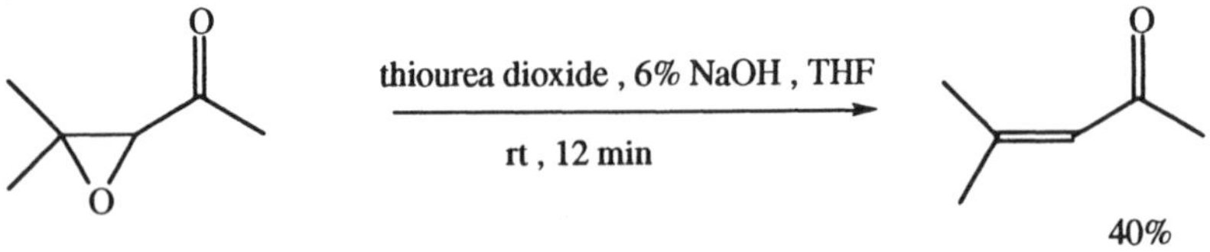
O
O
thiourea dioxide, 6% NaOH, THF
rt, 12 min
O
40%

dos Santos, R.B.; Brocksom, T.J.; Brocksom, U. *Tetrahedron Lett.*, ***1997***, *38*, 745.

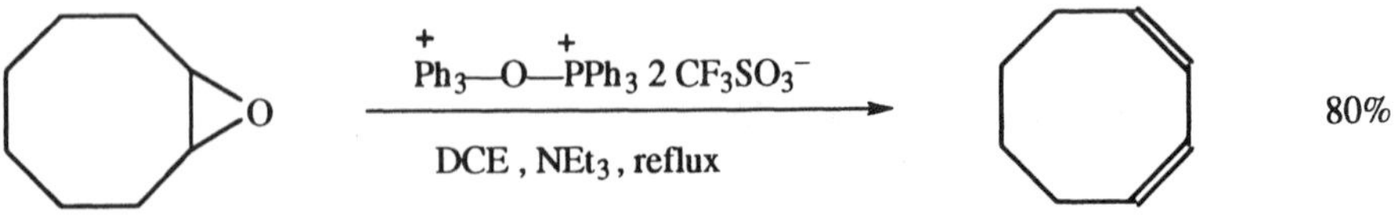
O
$\overset{+}{Ph_3}$—O—$\overset{+}{P}Ph_3$ 2 $CF_3SO_3^-$
DCE, NEt_3, reflux
80%

Hendrickson, J.B.; Walker, M.A.; Varvak, A.; Hussoin, Md.S. *SynLett*, ***1996***, 661.

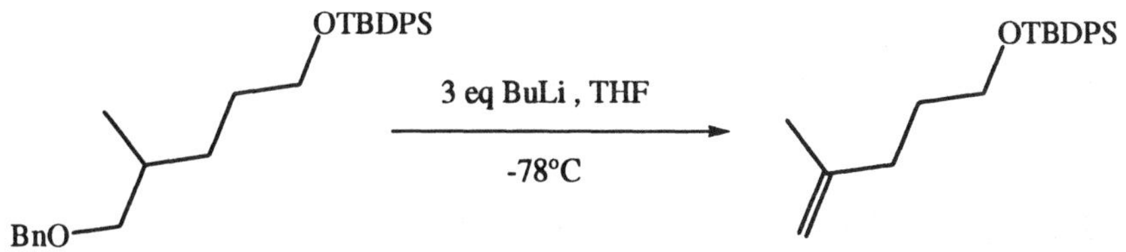
OTBDPS
3 eq BuLi , THF
-78°C
OTBDPS
BnO

Matsushita, M.; Nagaoka, Y.; Hioki, H.; Fukuyama, Y.; Kodama, M. *Chem. Lett.*, ***1996***, 1039

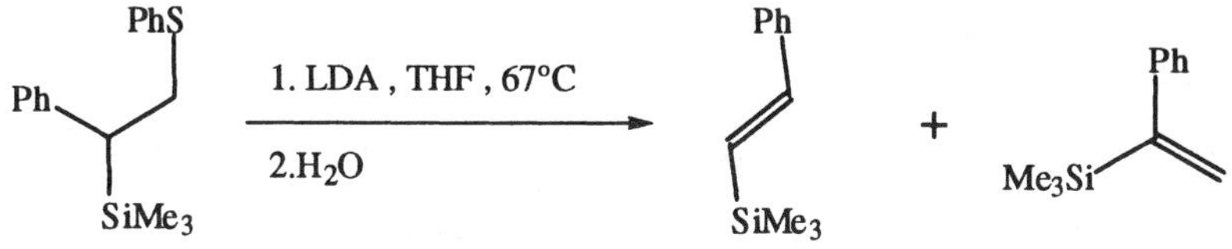
PhS
Ph
SiMe3
1. LDA , THF , 67°C
2.H2O
Ph
SiMe3
+
Ph
Me3Si

(10 : 1) 80%

Menichetti, S.; Stirling, C.J.M. *J. Chem. Soc., Perkin Trans. 1*, ***1996***, 1511.

REVIEW:

"Recent Advances in the Chemistry of Carbenoids Derived from Epoxides."
Doris, E.; Dechoux, L.; Mioskowski, C. *SynLett*, ***1998***, 337.

SECTION 205: ALKENES FROM HALIDES AND SULFONATES

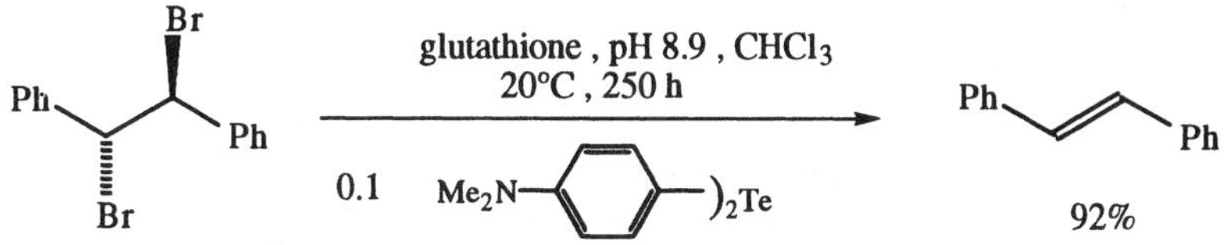
Br
Ph
Ph
Br
glutathione , pH 8.9 , CHCl3
20°C , 250 h
0.1
Me2N
)2Te
Ph
Ph
92%

Butcher, T.S.; Detty, M.R. *J. Org. Chem.*, ***1998***, ***63***, 177.

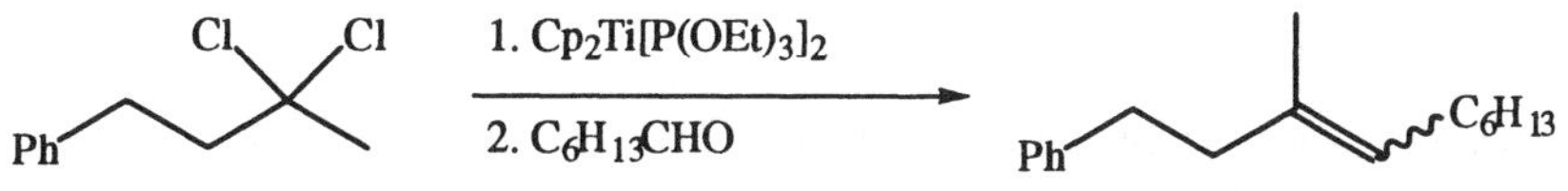
Cl
Cl
Ph
1. Cp2Ti[P(OEt)3]2
2. C6H13CHO
Ph
C6H13

57% (40:60 *E:Z*)

Takeda, T.; Sasaki, R.; Fujiwara, T. *J. Org. Chem.*, ***1998***, ***63***, 7286.

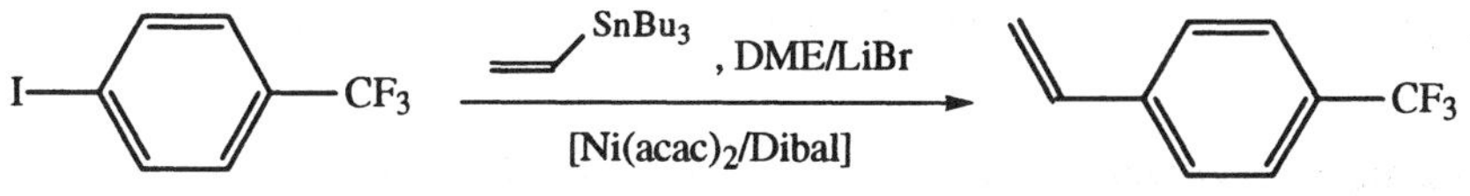
I
CF3
SnBu3
, DME/LiBr
[Ni(acac)2/Dibal]
CF3

91%

Shirakawa, E.; Yamasaki, K.; Hiyama, T. *Synthesis*, ***1998***, 1544.

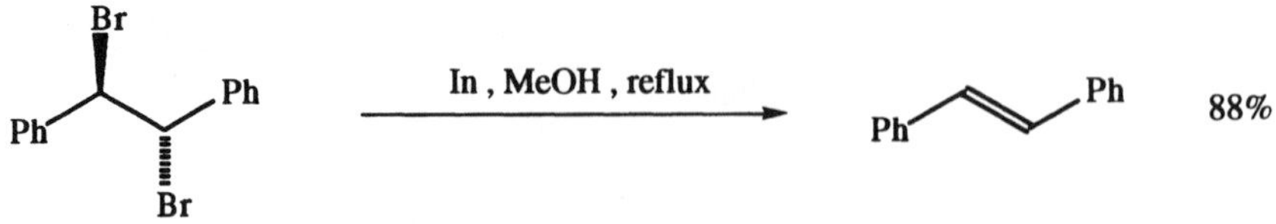
Br
Ph
Ph
Br
In , MeOH , reflux
Ph
Ph
88%

Ranu, B.C.; Guchhait, S.K.; Sarkar, A. *Chem. Commun.*, ***1998***, 2113.

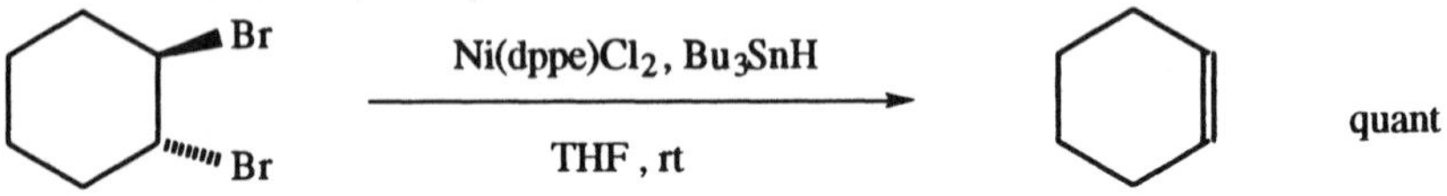
Br
Br
Ni(dppe)Cl2 , Bu3SnH
THF , rt
quant

Malanga, C.; Mannucci, S.; Lardicci, L. *Tetrahedron*, ***1998***, *54*, 1021.

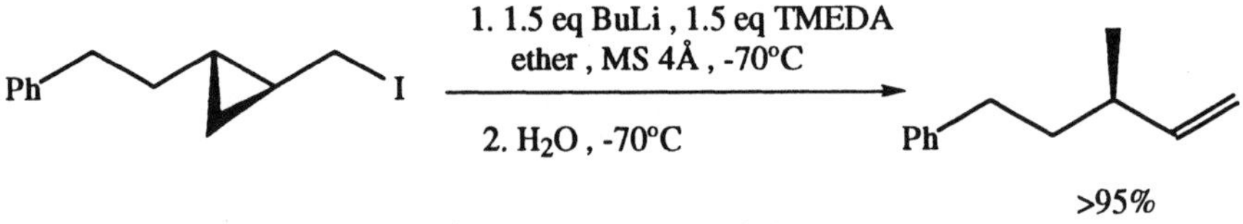
Ph
I
1. 1.5 eq BuLi , 1.5 eq TMEDA
ether , MS 4Å , -70°C
2. H2O , -70°C
Ph
>95%

Charette, A.B.; Naud, J. *Tetrahedron Lett*, ***1998***, *39*, 7259.

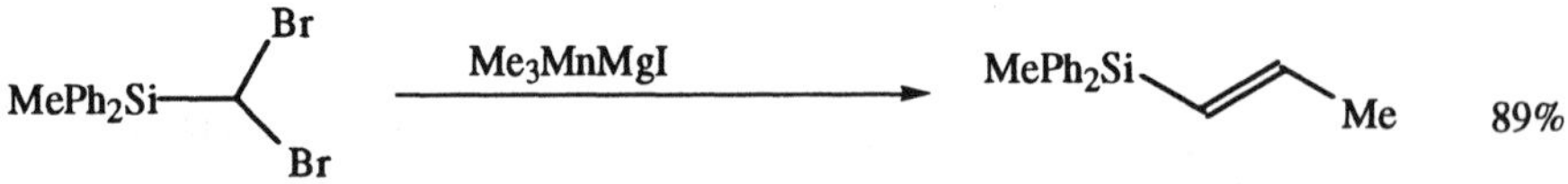
Br
MePh2Si
Br
Me3MnMgI
MePh2Si
Me
89%

Kakiya, H.; Inoue, R.; Shinokubo, H.; Oshima, K. *Tetrahedron Lett*, ***1997***, *38*, 3275.

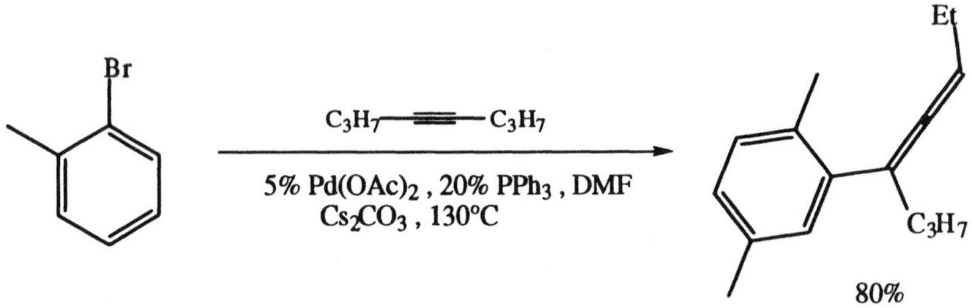
Br
C3H7 C3H7
5% Pd(OAc)2 , 20% PPh3 , DMF
Cs2CO3 , 130°C
Et
C3H7
80%

Pivsa-Art, S.; Satoh, T.; Miura, M.; Nomura, M. *Chem. Lett.*, ***1997***, 823.

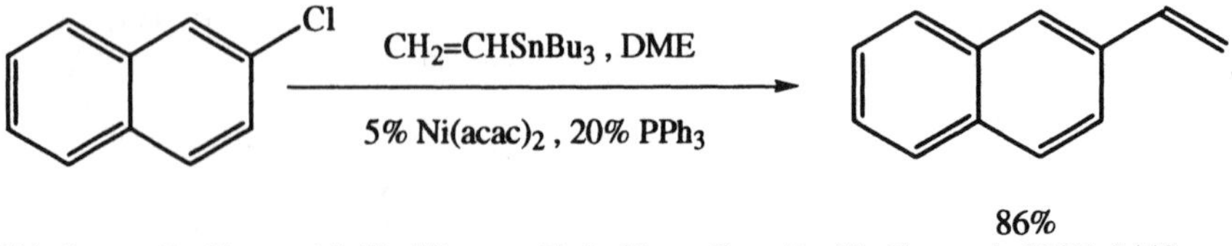
Cl
CH2=CHSnBu3 , DME
5% Ni(acac)2 , 20% PPh3
86%

Shirakawa, E.; Yamasaki, K.; Hiyama, T. *J. Chem. Soc., Perkin Trans. 1*, ***1997***, 2449.

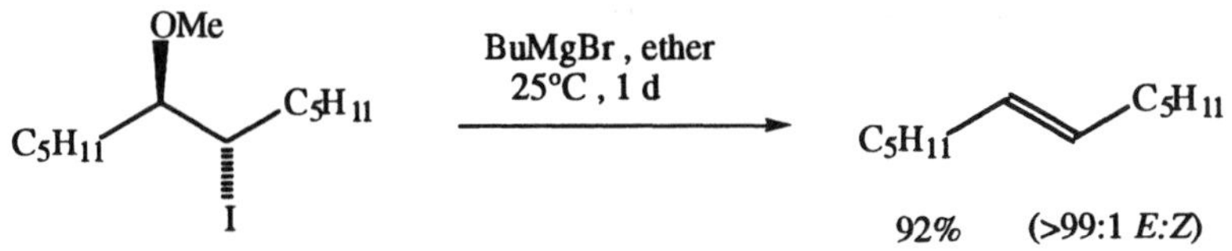
OMe
C_5H_{11}
C_5H_{11}
I
BuMgBr , ether
25°C , 1 d
C_5H_{11}
C_5H_{11}
92% (>99:1 *E:Z*)

Maeda, K.; Shinokubo, H.; Oshima, K. *J. Org. Chem.*, ***1996***, *61*, 6770.

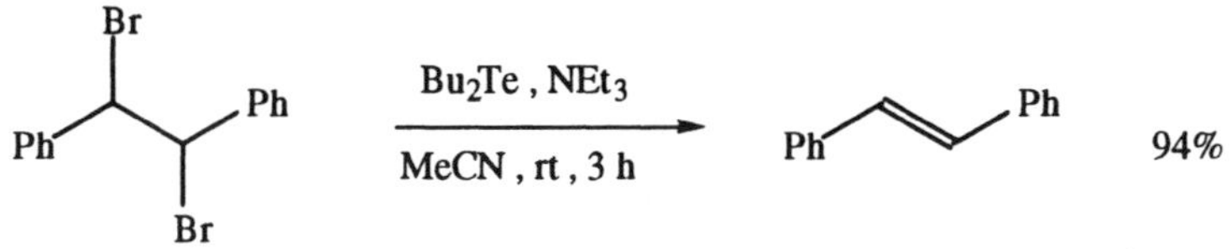
Br
Ph
Ph
Br
Bu_2Te , NEt_3
MeCN , rt , 3 h
Ph
Ph
94%

Zheng, M.; Zhang, M.–H.; Shao, J.–G.; Zhong, Q. *Org. Prep. Proceed. Int.*, ***1996***, *28*, 117.

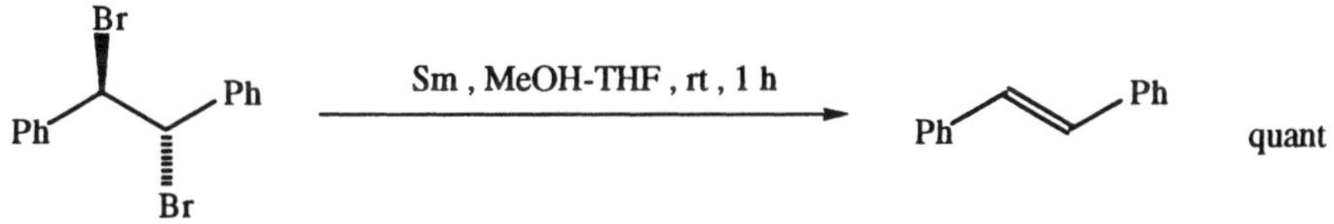
Br
Ph
Ph
Br
Sm , MeOH-THF , rt , 1 h
Ph
Ph
quant

Yanada, R.; Negoro, N.; Yanada, K.; Fujita, T. *Tetrahedron Lett.*, ***1996***, *37*, 9313.

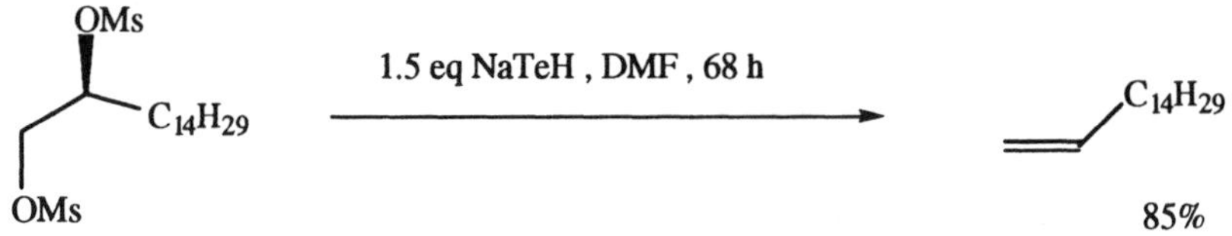
OMs
$C_{14}H_{29}$
OMs
1.5 eq NaTeH , DMF , 68 h
$C_{14}H_{29}$
85%

Bargues, V.; Blay, G.; Fernández, I.; Pedro, J.R. *SynLett*, ***1996***, 655.

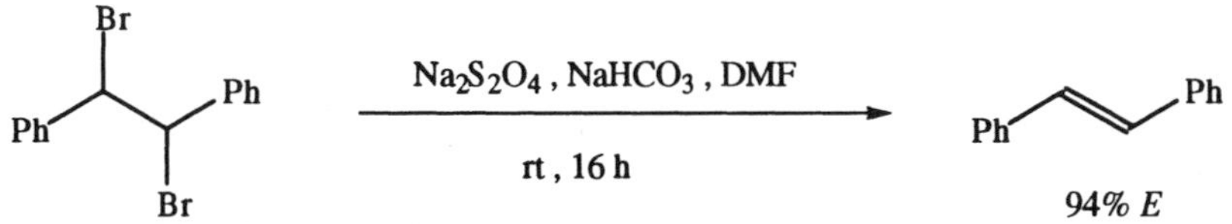
Br
Ph
Ph
Br
$Na_2S_2O_4$, $NaHCO_3$, DMF
rt , 16 h
Ph
Ph
94% *E*

Khurana, J.M.; Sehgal, A. *Synth. Commun.*, ***1996***, *26*, 3791.

SECTION 206: ALKENES FROM HYDRIDES

For conversions of methylenes to alkenes ($RCH_2R' \rightarrow RR'C{=}CH_2$), see Section 200 (Alkenes from Alkyls).

NO ADDITIONAL EXAMPLES

SECTION 207: ALKENES FROM KETONES

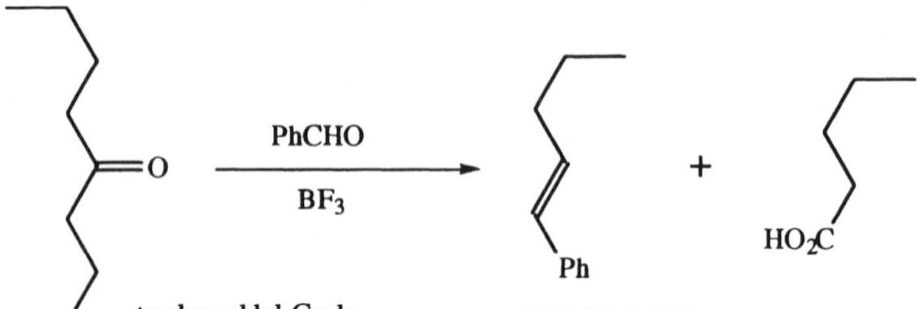

tandem aldol-Grob 78% (97:3 *E:Z*)

Kabalka, G.W.; Tejedor, D.; Li, N.–S.; Malladi, R.R.; Trotman, S.
J. Org. Chem., ***1998***, *63*, 6438.

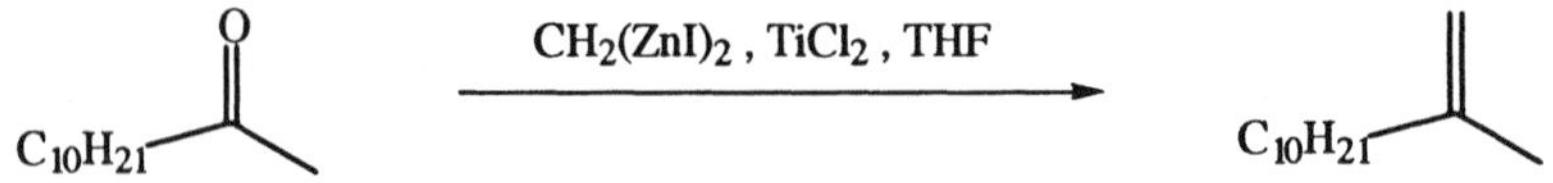

83%

Matsubara, S.; Mizuno, T.; Otake, T.; Kobata, M.; Utimoto, K.; Takai, K.
SynLett, ***1998***, 1369.

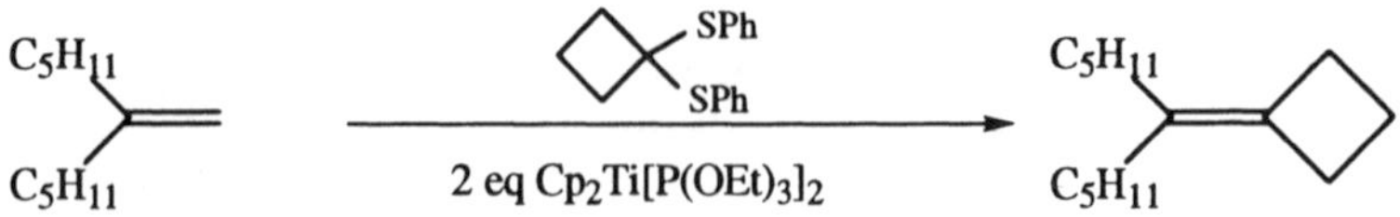

73%

Fujiwara, T.; Iwasaki, N.; Takeda, T. *Chem. Lett.*, ***1998***, 741.

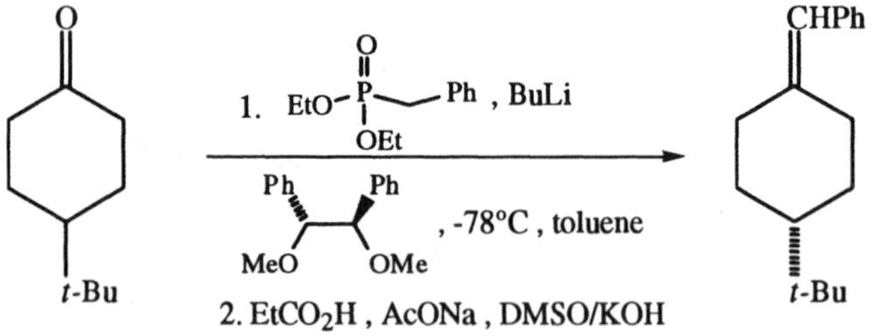

85% (84% ee , *S*)

Mizuno, M.; Fujii, K.; Tomioka, K. *Angew. Chem. Int. Ed.*, ***1998***, *37*, 515.

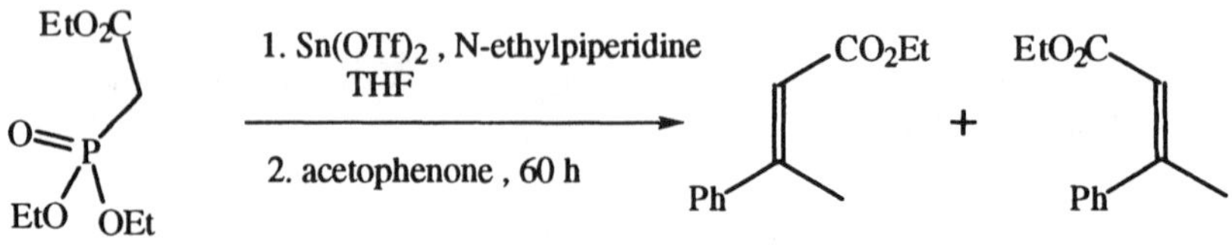

(85 : 15) quant

Sano, S.; Yokoyama, K.; Fukushima, M.; Yagi, T.; Nagao, Y. *Chem. Commun.*, ***1997***, 559.

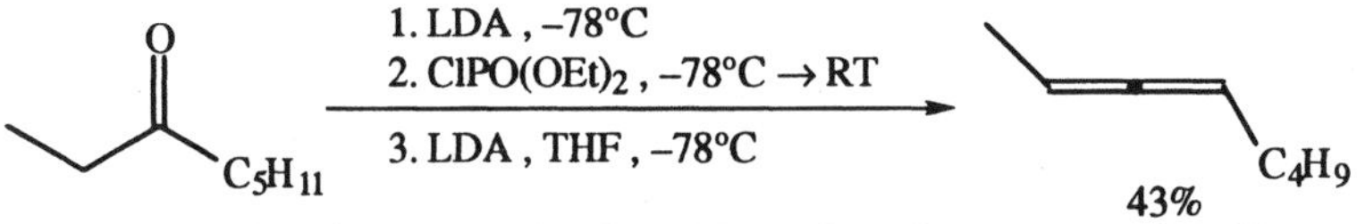
O
1. LDA , –78°C
2. ClPO(OEt)2 , –78°C → RT
3. LDA , THF , –78°C
C5H11
C4H9
43%

<u>Brummond, K.M.</u>; Dingess, E.A.; Kent, J.L. *J. Org. Chem.*, ***1996***, *61*, 6096.

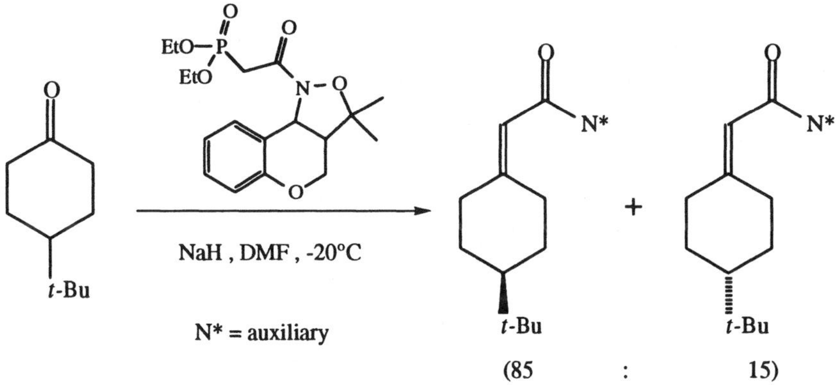
EtO—P
EtO
N–O
O
NaH , DMF , -20°C
N* = auxiliary
t-Bu
N*
+
(85 : 15)

<u>Abiko, A.</u>; Masamune, S. *Tetrahedron Lett.*, ***1996***, *37*, 1077.

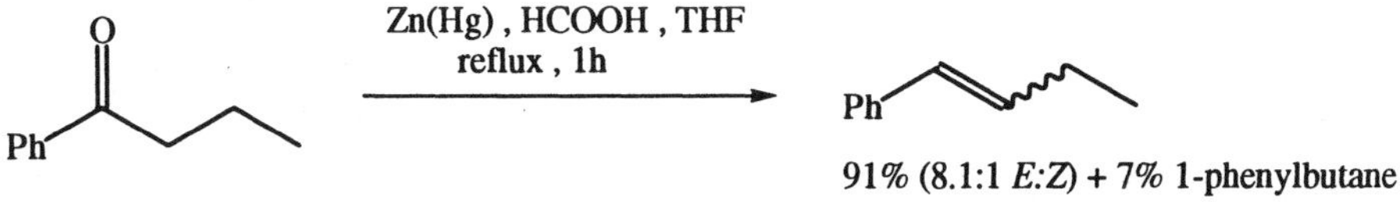
Zn(Hg) , HCOOH , THF
reflux , 1h
Ph
91% (8.1:1 E:Z) + 7% 1-phenylbutane

<u>Hiegel, G.A.</u>; Carney, J.R. *Synth. Commun.*, ***1996***, *26*, 2625.

Related Methods: Section 199 (Alkenes from Aldehydes).

SECTION 208: ALKENES FROM NITRILES

NO ADDITIONAL EXAMPLES

SECTION 209: ALKENES FROM ALKENES

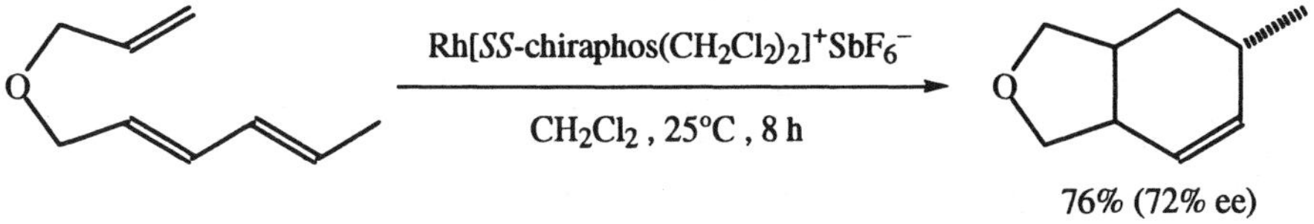
Rh[SS-chiraphos(CH2Cl2)2]+SbF6–
CH2Cl2 , 25°C , 8 h
O
76% (72% ee)

<u>Gilbertson, S.R.</u>; Hoge, G.S.; Genov, D.G. *J. Org. Chem.*, ***1998***, *63*, 10077.

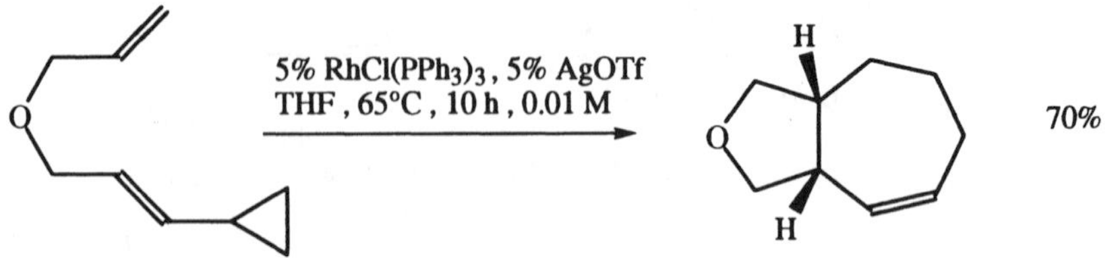
5% RhCl(PPh3)3 , 5% AgOTf
THF , 65°C , 10 h , 0.01 M
O
H
O
H
70%

Wender, P.A.; Husfeld, C.O.; Langkopf, E.; Love, J.A. *J. Am. Chem. Soc.*, ***1998***, *120*, 1940

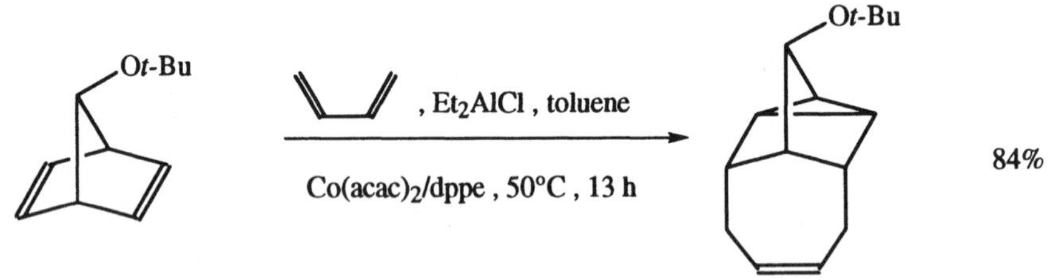
Ot-Bu
, Et2AlCl , toluene
Co(acac)2/dppe , 50°C , 13 h
Ot-Bu
84%

Chen, Y.; Snyder, J.K. *J. Org. Chem.*, ***1998***, *63*, 2060.

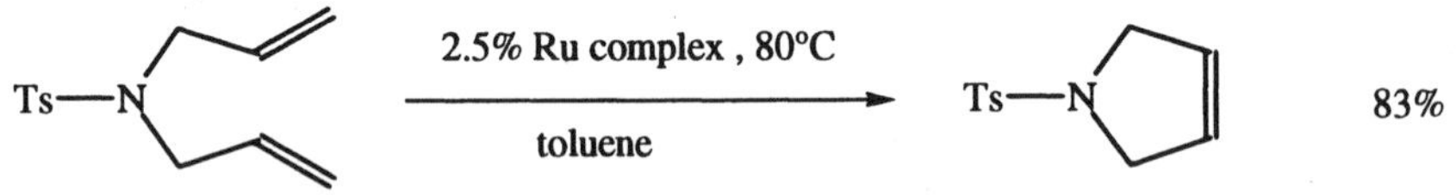
Ts—N
2.5% Ru complex , 80°C
toluene
Ts—N
83%

Fürstner, A.; Picquet, M.; Bruneau, C.; Dixneuf, P.H. *Chem. Commun.*, ***1998***, 1315.

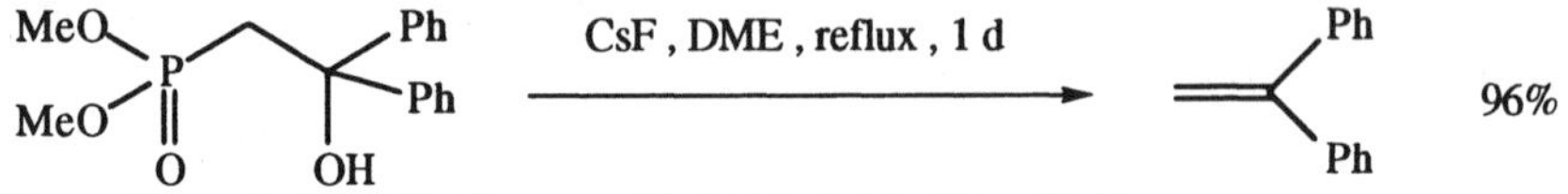
MeO
MeO
P
O
Ph
Ph
OH
CsF , DME , reflux , 1 d
Ph
Ph
96%

Kawashima, T.; Ishii, T.; Inamoto, N.; Tokitoh, N.; Okazaki, R.
Bull. Chem. Soc., Jpn., ***1998***, *71*, 209.

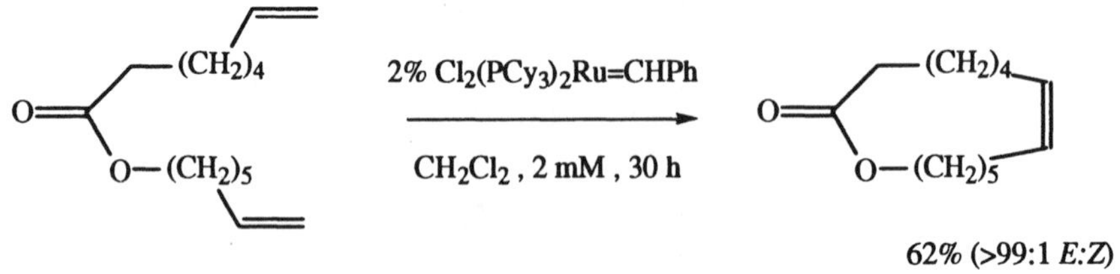
(CH2)4
O
O—(CH2)5
2% Cl2(PCy3)2Ru=CHPh
CH2Cl2 , 2 mM , 30 h
(CH2)4
O
O—(CH2)5
62% (>99:1 E:Z)

Goldring, W.P.D.; Hodder, A.S.; Weiler, L. *Tetrahedron Lett.*, ***1998***, *39*, 4955.

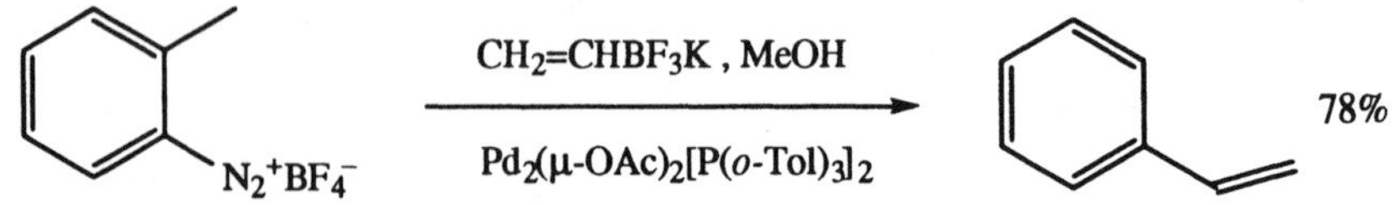
N2+BF4−
CH2=CHBF3K , MeOH
Pd2(μ-OAc)2[P(o-Tol)3]2
78%

Darses, S.; Michaud, G.; Genêt, J.-P. *Tetrahedron Lett.*, ***1998***, *39*, 5045.

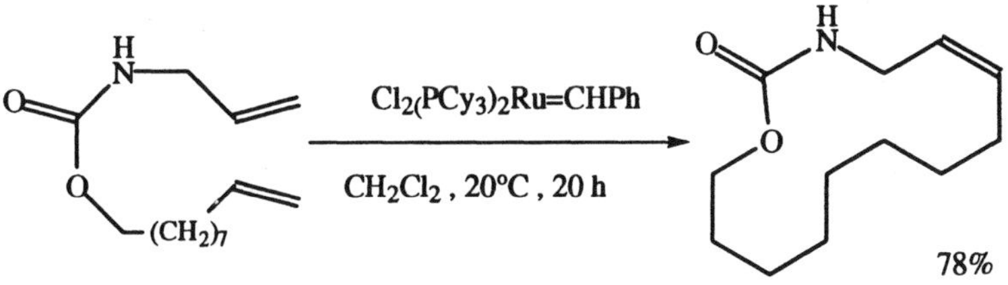
H
N
O
O
(CH2)7
Cl2(PCy3)2Ru=CHPh
CH2Cl2 , 20°C , 20 h
78%

Chosh, A.K.; Hussain, K.A. *Tetrahedron Lett.*, ***1998***, *39*, 1881.

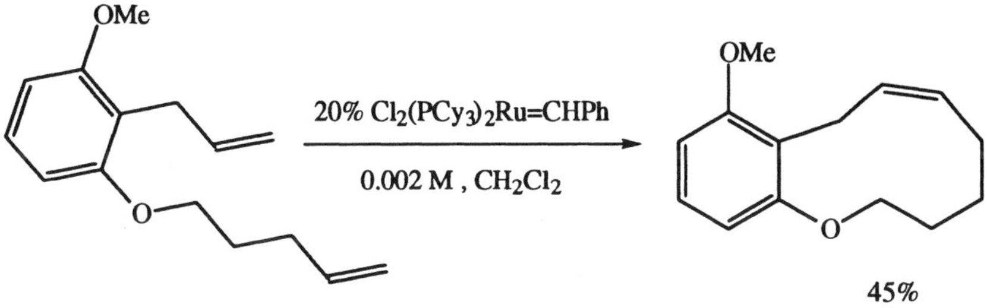
OMe
O
20% Cl2(PCy3)2Ru=CHPh
0.002 M , CH2Cl2
45%

Stefinovic, M.; Snieckus, V. *J. Org. Chem.*, ***1998***, *63*, 2808.

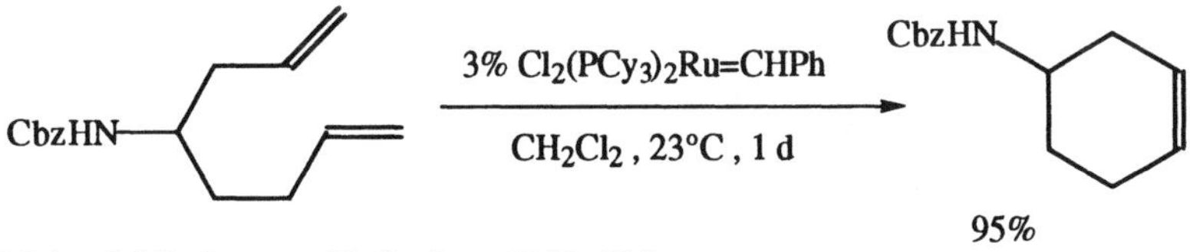
CbzHN
3% Cl2(PCy3)2Ru=CHPh
CH2Cl2 , 23°C , 1 d
95%

Maier, M.E.; Lapeva, T. *SynLett*, ***1998***, 891.

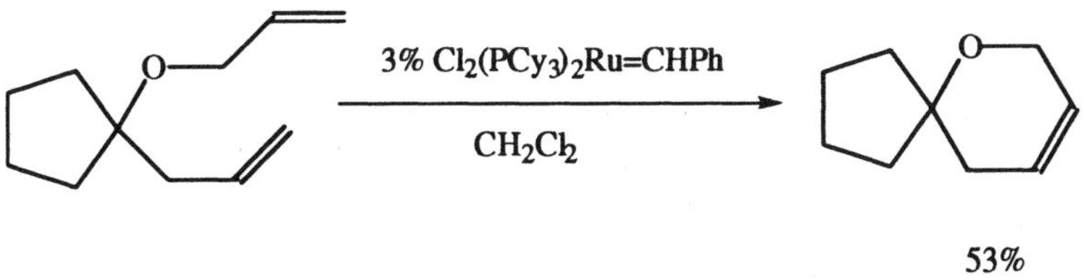
O
3% Cl2(PCy3)2Ru=CHPh
CH2Cl2
53%

Maier, M.E.; Bugl, M. *SynLett*, ***1998***, 1390.

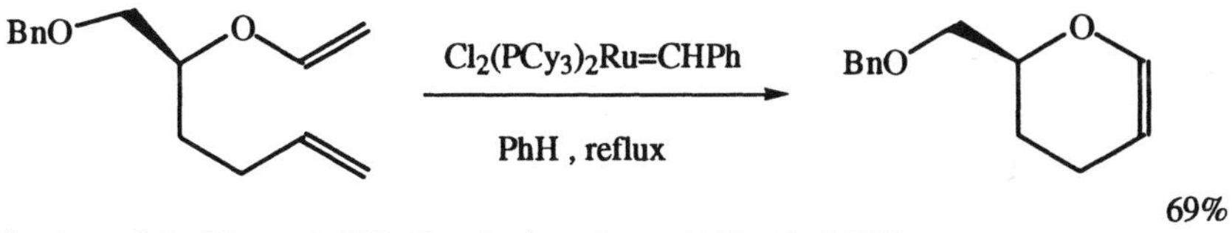
BnO
O
Cl2(PCy3)2Ru=CHPh
PhH , reflux
69%

Sturino, C.F.; Wong, J.C.Y. *Tetrahedron Lett.*, ***1998***, *39*, 9623.

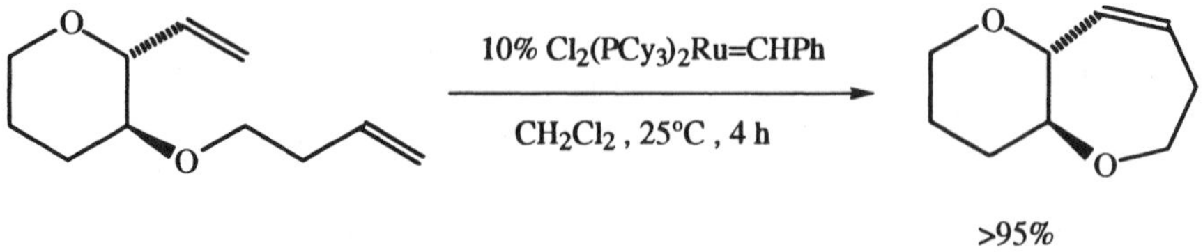
10% Cl2(PCy3)2Ru=CHPh
CH2Cl2, 25°C, 4 h
>95%

Delgado, M.; Martín, J.D. *Tetrahedron Lett*, ***1997***, *38*, 6299.

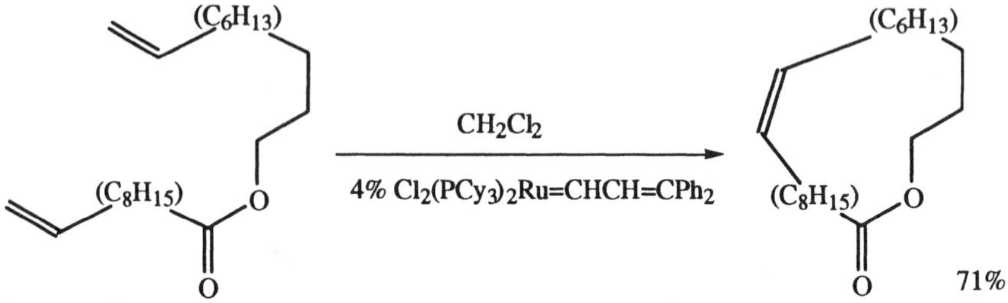
(C6H13)
(C8H15)
CH2Cl2
4% Cl2(PCy3)2Ru=CHCH=CPh2
71%

Fürstner, A.; Langemann, K. *J. Org. Chem.*, ***1996***, *61*, 3942.

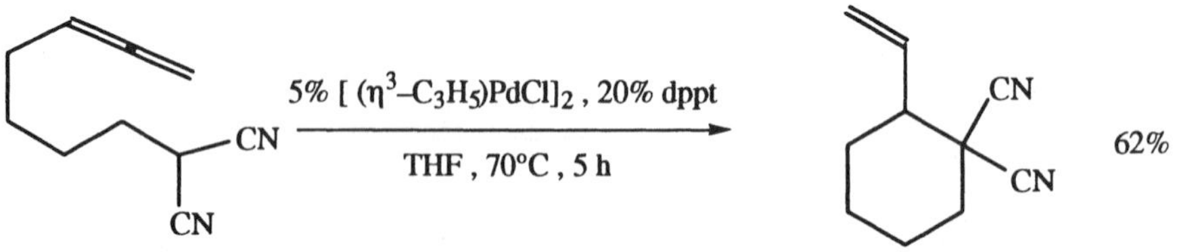
5% [(η3–C3H5)PdCl]2, 20% dppt
THF, 70°C, 5 h
62%

Meguro, M.; Kamijo, S.; Yamamoto, Y. *Tetrahedron Lett*, ***1996***, *37*, 7453.

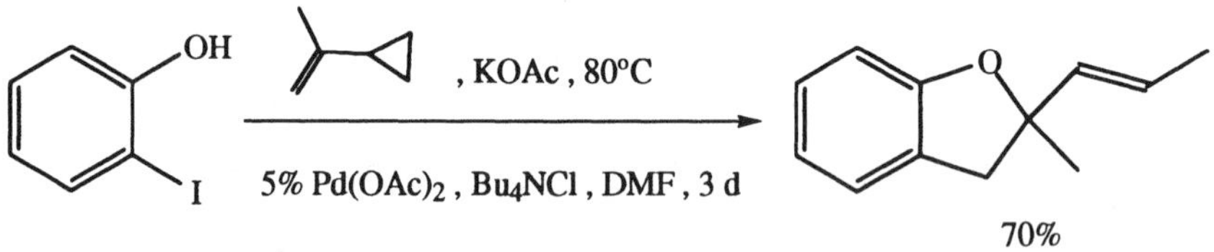
, KOAc, 80°C
5% Pd(OAc)2, Bu4NCl, DMF, 3 d
70%

Larock, R.C.; Yum, E.K. *Tetrahedron*, ***1996***, *52*, 2743.

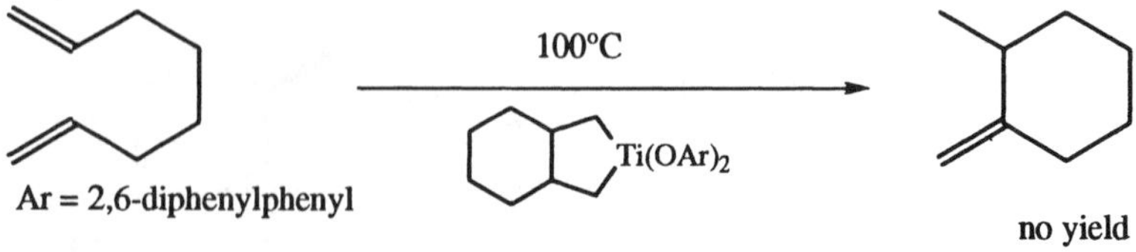
100°C
Ti(OAr)2
Ar = 2,6-diphenylphenyl
no yield

Wartuke, S.A.; Johnson, E.S.; Thorn, M.G.; Fanwick, P.E.; Rothwell, I.P. *Chem. Commun.*. ***1996***. 2617.

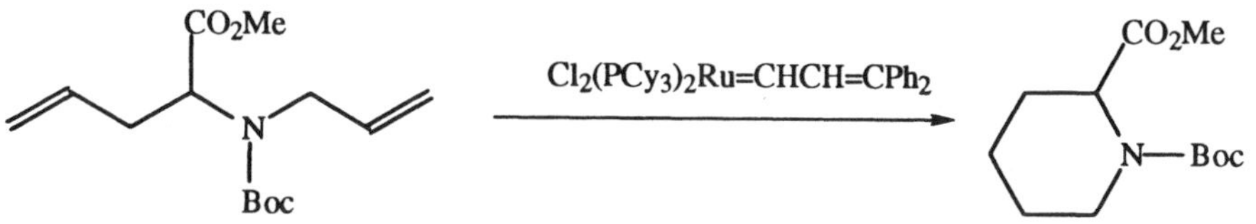

91%

Miller, S.J.; Blackwell, H.E.; Grubbs, R.H. *J. Am. Chem. Soc.*, ***1996***, *118*, 9606.

REVIEW:

"The Asymmetric Heck Reaction"

Shibasaki, M.; Boden, D.J.; Kojima, A. *Tetrahedron*, ***1997***, *53*, 7371.

SECTION 210: ALKENES FROM MISCELLANEOUS COMPOUNDS

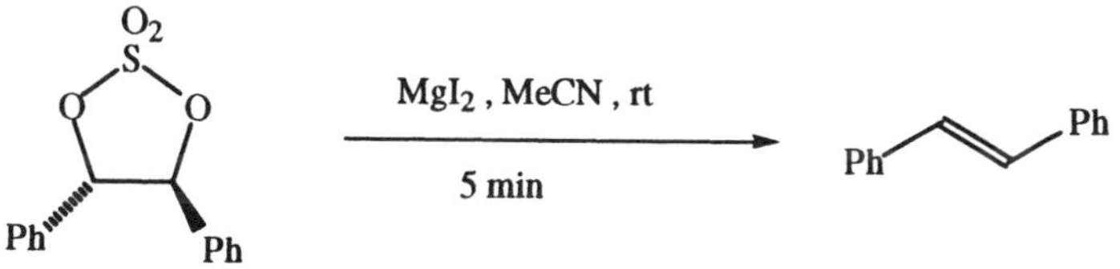

90%

Jang, D.O.; Joo, Y.H. *Synth. Commun.*, ***1998***, *28*, 871.

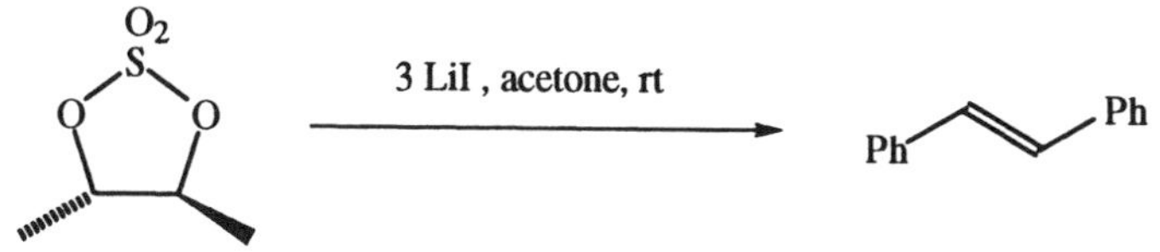

95%

Jang, D.O.; Joo, Y.H. *SynLett*, ***1997***, 279.

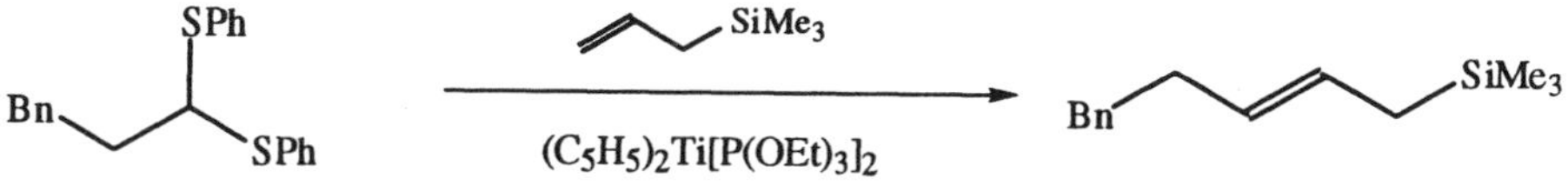

64% (13:87 *E:Z*)

Fujiwara, T.; Takamori, M.; Takeda, T. *Chem. Commun.*, ***1998***, 51.

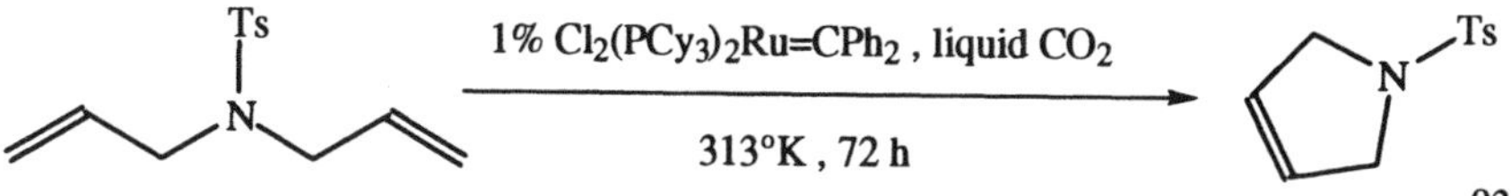

93%

Fürstner, A.; Koch, D.; Langemann, K.; Leitner, W.; Six, C.
Angew. Chem. Int. Ed., ***1997***, *36*, 2466.

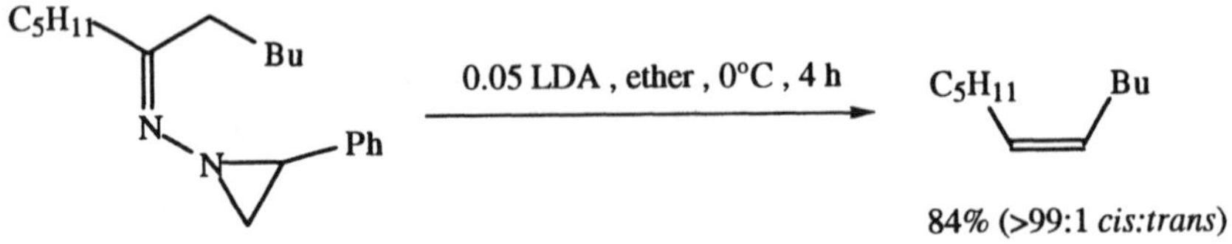

0.05 LDA , ether , 0°C , 4 h
84% (>99:1 cis:trans)

Maruoka, K.; Oishi, M.; Yamammoto, H. *J. Am. Chem. Soc.*, ***1996***, *118*, 2289.

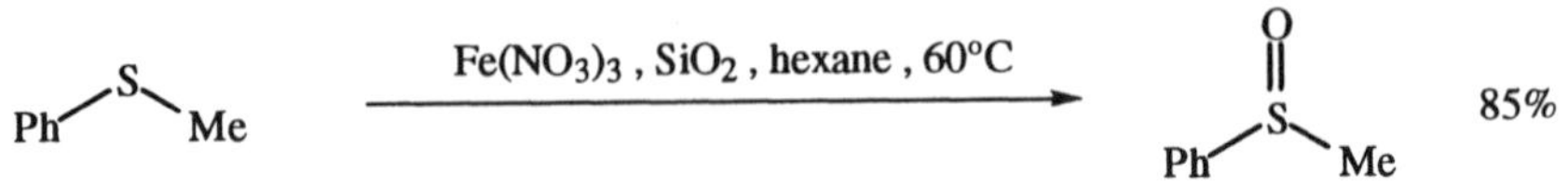

Fe(NO3)3 , SiO2 , hexane , 60°C
85%

Hiyano, M.; Komiya, K.; Yakabe, S.; Clark, J.H.; Morimoto, T.
Org. Prep. Proceed. Int., ***1996***, *28*, 705.

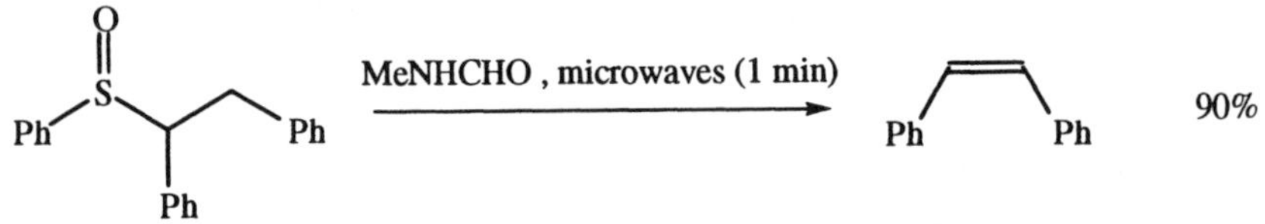

MeNHCHO , microwaves (1 min)
90%

Moghaddam, F.M.; Ghaffarzadeh, M. *Tetrahedron Lett.*, ***1996***, *37*, 1855.

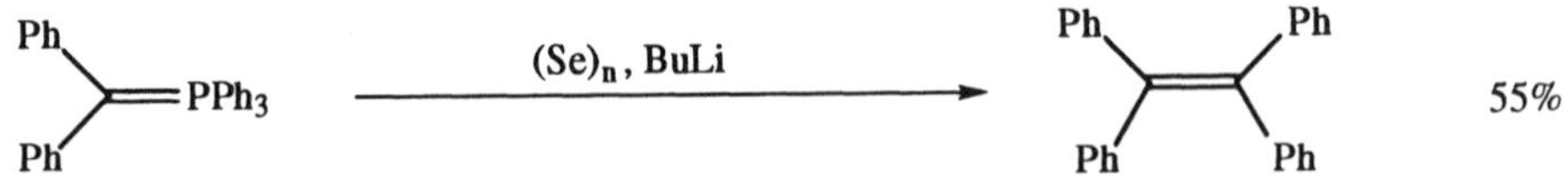

(Se)n , BuLi
55%

Okuma, K.; Koda, G.; Okumura,, S.; Ohno, A. *Chem. Lett.*, ***1996***, 609.

CHAPTER 15

PREPARATION OF OXIDES

This chapter contains reactions which prepare the oxides of nitrogen, sulfur and selenium. Included are *N*-oxides, nitroso and nitro compounds, nitrile oxides, sulfoxides, selenoxides and sulfones. Oximes are considered to be amines and appear in those sections. Preparation of sulfonic acid derivatives are found in Chapter Two and the preparation of sulfonates in Chapter Ten.

SECTION 211: OXIDES FROM ALKYNES

NO ADDITIONAL EXAMPLES

SECTION 212: OXIDES FROM ACID DERIVATIVES

$$PhSO_2Cl \xrightarrow[0°C\text{ , }3\text{ h}]{PhCH_2Br\text{ , }Zn\text{ , }THF\text{ , aq }NH_4Cl} PhSO_2CH_2Ph \qquad 80\%$$

Sun, X.; Wang, L.; Zhang, Y. *Synth. Commun.*, ***1998***, *28*, 1785.

SECTION 213: OXIDES FROM ALCOHOLS AND THIOLS

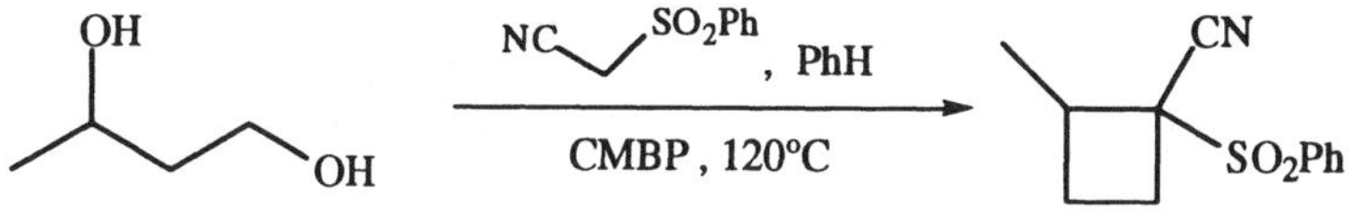
OH
OH
NC
SO2Ph
, PhH
CMBP , 120°C
CN
SO2Ph

99%

CMBP = cyanomethylenetributylphosphorane

Tsunoda, T.; Nagino, C.; Oguri, M.; Itô, S. *Tetrahedron Lett.*, ***1996***, *37*, 2459.
Tsunoda, T.; Ozaki, F.; Shiiakata, N.; Tamaoka, Y.; Yamamoto, H.; Itô, S. *Tetrahedron Lett.*, ***1996***, *37*, 2463.

SECTION 214: OXIDES FROM ALDEHYDES

NO ADDITIONAL EXAMPLES

SECTION 215: OXIDES FROM ALKYLS, METHYLENES AND ARYLS

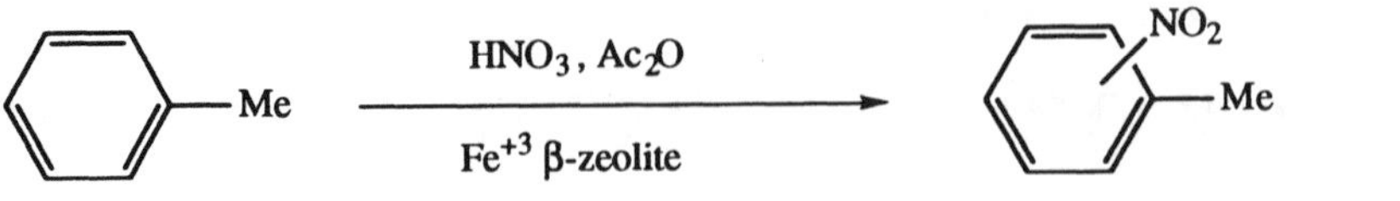
Me
HNO3 , Ac2O
Fe+3 β-zeolite
NO2
Me

>99% (*o:m:p* 25:3:72)

Smith, K.; Musson, A.; De Boos, G.A. *J. Org. Chem.*, ***1998***, *63*, 8448.

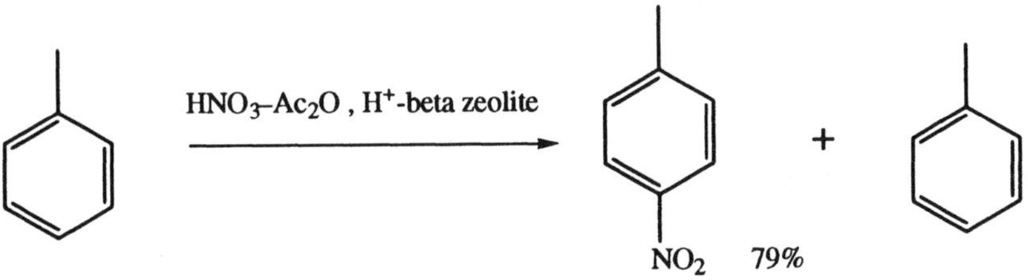
HNO3–Ac2O , H+-beta zeolite
+
NO2

79% 18%

Smith, K.; Musson, A.; DeBoos, G.A. *Chem. Commun.*, ***1996***, 469.

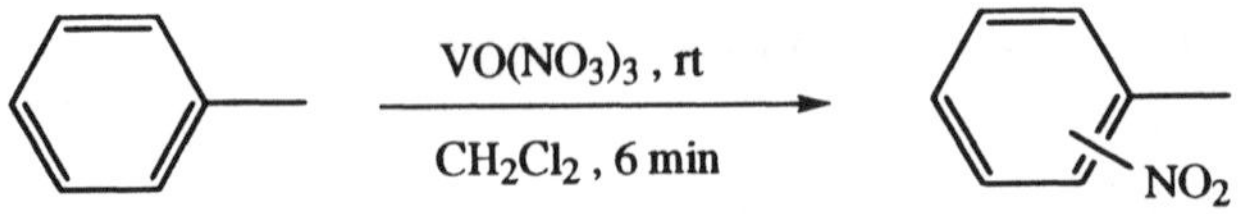
VO(NO3)3 , rt
CH2Cl2 , 6 min
NO2

>99% (35:2:41 *o:m:p*)
+ 19% 2,4-disubstituted

Dove, M.F.A.; Manz, B.; Montgomery, J.; Pattenden, G.; Wood, S.A. *J. Chem. Soc., Perkin Trans. 1*, ***1998***, 1589.

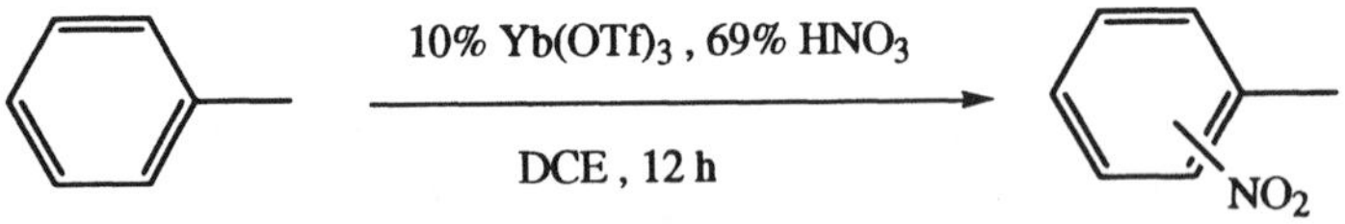
10% Yb(OTf)3 , 69% HNO3
DCE , 12 h
NO2

51% *o*/41% *p*

Waller, F.J.; Barrett, A.G.M.; Braddock, D.C.; Ramprasad, D. *Chem. Commun.*, ***1997***, 613.

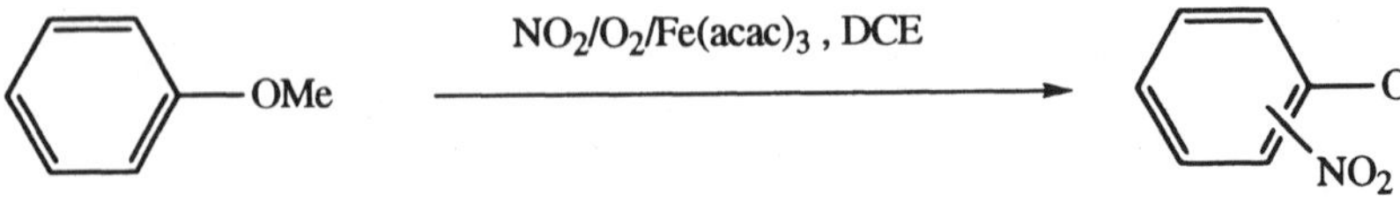
OMe
NO2/O2/Fe(acac)3 , DCE
OMe
NO2

55% (*o:m:p* = 40:<1:60)

Suzuki, H.; Yonezawa, S.; Nonoyama, N.; Mori, T. *J. Chem. Soc.. Perkin Trans. 1. **1996***. 2385.

SECTION 216: OXIDES FROM AMIDES

NO ADDITIONAL EXAMPLES

SECTION 217: OXIDES FROM AMINES

$PhNH_2$ —(H_2O_2, $MeReO_3$, rt, 2h)→ $PhNO_2$ 70%

Murray, R.W.; Iyanar, K.; Chen, J.; Wearing, J.T. *Tetrahedron Lett*, ***1996***, *37*, 805.

SECTION 218: OXIDES FROM ESTERS

NO ADDITIONAL EXAMPLES

SECTION 219: OXIDES FROM ETHERS, EPOXIDES AND THIOETHERS

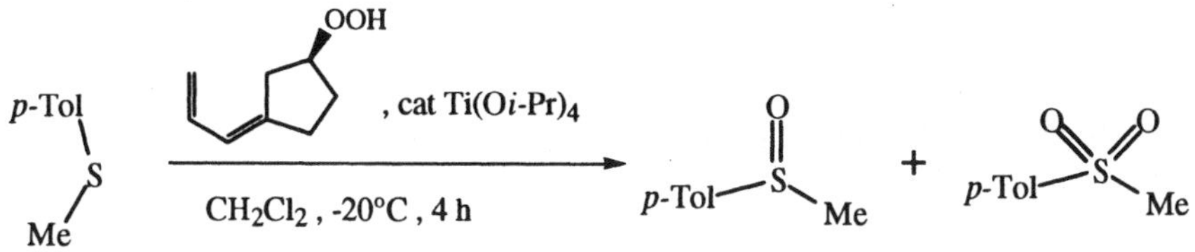

(79 : 21)

20% ee

95% conversion

Adam, W.; Korb, M.N.; Roschmann, K.J.; Saha–Möller, C.R. *J. Org. Chem.*, ***1998***, *63*, 3423

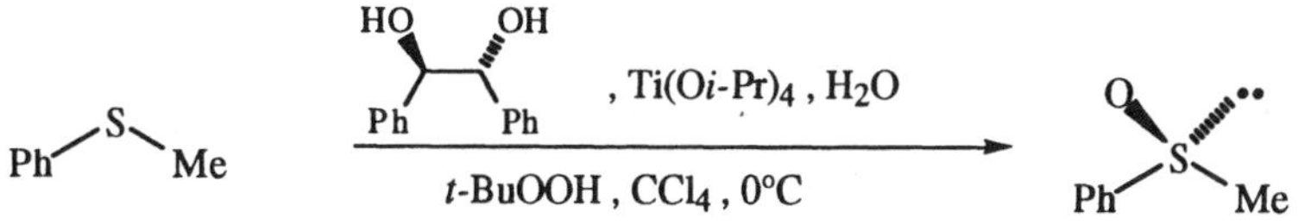

63% (80% ee, *S*)

Donnoli, M.I.; Superchi, S.; Rosini, C. *J. Org. Chem.*, ***1998***, *63*, 9392.

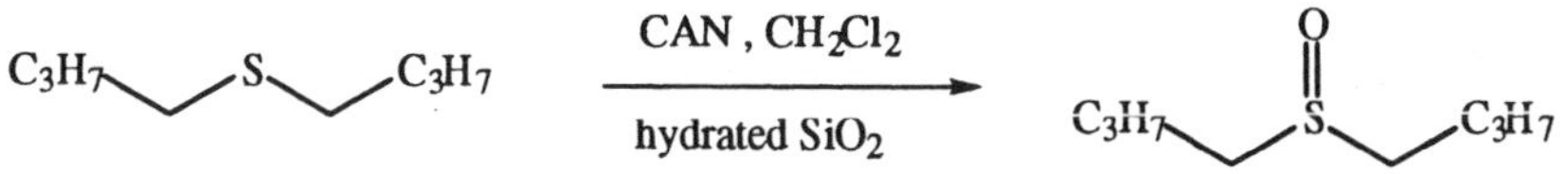

90%

Ali, M.H.; Leach, D.R.; Schmitz, C.E. *Synth. Commun.*, ***1998***, *28*, 2969.

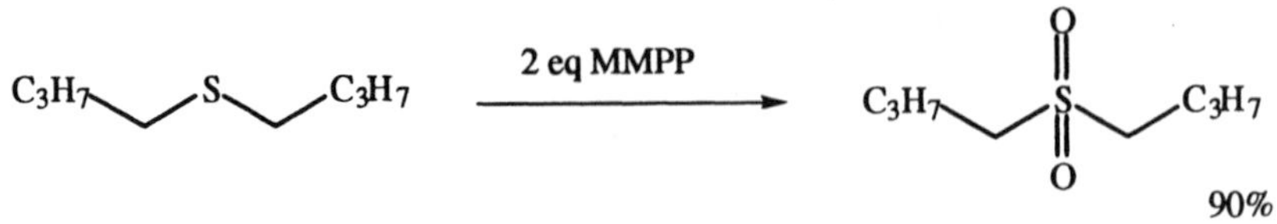
C_3H_7 S C_3H_7
2 eq MMPP
O
C_3H_7 S C_3H_7
O
90%

Ali, M.H.; Bohnert, C.J. *Synth. Commun.*, ***1998***, *28*, 2983.

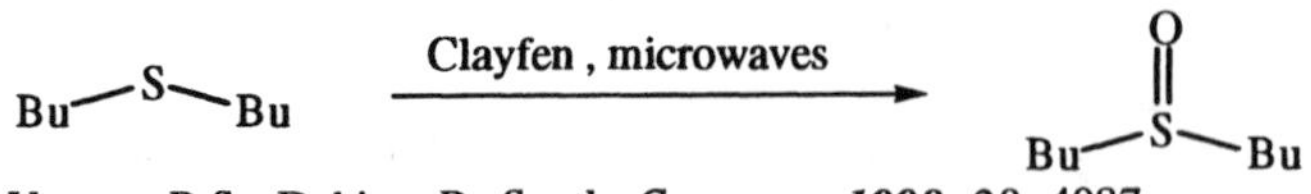
Bu S Bu
Clayfen , microwaves
O
Bu S Bu

Varma, R.S.; Dahiya, R. *Synth. Commun.*, ***1998***, *28*, 4087.

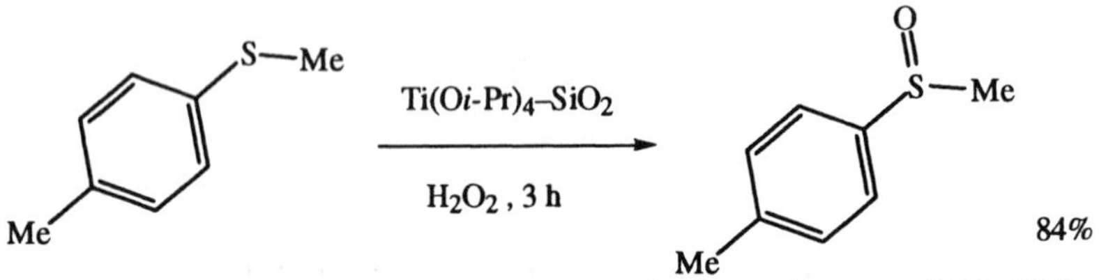
S—Me
Me
$Ti(Oi\text{-}Pr)_4\text{–}SiO_2$
H_2O_2 , 3 h
O
S—Me
Me
84%

Faile, J.M.; García, J.I.; Lázaro, B.; Mayoral, J.A. *Chem. Commun.*, ***1998***, 1807.

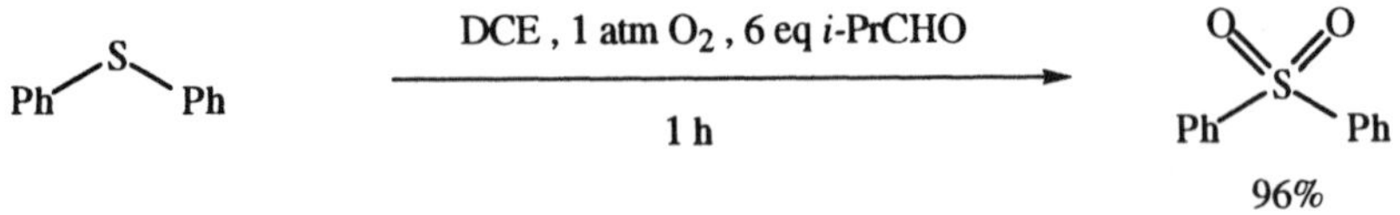
Ph S Ph
DCE , 1 atm O_2 , 6 eq *i*-PrCHO
1 h
O O
S
Ph Ph
96%

Rao, T.V.; Sain, B.; Kumar, K.; Murthy, P.S.; Rao, T.S.R.P.; Joshi, G.C. *Synth. Commun.*, ***1998***, *28*, 319.

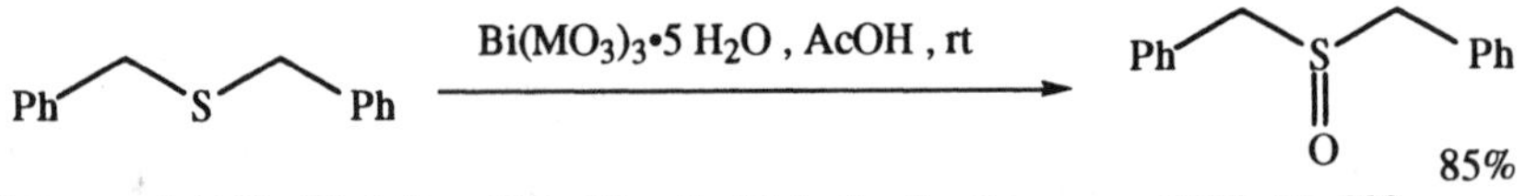
Ph S Ph
$Bi(MO_3)_3\cdot 5\ H_2O$, AcOH , rt
Ph S Ph
O
85%

Masraqui, S.H.; Mudaliar, C.D.; Karnik, M.A. *Synth. Commun.*, ***1998***, *28*, 939.

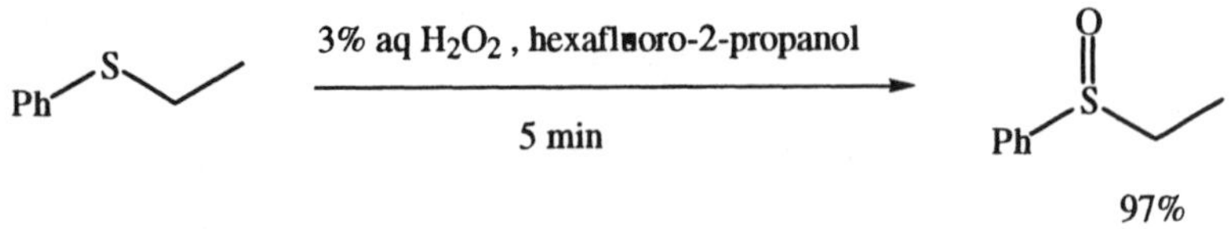
Ph S
3% aq H_2O_2 , hexafluoro-2-propanol
5 min
O
S
Ph
97%

Ravikumar, K.S.; Bégué, J.-P.; Bonnet-Delpon, D. *Tetrahedron Lett.*, ***1998***, *39*, 3141.

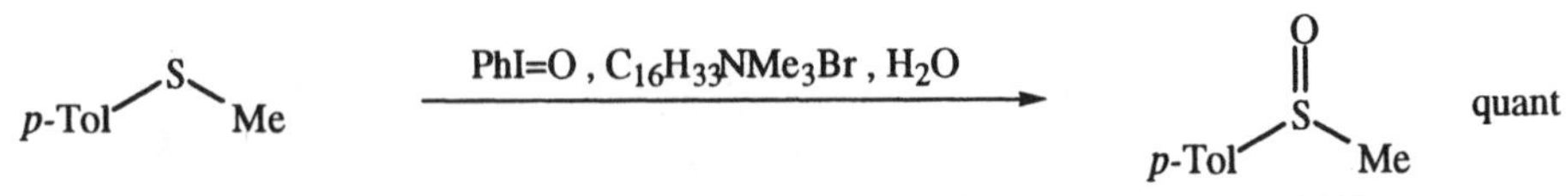
S
p-Tol Me
PhI=O , $C_{16}H_{33}NMe_3Br$, H_2O
O
S
p-Tol Me
quant

Tohma, H.; Takizawa, S.; Watanabe, H.; Kita, H. *Tetrahedron Lett.*, ***1998***, *39*, 4547.

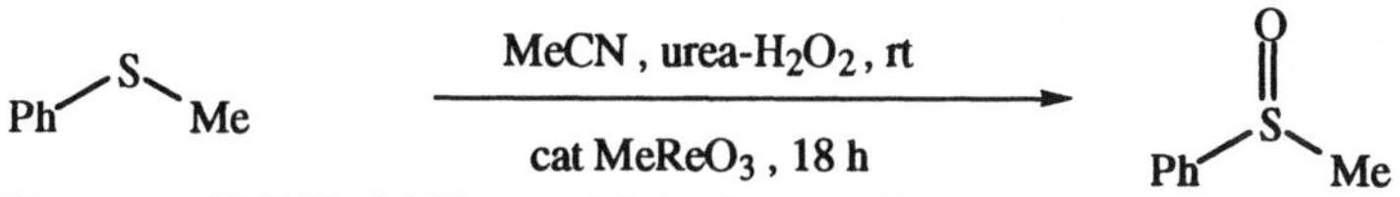
MeCN , urea-H2O2 , rt
cat MeReO3 , 18 h
Ph S Me
O

92%

Gunaratne, H.Q.N.; McKervey, M.A.; Feutren, S.; Finlay, J.; Boyd, J. *Tetrahedron Lett.*, ***1998***, *39*, 5655.

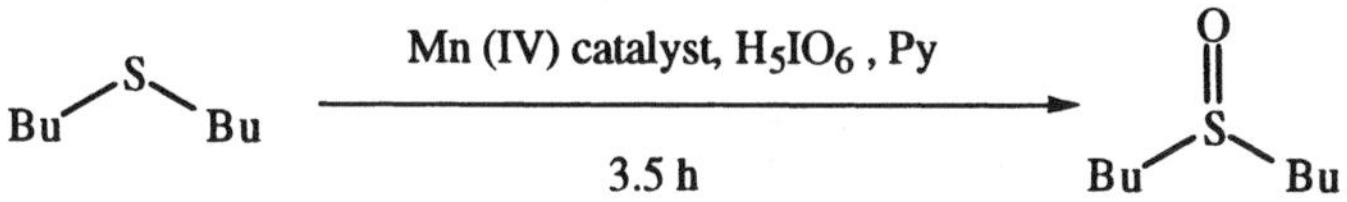
Mn (IV) catalyst, H5IO6 , Py
3.5 h
Bu S Bu
O

96%

reaction overnight gives sulfones

Barton, D.H.R.; Li, W.; Smith, J.A. *Tetrahedron Lett.*, ***1998***, *39*, 7055.

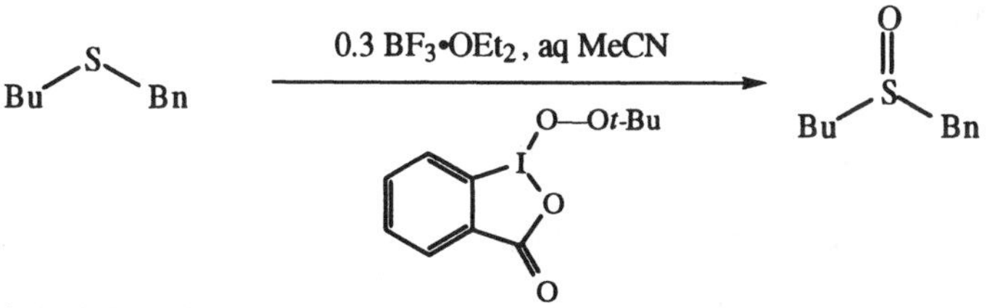
0.3 BF3•OEt2 , aq MeCN
O—Ot-Bu
I
O
O
Bu S Bn

quant

Ochiai, M.; Nakanishi, A.; Ito, T. *J. Org. Chem.*, ***1997***, *62*, 4253.

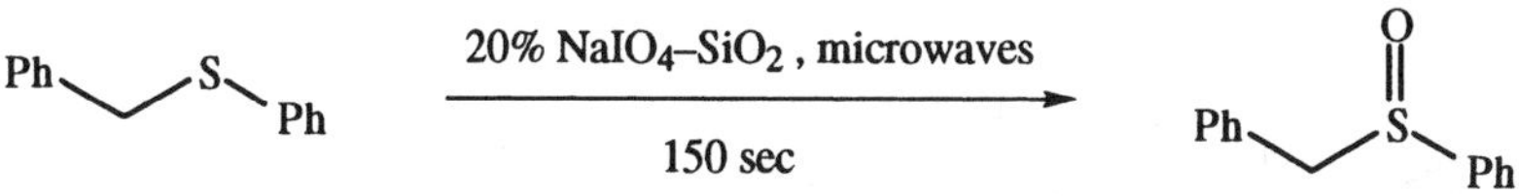
20% NaIO4–SiO2 , microwaves
150 sec
Ph S Ph
O

higher equivalents of reagent can lead to sulfones

83%

Varma, R.S.; Saini, R.K.; Meshram, H.M. *Tetrahedron Lett*, ***1997***, *38*, 6525.

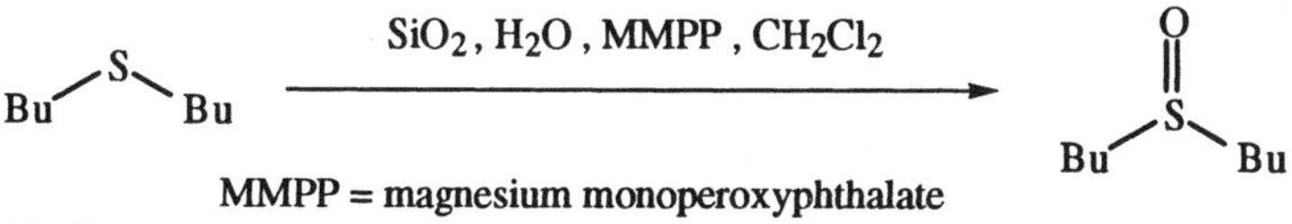
SiO2 , H2O , MMPP , CH2Cl2
Bu S Bu
O

99%

MMPP = magnesium monoperoxyphthalate

Ali, M.H.; Steven, W.C. *Synthesis*, ***1997***, 764.

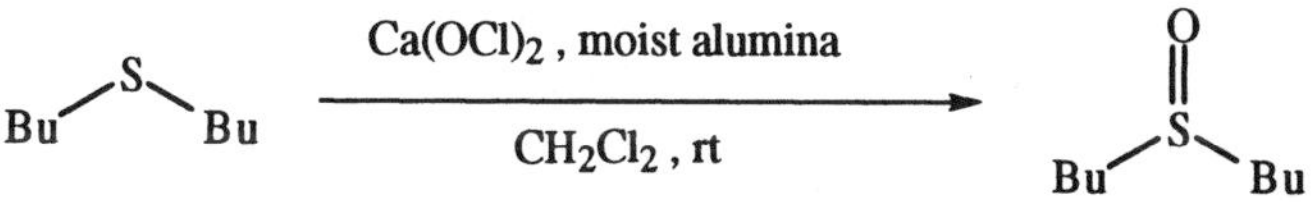
Ca(OCl)2 , moist alumina
CH2Cl2 , rt
Bu S Bu
O

90%

Hirano, M.; Yakabe, S.; Itoh, S.; Clark, J.H.; Morimotoa, T. *Synthesis*, ***1997***, 1161.

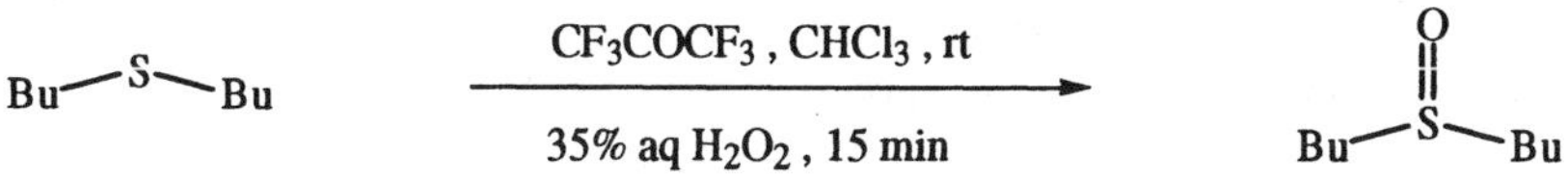
CF3COCF3 , CHCl3 , rt
35% aq H2O2 , 15 min
Bu S Bu
O

Lupattelli, P.; Ruzziconi, R.; Scafato, P. *Synth. Commun.*, ***1997***, *27*, 441.

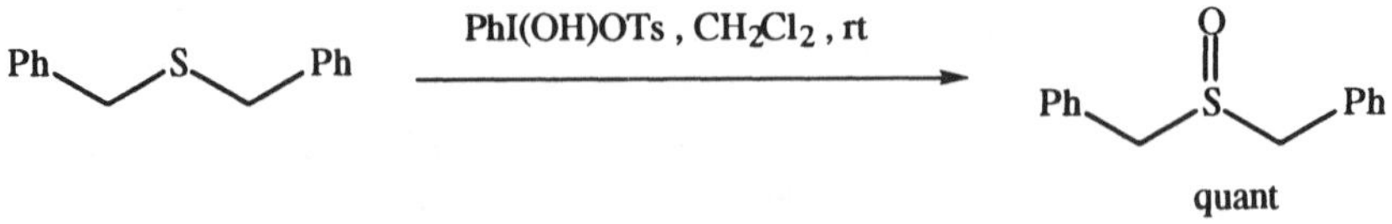
PhI(OH)OTs , CH2Cl2 , rt
quant

Xia, M.; Chen, Z.-C. *Synth. Commun.*, ***1997***, *27*, 1315.

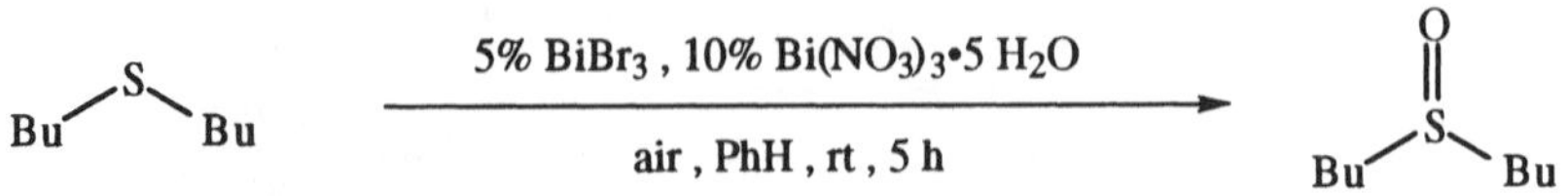
5% BiBr3 , 10% Bi(NO3)3•5 H2O
air , PhH , rt , 5 h

Komatsu, N.; Uda, M.; Suzuki, H. *Chem. Lett.*, ***1997***, 1229.

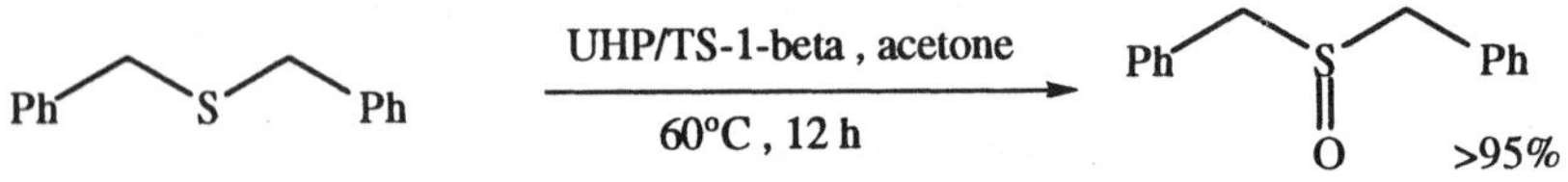
UHP/TS-1-beta , acetone
60°C , 12 h
>95%

TS-1 = microporous titanium silicate ; UHP = urea adduct of hydrogen peroxide

Reddy, T.I.; Varma, R.S. *Chem. Commun.*, ***1997***, 471.

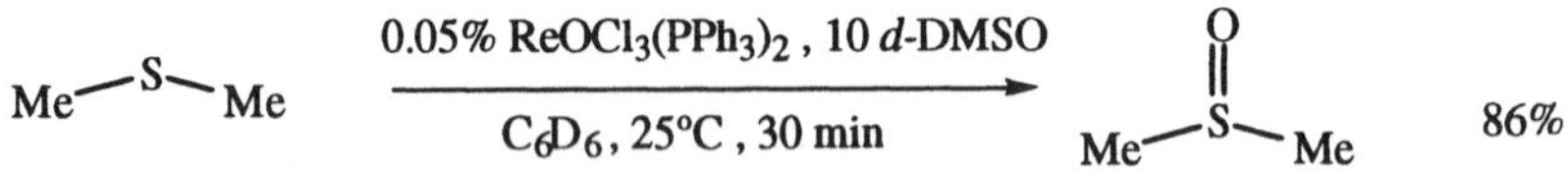
0.05% ReOCl3(PPh3)2 , 10 d-DMSO
C6D6 , 25°C , 30 min
86%

Arterburn, J.B.; Nelson, S.L. *J. Org. Chem.*, ***1996***, *61* 2260.

10 eq NaOCl , MeCN , rt
89%

Khurana, J.M.; Panda, A.K.; Rag, A.; Gogia, A. *Org. Prep. Proceed. Int.*, ***1996***, *28*, 234.

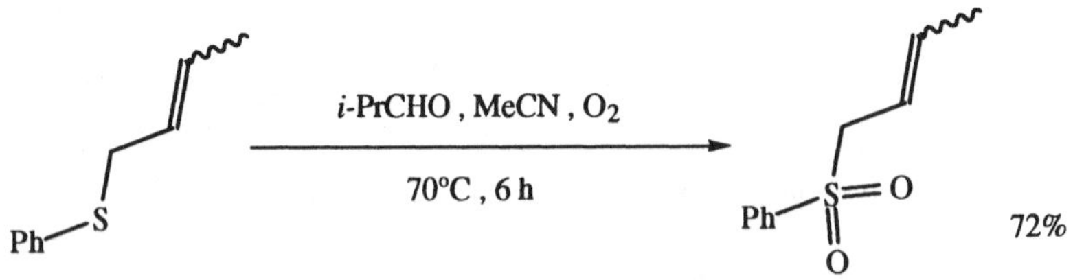
i-PrCHO , MeCN , O2
70°C , 6 h
72%

Khanna, V.; Maikap, G.C.; Iqbal, J. *Tetrahedron Lett.*, ***1996***, *37*, 3367.

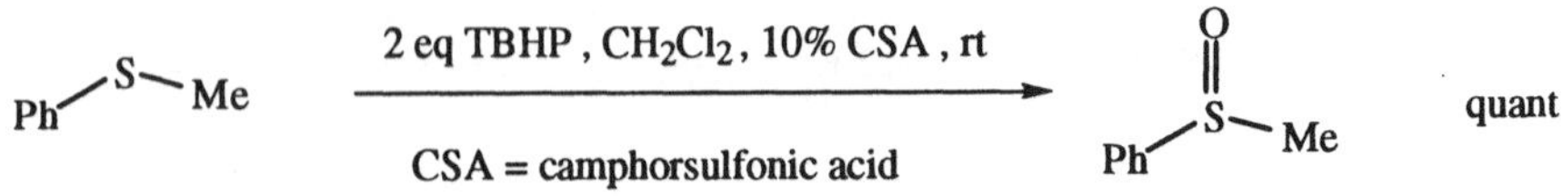
2 eq TBHP , CH2Cl2 , 10% CSA , rt
CSA = camphorsulfonic acid
quant

Bonadies, F.; De Angelis, F.; Locati, L.; Scettri, A. *Tetrahedron Lett.*, ***1996***, *37*, 7129.

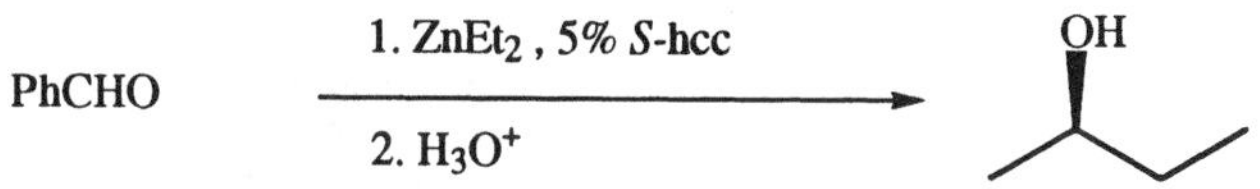
PhCHO
1. ZnEt2 , 5% S-hcc
2. H3O+
OH

80% (94:6 *S:R*)

hcc = hyperbranched chiral catalysts

Brunel, J.M.; Kagan, H.B. *SynLett*, ***1996***, 404.

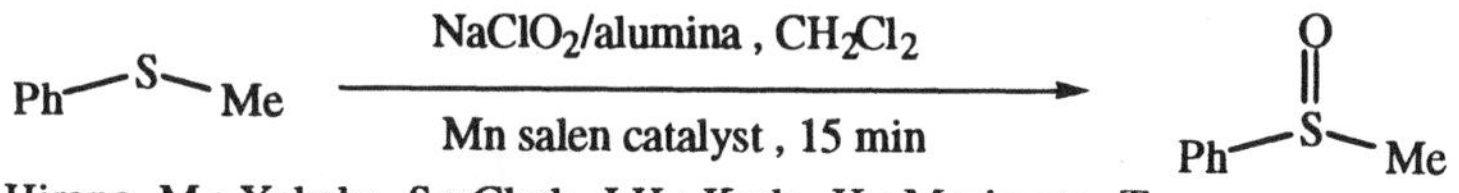
Ph S Me
NaClO2/alumina , CH2Cl2
Mn salen catalyst , 15 min
O
Ph S Me

89%

Hirano, M.; Yakabe, S.; Clark, J.H.; Kudo, H.; Morimoto, T. *Synth. Commun.*, ***1996***, *26*, 1875.

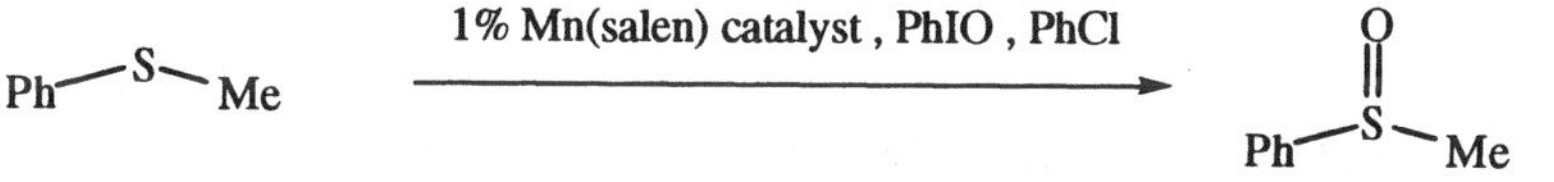
Ph S Me
1% Mn(salen) catalyst , PhIO , PhCl
O
Ph S Me

93% (77% ee , *S*)

Kokubo, C.; Katsuki, T. *Tetrahedron*, ***1996***, *52*, 13895.

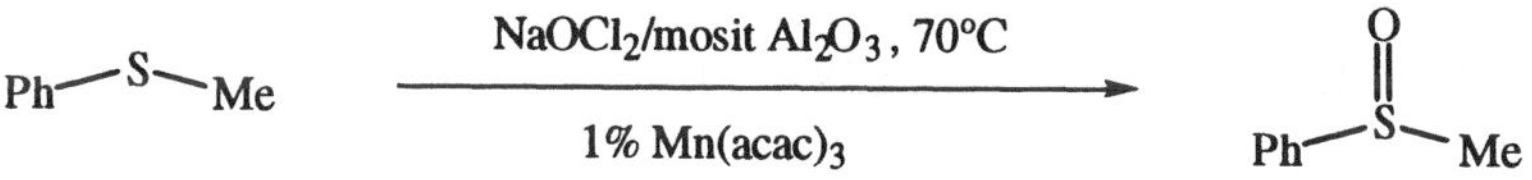
Ph S Me
NaOCl2/mosit Al2O3 , 70°C
1% Mn(acac)3
O
Ph S Me

93%

Hirano, M.; Yakabe, S.; Clark, J.H.; Morimoto, T. *J. Chem. Soc., Perkin Trans. 1*, ***1996***, 2693.

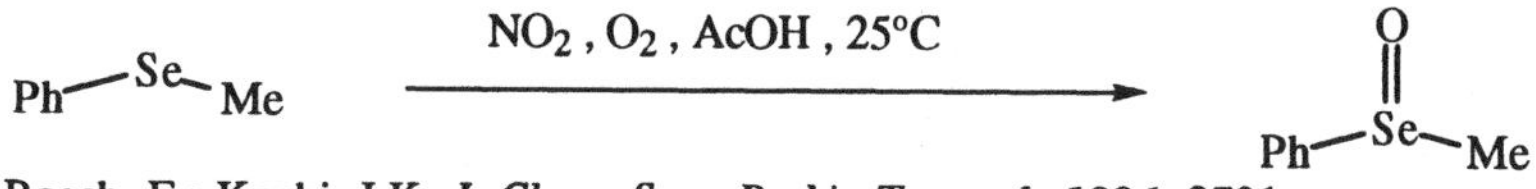
Ph Se Me
NO2 , O2 , AcOH , 25°C
O
Ph Se Me

96%

Bosch, E.; Kochi, J.K. *J. Chem. Soc., Perkin Trans. 1*, ***1996***, 2731.

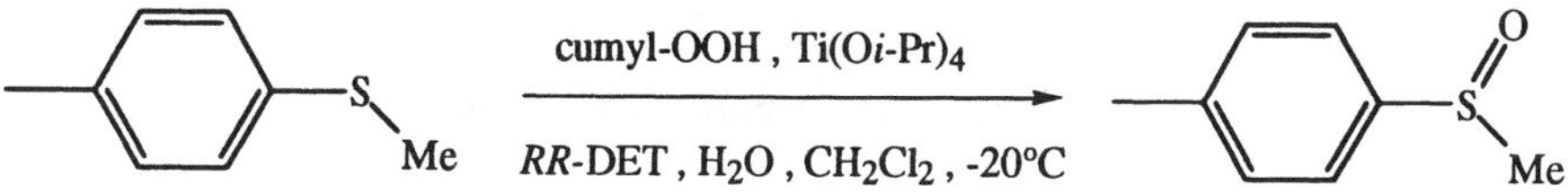
S
Me
cumyl-OOH , Ti(Oi-Pr)4
RR-DET , H2O , CH2Cl2 , -20°C
O
S
Me

76% (>99.5% ee)

Brunel, J.–M.; Kagan, H.B. *Bull. Soc. Chim. Fr.*, ***1996***, *133*, 1109.

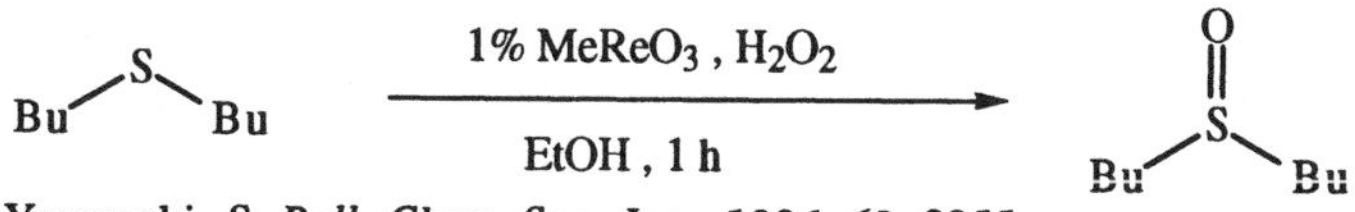
Bu S Bu
1% MeReO3 , H2O2
EtOH , 1 h
O
Bu S Bu

98%

Yamazaki, S. *Bull. Chem. Soc. Jpn.*, ***1996***, *69*, 2955.

REVIEW:

"Enantioselective Oxidation of Sulfides to Sulfoxides Catalyzed by Bacterial Cyclohexanone Monooxygenases."

Colonna, S.; Gaggero, N.; Pasta, P.; Ottolina, G. *Chem. Commun.*, ***1996***, 2303.

SECTION 220: OXIDES FROM HALIDES AND SULFONATES

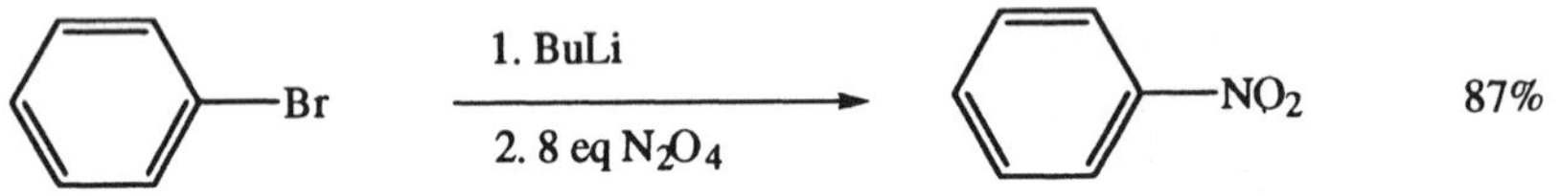

Br
1. BuLi
2. 8 eq N_2O_4
NO_2
87%

Tani, K.; Lukin, K.; Eaton, P.E. *J. Am. Chem. Soc.*, ***1997***, *119*, 1476.

SECTION 221: OXIDES FROM HYDRIDES

NO ADDITIONAL EXAMPLES

SECTION 222: OXIDES FROM KETONES

NO ADDITIONAL EXAMPLES

SECTION 223: OXIDES FROM NITRILES

NO ADDITIONAL EXAMPLES

SECTION 224: OXIDES FROM ALKENES

NO ADDITIONAL EXAMPLES

SECTION 225: OXIDES FROM MISCELLANEOUS COMPOUNDS

PhCH=N—OH —(Oxone , microwaves)→ $PhCH_2NO_2$ 72%

Bose, D.S.; Vanajatha, G. *Synth. Commun.*, ***1998***, *28*, 4531.

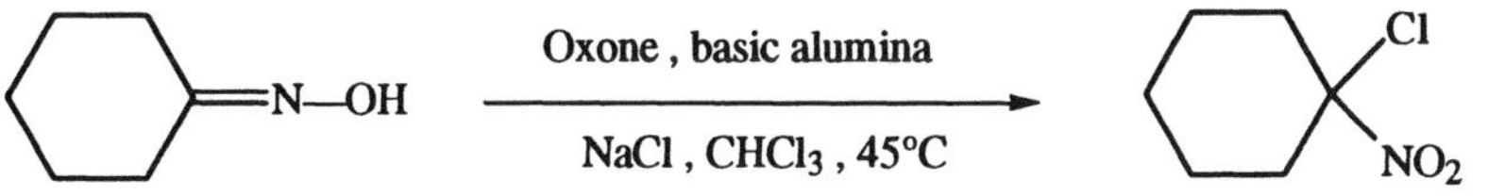

=N—OH
Oxone , basic alumina
NaCl , $CHCl_3$, 45°C
Cl
NO_2

79%

Ceccherelli, P.; Curini, M.; Epifano, F.; Marcotullio, M.C.; Rosati, O. *Tetrahedron Lett.*, ***1998***, *39*, 4385.

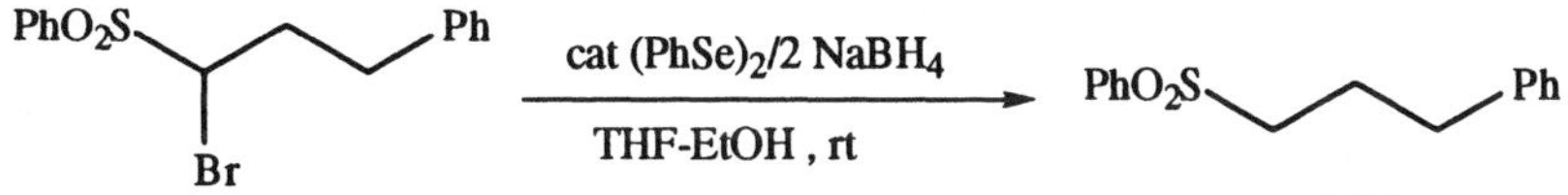

PhO_2S
Ph
Br
cat $(PhSe)_2$/2 $NaBH_4$
THF-EtOH , rt
PhO_2S
Ph

85%

Yoshimatsu, M.; Ohara, M. *Tetrahedron Lett.*, ***1997***, *38*, 5651.

CHAPTER 16

PREPARATION OF DIFUNCTIONAL COMPOUNDS

SECTION 300: ALKYNE - ALKYNE

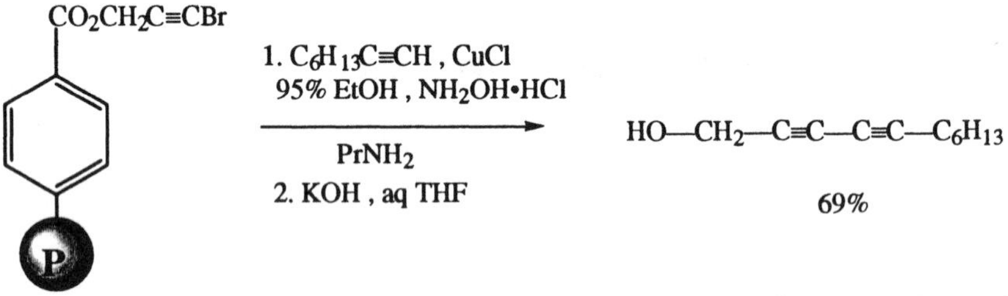
CO2CH2C≡CBr
1. C6H13C≡CH , CuCl
95% EtOH , NH2OH•HCl
PrNH2
2. KOH , aq THF
HO—CH2—C≡C—C≡C—C6H13
69%
P

Montierth, J.M.; DeMario, D.R.; Kurth, M.J.; Schore, N.E. *Tetrahedron*, ***1998***, *54*, 11745.

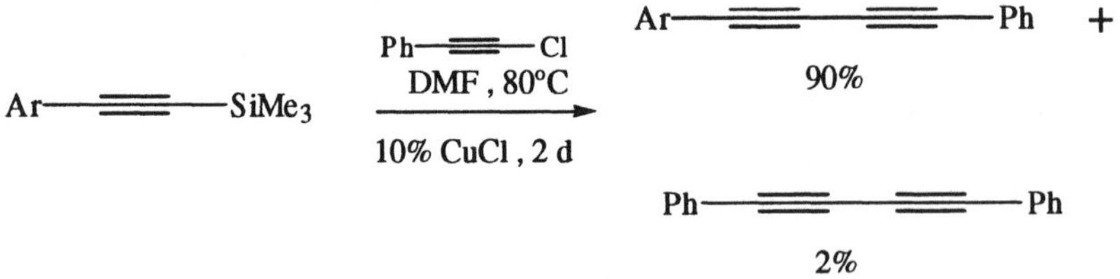
Ar
SiMe3
Ph
Cl
DMF , 80°C
10% CuCl , 2 d
Ar
Ph
+
90%
Ph
Ph
2%

Nishihara, Y.; Ikegashira, K.; Mori, A.; Hiyama, T. *Tetrahedron Lett.*, ***1998***, *39*, 4075.

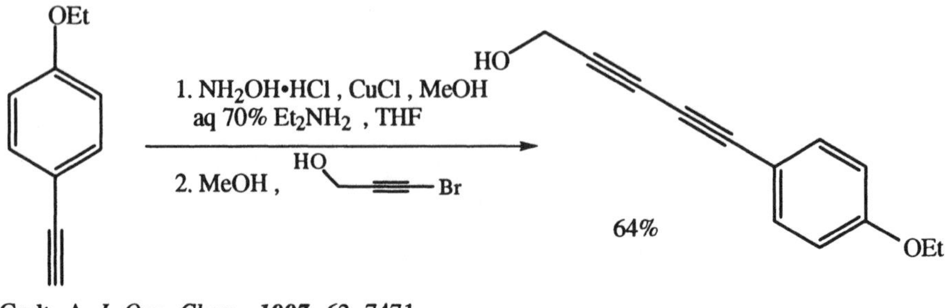
OEt
1. NH2OH•HCl , CuCl , MeOH
aq 70% Et2NH2 , THF
2. MeOH ,
HO
Br
HO
64%
OEt

Godt, A. *J. Org. Chem.*, ***1997***, *62*, 7471.

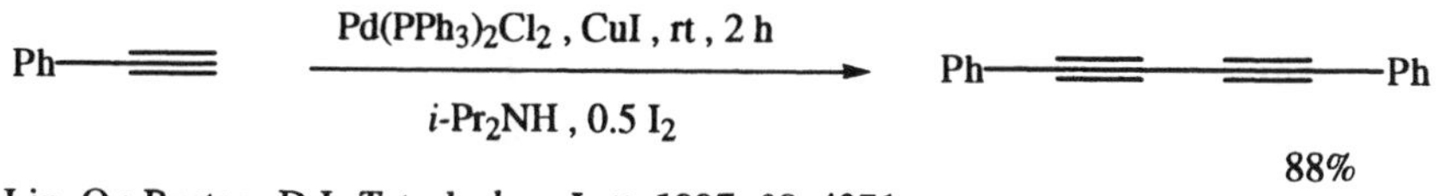
Ph
Pd(PPh3)2Cl2 , CuI , rt , 2 h
i-Pr2NH , 0.5 I2
Ph
Ph
88%

Liu, Q.; Burton, D.J. *Tetrahedron Lett*, ***1997***, *38*, 4371.

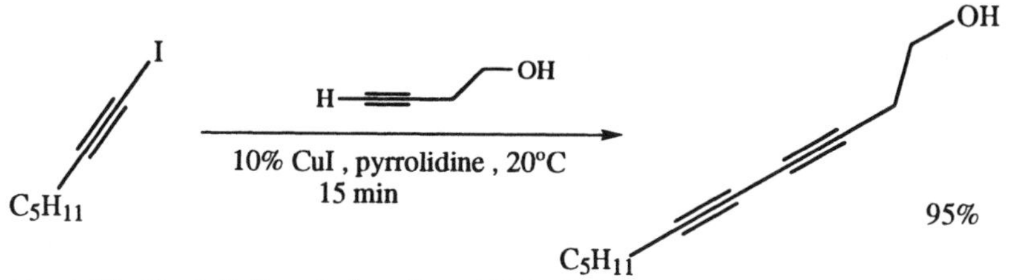
I
C5H11
H
OH
10% CuI , pyrrolidine , 20°C
15 min
OH
C5H11
95%

Alami, M.; Ferri, F. *Tetrahedron Lett*, ***1996***, *37*, 2763.

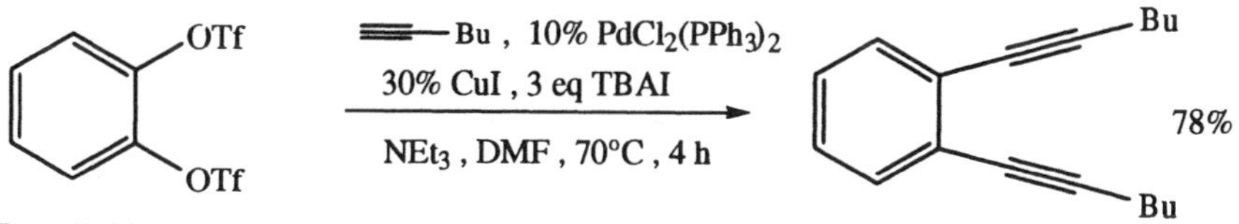
OTf
OTf
Bu , 10% PdCl2(PPh3)2
30% CuI , 3 eq TBAI
NEt3 , DMF , 70°C , 4 h
Bu
Bu
78%

Powell, N.A.; Rychnovsky, S.D. *Tetrahedron Lett.*, ***1996***, *37*, 7901.

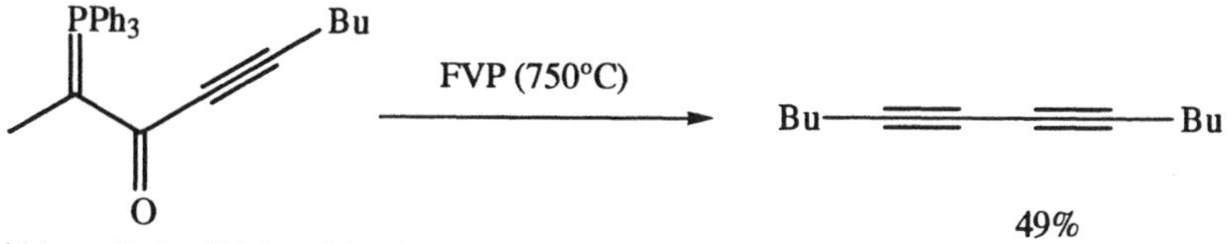
PPh3
Bu
O
FVP (750°C)
Bu
Bu
49%

Aitken, R.A.; Hérion, H.; Horsburgh, C.E.R.; Karodia, N.; Seth, S. *J. Chem. Soc., Perkin Trans. 1*, ***1996***, 485.

SECTION 301: ALKYNE - ACID DERIVATIVES

NO ADDITIONAL EXAMPLES

SECTION 302: ALKYNE - ALCOHOL, THIOL

O
Ph
C3H7
1. LDA
2. H-2O
OH
Ph
C3H7
85%

Dollinger, L.M.; Howell, A.R. *J. Org. Chem.*, ***1998***, *63*, 6782.

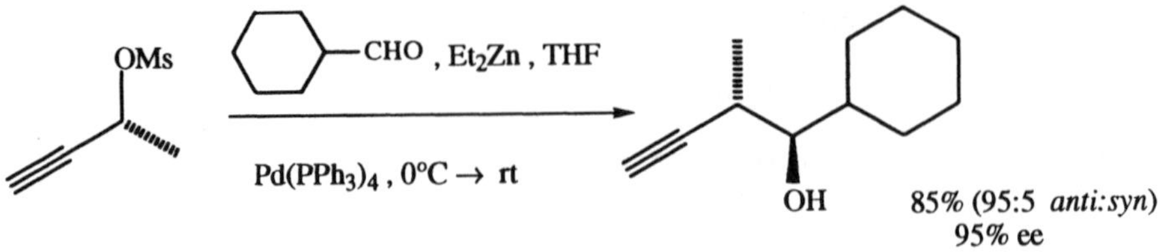
OMs
CHO , Et_2Zn , THF
$Pd(PPh_3)_4$, 0°C → rt
OH
85% (95:5 *anti:syn*)
95% ee

Marshall, J.A.; Adams, N.D. *J. Org. Chem.*, ***1998***, *63*, 3812.

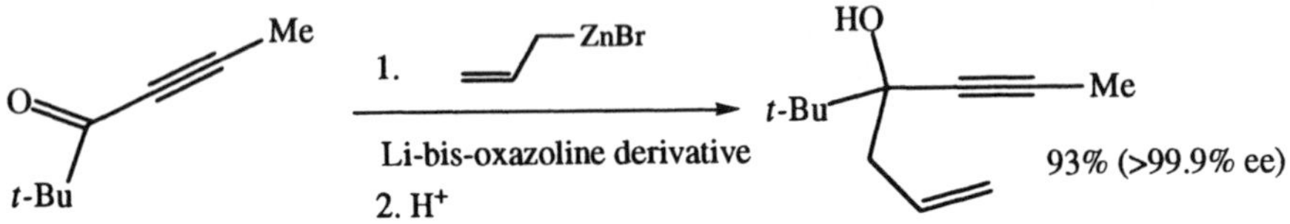
Me
O
t-Bu
1. ZnBr
Li-bis-oxazoline derivative
2. H^+
HO
t-Bu
Me
93% (>99.9% ee)

Nakamura, M.; Hirai, A.; Sogi, M.; Nakamura, E. *J. Am. Chem. Soc.*, ***1998***, *120*, 5846.

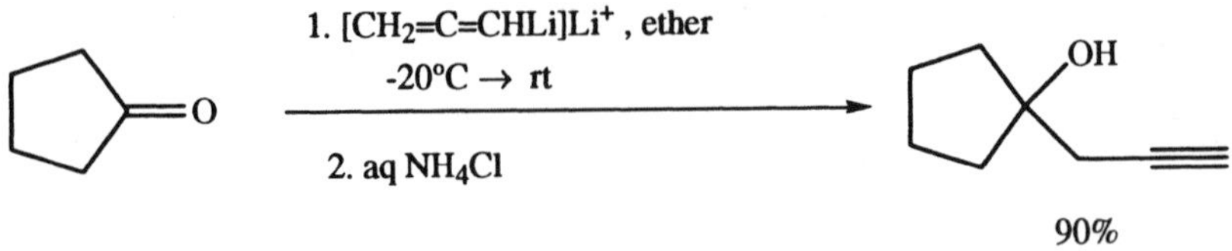
1. $[CH_2{=}C{=}CHLi]Li^+$, ether
-20°C → rt
O
2. aq NH_4Cl
OH
90%

Cabezas, J.A.; Alvarez, L.X. *Tetrahedron Lett.*, ***1998***, *39*, 3935.

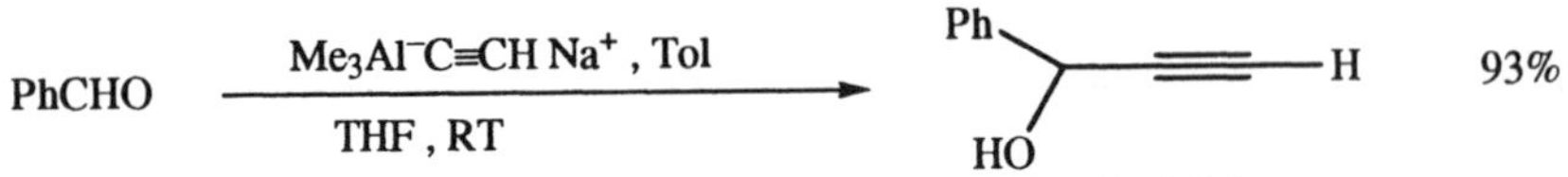
PhCHO
$Me_3Al^-C{\equiv}CH\ Na^+$, Tol
THF , RT
Ph
H
HO
93%

Joung, M.J.; Ahn, J.H.; Yoon, N.M. *J. Org. Chem.*, ***1996***, *61*, 4472.

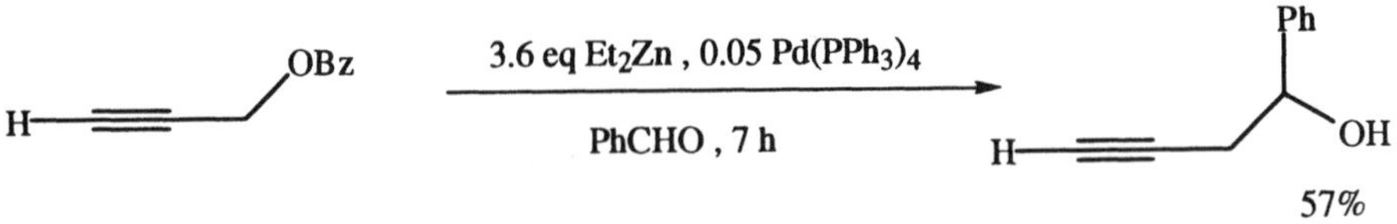
OBz
H
3.6 eq Et_2Zn , 0.05 $Pd(PPh_3)_4$
PhCHO , 7 h
Ph
OH
H
57%

Tamaru, Y.; Goto, S.; Tanaka, A.; Shimizu, M.; Kimura, M.
Angew. Chem. Int. Ed., ***1996***, *35*, 878.

SECTION 303: ALKYNE - ALDEHYDE

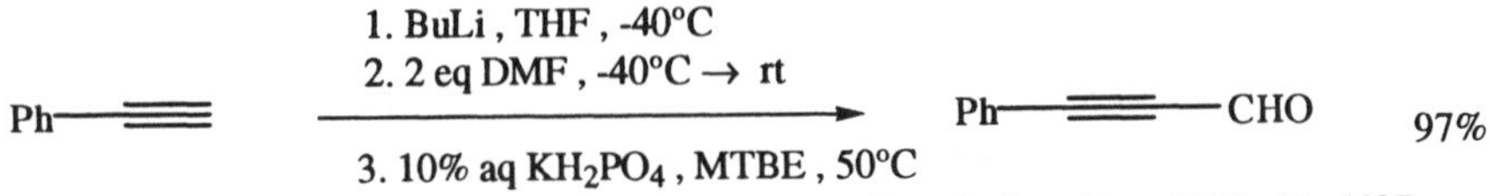
1. BuLi , THF , -40°C
2. 2 eq DMF , -40°C → rt
Ph
Ph
CHO
97%
3. 10% aq KH_2PO_4 , MTBE , 50°C

Journet, M.; Cai, D.; DiMichele, L.M.; Larsen, R.D. *Tetrahedron Lett*, ***1998***, *39*, 6427.

SECTION 304: ALKYNE - AMIDE

NO ADDITIONAL EXAMPLES

SECTION 305: ALKYNE - AMINE

NO ADDITIONAL EXAMPLES

SECTION 306: ALKYNE - ESTER

NO ADDITIONAL EXAMPLES

SECTION 307: ALKYNE - ETHER, EPOXIDE, THIOETHER

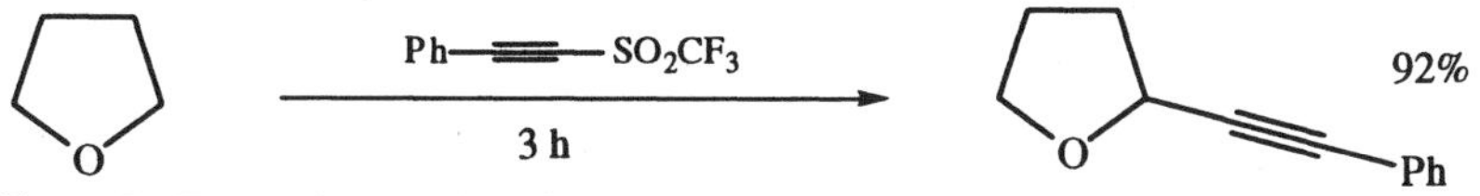
Ph—≡—SO2CF3
3 h
O
O
Ph
92%

Gong, J.; Fuchs, P.L. *J. Am. Chem. Soc.*, ***1996***, *118*, 4486.

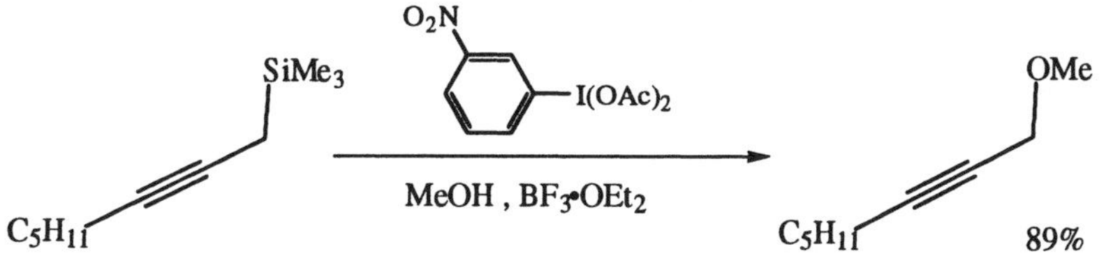
O2N
SiMe3
I(OAc)2
MeOH , BF3•OEt2
OMe
C5H11
C5H11
89%

Kida, M.; Sueda, T.; Goto, S.; Okuyama, T.; Ochiai, M. *Chem. Commun.*, ***1996***, 1933.

SECTION 308: ALKYNE - HALIDE

NO ADDITIONAL EXAMPLES

SECTION 309: ALKYNE - KETONE

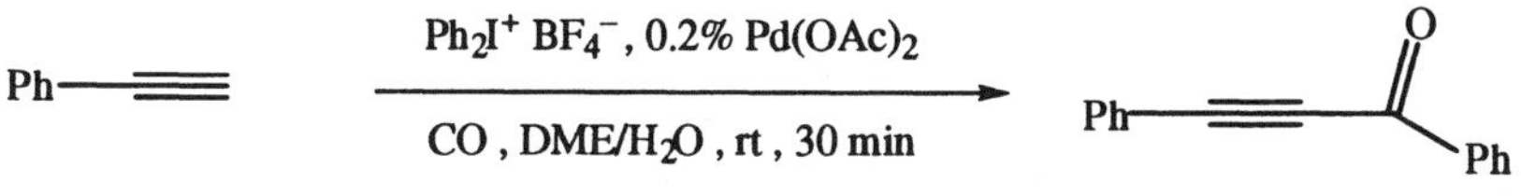
Ph
Ph2I+ BF4−, 0.2% Pd(OAc)2
CO , DME/H2O , rt , 30 min
O
Ph
Ph

89%

Kang, S.-K.; Lim, K.-H.; Ho, P.-S.; Kim, W.-Y. *Synthesis*, ***1997***, 874.

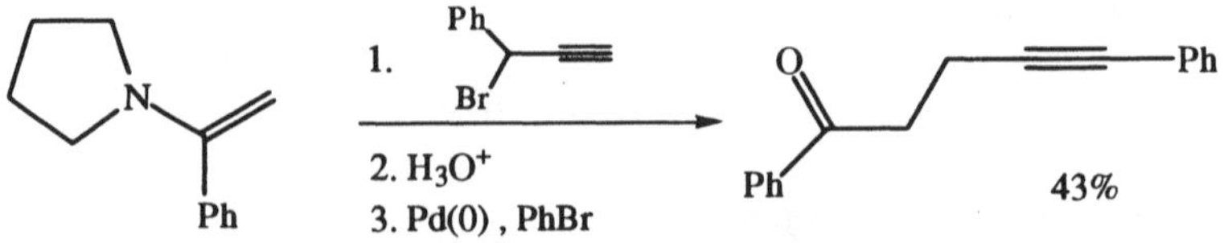
Ph
1.
Br
O
Ph
N
2. H3O+
3. Pd(0) , PhBr
Ph
Ph
43%

Arcadi, A.; Marinelli, F.; Pini, E.; Rossi, E. *Tetrahedron Lett*, ***1996***, *37*, 3387.

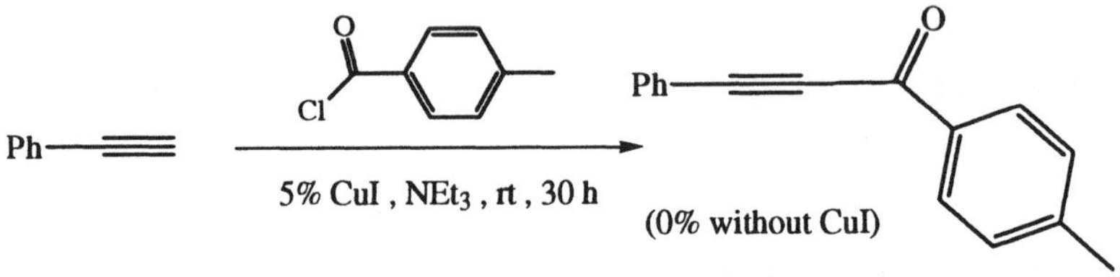
O
Cl
Ph
O
Ph
5% CuI , NEt3 , rt , 30 h
(0% without CuI)

83%

Choudhury, C.; Kundu, N.G. *Tetrahedron Lett.*, ***1996***, *37*, 7323.

SECTION 310: ALKYNE - NITRILE

NO ADDITIONAL EXAMPLES

SECTION 311: ALKYNE - ALKENE

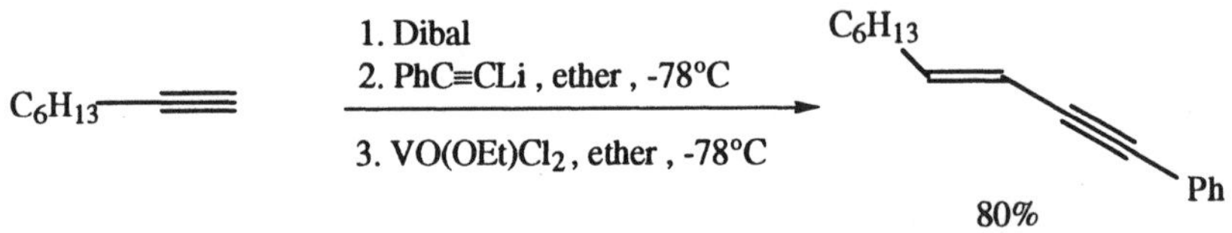
1. Dibal
2. PhC≡CLi , ether , -78°C
3. VO(OEt)Cl2 , ether , -78°C
C6H13
C6H13
Ph

80%

Ishikawa, T.; Ogawa, A.; Hirao, T. *J. Am. Chem. Soc.*, ***1998***, *120*, 5124.

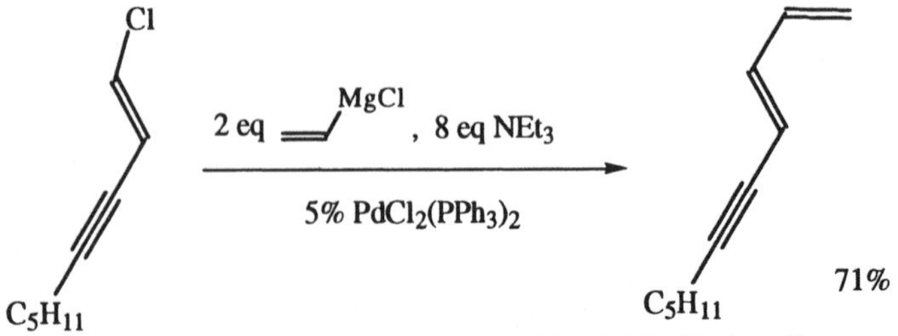
Cl
2 eq
MgCl
, 8 eq NEt3
5% PdCl2(PPh3)2
C5H11
C5H11
71%

Ramiandrasoa, P.; Bréhon, B.; Thivet, A.; Alami, M.; Chaiez, G. *Tetrahedron Lett.*. ***1997***. *38*. 2447.

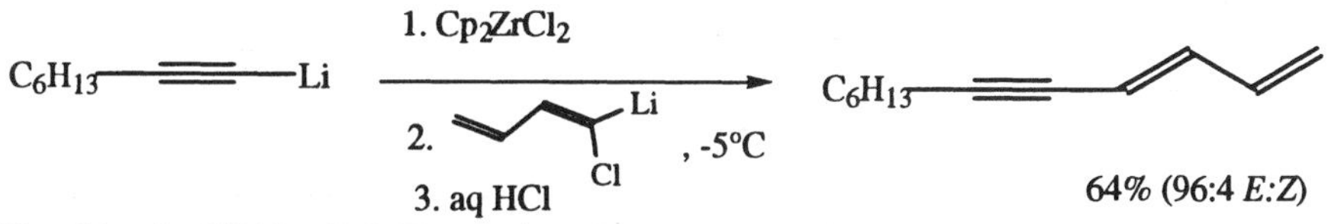
C6H13
Li
1. Cp2ZrCl2
2.
Li
Cl
, -5°C
3. aq HCl
C6H13
64% (96:4 E:Z)

Kasatkin, A.; Whitby, R.J. *Tetrahedron Lett.*, ***1997***, *38*, 4857.

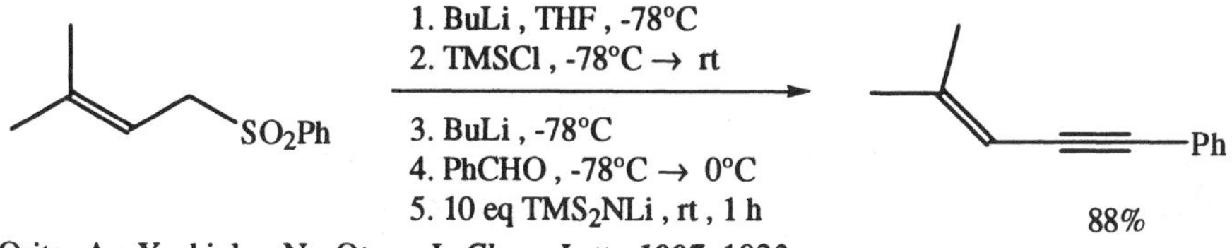
SO2Ph
1. BuLi, THF, -78°C
2. TMSCl, -78°C → rt
3. BuLi, -78°C
4. PhCHO, -78°C → 0°C
5. 10 eq TMS2NLi, rt, 1 h
Ph
88%

Orita, A.; Yoshioka, N.; Otera, J. *Chem. Lett.*, ***1997***, 1023.

SECTION 312: CARBOXYLIC ACID - CARBOXYLIC ACID

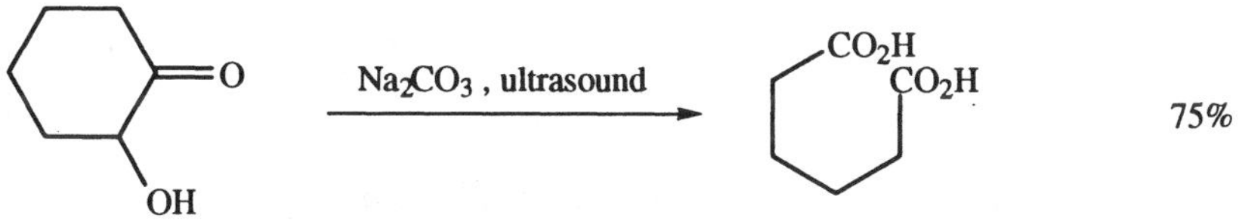
O
OH
Na2CO3, ultrasound
CO2H
CO2H
75%

Yang, D.T.C.; Cao, Y.H.; Evans, T.T.; Kabalka, G.W. *Synth. Commun.*, ***1996***, *26*, 4275.

Also, with NaOCl and ultrasound; 87%.
Yang, D.T.C.; Zhang, C.J.; Fu, P.P.; Kabalka, G.W. *Synth. Commun*, ***1997***, *27*, 1601.
Also, with NaOCl and microwaves; 86%.
Yang, D.T.C.; Zhang, C.J.; Haynie, B.C.; Fu, P.P.; Kabalka, G.W.
Synth. Commun., ***1997***, *27*, 3235.

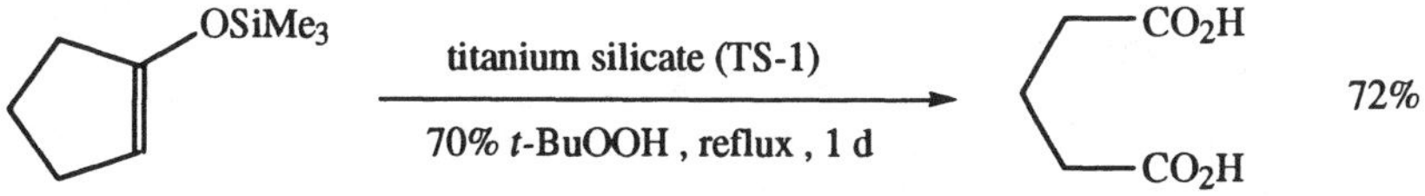
OSiMe3
titanium silicate (TS-1)
70% t-BuOOH, reflux, 1 d
CO2H
CO2H
72%

Raju, S.V.N.; Upadhya, T.T.; Ponrathnam, S.; Daniel, T.; Sudalai, A.
Chem. Commun., ***1996***, 1969.

SECTION 313: CARBOXYLIC ACID - ALCOHOL, THIOL

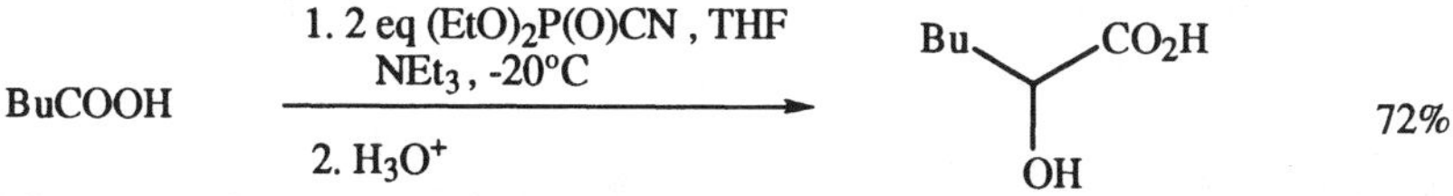
BuCOOH
1. 2 eq (EtO)2P(O)CN, THF
NEt3, -20°C
2. H3O+
Bu
CO2H
OH
72%

Mizuno, M.; Shioiri, T. *Tetrahedron Lett.*, ***1998***, *39*, 9209.

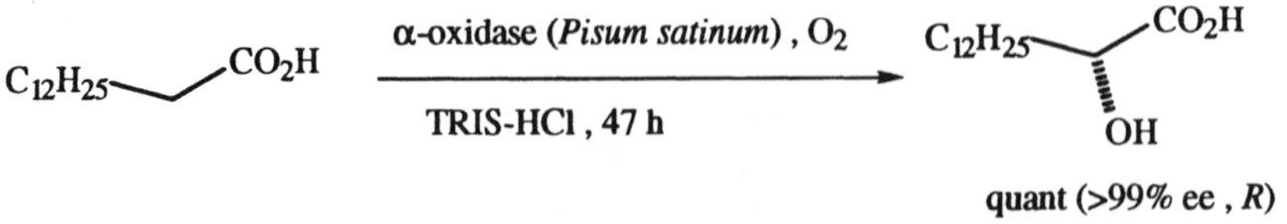
C12H25
CO2H
α-oxidase (Pisum satinum), O2
TRIS-HCl, 47 h
C12H25
CO2H
OH
quant (>99% ee, R)

Adam, W.; Boland, W.; Hartmann–Schreier, J.; Humpf, H.–U.; Lazarus, M.; Saffert, A.; Saha–Möller, C.R.; Schreier, P. *J. Am. Chem. Soc.*, ***1998***, *120*, 11044.

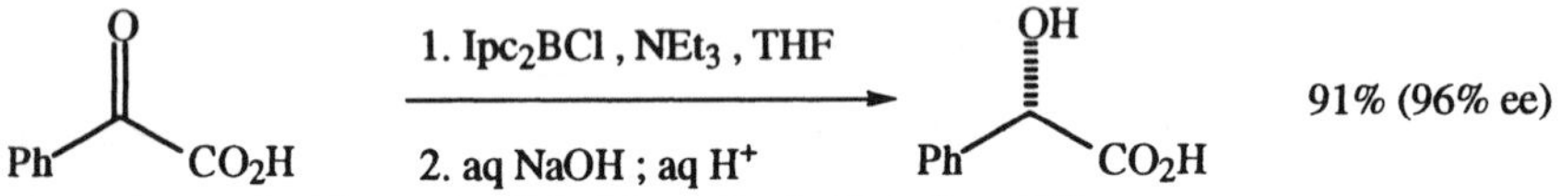
O
Ph
CO2H
1. Ipc2BCl, NEt3, THF
2. aq NaOH ; aq H+
OH
Ph
CO2H
91% (96% ee)

Wang, Z.; La, B.; Fortunak, J.M.; Meng, X.–J.; Kabalka, G.W. *Tetrahedron Lett*, ***1998***, *39*, 5501.

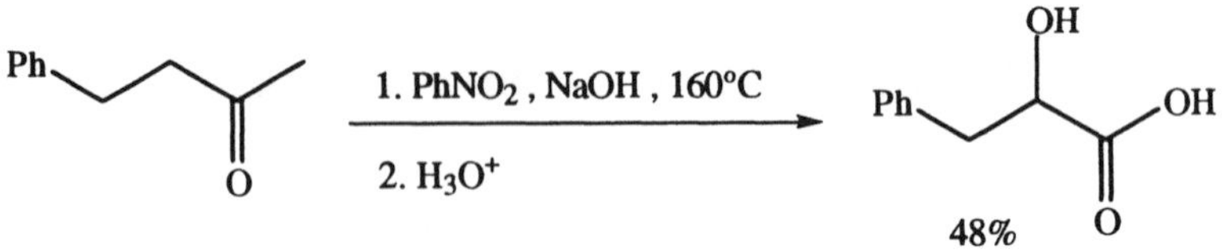
Ph
O
1. PhNO2, NaOH, 160°C
2. H3O+
OH
Ph
OH
O
48%

Srinivasan, P.S.; Mahesh, R.; Rao, G.V.; Kalyanam, N. *Synth. Commun.*, ***1996***, *26*, 2161.

SECTION 314: CARBOXYLIC ACID - ALDEHYDE

NO ADDITIONAL EXAMPLES

SECTION 315: CARBOXYLIC ACID - AMIDE

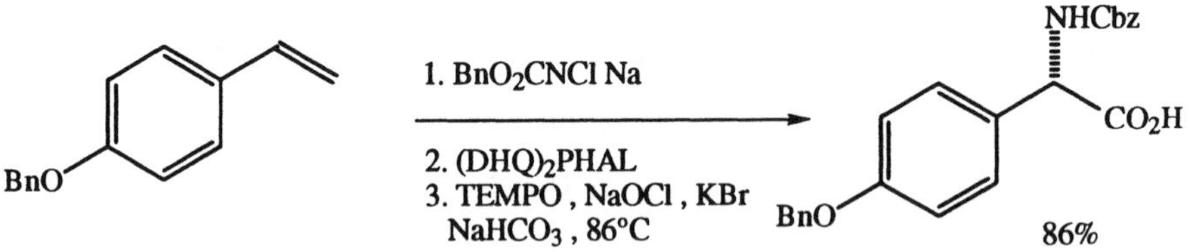
BnO
1. BnO2CNCl Na
2. (DHQ)2PHAL
3. TEMPO, NaOCl, KBr
NaHCO3, 86°C
NHCbz
CO2H
BnO
86%

Reddy, K.L; Sharpless, K.B. *J. Am. Chem. Soc.*, ***1998***, *120*, 1207.

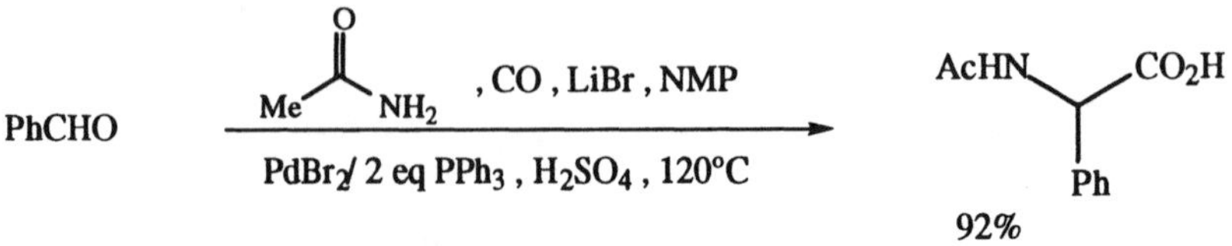
PhCHO
O
Me
NH2
, CO, LiBr, NMP
PdBr2/ 2 eq PPh3, H2SO4, 120°C
AcHN
CO2H
Ph
92%

Beller, M.; Eckert, M.; Holla, E.W. *J. Org. Chem.*, ***1998***, *63*, 5658.

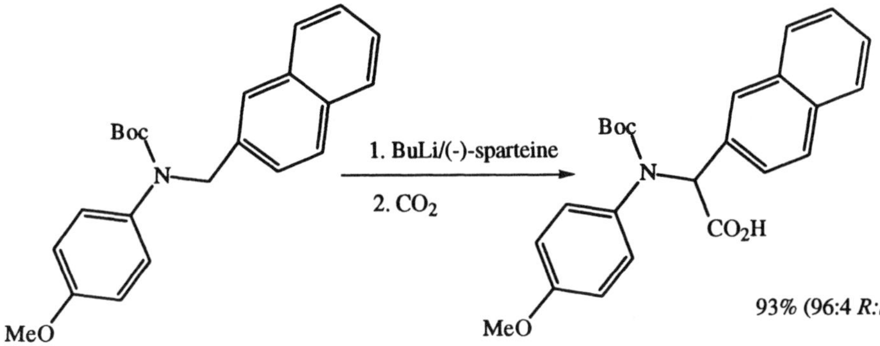

93% (96:4 *R:S*)

Park, Y.S.; Beak, P. *J. Org. Chem.*, ***1997***, *62*, 1574.

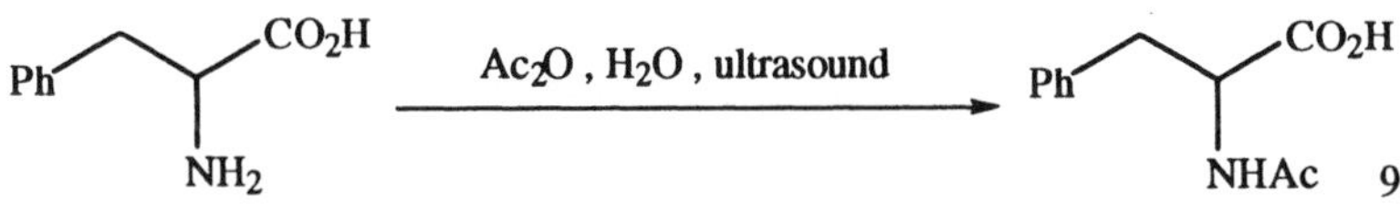

90%

Anuradha, M.V.; Ravindranath, B. *Tetrahedron*, ***1997***, *53*, 1123.

SECTION 316: CARBOXYLIC ACID - AMINE

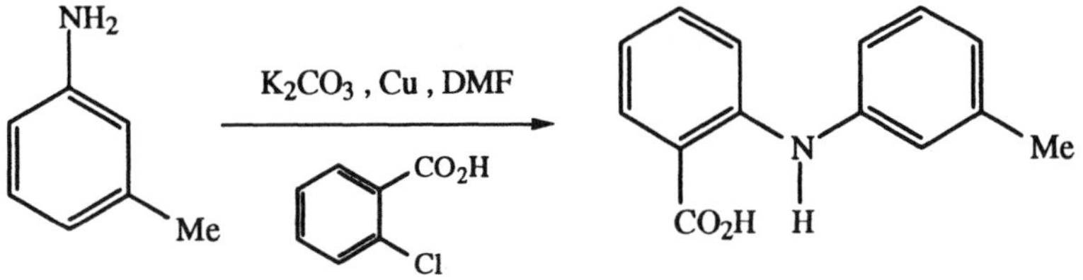

63%

Pellón, R.F.; Carrasco, R.; Márquez, T.; Mamposo, T. *Tetrahedron Lett.*, ***1997***, *38*, 5107.

Related Methods: Section 315 (Carboxylic Acid - Amide).
Section 344 (Amide - Ester).
Section 351 (Amine - Ester).

SECTION 317: CARBOXYLIC ACID - ESTER

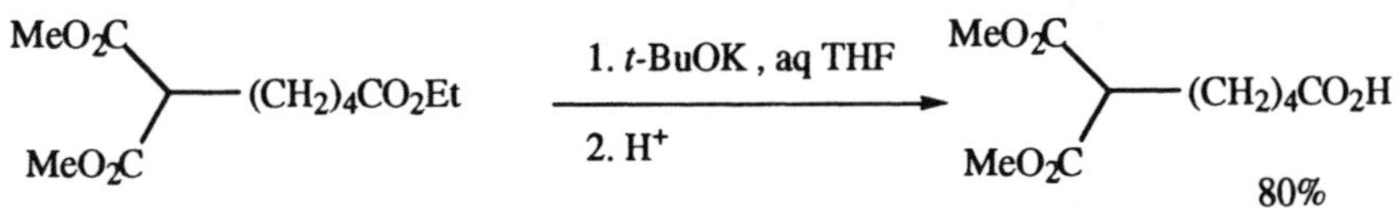

80%

Wilk, B.K. *Synth. Commun.*, ***1996***, *26*, 3859.

SECTION 318: CARBOXYLIC ACID - ETHER, EPOXIDE, THIOETHER

NO ADDITIONAL EXAMPLES

SECTION 319: CARBOXYLIC ACID - HALIDE, SULFONATE

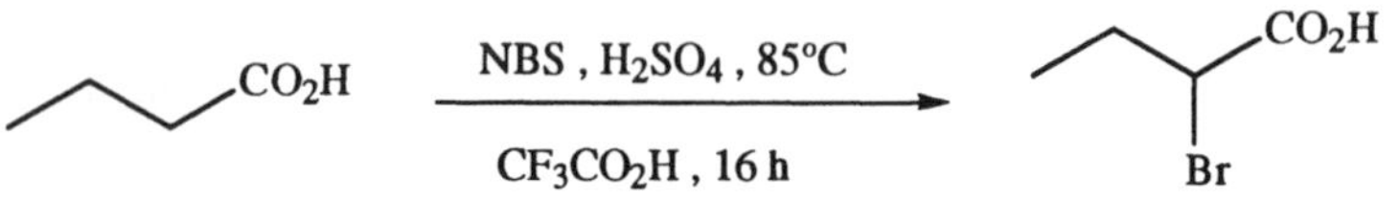
CO_2H
NBS , H_2SO_4 , 85°C
CF_3CO_2H , 16 h
CO_2H
Br

87%

Zhang, L.H.; Duan, J.; Xu, Y.; Dolbier Jr., W.R. *Tetrahedron Lett.*, ***1998***, *39*, 9621.

SECTION 320: CARBOXYLIC ACID - KETONE

OTf
e^- , CO_2 , Bu_4NBF_4–DMF , 5°C
O
CO_2H

75%

Kamekawa, H.; Senboku, H.; Tokuda, M. *Tetrahedron Lett.*, ***1998***, *39*, 1591.

Also via: Section 360 (Ketone - Ester).

SECTION 321: CARBOXYLIC ACID - NITRILE

NO ADDITIONAL EXAMPLES

Also via: Section 361 (Nitrile - Ester).

SECTION 322: CARBOXYLIC ACID - ALKENE

PhCHO —[$CH_2(COOH)_2$, NH_4OAc , microwaves; 3 min]→ $PhCH{=}CHCO_2H$

98%

Kumar, H.M.S.; Subbareddy, B.V.; Anjaneyulu, S.; Yadav, J.S. *Synth. Commun.*, ***1998***, *28*, 3811.

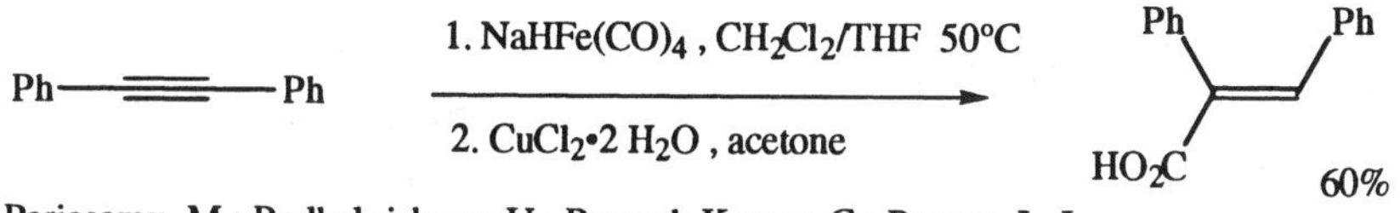
Ph
Ph
1. NaHFe(CO)4 , CH2Cl2/THF 50°C
2. CuCl2•2 H2O , acetone
Ph
Ph
HO2C
60%

Periasamy, M.; Radhakrishnan, U.; Ramesh Kumar, C.; Brunet, J.–J. *Tetrahedron Lett.*, ***1997***, *38*, 1623.

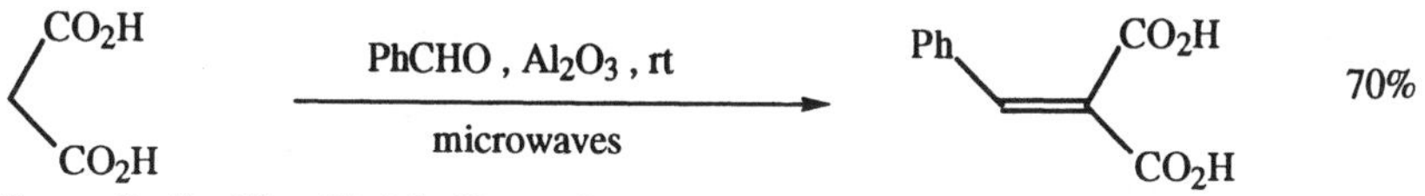
CO2H
CO2H
PhCHO , Al2O3 , rt
microwaves
Ph
CO2H
CO2H
70%

Kwon, P.–S.; Kim, Y.–M.; Kang, C.–J.; Kwon, T.–W. *Synth. Commun.*, ***1997***, *27*, 4091.

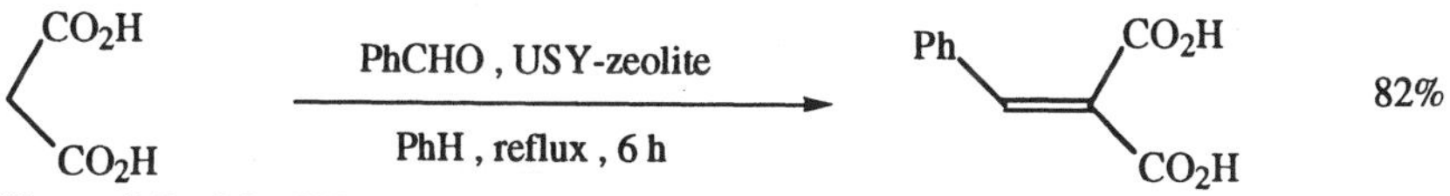
CO2H
CO2H
PhCHO , USY-zeolite
PhH , reflux , 6 h
Ph
CO2H
CO2H
82%

Wang, Q.L.; Ma, Y.D.; Zuo, B.J. *Synth. Commun.*, ***1997***, *27*, 4107.

Also via: Section 313 (Alcohol - Carboxylic Acids).
Section 349 (Amide - Alkene).
Section 362 (Ester - Alkene).
Section 376 (Nitrile - Alkene).

SECTION 323: ALCOHOL, THIOL - ALCOHOL, THIOL

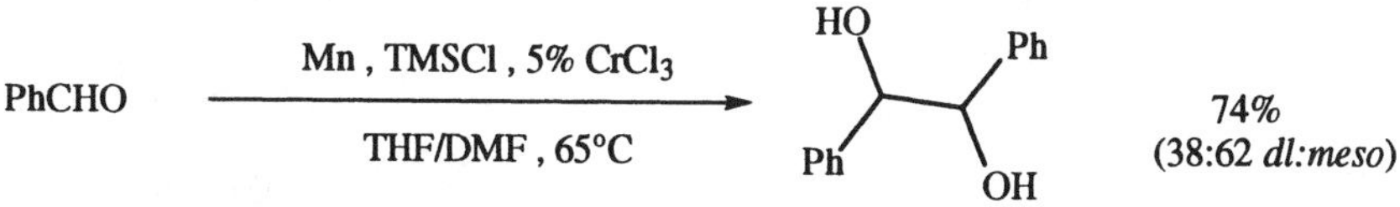
PhCHO
Mn , TMSCl , 5% CrCl3
THF/DMF , 65°C
HO
Ph
Ph
OH
74%
(38:62 dl:meso)

Svatoš, A.; Boland, W. *SynLett*, ***1998***, 549.

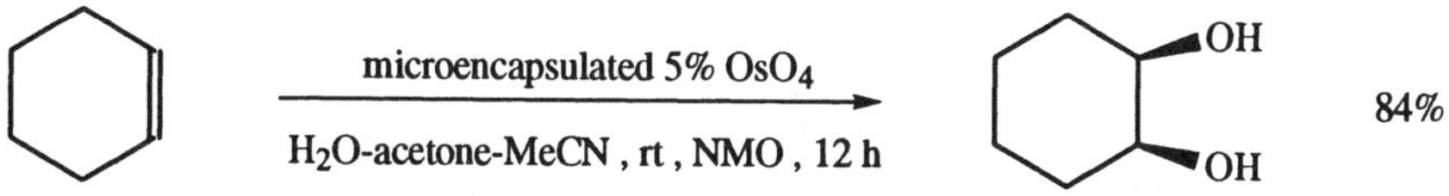
microencapsulated 5% OsO4
H2O-acetone-MeCN , rt , NMO , 12 h
OH
OH
84%

Nagayama, S.; Endo, M.; Kobayashi, S. *J. Org. Chem.*, ***1998***, *63*, 6094.

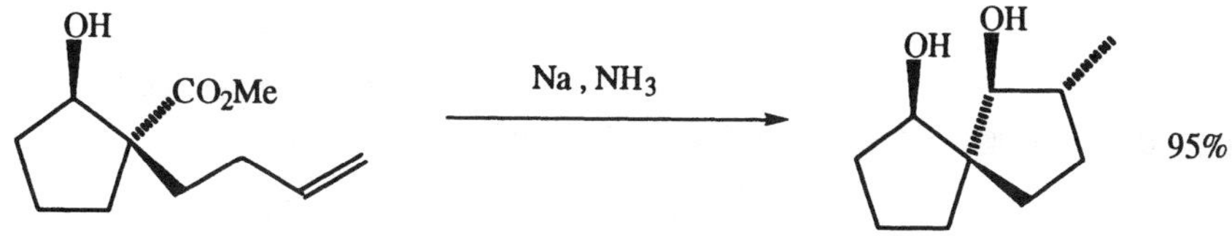
OH
CO2Me
Na , NH3
OH
OH
95%

Cossy, J.; Gille, B.; Bellosta, V. *J. Org. Chem.*, ***1998***, *63*, 3141.

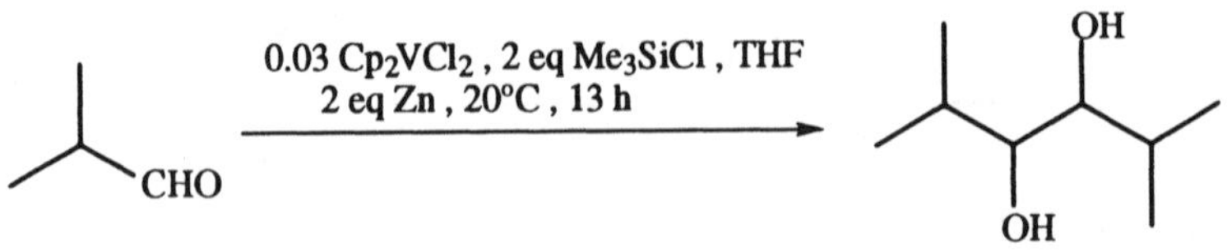

CHO
0.03 Cp_2VCl_2 , 2 eq Me_3SiCl , THF
2 eq Zn , 20°C , 13 h
OH
OH

89% (91:9 *dl:meso*)

Hirao, T.; Asahara, M.; Muguruma, Y.; Ogawa, A. *J. Org. Chem.*, ***1998***, *63*, 2812.

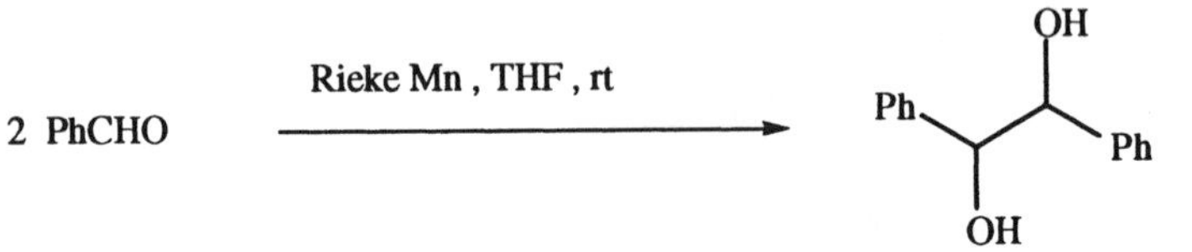

2 PhCHO
Rieke Mn , THF , rt
OH
Ph
Ph
OH

64% (63:37 *dl:meso*)

Rieke, R.D.; Kim, S.–H. *J. Org. Chem.*, ***1998***, *63*, 5235.

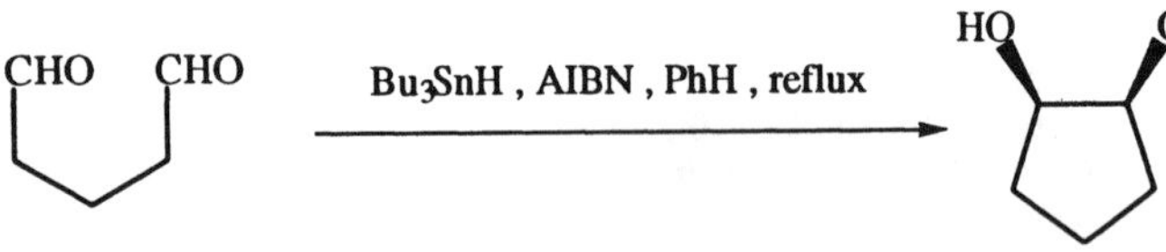

CHO CHO
Bu_3SnH , AIBN , PhH , reflux
HO OH

46% (98:1 *cis:trans*)

Hays, D.S.; Fu, G.C. *J. Org. Chem.*, ***1998***, *63*, 6375.

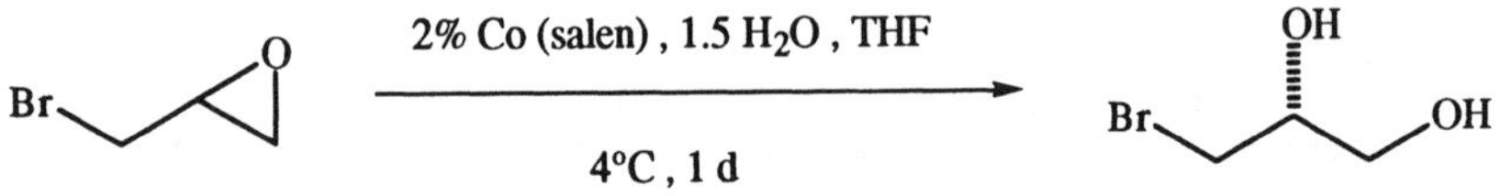

Br
O
2% Co (salen) , 1.5 H_2O , THF
4°C , 1 d
OH
Br
OH

93% (96% ee)

Furrow, M.E.; Schaus, S.E.; Jacobsen, E.N. *J. Org. Chem.*, ***1998***, *63*, 6776.

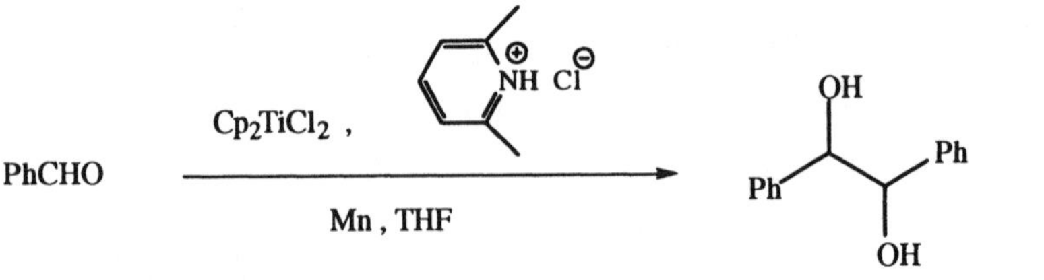

PhCHO
Cp_2TiCl_2 ,
NH Cl
Mn , THF
OH
Ph
Ph
OH

75% (82:18 *syn:anti*)

Gansäuer, A.; Bauer, D. *Eur. J. Org. Chem.*, ***1998***, 2673.

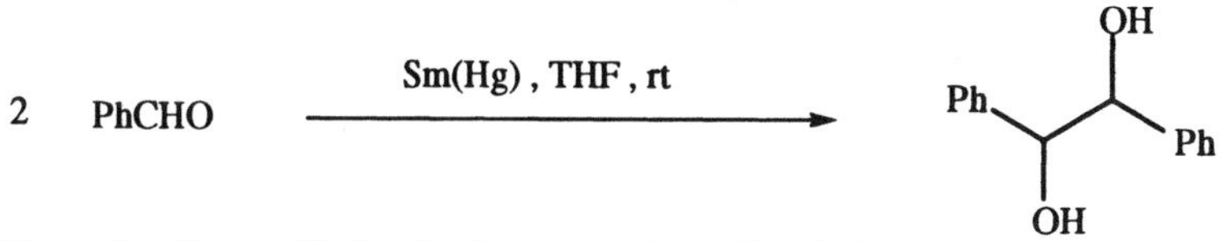
2 PhCHO
Sm(Hg) , THF , rt
OH
Ph
Ph
OH

67%

Wang, L.; Zhang, Y. *Synth. Commun.*, ***1998***, *28*, 3991.

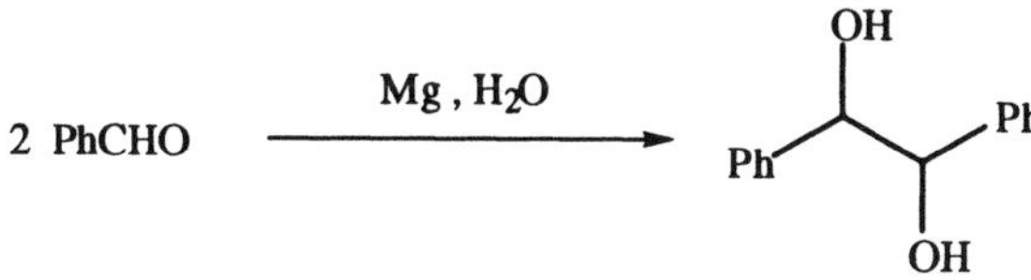
2 PhCHO
Mg , H2O
OH
Ph
Ph
OH

80% (59:41 *threo:erythro*)

Zhang, W.–C.; Li, C.–J. *J. Chem. Soc., Perkin Trans. 1*, ***1998***, 3131.

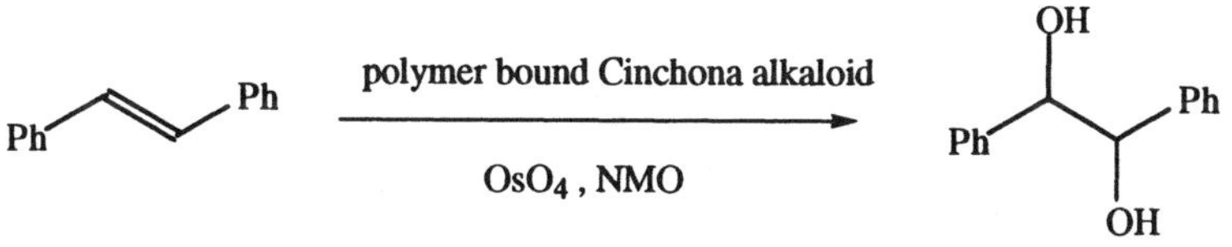
Ph
Ph
polymer bound Cinchona alkaloid
OsO4 , NMO
OH
Ph
Ph
OH

90% ee

Bolm, C.; Gerlach, A. *Eur. J. Org. Chem.*, ***1998***, 21.

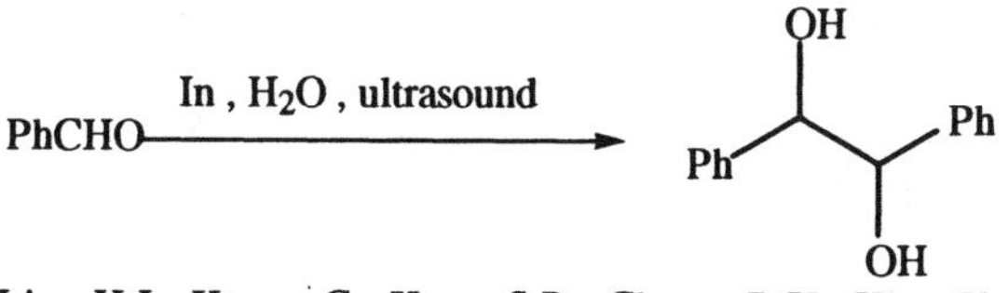
PhCHO
In , H2O , ultrasound
OH
Ph
Ph
OH

70% (5:3 *dl:meso*)

Lim, H.J.; Keum, G.; Kang, S.B.; Chung, B.Y.; Kim, Y. *Tetrahedron Lett.*, ***1998***, *39*, 4367.

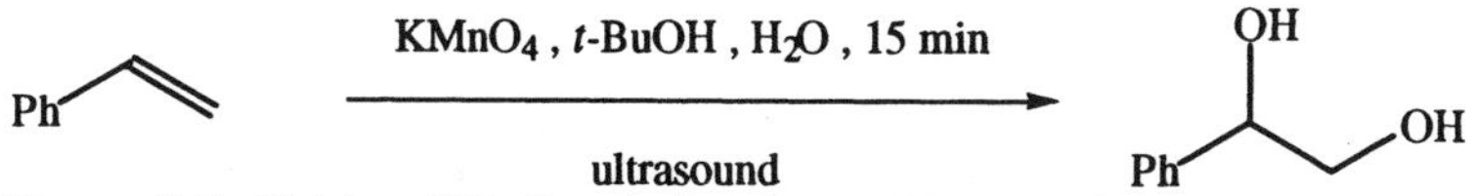
Ph
KMnO4 , t-BuOH , H2O , 15 min
ultrasound
OH
OH
Ph

80%

Varma, R.S.; Naicker, K.P. *Tetrahedron Lett.*, ***1998***, *39*, 7463.

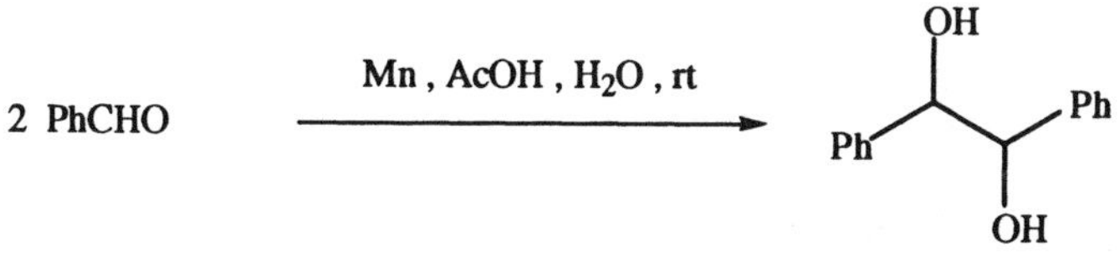
2 PhCHO
Mn , AcOH , H2O , rt
OH
Ph
Ph
OH

74% (44:56 *threo:erythro*)

Li, C.–J.; Meng, Y.; Yi, X.–H. *J. Org. Chem.*, ***1997***, *62*, 8632.

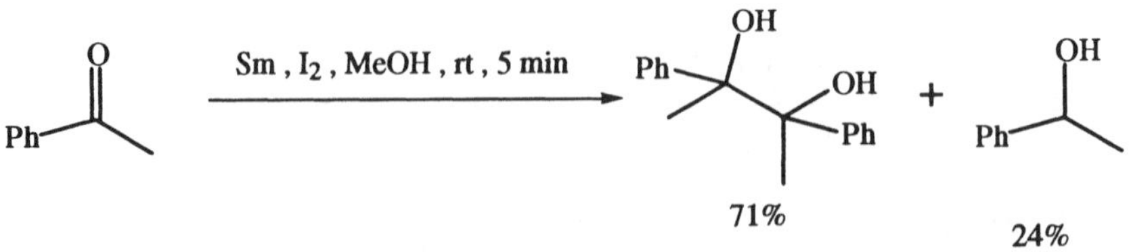
Sm , I2 , MeOH , rt , 5 min
71%
24%

Yanada, R.; Negoro, N.; Yanada, K.; Fujita, T. *Tetrahedron Lett.*, ***1997***, *38*, 3271.

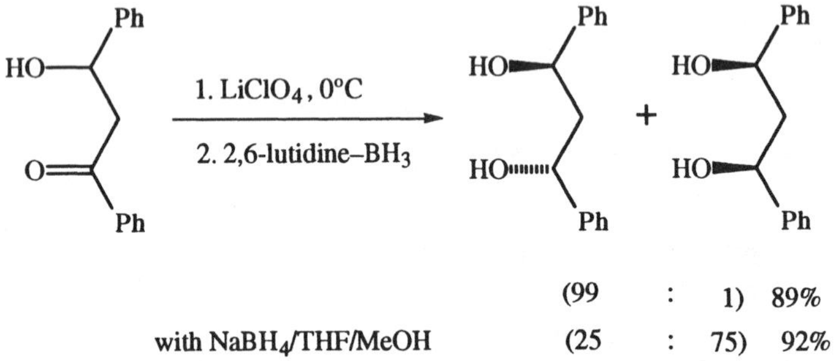
1. LiClO4 , 0°C
2. 2,6-lutidine–BH3
(99 : 1) 89%
with NaBH4/THF/MeOH (25 : 75) 92%

Narayana, C.; Reddy, M.R.; Hair, M.; Kabalka, G.W. *Tetrahedron Lett.*, ***1997***, *38*, 7705.

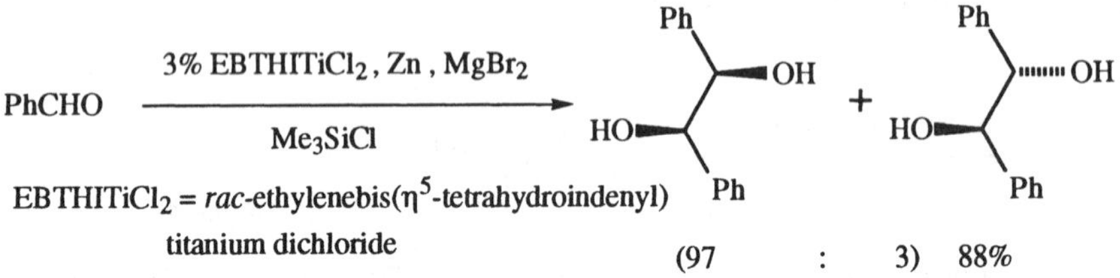
3% EBTHITiCl2 , Zn , MgBr2
Me3SiCl
(97 : 3) 88%

$EBTHITiCl_2$ = *rac*-ethylenebis(η^5-tetrahydroindenyl) titanium dichloride

Gansäuer, A. *SynLett*, ***1997***, 363.

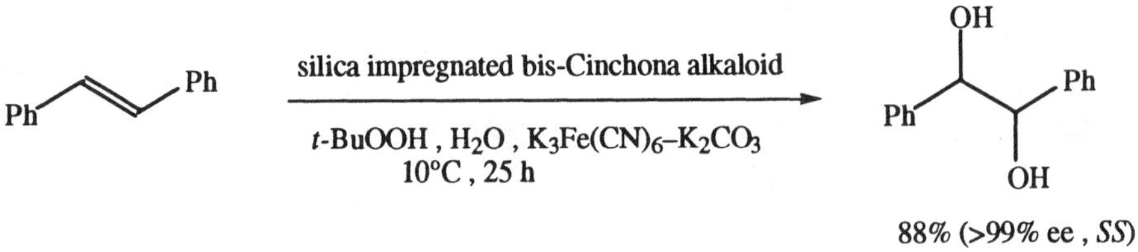
silica impregnated bis-Cinchona alkaloid
t-BuOOH , H2O , K3Fe(CN)6–K2CO3
10°C , 25 h
88% (>99% ee , SS)

Song, C.E.; Yang, J.W.; Ha, H.–J. *Tetrahedron Asymmetry*, ***1997***, *8*, 841.

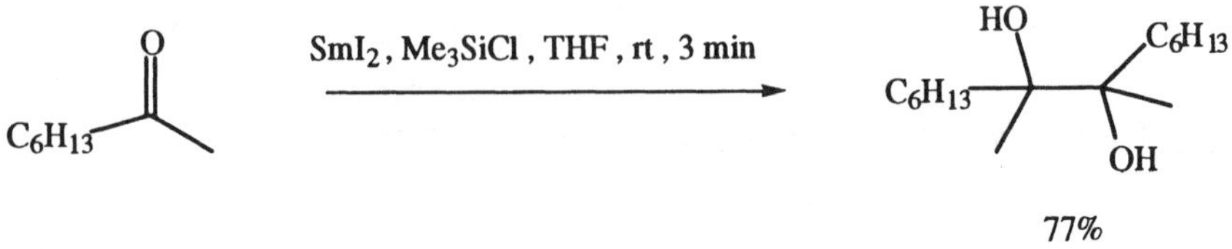
SmI2 , Me3SiCl , THF , rt , 3 min
77%

Honda, T.; Katoh, M. *Chem. Commun.*, ***1997***, 369.

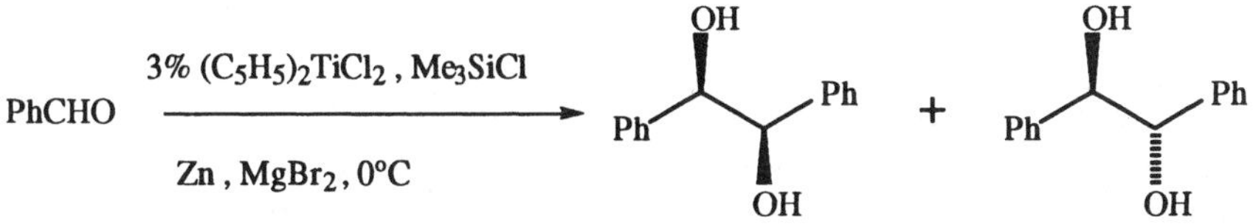
PhCHO
3% (C5H5)2TiCl2 , Me3SiCl
Zn , MgBr2 , 0°C
OH
Ph
OH
Ph
+
OH
Ph
OH
Ph

(86 : 14) 93%

Gansäuer, A. *Chem. Commun.*, ***1997***, 457.

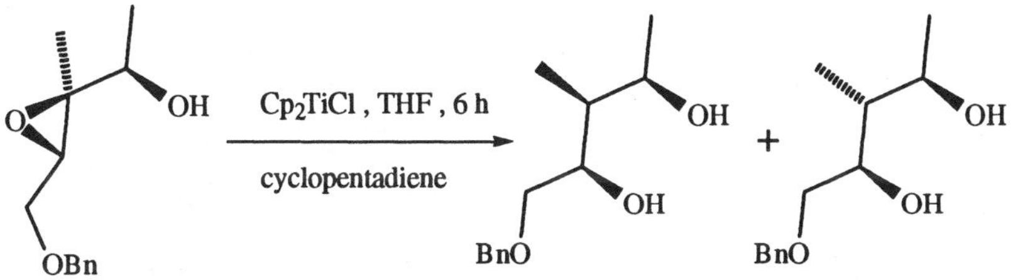
O
OH
OBn
Cp2TiCl , THF , 6 h
cyclopentadiene
OH
OH
BnO
+
OH
OH
BnO

(5 : 1) 84%

Chakraborty, T.K.; Dutta, S. *J. Chem. Soc., Perkin Trans. 1*, ***1997***, 1257.

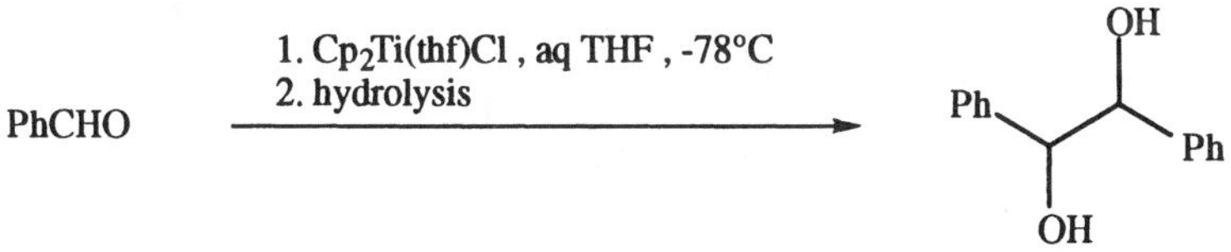
PhCHO
1. Cp2Ti(thf)Cl , aq THF , -78°C
2. hydrolysis
OH
Ph
Ph
OH

95% (98:2 *dl:meso*)

Barden, M.C.; Schwartz, J. *J. Am. Chem. Soc.*, ***1996***, *118*, 5484.

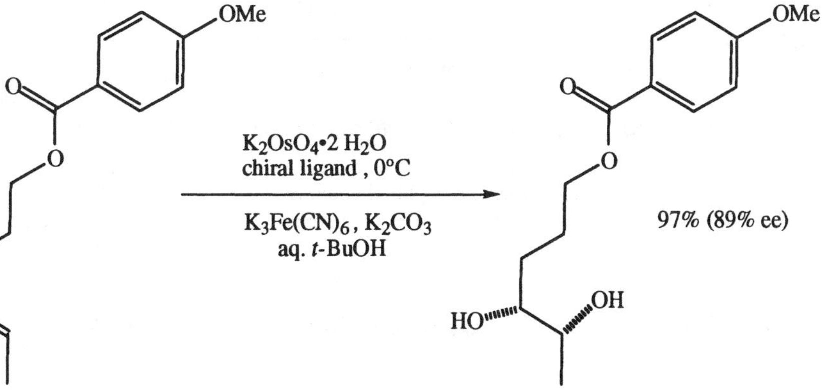
OMe
O
O
K2OsO4•2 H2O
chiral ligand , 0°C
K3Fe(CN)6 , K2CO3
aq. t-BuOH
OMe
O
O
HO
OH
97% (89% ee)

Corey, E.J.; Noe, M.C.; Ting, A.Y. *Tetrahedron Lett.*, ***1996***, *37*, 1735.
Noe, M.C.; Corey, E.J. *Tetrahedron Lett.*, ***1996***, *37*, 1739.

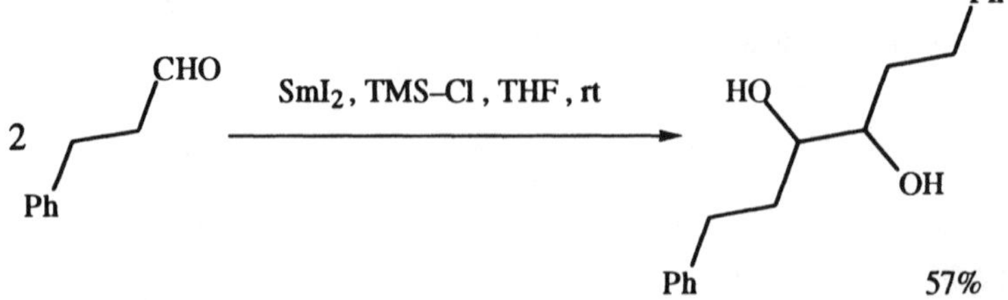
2
CHO
Ph
SmI_2, TMS–Cl, THF, rt
HO
OH
Ph
Ph
57%

Nomura, R.; Matsuno, T.; Endo, T. *J. Am. Chem. Soc.*, ***1996***, *118*, 11666.

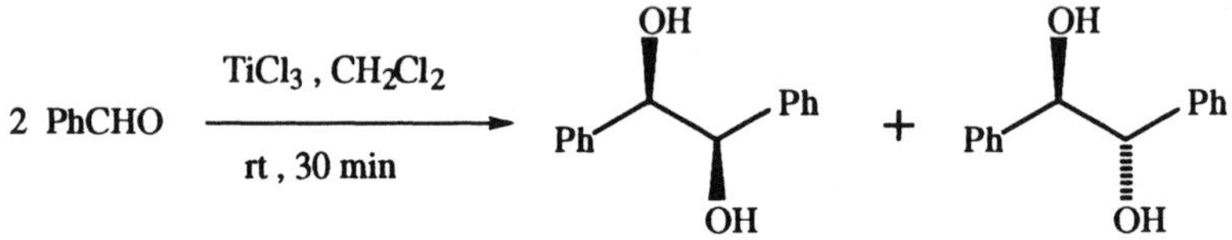
2 PhCHO
$TiCl_3$, CH_2Cl_2
rt, 30 min
OH
Ph
Ph
OH
+
OH
Ph
Ph
OH

(200 : 1) 65%

Clerici, A.; Clerici, L.; Porta, O. *Tetrahedron Lett*, ***1996***, *37*, 3035.

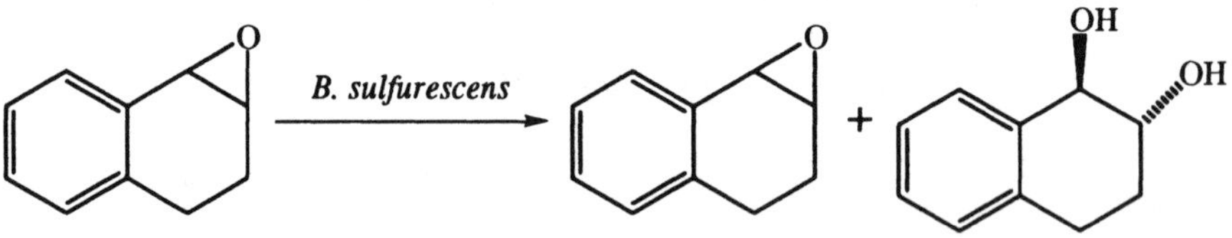
O
B. sulfurescens
O
+
OH
OH

20% (98% ee 1R,2S) 48% (69% ee 1R2R)

Pedragosa–Moreau, S.; Archlas, A.; Furstoss, R. *Tetrahedron Lett.*, ***1996***, *37*, 3319.

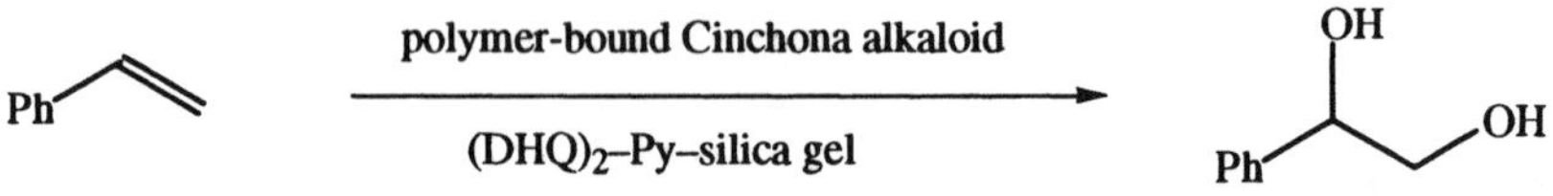
Ph
polymer-bound Cinchona alkaloid
$(DHQ)_2$–Py–silica gel
OH
OH
Ph

95%

Lohray, B.B.; Nandanan, E.; Bhushan, V. *Tetrahedron Asymmetry*, ***1996***, *7*, 2805.

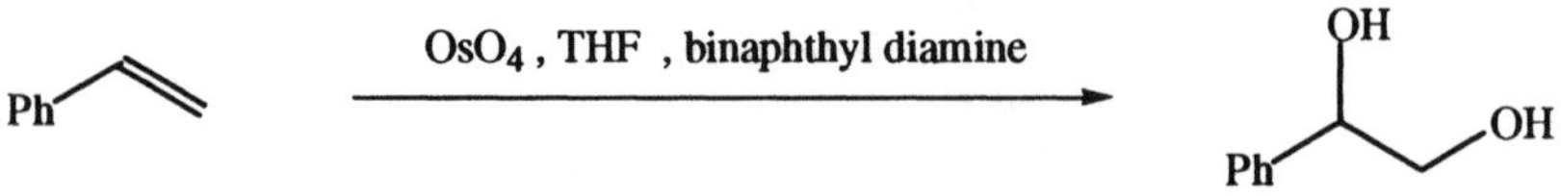
Ph
OsO_4, THF, binaphthyl diamine
OH
OH
Ph

83% (96% ee, *R*)

Rosini, C.; Tanturli, R.; Pertici, P.; Salvadori, P. *Tetrahedron Asymmetry*, ***1996***, *7*, 2971.

Also via: Section 327 (Alcohol - Ester). Section 357 (Ester - Ester).

SECTION 324: ALCOHOL, THIOL - ALDEHYDE

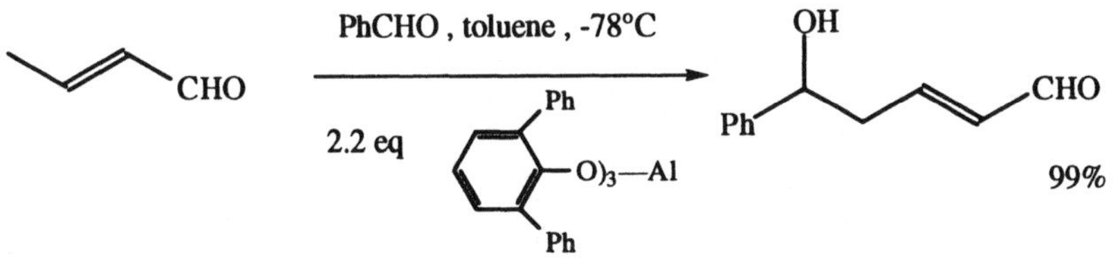

Saito, S.; Shiozawa, M.; Ito, M.; Yamamoto, H. *J. Am. Chem. Soc.*, ***1998***, *120*, 813.

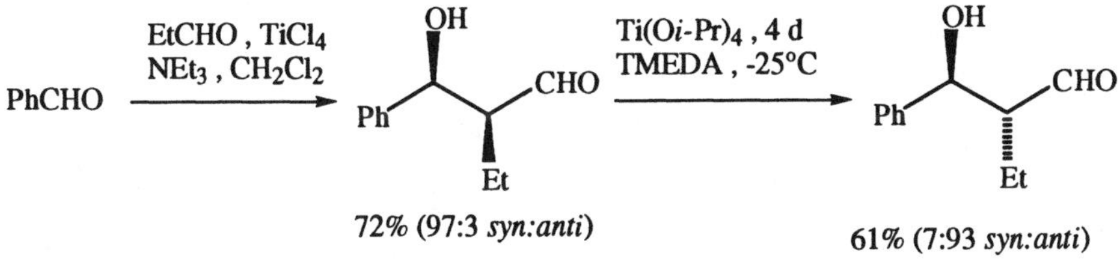

Mahrwald, R.; Costisella, B.; Gündogan, B. *Synthesis*, ***1998***, 262.

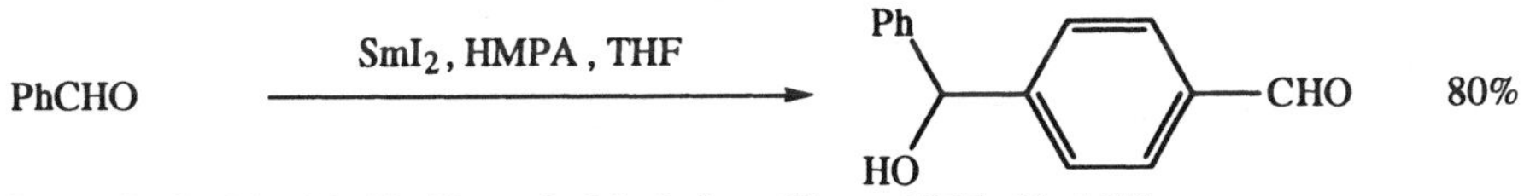

Shiue, J.–S.; Lin, M.–H.; Fang, J.–M. *J. Org. Chem.*, ***1997***, *62*, 4643.

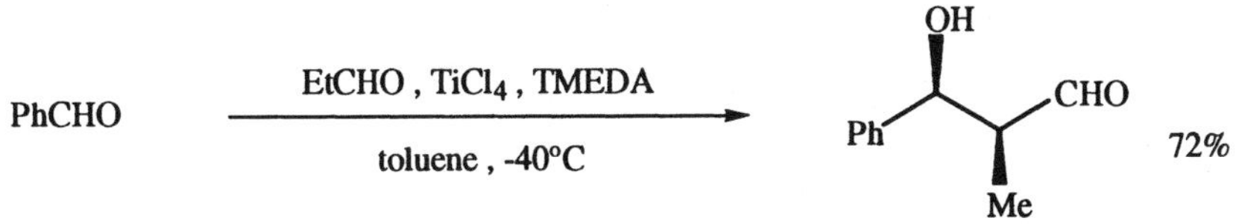

Mahrwald, R.; Costisella, B.; Gündogan, B. *Tetrahedron Lett.*, ***1997***, *38*, 4543.

Related Methods: Section 330 (Alcohol - Ketone).

SECTION 325: ALCOHOL, THIOL - AMIDE

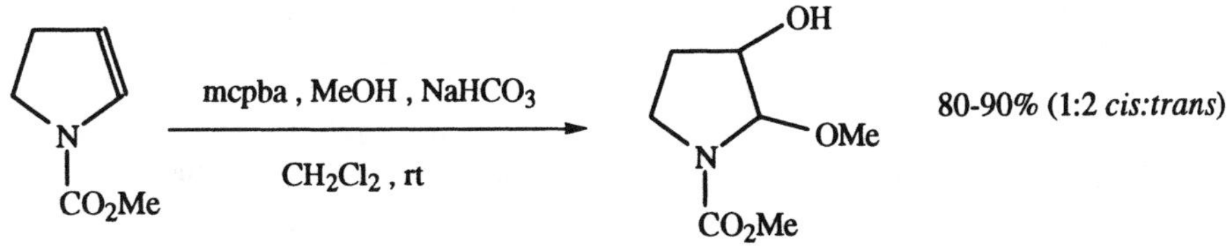

Sugisaki, C.H.; Carroll, P.J.; Correia, C.R.D. *Tetrahedron Lett.*, ***1998***, *39*, 3413.

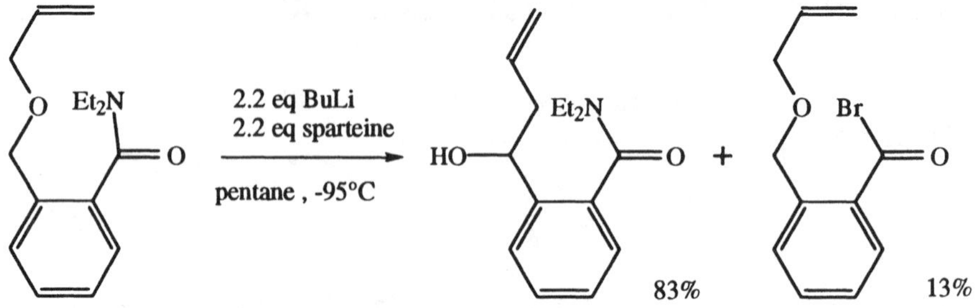

Et2N
O
2.2 eq BuLi
2.2 eq sparteine
pentane , -95°C
HO
Br
+
83%
13%

Kawasaki, T.; Kimachi, T. *SynLett*, ***1998***, 1429.

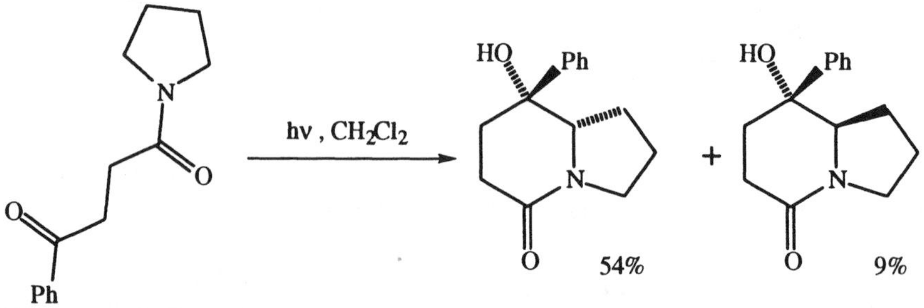

N
O
Ph
hv , CH2Cl2
HO
Ph
+
54%
9%

Lindemann, U.; Reck, G.; Wulff–Molder, D.; Wessig, P. *Tetrahedron*, ***1998***, *54*, 2529.

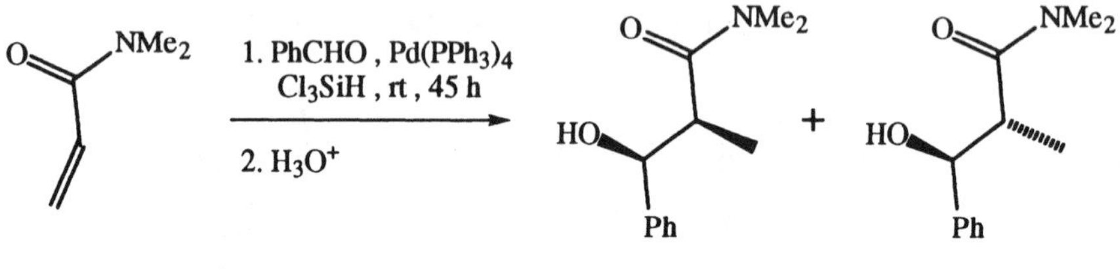

NMe2
O
1. PhCHO , Pd(PPh3)4
Cl3SiH , rt , 45 h
2. H3O+
HO
Ph
+

(32 : 68) 87%

Kiyooka, S.-i.; Shimizu, A.; Torii, S. *Tetrahedron Lett.*, ***1998***, *39*, 5237.

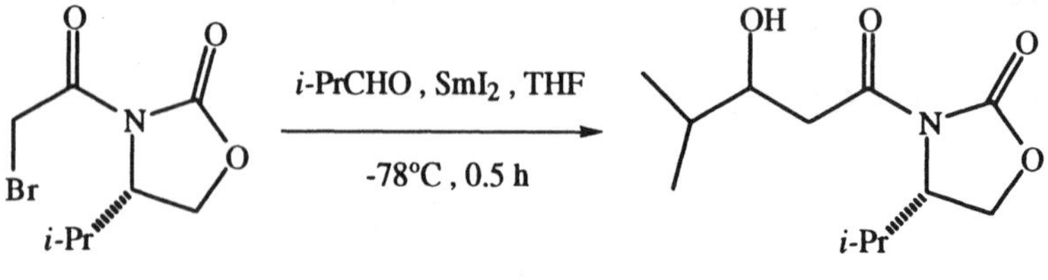

O
N
Br
i-Pr
i-PrCHO , SmI2 , THF
-78°C , 0.5 h
OH

92% (78% de)

Fukuzawa, S.-i.; Tatsuzawa, M.; Hirano, K. *Tetrahedron Lett.*, ***1998***, *39*, 6899.

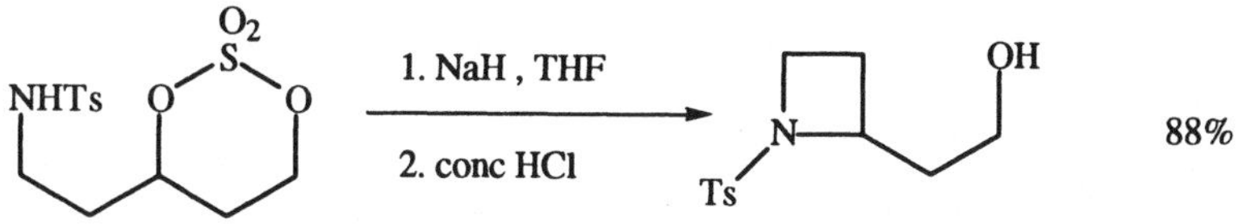

Littler, B.J.; Gallagher, T.; Boddy, I.K.; Riordan, P.D. *SynLett*, ***1997***, 22.

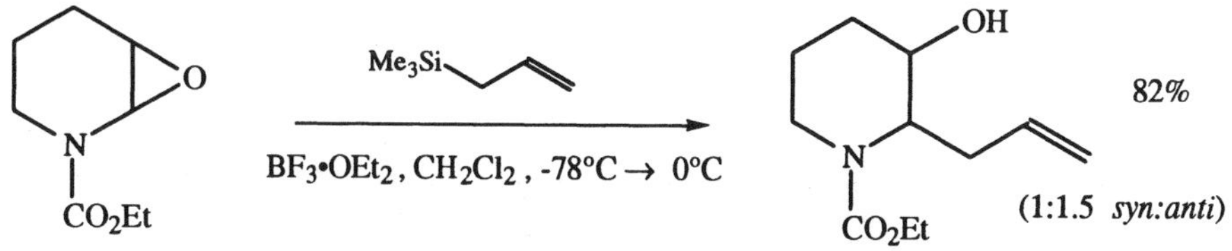

Brugess, L.E.; Gross, E.K.M.; Jurka, J. *Tetrahedron Lett,*. ***1996***, *37*, 3255.

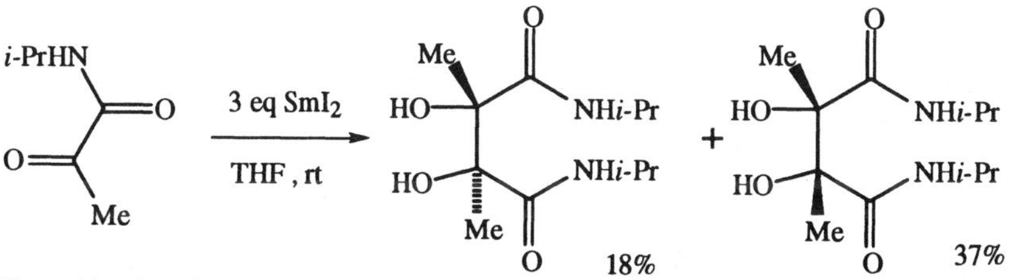

Yamashita, M.; Okuyama, K.; Kawasaki, I.; Ohta, S. *Tetrahedron Lett.*, ***1996***, *37*, 7755.

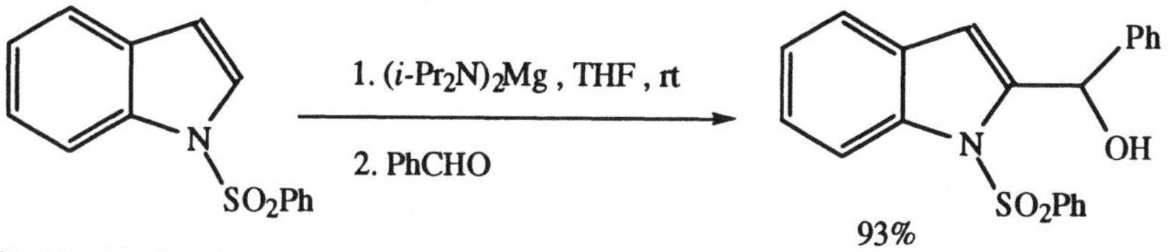

Kondo, Y.; Yoshida, A.; Sakamoto, T. *J. Chem. Soc., Perkin Trans. 1*, ***1996***, 2331.

SECTION 326: ALCOHOL, THIOL - AMINE

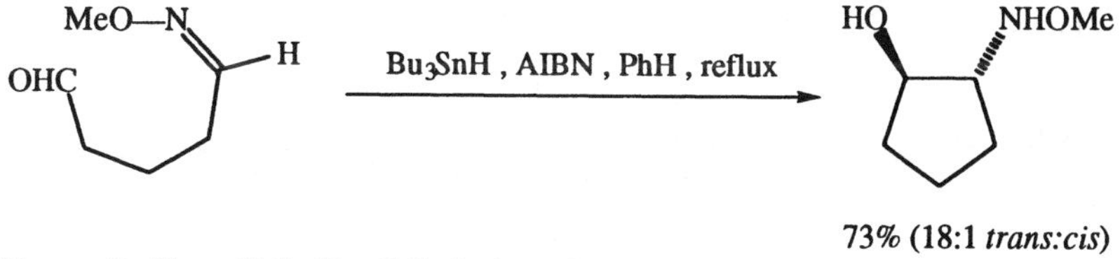

Tormo, J.; Hays, D.S.; Fu, G.C. *J. Org. Chem.*, ***1998***, *63*, 201.

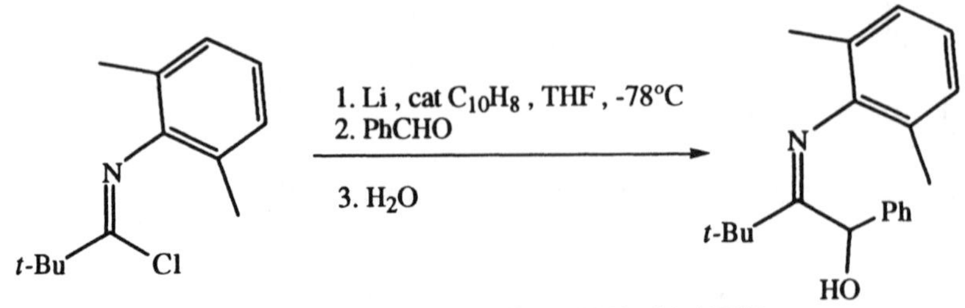

N
t-Bu
Cl
1. Li, cat $C_{10}H_8$, THF, -78°C
2. PhCHO
3. H_2O
N
t-Bu
Ph
HO

45%

Alonso, E.; Ramón, D.J.; Yus, M. *Tetrahedron*, ***1998***, *54*, 12007.

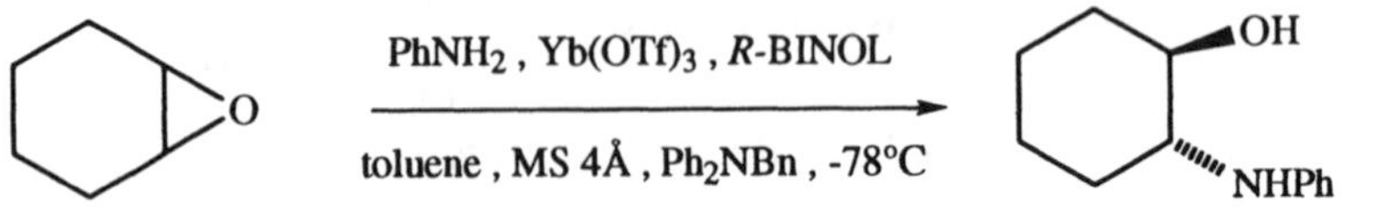

O
$PhNH_2$, $Yb(OTf)_3$, *R*-BINOL
toluene, MS 4Å, Ph_2NBn, -78°C
OH
NHPh

90% (80% ee)

Hou, X.-L.; Wu, J.; Dai, L.-X.; Xia, L.-J.; Tang, M.-H. *Tetrahedron Asymmetry*, ***1998***, *9*, 1747.

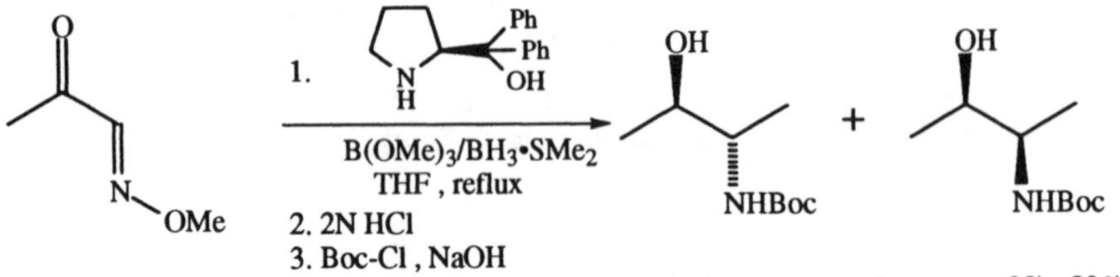

O
N
OMe
1.
Ph
Ph
OH
N
H
$B(OMe)_3/BH_3•SMe_2$
THF, reflux
2. 2N HCl
3. Boc-Cl, NaOH
OH
NHBoc
+
OH
NHBoc

(72 : 28) 89%

65% ee 36% ee

Masui, M.; Shioiri, T. *Tetrahedron Lett.*, ***1998***, *39*, 5193.

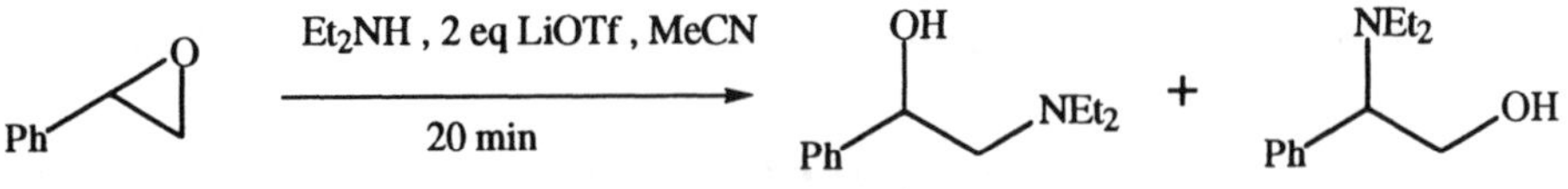

Ph
O
Et_2NH, 2 eq LiOTf, MeCN
20 min
OH
Ph
NEt_2
+
NEt_2
Ph
OH

(2.8 : 1) 98%

Augé, J.; Leroy, R. *Tetrahedron Lett.*, ***1996***, *37*, 7715.

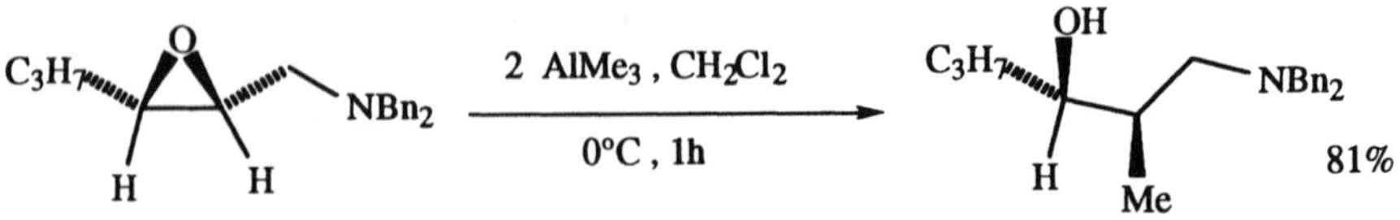

C_3H_7
O
NBn_2
H
H
2 $AlMe_3$, CH_2Cl_2
0°C, 1h
OH
C_3H_7
NBn_2
H
Me
81%

Liu, C.; Hashimoto, Y.; Saigo, K. *Tetrahedron Lett.*, ***1996***, *37*, 6177.

SECTION 327: ALCOHOL, THIOL - ESTER

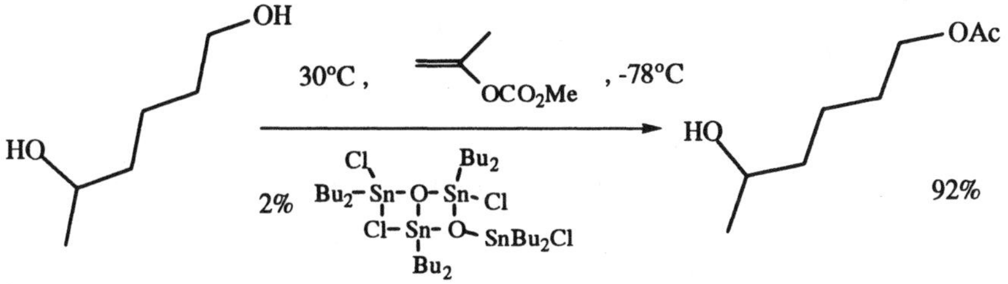
OH
30°C ,
OCO2Me
, -78°C
OAc
HO
HO
Cl
Bu2
2%
Bu2-Sn-O-Sn-Cl
Cl-Sn-O-SnBu2Cl
Bu2
92%

Orita, A.; Mitsutome, A.; Otera, J. *J. Org. Chem.*, ***1998***, *63*, 2420.

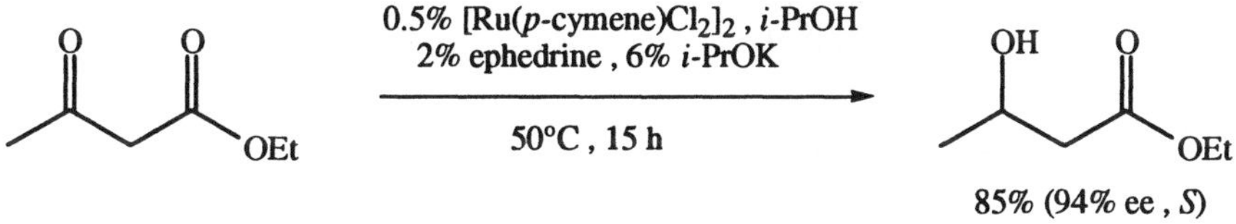
O
O
OEt
0.5% [Ru(p-cymene)Cl2]2 , i-PrOH
2% ephedrine , 6% i-PrOK
50°C , 15 h
OH
O
OEt
85% (94% ee , S)

Everaere, K.; Carpentier, J.–F.; Mortreux, A.; Bulliard, M. *Tetrahedron Asymmetry*, ***1998***, *9*, 2971.

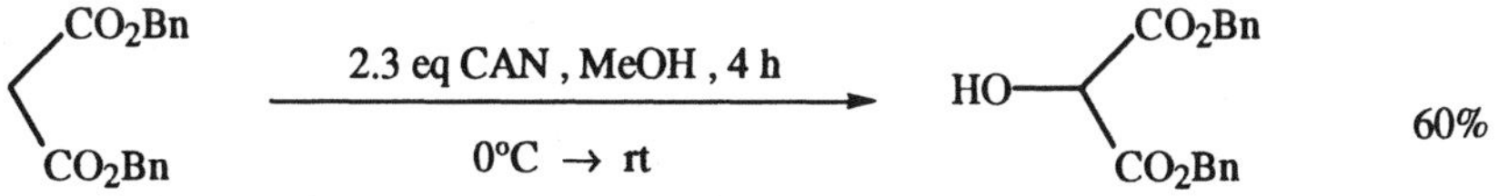
CO2Bn
CO2Bn
2.3 eq CAN , MeOH , 4 h
0°C → rt
HO
CO2Bn
CO2Bn
60%

Nair, V.; Nair, L.G.; Mathew, J. *Tetrahedron Lett.*, ***1998***, *39*, 2801.

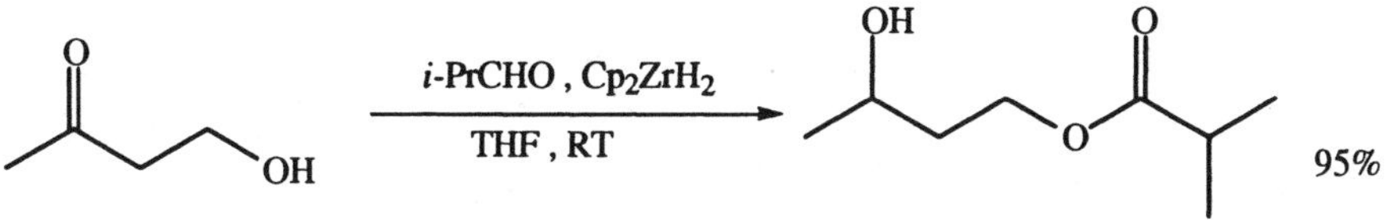
O
OH
i-PrCHO , Cp2ZrH2
THF , RT
OH
O
O
95%

Umekawa, Y.; Sakaguchi, S.; Nishiuama, Y.; Ishii, Y. *J. Org. Chem.*, ***1997***, *62*, 3409.

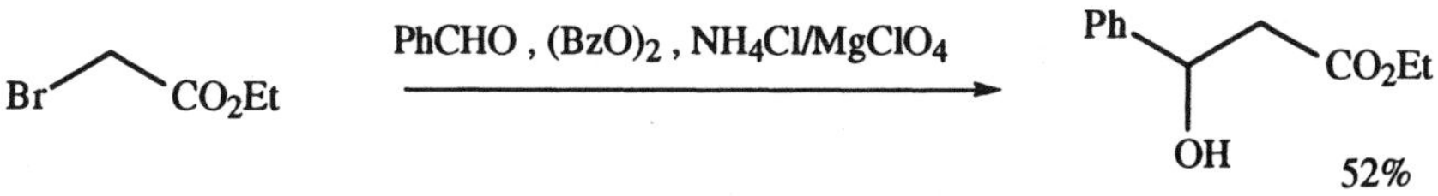
Br
CO2Et
PhCHO , (BzO)2 , NH4Cl/MgClO4
Ph
CO2Et
OH
52%

Bieber, L.W.; Malvestiti, I.; Storch, E.C. *J. Org. Chem.*, ***1997***, *62*, 904.

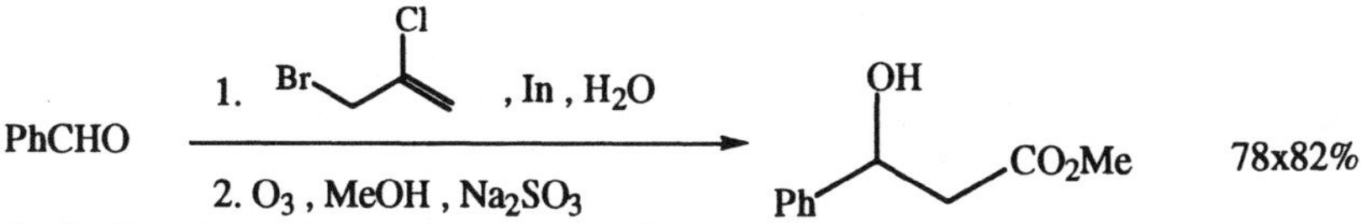
PhCHO
Cl
1. Br
, In , H2O
2. O3 , MeOH , Na2SO3
OH
Ph
CO2Me
78x82%

Yi, X.–H.; Meng, Y.; Li, C.–J. *Tetrahedron Lett.*, ***1997***, *38*, 4731.

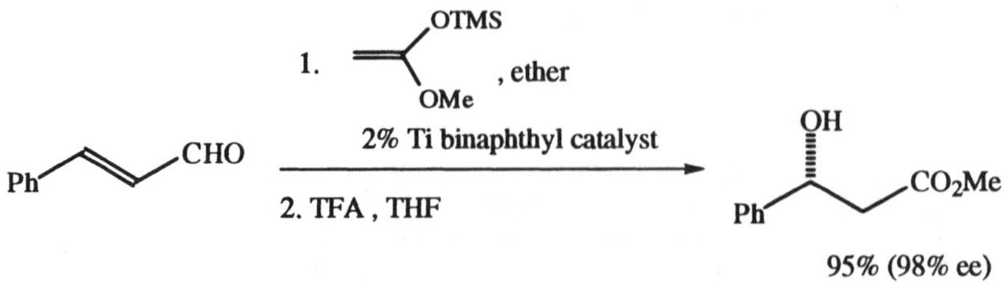
OTMS
1.
, ether
OMe
2% Ti binaphthyl catalyst
2. TFA , THF
Ph
CHO
OH
Ph
CO_2Me
95% (98% ee)

Singer, R.A.; Carreira, E.M. *Tetrahedron Lett.*, ***1997***, *38*, 927.

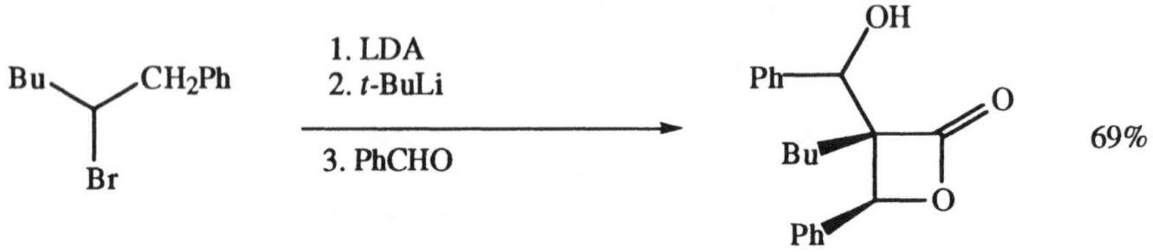
Bu
CH_2Ph
Br
1. LDA
2. *t*-BuLi
3. PhCHO
OH
Ph
O
Bu
O
Ph
69%

Shindo, M. *Tetrahedron Lett.*, ***1997***, *38*, 4433.

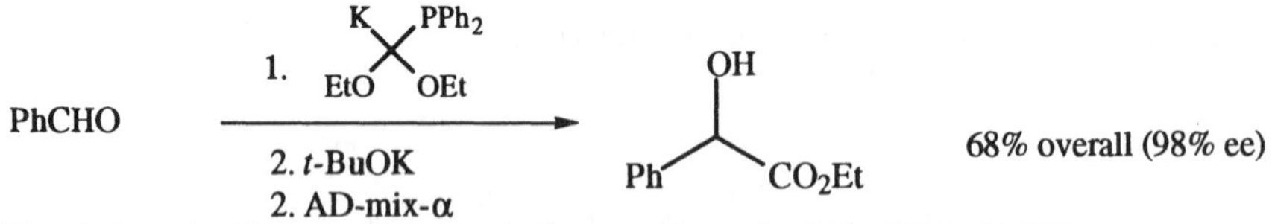
K
PPh_2
1.
EtO
OEt
PhCHO
2. *t*-BuOK
2. AD-mix-α
OH
Ph
CO_2Et
68% overall (98% ee)

Kirschning, A.; Dräger, G.; Jung, A. *Angew. Chem. Int. Ed.*, ***1997***, *36*, 253.

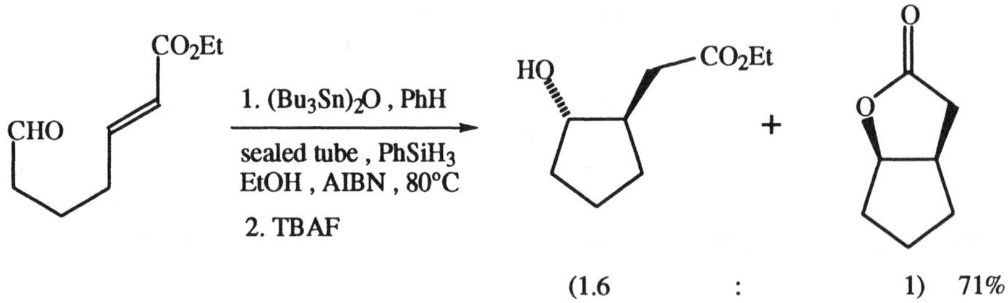
CO_2Et
CHO
1. $(Bu_3Sn)_2O$, PhH
sealed tube , $PhSiH_3$
EtOH , AIBN , 80°C
2. TBAF
HO
CO_2Et
+
O
O
(1.6 : 1) 71%

Hays, D.S.; Fu, G.C. *J. Org. Chem.*, ***1996***, *61*, 4.

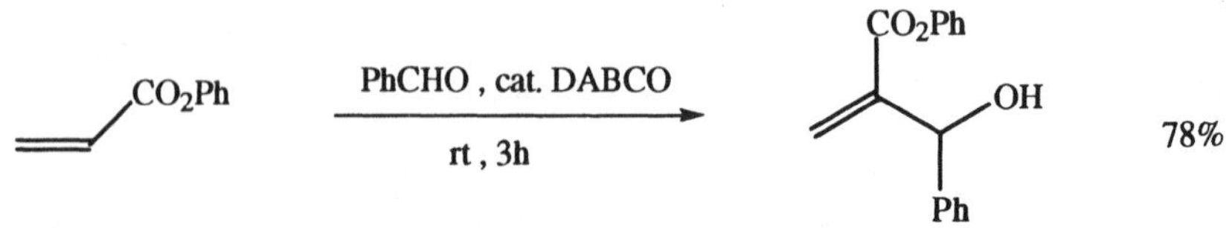
CO_2Ph
PhCHO , cat. DABCO
rt , 3h
CO_2Ph
OH
Ph
78%

Perlmutter, P.; Puniani, E.; Westman, G. *Tetrahedron Lett.*, ***1996***, *37*, 1715.

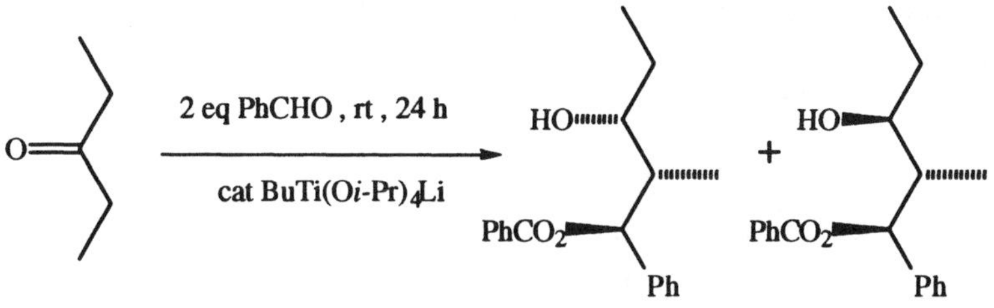
2 eq PhCHO , rt , 24 h
cat BuTi(O*i*-Pr)$_4$Li
HO
PhCO$_2$
Ph
+
HO
PhCO$_2$
Ph

(99 : 1) 63%

Mahrwald, R.; Costisella, B. *Synthesis*, ***1996***, 1087.

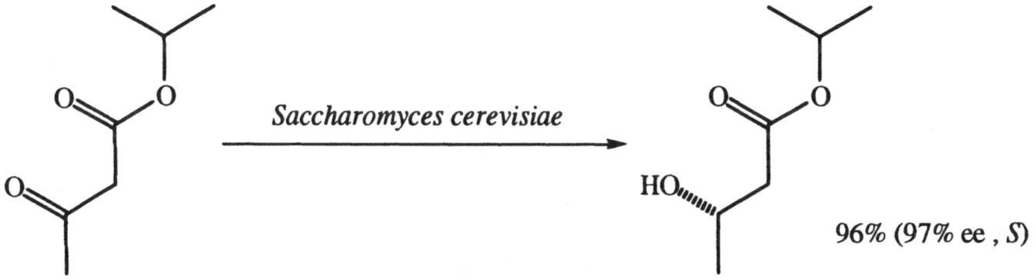
O
O
O
Saccharomyces cerevisiae
O
O
HO
96% (97% ee , *S*)

Medson, C.; Smallridge, A.J.; Trewhella, M.A. *Tetrahedron Asymmetry*, ***1997***, *8*, 1049.

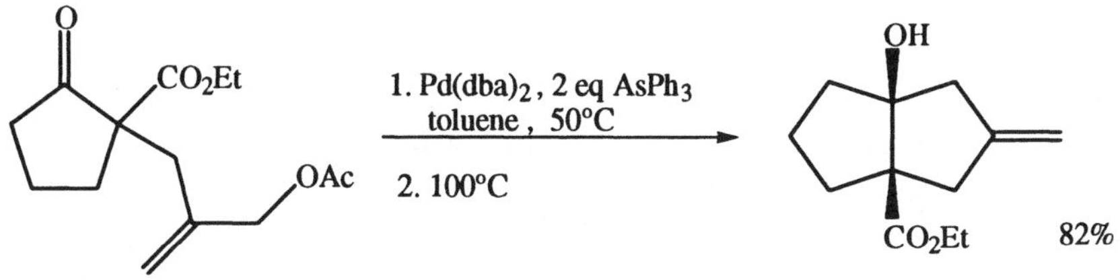
O
CO$_2$Et
OAc
1. Pd(dba)$_2$, 2 eq AsPh$_3$
toluene , 50°C
2. 100°C
OH
CO$_2$Et
82%

Ahiko, T.–a.; Ishiyama, T.; Miyara, N. *Chem. Lett*, ***1997***, 811.

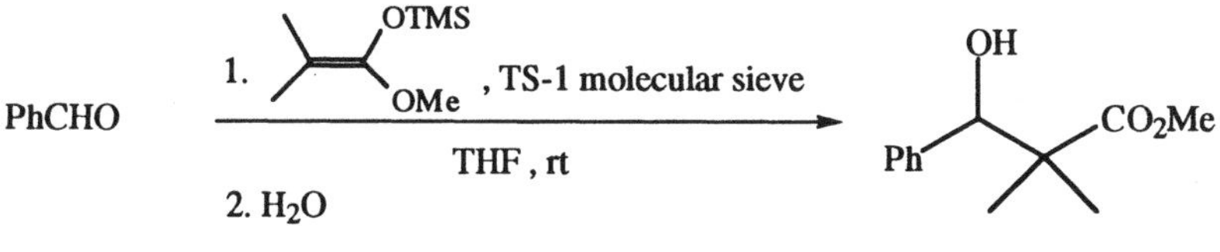
PhCHO
1.
OTMS
OMe
, TS-1 molecular sieve
THF , rt
2. H$_2$O
OH
Ph
CO$_2$Me

85%

Sasidharan, M.; Raju, S.V.N.; Srinivasan, K.V.; Paul, V.; Kumar, R. *Chem. Lett.*, ***1996***, 129

Also via: Section 313 (Alcohol - Carboxylic Acid).

SECTION 328: ALCOHOL, THIOL - ETHER, EPOXIDE, THIOETHER

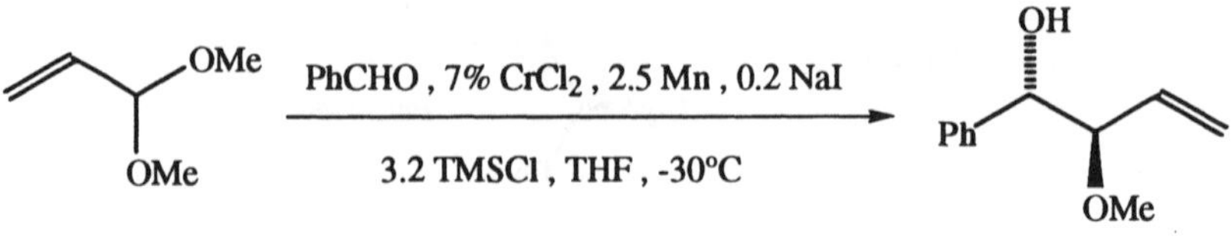
OMe
OMe
PhCHO , 7% $CrCl_2$, 2.5 Mn , 0.2 NaI
3.2 TMSCl , THF , -30°C
OH
Ph
OMe

88% (10.9:1 *anti:syn*)

Boeckman Jr., R.K.; Hudack Jr., R.A. *J. Org. Chem.*, ***1998***, *63*, 3524.

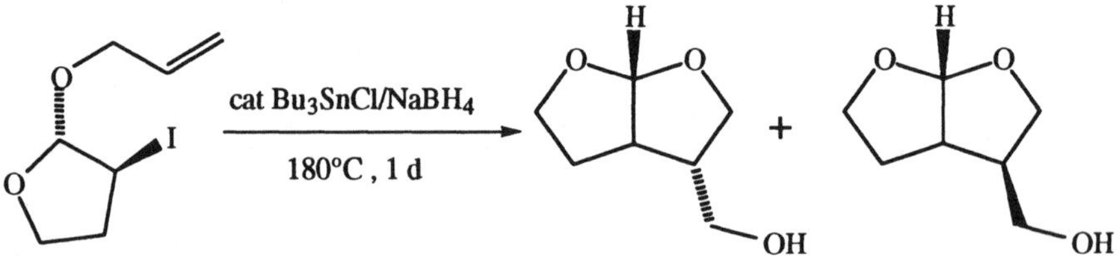
O
O
I
cat $Bu_3SnCl/NaBH_4$
180°C , 1 d
H
O
O
OH
+
H
O
O
OH

(15 : 1) 72%

Mayer, S.; Prandi, J.; Bamhaoud, T.; Bakkas, S.; Guillou, O. *Tetrahedron*, ***1998***, *54*, 8753.

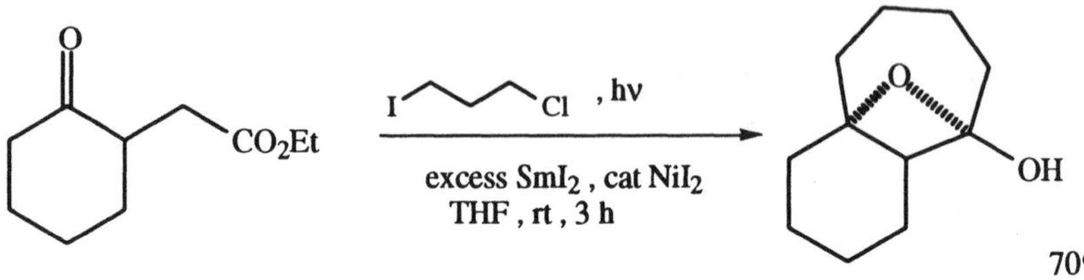
O
CO_2Et
I Cl , hν
excess SmI_2 , cat NiI_2
THF , rt , 3 h
O
OH

70%

Molander, G.A.; Sono, M. *Tetrahedron*, ***1998***, *54*, 9289.

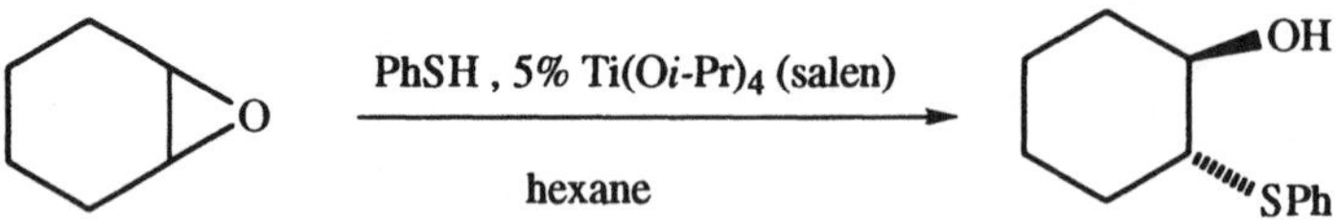
O
PhSH , 5% Ti(O*i*-Pr)$_4$ (salen)
hexane
OH
SPh

93% (63% ee , *1S,2S*)

Wu, J.; Hou, X.–L.; Dai, L.–X.; Xia, J.–J.; Tang, M.–H. *Tetrahedron Asymmetry*, ***1998***, *9*, 3431.

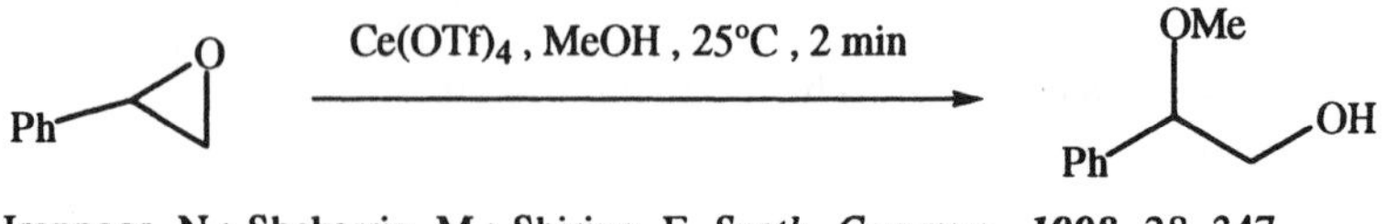
Ph
O
Ce(OTf)$_4$, MeOH , 25°C , 2 min
OMe
Ph
OH

92%

Iranpoor, N.; Shekarriz, M.; Shiring, F. *Synth. Commun.*, ***1998***, *28*, 347.

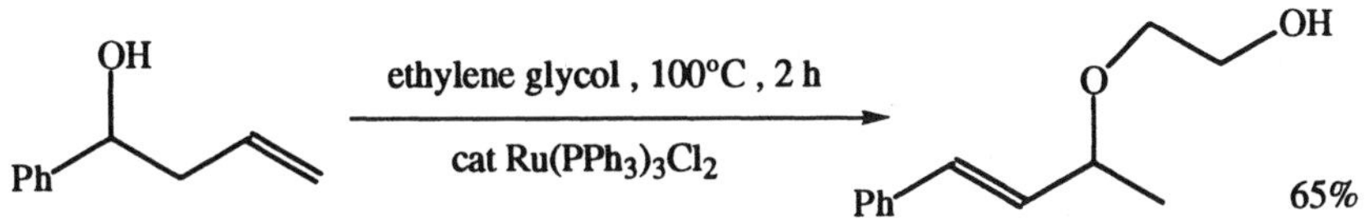
OH
Ph
ethylene glycol , 100°C , 2 h
cat Ru(PPh3)3Cl2
O
OH
Ph
65%

Wang, D.; Li, C.-J. *Synth. Commun.*, ***1998***, *28*, 507.

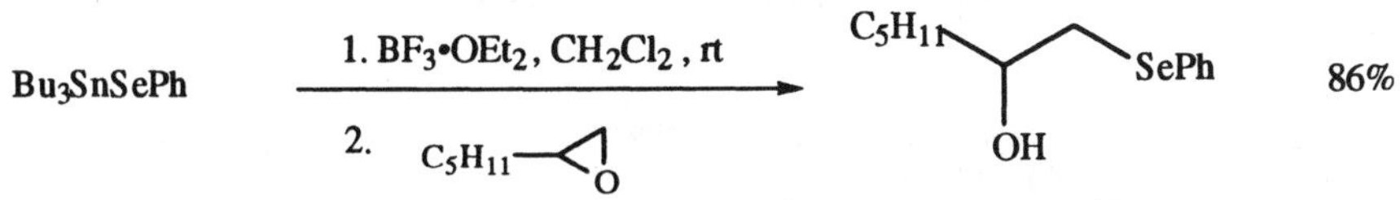
Bu3SnSePh
1. BF3•OEt2, CH2Cl2 , rt
2. C5H11 O
C5H11
SePh
OH
86%

Nishiyama, Y.; Ohashi, H.; Itoh, K.; Sonoda, N. *Chem. Lett*, ***1998***, 159.

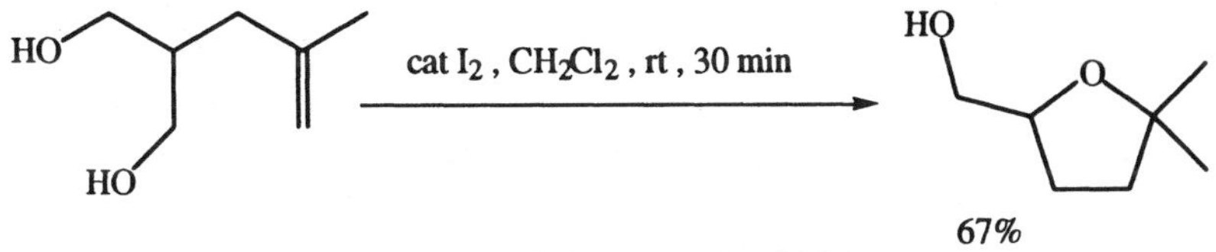
HO
HO
cat I2 , CH2Cl2 , rt , 30 min
HO
O
67%

Kim, K.M.; Jeon, D.J.; Ryu, E.K. *Synthesis*, ***1998***, 835.

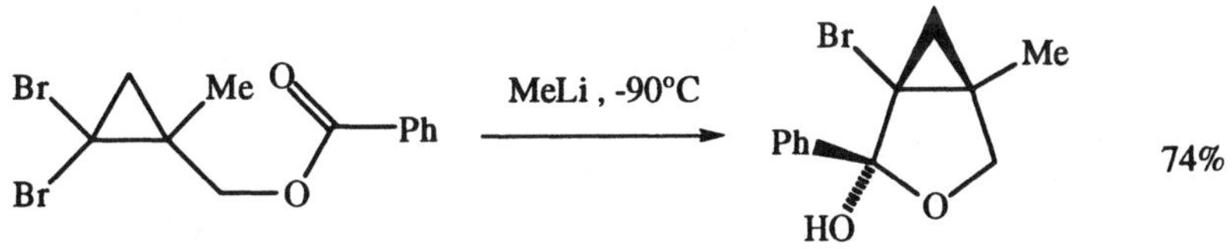
Br
Br
Me
O
Ph
O
MeLi , -90°C
Br
Me
Ph
HO
O
74%

Baird, M.S.; Huber, F.A.M. *Tetrahedron Lett.*, ***1998***, *39*, 9081.

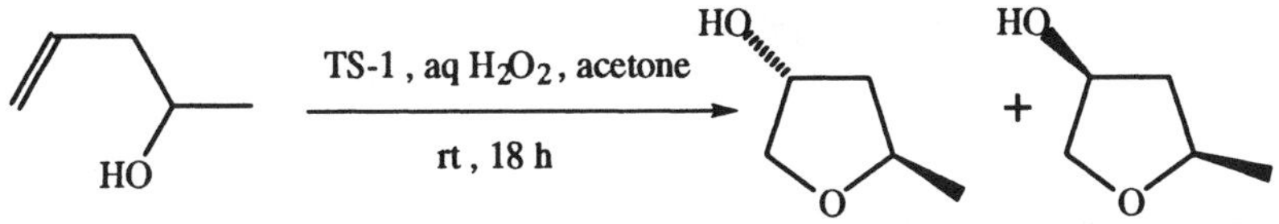
HO
TS-1 , aq H2O2 , acetone
rt , 18 h
HO
O
+
HO
O

(70 : 30) 80%

TS-1 = titanium silicate

Bhaumik, A.; Tatsumi, T. *Chem. Commun.*, ***1998***, 463.

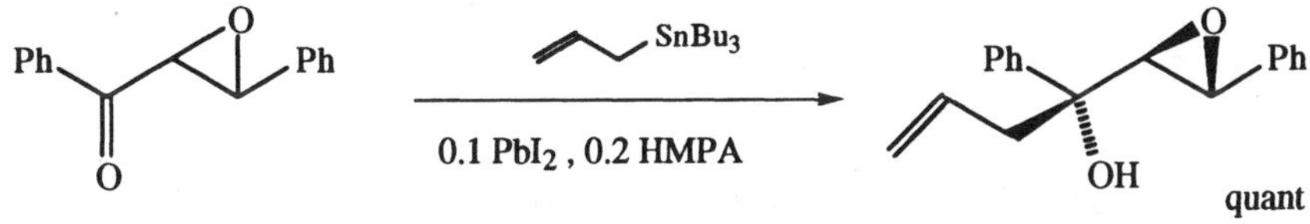
Ph
O
Ph
O
SnBu3
0.1 PbI2 , 0.2 HMPA
Ph
O
Ph
OH
quant

Shibata, I.; Fukuoka, S.; Yoshimura, N.; Matsuda, H.; Baba, A. *J. Org. Chem.*, ***1997***, *62*, 3790.

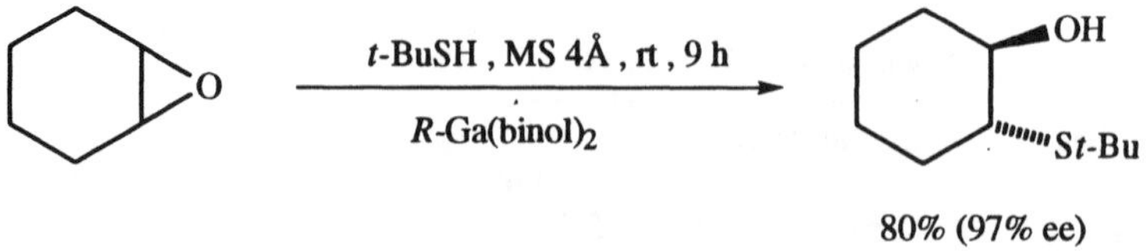
O
t-BuSH , MS 4Å , rt , 9 h
R-Ga(binol)2
OH
St-Bu
80% (97% ee)

Iida, T.; Yamamoto, N.; Sasai, H.; Shibasaki, M. *J. Am. Chem. Soc.*, ***1997***, *119*, 4783.

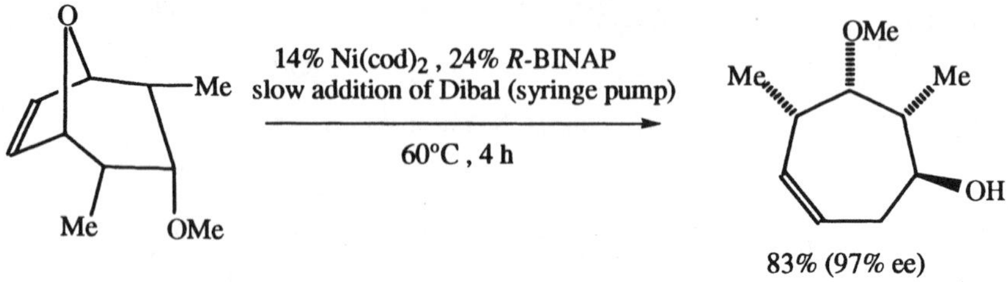
O
Me
Me
OMe
14% Ni(cod)2 , 24% R-BINAP
slow addition of Dibal (syringe pump)
60°C , 4 h
OMe
Me
Me
OH
83% (97% ee)

Lautens, M.; Rovis, T. *J. Am. Chem. Soc.*, ***1997***, *119*, 11090.

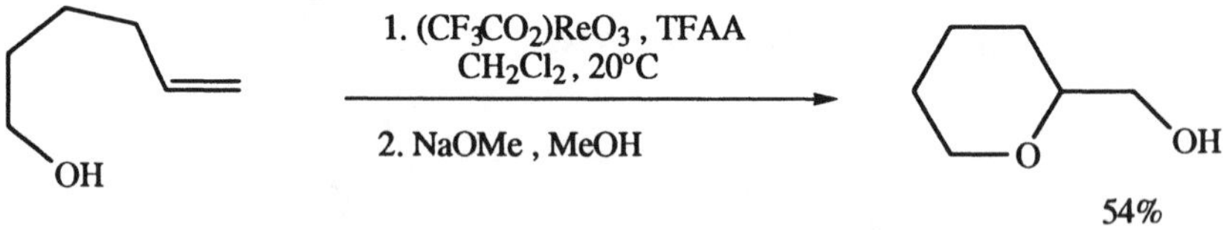
OH
1. (CF3CO2)ReO3 , TFAA
CH2Cl2 , 20°C
2. NaOMe , MeOH
O
OH
54%

McDonald, F.E.; Singhi, A.D. *Tetrahedron Lett.*, ***1997***, *38*, 7683.

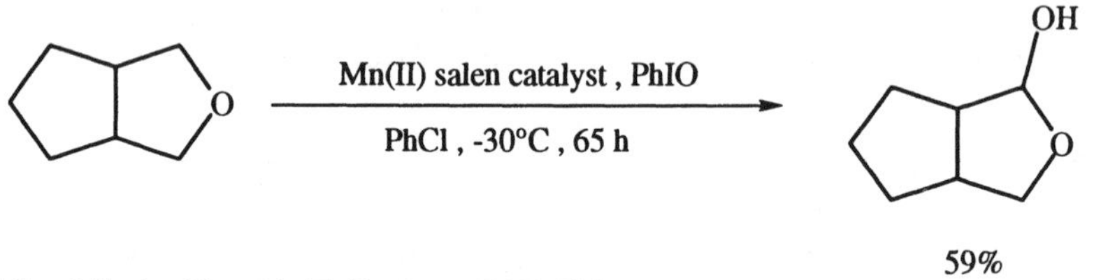
O
Mn(II) salen catalyst , PhIO
PhCl , -30°C , 65 h
OH
O
59%

Miyafuji, A.; Katsuki, T. *SynLett*, ***1997***, 836.

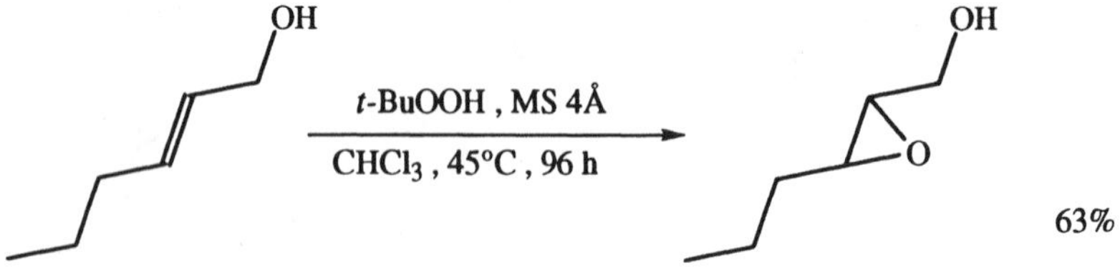
OH
t-BuOOH , MS 4Å
CHCl3 , 45°C , 96 h
OH
O
63%

Palombi, L.; Bonadies, F.; Scettri, A. *Tetrahedron*, ***1997***, *53*, 11369.

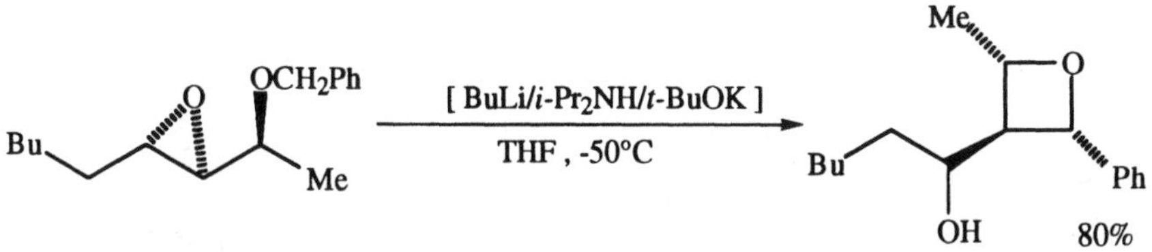
OCH2Ph
O
Bu
Me
[BuLi/i-Pr2NH/t-BuOK]
THF , -50°C
Me
O
Bu
Ph
OH
80%

Mordini, A.; Bindi, S.; Pecchi, S.; Capperucci, A.; Degl'Innocenti, A.; Reginato, G. *J. Org. Chem.*, ***1996***, *61*, 4466.

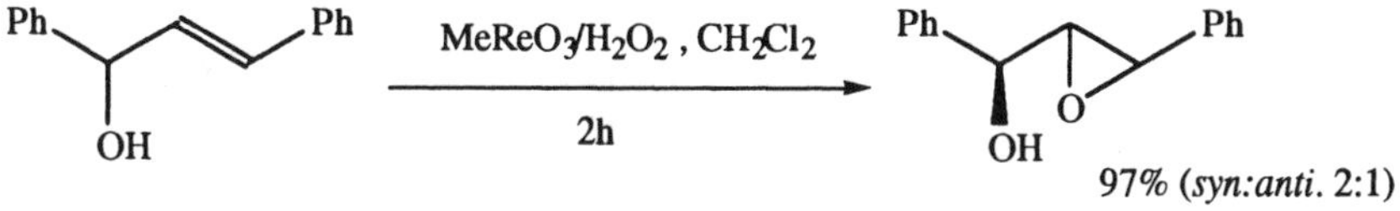
Ph
Ph
OH
MeReO3/H2O2 , CH2Cl2
2h
Ph
Ph
O
OH
97% (syn:anti. 2:1)

Boehlow, T.R.; Spinning, C.D. *Tetrahedron Lett.*, ***1996***, *37*, 2717.

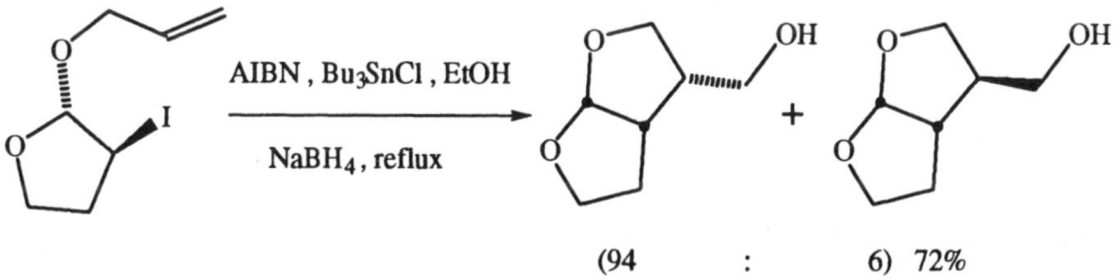
O
I
O
AIBN , Bu3SnCl , EtOH
NaBH4 , reflux
O
OH
O
+
O
OH
O
(94 : 6) 72%

Mayer, S.; Prandi, J. *Tetrahedron Lett.*, ***1996***, *37*, 3117.

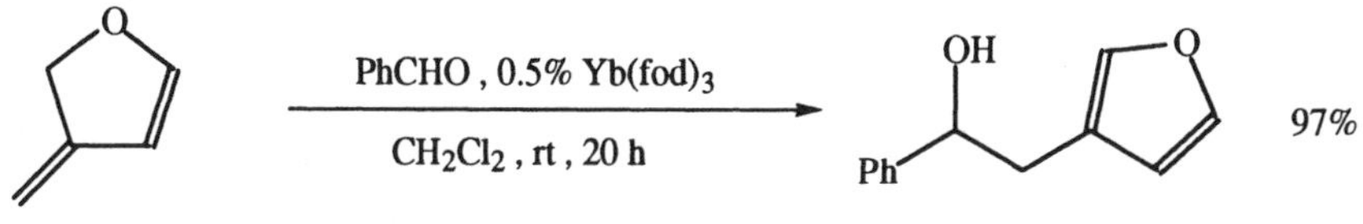
O
PhCHO , 0.5% Yb(fod)3
CH2Cl2 , rt , 20 h
OH
O
Ph
97%

Miles, W.H.; Berreth, C.L.; Anderton, C.A. *Tetrahedron Lett.*, ***1996***, *37*, 7893.

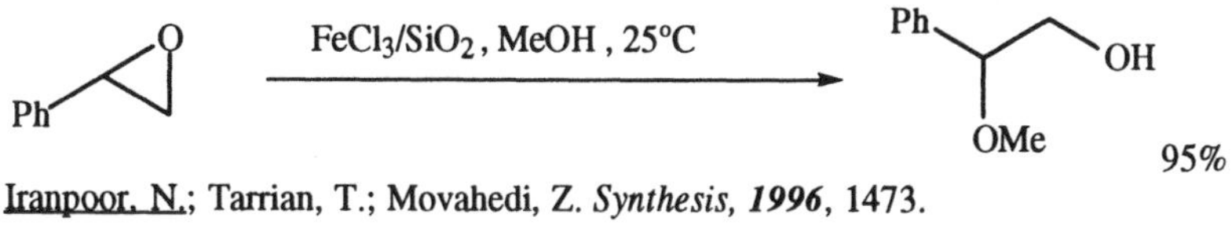
O
Ph
FeCl3/SiO2 , MeOH , 25°C
Ph
OH
OMe
95%

Iranpoor, N.; Tarrian, T.; Movahedi, Z. *Synthesis*, ***1996***, 1473.

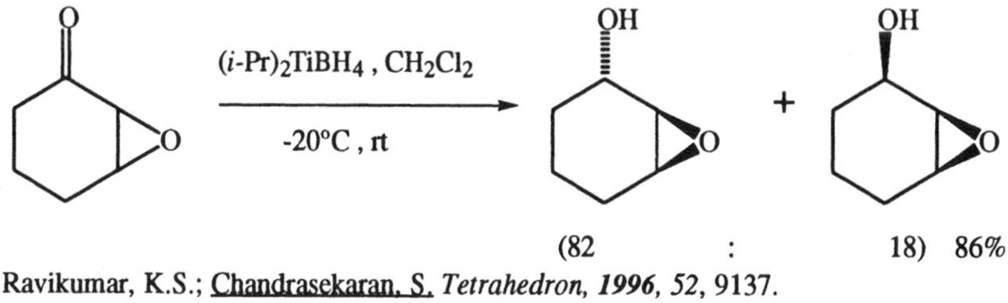
O
O
(i-Pr)2TiBH4 , CH2Cl2
-20°C , rt
OH
O
+
OH
O
(82 : 18) 86%

Ravikumar, K.S.; Chandrasekaran, S. *Tetrahedron*, ***1996***, *52*, 9137.

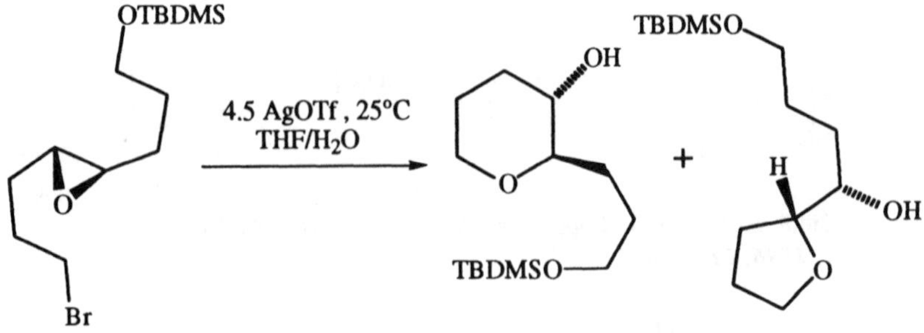

(3.8 : 1) 91%

Hayashi, N.; Hujiwara, K.; Murai, A. *Chem. Lett.*, ***1996***, 341.

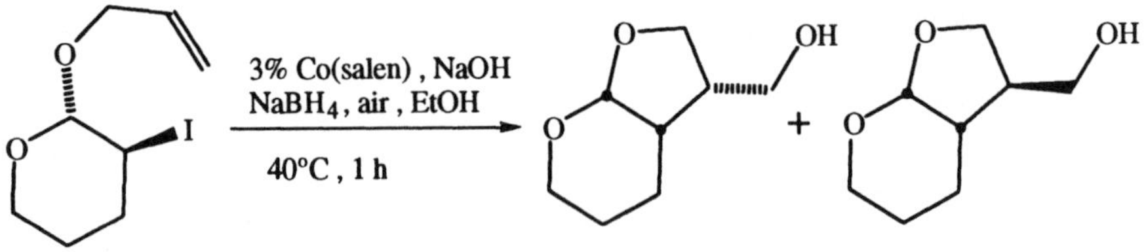

(7 : 1) 80%

Bamhaoud, T.; Prandi, J. *Chem. Commun.*, ***1996***, 1229.

SECTION 329: ALCOHOL, THIOL - HALIDE, SULFONATE

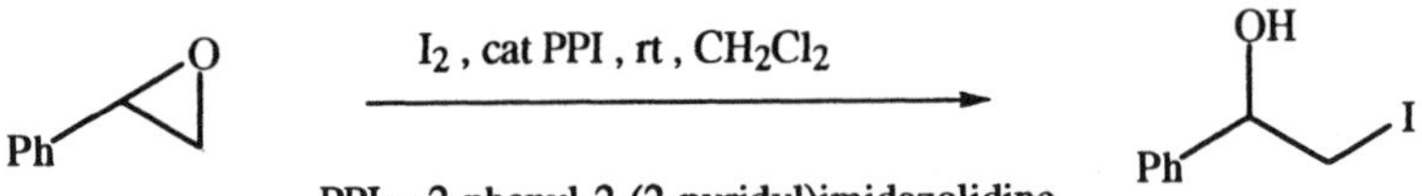

97%

PPI = 2-phenyl-2-(2-pyridyl)imidazolidine

Sharghi, H.; Naeimi, H. *SynLett*, ***1998***, 1343.

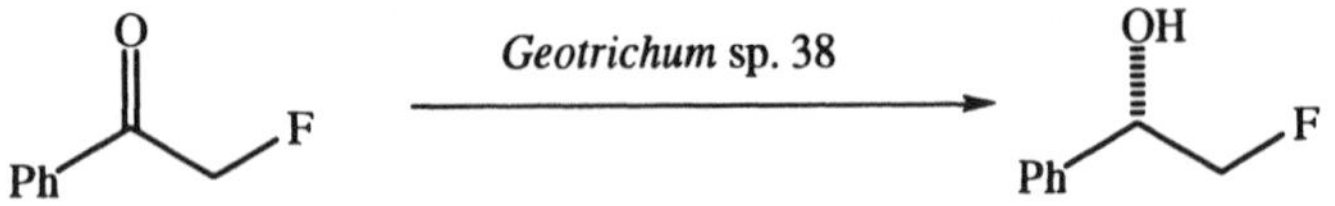

80% (70% ee , *S*)

Wei, Z.-L.; Li, Z.-Y.; Lin, G.-Q. *Tetrahedron*, ***1998***, *54*, 13059.

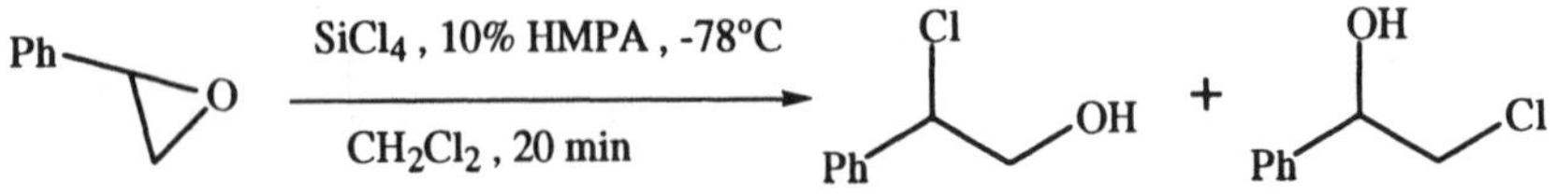

(17 : 1) 94%

Denmark, S.E.; Barsanti, P.A.; Wong, K.-T.; Stavenger, R.A. *J. Org. Chem.* ***1998***, *63*, 2428

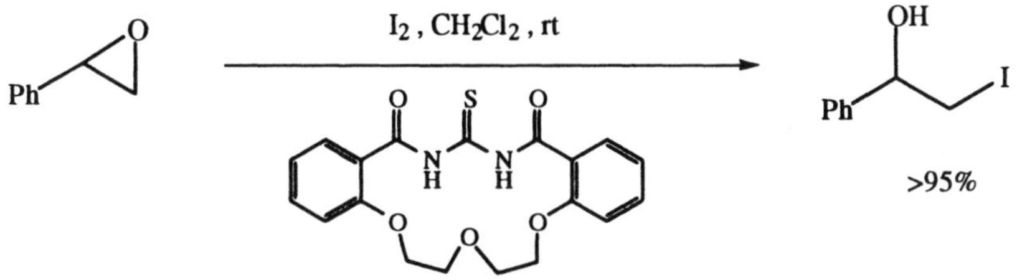
I2 , CH2Cl2 , rt
Ph
O
OH
I
>95%

Sarghi, H.; Massah, A.R.; Eshghi, H.; Niknam, K. *J. Org. Chem.*, ***1998***, *63*, 1455.

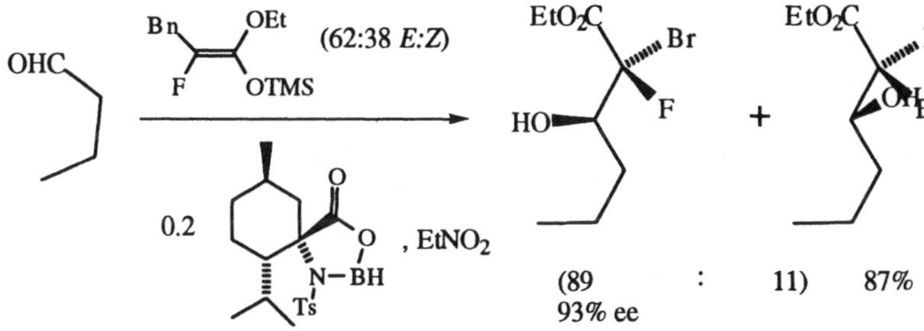
OHC
Bn
OEt
F
OTMS
(62:38 E:Z)
0.2
, EtNO2
N–BH
Ts
EtO2C
Br
F
HO
OH
+
(89 : 11) 87%
93% ee

Iseki, K.; Kuroki, Y.; Kobayashi, Y. *SynLett*, ***1998***, 437.

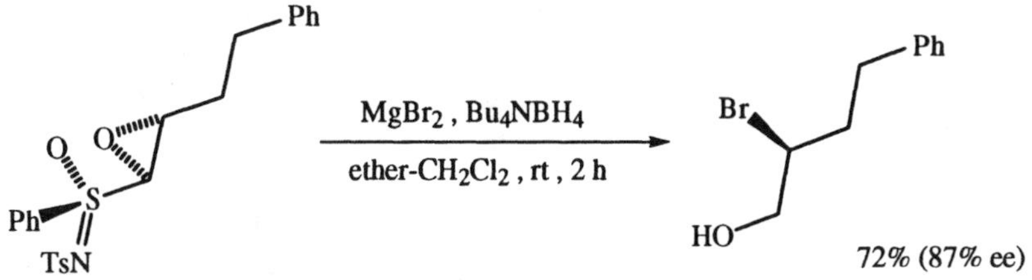
Ph
MgBr2 , Bu4NBH4
ether-CH2Cl2 , rt , 2 h
TsN
Br
HO
72% (87% ee)

Bailey, P.L.; Briggs, A.D.; Jackson, R.F.W.; Pietruszka, J. *J. Chem. Soc., Perkin Trans. 1*, ***1998***, 3359.

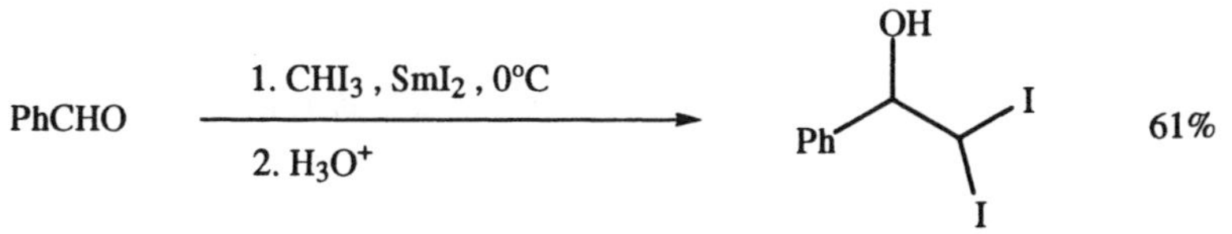
PhCHO
1. CHI3 , SmI2 , 0°C
2. H3O+
OH
Ph
I
I
61%

Concellón, J.M.; Bernad, P.L.; Perez–Andrés, J.A. *Tetrahedron Lett.*, ***1998***, *39*, 1409.

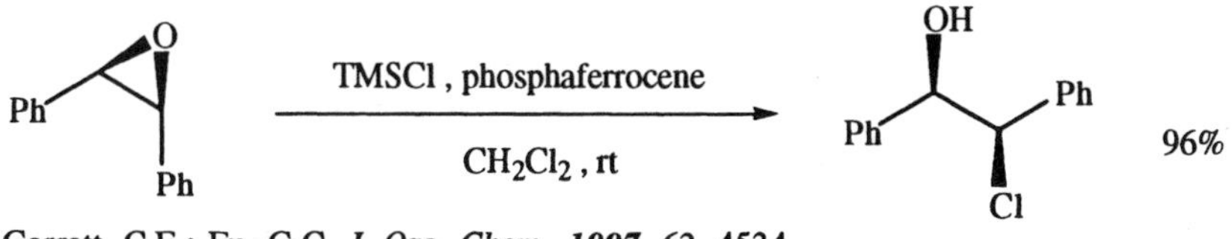
TMSCl , phosphaferrocene
CH2Cl2 , rt
Ph
OH
Cl
96%

Garrett, C.E.; Fu, G.C. *J. Org. Chem.*, ***1997***, *62*, 4534.

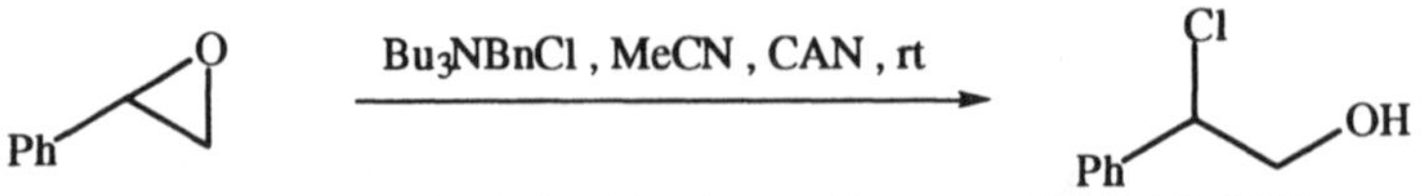
Ph
O
Bu_3NBnCl , MeCN , CAN , rt
Cl
OH
Ph

85%

Iranpoor, N.; Kazemi, F.; Salehi, P. *Synth. Commun.*, ***1997***, *37*, 1247.

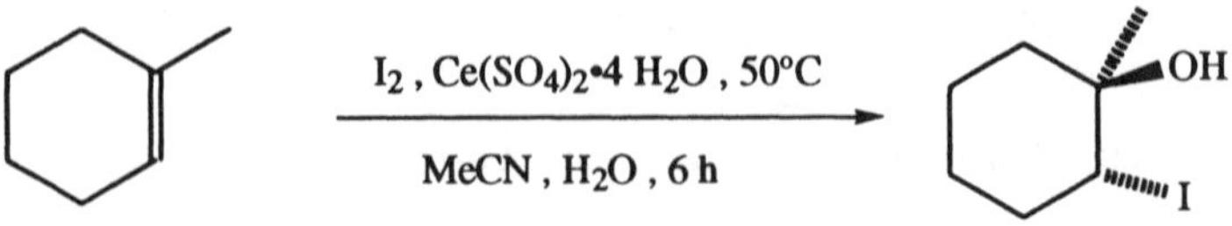
I_2 , $Ce(SO_4)_2 \cdot 4\ H_2O$, 50°C
MeCN , H_2O , 6 h
OH
I

35%

Horiuchi, C.A.; Ikeda, A.; Kanamori, M.; Hosokawa, H.; Sugiyama, T.; Takahashi, T.T. *J. Chem. Res. (S)*, ***1997***, 60.

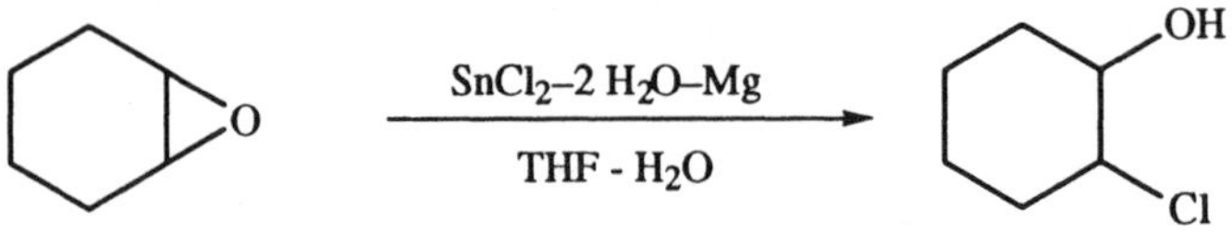
O
$SnCl_2$–2 H_2O–Mg
THF - H_2O
OH
Cl

64%

Sarangi, C.; Das, N.B.; Nanda, B.; Nayak, A.; Sharma, R.P. *J. Chem. Res. (S)*, ***1997***, 180.

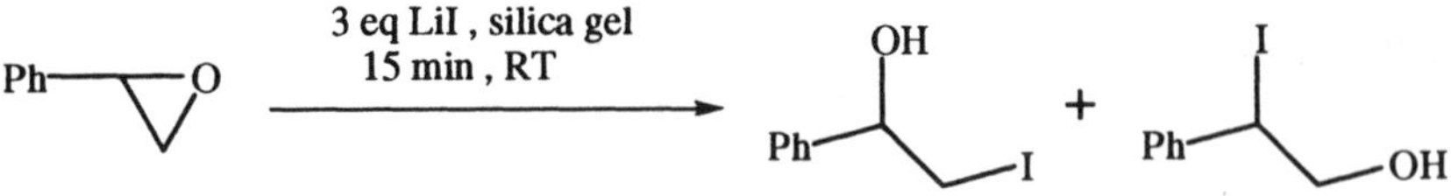
Ph
O
3 eq LiI , silica gel
15 min , RT
OH
Ph
I
+
I
Ph
OH

(20 : 80) 83%

suki, H.; Shimanduchi, T. *Tetrahedron Lett.*, ***1996***, *36*, 1845.

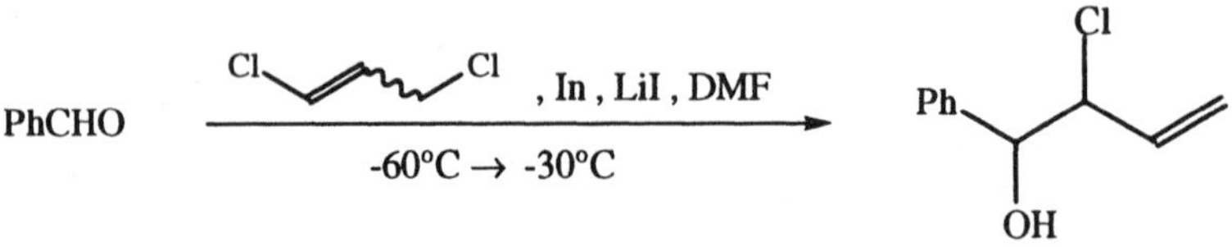
PhCHO
Cl
Cl
, In , LiI , DMF
-60°C → -30°C
Cl
Ph
OH

83% (92:8 *syn:anti*)

Araki, S.; Hirashita, T.; Shimizu, H.; Yamamura, H.; Kawai, M.; Butsugan, Y. *Tetrahedron Lett.*, ***1996***, *37*, 8417.

SECTION 330: ALCOHOL, THIOL - KETONE

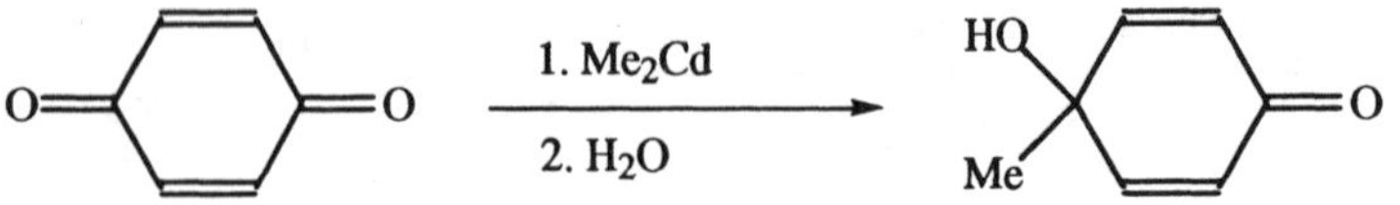
O
O
1. Me_2Cd
2. H_2O
HO
Me
O

71%

Aponick, A.; McKinley, J.D.; Raber, J.C.; Wigal, C.T. *J. Org. Chem.*, ***1998***, *63*, 2676.

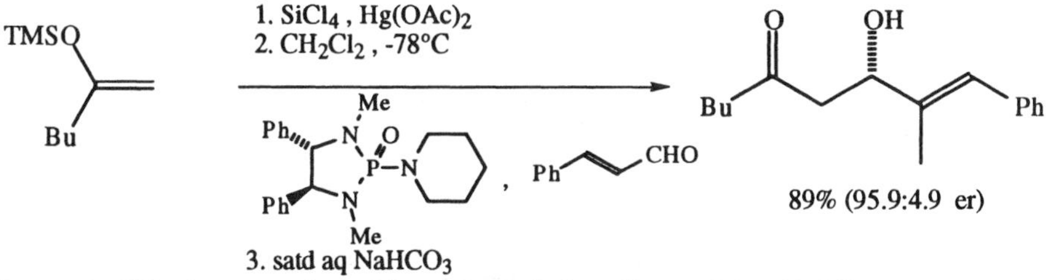
TMSO
Bu
1. $SiCl_4$, $Hg(OAc)_2$
2. CH_2Cl_2 , -78°C
Me
Ph
N
O
P
N
Ph
N
Me
Ph
CHO
3. satd aq $NaHCO_3$
O
OH
Bu
Ph
89% (95.9:4.9 er)

Denmark, S.E.; Stavenger, R.A.; Wong, K.–T. *J. Org. Chem.*, ***1998***, *63*, 918.

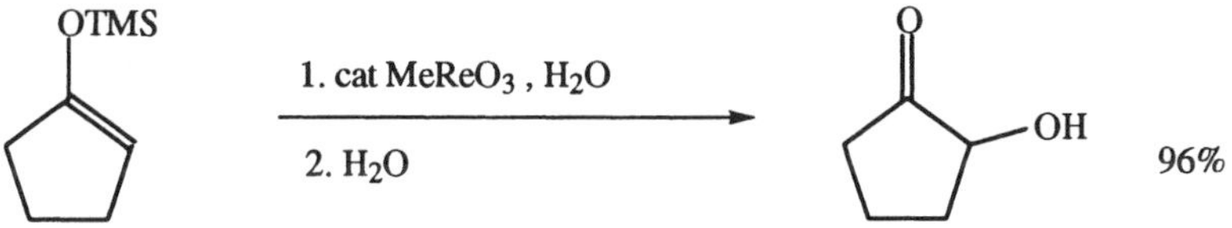
OTMS
1. cat $MeReO_3$, H_2O
2. H_2O
O
OH
96%

Stanković, S.; Espenson, J.H. *J. Org. Chem.*, ***1998***, *63*, 4129.

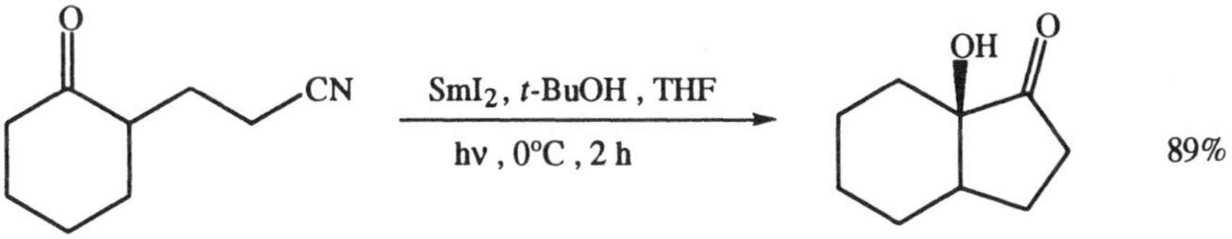
O
CN
SmI_2 , *t*-BuOH , THF
hν , 0°C , 2 h
OH
O
89%

Molander, G.A.; Wolfe, C.N. *J. Org. Chem.*, ***1998***, *63*, 9031.

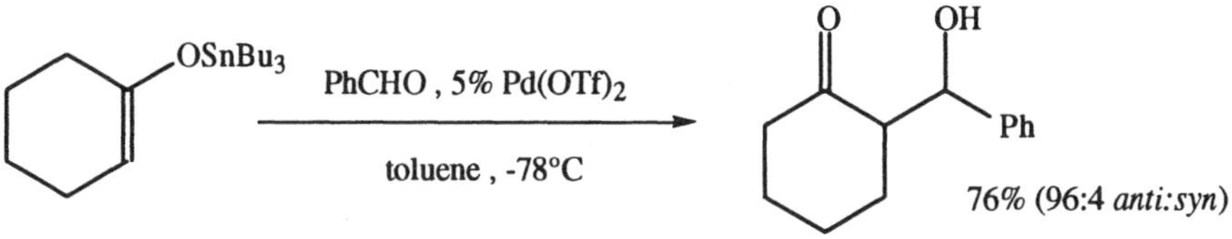
$OSnBu_3$
PhCHO , 5% $Pd(OTf)_2$
toluene , -78°C
O
OH
Ph
76% (96:4 *anti:syn*)

Yanagisawa, A.; Kimura, K.; Nakatsuka, Y.; Yamamoto, H. *SynLett*, ***1998***, 958.

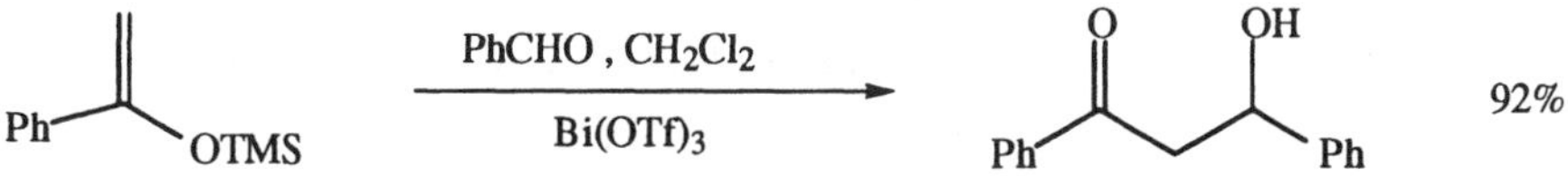
Ph
OTMS
PhCHO , CH_2Cl_2
$Bi(OTf)_3$
O
OH
Ph
Ph
92%

LeRaex, C.; Ciliberti, L.; Laurent–Robert, H.; Laporerie, A.; Dubac, J. *SynLett*, ***1998***, 1249.

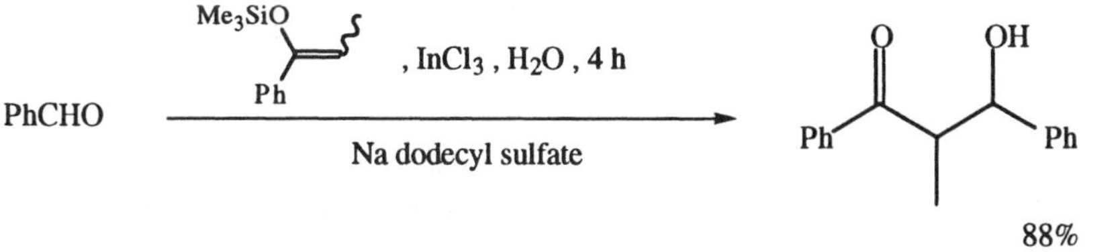
Me_3SiO
Ph
, $InCl_3$, H_2O , 4 h
PhCHO
Na dodecyl sulfate
O
OH
Ph
Ph
88%

Kobayashi, S.; Busujima, T.; Nagayama, S. *Tetrahedron Lett.*, ***1998***, *39*, 1579.

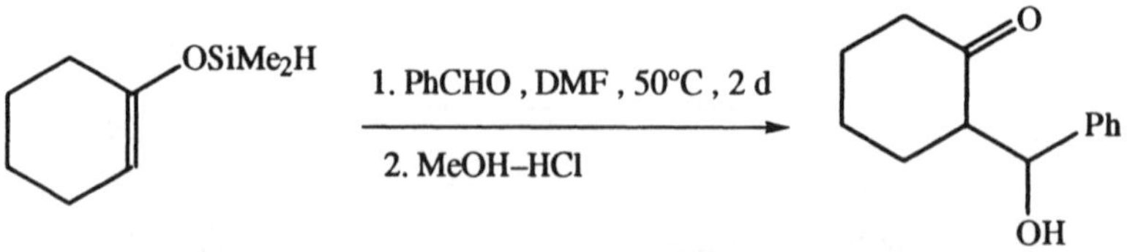
OSiMe2H
1. PhCHO , DMF , 50°C , 2 d
2. MeOH–HCl
O
Ph
OH

79% (58:42 *syn:anti*)

Miura, K.; Sato, H.; Tamaki, K.; Ito, H.; Hosomi, A. *Tetrahedron Lett.*, ***1998***, *39*, 2585.

O
Ph
OAc
1. Bu3MnLi , THF
2. PhCHO
O
OH
Ph
Ph

69%

Hojo, M.; Harada, H.; Ito, H.; Hosomi, A. *J. Am. Chem. Soc.*, ***1997***, *119*, 5459.

O
SnBu3
PhCHO , cat R-BINAP-AgOTf
THF
O
OH
Ph

73% (77% ee)

Yanigisawa, A.; Matsumoto, Y.; Nakashima, H.; Asakawa, K.; Yamamoto, H. *J. Am. Chem. Soc.*, ***1997***, *119*, 9319.

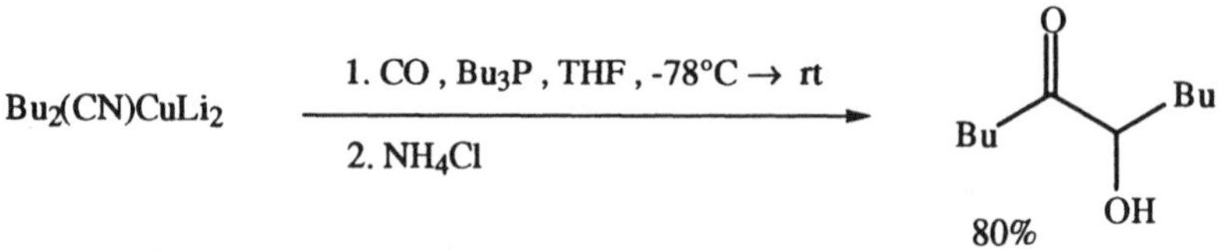
Bu2(CN)CuLi2
1. CO , Bu3P , THF , -78°C → rt
2. NH4Cl
O
Bu
Bu
OH

80%

Kabalka, G.W.; Lin, N.–S.; Yu, S. *Tetrahedron Lett.*, ***1997***, *38*, 2203.

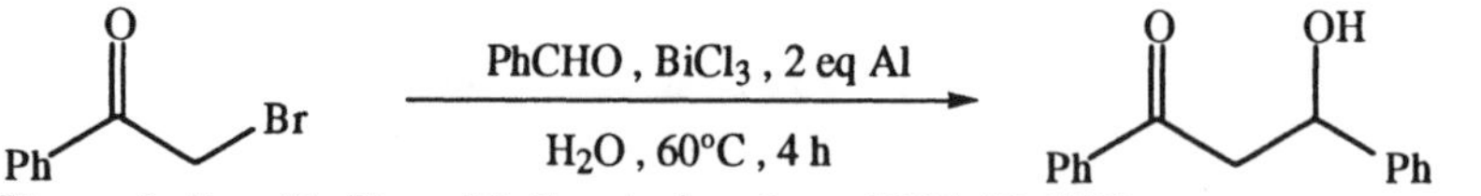
O
Ph
Br
PhCHO , BiCl3 , 2 eq Al
H2O , 60°C , 4 h
O
OH
Ph
Ph

63%

Zhang, J.; Zou, H.; Yang, M. *Tetrahedron Lett.*, ***1997***, *38*, 2733.

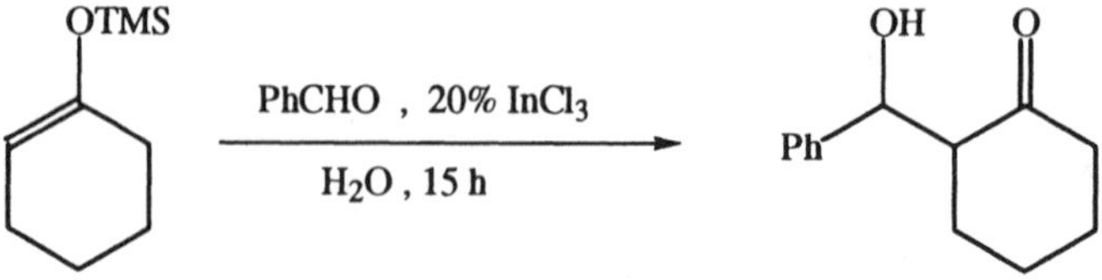
OTMS
PhCHO , 20% InCl3
H2O , 15 h
OH
O
Ph

51% (48:52 *anti:syn*)

Loh, T.–P.; Pei, J.; Koh, K.S.–V.; Cao, G.–Q.; Li, X.–R. *Tetrahedron Lett.*, ***1997***, *38*, 3465

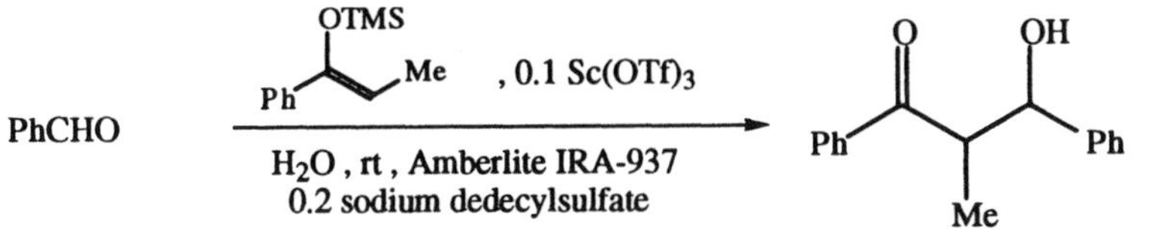
OTMS
Ph
Me
, 0.1 Sc(OTf)3
PhCHO
H2O , rt , Amberlite IRA-937
0.2 sodium dedecylsulfate
O
OH
Ph
Ph
Me

88% (1:1 *syn:anti*)

Kobayashi, S.; Wakabayashi, T.; Nagayama, S.; Oyamada, H. *Tetrahedron Lett.*, ***1997***, *38*, 4559.

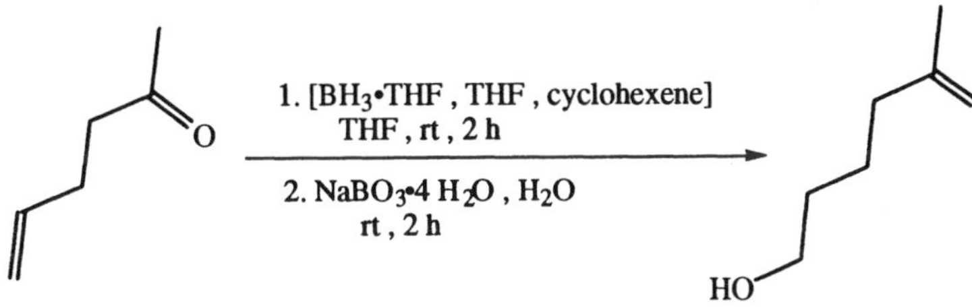
O
1. [BH3•THF , THF , cyclohexene]
THF , rt , 2 h
2. NaBO3•4 H2O , H2O
rt , 2 h
O
HO

76%

Kabalka, G.W.; Yu, S.; Li, N.–S. *Tetrahedron Lett.*, ***1997***, *38*, 5455.

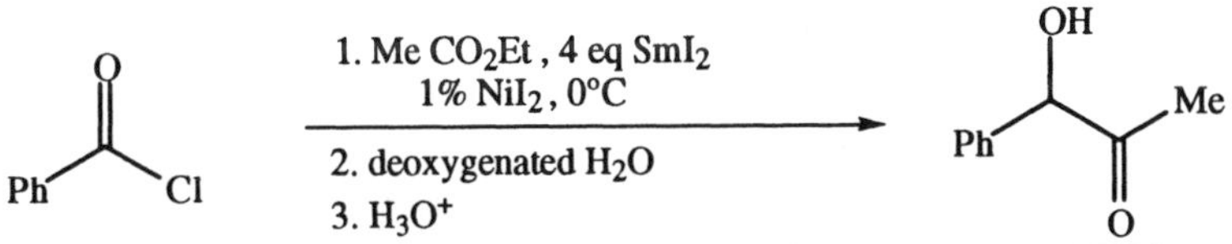
O
Ph
Cl
1. Me CO2Et , 4 eq SmI2
1% NiI2 , 0°C
2. deoxygenated H2O
3. H3O+
OH
Ph
Me
O

81%

Machrouhi, F.; Namy, J.–L.; Kagan, H.B. *Tetrahedron Lett*, ***1997***, *38*, 7183.

CO2Me
CO2Me
TiCl3/NBu3 cat TMSOTf
O
CO2Me

91%

Yoshida, Y.; Hayashi, R.; Sumihara, H.; Tanabe, Y. *Tetrahedron Lett*, ***1997***, *38*, 8727.

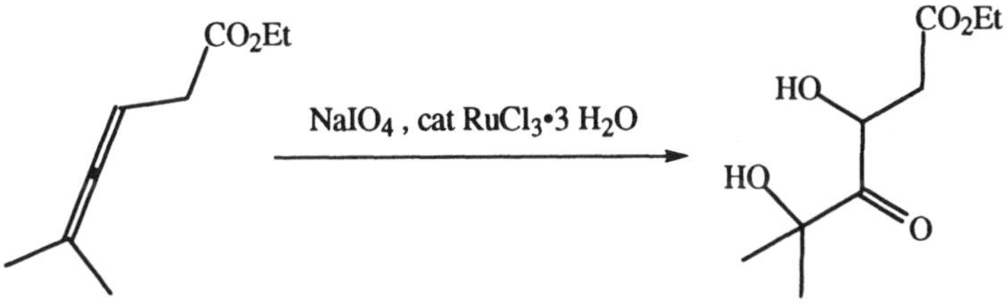
CO2Et
NaIO4 , cat RuCl3•3 H2O
CO2Et
HO
HO
O

32%

Laux, M.; Krause, N. *SynLett*, ***1997***, 765.

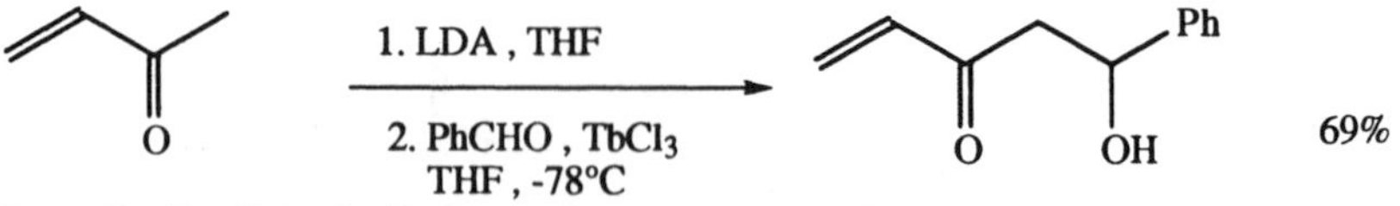
1. LDA , THF
2. PhCHO , TbCl3
THF , -78°C
Ph
O
O
OH
69%

Hong, B.-C.; Chin, S.-F. *Synth. Commun.*, ***1997***, *27*, 1191.

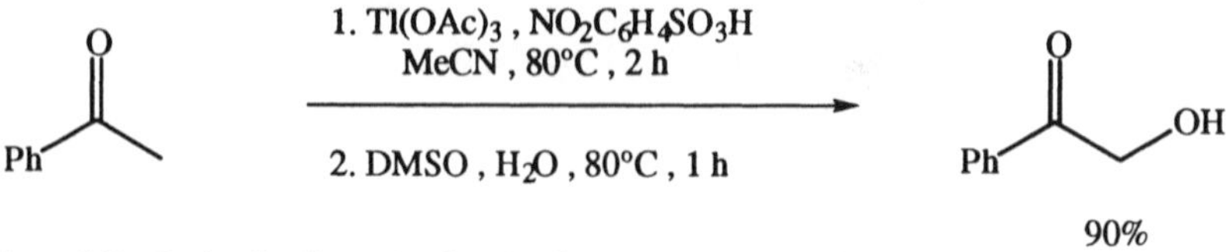
1. Tl(OAc)3 , NO2C6H4SO3H
MeCN , 80°C , 2 h
2. DMSO , H2O , 80°C , 1 h
O
Ph
O
OH
Ph
90%

Lee, J.C.; Park, C.; Choi, Y. *Synth. Commun.*, ***1997***, *27*, 4079.

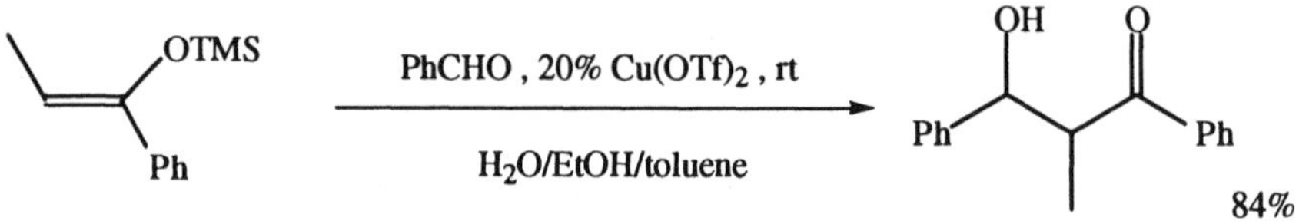
OTMS
Ph
PhCHO , 20% Cu(OTf)2 , rt
H2O/EtOH/toluene
OH
O
Ph
Ph
84%

Kobayashi, S.; Nagayama, S.; Busujima, T. *Chem. Lett.*, ***1997***, 959.

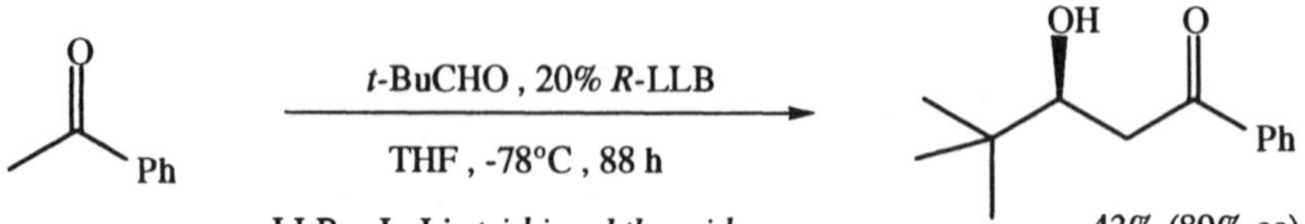
O
Ph
t-BuCHO , 20% R-LLB
THF , -78°C , 88 h
OH
O
Ph

LLB = $LnLi_3$ *tri-binaphthaoxide* 43% (89% ee)

Yamada, Y.M.A.; Yoshikawa, N.; Sasai, H.; Shibasaki, M. *Angew. Chem. Int. Ed.*, ***1997***, *36*, 1871.

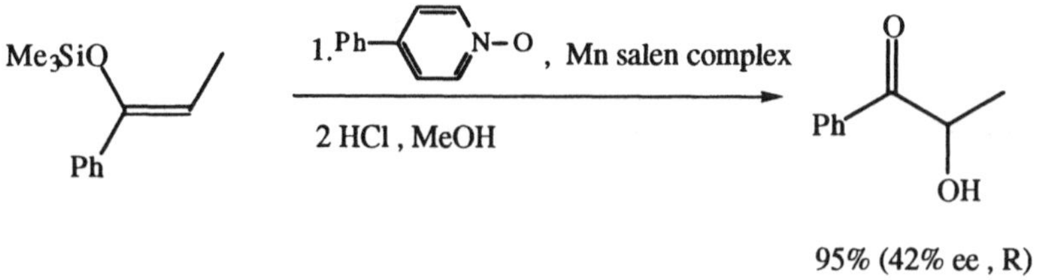
Me3SiO
Ph
1. Ph N–O , Mn salen complex
2 HCl , MeOH
O
Ph
OH
95% (42% ee , R)

Adam, W.; Fell, R.T.; Mock-Knoblauch, C.; Saha-Möller, C.R. *Tetrahedron Lett.*, ***1996***, *37*, 6531.

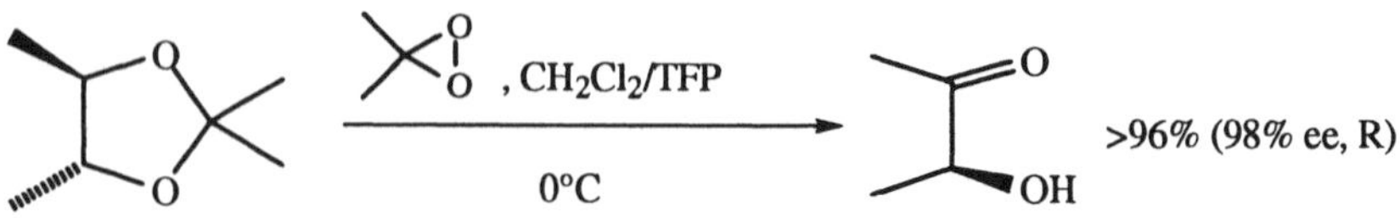
O
O
O
O
, CH2Cl2/TFP
0°C
O
OH
>96% (98% ee, R)

TFP = 1,1,1trifluoro-2-propanone

Curci, R.; D'Accolti, L.; Dinoi, A.; Fusco, C.; Rosa, A. *Tetrahedron Lett*, ***1996***, *37*, 115.

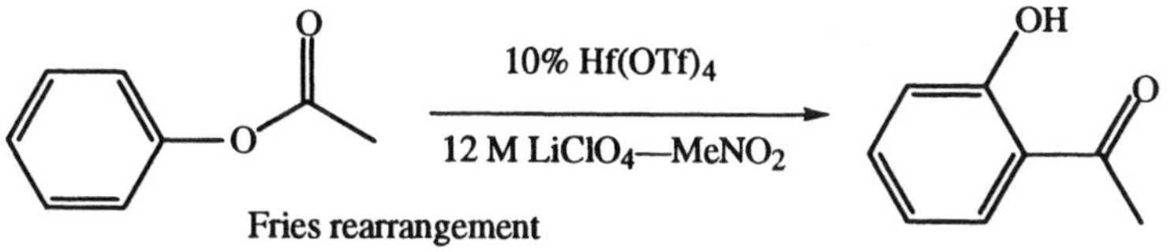
O
10% Hf(OTf)$_4$
12 M LiClO$_4$—MeNO$_2$
O
OH
O
Fries rearrangement

57%

Kobayshi, S.; Moriwaki, M.; Hachiya, I. *Tetrahedron Lett.*, ***1996***, *37*, 2053.

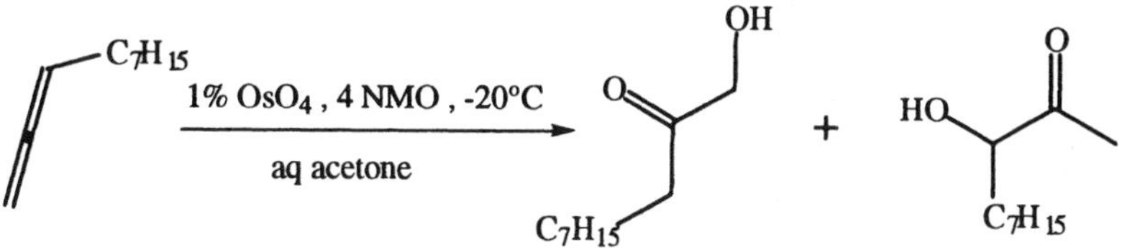
C$_7$H$_{15}$
1% OsO$_4$, 4 NMO , -20°C
aq acetone
OH
O
C$_7$H$_{15}$
+
O
HO
C$_7$H$_{15}$

(76 : 24) 60%

David, K.; Ariente, C.; Greiner, A.; Goré, J.; Cazes, B. *Tetrahedron Lett*, ***1996***, *37*, 3335.

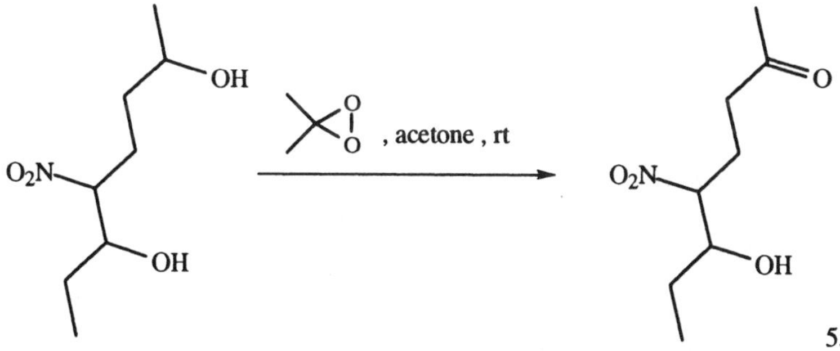
OH
O$_2$N
OH
, acetone , rt
O
O$_2$N
OH

55%

Ballini, R.; Papa, F.; Bovicelli, P. *Tetrahedron Lett.*, ***1996***, *37*, 3507.

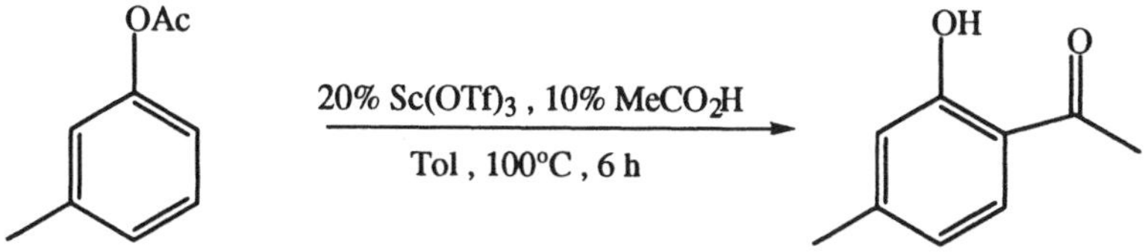
OAc
20% Sc(OTf)$_3$, 10% MeCO$_2$H
Tol , 100°C , 6 h
OH
O

60%

Kobayashi, S.; Moriwaki, M.; Hachiya, I. *Tetrahedron Lett.*, ***1996***, *37*, 4183.

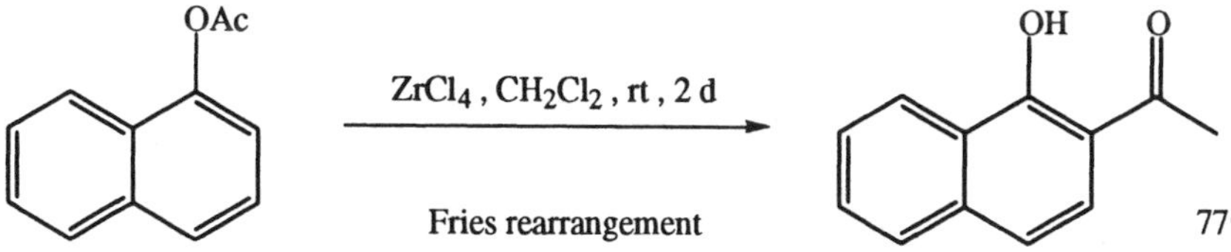
OAc
ZrCl$_4$, CH$_2$Cl$_2$, rt , 2 d
Fries rearrangement
OH
O

77%

Harrowven, D.C.; Dainty, R.F. *Tetrahedron Lett.*, ***1996***, *37*, 7659.

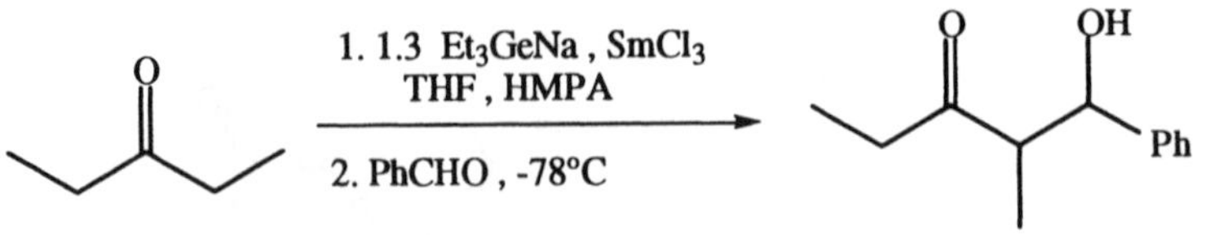
1. 1.3 Et_3GeNa , $SmCl_3$
THF , HMPA
2. PhCHO , -78°C
O
OH
Ph

98% (96:4 *syn:anti*)

Yokoyama, Y.; Mochida, K. *SynLett,* ***1996***, 445.

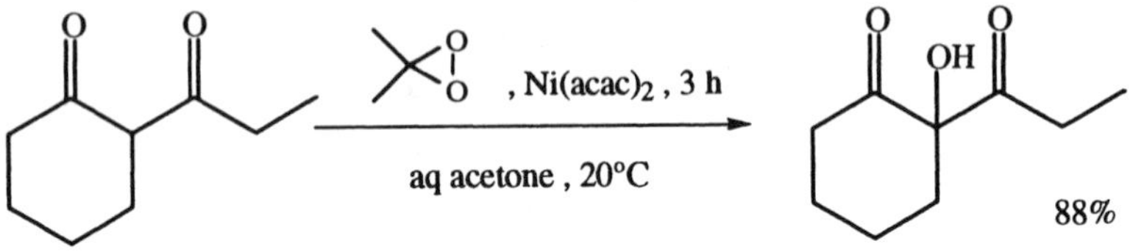
, $Ni(acac)_2$, 3 h
aq acetone , 20°C
O
OH
88%

Adam, W.; Smerz, A.K. *Tetrahedron,* ***1996***, *52*, 5799.

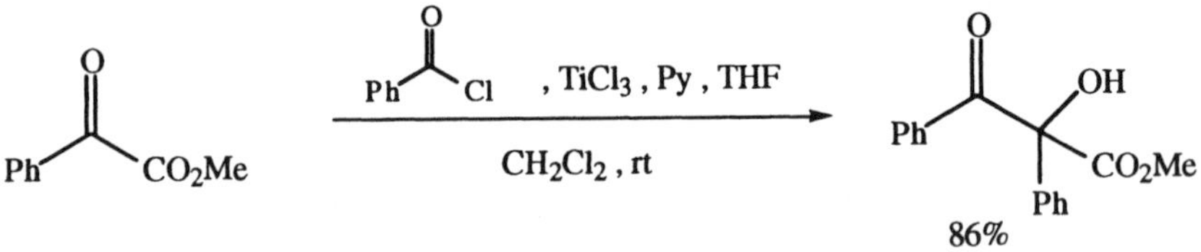
Ph
Cl
, $TiCl_3$, Py , THF
CH_2Cl_2 , rt
O
Ph
CO_2Me
OH
86%

Clerici, A.; Clerici, L.; Porta, O. *Tetrahedron,* ***1996***, *52*, 11037.

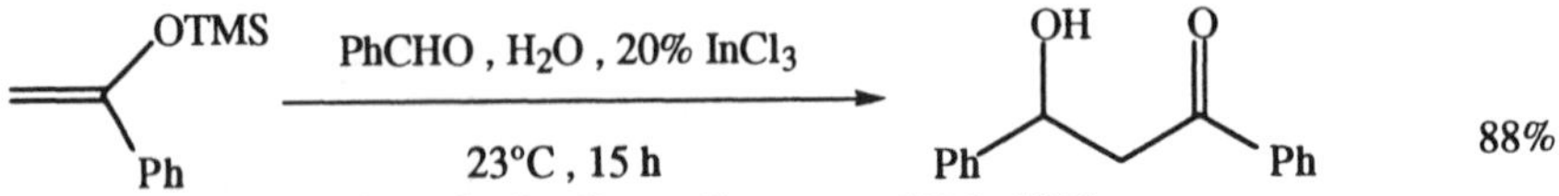
OTMS
Ph
PhCHO , H_2O , 20% $InCl_3$
23°C , 15 h
OH
O
Ph
88%

Loh, T.-P.; Pei, J.; Cao, G.-Q. *Chem. Commun.,* ***1996***, 1819.

SECTION 331: ALCOHOL, THIOL - NITRILE

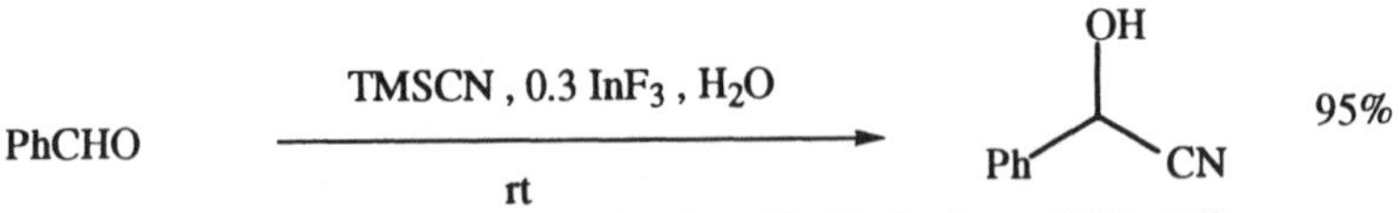
PhCHO
TMSCN , 0.3 InF_3 , H_2O
rt
OH
Ph
CN
95%

Loh, T.-P.; Xu, K.-C.; Ho, D.S.-C.; Sim, K.-Y. *SynLett,* ***1998***, 369.

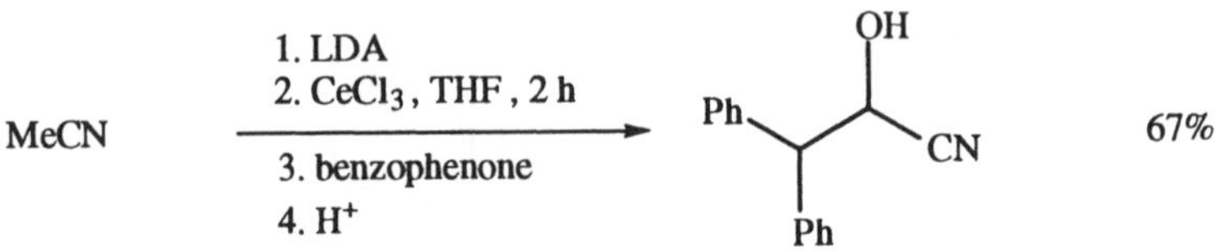
MeCN
1. LDA
2. $CeCl_3$, THF , 2 h
3. benzophenone
4. H^+
OH
Ph
CN
Ph
67%

Xiao, Z.; Timberlake, J.W. *Tetrahedron,* ***1998***, *54*, 4211.

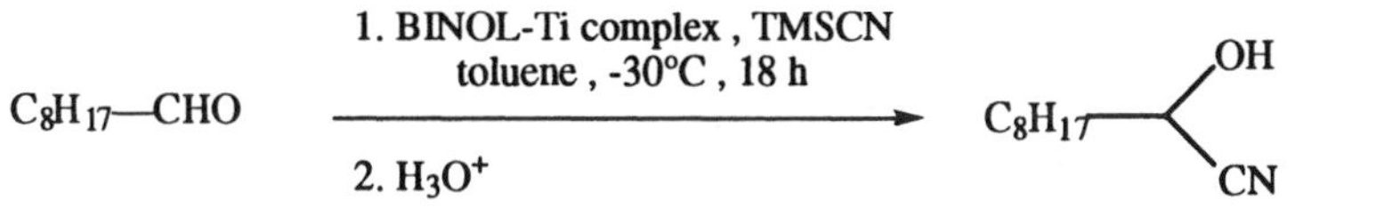
1. BINOL-Ti complex , TMSCN
toluene , -30°C , 18 h
2. H3O+
C8H17—CHO
OH
C8H17
CN

72% (78% ee , *S*)

Mori, M.; Imma, H.; Nakai, T. *Tetrahedron Lett.*, ***1997***, *38*, 6229.

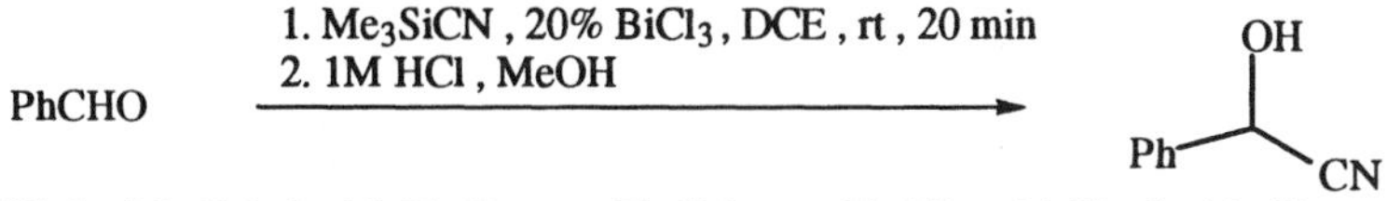
1. Me3SiCN , 20% BiCl3 , DCE , rt , 20 min
2. 1M HCl , MeOH
PhCHO
OH
Ph
CN

quant

Wada, M.; Takahashi, T.; Domae, T.; Fukuma, T.; Miyoshi, N.; Smith, K. *Tetrahedron Asymmetry*, ***1997***, *8*, 3939.

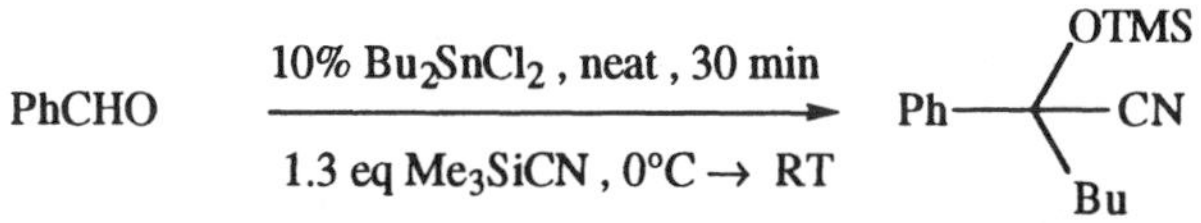
10% Bu2SnCl2 , neat , 30 min
1.3 eq Me3SiCN , 0°C → RT
PhCHO
OTMS
Ph
CN
Bu

90%

Whitesell, J.K.; Apodaca, R. *Tetrahedron Lett.*, ***1996***, *37*, 2525.

SECTION 332: ALCOHOL, THIOL - ALKENE

Allylic and benzylic hydroxylation (C=C-C-H → C=C-C-OH, etc.) is listed in Section 41 (Alcohols from Hydrides).

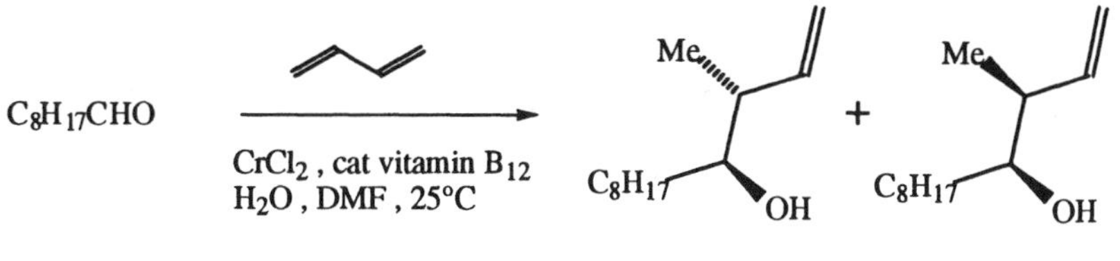
C8H17CHO
CrCl2 , cat vitamin B12
H2O , DMF , 25°C
Me
C8H17
OH
+
Me
C8H17
OH

(72 : 28) 88%

Takai, K.; Toratsu, C. *J. Org. Chem.*, ***1998***, *63*, 6450.

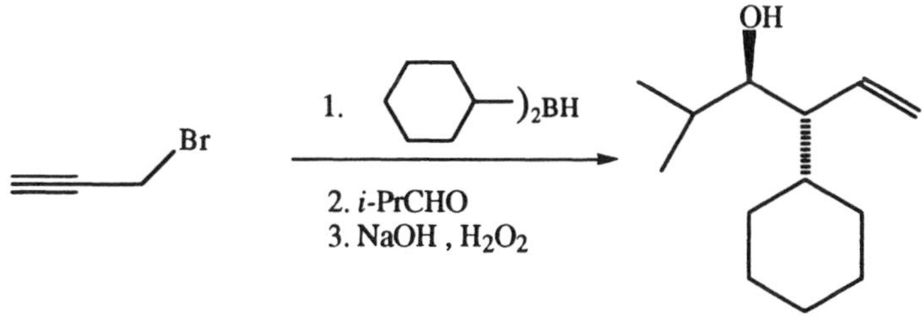
Br
1.
)2BH
2. i-PrCHO
3. NaOH , H2O2
OH

72%

Gaddoni, L.; Lombardo, M.; Trombini, C. *Tetrahedron Lett.*, ***1998***, *39*, 7571.

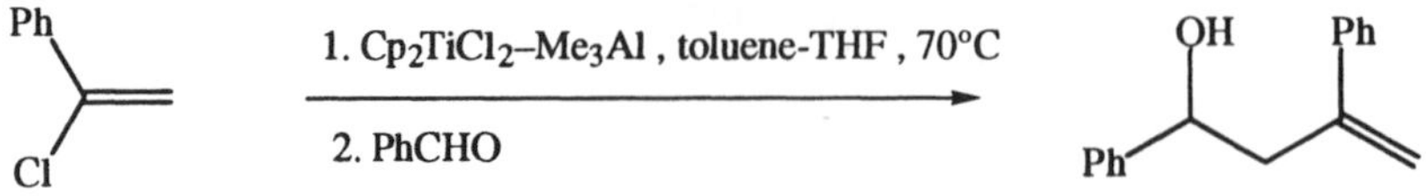

50%

Hanzawa, Y.; Kowase, N.; Taguchi, T. *Tetrahedron Lett.*, ***1998***, *39*, 583.

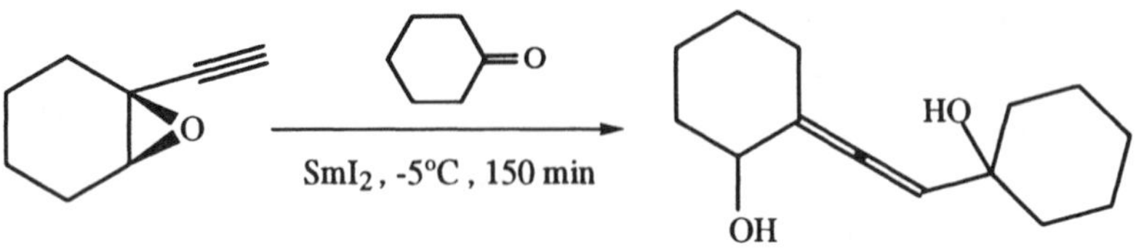

74% (4.9:1 *anti:syn*)

Aurrecoechea, J.M.; Alonso, E.; Solay, M. *Tetrahedron*, ***1998***, *54*, 3833.

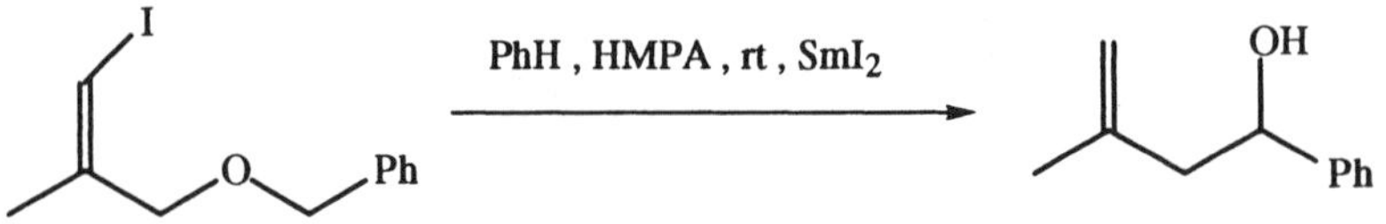

81%

Kunishima, M.; Hioki, K.; Kono, K.; Kato, A.; Tani, S. *J. Org. Chem.*, ***1997***, *62*, 7542.

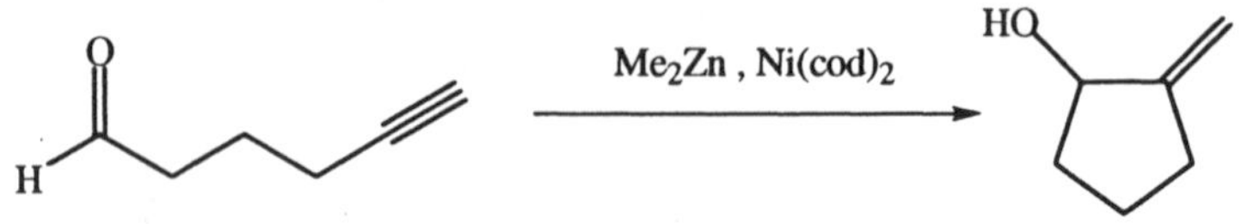

70%

Oblinger, E.; Montogomery, J. *J. Am. Chem. Soc.*, ***1997***, *119*, 9065.

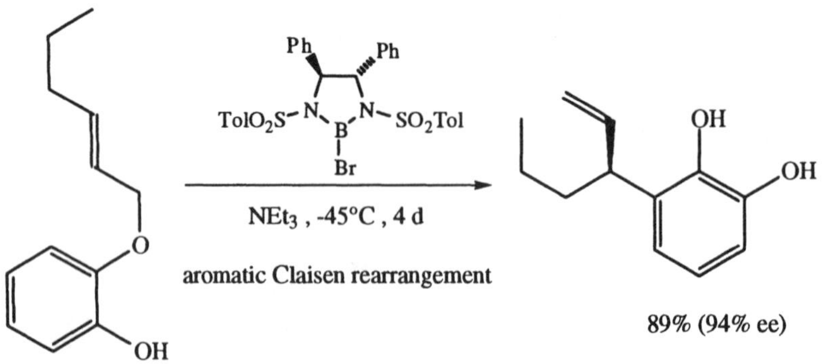

89% (94% ee)

Ito, H.; Sato, A.; Taguchi, T. *Tetrahedron Lett.*, ***1997***, *38*, 4815.

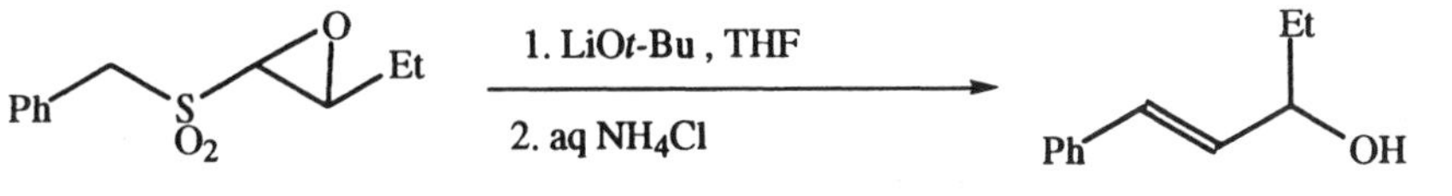
Ph
S
O2
O
Et
1. LiOt-Bu , THF
2. aq NH4Cl
Et
Ph
OH

75% (9:1 *E:Z*)

Evans, P.; Taylor, R.J.K. *Tetrahedron Lett.*, ***1997***, *38*, 3055.

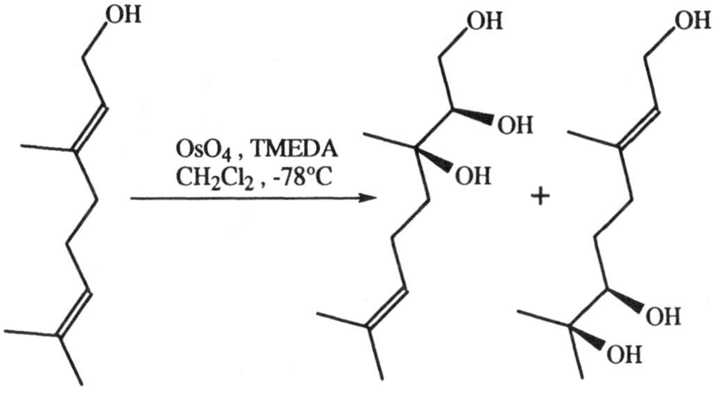
OH
OsO4 , TMEDA
CH2Cl2 , -78°C
OH
OH
OH
+
OH
OH
OH

(>25 : 1) 74%

Donohoe, T.J.; Moore, P.R.; Waring, M.J.; Newcombe, N.J. *Tetrahedron Lett.*, ***1997***, *38*, 5027

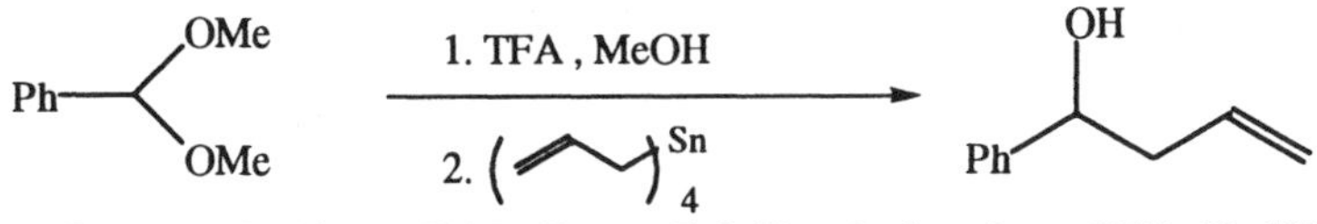
OMe
Ph
OMe
1. TFA , MeOH
2.
Sn
4
OH
Ph

quant

McCluskey, A.; Mayer, D.M.; Young, D.J. *Tetrahedron Lett.*, ***1997***, *38*, 5217.

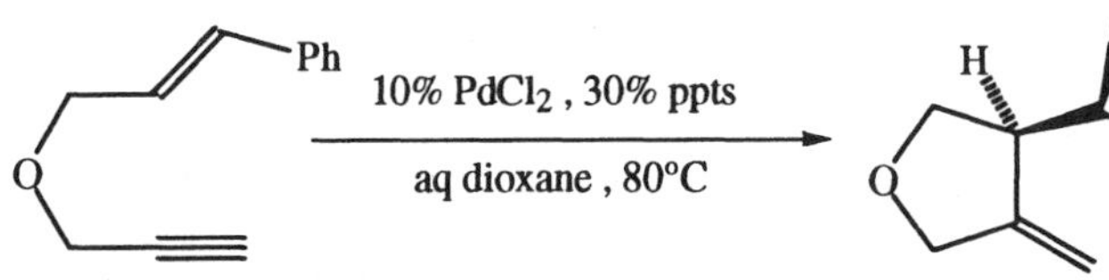
Ph
O
10% PdCl2 , 30% ppts
aq dioxane , 80°C
H
OH
Ph
O

85%

Galland, J.–C.; Savignac, M.; Genêt, J.–P. *Tetrahedron Lett.*, ***1997***, *38*, 8695.

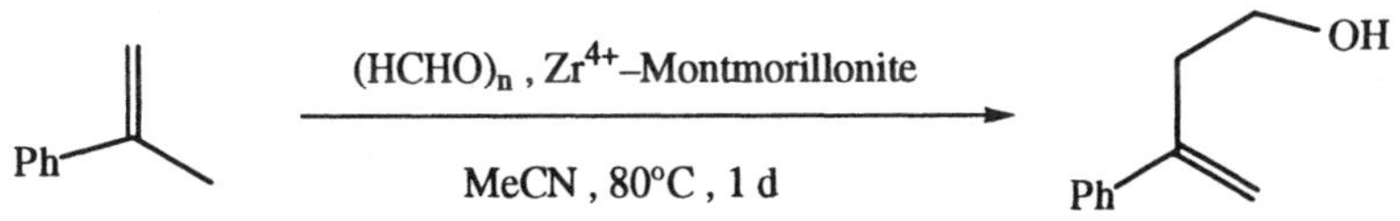
Ph
(HCHO)n , Zr4+–Montmorillonite
MeCN , 80°C , 1 d
OH
Ph

47%

Tateiwa, J.–i.; Kimura, A.; Takasuka, M.; Uemura, S.
J. Chem. Soc.. Perkin Trans. 1. ***1997***. 2169.

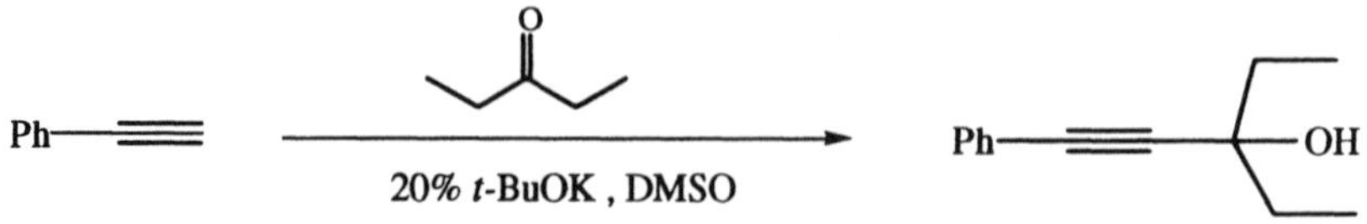
Ph
O
20% t-BuOK , DMSO
Ph
OH

70%

Babler, J.H.; Liptak, V.P.; Phan, N. *J. Org. Chem.*, ***1996***, *61*, 416.

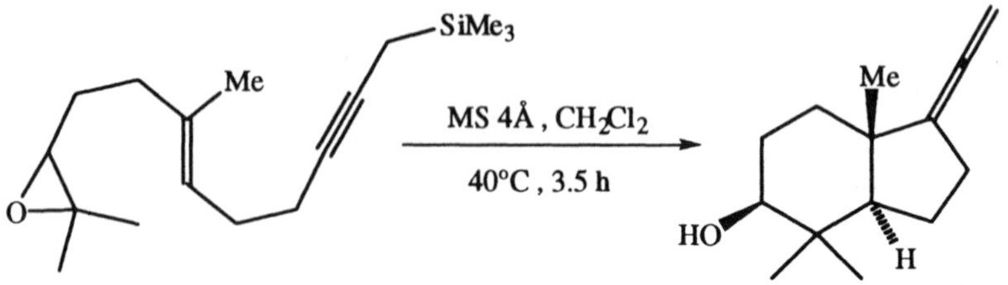
SiMe3
Me
O
MS 4Å , CH2Cl2
40°C , 3.5 h
Me
HO
H

quant

Sen, S.E.; Zhang, Y.Z.; Roach, S.L. *J. Org. Chem.*, ***1996***, *61*, 9534.

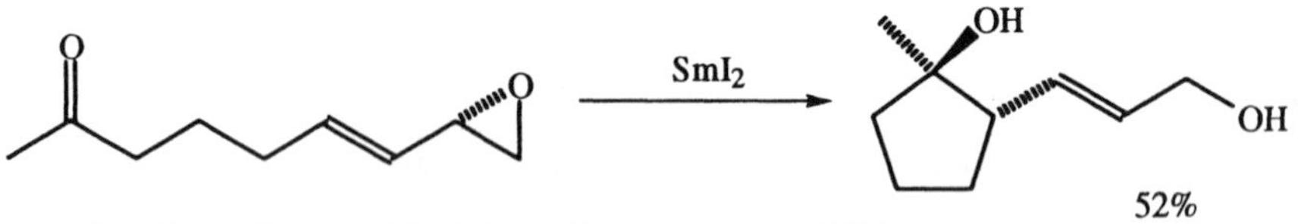
O
O
SmI2
OH
OH

52%

Molander, G.A.; Shakya, S.R. *J. Org. Chem.*, ***1996***, *61*, 5885.

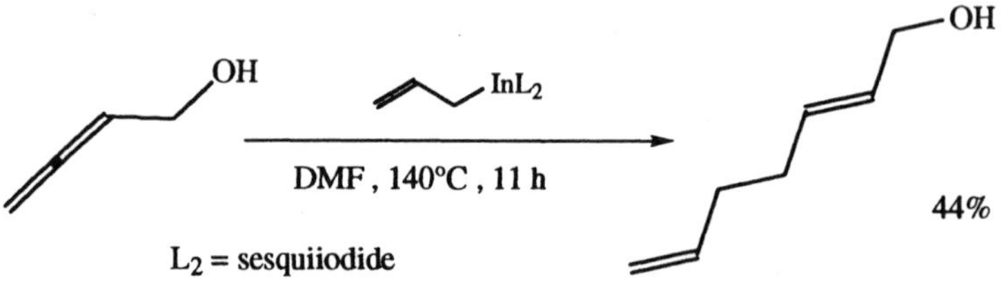
OH
InL2
DMF , 140°C , 11 h
OH

44%

L_2 = sesquiiodide

Araki, S.; Usui, H.; Kato, M.; Butsugan, Y. *J. Am. Chem. Soc.*, ***1996***, *118*, 4699.

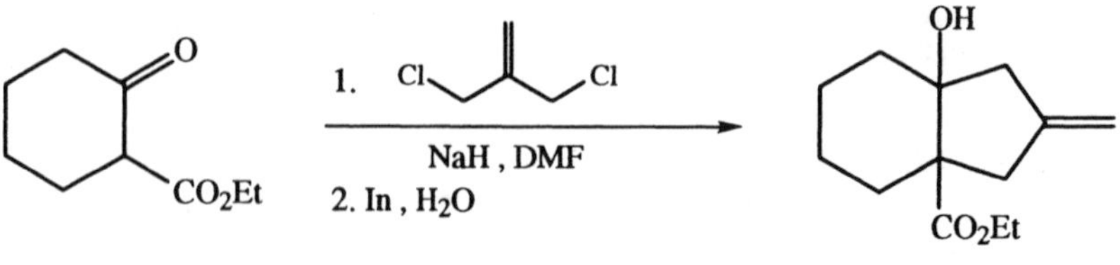
O
CO2Et
1. Cl Cl
NaH , DMF
2. In , H2O
OH
CO2Et

65x98%

Lu, Y.–Q.; Li, C.–J. *Tetrahedron Lett*, ***1996***, *37*, 471.

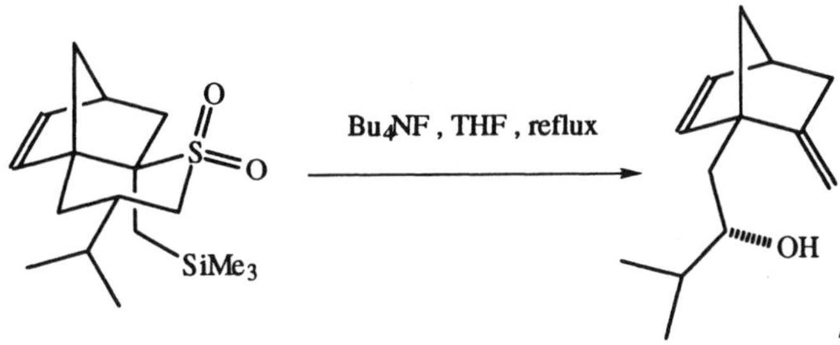
O
S
O
SiMe3
Bu4NF, THF, reflux
OH

53%

Metz, P.; Seng, D.; Plietker, B. *Tetrahedron Lett.*, ***1996***, *37*, 3841.

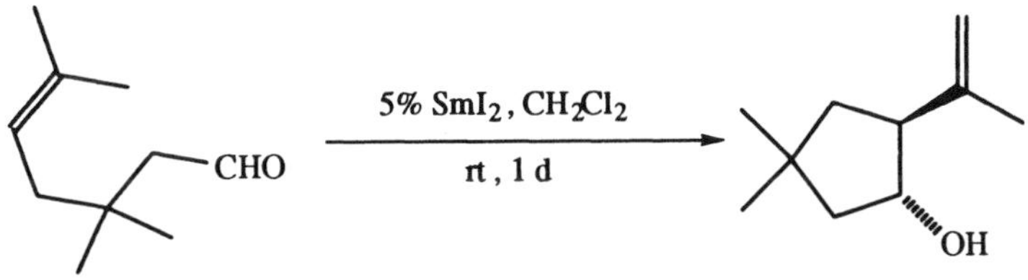
CHO
5% SmI2, CH2Cl2
rt, 1 d
OH

60%

Sarkar, T.K.; Nandy, S.K. *Tetrahedron Lett.*, ***1996***, *37*, 5195.

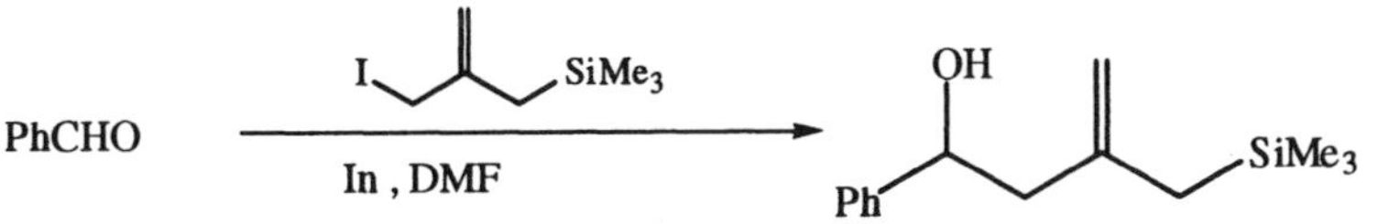
PhCHO
I
SiMe3
In, DMF
OH
Ph
SiMe3

90%

Bardot, V.; Remuson, R.; Gelas–Mialhe, Y.; Gramain, J.–C. *SynLett*, ***1996***, 37.

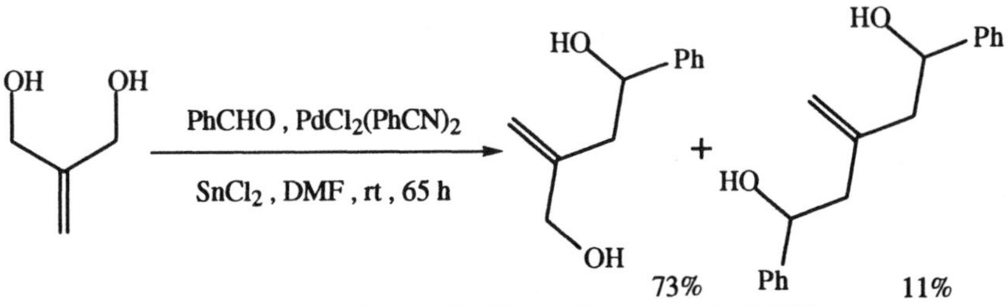
OH
OH
PhCHO, PdCl2(PhCN)2
SnCl2, DMF, rt, 65 h
HO
Ph
OH
73%
+
HO
Ph
HO
Ph
11%

Masuyama, Y.; Kagawa, M.; Kurusu, Y. *Chem. Commun.*, ***1996***, 585.

REVIEW:

"The Baylis-Hillman Reaction. A Novel Carbon–Carbon Bond Forming Reaction."
Basavaiah, D.; Rao, P.D.; Hyma, R.S. *Tetrahedron*, ***1996***, *52*, 8001.

Also via: Section 302 (Alkyne - Alcohol).

SECTION 333: ALDEHYDE - ALDEHYDE

NO ADDITIONAL EXAMPLES

SECTION 334: ALDEHYDE - AMIDE

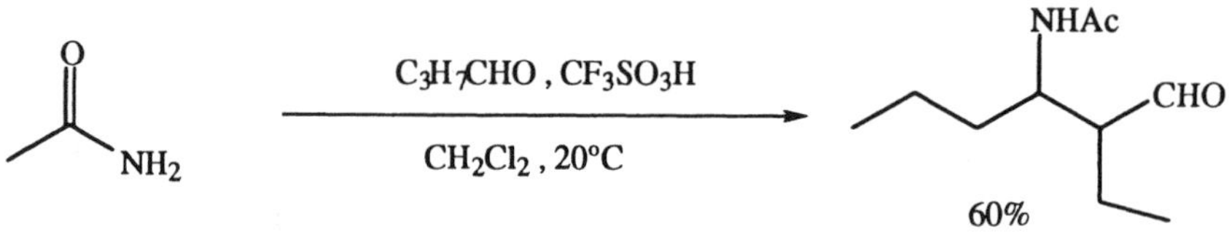

NH2
C_3H_7CHO , CF_3SO_3H
CH_2Cl_2 , 20°C
NHAc
CHO
60%

Marson, C.M.; Fallah, A. *Chem. Commun.*, ***1998***, 83.

SECTION 335: ALDEHYDE - AMINE

NO ADDITIONAL EXAMPLES

SECTION 336: ALDEHYDE - ESTER

NO ADDITIONAL EXAMPLES

SECTION 337: ALDEHYDE - ETHER, EPOXIDE, THIOETHER

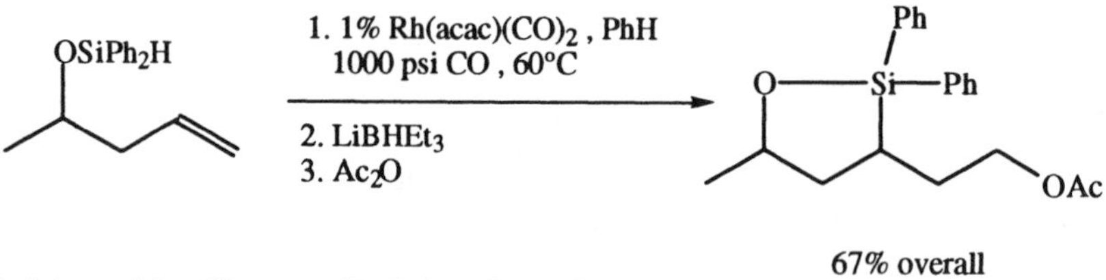

OSiPh2H
1. 1% Rh(acac)(CO)2 , PhH
1000 psi CO , 60°C
2. LiBHEt3
3. Ac2O
Ph
Si
Ph
O
OAc
67% overall

Leighton, J.L.; Chapman, E. *J. Am. Chem. Soc.*, ***1997***, *119*, 12416.

SECTION 338: ALDEHYDE - HALIDE, SULFONATE

NO ADDITIONAL EXAMPLES

SECTION 339: ALDEHYDE - KETONE

NO ADDITIONAL EXAMPLES

SECTION 340: ALDEHYDE - NITRILE

NO ADDITIONAL EXAMPLES

SECTION 341: ALDEHYDE - ALKENE

For the oxidation of allylic alcohols to alkene aldehydes, also see Section 48 (Aldehydes from Alcohols).

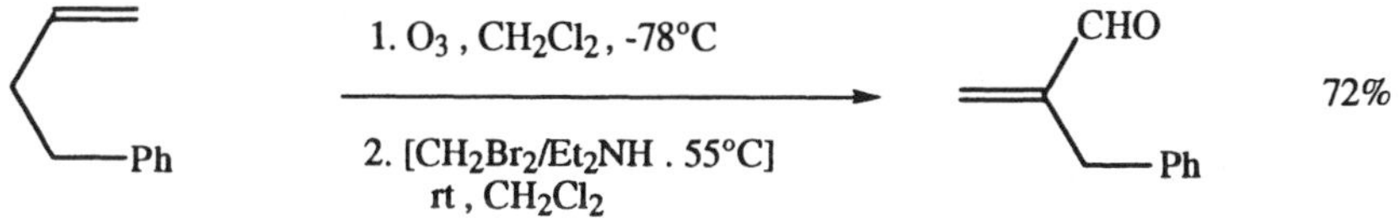
Ph
1. O_3 , CH_2Cl_2 , -78°C
2. [CH_2Br_2/Et_2NH . 55°C]
rt , CH_2Cl_2
CHO
Ph
72%

Hon, Y.-S.; Chang, F.-J.; Lu, L.; Lin, W.-C. *Tetrahedron*, ***1998***, *54*, 5233.

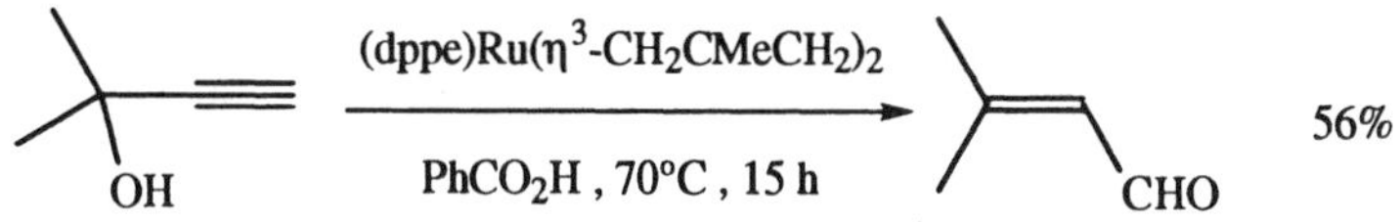
OH
(dppe)Ru(η^3-CH_2CMeCH_2)$_2$
$PhCO_2H$, 70°C , 15 h
CHO
56%

Picquet, M.; Bruneau, C.; Dixneuf, P.H. *Chem. Commun.*, ***1997***, 1201.

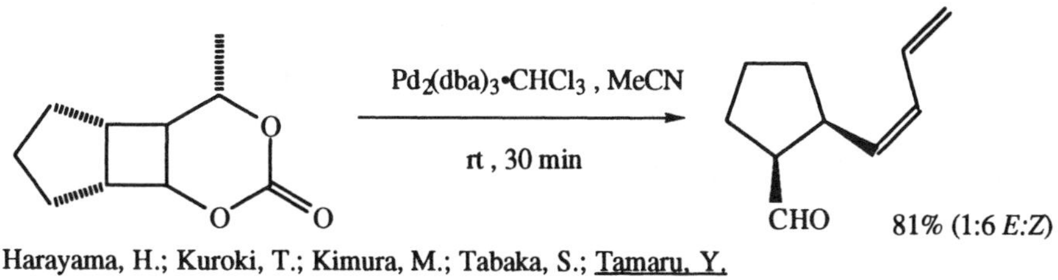
O
O
O
$Pd_2(dba)_3$•$CHCl_3$, MeCN
rt , 30 min
CHO
81% (1:6 *E:Z*)

Harayama, H.; Kuroki, T.; Kimura, M.; Tabaka, S.; Tamaru, Y. *Angew. Chem. Int. Ed.*, ***1997***, *36*, 2352.

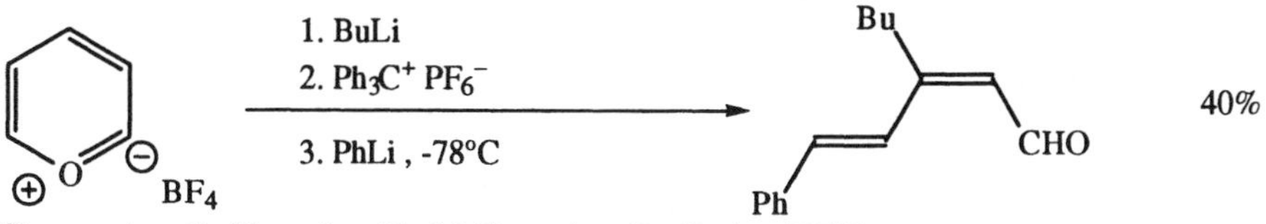
O
BF_4
1. BuLi
2. Ph_3C^+ PF_6^-
3. PhLi , -78°C
Bu
CHO
Ph
40%

Charoenying, P.; Hemming, K.; McKerrecher, D.; Taylor, T.J.K. *J. Heterocyclic Chem.*, ***1996***, *33*, 1083.

Also via β-Hydroxy aldehydes: Section 324 (Alcohols - Aldehyde).

SECTION 342: AMIDE - AMIDE

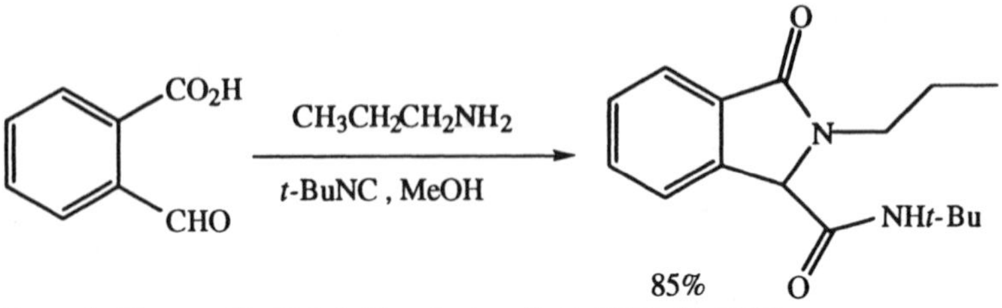
CO2H
CHO
CH3CH2CH2NH2
t-BuNC , MeOH
NHt-Bu

85%

Hanusch–Kompa, C.; Ugi, I. *Tetrahedron Lett*, ***1998***, *39*, 2725.

Also via Dicarboxylic Acids:	Section 312 (Carboxylic Acid - Carboxylic Acid)
Diamines	Section 350 (Amines - Amines)

SECTION 343: AMIDE - AMINE

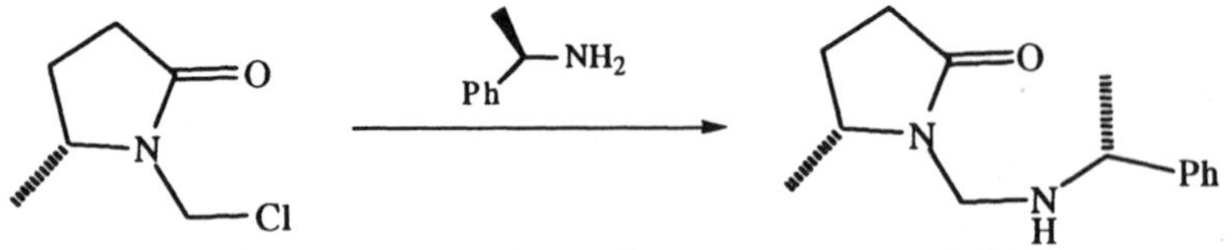
NH2
Ph
Cl
Ph

75%

Chen, P.; Tao, S.; Smith, M.B. *Synth. Commun.*, ***1998***, *28*, 1641.

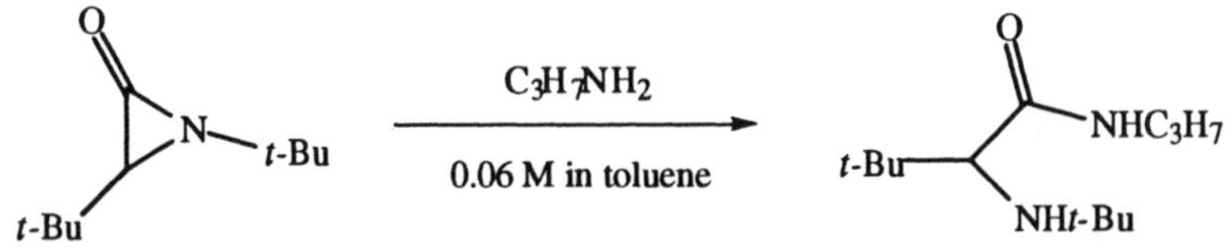
t-Bu
t-Bu
C3H7NH2
0.06 M in toluene
NHC3H7
t-Bu
NHt-Bu

88%

Talaty, E.R.; Yusoff, M.M. *Chem. Commun.*, ***1998***, 985.

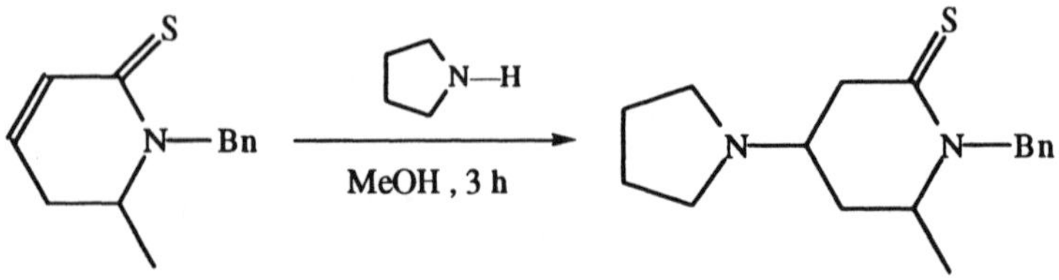
Bn
N—H
MeOH , 3 h
Bn

73% (85:15 *trans:cis*)

Sośnicki, J.G.; Jagodziński, T.S.; Liebscher, J. *J. Heterocyclic Chem.*, ***1997***, *34*, 643.

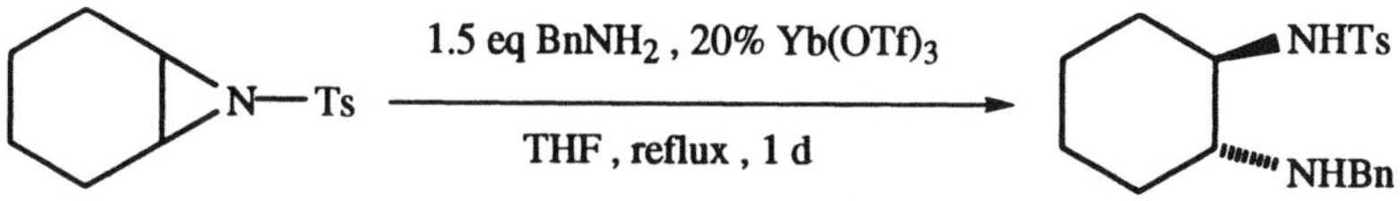
N—Ts
1.5 eq BnNH2 , 20% Yb(OTf)3
THF , reflux , 1 d
NHTs
NHBn

quant

Meguro, M.; Yamamoto, Y. *Heterocycles*, ***1996***, *43*, 2473.

SECTION 344: AMIDE - ESTER

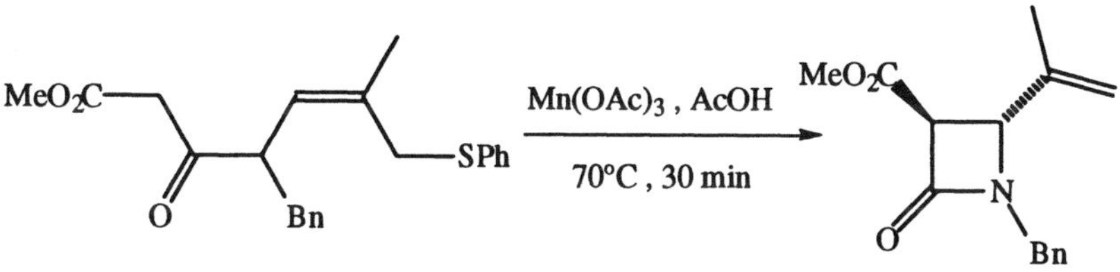
MeO2C
O
Bn
SPh
Mn(OAc)3 , AcOH
70°C , 30 min
MeO2C
N
O
Bn

43%

Attenni, B.; Cerreti, A.; D'Annibale, A.; Resta, S.; Trogolo, C. *Tetrahedron*, ***1998***, *54*, 12029

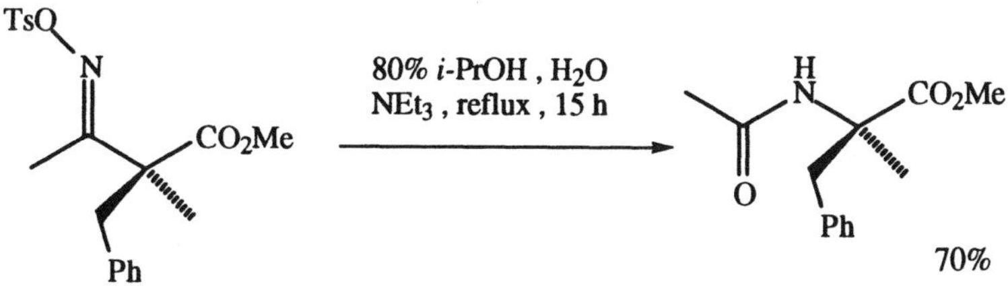
TsO
N
CO2Me
Ph
80% i-PrOH , H2O
NEt3 , reflux , 15 h
H
N
CO2Me
O
Ph

70%

Frutos, R.P.; Spero, D.M. *Tetrahedron Lett.*, ***1998***, *39*, 2475.

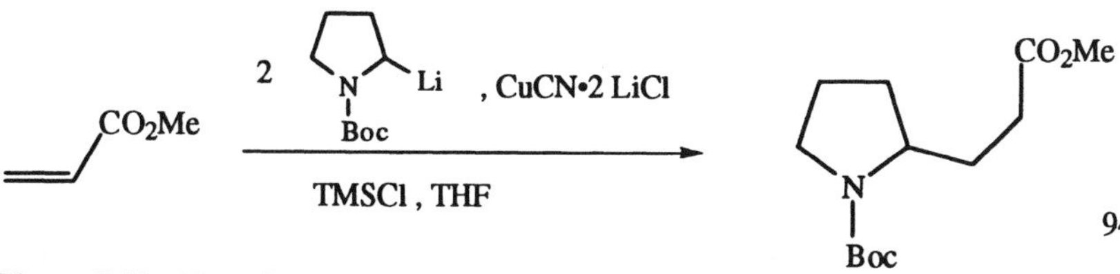
CO2Me
2
N
Li
Boc
, CuCN•2 LiCl
TMSCl , THF
CO2Me
N
Boc

94%

Dieter, R.K.; Velu, S.E. *J. Org. Chem.*, ***1997***, *62*, 3798.

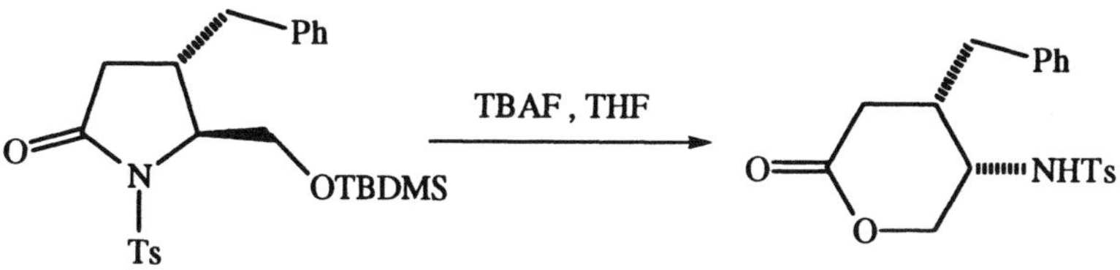
Ph
O
N
Ts
OTBDMS
TBAF , THF
Ph
O
NHTs
O

98%

Cossy, J.; Cases, M.; Pardo, D.G. *Tetrahedron Lett.*, ***1996***, *37*, 8173.

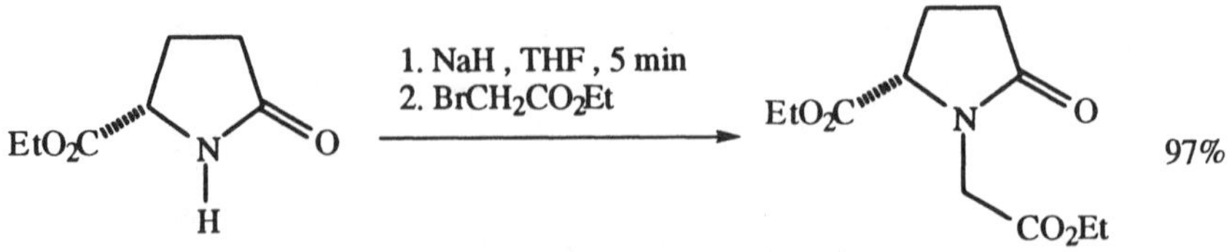
EtO2C
N
H
O
1. NaH , THF , 5 min
2. BrCH2CO2Et
EtO2C
N
O
CO2Et
97%

Simandan, T.; Smith, M.B. *Synth. Commun.*, ***1996***, *26*, 1827.

Related Methods:

Section 315 (Carboxylic Acid - Amide)
Section 316 (Carboxylic Acid - Amine)
Section 351 (Amine - Ester)

SECTION 345: AMIDE - ETHER, EPOXIDE, THIOETHER

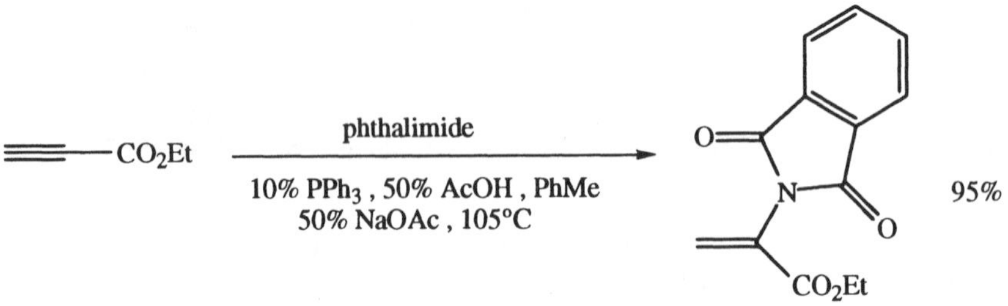
CO2Et
phthalimide
10% PPh3 , 50% AcOH , PhMe
50% NaOAc , 105°C
O
N
O
CO2Et
95%

Trost, B.M.; Dake, G.R. *J. Am. Chem. Soc.*, ***1997***, *119*, 7595.

SECTION 346: AMIDE - HALIDE, SULFONATE

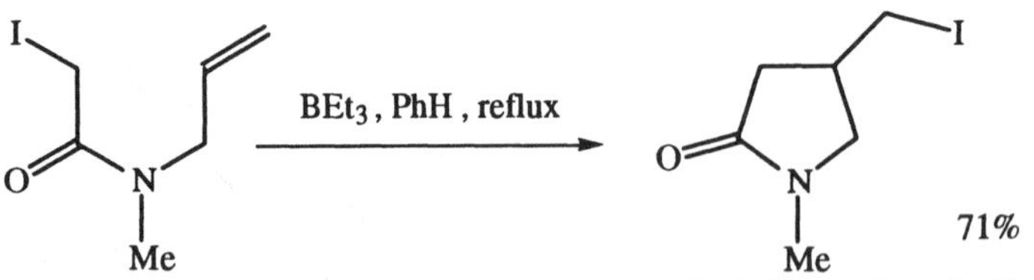
I
O
N
Me
BEt3 , PhH , reflux
I
O
N
Me
71%

Ikeda, M.; Teranishi, H.; Nozaki, K.; Ishibashi, H. *J. Chem. Soc., Perkin Trans. 1*, ***1998***, 1691

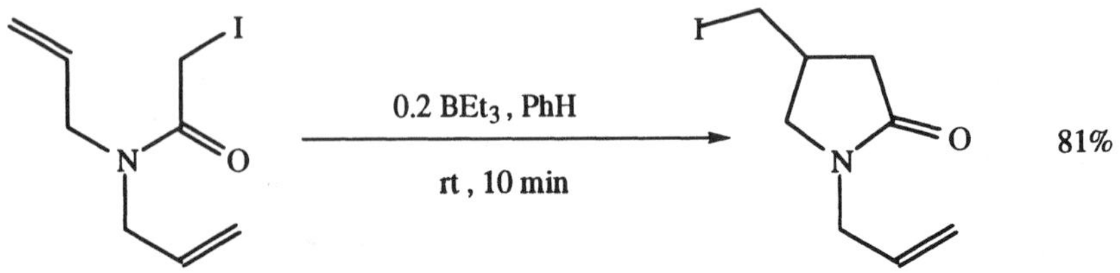
I
N
O
0.2 BEt3 , PhH
rt , 10 min
I
N
O
81%

Ikeda, M.; Terahishi, H.; Iwamura, N.; Ishibashi, H. *Heterocycles*, ***1997***, *45*, 863.

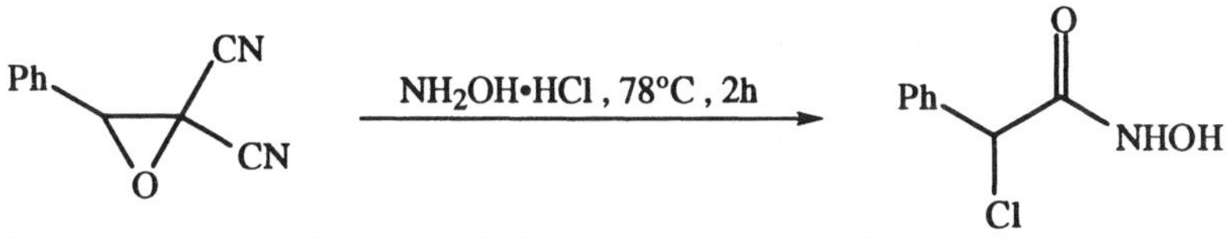
Ph
CN
CN
O
$NH_2OH•HCl$, 78°C , 2h
Ph
O
NHOH
Cl

75%

Said, B.; Abdelaziz, S.; Albert, R. *Tetrahedron Lett.*, ***1996***, *37*, 179.

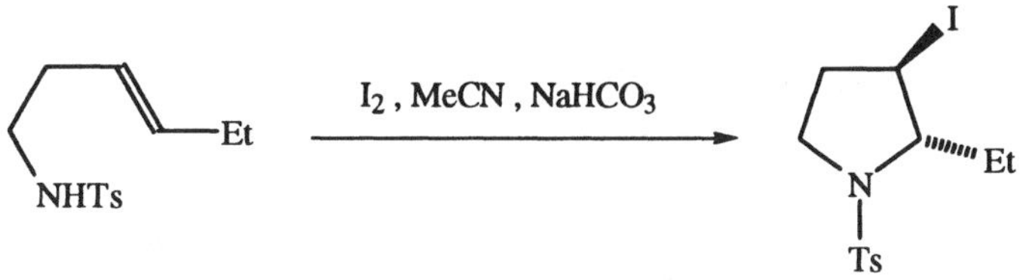
Et
NHTs
I_2 , MeCN , $NaHCO_3$
I
Et
N
Ts

95%

Jones, A.D.; Knight, D.W. *Chem. Commun.*, ***1996***, 915.

SECTION 347: AMIDE - KETONE

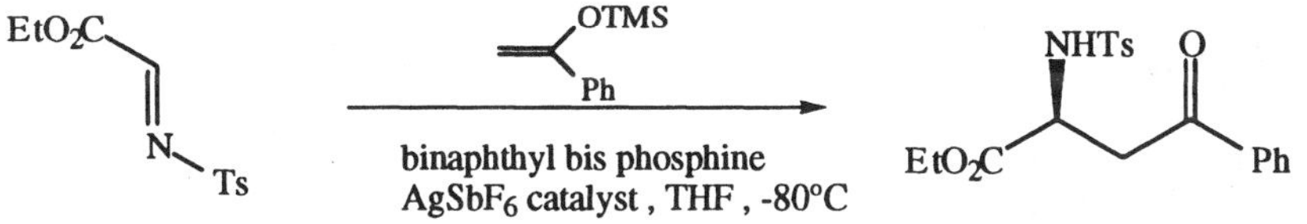
EtO_2C
N
Ts
OTMS
Ph
binaphthyl bis phosphine
$AgSbF_6$ catalyst , THF , -80°C
NHTs
O
EtO_2C
Ph

95% (90% ee)

Ferraris, D.; Young, B.; Dudding, T.; Kectka, T. *J. Am. Chem. Soc.*, ***1998***, *120*, 4548.

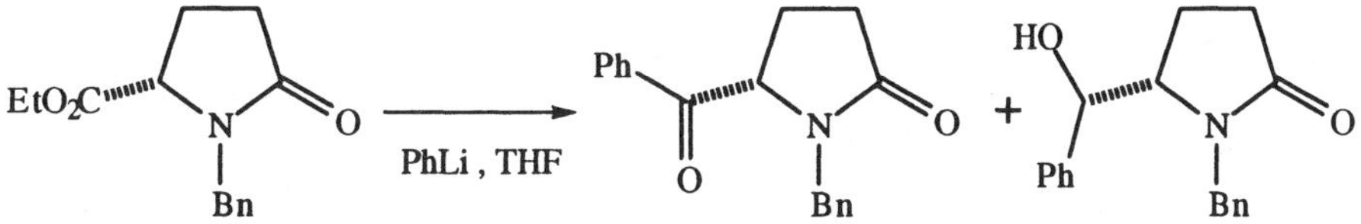
EtO_2C
N
O
Bn
PhLi , THF
Ph
O
N
O
Bn
+
HO
Ph
N
O
Bn

40% 10%

Deskus, J.; Fan, D.; Smith, M.B. *Synth. Commun.*, ***1998***, *28*, 1649.

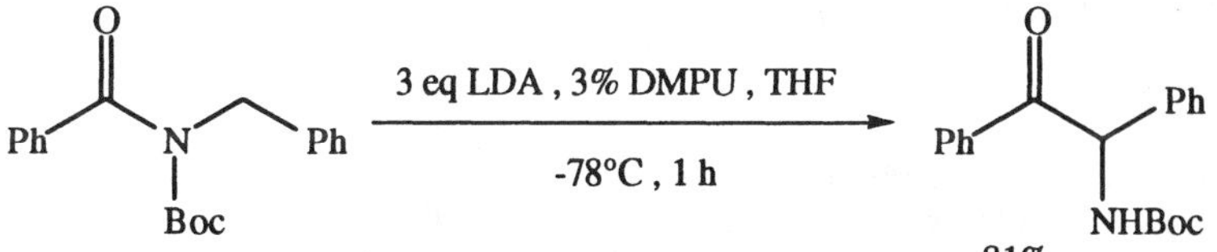
O
Ph
N
Ph
Boc
3 eq LDA , 3% DMPU , THF
-78°C , 1 h
O
Ph
Ph
NHBoc

81%

Hara, O.; Ito, M.; Hamada, Y. *Tetrahedron Lett*, ***1998***, *39*, 5537.

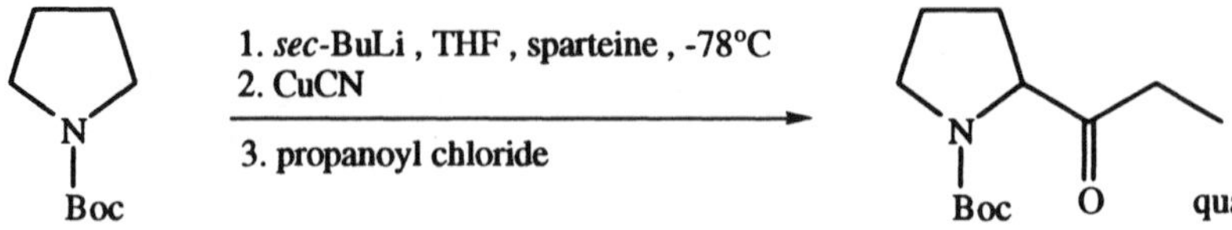

quant

Dieter, R.K.; Sharma, R.R.; Ryan, W. *Tetrahedron Lett.*, ***1997***, *38*, 783.

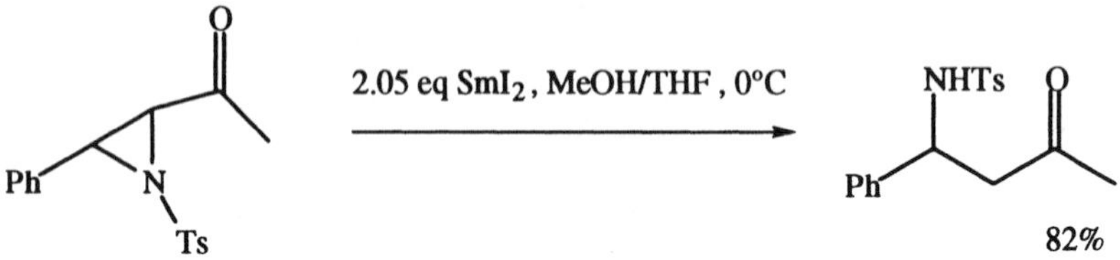

82%

Molander, G.A.; Stengel, P.J. *Tetrahedron*, ***1997***, *53*, 8887.

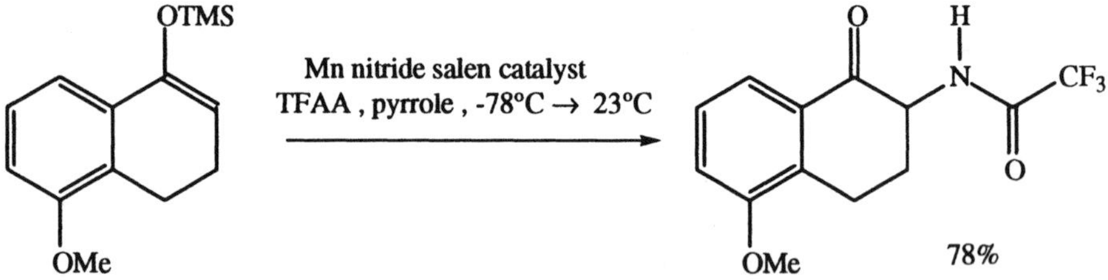

78%

DuBois, J.; Hong, J.; Carreira, E.M.; Day, M.W. *J. Am. Chem. Soc.*, ***1996***, *118*, 915.

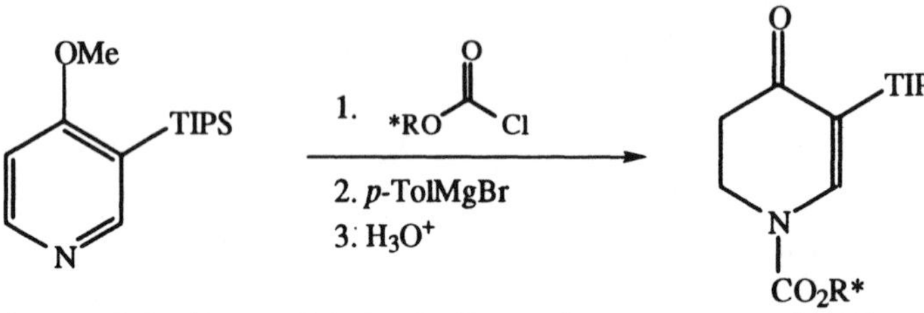

89% (89% de)

Comins, D.L.; Guerra–Weltzien, L. *Tetrahedron Lett*, ***1996***, *37*, 3807.

SECTION 348: AMIDE - NITRILE

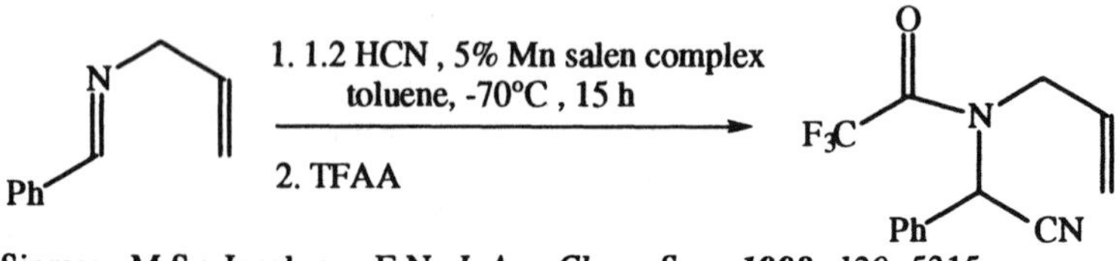

95%

Sigman, M.S.; Jacobsen, E.N. *J. Am. Chem. Soc.*, ***1998***, *120*, 5315.

SECTION 349: AMIDE - ALKENE

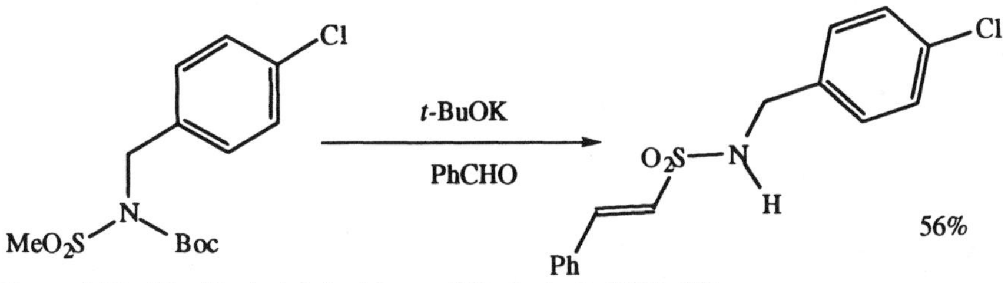
Cl
t-BuOK
PhCHO
MeO2S
N
Boc
O2S
N
H
Ph
Cl
56%

Tozer, M.J.; Woolford, A.J.A.; Linney, I.D. *SynLett, **1998***, 186.

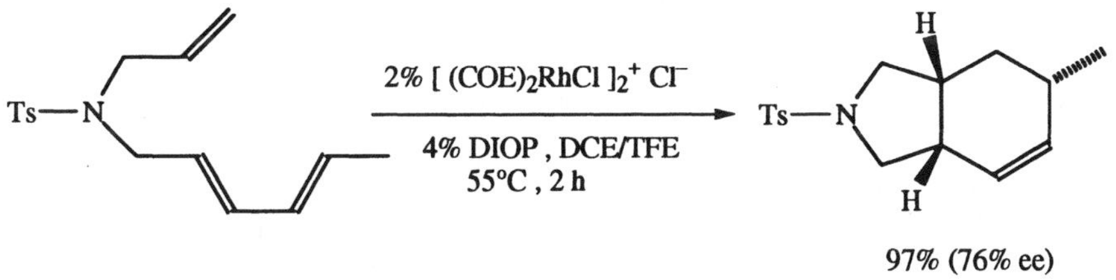
Ts
N
2% [(COE)2RhCl]2+ Cl−
4% DIOP, DCE/TFE
55°C, 2 h
Ts
N
H
H
97% (76% ee)

O'Mahony, D.J.R.; Belanger, D.B.; Livinghouse, T. *SynLett, **1998***, 443.

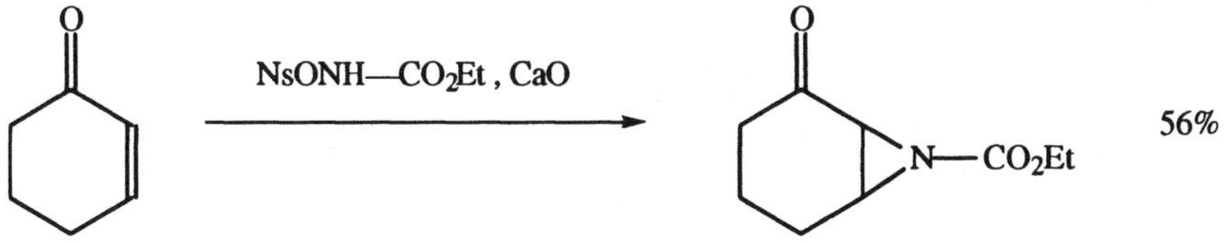
O
NsONH—CO2Et, CaO
O
N—CO2Et
56%

Fioravanti, S.; Pellacani, L.; Tabanella, S.; Tardella, P.A. *Tetrahedron, **1998**, 54*, 14105.

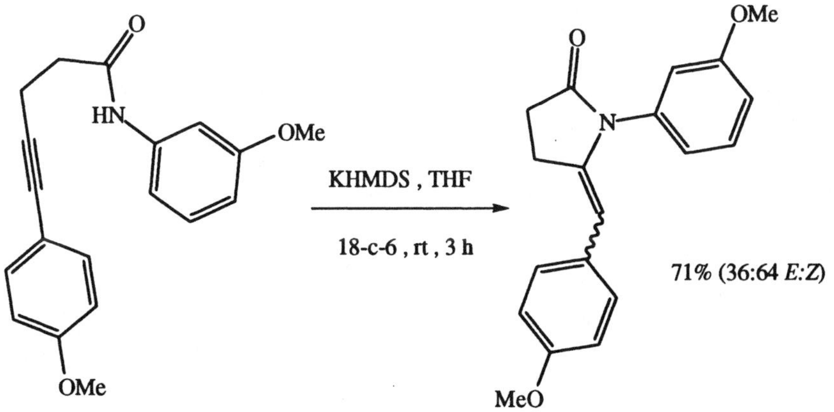
O
HN
OMe
OMe
KHMDS, THF
18-c-6, rt, 3 h
O
N
OMe
MeO
71% (36:64 E:Z)

Koseki, Y.; Kusano, S.; Nagasaka, T. *Tetrahedron Lett., **1998**, 39*, 3517.

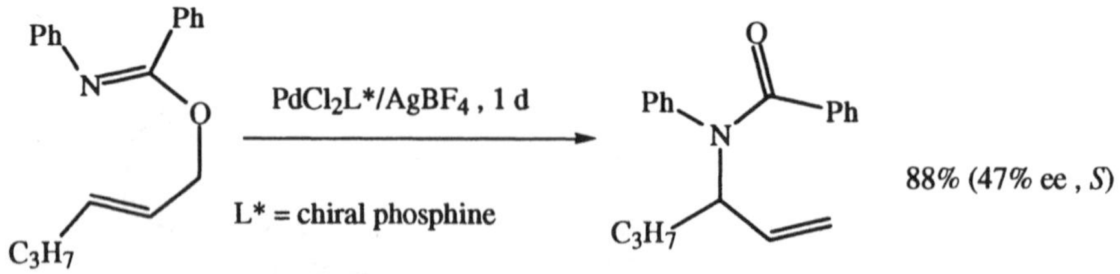
Ph
Ph
N
O
C_3H_7
$PdCl_2L^*/AgBF_4$, 1 d
L* = chiral phosphine
O
Ph
N
Ph
C_3H_7
88% (47% ee, S)

Uozumi, Y.; Kato, K.; Hayashi, T. *Tetrahedron Asymmetry*, ***1998***, *9*, 1065.

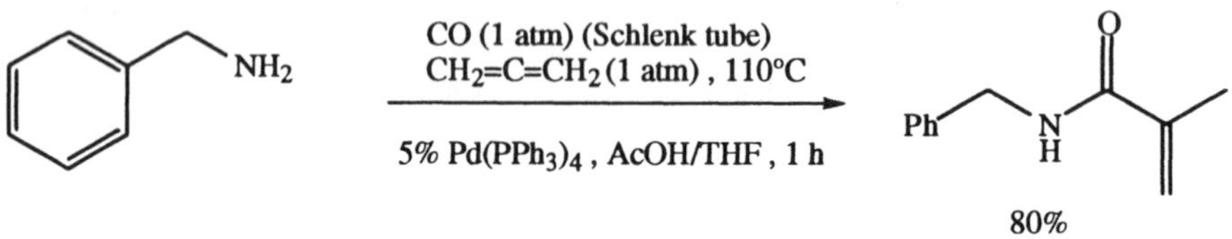
NH_2
CO (1 atm) (Schlenk tube)
$CH_2=C=CH_2$ (1 atm), 110°C
5% $Pd(PPh_3)_4$, AcOH/THF, 1 h
O
Ph
N
H
80%

Grigg, R.; Montieth, M.; Sridharan, V.; Terrier, C. *Tetrahedron*, ***1998***, *54*, 3885.

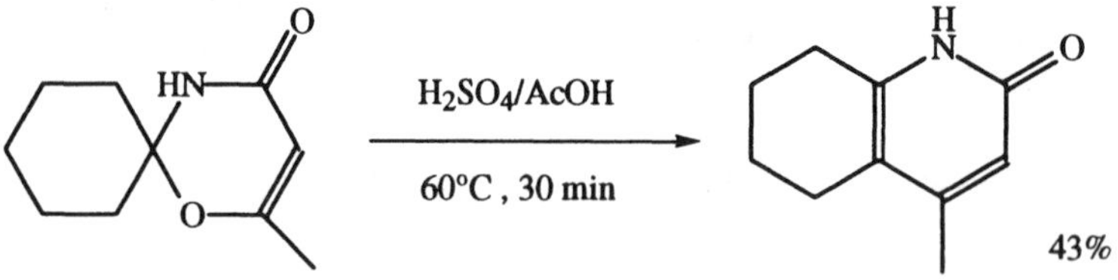
O
HN
O
H_2SO_4/AcOH
60°C, 30 min
H
N
O
43%

Stambach, J.F.; Jung, L.; Hug, R. *Synthesis*, ***1998***, 265.

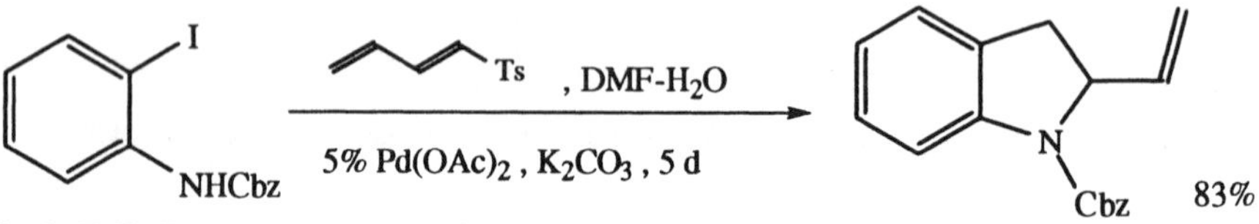
I
NHCbz
Ts
, DMF-H_2O
5% $Pd(OAc)_2$, K_2CO_3, 5 d
N
Cbz
83%

Back, T.G.; Bethell, R.J. *Tetrahedron Lett*, ***1998***, *39*, 5463.

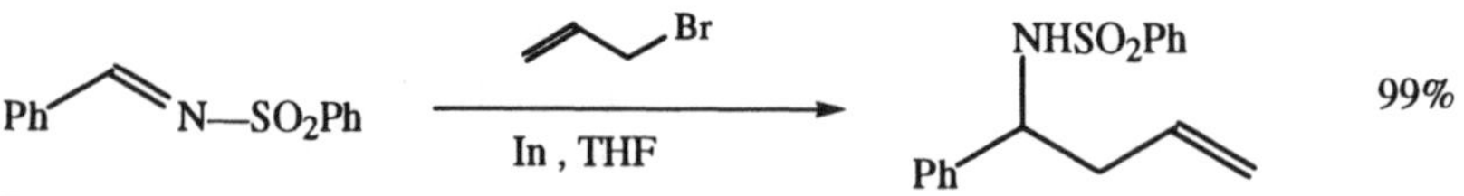
Ph
N—SO_2Ph
Br
In, THF
$NHSO_2Ph$
Ph
99%

Chan, T.H.; Lu, W. *Tetrahedron Lett.*, ***1998***, *39*, 8605.

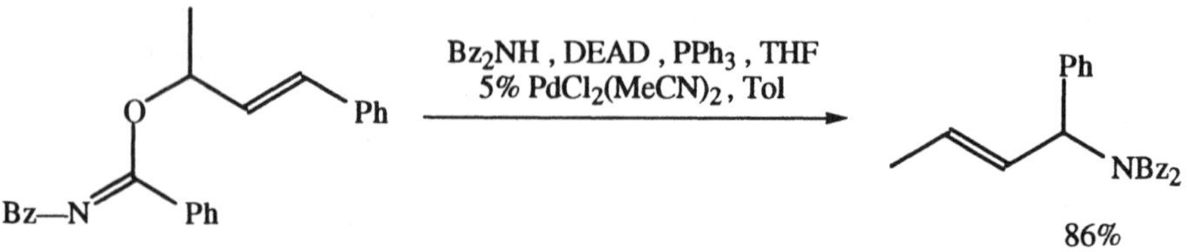
O
Ph
Bz—N
Ph
Bz_2NH, DEAD, PPh_3, THF
5% $PdCl_2(MeCN)_2$, Tol
Ph
NBz_2
86%

Overman, L.E.; Zipp, G.G. *J. Org. Chem.*, ***1997***, *62*, 2288.

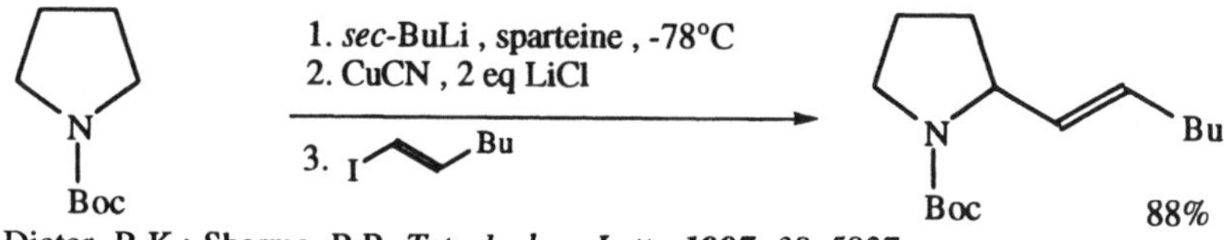
1. *sec*-BuLi , sparteine , -78°C
2. CuCN , 2 eq LiCl
3. I Bu
N
Boc
Bu
88%

Dieter, R.K.; Sharma, R.R. *Tetrahedron Lett.*, ***1997***, *38*, 5937.

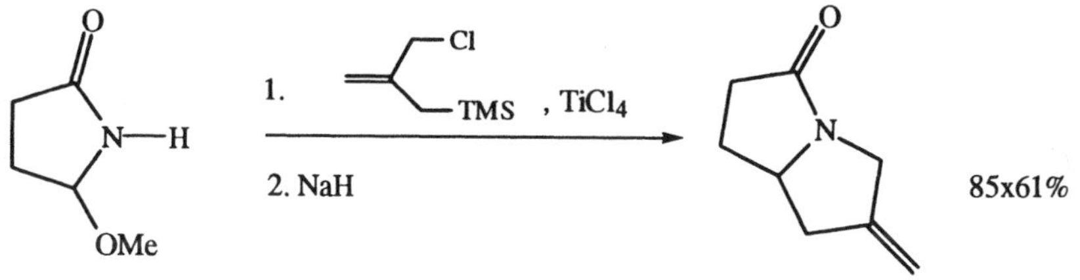
O
N—H
OMe
1. Cl TMS , $TiCl_4$
2. NaH
85x61%

Sadakane, M.; Vahle, R.; Schierle, K.; Kolter, D.; Steckhan, E. *SynLett*, ***1997***, 95.

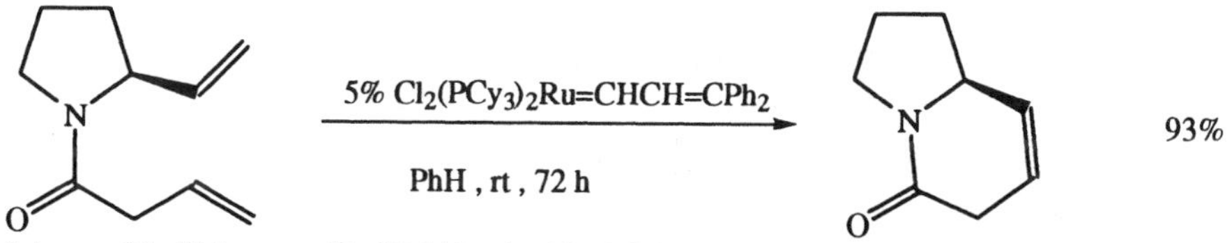
5% $Cl_2(PCy_3)_2Ru$=CHCH=CPh_2
PhH , rt , 72 h
93%

Arisawa, M.; Takezawa, E.; Nishida, A.; Mori, M.; Nakagawa, M. *SynLett*, ***1997***, 1179.

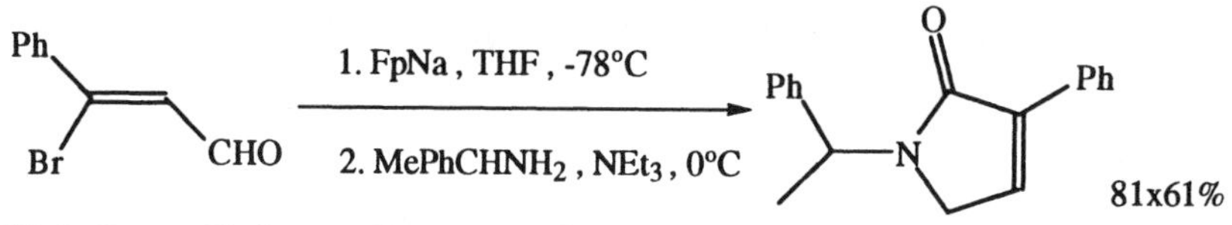
Ph
Br
CHO
1. FpNa , THF , -78°C
2. $MePhCHNH_2$, NEt_3 , 0°C
81x61%

Rück–Braun, K. *Angew. Chem. Int. Ed.*, ***1997***, *36*, 509.

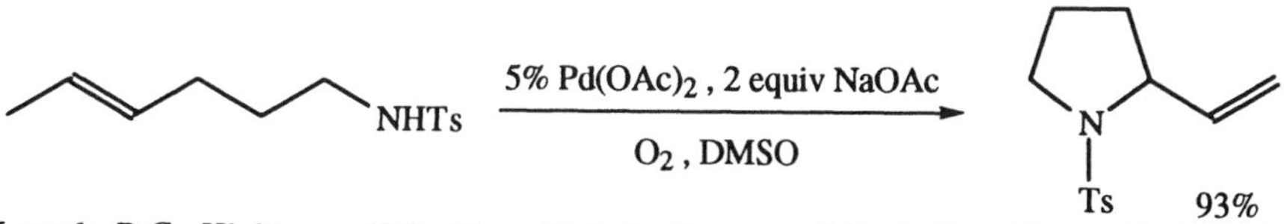
NHTs
5% $Pd(OAc)_2$, 2 equiv NaOAc
O_2 , DMSO
Ts
93%

Larock, R.C.; Hightower, T.R.; Hasvold, L.A.; Peterson, K.P. *J. Org. Chem.*, ***1996***, *61*, 3584

NHBn
Bu_4NF , THF , 66°C
Bn
84%

Jacobi, P.A.; Brielmann, H.L.; Hauck, S.I. *J. Org. Chem.*, ***1996***, *61*, 5013.

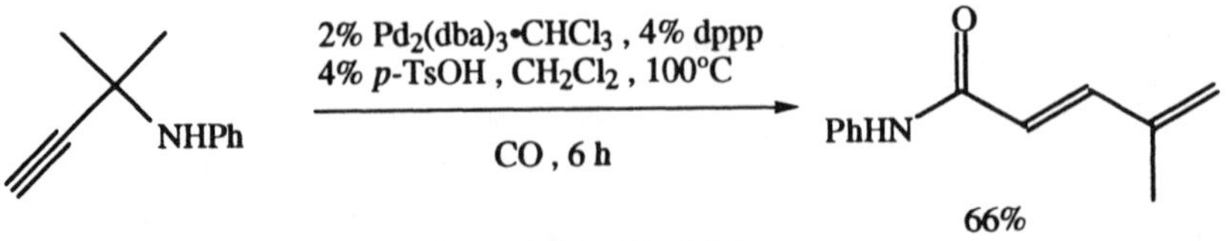
NHPh
2% Pd2(dba)3•CHCl3 , 4% dppp
4% p-TsOH , CH2Cl2 , 100°C
CO , 6 h
O
PhHN
66%

Imada, Y.; Alper, H. *J. Org. Chem.*, ***1996***, *61*, 6766.

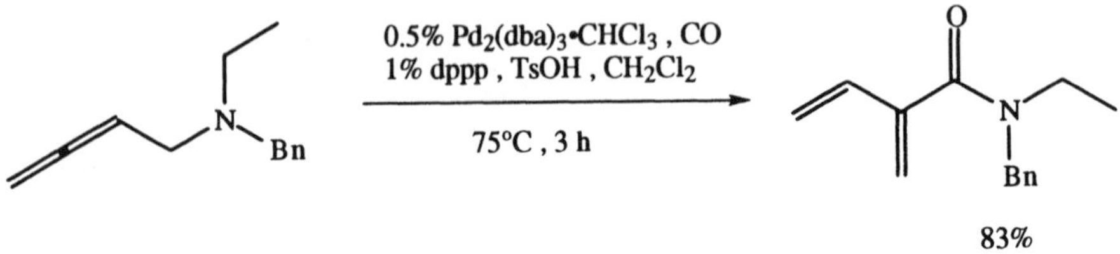
N
Bn
0.5% Pd2(dba)3•CHCl3 , CO
1% dppp , TsOH , CH2Cl2
75°C , 3 h
O
N
Bn
83%

Imada, Y.; Vasapollo, G.; Alper, H. *J. Org. Chem.*, ***1996***, *61*, 7982.

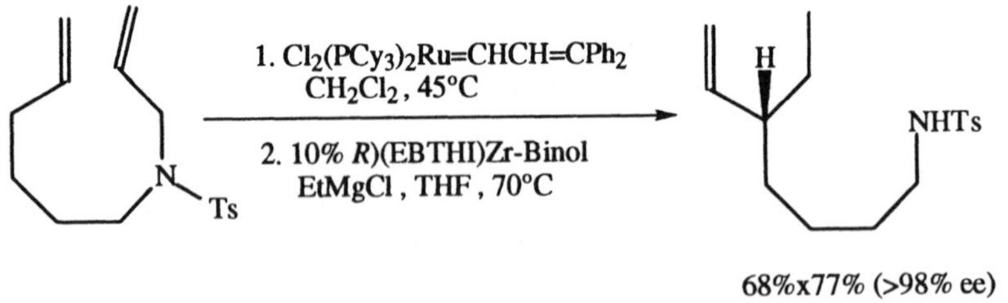
N
Ts
1. Cl2(PCy3)2Ru=CHCH=CPh2
CH2Cl2 , 45°C
2. 10% R)(EBTHI)Zr-Binol
EtMgCl , THF , 70°C
H
NHTs
68%x77% (>98% ee)

Visser, M.S.; Heron, N.M.; Didiuk, M.T.; Sagal, J.F.; Hoveyda, A.H. *J. Am. Chem. Soc.*, ***1996***, *118*, 4291.

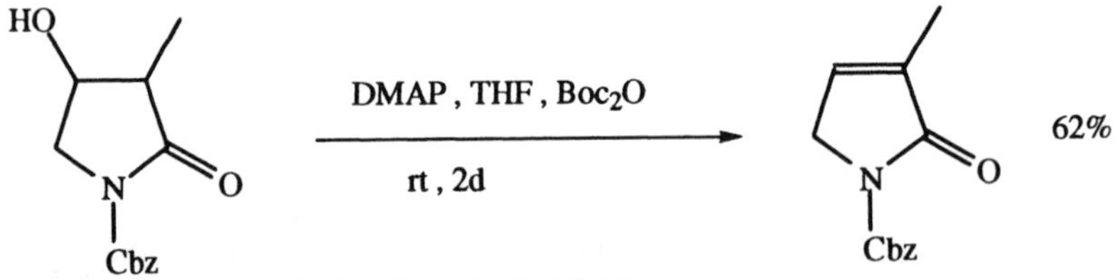
HO
N
O
Cbz
DMAP , THF , Boc2O
rt , 2d
N
O
Cbz
62%

Mattern, R.-H. *Tetrahedron Lett*, ***1996***, *37*, 291.

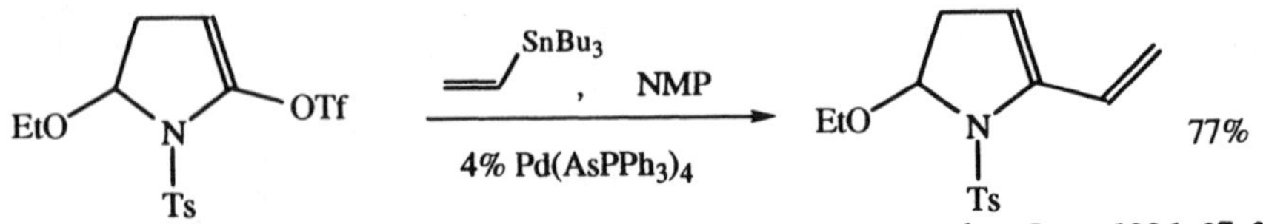
EtO
N
OTf
Ts
SnBu3
, NMP
4% Pd(AsPPh3)4
EtO
N
Ts
77%

Bernabé, P.; Rutjes, P.J.T.; Hiemstra, H.; Speckamp, W.N. *Tetrahedron Lett.*, ***1996***, *37*, 3561

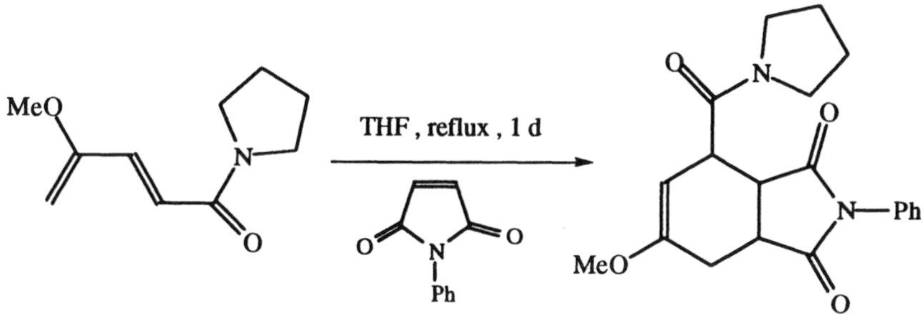
MeO
THF, reflux, 1 d
O
N
Ph
N—Ph
MeO

56% (10:1 *endo:exo*)

Bernabeu, M.C.; Chinchilla, R.; Nájera, C.; Rodríquez, M.A. *Tetrahedron Lett.*, ***1996***, *37*, 3595

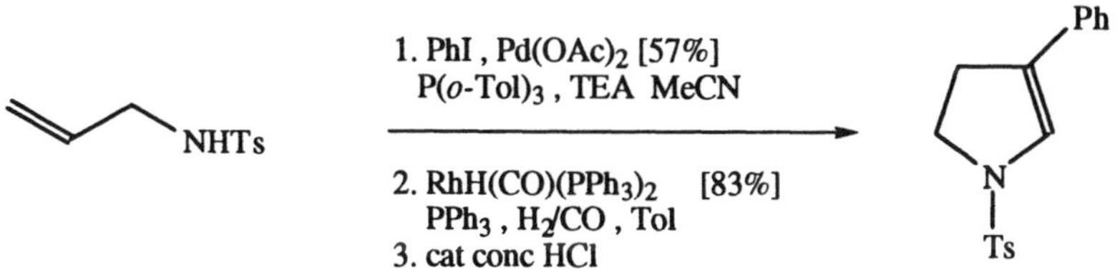
NHTs
1. PhI, Pd(OAc)2 [57%]
P(o-Tol)3, TEA MeCN
2. RhH(CO)(PPh3)2 [83%]
PPh3, H2/CO, Tol
3. cat conc HCl
Ph
N
Ts

57x83%

Busacca, C.A.; Dong, Y. *Tetrahedron Lett.*, ***1996***, *37*, 3947.

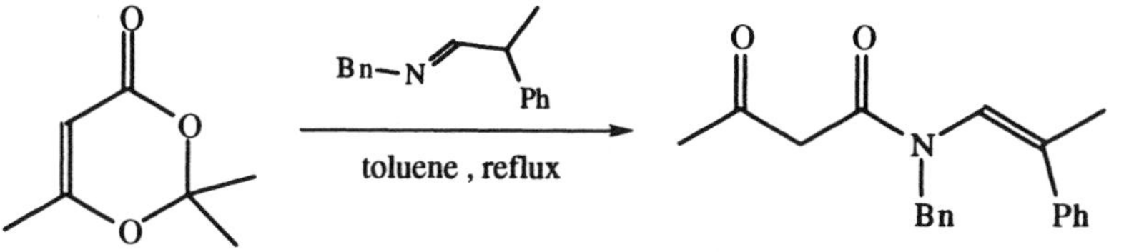
Bn—N
Ph
toluene, reflux
N
Bn
Ph

65%

D'Annibale, A.; Pesce, A.; Resta, S.; Trogolo, C. *Tetrahedron Lett.*, ***1996***, *37*, 7429.

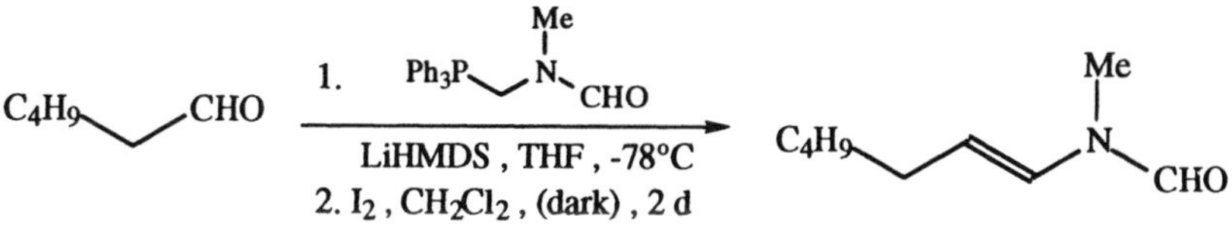
C4H9
CHO
1. Ph3P N CHO
Me
LiHMDS, THF, -78°C
2. I2, CH2Cl2, (dark), 2 d
Me
N
CHO

79%

Paterson, I.; Cowden, C.; Watson, C. *SynLett*, ***1996***, 209.

Also via Alkenyl Acids: Section 322 (Carboxylic Acid -Alkene)

SECTION 350: AMINE - AMINE

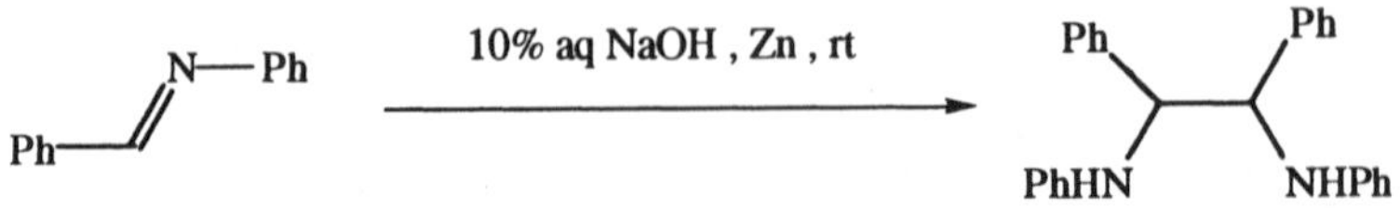
10% aq NaOH , Zn , rt
N—Ph
Ph
Ph
Ph
PhHN
NHPh

90% (80:20 *dl:meso*)

Dutta, M.P.; Baruah, B.; Boruah, A.; Prajapati, D.; Sandhu, J.S. *SynLett*, ***1998***, 857.

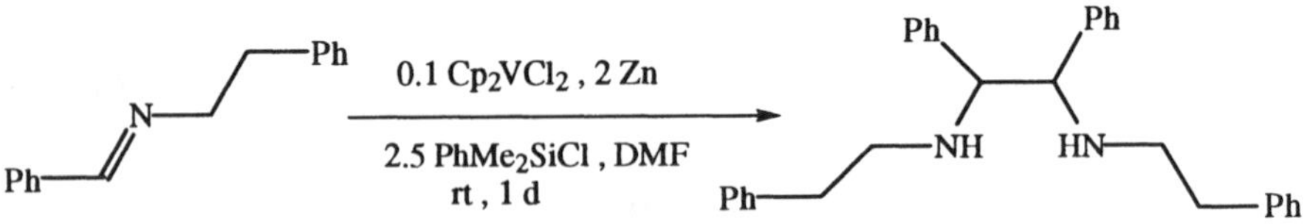
0.1 Cp_2VCl_2 , 2 Zn
2.5 $PhMe_2SiCl$, DMF
rt , 1 d
Ph
N
Ph
Ph
Ph
NH
HN
Ph
Ph

64% (83:17 *meso:dl*)

Hatano, B.; Ogawa, A.; Hirao, T. *J. Org. Chem.*, ***1998***, *63*, 9421.

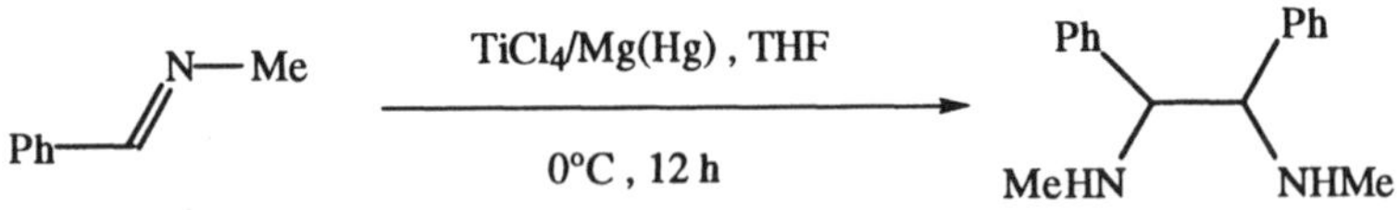
$TiCl_4$/Mg(Hg) , THF
0°C , 12 h
N—Me
Ph
Ph
Ph
MeHN
NHMe

63% (80:20 *dl:meso*)

+ 24% BnNHMe

Alexakis, A.; Aujard, I.; Mangeney, P. *SynLett*, ***1998***, 873, 875.

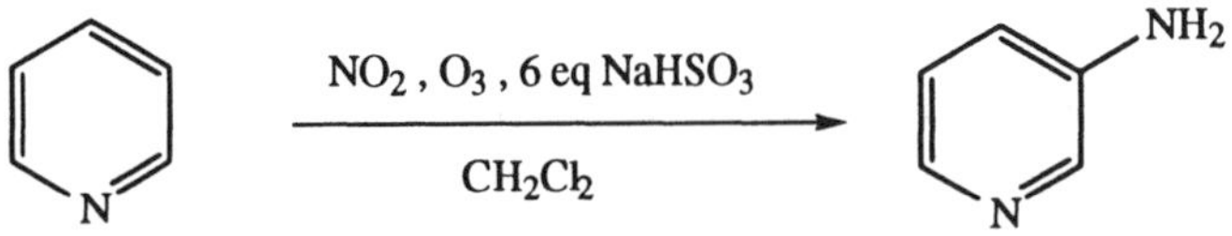
NO_2 , O_3 , 6 eq $NaHSO_3$
CH_2Cl_2
N
N
NH_2

56%

Suzuki, H.; Iwaya, M.; Mori, T. *Tetrahedron Lett.*, ***1997***, *38*, 5647.

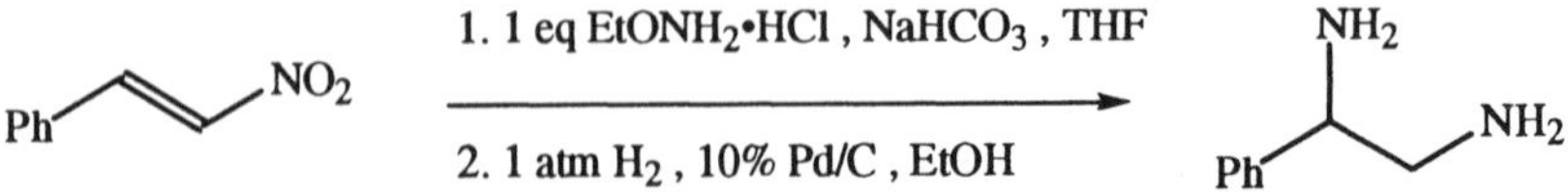
1. 1 eq $EtONH_2$•HCl , $NaHCO_3$, THF
2. 1 atm H_2 , 10% Pd/C , EtOH
Ph
NO_2
NH_2
NH_2
Ph

88%

Imagawa, K.; Hata, E.; Yamada, T.; Mukaiyama, T. *Chem. Lett.*, ***1996***, 291.

SECTION 351: AMINE - ESTER

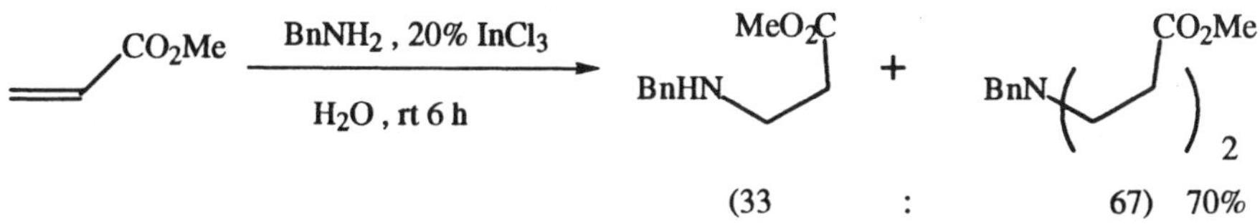
CO2Me
BnNH2 , 20% InCl3
H2O , rt 6 h
MeO2C
BnHN
+
CO2Me
BnN
2
(33 : 67) 70%

Loh, T.–P.; Wei, L.–L. *SynLett, **1998***, 975.

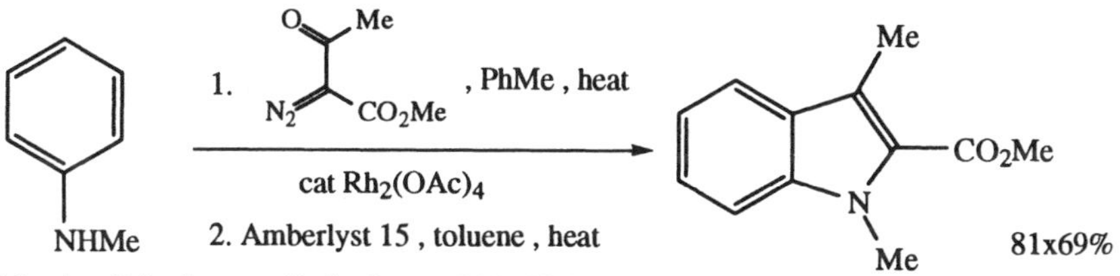
NHMe
1.
Me
N2
CO2Me
, PhMe , heat
cat Rh2(OAc)4
2. Amberlyst 15 , toluene , heat
Me
CO2Me
N
Me
81x69%

Moody, C.J.; Swann, E. *SynLett, **1998***, 135.

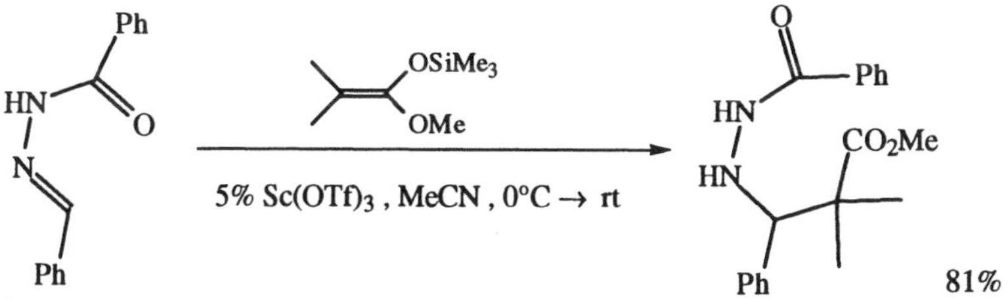
Ph
HN
O
N
Ph
OSiMe3
OMe
5% Sc(OTf)3 , MeCN , 0°C → rt
O
Ph
HN
HN
CO2Me
Ph
81%

Oyamada, H.; Kobayashi, S. *SynLett, **1998***, 249.

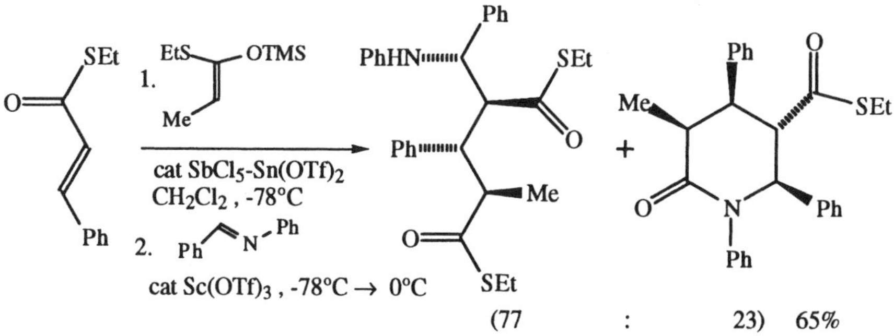
SEt
O
Ph
1.
EtS
OTMS
Me
cat SbCl5-Sn(OTf)2
CH2Cl2 , -78°C
2.
Ph
N
Ph
cat Sc(OTf)3 , -78°C → 0°C
Ph
PhHN
SEt
O
Ph
Me
O
SEt
+
Ph
O
Me
SEt
O
N
Ph
Ph
(77 : 23) 65%

Kobayashi, S.; Akiyama, R.; Moriwaki, M. *Tetrahedron Lett, **1997**, 38*, 4819.

$$N_2CHCO_2Et \xrightarrow{Et_2NH\ ,\ Ru(tmp)CO} EtO_2CCH_2NEt_2 \quad 81\%$$

tmp = 5,10,15,20-tetramesitylporphyrin dianion

Galardon, E.; Le Maux, P.; Simonneaux, G. *J. Chem. Soc., Perkin Trans. 1*, ***1997***, 2455.

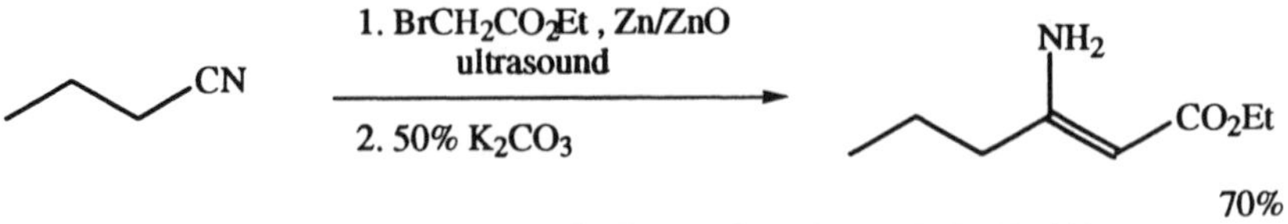
CN
1. $BrCH_2CO_2Et$, Zn/ZnO
ultrasound
2. 50% K_2CO_3
NH_2
CO_2Et
70%

Lee, A.S.–Y.; Cheng, R.–Y.; Pan, O.–G. *Tetrahedron Lett.*, ***1997***, *38*, 443.

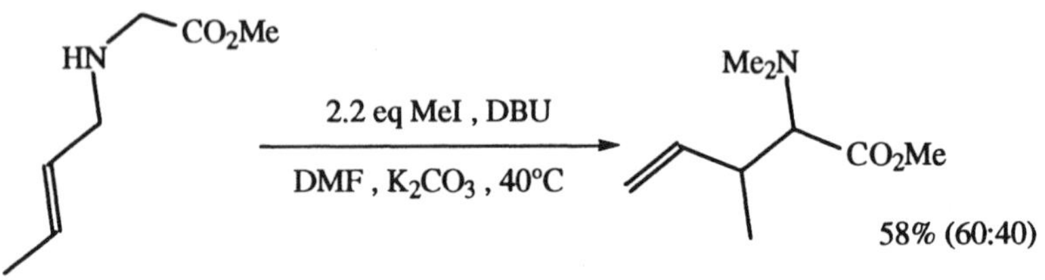
HN
CO_2Me
2.2 eq MeI, DBU
DMF, K_2CO_3, 40°C
Me_2N
CO_2Me
58% (60:40)

Coldham, I.; Middleton, M.L.; Taylor, P.L. *J. Chem. Soc., Perkin Trans. 1*, ***1997***, 2951.

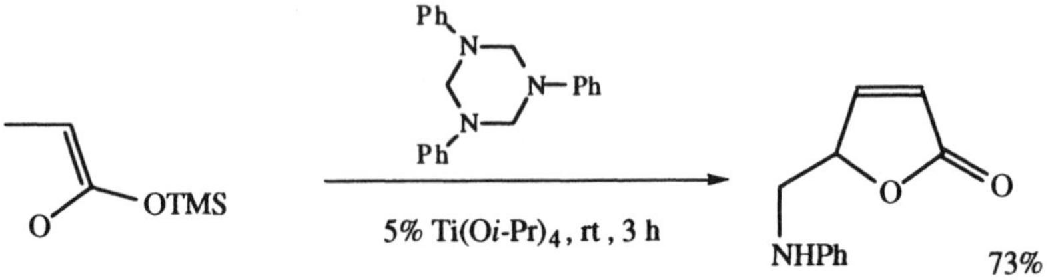
OTMS
O
Ph
N
N–Ph
5% $Ti(Oi\text{-}Pr)_4$, rt, 3 h
O
O
NHPh
73%

Ha, H.–J.; Kang, K.–H.; Ahn, Y.–G.; Oh, S.–J. *Heterocycles*, ***1997***, *45*, 277.

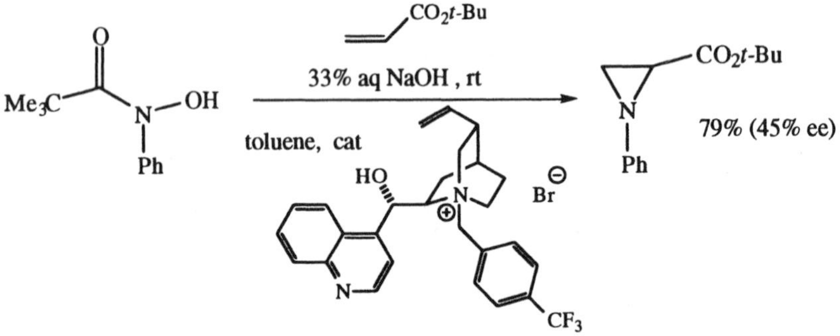
Me_3C
O
N
OH
Ph
CO_2t-Bu
33% aq NaOH, rt
toluene, cat
HO
N
Br
N
CF_3
CO_2t-Bu
N
Ph
79% (45% ee)

Aires–de–Sousa, J.; Labo, A.M.; Prabhakar, S. *Tetrahedron Lett.*, ***1996***, *37*, 3183.

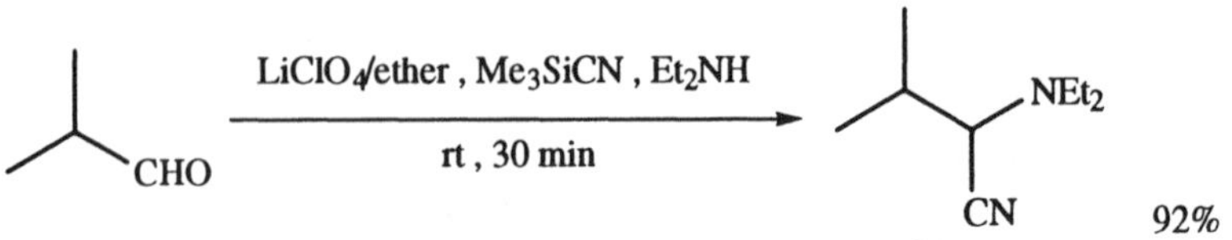
CHO
$LiClO_4$/ether, Me_3SiCN, Et_2NH
rt, 30 min
NEt_2
CN
92%

Shimizu, M.; Kume, K.; Fujisawa, T. *Chem. Lett.*, ***1996***, 545.

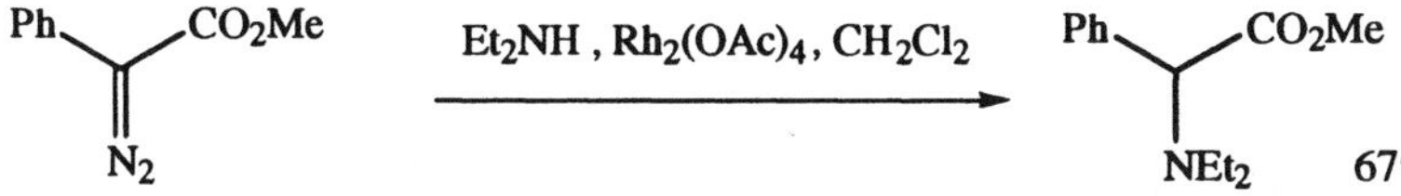

67%

Aller, E.; Buck, R.T.; Drysdale, M.J.; Ferris, L.; Haigh, D.; Moody, C.J.; Pearson, N.D.; Sanghera, J.B. *J. Chem. Soc., Perkin Trans. 1*, ***1996***, 2879.

Related Methods: Section 315 (Carboxylic Acid - Amide)
Section 316 (Carboxylic Acid - Amine)
Section 344 (Amide - Ester)

SECTION 352: AMINE - ETHER, EPOXIDE, THIOETHER

NO ADDITIONAL EXAMPLES

SECTION 353: AMINE - HALIDE, SULFONATE

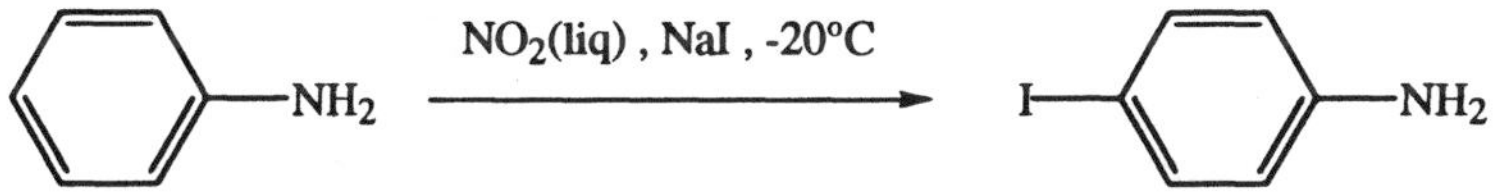

85%

Suzuki, H.; Nonoyama, N. *Tetrahedron Lett*, ***1998***, *39*, 4533.

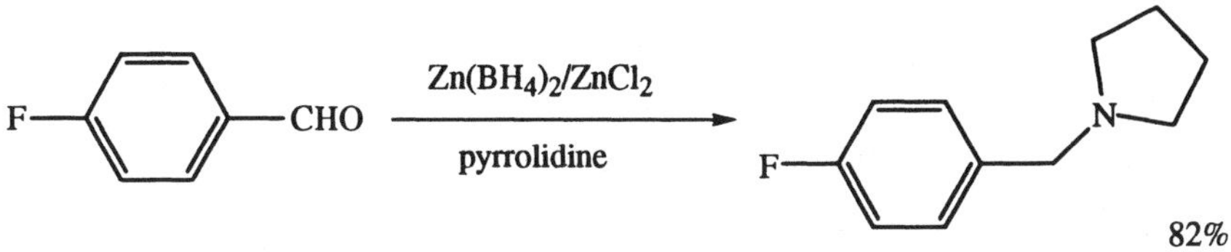

82%

Bhattacharyya, S.; Chatterjee, A.; Williamson, J.S. *Synth. Commun.*, ***1997***, *27*, 4265.

SECTION 354: AMINE - KETONE

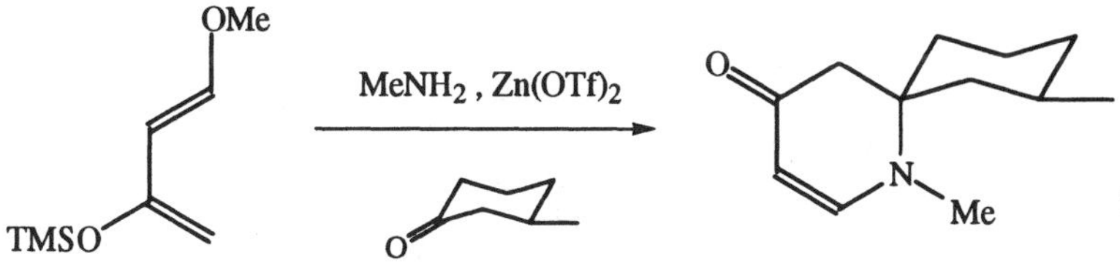

76%

Huang, P.; Isayan, K.; Sarkissian, A.; Oh, T. *J. Org. Chem.*, ***1998***, *63*, 4500.

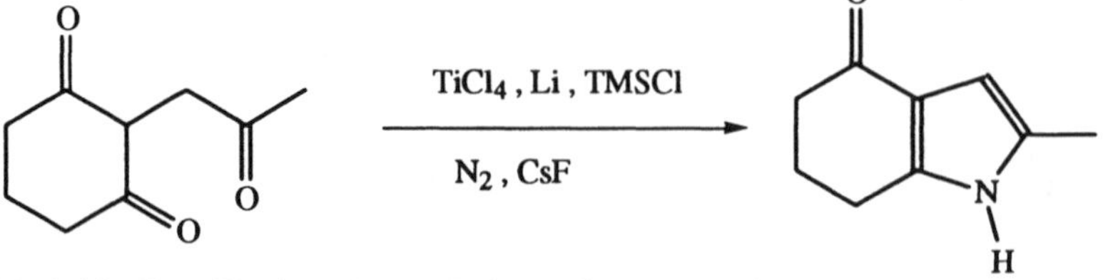
TiCl4 , Li , TMSCl
N2 , CsF

86%

Mori, M.; Hori, K.; Akashi, M.; Hori, M.; Sato, Y.; Nishida, M.
Angew. Chem. Int. Ed., ***1998***, *37*, 636.

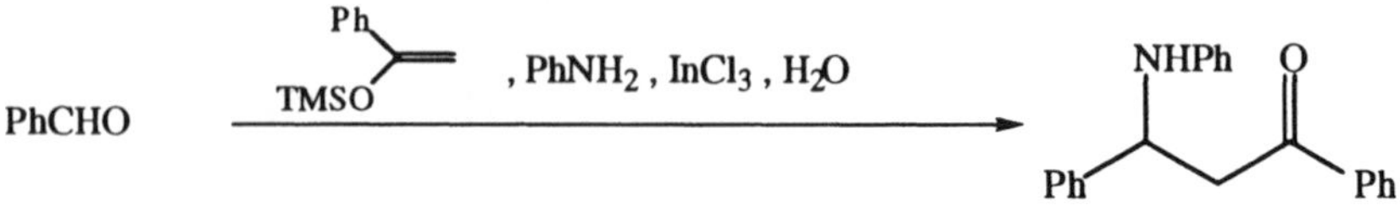
PhCHO
, PhNH2 , InCl3 , H2O

54%

Loh, T.-P.; Wei, L.-L. *Tetrahedron Lett.*, ***1998***, *39*, 323.

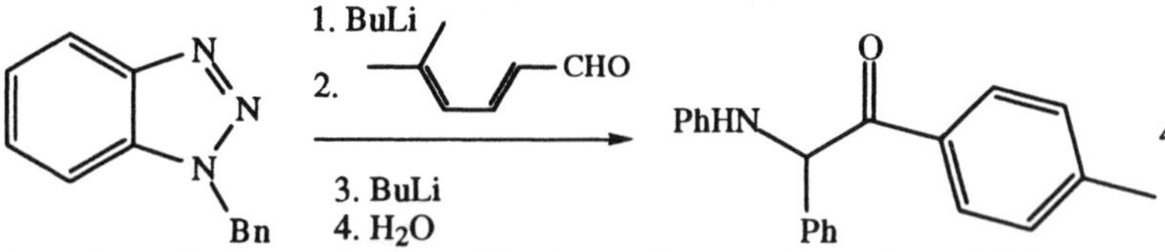
1. BuLi
2.
3. BuLi
4. H2O

40%

Katritzky, A.R.; Zhang, Z.; Lang, H.; Xie, L. *Heterocycles*, ***1998***, *47*, 187.

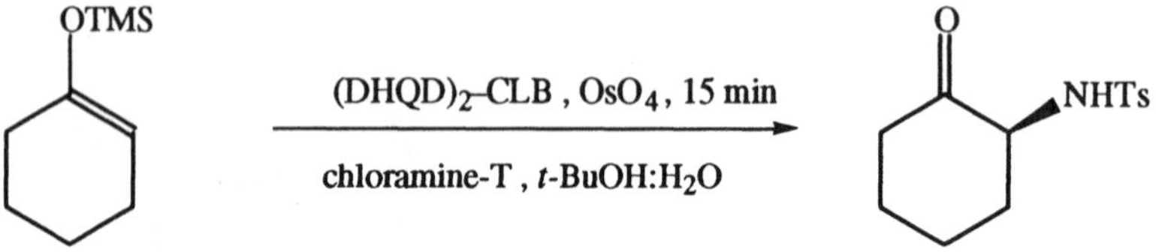
(DHQD)2-CLB , OsO4, 15 min
chloramine-T , t-BuOH:H2O

$(DHQD)_2$-CLB = dihydroquinidine-*p*-chlorobenzoate 35% (92% ee)

Phukan, P.; Sudalai, A. *Tetrahedron Asymmetry*, ***1998***, *9*, 1001.

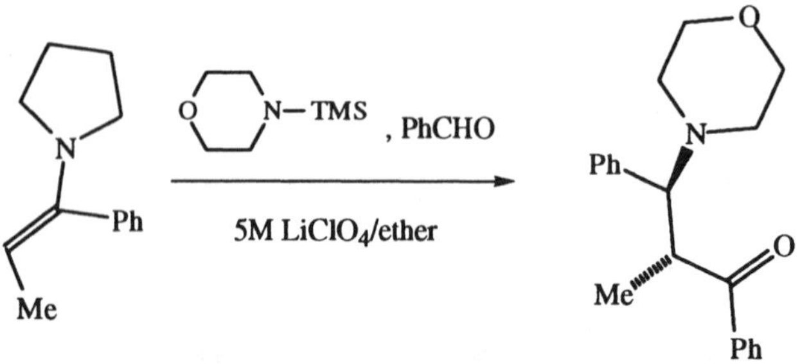
, PhCHO
5M LiClO4/ether

91% (100% *anti*)

Zarghi, A.; Naimi-Jamal, M.R.; Webb, S.A.; Balalaie, S.; Saidi, M.R.; Ipaktschi, J.
Eur. J. Org. Chem., ***1998***, 197.

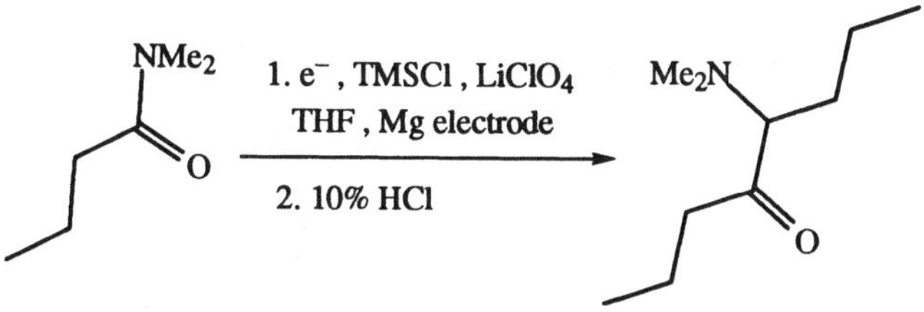
NMe2
1. e⁻ , TMSCl , LiClO4
THF , Mg electrode
2. 10% HCl
O
Me2N
O

70x98%

Kashimura, S.; Ishifune, M.; Murai, Y.; Murase, H.; Shimomura, M.; Shono, T. *Tetrahedron Lett.*, ***1998***, *39*, 6199.

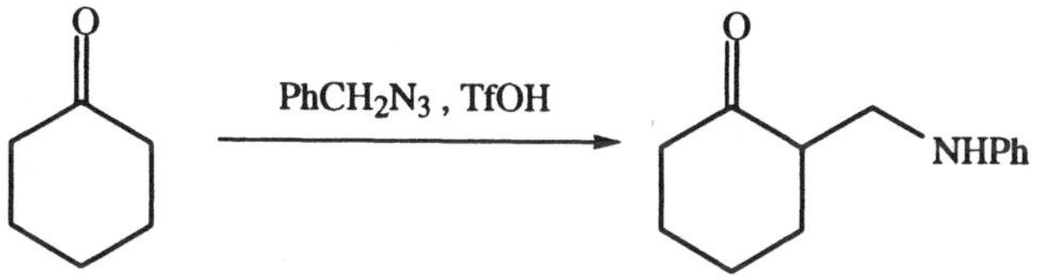
O
PhCH2N3 , TfOH
O
NHPh

63%

Schildknegt, K.; Agrios, K.A.; Aubé, J. *Tetrahedron Lett.*, ***1998***, *39*, 7687.

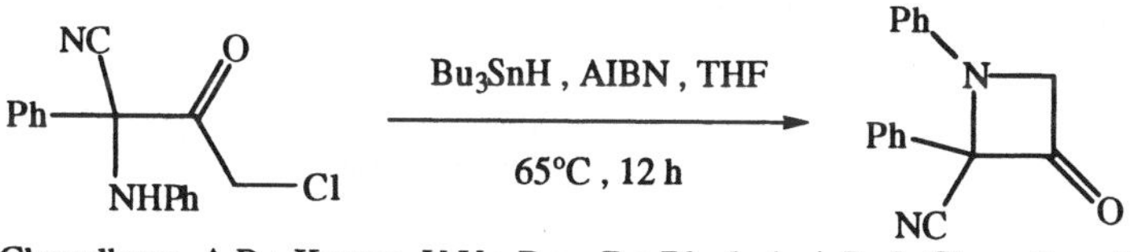
NC
O
Ph
NHPh
Cl
Bu3SnH , AIBN , THF
65°C , 12 h
Ph
N
Ph
NC
O

77%

Chowdhury, A.R.; Kumar, V.V.; Roy, R.; Bhaduri, A.P. *J. Chem. Res. (S)*, ***1997***, 254.

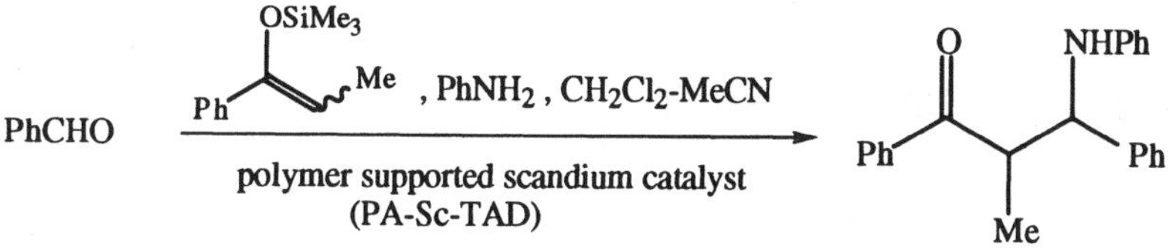
OSiMe3
Ph
Me
, PhNH2 , CH2Cl2-MeCN
PhCHO
polymer supported scandium catalyst
(PA-Sc-TAD)
O
NHPh
Ph
Ph
Me

91% (1:1)

Kobayashi, S.; Nagayama, S.; Busujima, T. *Tetrahedron Lett.*, ***1996***, *37*, 9221.

SECTION 355: AMINE - NITRILE

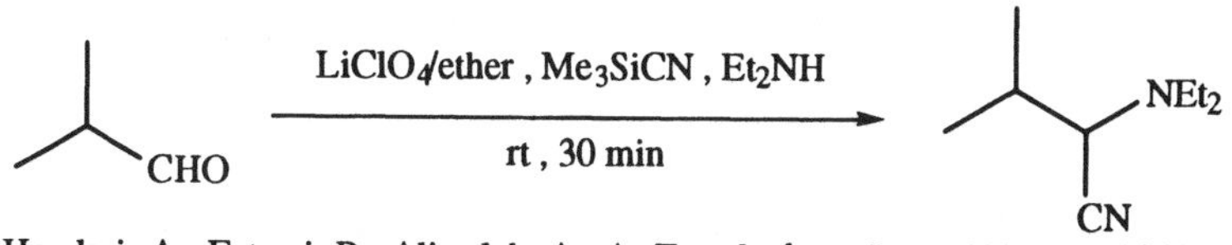
CHO
LiClO4/ether , Me3SiCN , Et2NH
rt , 30 min
NEt2
CN

92%

Heydari, A.; Fatemi, P.; Alizadeh, A.–A. *Tetrahedron Lett.*, ***1998***, *39*, 3049.

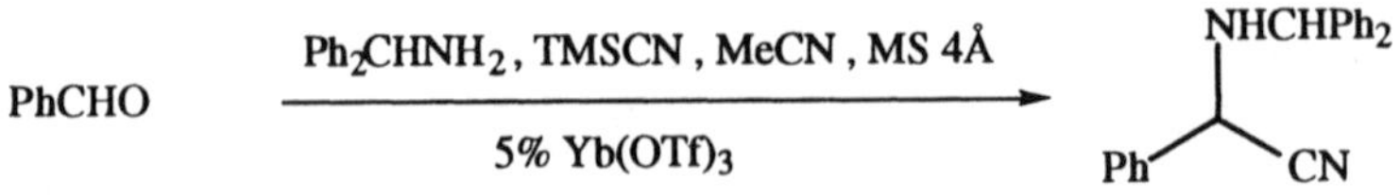
PhCHO
Ph2CHNH2, TMSCN, MeCN, MS 4Å
5% Yb(OTf)3
NHCHPh2
Ph
CN

88%

Kobayashi, S.; Ishitani, H.; Ueno, M. *SynLett*, ***1997***, 115.

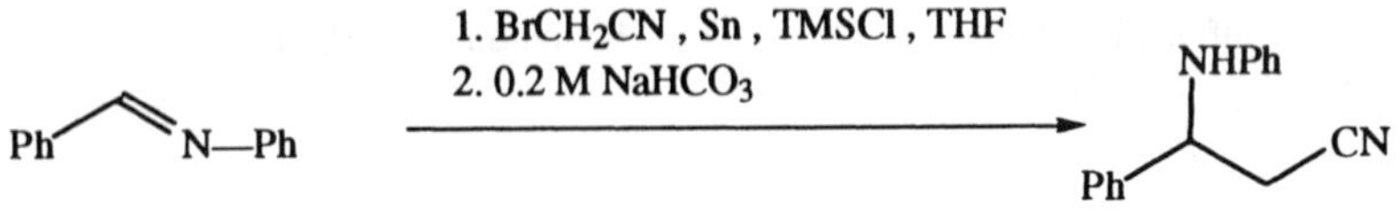
1. BrCH2CN, Sn, TMSCl, THF
2. 0.2 M NaHCO3
Ph
N—Ph
NHPh
Ph
CN

68%

Sun, P.; Zhang, Y. *Synth. Commun.*, ***1997***, *27*, 3175.

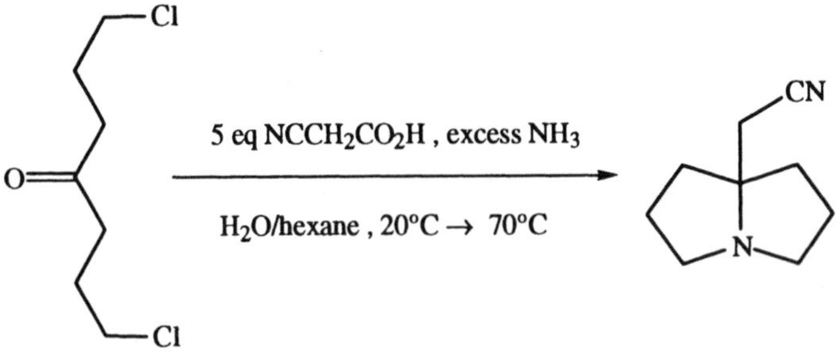
Cl
O
Cl
5 eq NCCH2CO2H, excess NH3
H2O/hexane, 20°C → 70°C
CN
N

54%

Oka, M.; Baba, K.; Suzuki, T.; Matsumoto, Y. *Heterocycles*, ***1997***, *45*, 2317.

SECTION 356: AMINE - ALKENE

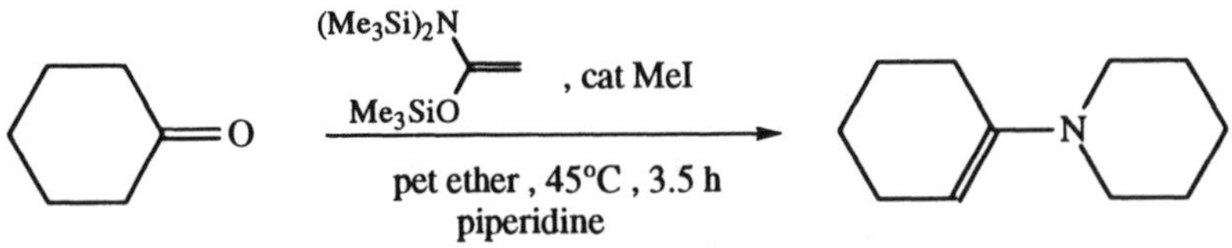
O
(Me3Si)2N
Me3SiO
, cat MeI
pet ether, 45°C, 3.5 h
piperidine
N

80%

Yamamoto, Y.; Matui, C. *J. Org. Chem.*, ***1998***, *63*, 377.

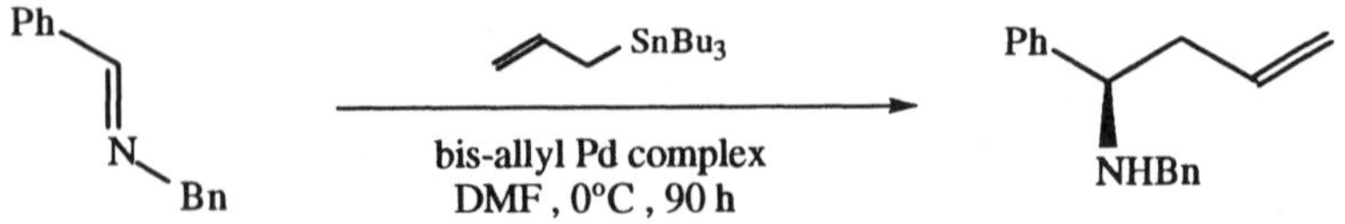
Ph
N
Bn
SnBu3
bis-allyl Pd complex
DMF, 0°C, 90 h
Ph
NHBn

62% (81% ee)

Nakamura, H.; Nakamura, K.; Yamamoto, Y. *J. Am. Chem. Soc.*, ***1998***, *120*, 4242.

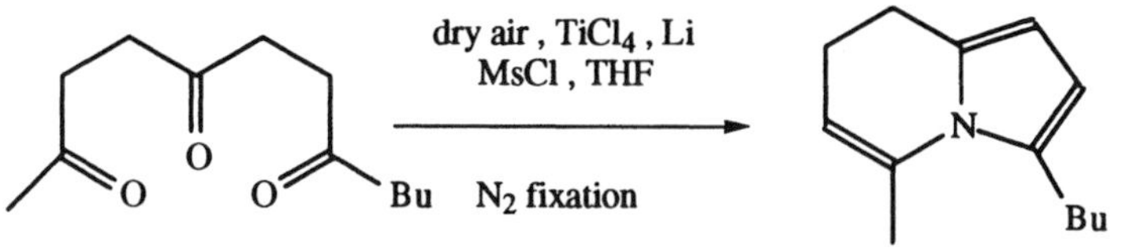
dry air , $TiCl_4$, Li
MsCl , THF
N_2 fixation
O
O
O
Bu
N
Bu

22%

Mori, M.; Hori, M.; Sato, Y. *J. Org. Chem.*, ***1998***, *63*, 4832.

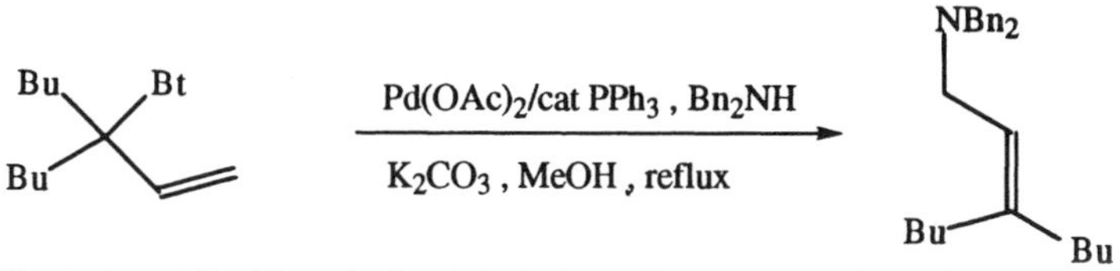
Bu
Bt
Bu
$Pd(OAc)_2$/cat PPh_3 , Bn_2NH
K_2CO_3 , MeOH , reflux
NBn_2
Bu
Bu

70%

Katritzky, A.R.; Yao, J.; Qui, M. *J. Org. Chem.*, ***1998***, *63*, 5232.

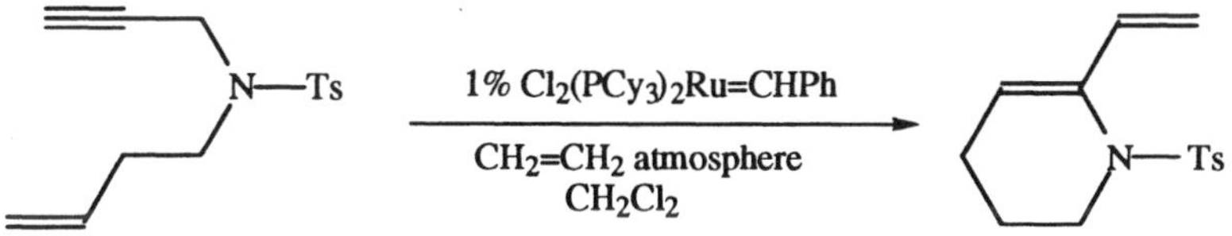
N—Ts
1% $Cl_2(PCy_3)_2Ru$=CHPh
CH_2=CH_2 atmosphere
CH_2Cl_2
N—Ts

99%(19% in Ar atmosphere)

Mori, M.; Sakakibara, N.; Kinoshita, A. *J. Org. Chem.*, ***1998***, *63*, 6082.

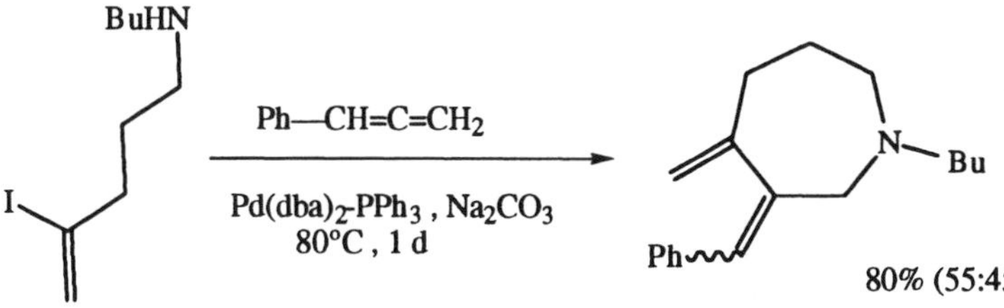
BuHN
I
Ph—CH=C=CH_2
$Pd(dba)_2$-PPh_3 , Na_2CO_3
80°C , 1 d
N
Bu
Ph

80% (55:45 *E:Z*)

Larock, R.C.; Tu, C.; Pace, P. *J. Org. Chem.*, ***1998***, *63*, 6859.

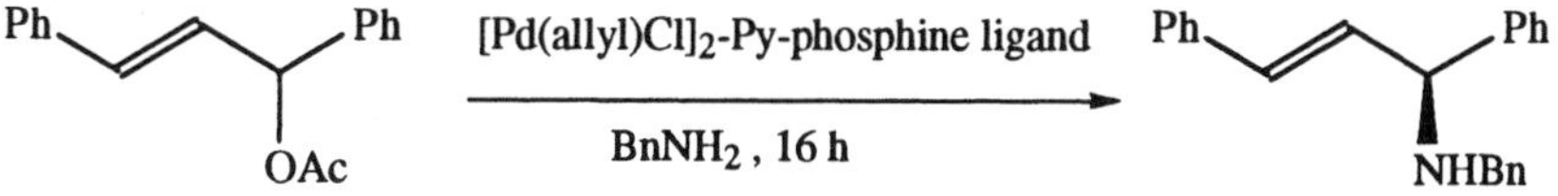
Ph
Ph
OAc
$[Pd(allyl)Cl]_2$-Py-phosphine ligand
$BnNH_2$, 16 h
Ph
Ph
NHBn

95% (93% ee , *S*)

Constantieux, T.; Brunel, J.–M.; Labande, A.; Buono, G. *SynLett.*, ***1998***, 49.

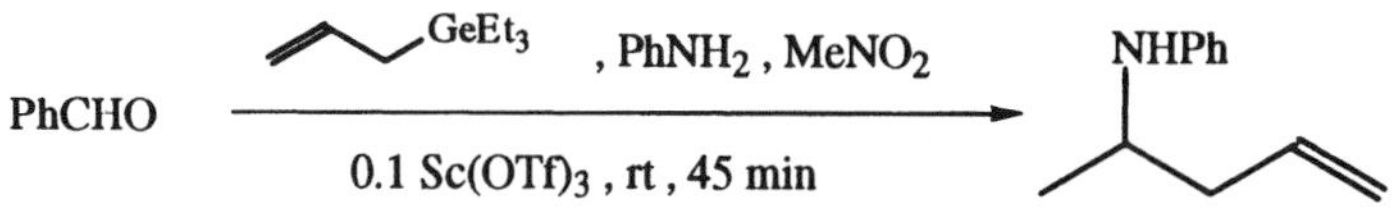
$GeEt_3$
, $PhNH_2$, $MeNO_2$
PhCHO
0.1 $Sc(OTf)_3$, rt , 45 min
NHPh

81%

Akiyama, T.; Iwai, J. *SynLett.*, ***1998***, 273.

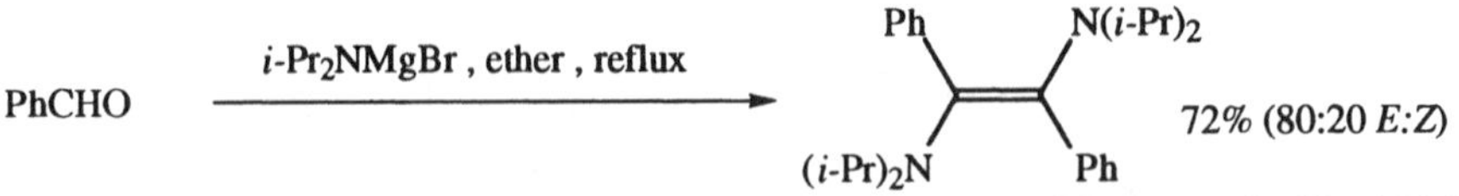
PhCHO
i-Pr_2NMgBr , ether , reflux
Ph
$N(i\text{-}Pr)_2$
$(i\text{-}Pr)_2N$
Ph
72% (80:20 E:Z)

Kobayashi, K.; Uneda, T.; Uchida, M.; Furuta, Y.; Tanmatsu, M.; Morikawa, O.; Konishi, H *Chem. Lett.*, ***1998***, 87.

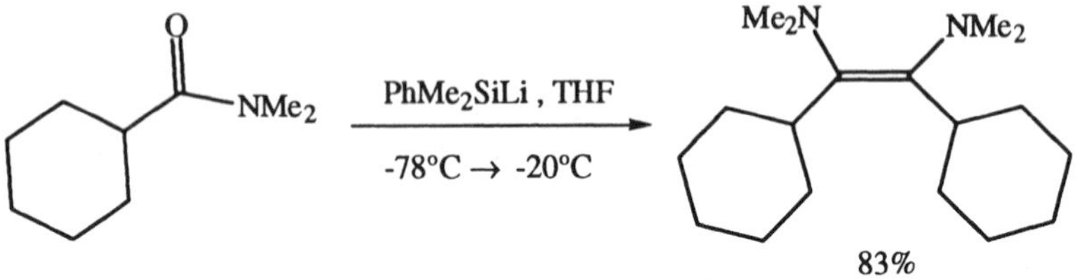
O
NMe_2
$PhMe_2SiLi$, THF
-78°C → -20°C
Me_2N
NMe_2
83%

Fleming, I.; Ghosh, U.; Mack, S.R.; Clark, B.P. *Chem. Commun.*, ***1998***, 74.

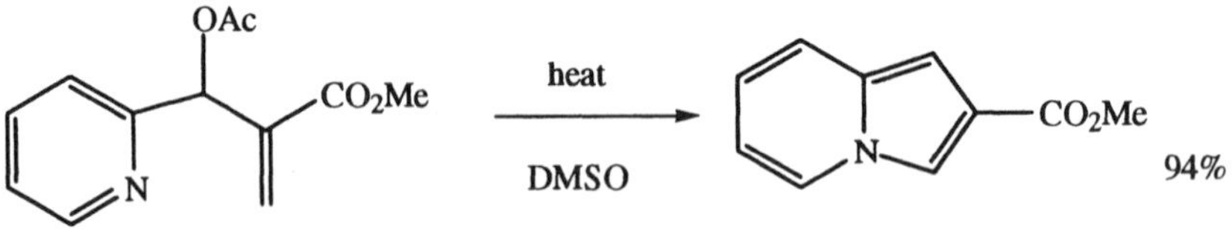
OAc
CO_2Me
N
heat
DMSO
N
CO_2Me
94%

Deane, P.O.; George, R.; Kaye, P.T. *Tetrahedron* , ***1998***, *54*, 3871.

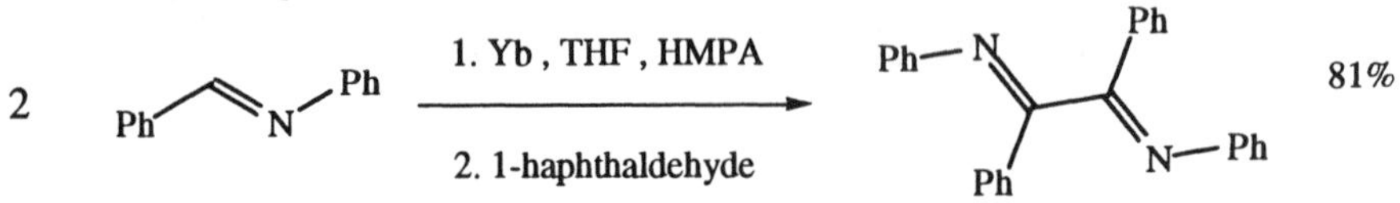
2
Ph
N
Ph
1. Yb , THF , HMPA
2. 1-haphthaldehyde
Ph—N
Ph
Ph
N—Ph
81%

Jin, W.–S.; Makioka, Y.; Tankguchi, Y.; Kitamura, T.; Fujiwara, Y. *Chem. Commun.*, ***1998***, 1101.

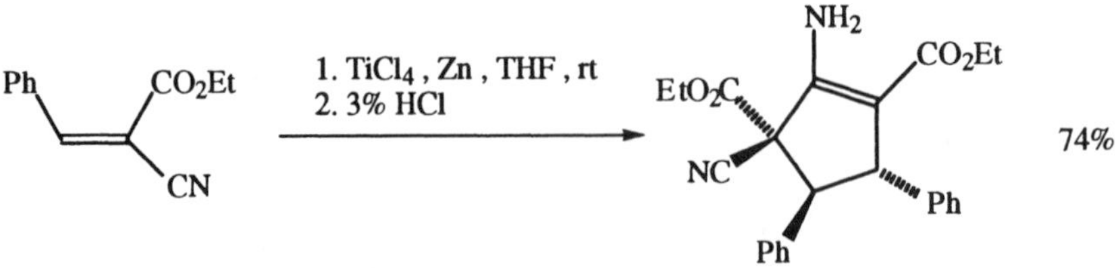
Ph
CO_2Et
CN
1. $TiCl_4$, Zn , THF , rt
2. 3% HCl
NH_2
CO_2Et
EtO_2C
NC
Ph
Ph
74%

Zhou, L.–L.; Tu, S.–j.; Shi, D.–q.; Dai, G.–y.; Chen, W.–x. *Synthesis*, ***1998***, 851.

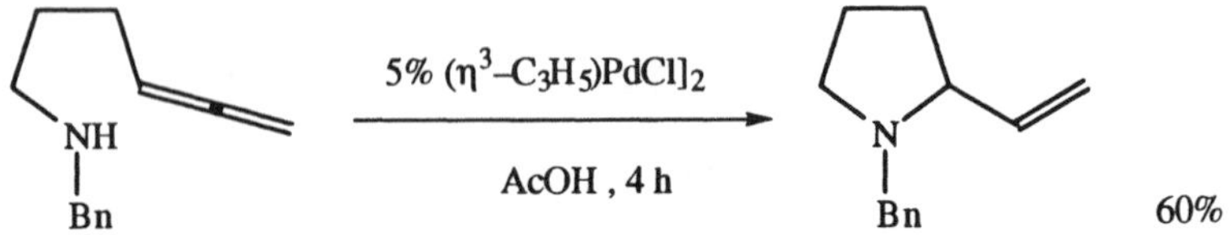
NH
Bn
5% $(\eta^3\text{-}C_3H_5)PdCl]_2$
AcOH , 4 h
N
Bn
60%

Meguro, M.; Yamamoto, Y. *Tetrahedron Lett.*, ***1998***, *39*, 5421.

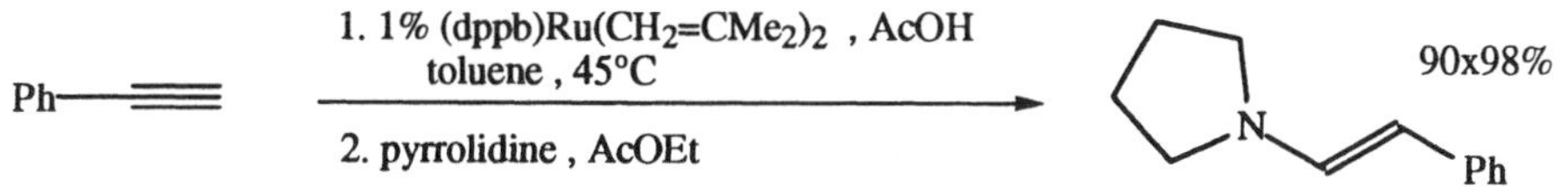

Ph
1. 1% (dppb)Ru(CH2=CMe2)2 , AcOH
toluene , 45°C
2. pyrrolidine , AcOEt
N
Ph
90x98%

Doucet, H.; Bruneau, C.; Dixneuf, P.H. *SynLett.*, ***1997***, 807.

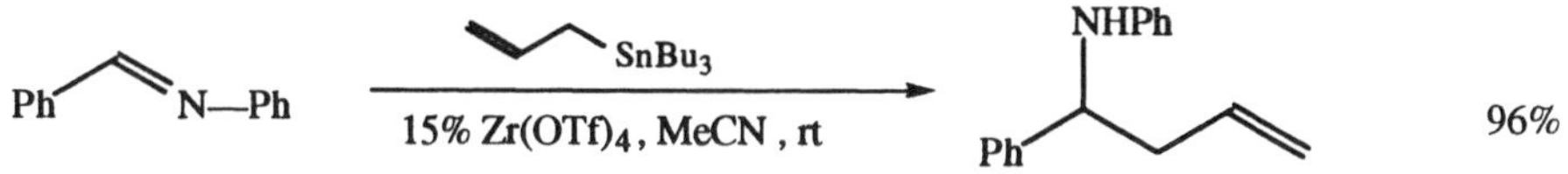

Ph
N—Ph
SnBu3
15% Zr(OTf)4 , MeCN , rt
NHPh
Ph
96%

Kobayashi, S.; Inamoto, S.; Nagayama, S. *SynLett.*, ***1997***, 1099.

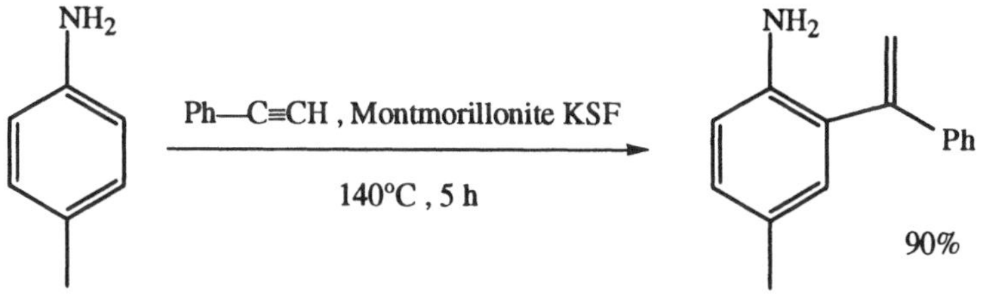

NH2
Ph—C≡CH , Montmorillonite KSF
140°C , 5 h
NH2
Ph
90%

Arienti, A.; Bigi, F.; Maggi, R.; Marzi, E.; Moggi, P.; Rastelli, M.; Sartori, G.; Tarantola, F. *Tetrahedron*, ***1997***, *53*, 3795.

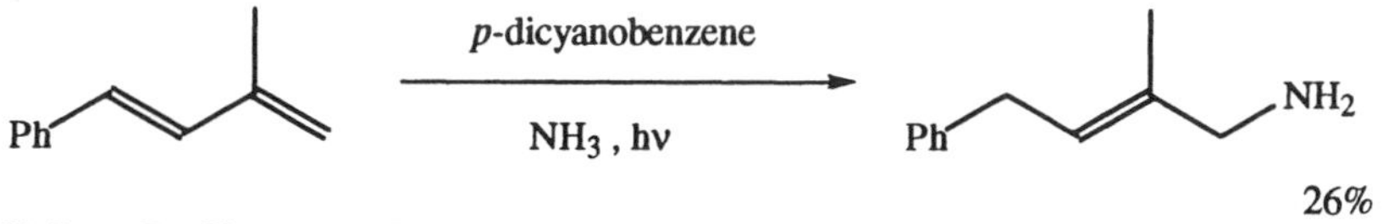

Ph
p-dicyanobenzene
NH3 , hν
Ph
NH2
26%

Kojima, R.; Yamashita, T.; Tanabe, K.; Shiragami, T.; Yasuda, M.; Shima, K. *J. Chem. Soc., Perkin Trans. 1*, ***1997***, 217.

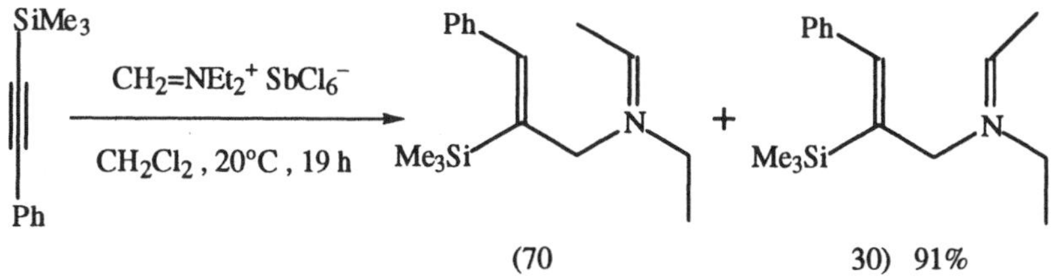

SiMe3
Ph
CH2=NEt2+ SbCl6−
CH2Cl2 , 20°C , 19 h
Ph
Me3Si
N
+
Ph
Me3Si
N
(70
30) 91%

Ofial, A.R.; Mayr, H. *Angew. Chem. Int. Ed.*, ***1997***, *36*, 143.

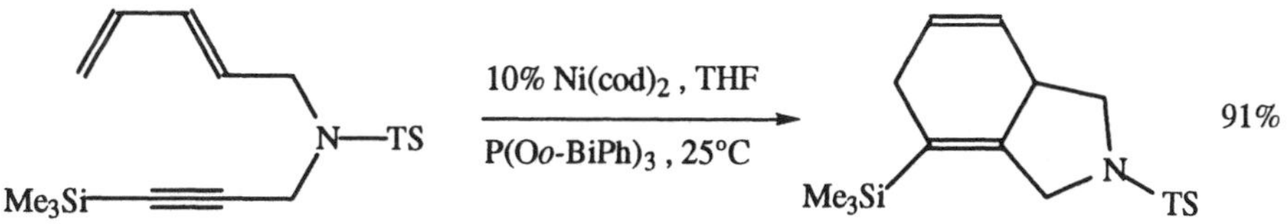

N—TS
Me3Si
10% Ni(cod)2 , THF
P(Oo-BiPh)3 , 25°C
Me3Si
N
TS
91%

Wender, P.A.; Smith, T.E. *J. Org. Chem.*, ***1996***, *61*, 824.

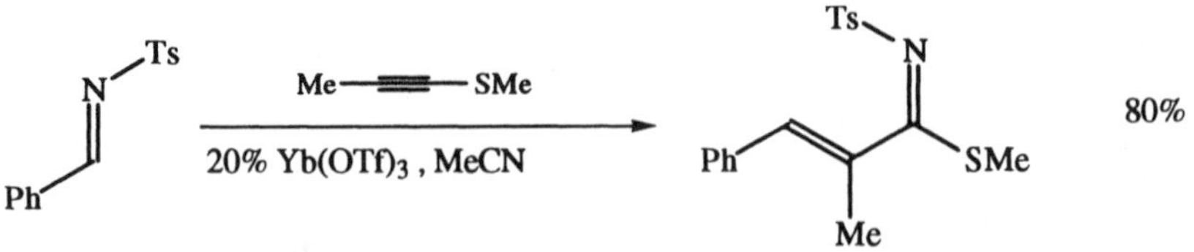
Ts
N
Ph
Me—≡—SMe
20% Yb(OTf)3 , MeCN
Ts
N
Ph
SMe
Me
80%

Ishitani, H.; Nagayama, S.; Kobayashi, S. *J. Org. Chem.*, ***1996***, *61*, 1902.

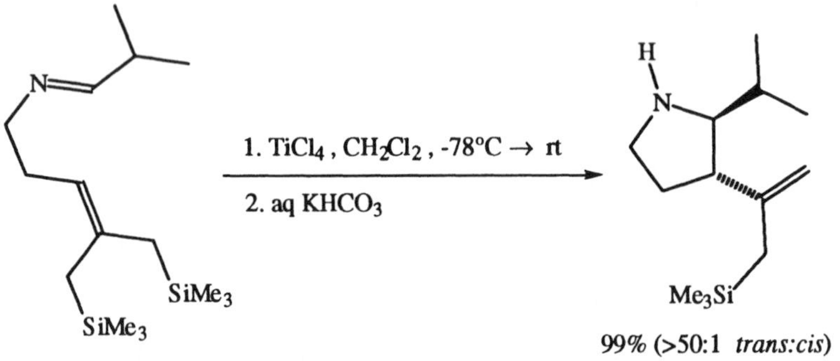
N
SiMe3
SiMe3
1. TiCl4 , CH2Cl2 , -78°C → rt
2. aq KHCO3
H
N
Me3Si
99% (>50:1 trans:cis)

Kercher, T.; Livinghouse, T. *J. Am. Chem. Soc.*, ***1996***, *118*, 4200.

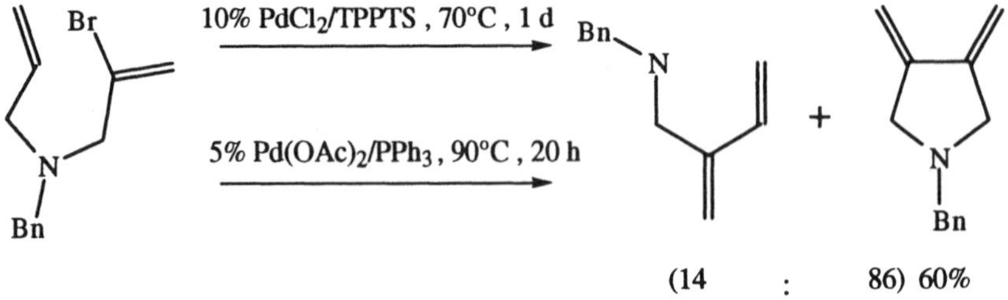
Br
N
Bn
10% PdCl2/TPPTS , 70°C , 1 d
5% Pd(OAc)2/PPh3 , 90°C , 20 h
Bn
N
+
N
Bn
(14 : 86) 60%

Lemaiare–Audoire, S.; Savignac, M.; Dupuis, C.; Genêt, J.–P. *Tetrahedron Lett.*, ***1996***, *37*, 2003.

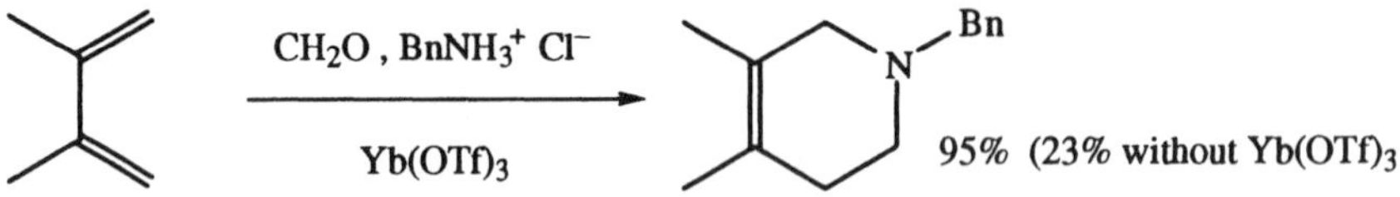
CH2O , BnNH3+ Cl−
Yb(OTf)3
Bn
N
95% (23% without Yb(OTf)3

Yu, L.; Chen, D.; Wang, P.G. *Tetrahedron Lett.*, ***1996***, *37*, 2169.

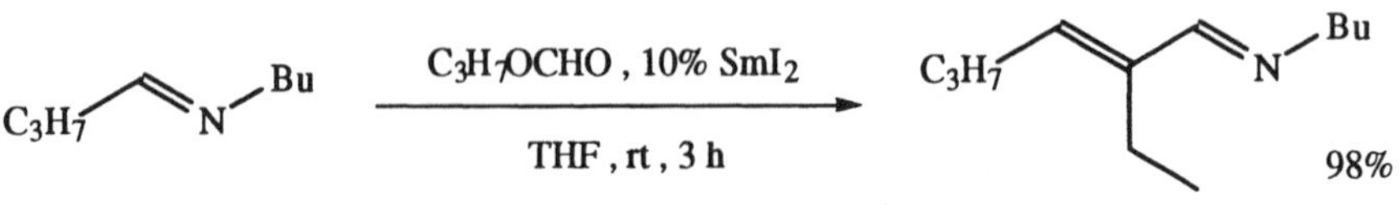
C3H7
N
Bu
C3H7OCHO , 10% SmI2
THF , rt , 3 h
C3H7
N
Bu
98%

Shiraishi, H.; Kawasaki, Y.; Sakaguchi, S.; Nishiyama, Y.; Ishii, Y. *Tetrahedron Lett.*. ***1996***. *37*. 7291.

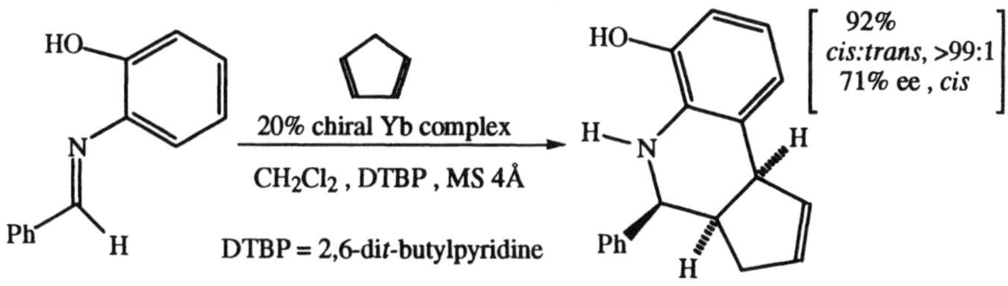
HO
Ph
H
N
20% chiral Yb complex
CH2Cl2 , DTBP , MS 4Å
DTBP = 2,6-di*t*-butylpyridine
HO
H
N
H
Ph
H
92%
cis:trans, >99:1
71% ee , *cis*

Ishitani, H.; Kobayashi, S. *Tetrahedron Lett.*, ***1996***, *37*, 7357.

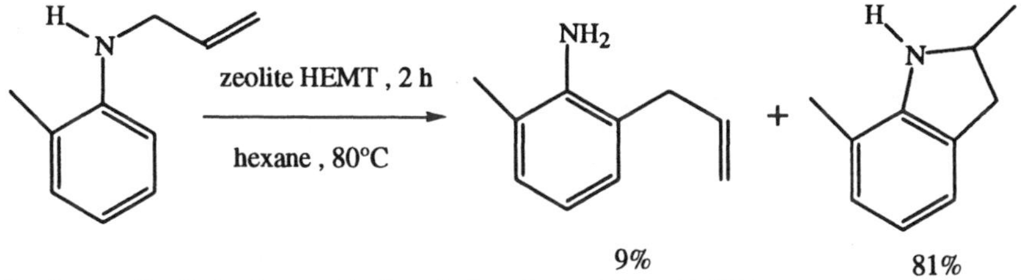
H
N
zeolite HEMT , 2 h
hexane , 80°C
NH2
+
H
N
9%
81%

Sreekumar, R.; Padmakumar, R. *Tetrahedron Lett.*, ***1996***, *37*, 5281.

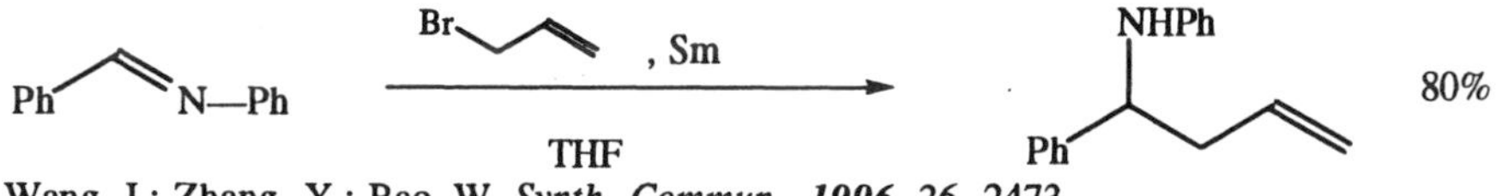
Ph
N—Ph
Br
, Sm
THF
NHPh
Ph
80%

Wang, J.; Zhang, Y.; Bao, W. *Synth. Commun.*, ***1996***, *26*, 2473.

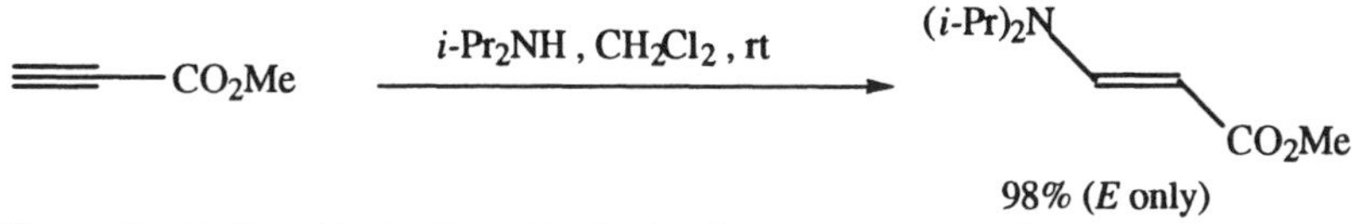
CO2Me
i-Pr2NH , CH2Cl2 , rt
(*i*-Pr)2N
CO2Me
98% (*E* only)

Cossu, S.; De Lucchi, O.; Durr, R. *Synth. Commun.*, ***1996***, *26*, 4597.

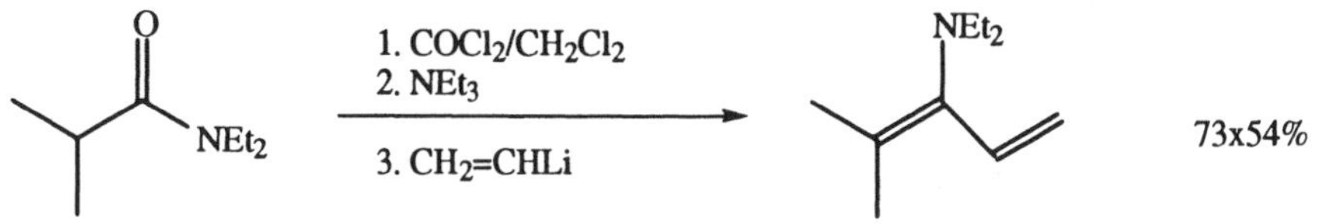
O
NEt2
1. COCl2/CH2Cl2
2. NEt3
3. CH2=CHLi
NEt2
73x54%

Enders, D.; Hecker, P.; Meyer, O. *Tetrahedron*, ***1996***, *52*, 2909.

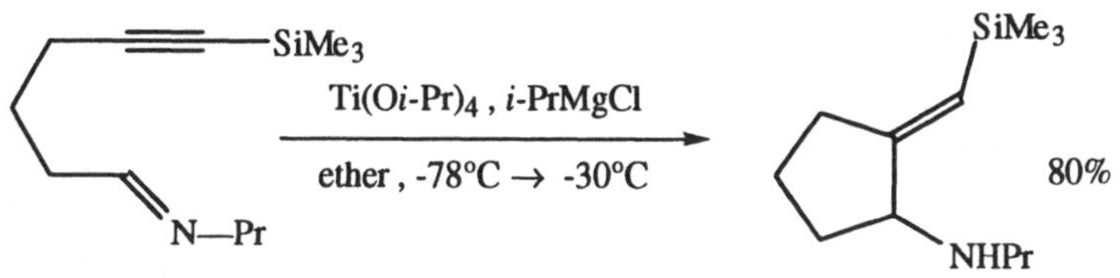
SiMe3
N—Pr
Ti(O*i*-Pr)4 , *i*-PrMgCl
ether , -78°C → -30°C
SiMe3
NHPr
80%

Gao, Y.; Harada, K.; Sato, F. *Chem. Commun.*, ***1996***, 533.

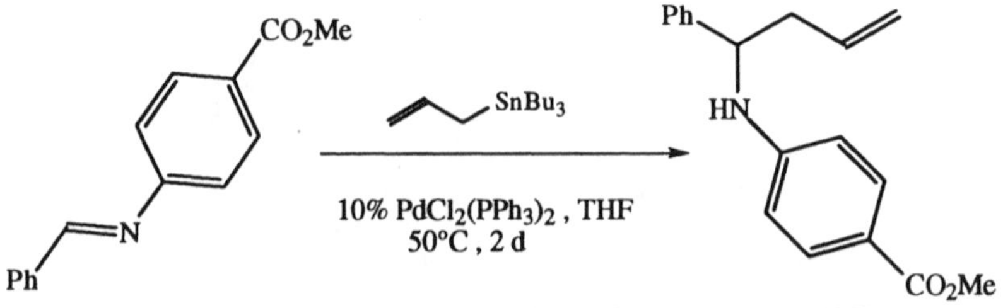
CO2Me
SnBu3
10% PdCl2(PPh3)2 , THF
50°C , 2 d
Ph
N
Ph
HN
CO2Me
98%

Nakamura, H.; Iwama, H.; Yamamoto, Y. *Chem. Commun.*, ***1996***, 1459.

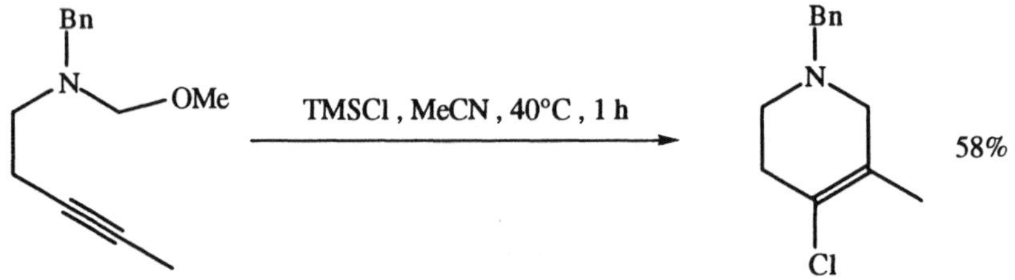
Bn
N
OMe
TMSCl , MeCN , 40°C , 1 h
Bn
N
58%
Cl

Murata, Y.; Overman, L.E. *Heterocycles, **1996**, 42,* 549.

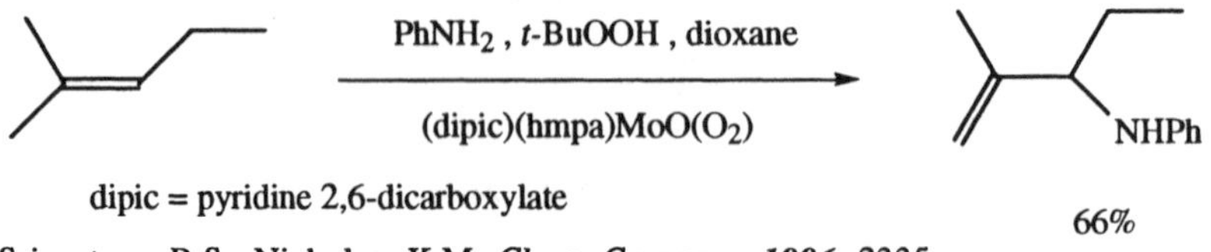
PhNH2 , t-BuOOH , dioxane
(dipic)(hmpa)MoO(O2)
NHPh
dipic = pyridine 2,6-dicarboxylate
66%

Srivastava, R.S.; Nicholas, K.M. *Chem. Commun.*, ***1996***, 2335.

SECTION 357: ESTER - ESTER

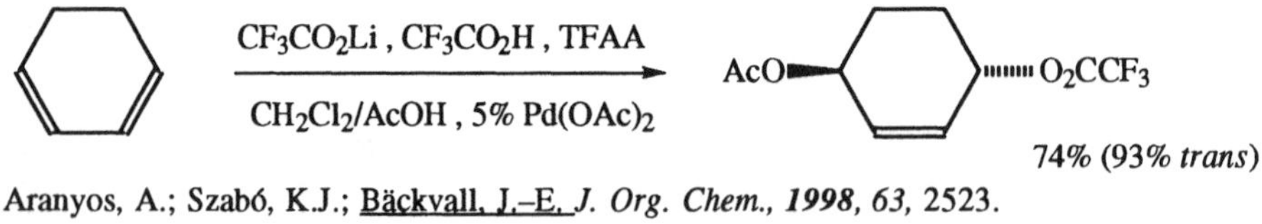
CF3CO2Li , CF3CO2H , TFAA
CH2Cl2/AcOH , 5% Pd(OAc)2
AcO
O2CCF3
74% (93% trans)

Aranyos, A.; Szabó, K.J.; Bäckvall, J.-E. *J. Org. Chem.*, ***1998***, *63,* 2523.

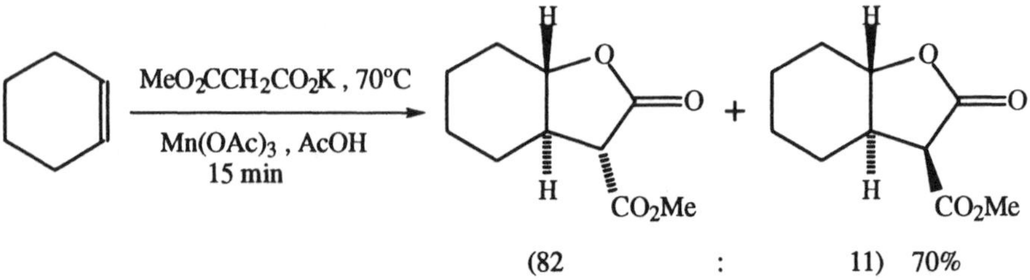
MeO2CCH2CO2K , 70°C
Mn(OAc)3 , AcOH
15 min
H
O
O
H
CO2Me
+
H
O
O
H
CO2Me
(82 : 11) 70%

Lamarque, L.; Méou, A.; Brun, P. *Tetrahedron, **1998**, 54,* 6497.

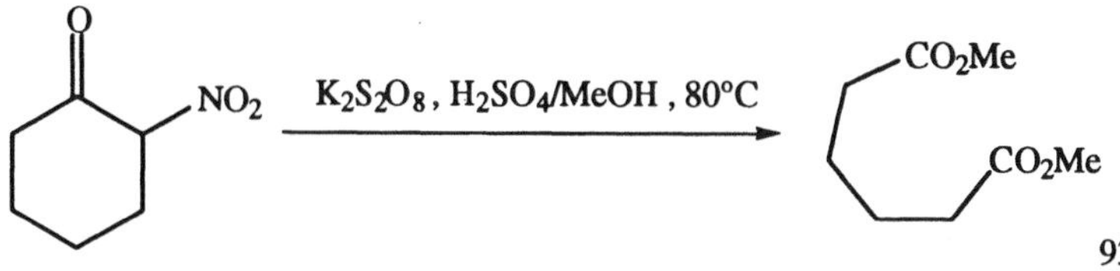
NO2
K2S2O8 , H2SO4/MeOH , 80°C
CO2Me
CO2Me

93%

Ballini, R.; Bosica, G. *Tetrahedron*, ***1997***, *53*, 16131.

Also via Dicarboxylic Acids:	Section 312 (Carboxylic Acids - Carboxylic Acids)
Hydroxy-esters	Section 327 (Alcohol - Ester)
Diols	Section 323 (Alcohol - Alcohol)

SECTION 358: ESTER - ETHER, EPOXIDE, THIOETHER

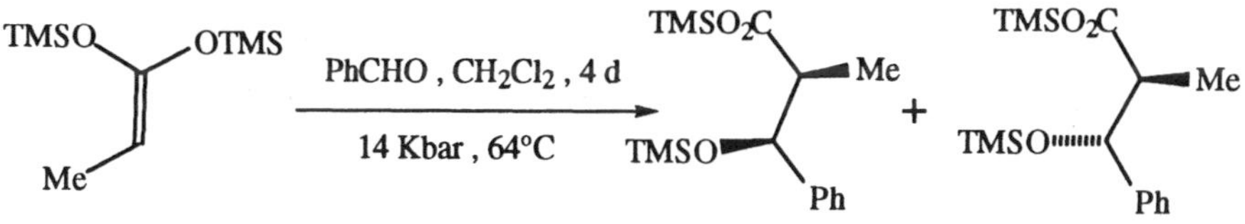
TMSO
OTMS
Me
PhCHO , CH2Cl2 , 4 d
14 Kbar , 64°C
TMSO2C
Me
TMSO
Ph
+
TMSO2C
Me
TMSO
Ph

(63 : 37) 85%

Bellassoued, M.; Reboul, E.; Dumas, F. *Tetrahedron Lett.*, ***1997***, *38*, 5631.

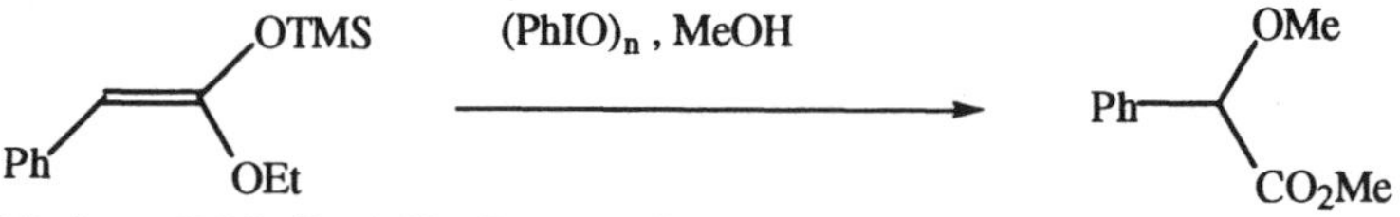
OTMS
Ph
OEt
(PhIO)n , MeOH
OMe
Ph
CO2Me

65%

Moriarty, R.M.; Rani, N.; Condeiu, C.; Duncan, M.P.; Prakash, O. *Synth. Commun.*, ***1997***, *27*, 3273.

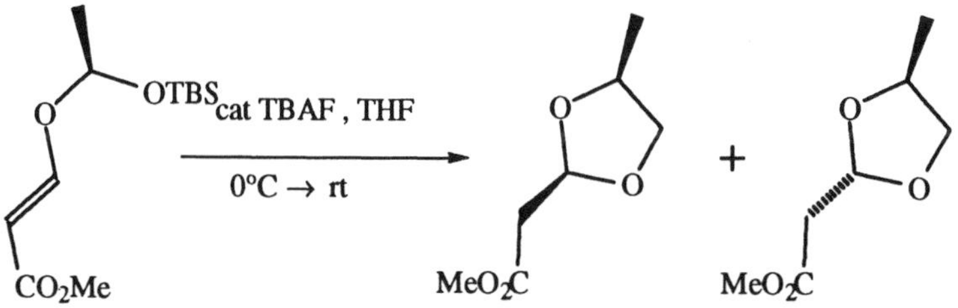
O
OTBS
CO2Me
cat TBAF , THF
0°C → rt
O
O
MeO2C
+
O
O
MeO2C

(1.2 : 1) 86%

Evans, P.A.; Garber, L.T. *Tetrahedron Lett.*, ***1996***, *37*, 2927.

SECTION 359: ESTER - HALIDE, SULFONATE

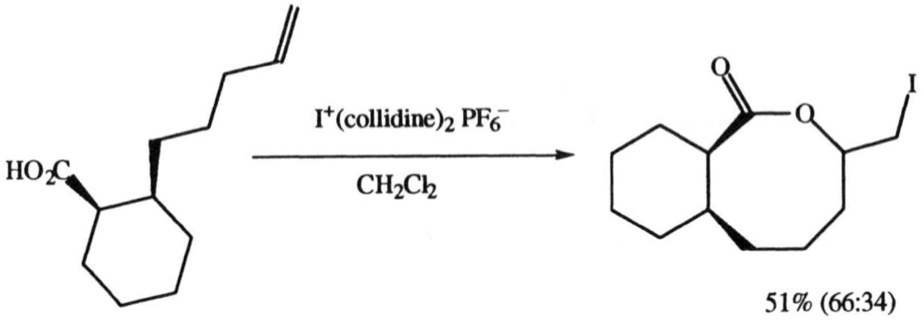
HO2C
I+(collidine)2 PF6−
CH2Cl2
O
O
I
51% (66:34)

Homsi, F.; Rousseau, G. *J. Org. Chem.*, ***1998***, *63*, 5255.

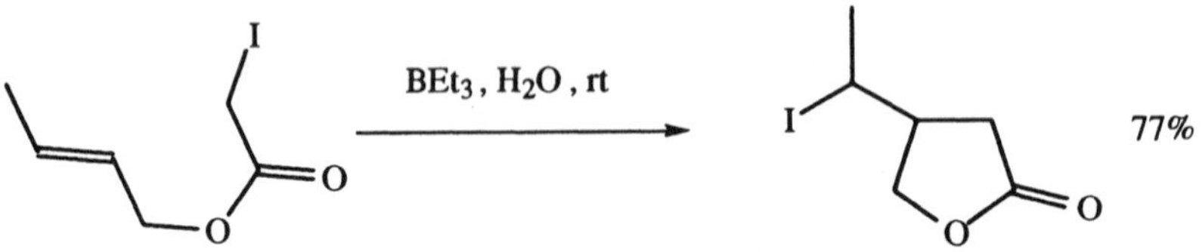
I
O
O
BEt3, H2O, rt
I
O
O
77%

Yorimitsu, H.; Nakamura, T.; Shinokubo, H.; Oshima, K. *J. Org. Chem.*, ***1998***, *63*, 8604.

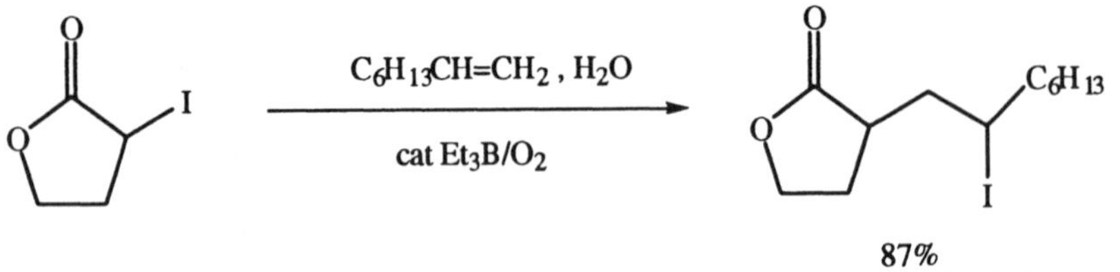
O
O
I
C6H13CH=CH2, H2O
cat Et3B/O2
O
O
C6H13
I
87%

Nakamura, T.; Yorimitsu, H.; Shinokubo, H.; Oshima, K. *SynLett.*, ***1998***, 1351.

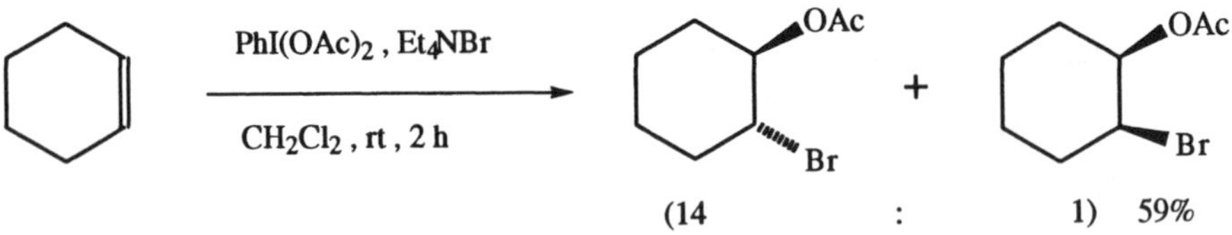
PhI(OAc)2, Et4NBr
CH2Cl2, rt, 2 h
OAc
Br
+
OAc
Br
(14 : 1) 59%

Hashem, Md.A.; Jung, A.; Ries, M.; Kirschning, A. *SynLett.*, ***1998***, 195.

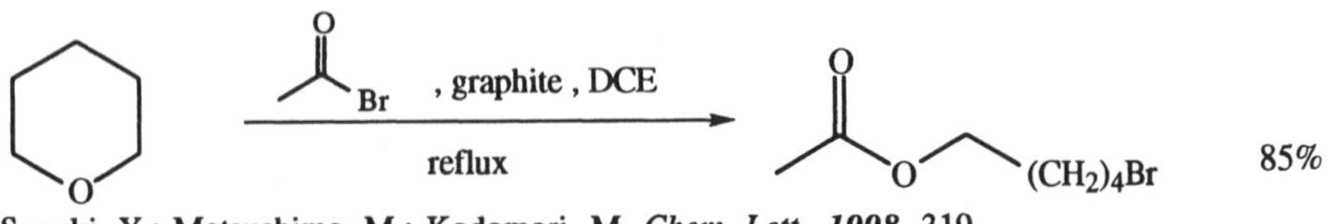
O
O
Br
, graphite, DCE
reflux
O
O
(CH2)4Br
85%

Suzuki, Y.; Matsushima, M.; Kodomari, M. *Chem. Lett.*, ***1998***, 319.

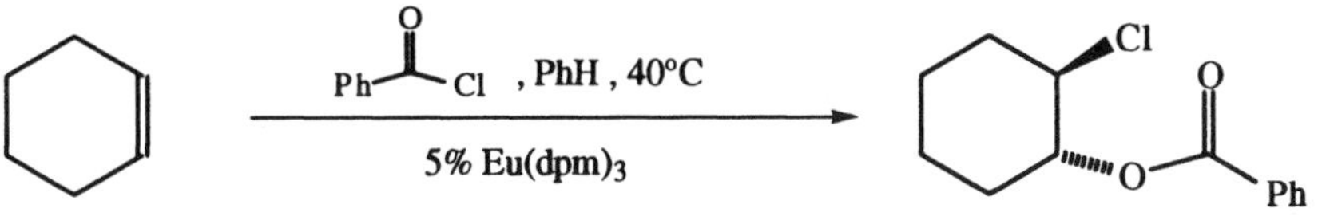

Ph Cl , PhH , 40°C
5% Eu(dpm)3
Cl
O
O
Ph

92%

Tankguchi, Y.; Tanbaka, S.; Kitamura, T.; Fujiwara, Y. *Tetrahedron Lett.*, ***1998***, *39*, 4559.

SECTION 360: ESTER - KETONE

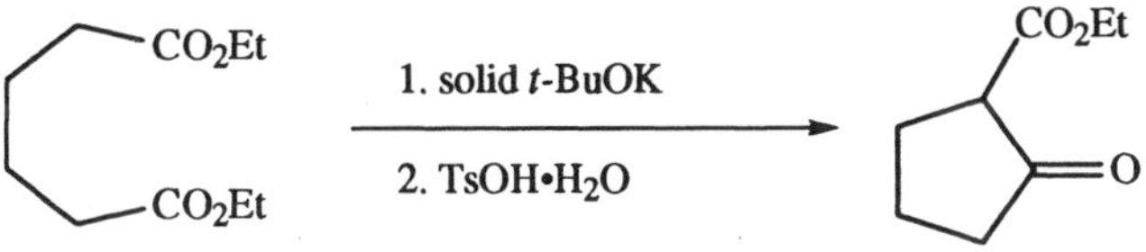

CO2Et
CO2Et
1. solid t-BuOK
2. TsOH•H2O
CO2Et
O

69%

solvent free Dieckmann

Toda, F.; Suzuki, Y.; Higa, S. *J. Chem. Soc., Perkin Trans. 1*, ***1998***, 3521.

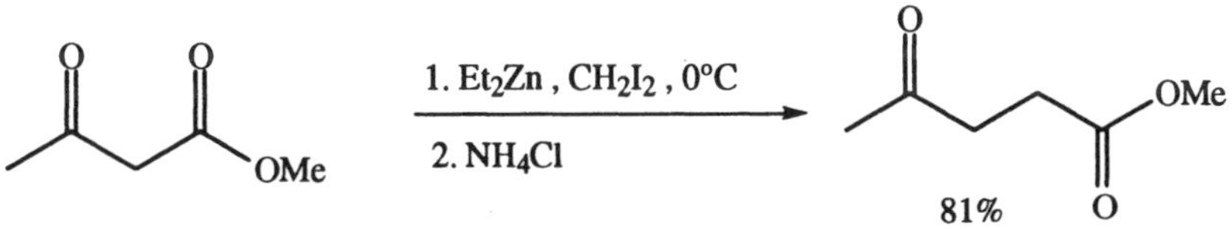

O
O
OMe
1. Et2Zn , CH2I2 , 0°C
2. NH4Cl
O
OMe
O

81%

Brogan, J.B.; Zercher, C.K. *J. Org. Chem.*, ***1997***, *62*, 6444.

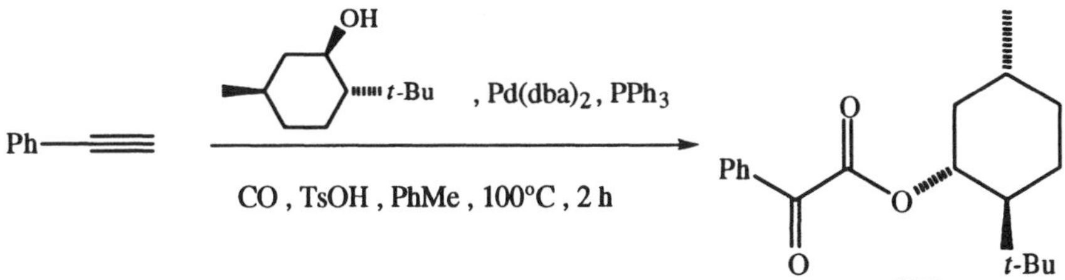

OH
t-Bu
, Pd(dba)2 , PPh3
Ph
CO , TsOH , PhMe , 100°C , 2 h
O
Ph
O
O
t-Bu

88%

Monteiro, A.L.; Lando, V.R.; Gasparini, V. *Synth. Commun.*, ***1997***, *27*, 3605.

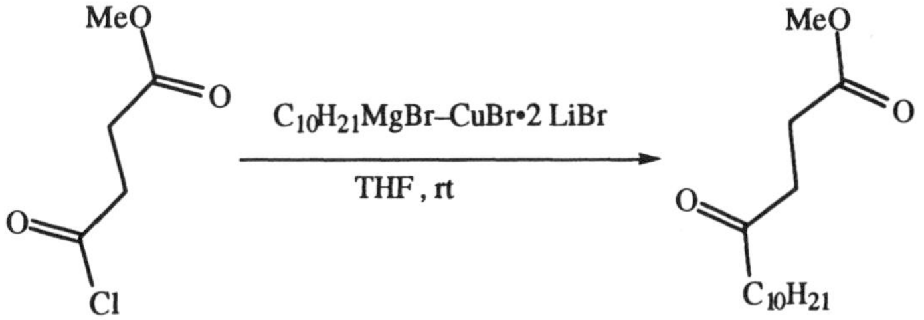

MeO
O
O
Cl
C10H21MgBr–CuBr•2 LiBr
THF , rt
MeO
O
O
C10H21

80%

Babudri, F.; Fiandanese, V.; Marchese, G.; Punzi, A. *Tetrahedron*, ***1996***, *52*, 13513.

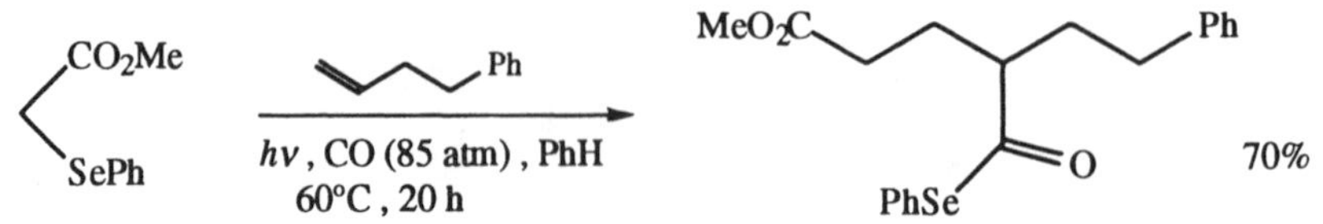

Ryu, I.; Muraoka, H.; Kambe, N.; Komatsu, M.; Sonoda, N. *J. Org. Chem.*, ***1996***, *61*, 6396

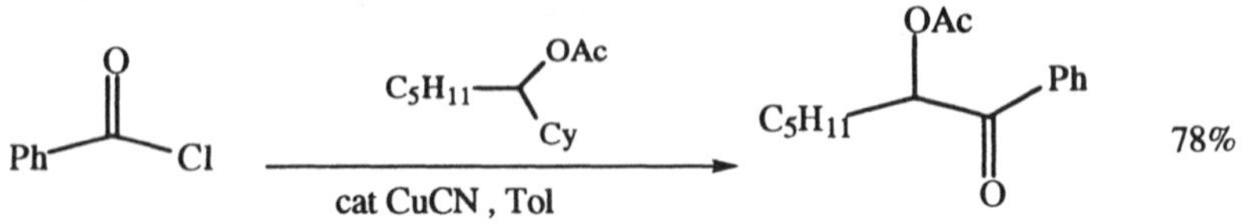

Linderman, R.J.; Siedlecki, J.M. *J. Org. Chem.*, ***1996***, *61*, 6492.

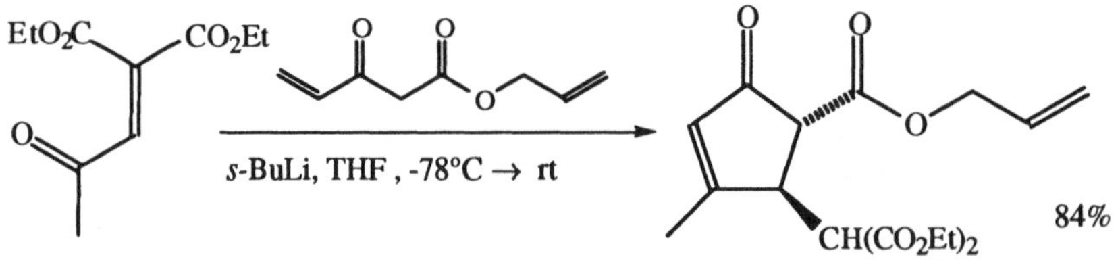

Hatanaka, M.; Ishida, A.; Tanaka, Y.; Ueda, I. *Tetrahedron Lett.*, ***1996***, *37*, 401.

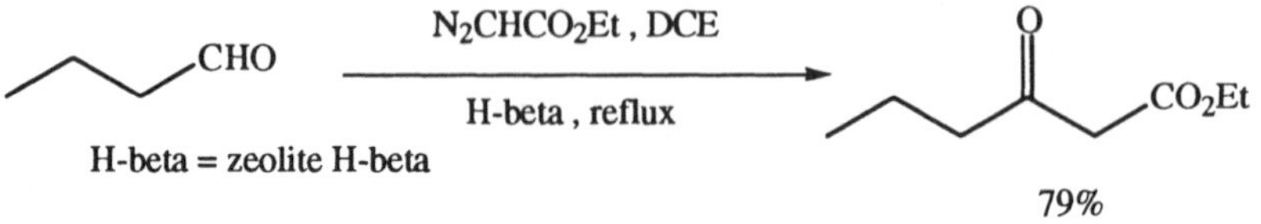

Sudrik, S.Gtra, R.B.; Sonawane, H.R. *SynLett.*, ***1996***, 369.

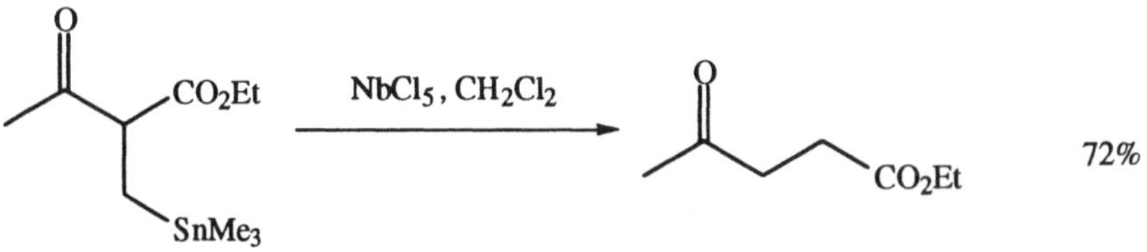

Yamamoto, M.; Nakazawa, M.; Kishikawa, K.; Kohmoto, S. *Chem. Commun.*, ***1996***, 2353.

Also via Keto acids	Section 320 (Carboxylic Acid - Ketone)
Hydroxy ketones	Section 330 (Alcohol - Ketone)

SECTION 361: ESTER - NITRILE

NO ADDITIONAL EXAMPLES

SECTION 362: ESTER - ALKENE

This section contains syntheses of enol esters and esters of unsaturated acids as well as ester molecules bearing a remote alkenyl unit.

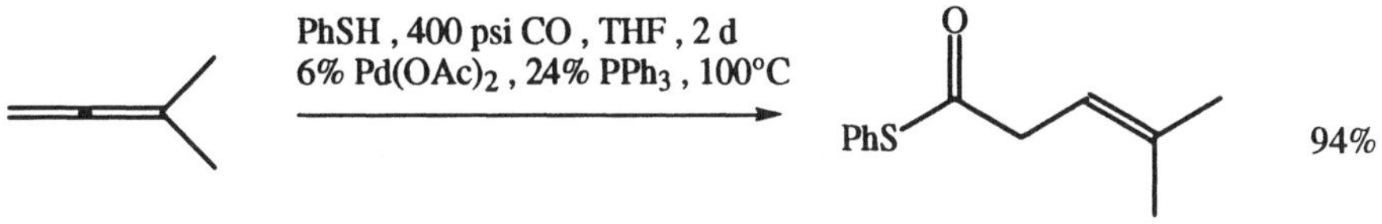
PhSH , 400 psi CO , THF , 2 d
6% Pd(OAc)2 , 24% PPh3 , 100°C
O
PhS
94%

Xiao, W.-J.; Vasapollo, G.; Alper, H. *J. Org. Chem.*, ***1998***, *63*, 2609.

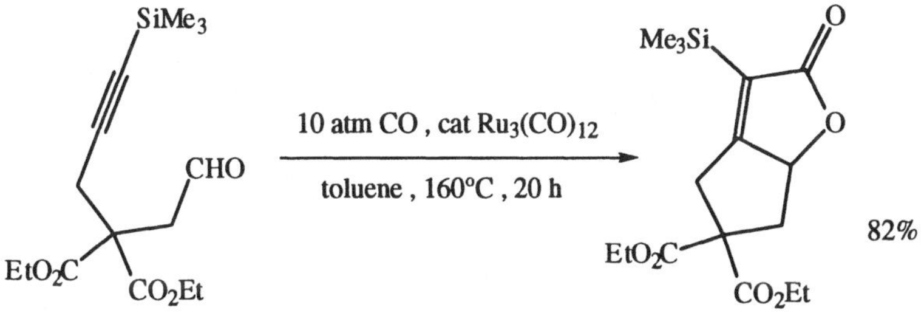
SiMe3
CHO
EtO2C
CO2Et
10 atm CO , cat Ru3(CO)12
toluene , 160°C , 20 h
Me3Si
O
O
EtO2C
CO2Et
82%

Chatani, N.; Morimoto, T.; Fukumoto, Y.; Murai, S. *J. Am. Chem. Soc.*, ***1998***, *120*, 5335.

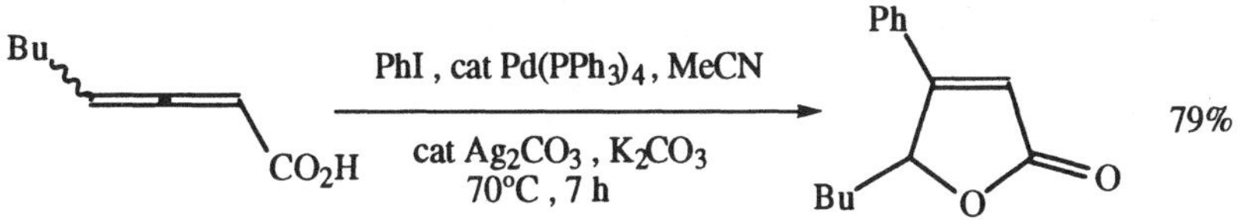
Bu
CO2H
PhI , cat Pd(PPh3)4 , MeCN
cat Ag2CO3 , K2CO3
70°C , 7 h
Ph
Bu
O
O
79%

Ma, S.; Shi, Z. *J. Org. Chem.*, ***1998***, *63*, 6387.

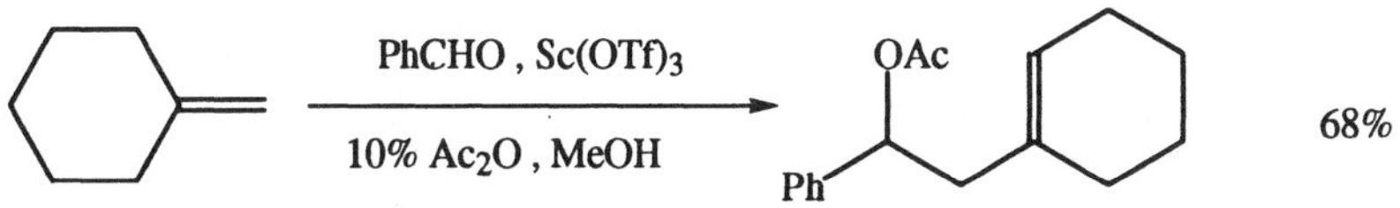
PhCHO , Sc(OTf)3
10% Ac2O , MeOH
OAc
Ph
68%

Aggarwal, V.K.; Vennall, G.P.; Davey, P.N.; Newman, C. *Tetrahedron Lett.*, ***1998***, *39*, 1997.

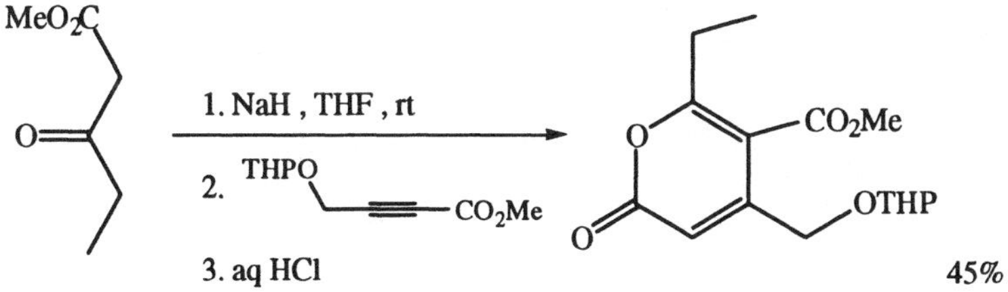
MeO2C
O
1. NaH , THF , rt
2. THPO
CO2Me
3. aq HCl
O
CO2Me
OTHP
O
45%

Covarrubias-Zúñiga, A.; Maldonado, L.A.; Díaz-Domínguez, J.
Synth. Commun., ***1998***, *28*, 1531.

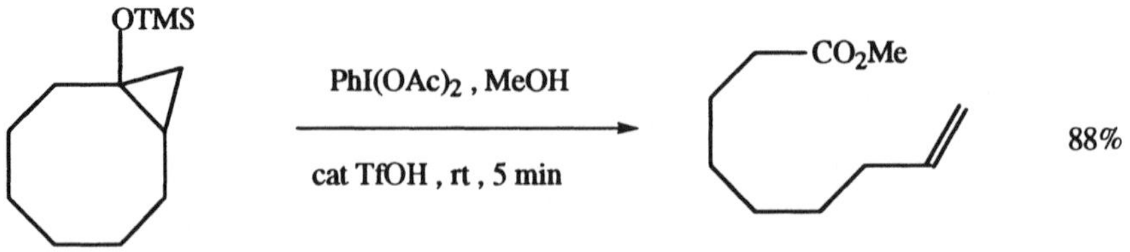
OTMS
PhI(OAc)2 , MeOH
cat TfOH , rt , 5 min
CO2Me
88%

Kirihara, M.; Yokoyama, S.; Momose, T. *Synth. Commun.*, ***1998***, *28*, 1947.

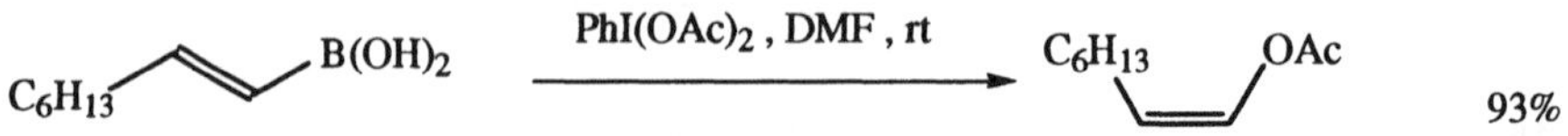
C6H13
B(OH)2
PhI(OAc)2 , DMF , rt
C6H13
OAc
93%

Murata, M.; Satoh, K.; Watanabe, S.; Masuda, Y. *J. Chem. Soc., Perkin Trans. 1*, ***1998***, 1465

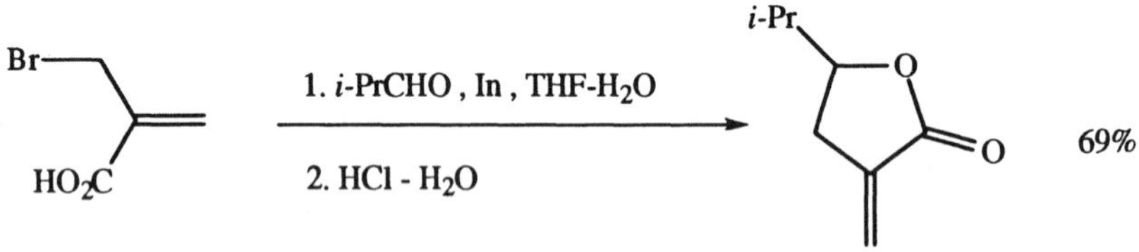
Br
HO2C
1. i-PrCHO , In , THF-H2O
2. HCl - H2O
i-Pr
O
O
69%

Choudhary, R.K.; Foubelo, F.; Yus, M. *Tetrahedron Lett.*, ***1998***, *39*, 3581.

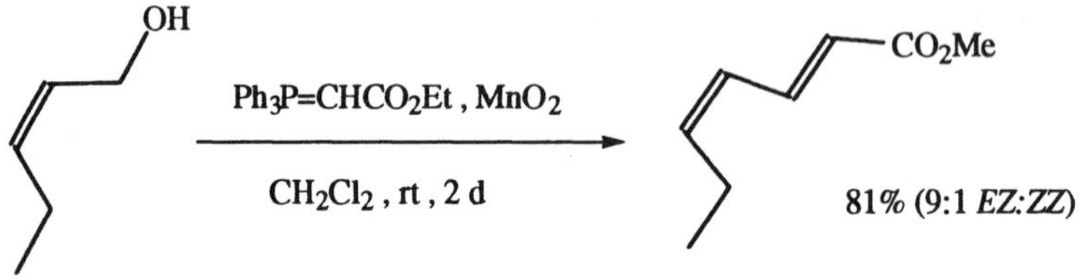
OH
Ph3P=CHCO2Et , MnO2
CH2Cl2 , rt , 2 d
CO2Me
81% (9:1 EZ:ZZ)

Wei, X.; Taylor, R.J.K. *Tetrahedron Lett.*, ***1998***, *39*, 3815.

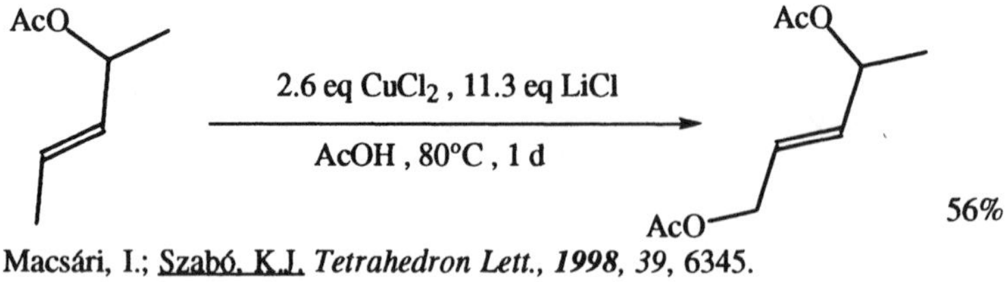
AcO
2.6 eq CuCl2 , 11.3 eq LiCl
AcOH , 80°C , 1 d
AcO
AcO
56%

Macsári, I.; Szabó, K.J. *Tetrahedron Lett.*, ***1998***, *39*, 6345.

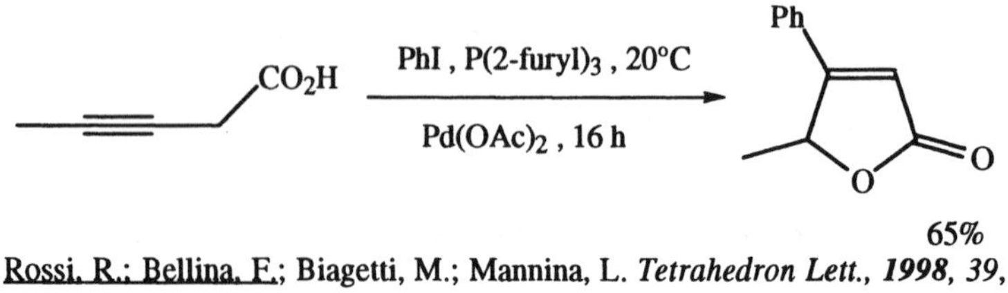
CO2H
PhI , P(2-furyl)3 , 20°C
Pd(OAc)2 , 16 h
Ph
O
O
65%

Rossi, R.; Bellina, F.; Biagetti, M.; Mannina, L. *Tetrahedron Lett.*, ***1998***, *39*, 7599.

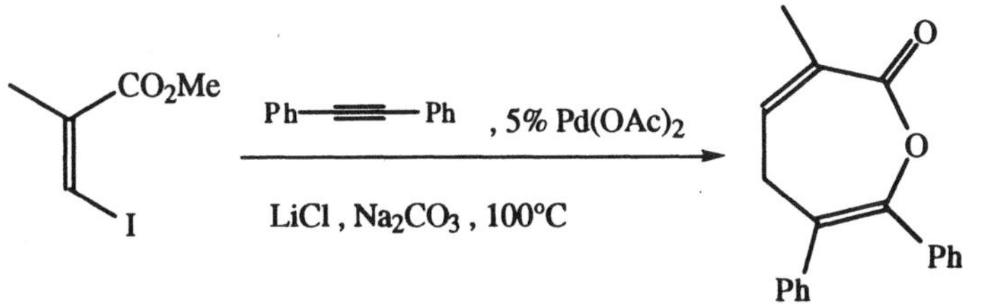
CO2Me
I
Ph——Ph , 5% Pd(OAc)2
LiCl , Na2CO3 , 100°C
O
O
Ph
Ph

40%

Larock, R.C.; Han, X.; Doty, M.J. *Tetrahedron Lett.*, ***1998***, *39*, 5713.

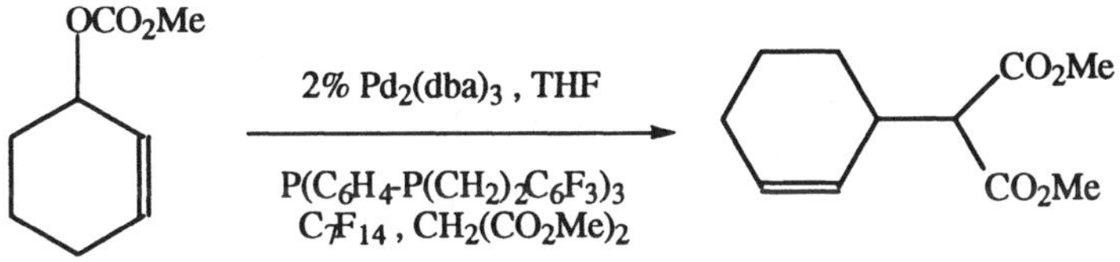
OCO2Me
2% Pd2(dba)3 , THF
P(C6H4-P(CH2)2C6F3)3
C7F14 , CH2(CO2Me)2
CO2Me
CO2Me

quant

Kling, R.; Sinou, D.; Pozzi, G.; Choplin, A.; Quignard, F.; Busch, S.; Kainz, S.; Koch, D.; Leitner, W. *Tetrahedron Lett.*, ***1998***, *39*, 9439.

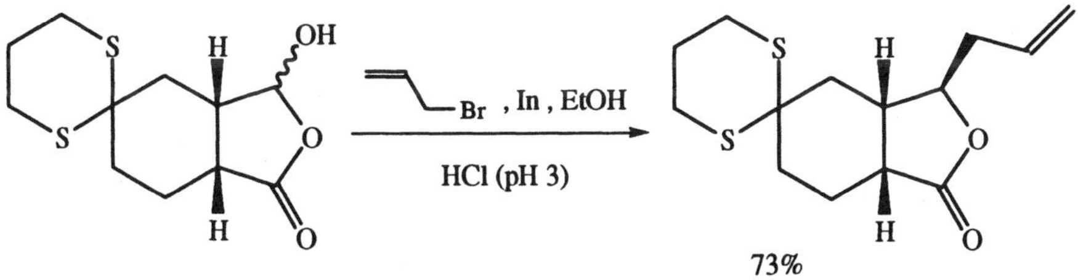
S
S
H
OH
O
H
O
Br , In , EtOH
HCl (pH 3)
S
S
H
O
H
O

73%
(6.5:1 α:β)

Bernardelli, P.; Paquette, L.A. *J. Org. Chem.*, ***1997***, *62*, 8284.

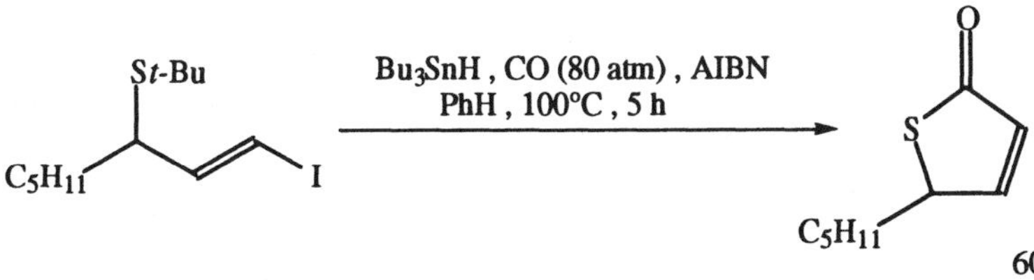
St-Bu
C5H11
I
Bu3SnH , CO (80 atm) , AIBN
PhH , 100°C , 5 h
O
S
C5H11

60%

Ryu, I.; Okuda, T.; Nagahara, K.; Kambe, N.; Komatsu, M.; Sonoda, N. *J. Org. Chem.*, ***1997***, *62*, 7550.

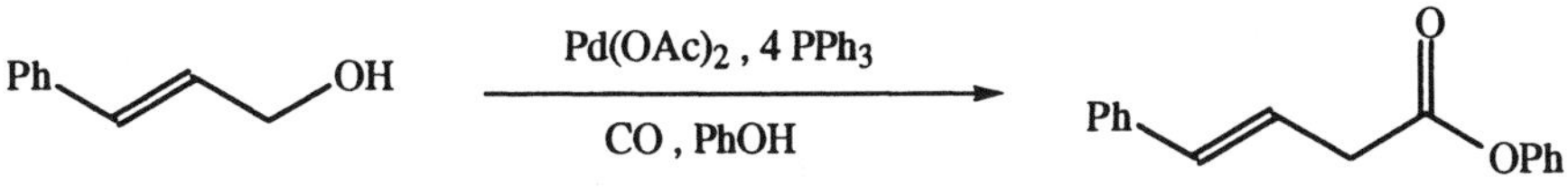
Ph
OH
Pd(OAc)2 , 4 PPh3
CO , PhOH
Ph
O
OPh

76%

Satoh, T.; Ikeda, M.; Kushino, Y.; Miura, M.; Nomura, M. *J. Org. Chem.*, ***1997***, *62*, 2662.

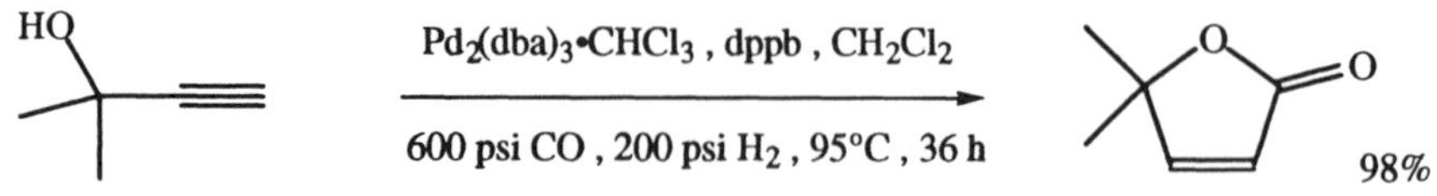
HO
Pd2(dba)3•CHCl3 , dppb , CH2Cl2
600 psi CO , 200 psi H2 , 95°C , 36 h
O
O
98%

Yu, W.–Y.; Alper, H. *J. Org. Chem.*, ***1997***, *62*, 5684.

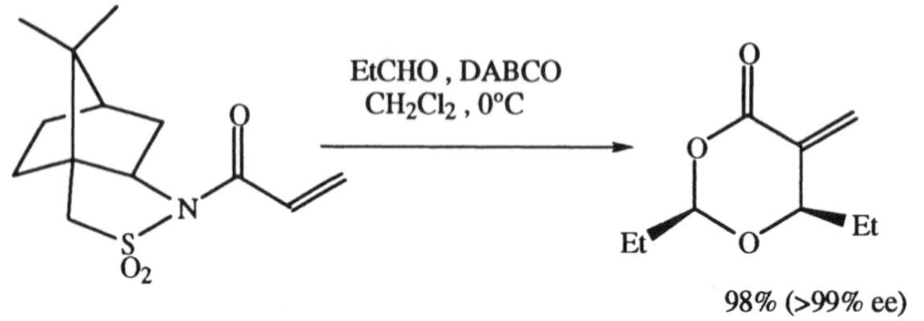
O
N
S
O2
EtCHO , DABCO
CH2Cl2 , 0°C
O
O
O
Et
Et
98% (>99% ee)

Brzezinski,. L.J.; Rafel, S.; Leahy, J.W. *J. Am. Chem. Soc.*, ***1997***, *119*, 4317.

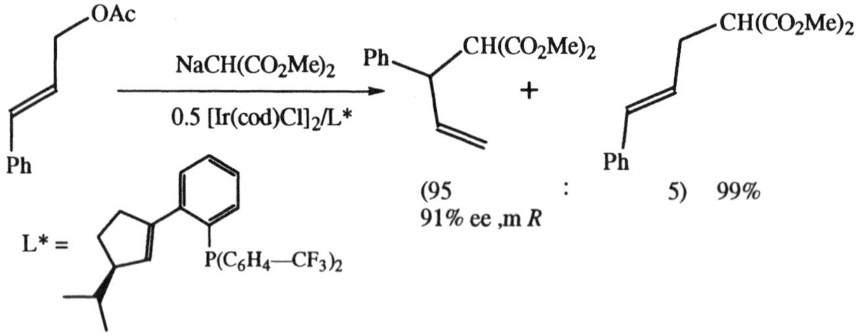
OAc
Ph
NaCH(CO2Me)2
0.5 [Ir(cod)Cl]2/L*
Ph
CH(CO2Me)2
+
CH(CO2Me)2
Ph
(95 : 5) 99%
91% ee ,m R
L* =
P(C6H4—CF3)2

Janssen, J.P.; Helmchen, G. *Tetrahedron Lett.*, ***1997***, *38*, 8025.

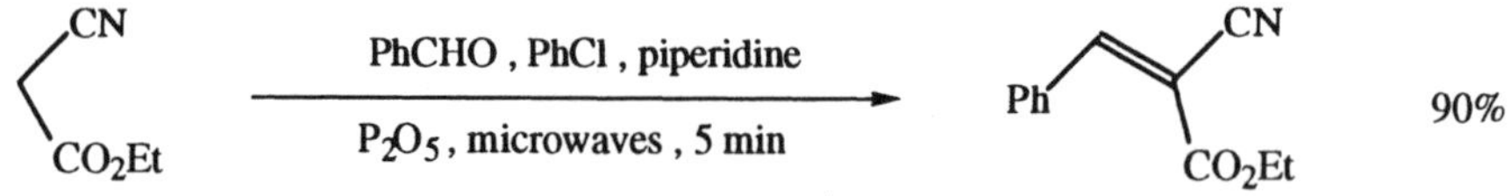
CN
CO2Et
PhCHO , PhCl , piperidine
P2O5 , microwaves , 5 min
Ph
CN
CO2Et
90%

Kim, S.–Y.; Kwon, P.S.; Kwon, T.–W.; Chung, S.–K.; Chang, Y.–T. *Synth. Commun.*, ***1997***, *27*, 533.

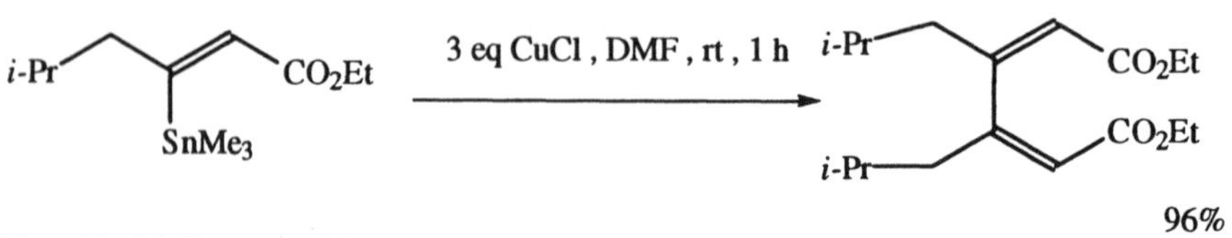
i-Pr
CO2Et
SnMe3
3 eq CuCl , DMF , rt , 1 h
i-Pr
CO2Et
CO2Et
i-Pr
96%

Piers, E.; McEachern, E.J.; Romero, M.A.; Gladstone, P.L. *Can. J. Chem.*, ***1997***, *75*, 694.

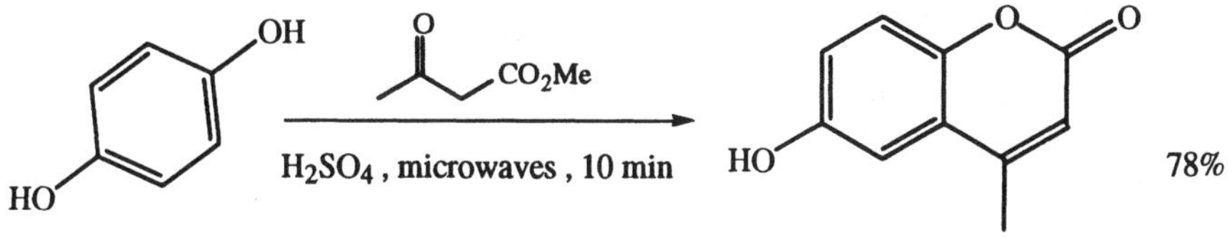
OH
O
CO_2Me
H_2SO_4, microwaves, 10 min
HO
O
O
HO
78%

Singh, V.; Singh, J.; Kaur, K.P.; Kad, G.L. *J. Chem. Res. (S)*, ***1997***, 58.

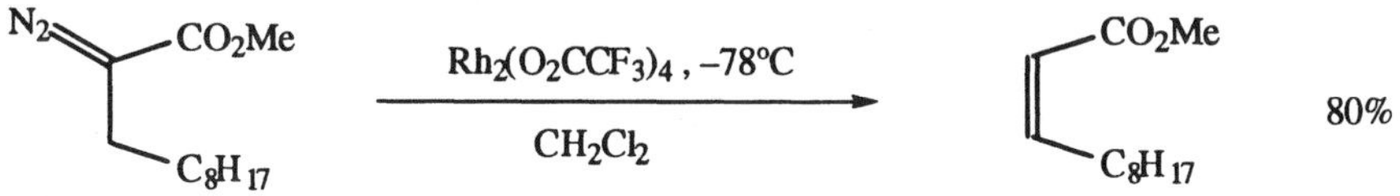
N_2
CO_2Me
C_8H_{17}
$Rh_2(O_2CCF_3)_4$, –78°C
CH_2Cl_2
CO_2Me
C_8H_{17}
80%

Taber, D.F.; Herr, R.J.; Pack, S.K.; Geremia, J.M. *J. Org. Chem.*, ***1996***, *61*, 2908.

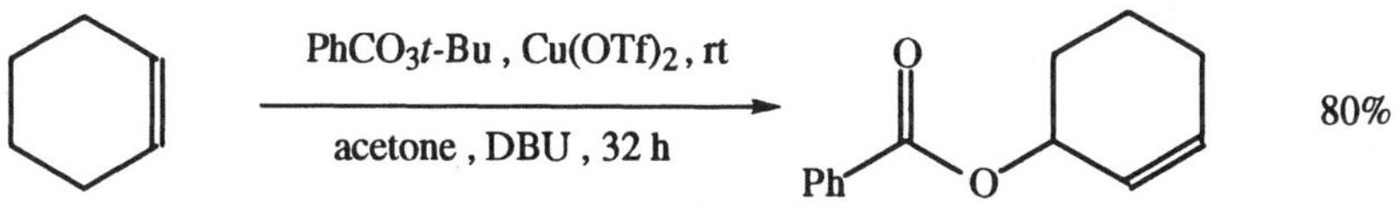
$PhCO_3t$-Bu, $Cu(OTf)_2$, rt
acetone, DBU, 32 h
O
Ph
O
80%

Sekar, G.; Datta Gupta, A.; Singh, V.K. *Tetrahedron Lett.*, ***1996***, *37*, 8435.

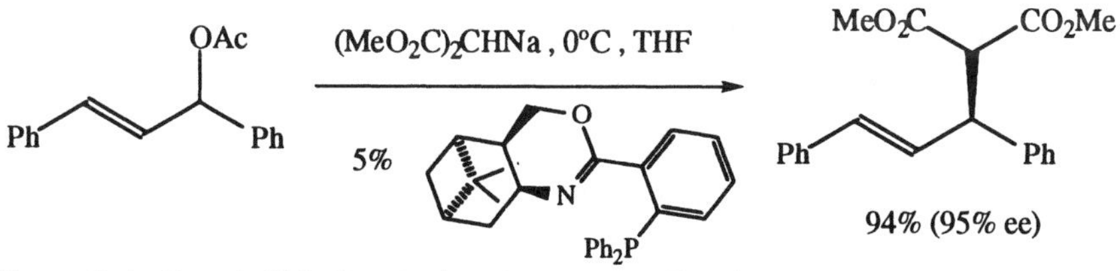
OAc
Ph
Ph
$(MeO_2C)_2CHNa$, 0°C, THF
5%
O
N
Ph_2P
MeO_2C
CO_2Me
Ph
Ph
94% (95% ee)

Evans, P.A.; Brandt, T.A. *Tetrahedron Lett.*, ***1996***, *37*, 9143.

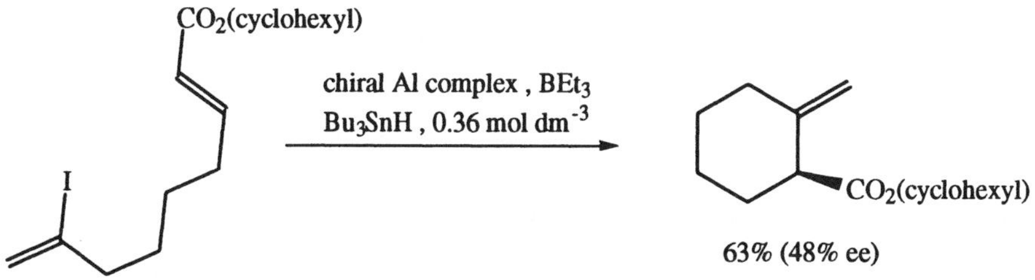
CO_2(cyclohexyl)
I
chiral Al complex, BEt_3
Bu_3SnH, 0.36 mol dm^{-3}
CO_2(cyclohexyl)
63% (48% ee)

Nishida, M.; Hayashi, H.; Nishida, A.; Kawahara, N. *Chem. Commun.*, ***1996***, 579.

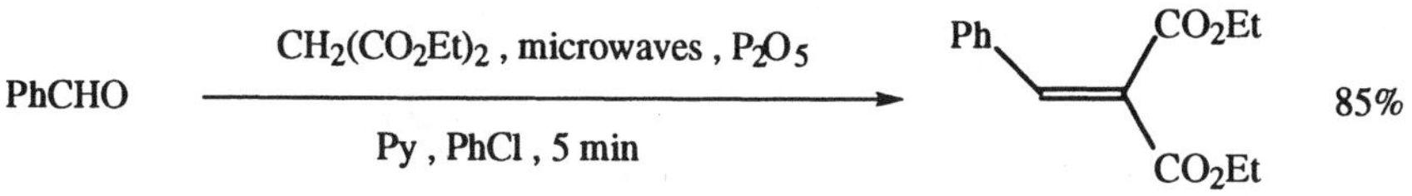
PhCHO
$CH_2(CO_2Et)_2$, microwaves, P_2O_5
Py, PhCl, 5 min
Ph
CO_2Et
CO_2Et
85%

Kim, J.–K.; Kwon, P.–S.; Kwon, T.–W.; Chung, S.–K.; Lee, J.–W. *Synth. Commun.*, ***1996***, *26*, 535.

Related Methods:	Section 60A (Protection of Aldehydes).
	Section 180A (Protection of Ketones).
Also via Acetylenic Esters:	Section 306 (Alkyne - Ester).
Alkenyl Acids:	Section 322 (Carboxylic Acid - Alkene).
β-Hydroxy-esters:	Section 327 (Alcohol - Ester).

SECTION 363: ETHER, EPOXIDE, THIOETHER - ETHER, EPOXIDE, THIOETHER

See Section 60A (Protection of Aldehydes) and Section 180A (Protection of Ketones) for reactions involving formation of Acetals and Ketals.

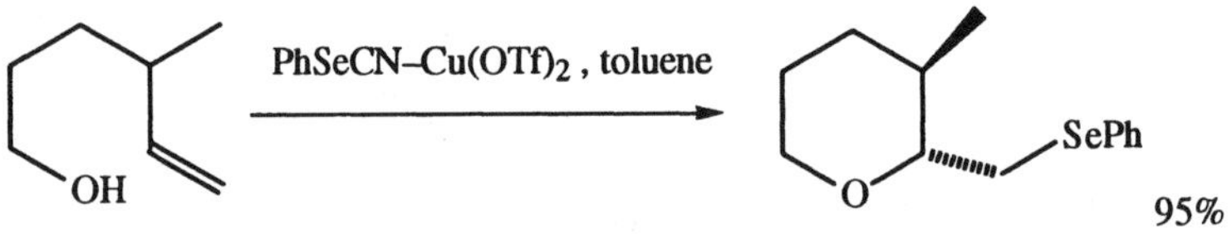
OH
PhSeCN–Cu(OTf)2 , toluene
SePh
O
95%

Inoue, H.; Murata, S. *Heterocycles*, ***1997***, *45*, 847.

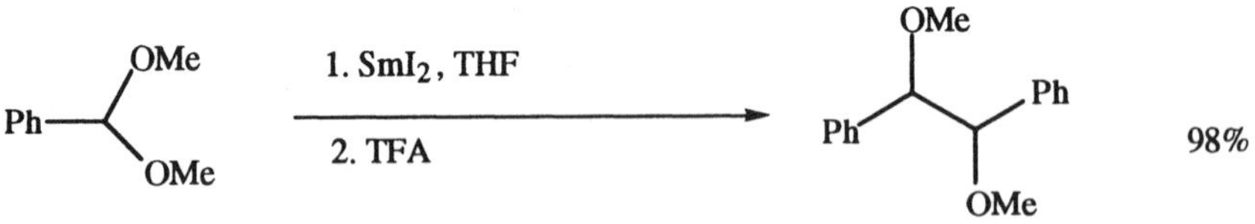
OMe
Ph
OMe
1. SmI2, THF
2. TFA
OMe
Ph
Ph
OMe
98%

Studer, A.; Curran, D.P. *SynLett.*, ***1996***, 255.

SECTION 364: ETHER, EPOXIDE, THIOETHER - HALIDE, SULFONATE

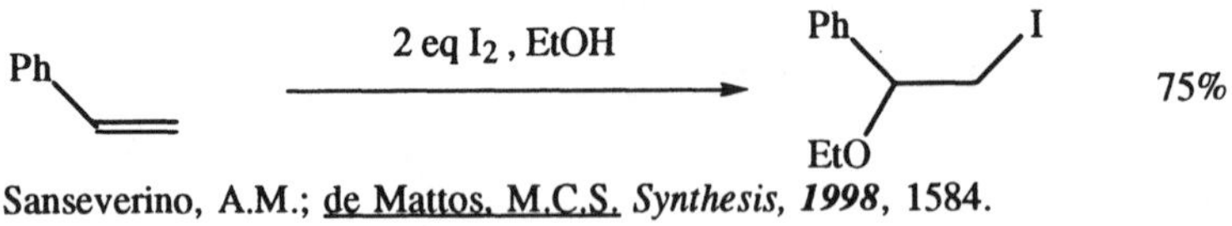
Ph
2 eq I2 , EtOH
Ph
I
EtO
75%

Sanseverino, A.M.; de Mattos, M.C.S. *Synthesis*, ***1998***, 1584.

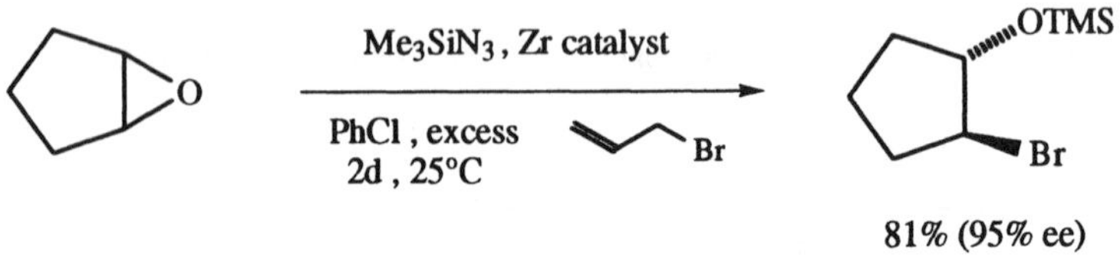
O
Me3SiN3 , Zr catalyst
PhCl , excess
2d , 25°C
Br
OTMS
Br
81% (95% ee)

Nugent, W.A. *J. Am. Chem. Soc.*, ***1998***, *120*, 7139.

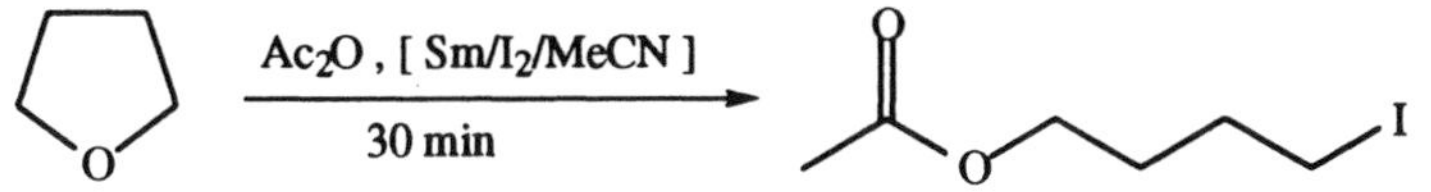
O
Ac_2O , [$Sm/I_2/MeCN$]
30 min
O
O
I

90%

Bao, W.; Ying, T.; Zhang, Y.; Yu, M. *Org. Prep. Proceed. Int.*, ***1997***, *29*, 335.

SECTION 365: ETHER, EPOXIDE, THIOETHER - KETONE

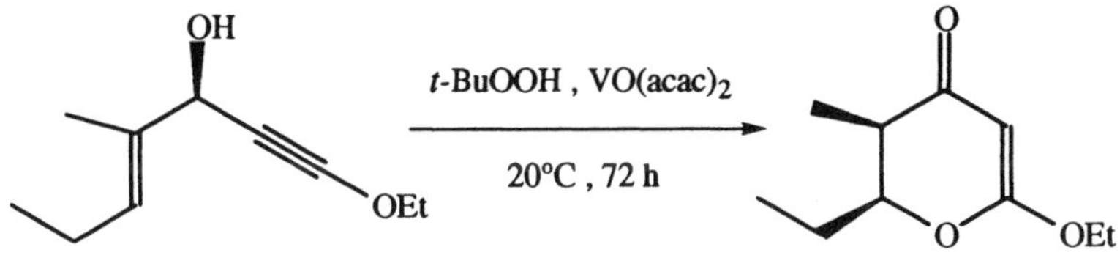
OH
OEt
t-BuOOH , $VO(acac)_2$
20°C , 72 h
O
O
OEt

55%

Marson, C.M.; Harper, S.; Oare, C.A. *J. Org. Chem.*, ***1998***, *63*, 3798.

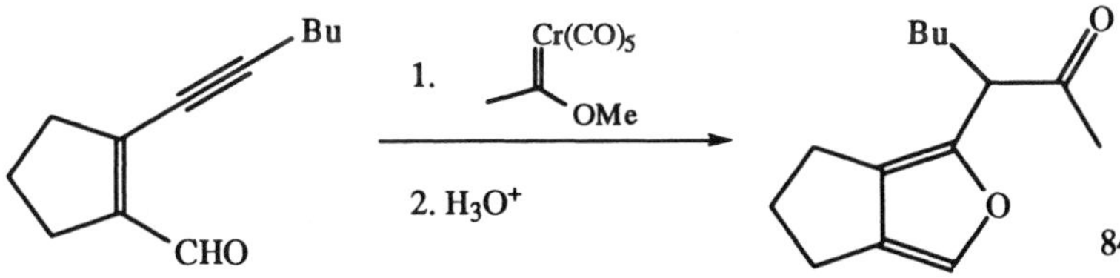
Bu
CHO
1.
$Cr(CO)_5$
OMe
2. H_3O^+
Bu
O
O

84%

Herndon, J.W.; Wang, H. *J. Org. Chem.*, ***1998***, *63*, 4564.

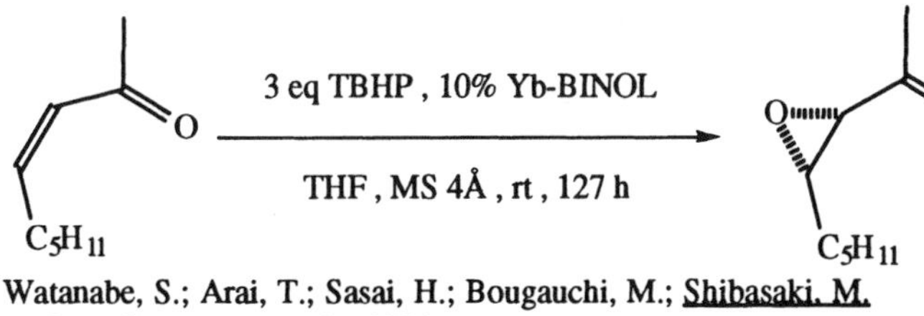
O
C_5H_{11}
3 eq TBHP , 10% Yb-BINOL
THF , MS 4Å , rt , 127 h
O
O
C_5H_{11}

80% (96% ee)

Watanabe, S.; Arai, T.; Sasai, H.; Bougauchi, M.; Shibasaki, M.
J. Org. Chem., ***1998***, *63*, 8090.

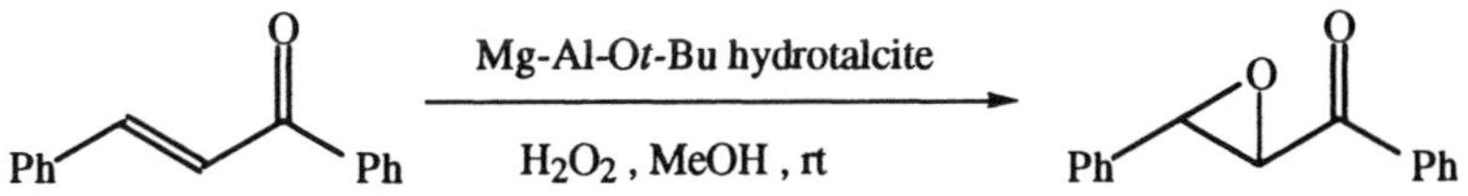
O
Ph
Ph
Mg-Al-O*t*-Bu hydrotalcite
H_2O_2 , MeOH , rt
O
O
Ph
Ph

quant

Choudary, B.M.; Kantam, M.L.; Bharathi, B.; Reddy, Ch.V. *SynLett.*, ***1998***, 1203.

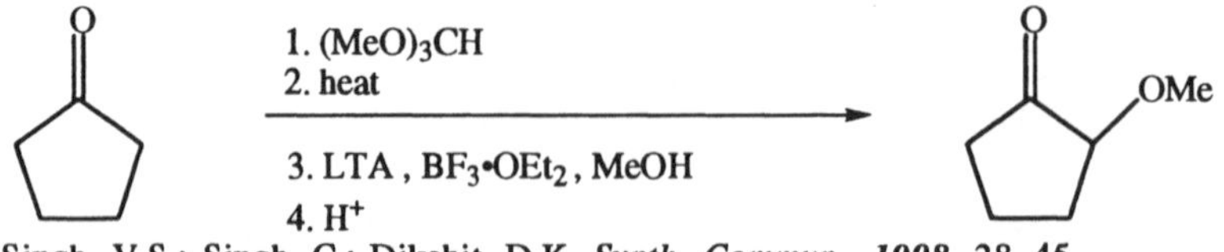
1. (MeO)3CH
2. heat
3. LTA , BF3•OEt2 , MeOH
4. H+
OMe

Singh, V.S.; Singh, C.; Dikshit, D.K. *Synth. Commun.*, ***1998***, *28*, 45.

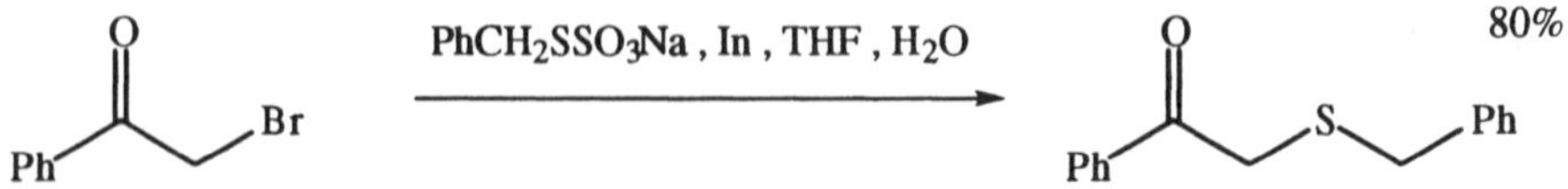
PhCH2SSO3Na , In , THF , H2O
Br
Ph
S
Ph
80%

Zhan, Z.; Zhang, Y.; Ma, Z. *J. Chem. Res. (S)*, ***1998***, 130.

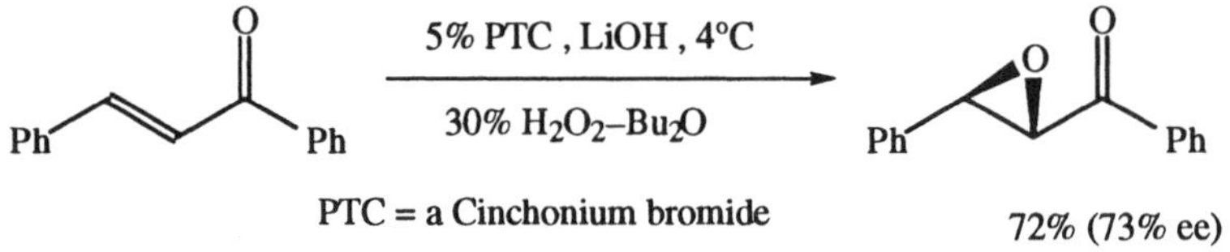
5% PTC , LiOH , 4°C
30% H2O2–Bu2O
Ph
Ph
PTC = a Cinchonium bromide
72% (73% ee)

Arai, S.; Tsuge, H.; Shioiri, T. *Tetrahedron Lett.*, ***1998***, *39*, 7567.

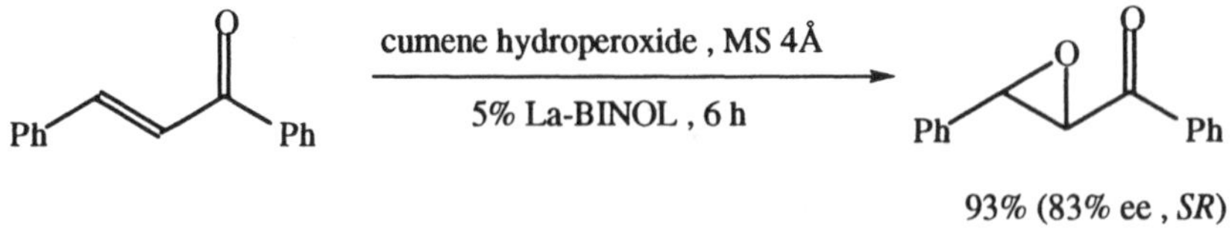
cumene hydroperoxide , MS 4Å
5% La-BINOL , 6 h
Ph
Ph
93% (83% ee , SR)

Bougauchi, M.; Watanabe, S.; Arai, T.; Sasai, H.; Shibasaki, M. *J. Am. Chem. Soc.*, ***1997***, *119*, 2329.

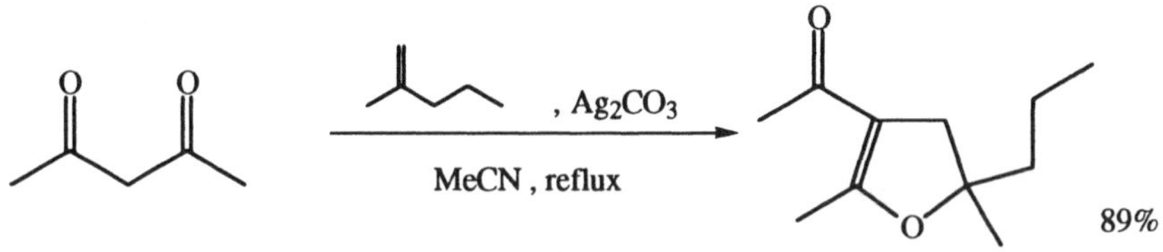
, Ag2CO3
MeCN , reflux
89%

Lee, Y.R.; Kim, B.S. *Tetrahedron Lett.*, ***1997***, *38*, 2095.

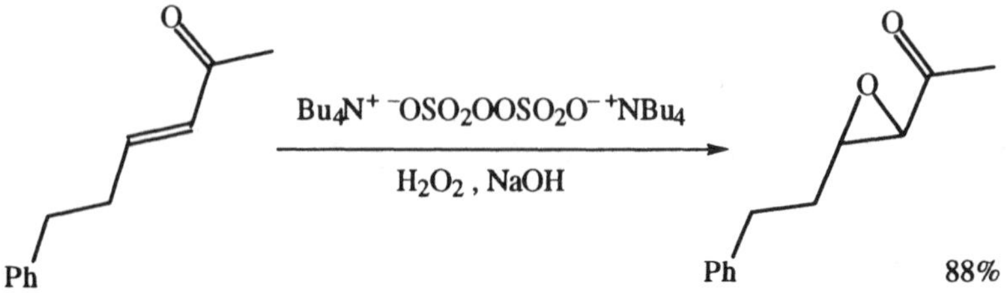
Bu4N+ −OSO2OOSO2O− +NBu4
H2O2 , NaOH
Ph
Ph
88%

Kim, Y.H.; Hwang, J.P.; Yang, S.G. *Tetrahedron Lett.*, ***1997***, *38*, 3009.

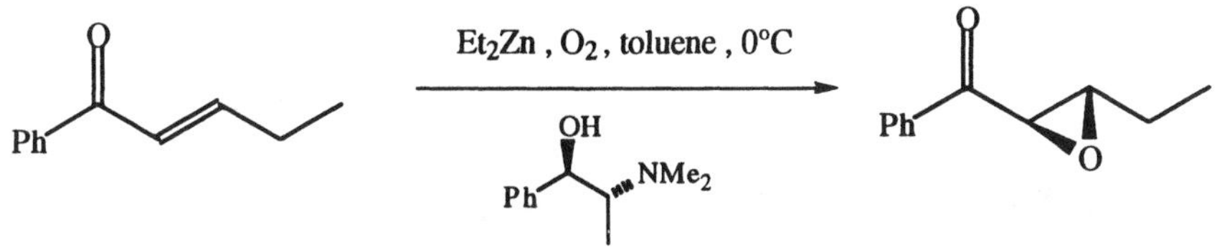
Ph
O
Et_2Zn , O_2 , toluene , 0°C
OH
Ph
NMe_2
Ph
O
O

99% (>94% de , 91% ee , *RS*)

Enders, D.; Zhu, J.; Kramps, L. *Liebigs Ann. Chem.*, ***1997***, 1101.

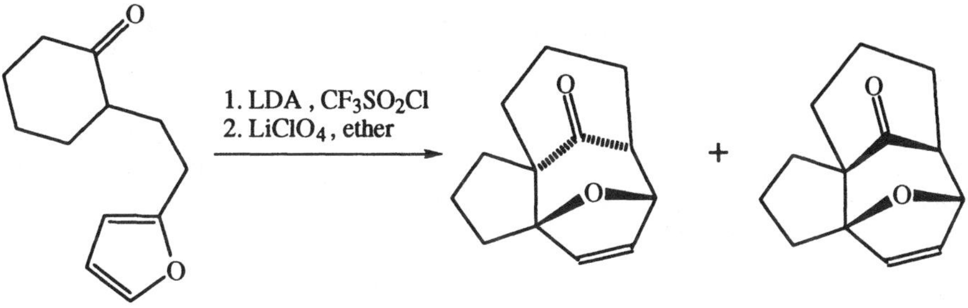
O
O
1. LDA , CF_3SO_2Cl
2. $LiClO_4$, ether
O
O
+
O
O

(1 : 2.5) 54%

Harmata, M.; Elomari, S.; Barnes, C.L. *J. Am. Chem. Soc.*, ***1996***, *118*, 2860.

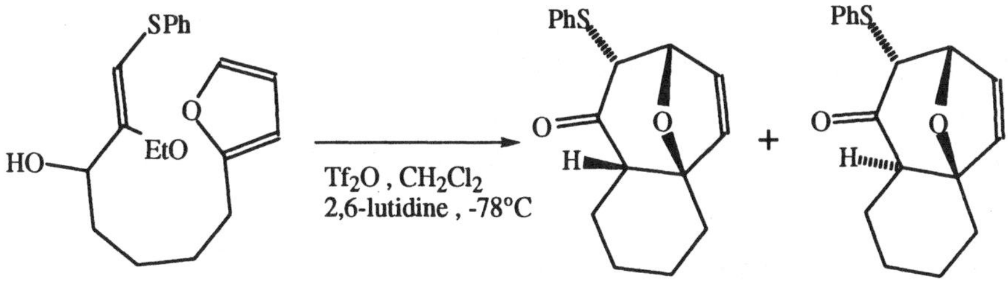
SPh
HO
EtO
O
Tf_2O , CH_2Cl_2
2,6-lutidine , -78°C
PhS
O
H
O
+
PhS
O
H
O

(2 : 1) 59%

Harmata, M.; Jones, D.E. *Tetrahedron Lett.*, ***1996***, *37*, 783.

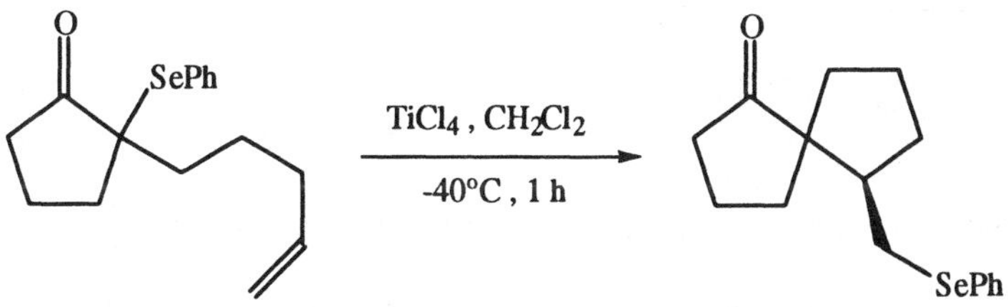
O
SePh
$TiCl_4$, CH_2Cl_2
-40°C , 1 h
O
SePh

77 % (97:3 *cis:trans*)

Toru, T.; Kawai, S.; Ueno, Y. *SynLett.*, ***1996***, 539.

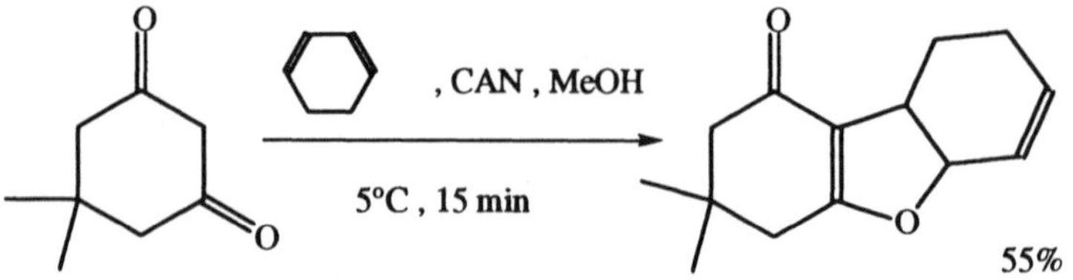
O
, CAN , MeOH
5°C , 15 min
O
O
O
55%

Nair, V.; Mathew, J.; Nair, L.G. *Synth. Commun.*, ***1996***, *26*, 4531.

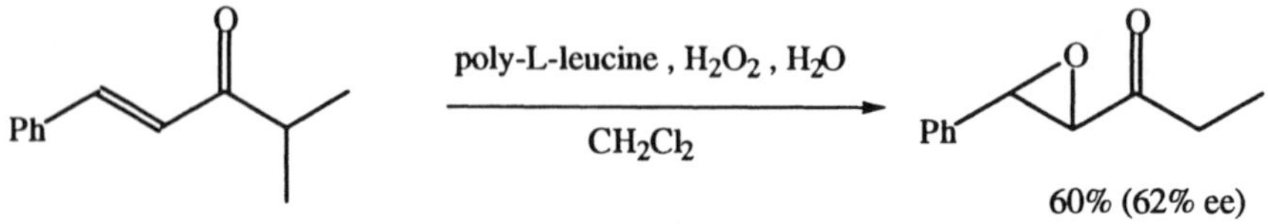
O
Ph
poly-L-leucine , H_2O_2 , H_2O
CH_2Cl_2
O
O
Ph
60% (62% ee)

Kroutil, W.; Lasterra-Sánchez, M.; Maddrell, S.J.; Mayon, P.; Morgan, P.; Roberts, S.M.; Thornton, S.R.; Todd, C.J.; Tüter, M. *J. Chem. Soc., Perkin Trans. 1*, ***1996***, 2837.

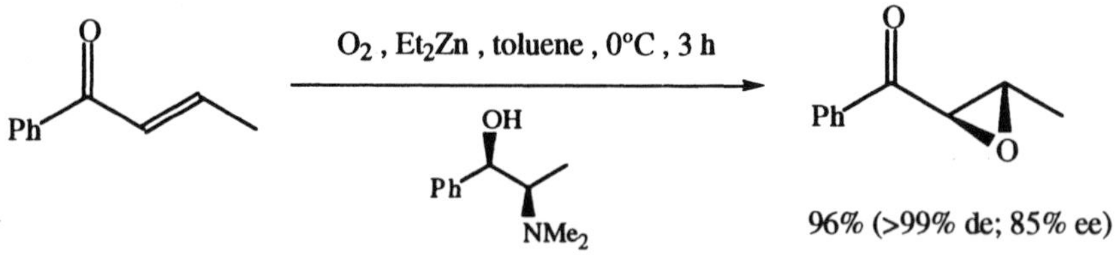
O
Ph
O_2 , Et_2Zn , toluene , 0°C , 3 h
OH
Ph
NMe_2
O
Ph
O
96% (>99% de; 85% ee)

Enders, D.; Zhu, J.; Raabe, G. *Angew. Chem. Int. Ed.*, ***1996***, *35*, 1725.

SECTION 366: ETHER, EPOXIDE, THIOETHER - NITRILE

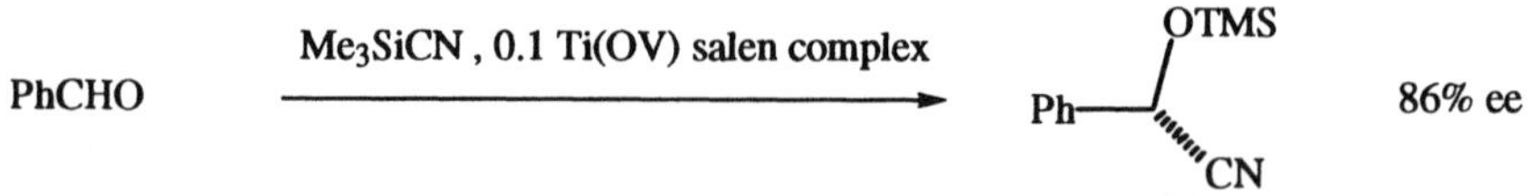
PhCHO
Me_3SiCN , 0.1 Ti(OV) salen complex
OTMS
Ph
CN
86% ee

Tararov, V.I.; Hibbs, D.E.; Hursthouse, M.B.; Ikonnikov, N.S.; Malik, K.M.A.; North, M.; Olrizu, C.; Belokon, Y.N. *Chem. Commun.*, ***1998***, 387.

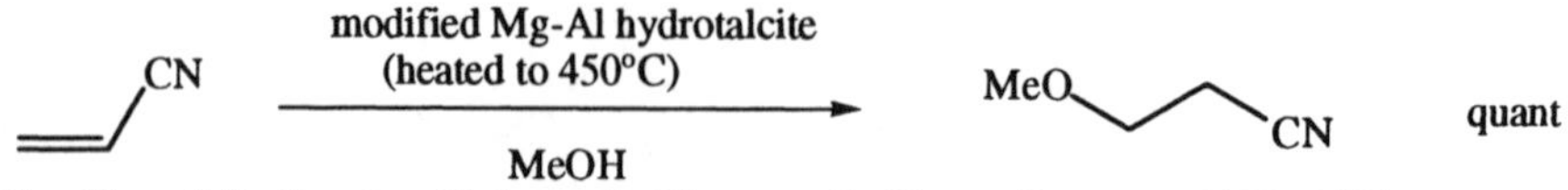
CN
modified Mg-Al hydrotalcite
(heated to 450°C)
MeOH
MeO
CN
quant

Kumbhar, P.S.; Sanchez-Valente, J.; Figueras, F. *Chem. Commun.*, ***1998***, 1091.

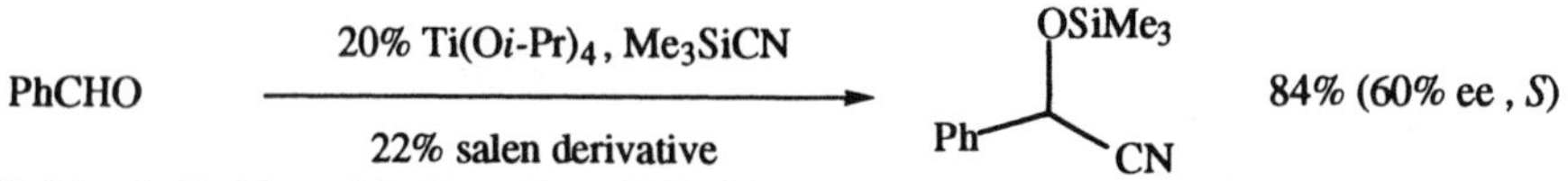
PhCHO
20% Ti(O*i*-Pr)$_4$, Me_3SiCN
22% salen derivative
$OSiMe_3$
Ph
CN
84% (60% ee , *S*)

Belokon', Y.; Flego, M.; Ikonnikov, N.S.; Moscalenko, M.; North, M.; Orizu, C.; Tararov, V Tasinazzo, M. *J. Chem. Soc.. Perkin Trans. 1*. ***1997***. 1293.

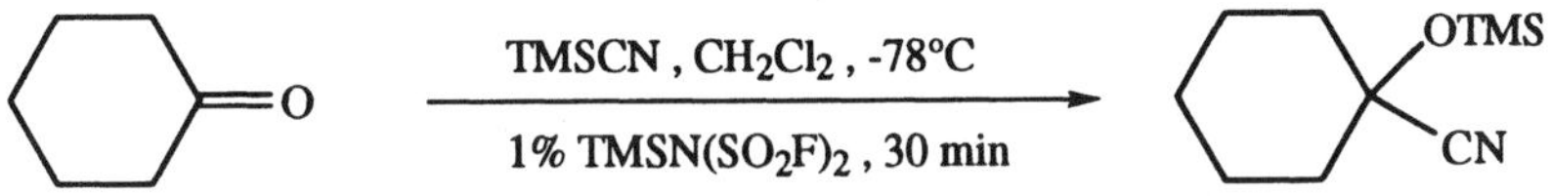
O
TMSCN, CH_2Cl_2, -78°C
1% $TMSN(SO_2F)_2$, 30 min
OTMS
CN

94%

Kaur, H.; Kaur, G.; Trehan, S. *Synth. Commun.*, ***1996***, *26*, 1925.

REVIEW:

"Sulfoximine–Titanium Reagents in Enantioselective Trimethylsilylcyanations of Aldehydes
Bolm, C.; Müller, P.; Harms, K. *Acta Chem. Scand. B*, ***1996***, *50*, 305.

SECTION 367: ETHER, EPOXIDE, THIOETHER - ALKENE

Enol ethers are found in this section as well as alkenyl ethers.

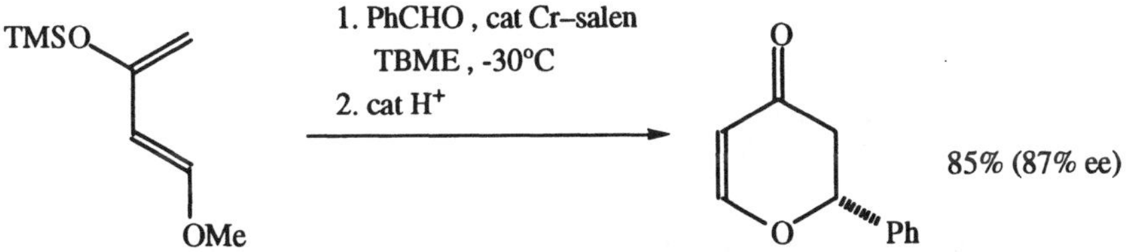
TMSO
OMe
1. PhCHO, cat Cr–salen
TBME, -30°C
2. cat H^+
O
O
Ph
85% (87% ee)

Schaus, S.E.; Brånalt, J.; Jacobsen, E.N. *J. Org. Chem.*, ***1998***, *63*, 403.

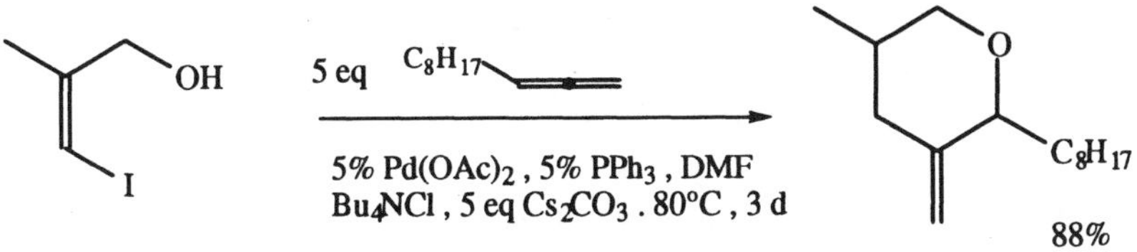
OH
I
5 eq C_8H_{17}
5% $Pd(OAc)_2$, 5% PPh_3, DMF
Bu_4NCl, 5 eq Cs_2CO_3. 80°C, 3 d
O
C_8H_{17}

88%

Larock, R.C.; He, Y.; Leong, W.W.; Han, X.; Refvik, M.D.; Zenner, J.M.
J. Org. Chem., ***1998***, *63*, 2154.

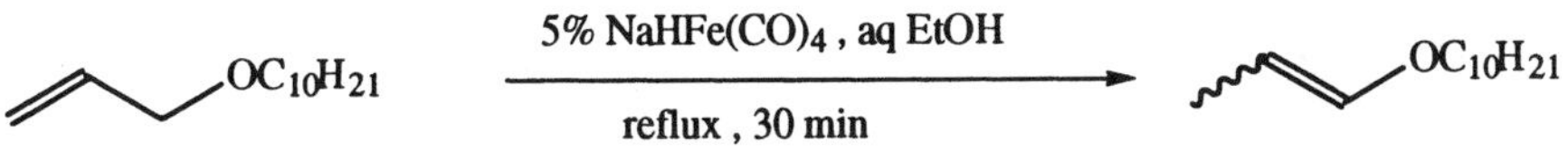
$OC_{10}H_{21}$
5% $NaHFe(CO)_4$, aq EtOH
reflux, 30 min
$OC_{10}H_{21}$

96% (42:58 *E:Z*)

Crivello, J.V.; Kong, S. *J. Org. Chem.*, ***1998***, *63*, 6745.

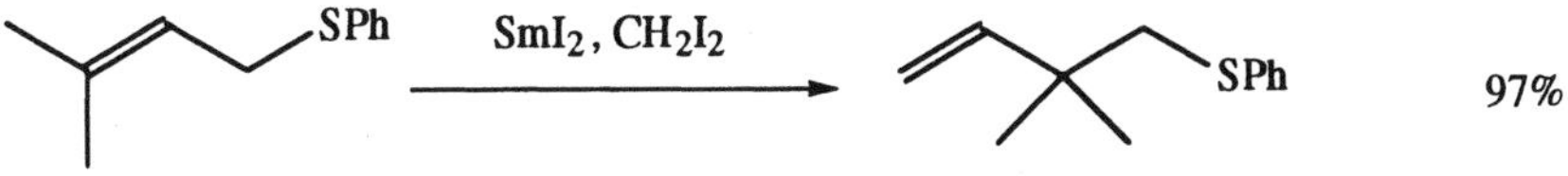
SPh
SmI_2, CH_2I_2
SPh
97%

Kunishima, M.; Nakata, D.; Goto, C.; Hioki, K.; Tani, S. *SynLett.*, ***1998***, 1366.

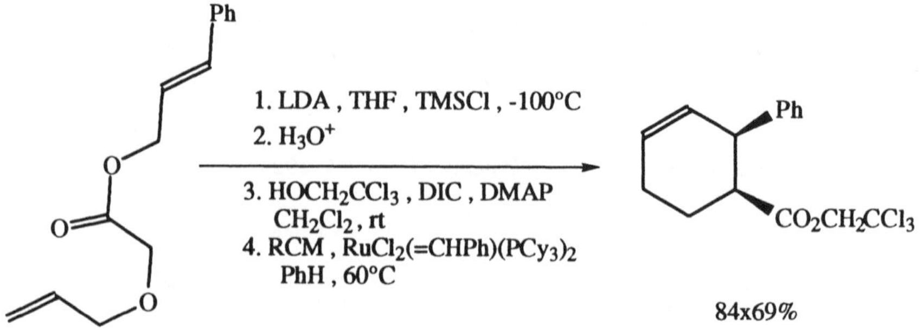
Ph
O
O
O
1. LDA , THF , TMSCl , -100°C
2. H_3O^+
3. $HOCH_2CCl_3$, DIC , DMAP
CH_2Cl_2 , rt
4. RCM , $RuCl_2(=CHPh)(PCy_3)_2$
PhH , 60°C
Ph
$CO_2CH_2CCl_3$
84x69%

tandem Claisen rearrangement

Burke, S.D.; Ng, R.A.; Morrison, J.A.; Alberti, M.J. *J. Org. Chem.*, ***1998***, *63*, 3160.

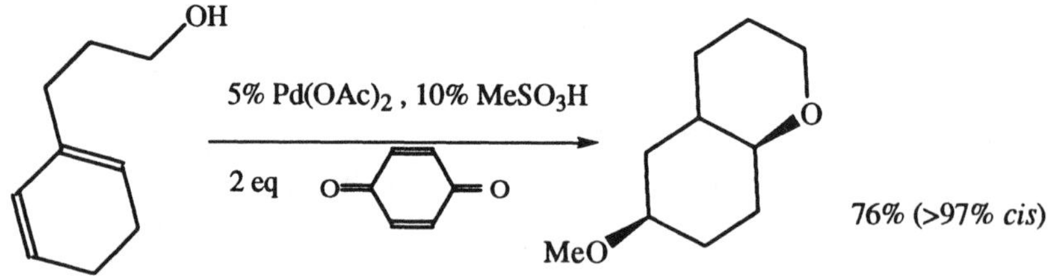
OH
5% $Pd(OAc)_2$, 10% $MeSO_3H$
2 eq
O
O
O
MeO
76% (>97% *cis*)

Itami, K.; Palmgren, A.; Bäckvall, J.-E. *Tetrahedron Lett.*, ***1998***, *39*, 1223.

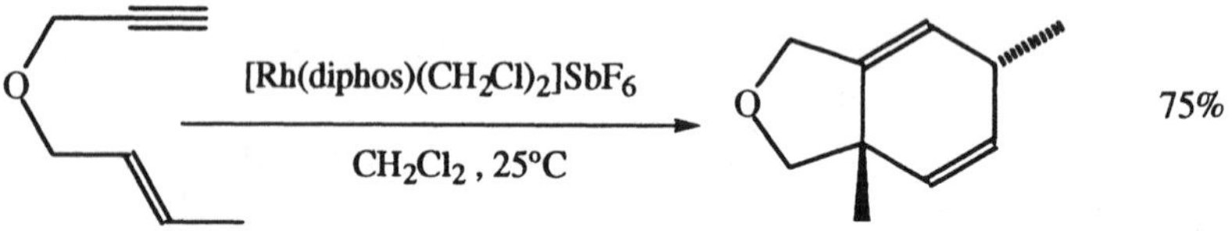
O
$[Rh(diphos)(CH_2Cl)_2]SbF_6$
CH_2Cl_2 , 25°C
O
75%

Gilbertson, S.R.; Hoge, G.S. *Tetrahedron Lett.*, ***1998***, *39*, 2075.

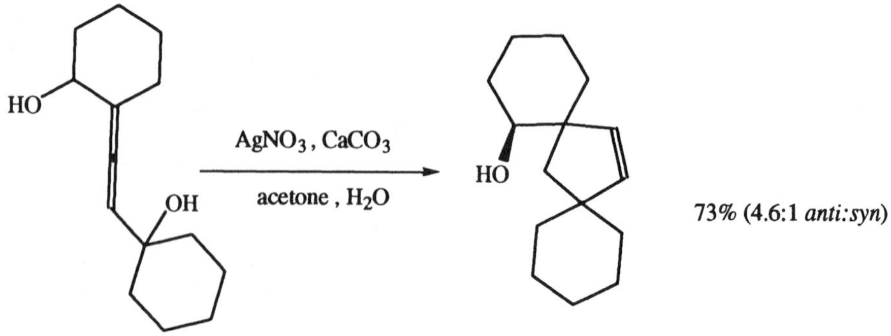
HO
OH
$AgNO_3$, $CaCO_3$
acetone , H_2O
HO
73% (4.6:1 *anti:syn*)

Aurrecoechea, J.M.; Solay, M. *Tetrahedron*, ***1998***, *54*, 3851.

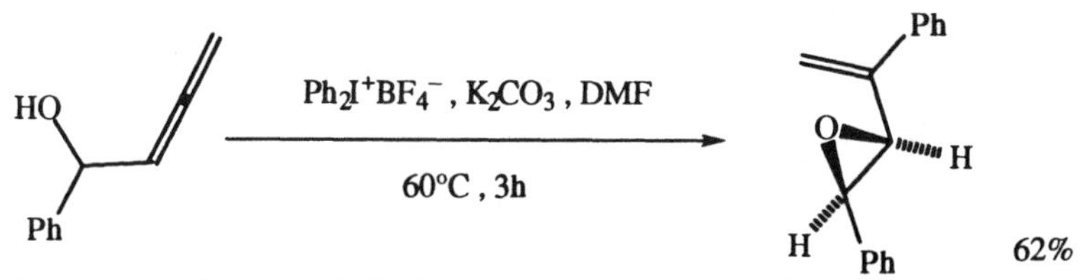
HO
Ph
$Ph_2I^+BF_4^-$, K_2CO_3 , DMF
60°C , 3h
Ph
O
H
H
Ph
62%

Kang, S.-K.; Yamaguchi, T.; Pyun, S.-J.; Lee, Y.-T.; Baik, T.-G. *Tetrahedron Lett.*, ***1998***, *39*, 2127.

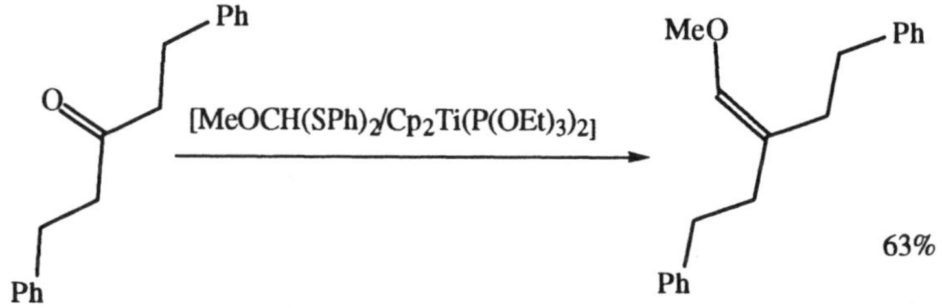
Ph
O
Ph
$[MeOCH(SPh)_2/Cp_2Ti(P(OEt)_3)_2]$
MeO
Ph
Ph
63%

Rahim, Md.A.; Taguchi, H.; Watanabe, M.; Fujiwara, T.; Takeda, T. *Tetrahedron Lett.*, ***1998***, *39*, 2153.

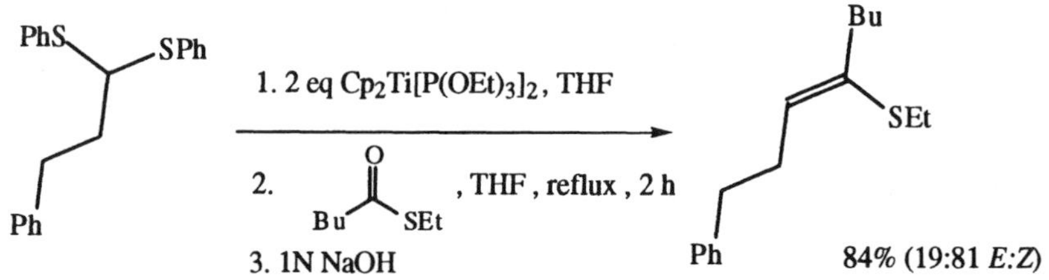
PhS
SPh
Ph
1. 2 eq $Cp_2Ti[P(OEt)_3]_2$, THF
2. , THF , reflux , 2 h
O
Bu
SEt
3. 1N NaOH
Bu
SEt
Ph
84% (19:81 *E:Z*)

Takeda, T.; Watanabe, M.; Nozaki, N.; Fujiwara, T. *Chem. Lett.*, ***1998***, 115.

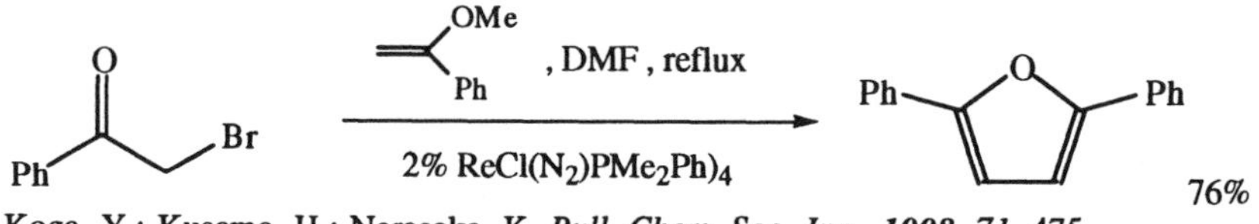
O
Ph
Br
OMe
Ph
, DMF , reflux
2% $ReCl(N_2)PMe_2Ph)_4$
Ph
O
Ph
76%

Koga, Y.; Kusama, H.; Narasaka, K. *Bull. Chem. Soc. Jpn.*, ***1998***, *71*, 475.

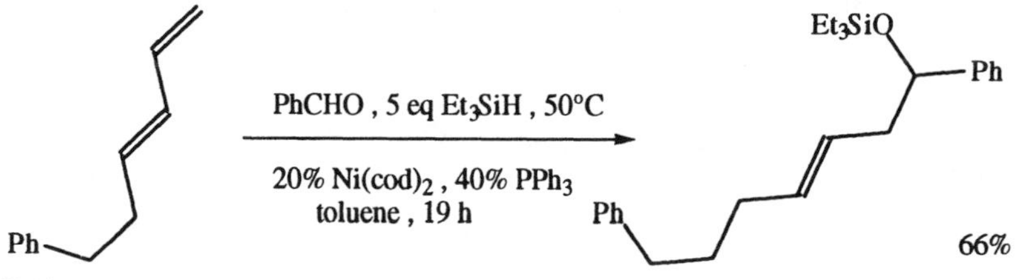
Ph
PhCHO , 5 eq Et_3SiH , 50°C
20% $Ni(cod)_2$, 40% PPh_3
toluene , 19 h
Et_3SiO
Ph
Ph
66%

Takimoto, M.; Hiraga, Y.; Sato, Y.; Mori, M. *Tetrahedron Lett.*, ***1998***, *39*, 4543.

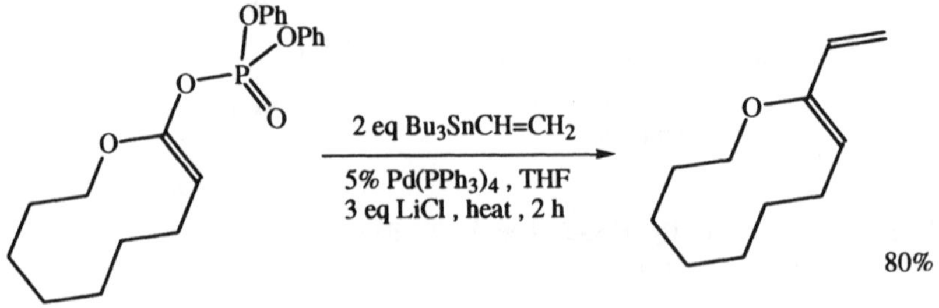
OPh
OPh
O
O
O
2 eq $Bu_3SnCH=CH_2$
5% $Pd(PPh_3)_4$, THF
3 eq LiCl, heat, 2 h
O
80%

Nicolaou, K.C.; Shi, G.–Q.; Gunzner, J.L.; Gärtner, P.; Yang, Z.
J. Am. Chem. Soc., ***1997***, *119*, 5467.

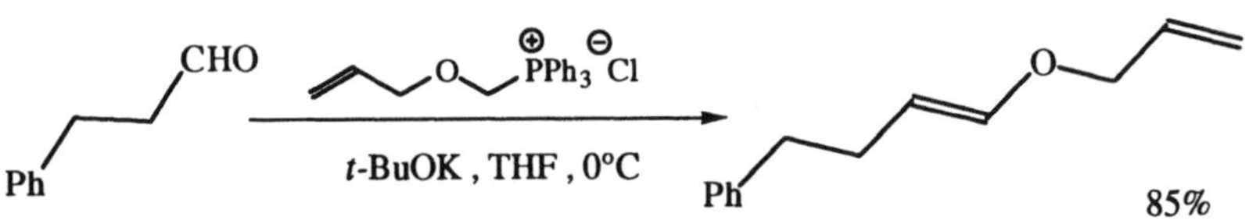
CHO
Ph
O
PPh_3 Cl
t-BuOK, THF, 0°C
O
Ph
85%

Kulkarni, M.G.; Pendharkar, D.S.; Rasne, R.M. *Tetrahedron Lett.*, ***1997***, *38*, 1459.

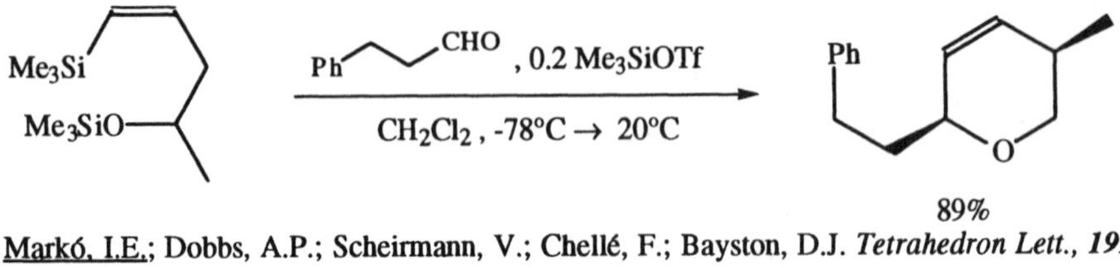
Me_3Si
Me_3SiO
Ph
CHO
, 0.2 Me_3SiOTf
CH_2Cl_2, -78°C → 20°C
Ph
O
89%

Markó, I.E.; Dobbs, A.P.; Scheirmann, V.; Chellé, F.; Bayston, D.J. *Tetrahedron Lett.*, ***1997***
38, 2899.

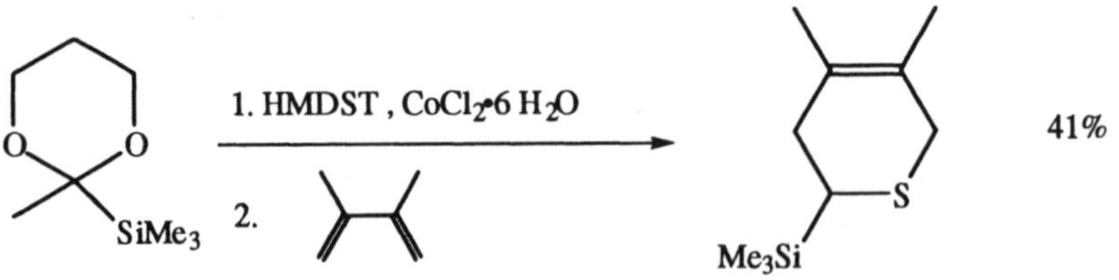
O
O
$SiMe_3$
1. HMDST, $CoCl_2$•6 H_2O
2.
S
Me_3Si
41%

HMDST = hexamethyldisilathiane

Degl'Innocenti, A.; Scafato, P.; Capperucci, A.; Bartoletti, L.; Spezzacatena, C.; Ruzziconi, R.
SynLett., ***1997***, 361.

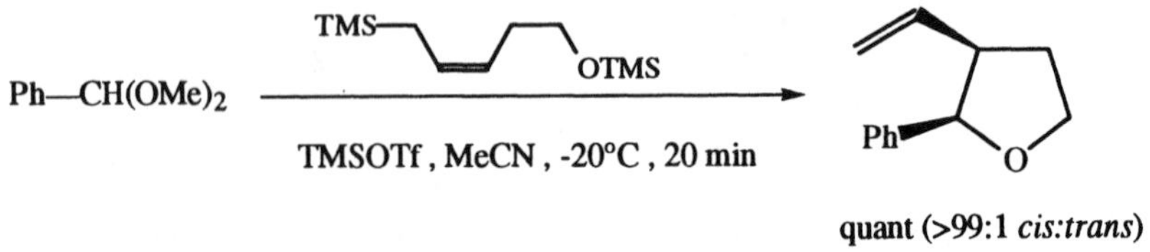
Ph—CH(OMe)$_2$
TMS
OTMS
TMSOTf, MeCN, -20°C, 20 min
Ph
O
quant (>99:1 *cis:trans*)

Sano, T.; Oriyama, T. *SynLett.*, ***1997***, 716.

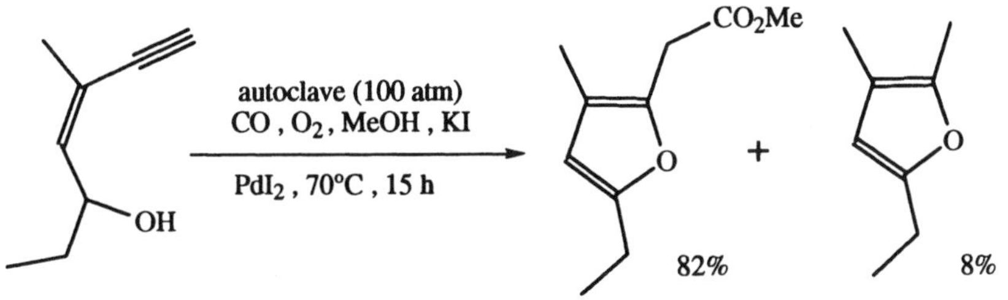
autoclave (100 atm)
CO , O2 , MeOH , KI
PdI2 , 70°C , 15 h
OH
CO2Me
O
+
O
82%
8%

Gabriele, B.; Salerno, G.; DePascali, F.; Scianò, G.T.; Costa, M.; Chiusoli, G.P. *Tetrahedron Lett.*, ***1997***, *38*, 6877.

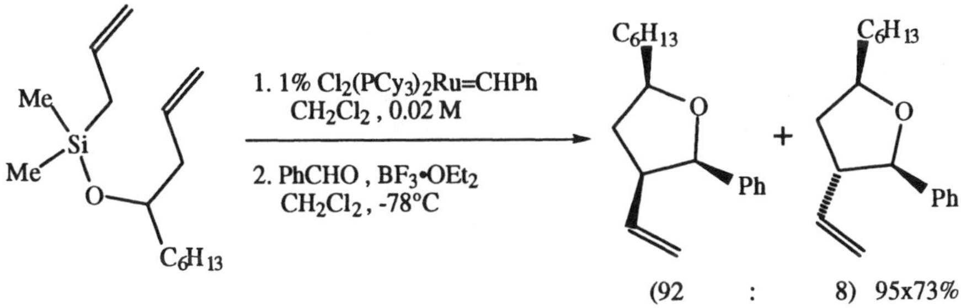
Me
Si
Me
O
C6H13
1. 1% Cl2(PCy3)2Ru=CHPh
CH2Cl2 , 0.02 M
2. PhCHO , BF3•OEt2
CH2Cl2 , -78°C
C6H13
O
Ph
+
C6H13
O
Ph
(92 : 8) 95x73%

Cassidy, J.H.; Marsden, S.P.; Stemp, G. *SynLett.*, ***1997***, 1411.

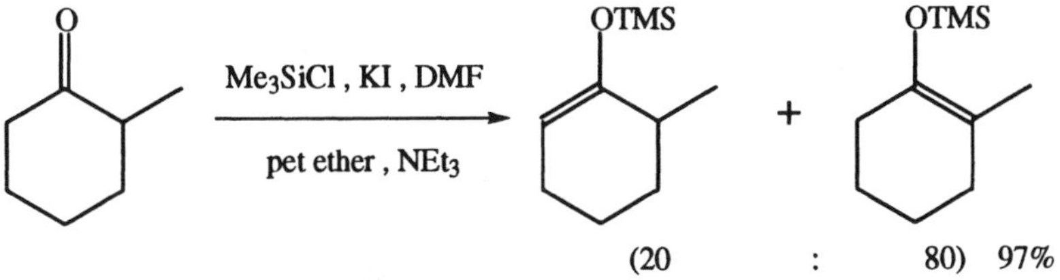
O
Me3SiCl , KI , DMF
pet ether , NEt3
OTMS
+
OTMS
(20 : 80) 97%

Lin, J.-M.; Liu, B.-S. *Synth. Commun.*, ***1997***, *27*, 739.

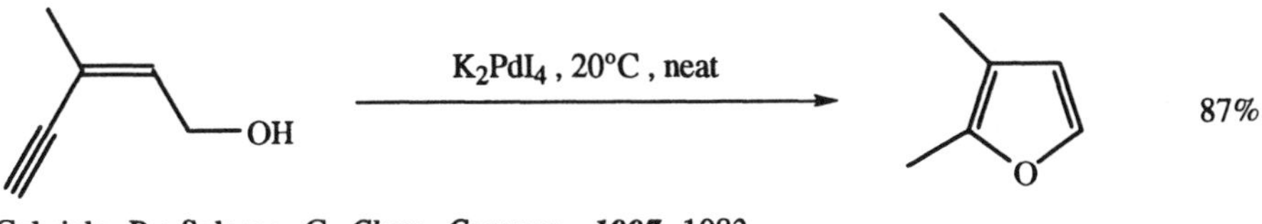
OH
K2PdI4 , 20°C , neat
O
87%

Gabriele, B.; Salerno, G. *Chem. Commun.*, ***1997***, 1083.

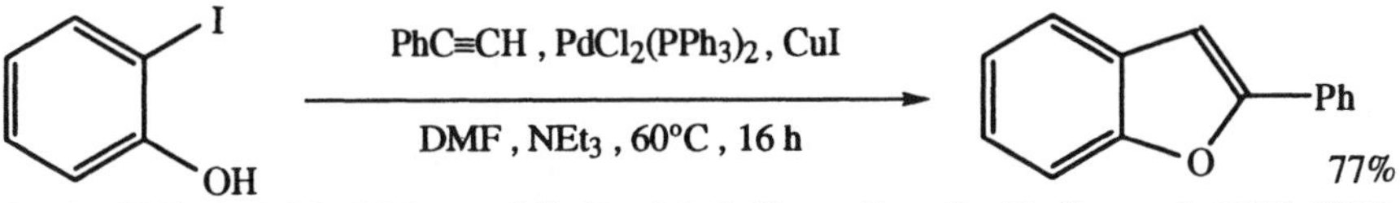
I
OH
PhC≡CH , PdCl2(PPh3)2 , CuI
DMF , NEt3 , 60°C , 16 h
Ph
O
77%

Kundu, N.G.; Pal, M.; Mahanty, J.S.; De, M. *J. Chem. Soc., Perkin Trans. 1*, ***1997***, 2815.

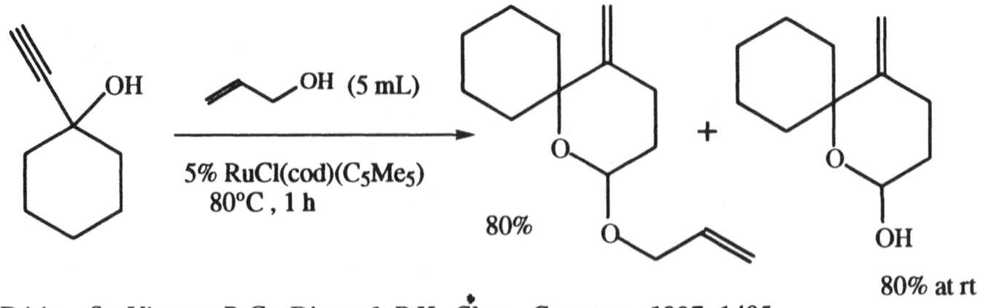
OH
OH (5 mL)
5% RuCl(cod)(C5Me5)
80°C , 1 h
O
+
O
OH
80%
80% at rt

Dérien, S.; Vicente, B.G.; Dixneuf, P.H. *Chem. Commun.*, ***1997***, 1405.

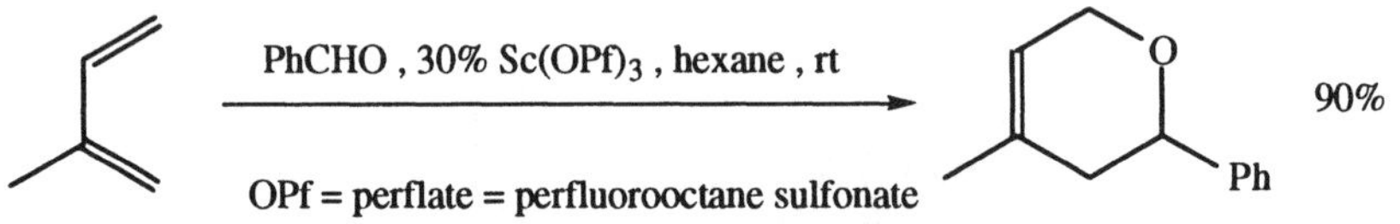
PhCHO , 30% Sc(OPf)3 , hexane , rt
OPf = perflate = perfluorooctane sulfonate
O
Ph
90%

Hanamoto, T.; Sugimoto, Y.; Jin, Y.Z.; Inanaga, J. *Bull. Chem. Soc. Jpn.*, ***1997***, *70*, 1421.

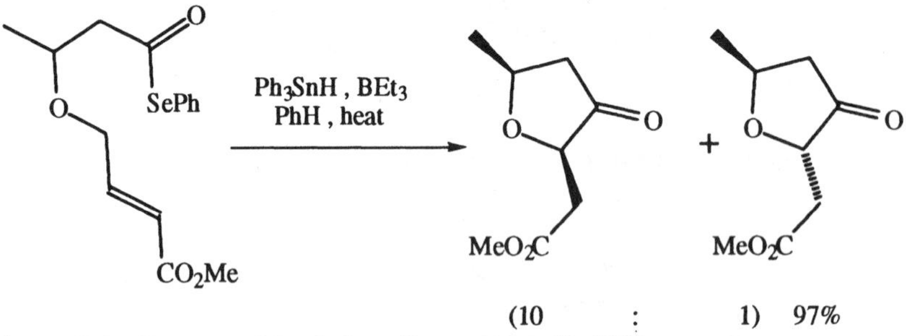
O
SePh
O
CO2Me
Ph3SnH , BEt3
PhH , heat
O
O
+
MeO2C
MeO2C
(10 : 1) 97%

Evans, P.A.; Roseman, J.D. *J. Org. Chem.*, ***1996***, *61*, 2252.

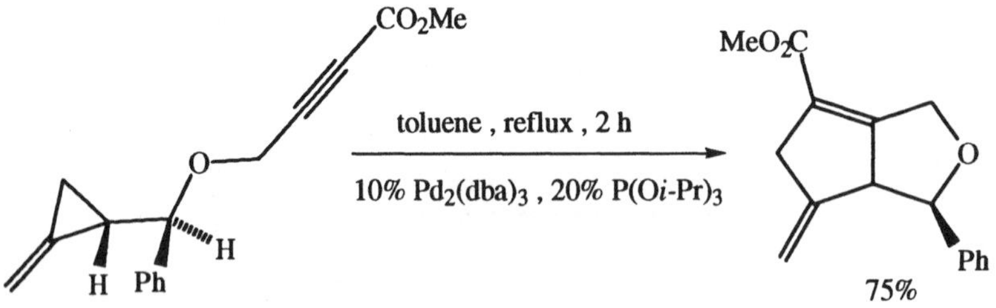
CO2Me
O
H
H Ph
toluene , reflux , 2 h
10% Pd2(dba)3 , 20% P(Oi-Pr)3
MeO2C
O
Ph
75%

Lautens, M.; Ren, Y. *J. Am. Chem. Soc.*, ***1996***, *118*, 9597.

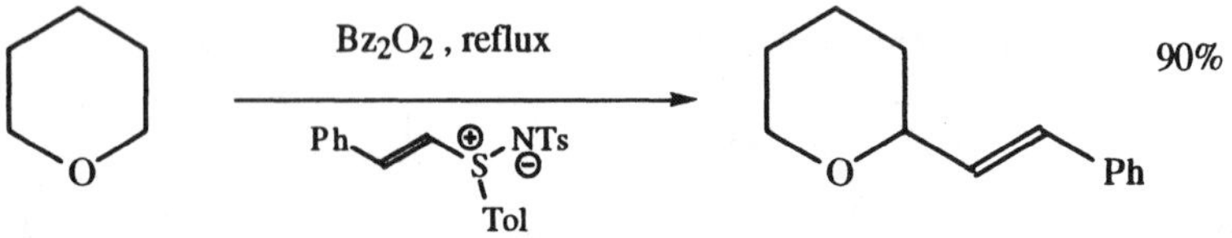
Bz2O2 , reflux
Ph
NTs
S
Tol
O
O
Ph
90%

Clark, A.J.; Rooke, S.; Sparey, T.J.; Taylor, P.C. *Tetrahedron Lett.*, ***1996***, *37*, 909.

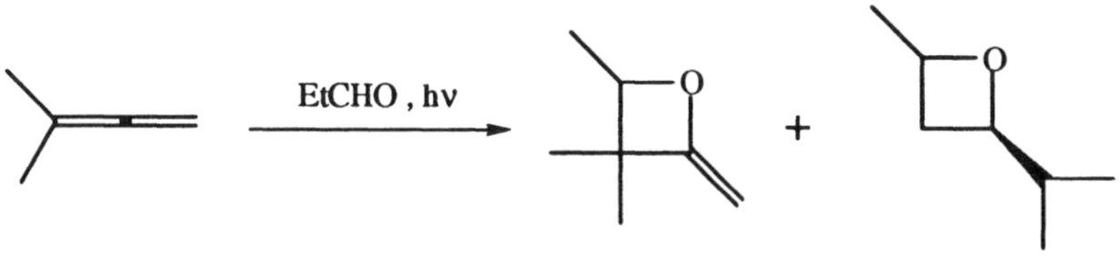
EtCHO , hν
O
O
+

(2 : 1) 69%

Howell, A.R.; Fan, R.; Truong, A. *Tetrahedron Lett.*, ***1996***, *37*, 8651.

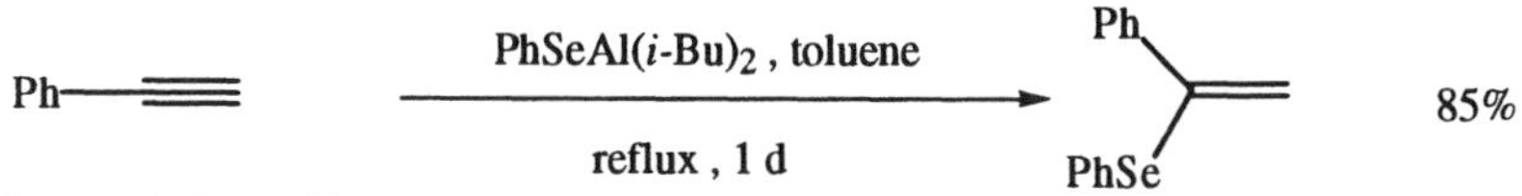
Ph
PhSeAl(i-Bu)2 , toluene
reflux , 1 d
Ph
PhSe
85%

Dabdoub, M.J.; Cassol, T.M.; Batista, A.C.F. *Tetrahedron Lett.*, ***1996***, *37*, 9005.

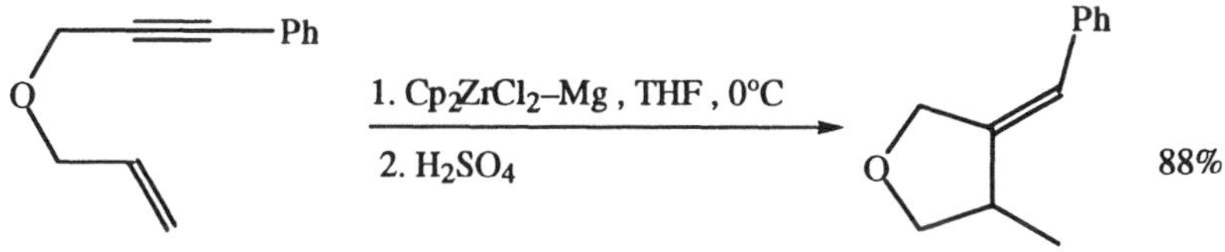
Ph
O
1. Cp2ZrCl2–Mg , THF , 0°C
2. H2SO4
Ph
O
88%

Miura, K.; Funatsu, M.; Saito, H.; Ito, H.; Hosomi, A. *Tetrahedron Lett.*, ***1996***, *37*, 9059.

Related Methods: Section 180A (Protection of Ketones)

SECTION 368: HALIDE, SULFONATE - HALIDE, SULFONATE

Halocyclopropanations are found in Section 74F (Alkyls from Alkenes).

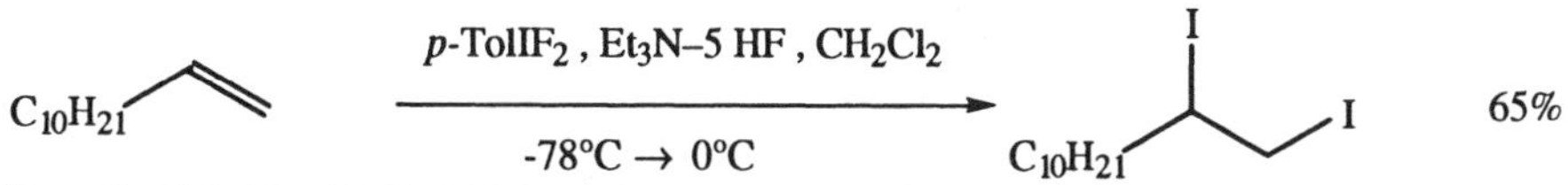
C10H21
p-TolIF2 , Et3N–5 HF , CH2Cl2
-78°C → 0°C
I
I
C10H21
65%

Hara, S.; Nakahigashi, J.; Ishi–i, K.; Sawaguchi, M.; Sakai, H.; Fukuhara, T.; Yoneda, N. *SynLett.*, ***1998***, 495.

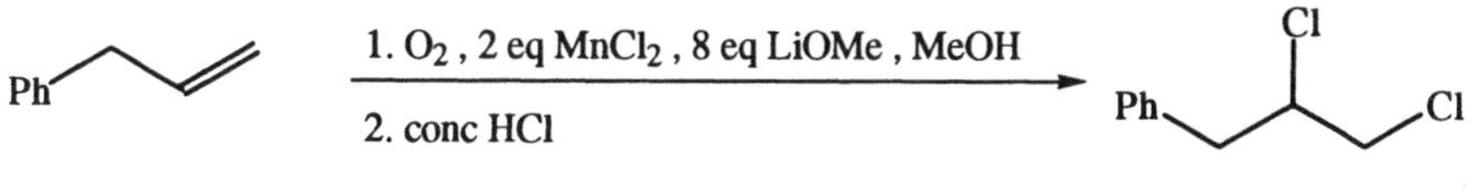
Ph
1. O2 , 2 eq MnCl2 , 8 eq LiOMe , MeOH
2. conc HCl
Cl
Ph
Cl

67%

Hojo, M.; Murakami, C.; Ohno, K.; Kuboyama, J.; Nakamura, S.–Y.; Ito, H.; Hosomi, A. *Heterocycles*, ***1998***, *47*, 97.

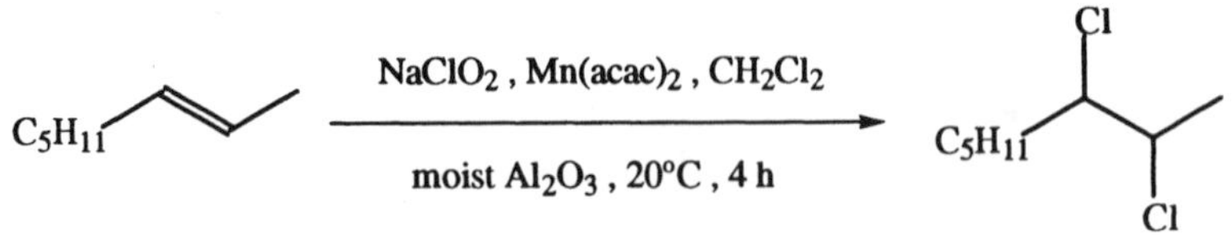

80%

Yakabe, S.; Hirano, M.; Morimoto, T. *Synth. Commun.*, ***1998***, *28*, 1871.

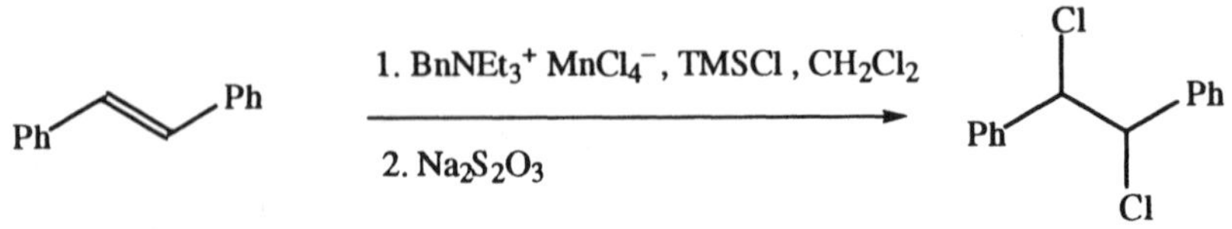

95%

Markó, I.E.; Richardson, P.R.; Bailey, M.; Maguire, A.R.; Coughlan, N. *Tetrahedron Lett.*, ***1997***, *38*, 2339.

SECTION 369: HALIDE, SULFONATE - KETONE

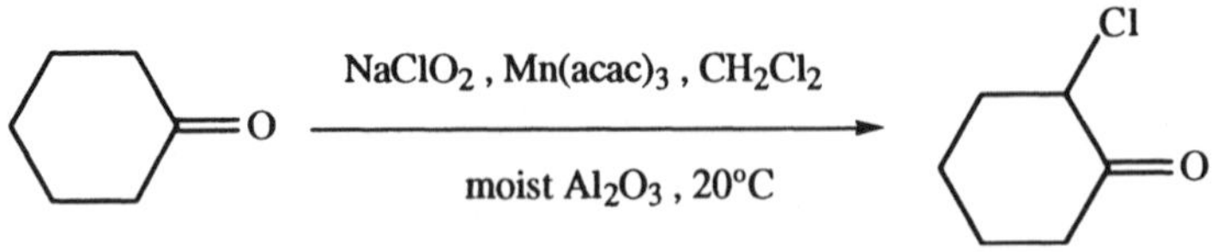

75%

Yakabe, S.; Hirano, M.; Morimoto, T. *Synth. Commun.*, ***1998***, *28*, 131.

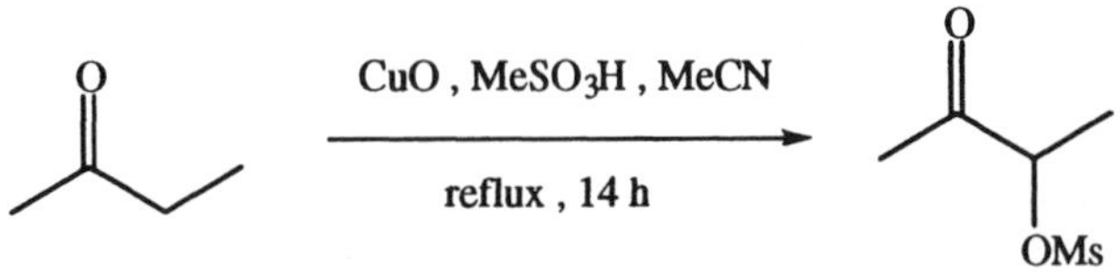

71%

Lee, J.C.; Choi, Y. *Tetrahedron Lett.*, ***1998***, *39*, 3171.

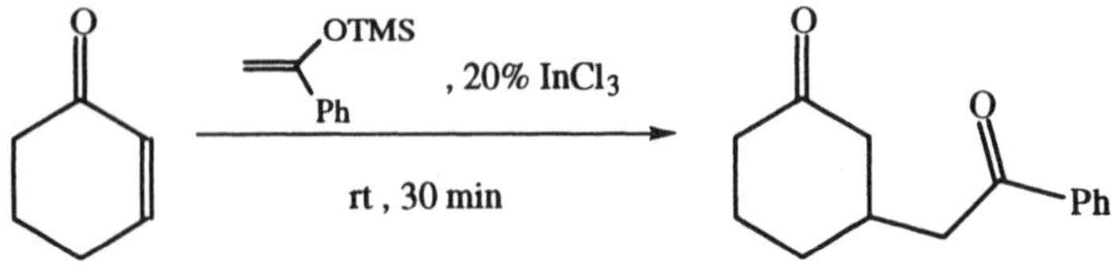

67%

Diwu, Z.; Beachdel, C.; Klaubert, D.H. *Tetrahedron Lett.*, ***1998***, *39*, 4987.

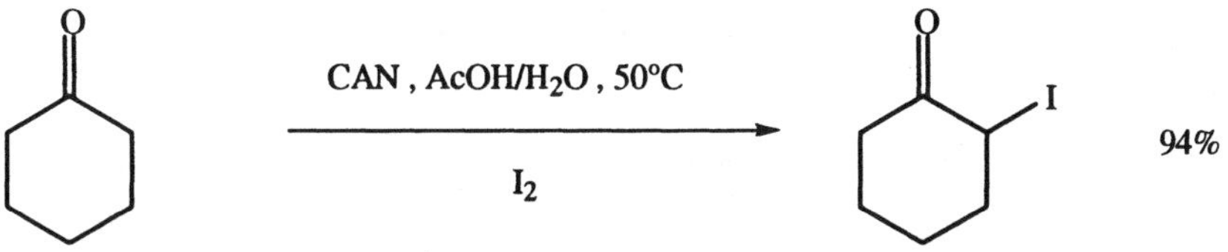
CAN , AcOH/H2O , 50°C
I2
I
94%

<u>Horiuchi, C.A.</u>; Kiji, S. *Bull. Chem. Soc. Jpn.*, ***1997***, *70*, 421.

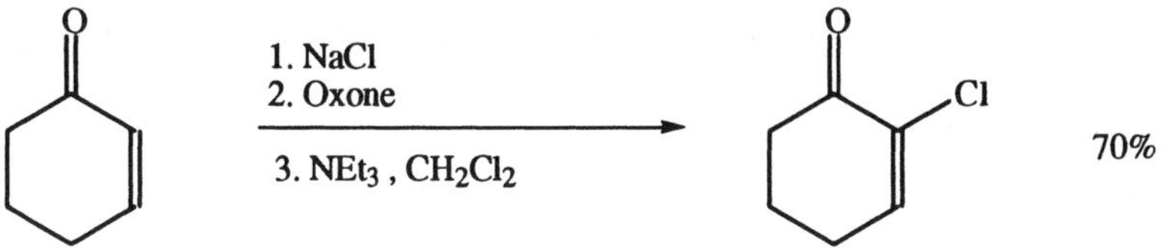
1. NaCl
2. Oxone
3. NEt3 , CH2Cl2
Cl
70%

<u>Dieter, R.K.</u>; Nice, L.E.; Velu, S.E. *Tetrahedron Lett.*, ***1996***, *37*, 2377.

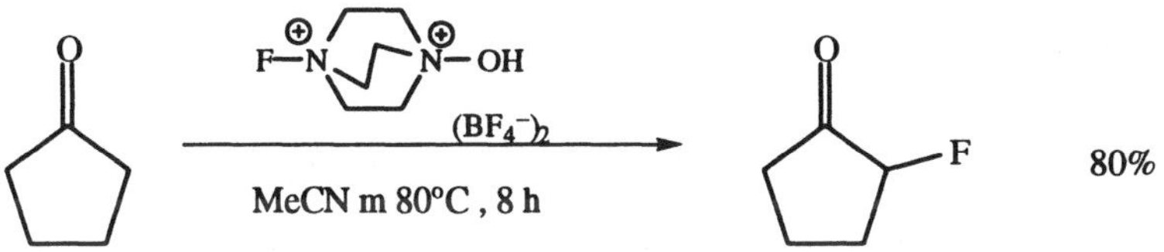
F—N N—OH
(BF4⁻)2
MeCN m 80°C , 8 h
F
80%

<u>Stavber, S.</u>; Zupan, M. *Tetrahedron Lett.*, ***1996***, *37*, 3591.

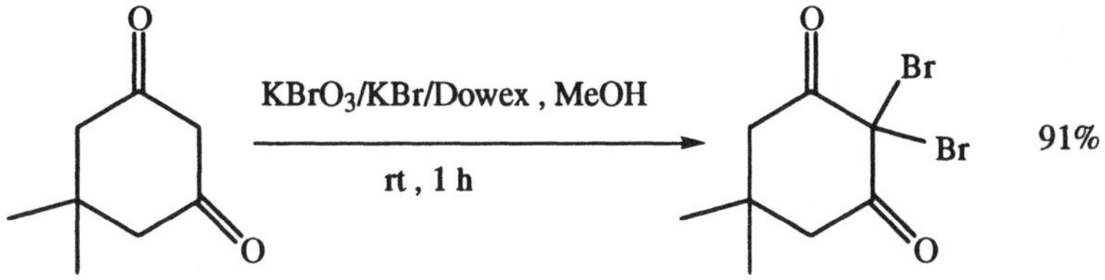
KBrO3/KBr/Dowex , MeOH
rt , 1 h
Br
Br
91%

Košmrlj, J.; Kočevar, M.; <u>Polanc, S.</u> *Synth. Commun.*, ***1996***, *27*, 3583.

SECTION 370: HALIDE, SULFONATE - NITRILE

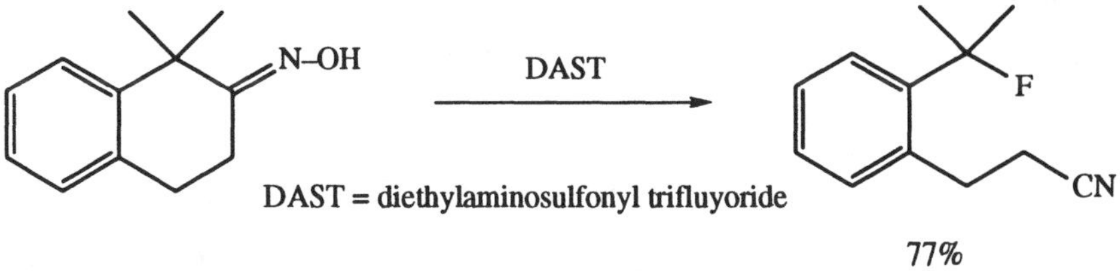
N–OH
DAST
DAST = diethylaminosulfonyl trifluyoride
F
CN

77%

Kirihara, M.; Niimi, K.; <u>Momose, T.</u> *Chem. Commun.*, ***1997***, 599.

SECTION 371: HALIDE, SULFONATE - ALKENE

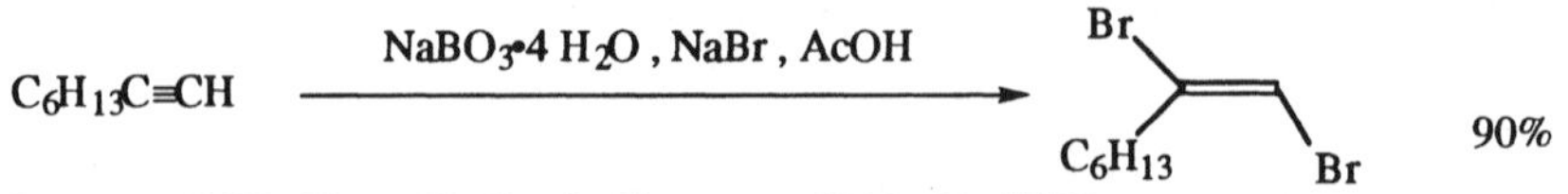
NaBO$_3$•4 H$_2$O , NaBr , AcOH
C$_6$H$_{13}$C≡CH
Br
C$_6$H$_{13}$
Br
90%

Kabalka, G.W.; Yang, K. *Synth. Commun.*, ***1998***, *28*, 3807.

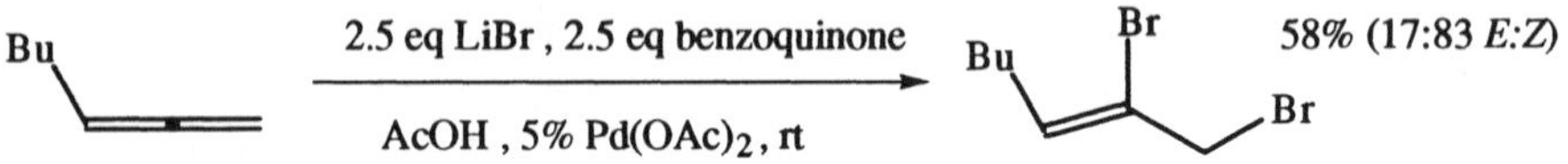
2.5 eq LiBr , 2.5 eq benzoquinone
AcOH , 5% Pd(OAc)$_2$, rt
Bu
Bu
Br
Br
58% (17:83 *E:Z*)

Bäckvall, J.-E.; Jonasson, C. *Tetrahedron Lett.*, ***1997***, *38*, 291.

SECTION 372: KETONE - KETONE

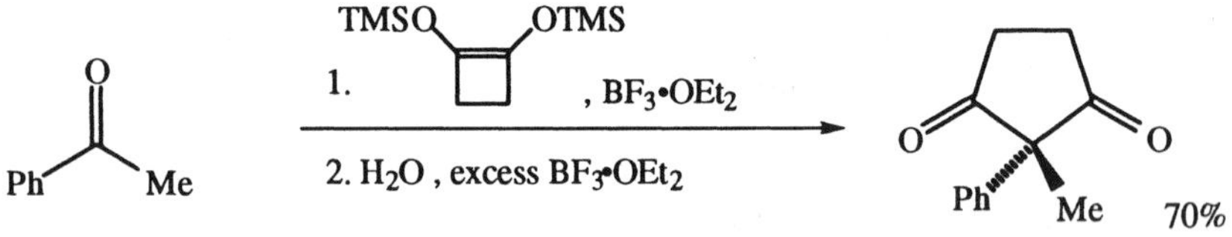
TMSO OTMS
1. , BF$_3$•OEt$_2$
2. H$_2$O , excess BF$_3$•OEt$_2$
O
Ph Me
O O
Ph Me
70%

Crane, S.N.; Burnell, D.J. *J. Org. Chem.*, ***1998***, *63*, 1352.

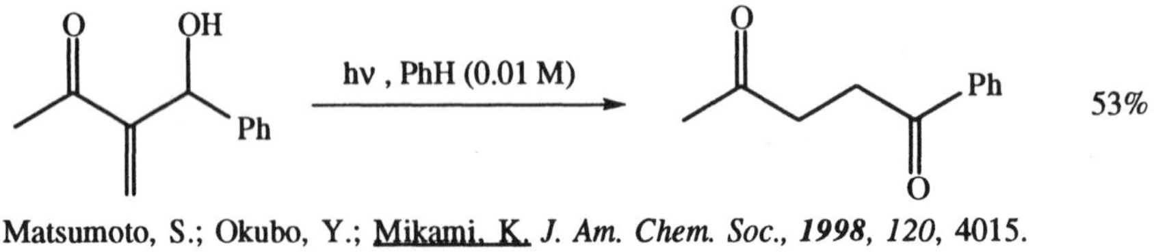
O OH
Ph
hν , PhH (0.01 M)
O
Ph
O
53%

Matsumoto, S.; Okubo, Y.; Mikami, K. *J. Am. Chem. Soc.*, ***1998***, *120*, 4015.

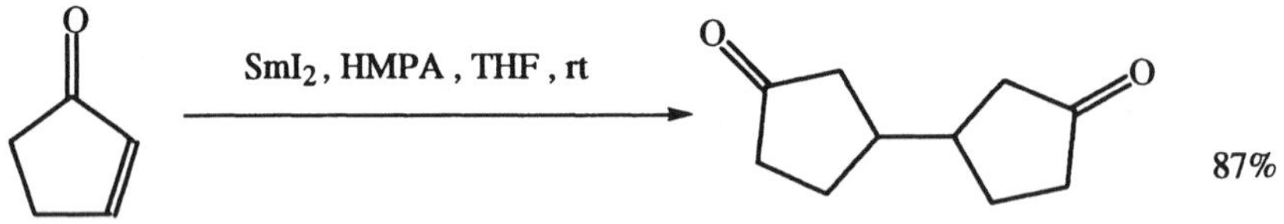
O
SmI$_2$, HMPA , THF , rt
O
O
87%

Cabrera, A.; Rosas, N.; Sharma, P.; LeLagadec, R.; Velasco, L.; Salmón, M. *Synth. Commun.*, ***1998***, *28*, 1103.

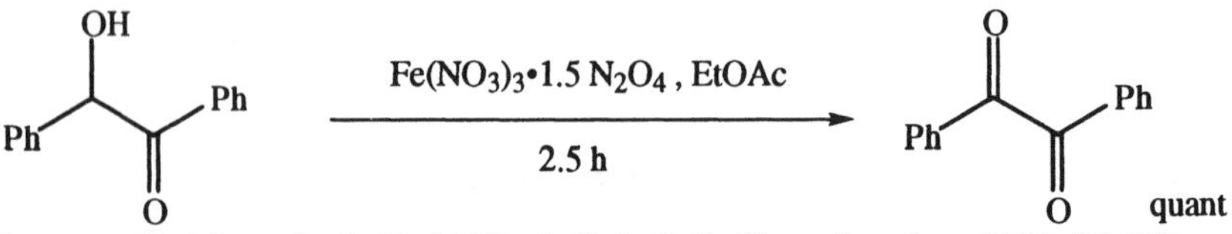
OH
Ph
Ph
O
Fe(NO$_3$)$_3$•1.5 N$_2$O$_4$, EtOAc
2.5 h
O
Ph
Ph
O
quant

Iranpoor, N.; Firouzabadi, H.; Zolfigol, M.A. *Bull. Chem. Soc. Jpn.*, ***1998***, *71*, 905.

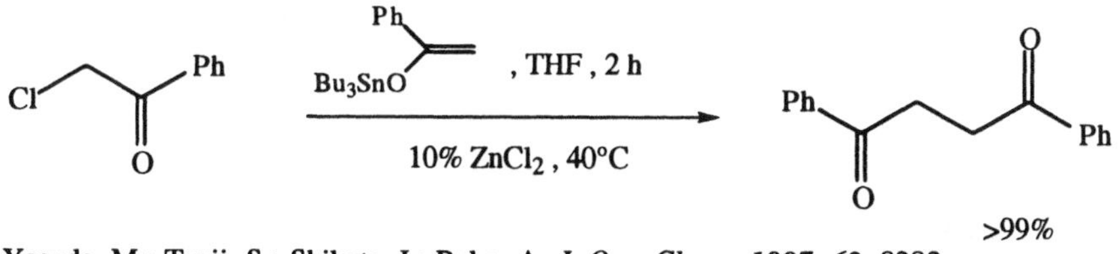

Yasuda, M.; Tsuji, S.; Shibata, I.; Baba, A. *J. Org. Chem.*, ***1997***, *62*, 8282.

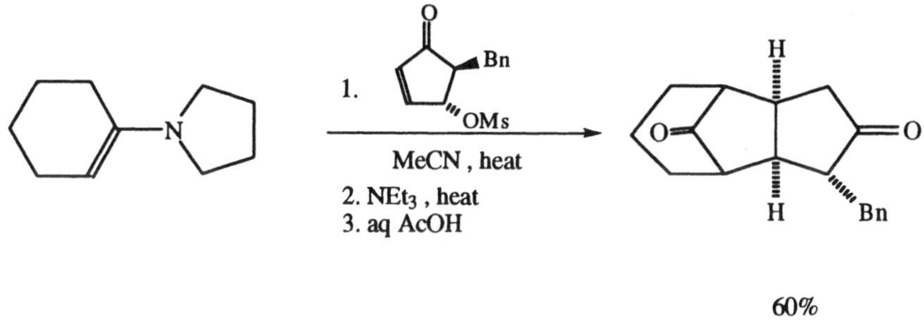

Gunawardena, G.U.; Arif, A.M.; West, F.G. *J. Am. Chem. Soc.*, ***1997***, *119*, 2066.

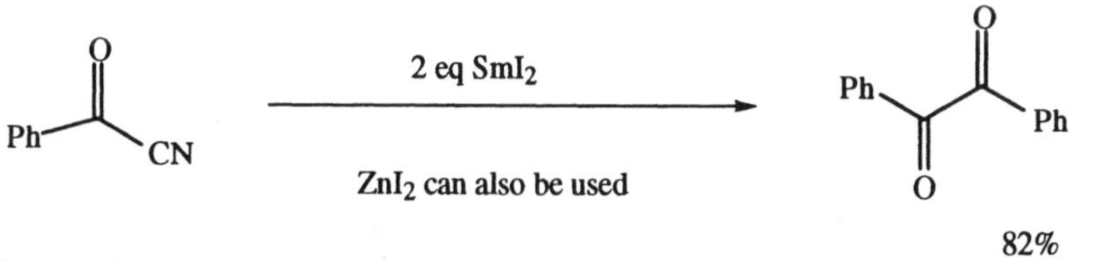

Baruah, B.; Boruah, A.; Prajapati, D.; Sandhu, J.S. *Tetrahedron Lett.*, ***1997***, *38*, 7603.

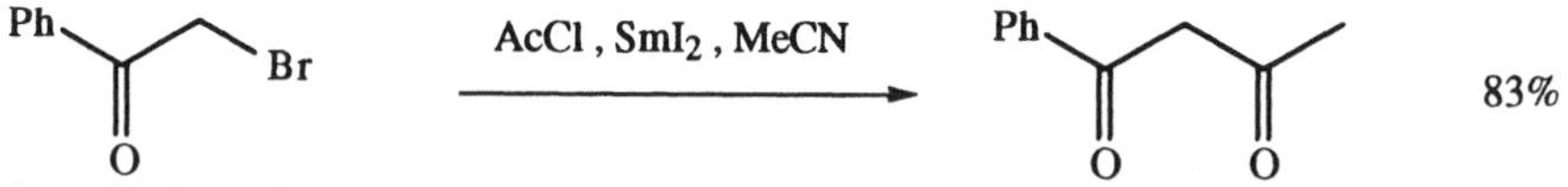

Ying, T.; Bao, W.; Zhang, Y.; Xu, W. *Tetrahedron Lett.*, ***1996***, *37*, 3885.

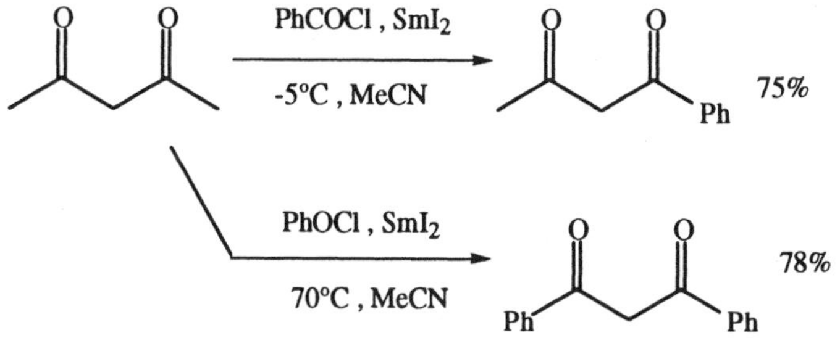

Hao, W.; Zhang, Y.; Ying, T.; Lu, P. *Synth. Commun.*, ***1996***, *26*, 2421.

SECTION 373: KETONE - NITRILE

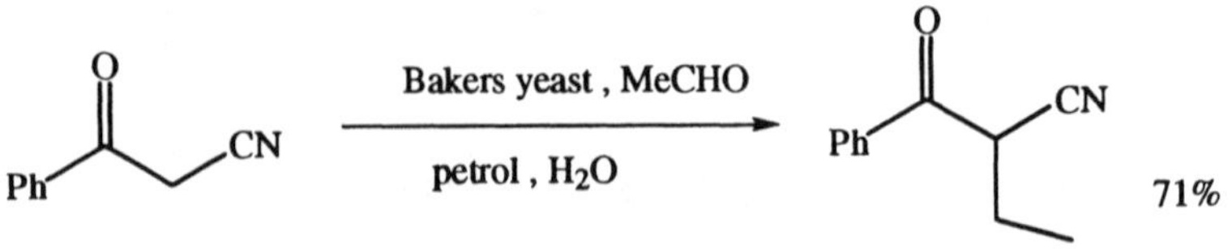
O
Ph
CN
Bakers yeast , MeCHO
petrol , H_2O
O
Ph
CN
71%

Smallridge, A.J.; Ten, A.; Trewhella, M.A. *Tetrahedron Lett.*, ***1998***, *39*, 5121.

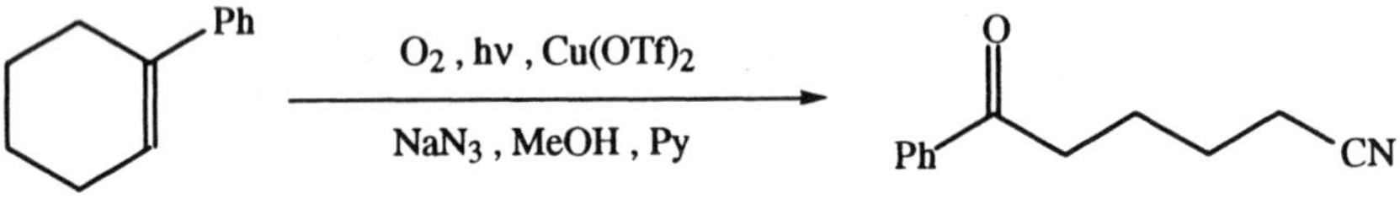
Ph
O_2 , hν , $Cu(OTf)_2$
NaN_3 , MeOH , Py
O
Ph
CN

89%

Shimizu, I.; Fujita, M.; Nakajima, T.; Sato, T. *SynLett.*, ***1997***, 887.

SECTION 374: KETONE - ALKENE

For the oxidation of allylic alcohols to alkene ketones, see Section 168 (Ketones from Alcohols and Phenols)

For the oxidation of allylic methylene groups (C=C-CH_2 → C=C-C=O), see Section 170 (Ketones from Alkyls and Methylenes).

For the alkylation of alkene ketones, also see Section 177 (Ketones from Ketones) and for conjugate alkylations see Section 74E (Alkyls form Alkenes).

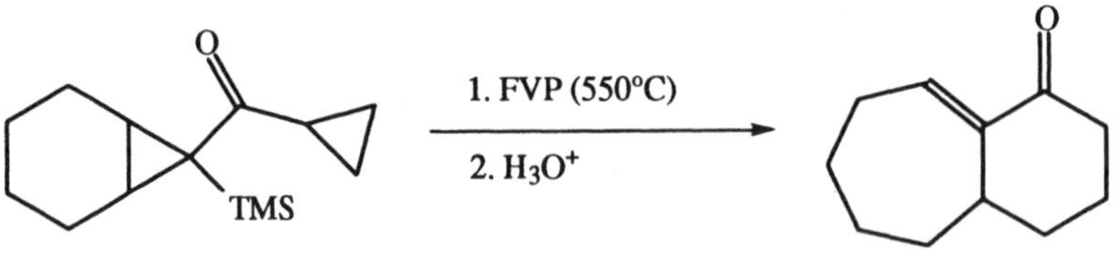
O
TMS
1. FVP (550°C)
2. H_3O^+
O

89%

Liu, H.; Shook, C.A.; Jamison, J.A.; Thiruvazhi, M.; Cohen, T.
J. Am. Chem. Soc., ***1998***, *120*, 605.

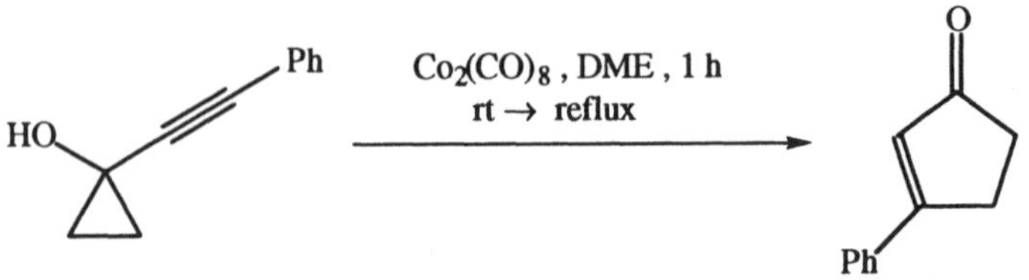
Ph
HO
$Co_2(CO)_8$, DME , 1 h
rt → reflux
O
Ph

91%

Iwasawa, N.; Matsuo, T.; Iwamoto, M.; Ikeno, T. *J. Am. Chem. Soc.*, ***1998***, *120*, 3903.

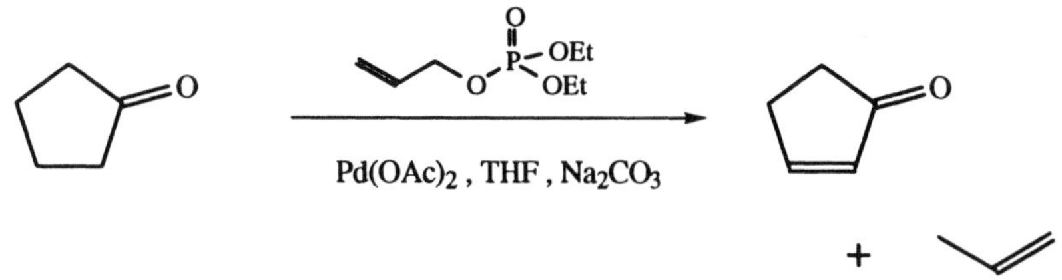
O
P
OEt
O
OEt
O
O
Pd(OAc)2 , THF , Na2CO3
+

Shvo, Y.; Arisha, A.H.I. *J. Org. Chem.*, ***1998***, *63*, 5640.

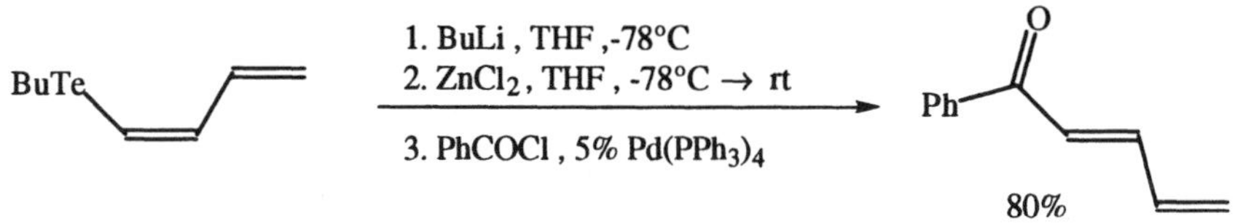
BuTe
1. BuLi , THF ,-78°C
2. ZnCl2 , THF , -78°C → rt
3. PhCOCl , 5% Pd(PPh3)4
O
Ph
80%

Huang, Y.-Z.; Mo, X.-S. *Tetrahedron Lett.*, ***1998***, *39*, 1945.

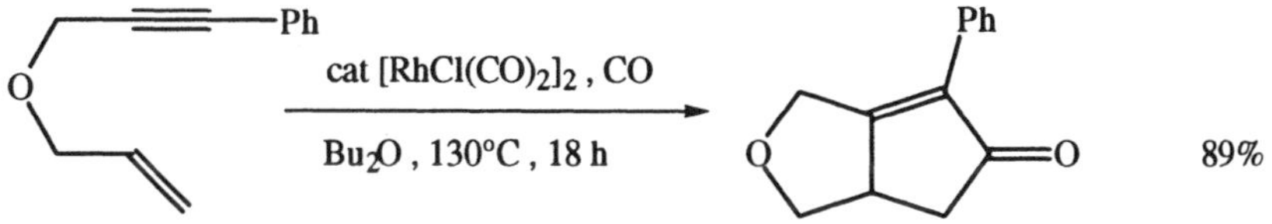
Ph
O
cat [RhCl(CO)2]2 , CO
Bu2O , 130°C , 18 h
Ph
O
O
89%

Koga, Y.; Kobayashi, T.; Narasaka, K. *Chem. Lett.*, ***1998***, 249.

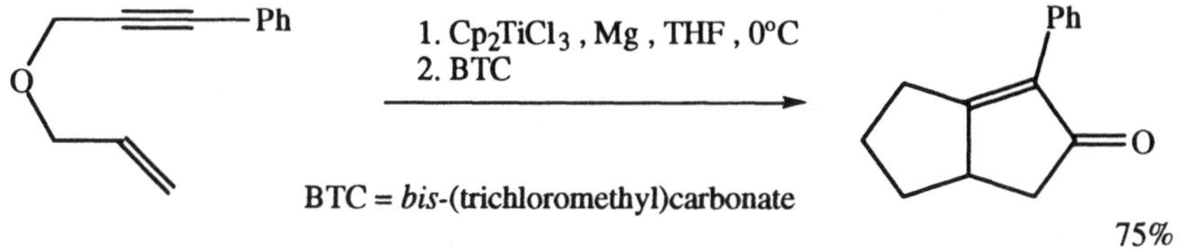
Ph
O
1. Cp2TiCl3 , Mg , THF , 0°C
2. BTC
BTC = bis-(trichloromethyl)carbonate
Ph
O
75%

Zhao, Z.; Ding, Y. *J. Chem. Soc., Perkin Trans. 1*, ***1998***, 171.

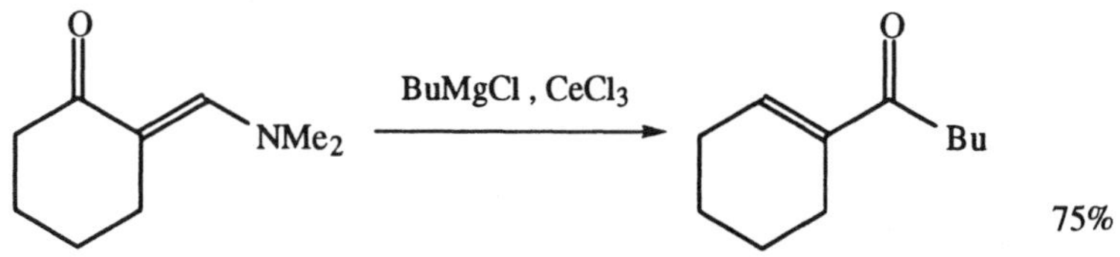
O
NMe2
BuMgCl , CeCl3
O
Bu
75%

Fonenas, C.; Aït-Haddou, H.; Balavoine, G.G.A. *Synth. Commun.*, ***1998***, *28*, 1743.

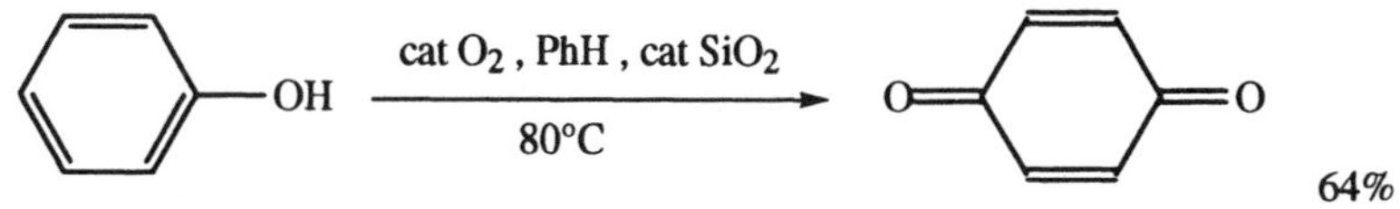
OH
cat O2 , PhH , cat SiO2
80°C
O
O
64%

Hashemi, M.M.; Beni, Y.A. *J. Chem. Res. (S)*, ***1998***, 138.

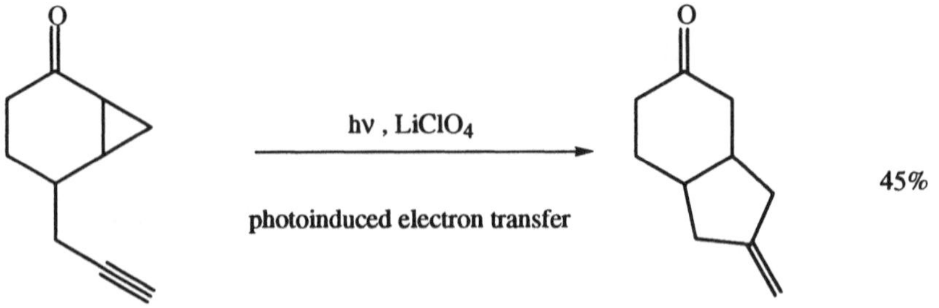
O
hv , LiClO4
photoinduced electron transfer
O
45%

Fagnoni, M.; Schmoldt, P.; Kirschberg, T.; Mattay, J. *Tetrahedron*, ***1998***, *54*, 6427.

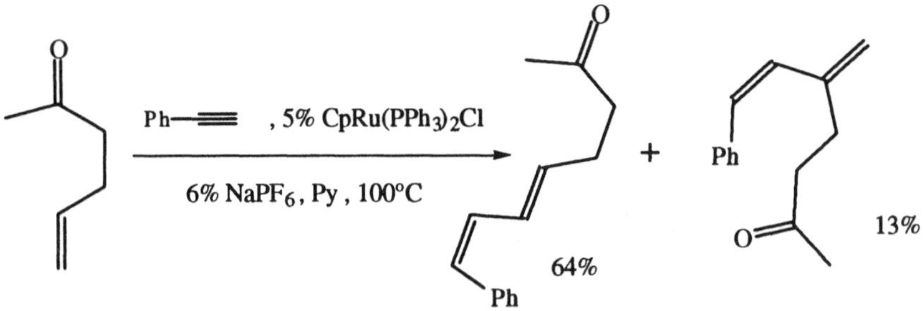
O
Ph
, 5% CpRu(PPh3)2Cl
6% NaPF6, Py , 100°C
O
+
Ph
64%
Ph
O
13%

Murakami, M.; Ubukata, M.; Ito, Y. *Tetrahedron Lett.*, ***1998***, *39*, 7361.

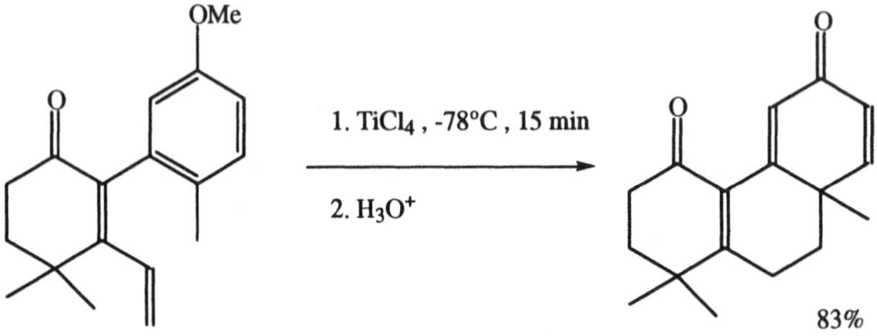
OMe
O
1. TiCl4 , -78°C , 15 min
2. H3O+
O
O
83%

Majetich, G.; Fang, J. *Tetrahedron Lett.*, ***1998***, *39*, 8381.

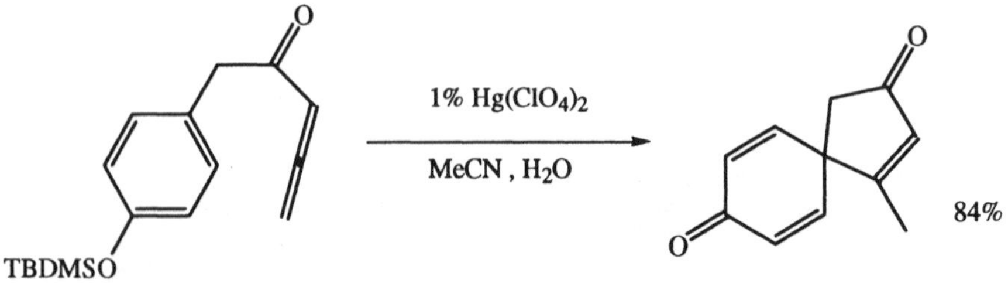
O
1% Hg(ClO4)2
MeCN , H2O
O
O
TBDMSO
84%

Hashmi, A.S.K.; Schwarz, L; Bolte, M. *Tetrahedron Lett.*, ***1998***, *39*, 8969.

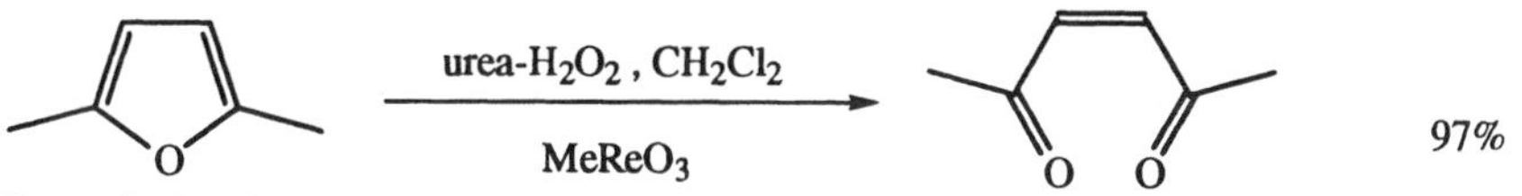
urea-H_2O_2 , CH_2Cl_2
$MeReO_3$
O
O O
97%

Finlay, J.; McKervey, M.A.; Gunaratne, H.Q.N. *Tetrahedron Lett.*, ***1998***, *39*, 5651.

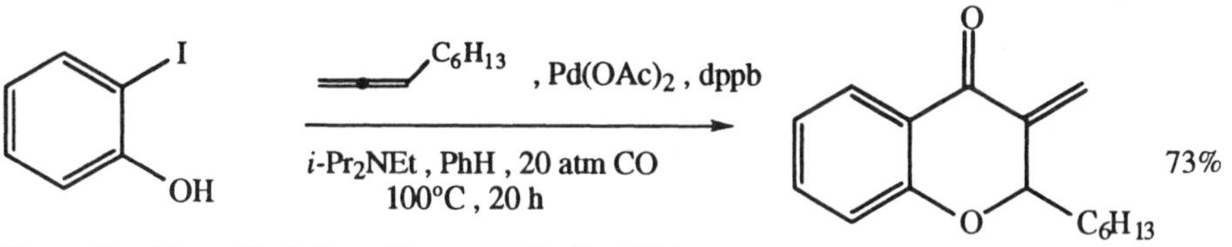
I
OH
C_6H_{13} , $Pd(OAc)_2$, dppb
i-Pr_2NEt , PhH , 20 atm CO
100°C , 20 h
O
O
C_6H_{13}
73%

Okuro, K.; Alper, H. *J. Org. Chem.*, ***1997***, *62*, 1566.

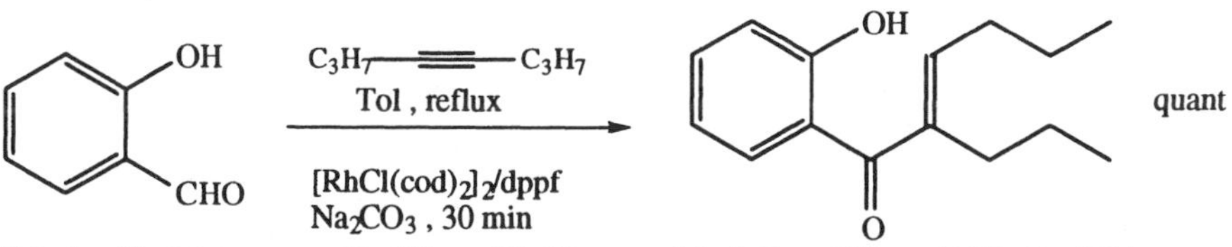
OH
CHO
C_3H_7 C_3H_7
Tol , reflux
$[RhCl(cod)_2]_2$/dppf
Na_2CO_3 , 30 min
OH
O
quant

Kokubo, K.; Matsumasa, K.; Miura, M.; Nomura, M. *J. Org. Chem.*, ***1997***, *62*, 4564.

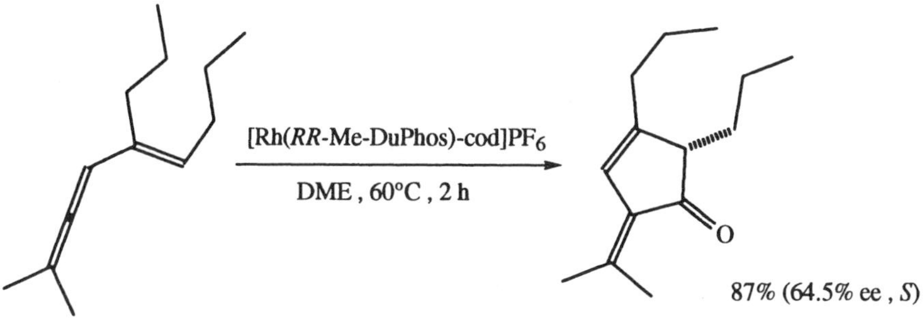
[Rh(*RR*-Me-DuPhos)-cod]PF_6
DME , 60°C , 2 h
O
87% (64.5% ee , *S*)

Murakami, M.; Itami, K.; Ito, Y. *J. Am. Chem. Soc.*, ***1997***, *119*, 2950.

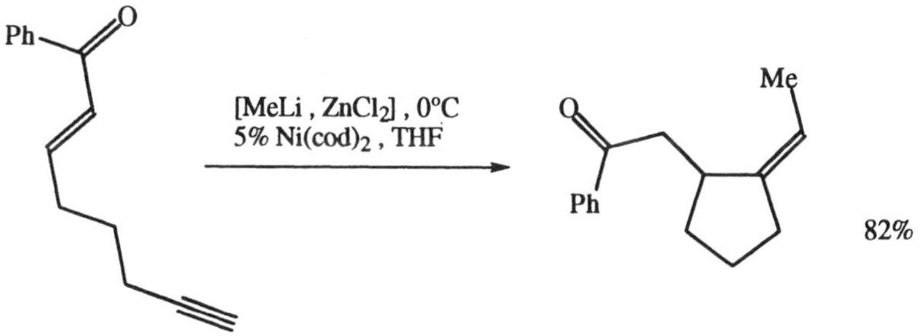
Ph
O
[MeLi , $ZnCl_2$] , 0°C
5% $Ni(cod)_2$, THF
Me
O
Ph
82%

Montogomery, J.; Oblinger, E.; Savchenko, A.V. *J. Am. Chem. Soc.*, ***1997***, *119*, 4911.

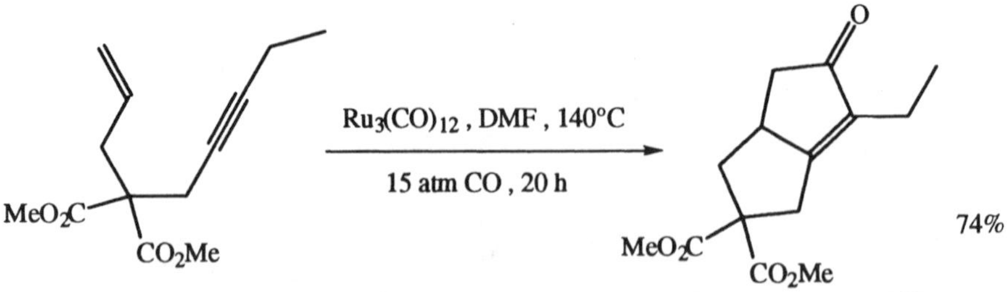

MeO2C
CO2Me
Ru3(CO)12, DMF, 140°C
15 atm CO, 20 h
O
MeO2C
CO2Me
74%

Kondo, T.; Suzuki, N.; Okada, T.; Mitsudo, T. *J. Am. Chem. Soc.*, ***1997***, *119*, 6187.

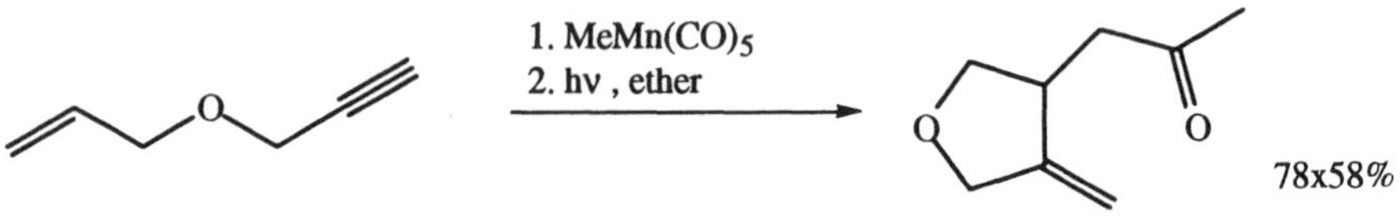

O
1. MeMn(CO)5
2. hν, ether
O
O
78x58%

Lee, J.E.; Hong, S.H.; Chung, Y.K. *Tetrahedron Lett.*, ***1997***, *38*, 1781.

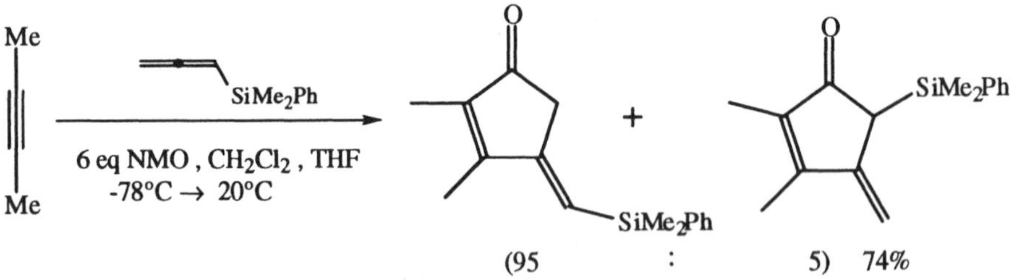

Me
Me
SiMe2Ph
6 eq NMO, CH2Cl2, THF
-78°C → 20°C
O
SiMe2Ph
+
O
SiMe2Ph
(95 : 5) 74%

Ahmar, M.; Chabanis, O.; Gauthier, J.; Cazes, B. *Tetrahedron Lett.*, ***1997***, *38*, 5277.
Ahmar, M.; Locatelli, C.; Colombier, D.; Cazes, B. *Tetrahedron Lett.*, ***1997***, *38*, 5281.

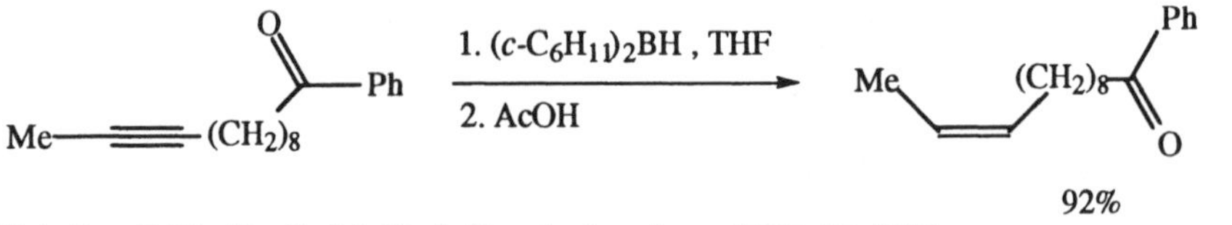

O
Ph
Me
(CH2)8
1. (c-C6H11)2BH, THF
2. AcOH
Ph
Me
(CH2)8
O
92%

Kabalka, G.W.; Yu, S.; Li, N.-S. *Tetrahedron Lett.*, ***1997***, *38*, 7681.

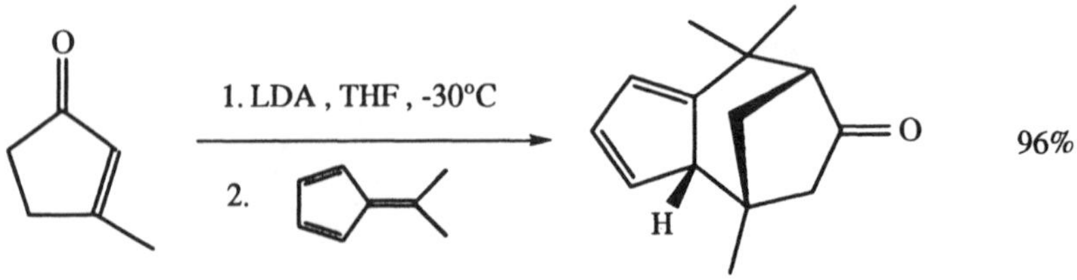

O
1. LDA, THF, -30°C
2.
O
H
96%

Hong, B.-C.; Hong, J.-H. *Tetrahedron Lett.*, ***1997***, *38*, 255.

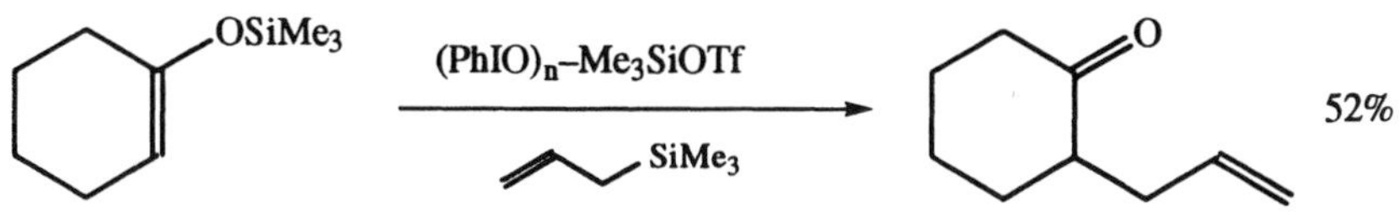
OSiMe3
(PhIO)n–Me3SiOTf
SiMe3
O
52%

Moriarty, R.M.; Epa, W.R.; Prakash, O. *J. Chem. Res (S)*, ***1997***, 262.

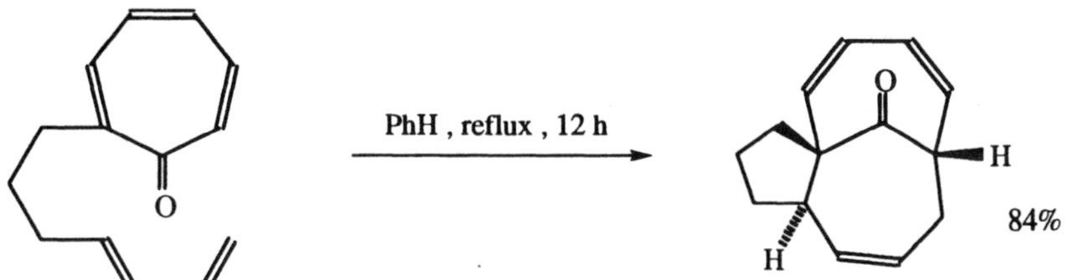
O
PhH , reflux , 12 h
O
H
H
84%

Rigby, J.H.; Rege, S.D.; Sandanayaka, V.P.; Kirova, M. *J. Org. Chem.*, ***1996***, *61*, 842.

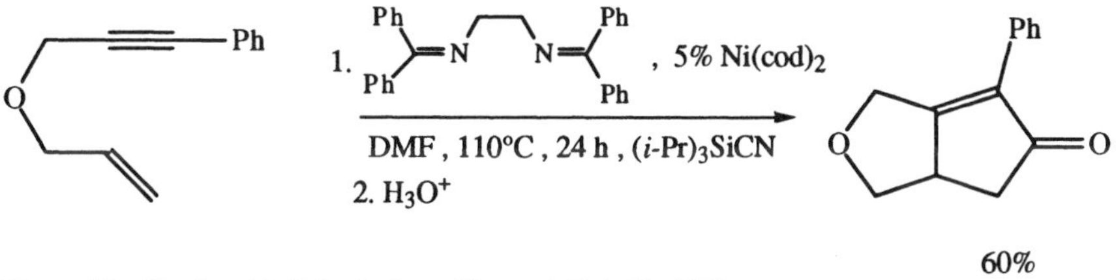
Ph
O
1.
Ph
Ph
N
N
Ph
Ph
, 5% Ni(cod)2
DMF , 110°C , 24 h , (i-Pr)3SiCN
2. H3O+
Ph
O
O
60%

Zhang, M.; Buchwald, S.L. *J. Org. Chem.*, ***1996***, *61*, 4498.

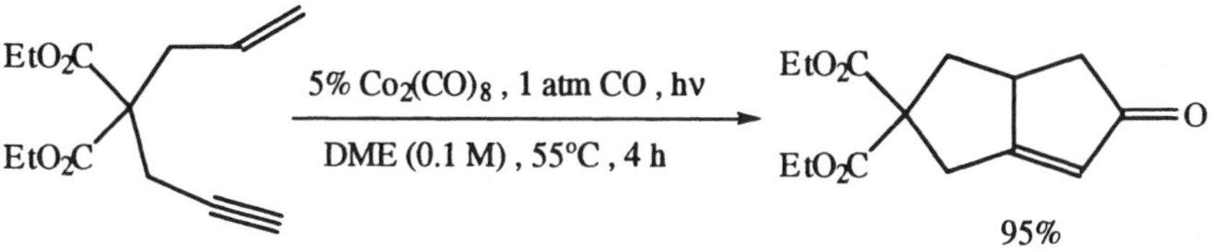
EtO2C
EtO2C
5% Co2(CO)8 , 1 atm CO , hv
DME (0.1 M) , 55°C , 4 h
EtO2C
EtO2C
O
95%

Pagenkopf, B.L.; Livinghouse, T. *J. Am. Chem. Soc.*, ***1996***, *118*, 2285.

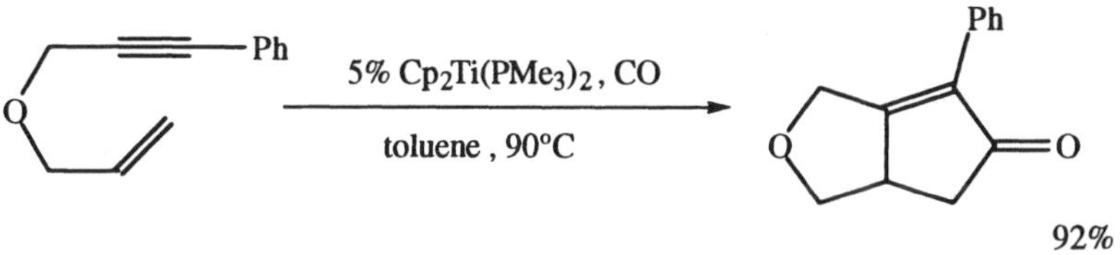
Ph
O
5% Cp2Ti(PMe3)2 , CO
toluene , 90°C
Ph
O
O
92%

Hicks, F.A.; Kablaoui, N.M.; Buchwald, S.L. *J. Am. Chem. Soc.*, ***1996***, *118*, 9450.

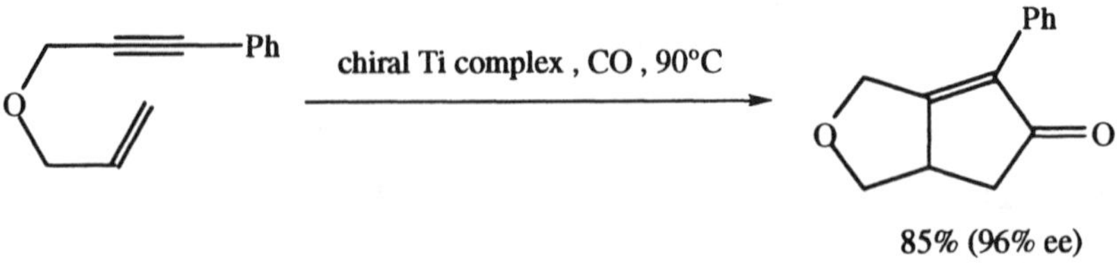
Ph
chiral Ti complex , CO , 90°C
Ph
O
O
85% (96% ee)

Hicks, F.A.; Buchwald, S.L. *J. Am. Chem. Soc.*, ***1996***, *118*, 11688.

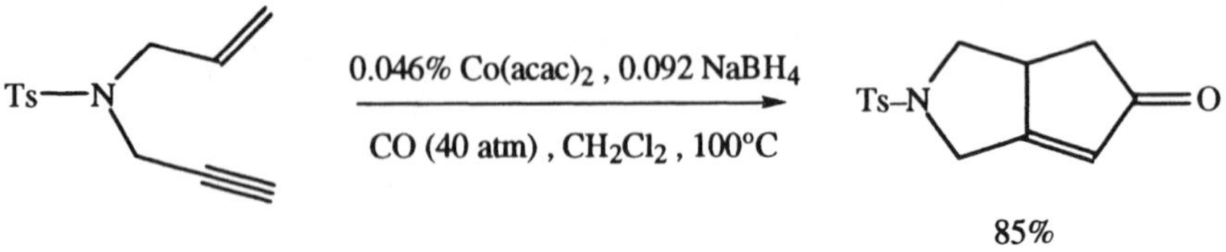
Ts—N
0.046% Co(acac)2 , 0.092 NaBH4
CO (40 atm) , CH2Cl2 , 100°C
Ts–N
O
85%

Lee, N.Y.; Chung, Y.K. *Tetrahedron Lett.*, ***1996***, *37*, 3145.

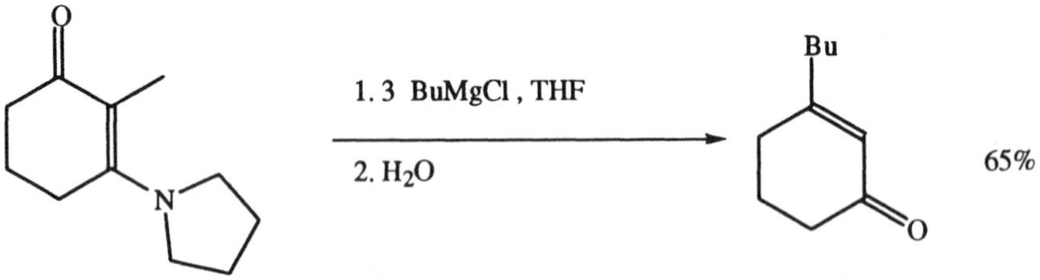
O
N
1. 3 BuMgCl , THF
2. H2O
Bu
O
65%

Shawe, T.T.; Landino, L.M.; Ross, A.A.; Prokopowicz, A.S.; Robinson, P.M.; Cannon, A. *Tetrahedron Lett.*, ***1996***, *37*, 3823.

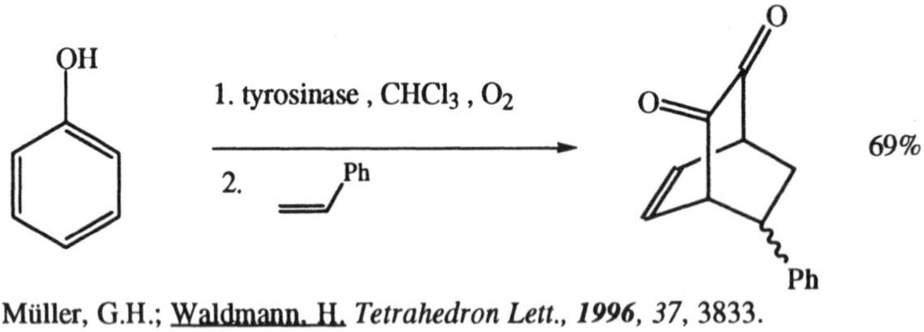
OH
1. tyrosinase , CHCl3 , O2
2. Ph
O
O
Ph
69%

Müller, G.H.; Waldmann, H. *Tetrahedron Lett.*, ***1996***, *37*, 3833.

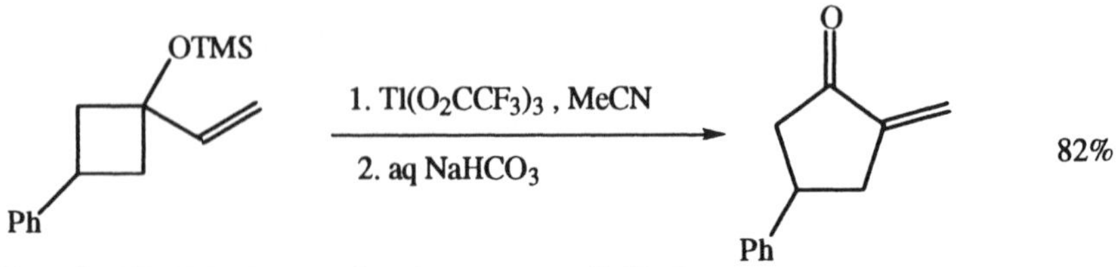
OTMS
Ph
1. Tl(O2CCF3)3 , MeCN
2. aq NaHCO3
O
Ph
82%

Kim, S.; Uh, K.H. *Tetrahedron Lett.*, ***1996***, *37*, 3865.

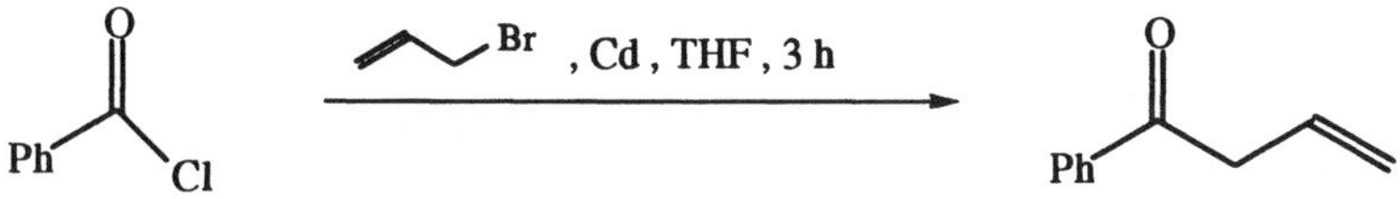
O
Ph
Cl
Br
, Cd , THF , 3 h
O
Ph

90%

Baruah, B.; Boruah, A.; Prajapati, D.; Sandhu, J.S. *Tetrahedron Lett.*, ***1996***, *37*, 9087.

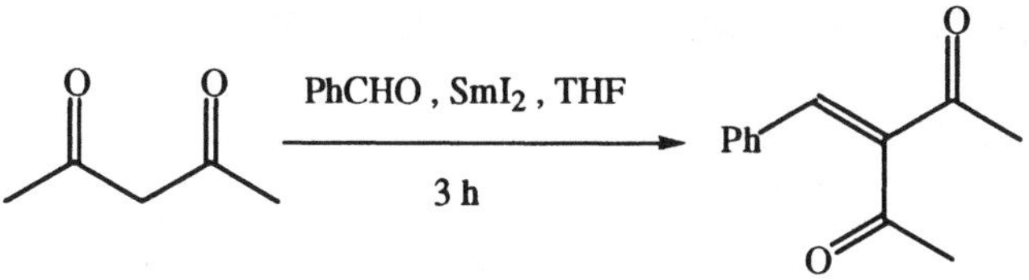
O
O
PhCHO , SmI2 , THF
3 h
O
Ph
O

78%

Bao, W.; Zhang, Y.; Wang, J. *Synth. Commun.*, ***1996***, *26*, 3025.

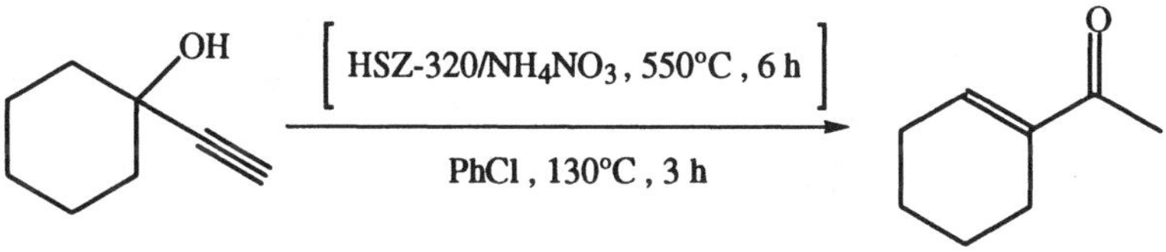
OH
HSZ-320/NH4NO3 , 550°C , 6 h
PhCl , 130°C , 3 h
O

quant

Sartori, G.; Pastorio, A.; Maggi, R.; Bigi, F. *Tetrahedron*, ***1996***, *52*, 8287.

SECTION 375: NITRILE - NITRILE

NO ADDITIONAL EXAMPLES

SECTION 376: NITRILE - ALKENE

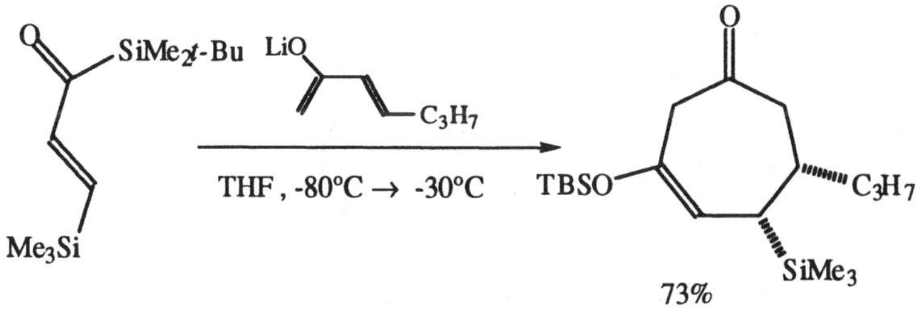
O
SiMe2t-Bu
LiO
C3H7
Me3Si
THF , -80°C → -30°C
O
TBSO
C3H7
SiMe3

73%

Takeda, K.; Nakajima, A.; Takeda, M.; Okamoto, Y.; Sato, T.; Yoshii, E.; Koizumi, T.; Shiro M. *J. Am. Chem. Soc.*, ***1998***, *120*, 4947.

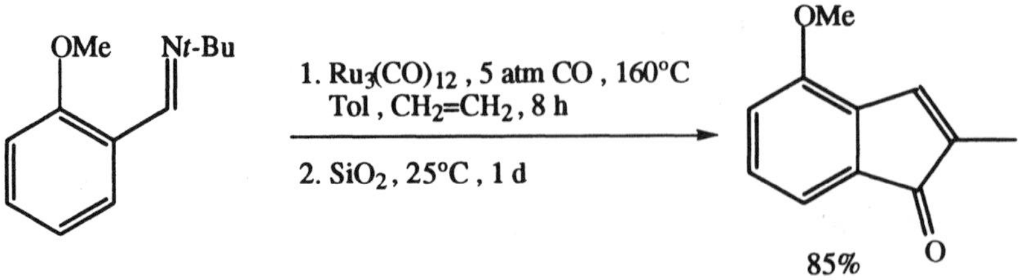
OMe
Nt-Bu
1. Ru3(CO)12 , 5 atm CO , 160°C
Tol , CH2=CH2 , 8 h
2. SiO2 , 25°C , 1 d
OMe
O
85%

Fukuyama, T.; Chatani, N.; Kakiuchi, F.; Murai, S. *J. Org. Chem.*, ***1997***, *62*, 5647.

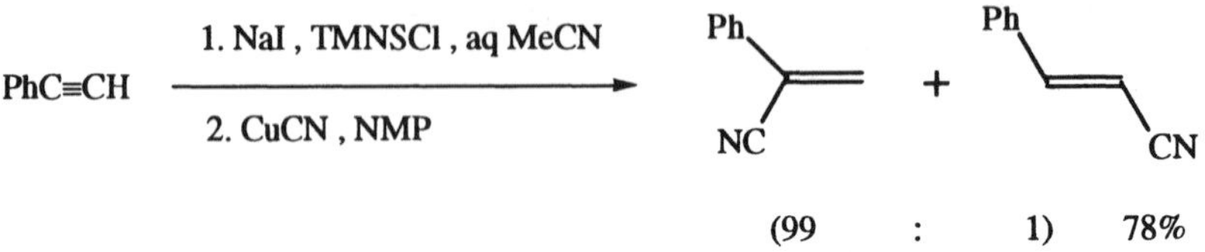
PhC≡CH
1. NaI , TMNSCl , aq MeCN
2. CuCN , NMP
Ph
NC
Ph
CN
(99 : 1) 78%

Luo, F.-T.; Ko, S.-L.; Chao, D.-Y. *Tetrahedron Lett.*, ***1997***, *38*, 8061.

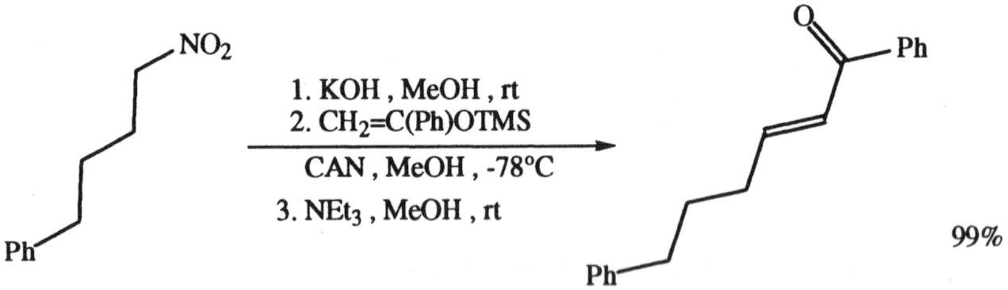
NO2
Ph
1. KOH , MeOH , rt
2. CH2=C(Ph)OTMS
CAN , MeOH , -78°C
3. NEt3 , MeOH , rt
O
Ph
Ph
99%

Arai, N.; Narasaka, K. *Bull. Chem. Soc. Jpn.*, ***1997***, *70*, 2525.

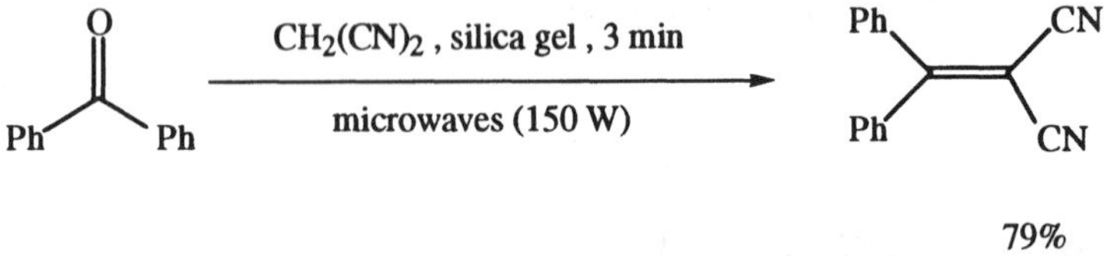
O
Ph
Ph
CH2(CN)2 , silica gel , 3 min
microwaves (150 W)
Ph
Ph
CN
CN
79%

de la Cruz, P.; Díez-Barra, E.; Loupy, A.; Langa, F. *Tetrahedron Lett.*, ***1996***, *37*, 1113.

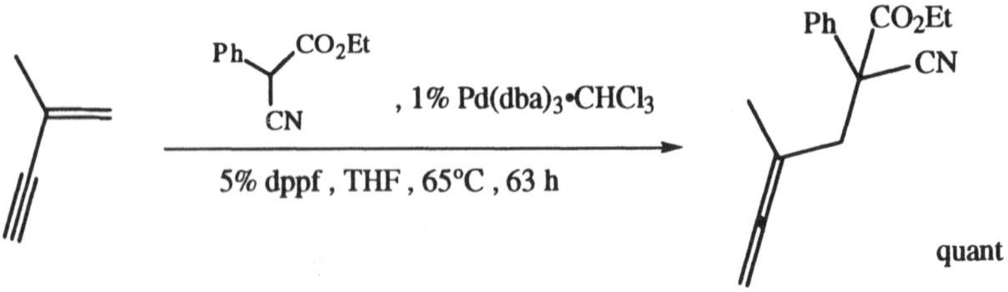
Ph
CO2Et
CN
, 1% Pd(dba)3•CHCl3
5% dppf , THF , 65°C , 63 h
Ph
CO2Et
CN
quant

Salter, M.M.; Gevorgyan, V.; Saito, S.; Yamamoto, Y. *Chem. Commun.*, ***1996***, 17.

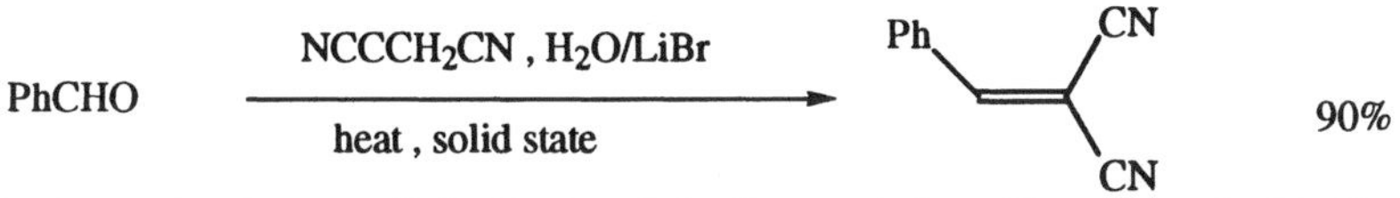
PhCHO
NCCCH2CN , H2O/LiBr
heat , solid state
Ph
CN
CN
90%

Prajapati, D.; Lekhok, K.C.; Sandhu, J.S.; Ghosh, A.C. *J. Chem. Soc., Perkin Trans. 1*, ***1996*** 959.

SECTION 377: ALKENE - ALKENE

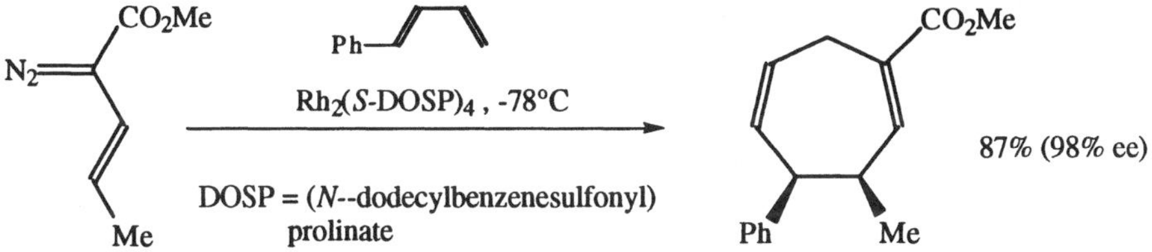
CO2Me
N2
Me
Ph
Rh2(S-DOSP)4 , -78°C
DOSP = (N--dodecylbenzenesulfonyl)
prolinate
CO2Me
Ph
Me
87% (98% ee)

Davies, H.M.L.; Stafford, D.G.; Doan, B.D.; Houser, J.H. *J. Am. Chem. Soc.*, ***1998***, *120*, 3326.

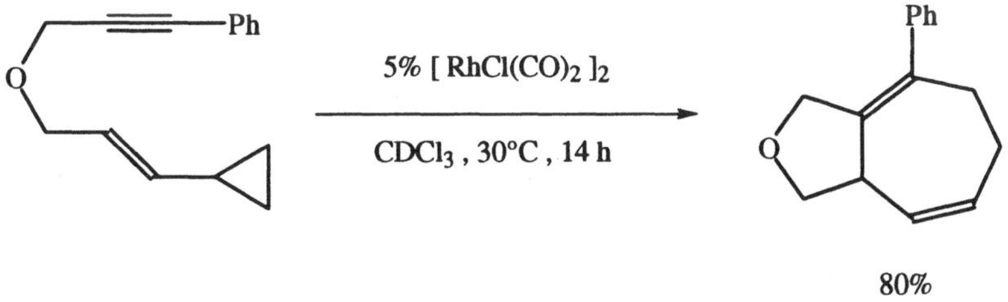
Ph
O
5% [RhCl(CO)2]2
CDCl3 , 30°C , 14 h
Ph
O
80%

Wender, P.A.; Sperandio, D. *J. Org. Chem.*, ***1998***, *63*, 4164.

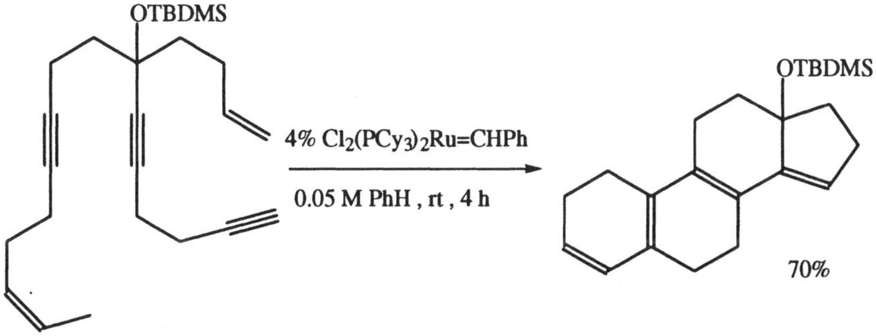
OTBDMS
4% Cl2(PCy3)2Ru=CHPh
0.05 M PhH , rt , 4 h
OTBDMS
70%

Zuercher, W.J.; Scholl, M.; Grubbs, R.H. *J. Org. Chem.*, ***1998***, *63*, 4291.

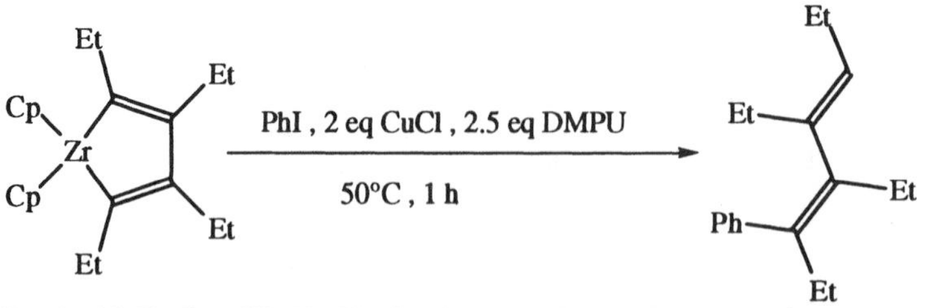

74%

Takahashi, T.; Sun, W.–H.; Xi, C.; Ubayama, H.; Xi, Z. *Tetrahedron*, ***1998***, *54*, 715.

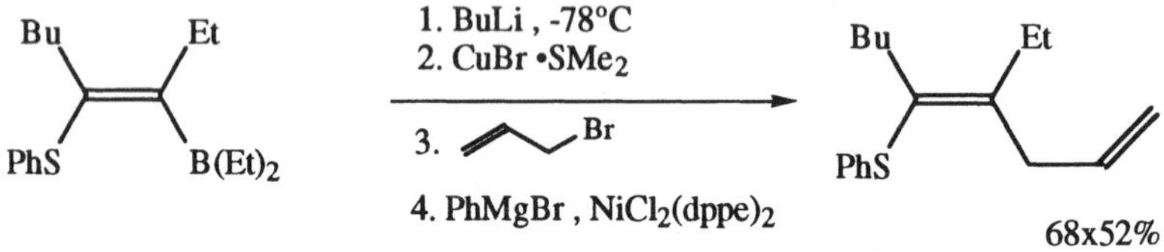

68x52%

Gerard, J.; Bietlot, E.; Hevesi, L. *Tetrahedron Lett.*, ***1998***, *39*, 8735.

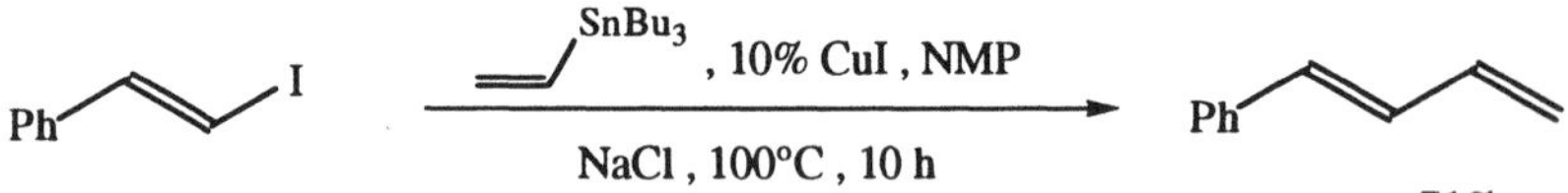

71%

Kang, S.–K.; Kim, J.–S.; Choi, S.–C. *J. Org. Chem.*, ***1997***, *62*, 4208.

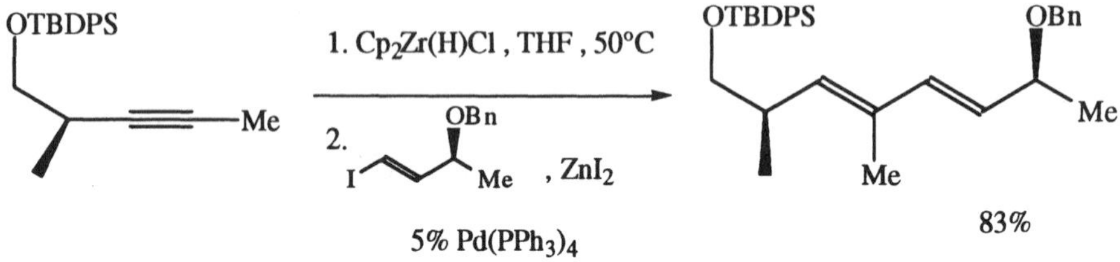

83%

Panek, J.S.; Hu, T. *J. Org. Chem.*, ***1997***, *62*, 4912.

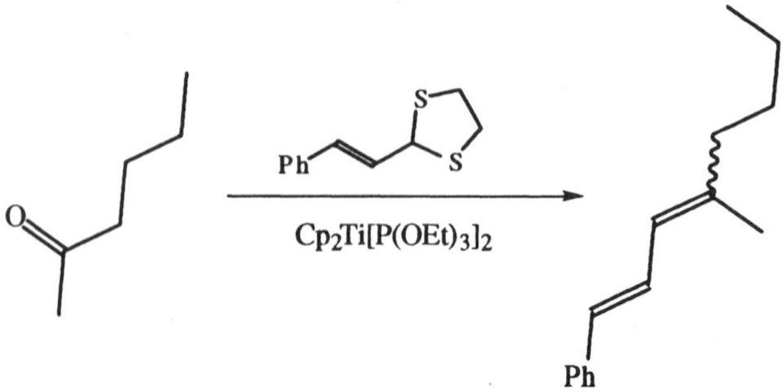

74% (60:40 *E:Z*)

Horikawa, Y.; Watanabe, M.; Fujiwara, T.; Takeda, T. *J. Am. Chem. Soc.*, ***1997***, *119*, 1127.

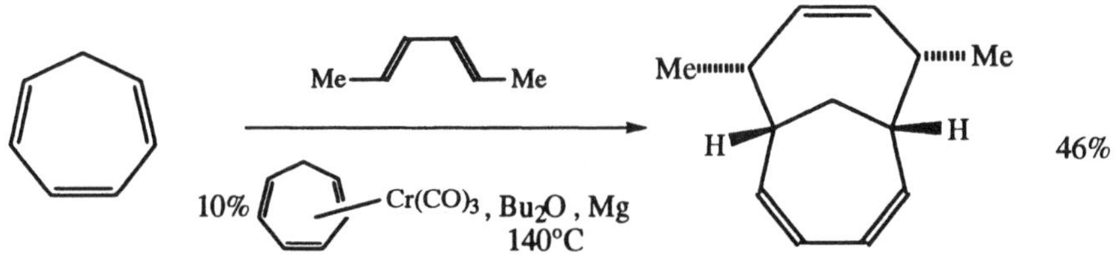
Me
Me
10%
Cr(CO)3 , Bu2O , Mg
140°C
Me
Me
H
H
46%

Rigby, J.H.; Fiedler, C. *J. Org. Chem.*, ***1997***, *62*, 6106.

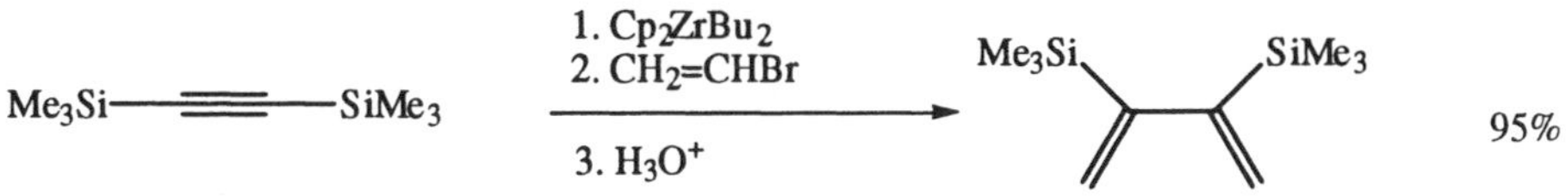
Me3Si
SiMe3
1. Cp2ZrBu2
2. CH2=CHBr
3. H3O+
Me3Si
SiMe3
95%

Takahashi, T.; Xi, Z.; Fischer, R.; Huo, S.; Xi, C.; Nakajima, K. *J. Am. Chem. Soc.*, ***1997***, *119*, 4561.

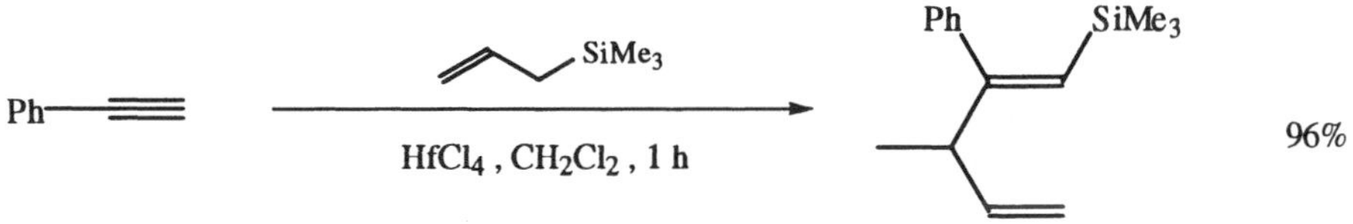
Ph
SiMe3
HfCl4 , CH2Cl2 , 1 h
Ph
SiMe3
96%

Yoshikawa, E.; Gevorgyan, V.; Asao, N.; Yamamoto, Y. *J. Am. Chem. Soc.*, ***1997***, *119*, 6781.

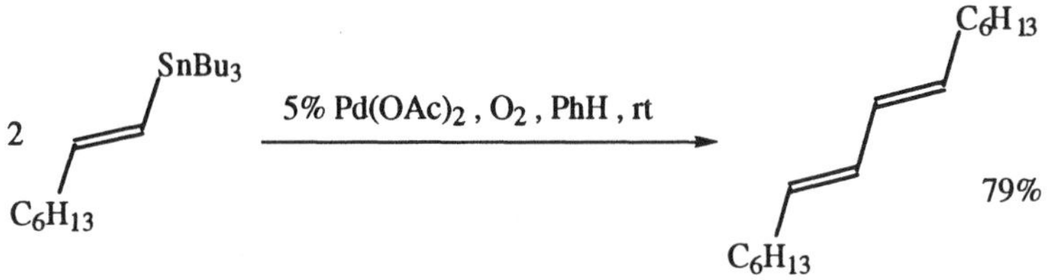
2
SnBu3
C6H13
5% Pd(OAc)2 , O2 , PhH , rt
C6H13
C6H13
79%

Alcaraz, L.; Taylor, R.J.K. *SynLett.*, ***1997***, 791.

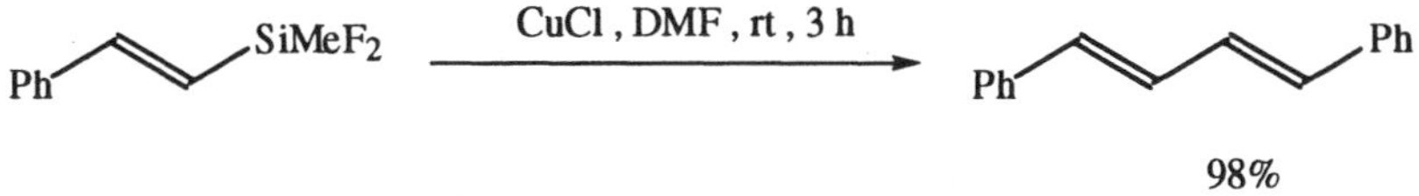
Ph
SiMeF2
CuCl , DMF , rt , 3 h
Ph
Ph
98%

Ikegashira, K.; Nishihara, Y.; Hirabayashi, K.; Mori, A.; Hiyama, T. *Chem. Commun.*, ***1997***, 1039.

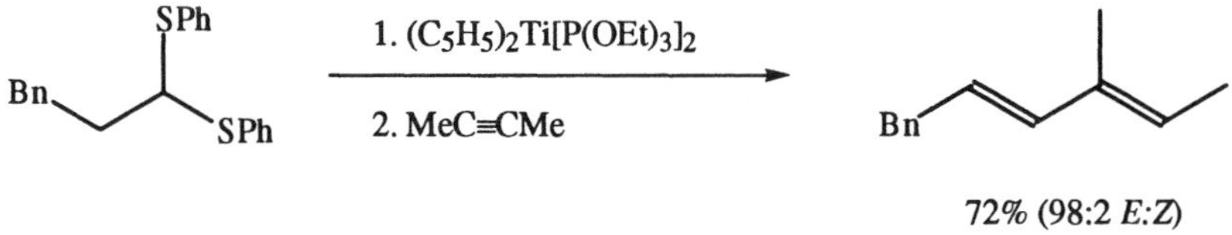
SPh
Bn
SPh
1. (C5H5)2Ti[P(OEt)3]2
2. MeC≡CMe
Bn
72% (98:2 E:Z)

Takeda, T.; Shimokawa, H.; Miyachi, Y.; Fujiwara, T. *Chem. Commun.*, ***1997***, 105.

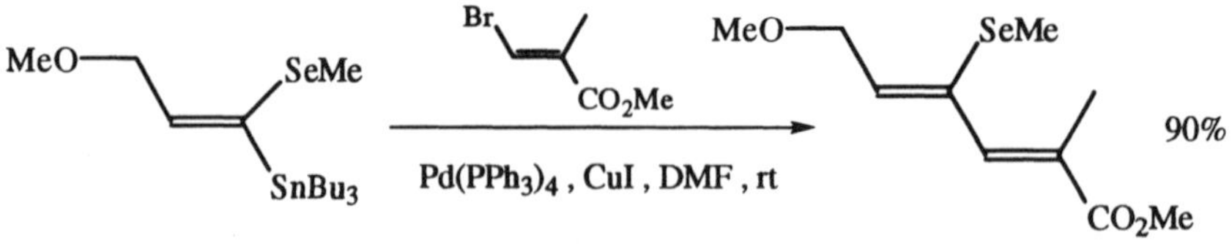
MeO
SeMe
SnBu3
Br
CO2Me
Pd(PPh3)4 , CuI , DMF , rt
MeO
SeMe
CO2Me
90%

Ma, Y.; Huang, X. *J. Chem. Soc., Perkin Trans., 1*, ***1997***, 2953.

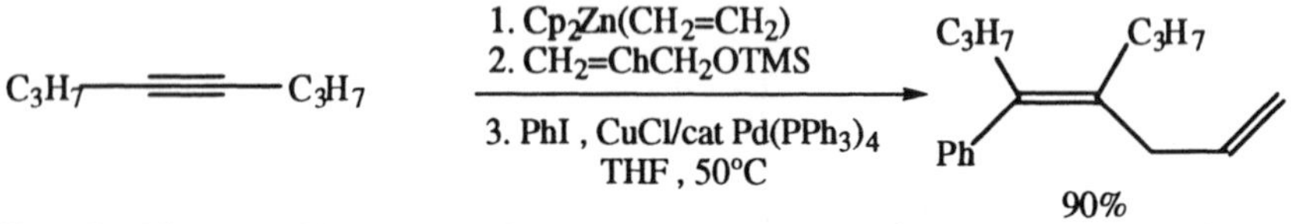
C3H7
C3H7
1. Cp2Zn(CH2=CH2)
2. CH2=ChCH2OTMS
3. PhI , CuCl/cat Pd(PPh3)4
THF , 50°C
C3H7
C3H7
Ph
90%

Hara, R.; Nishihara, Y.; Landré, P.D.; Takahashi, T. *Tetrahedron Lett.*, ***1997***, *38*, 447.

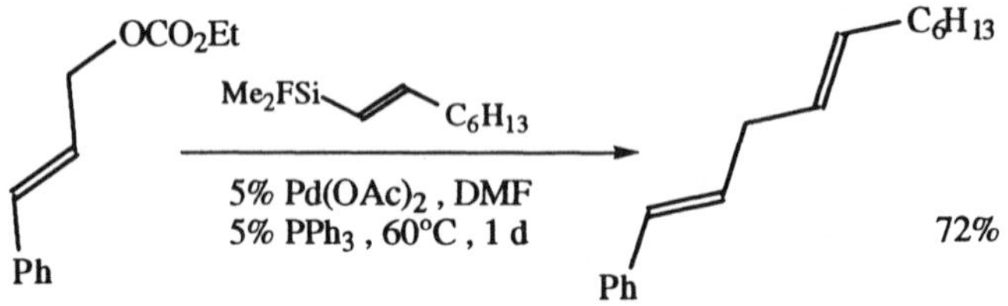
OCO2Et
Ph
Me2FSi
C6H13
5% Pd(OAc)2 , DMF
5% PPh3 , 60°C , 1 d
C6H13
Ph
72%

Matsuhashi, H.; Asai, S.; Hirabayashi, K.; Hatanaka, Y.; Mori, A.; Hiyama, T. *Bull. Chem. Soc. Jpn.*, ***1997***, *70*, 1943.

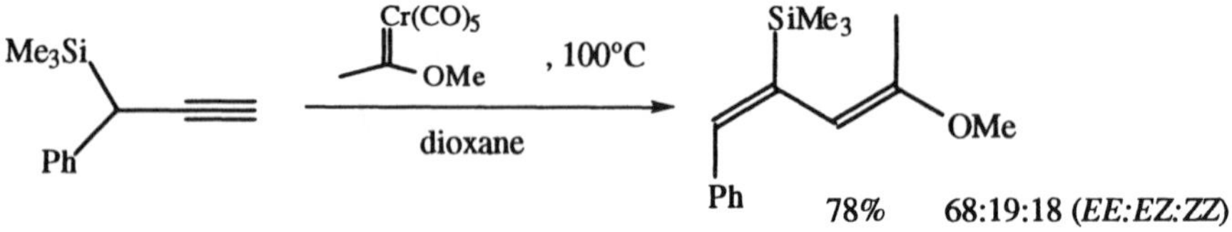
Me3Si
Ph
Cr(CO)5
OMe
, 100°C
dioxane
SiMe3
OMe
Ph
78%
68:19:18 (*EE:EZ:ZZ*)

Herndon, J.W.; Patel, P.P. *J. Org. Chem.*, ***1996***, *61*, 4500.

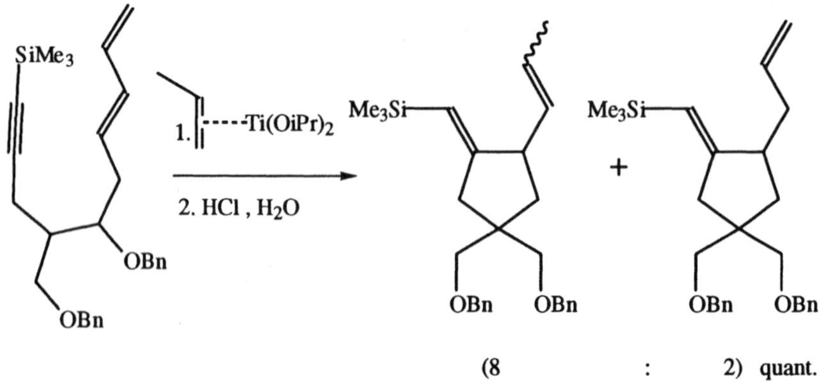
SiMe3
OBn
OBn
1.
Ti(OiPr)2
2. HCl , H2O
Me3Si
OBn
OBn
+
Me3Si
OBn
OBn
(8
:
2) quant.

Urabe, H.; Takeda, T.; Sato, F. *Tetrahedron Lett.*, ***1996***, *37*, 1253.

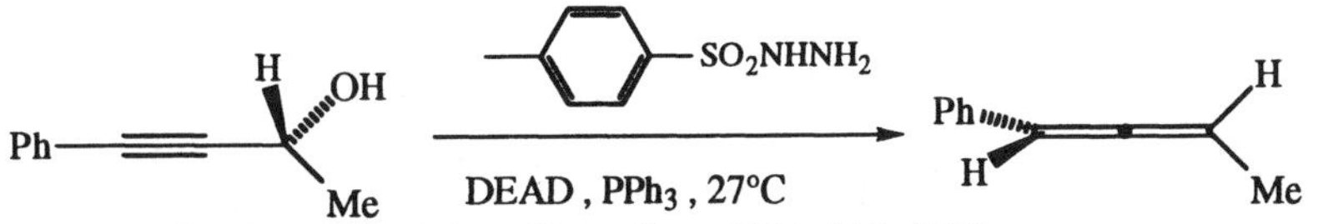
SO2NHNH2
DEAD, PPh3, 27°C

85%

Myers, A.G.; Zheng, B. *J. Am. Chem. Soc.*, ***1996***, *118*, 4492.

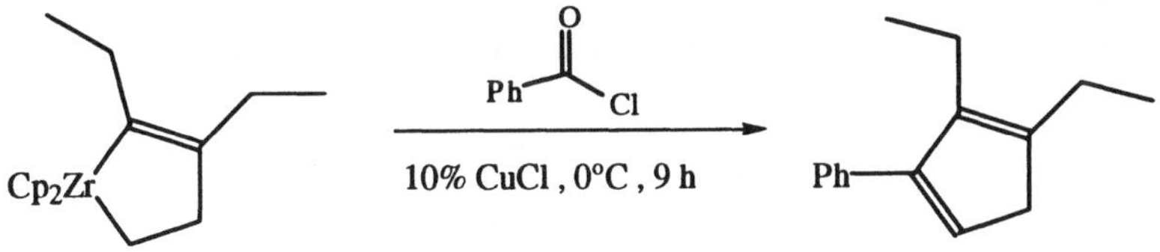
10% CuCl, 0°C, 9 h

51% (20:10:1 positional isomers)

Takahashi, T.; Xi, Z.; Kotora, M.; Xi, C.; Nakajima, K. *Tetrahedron Lett.*, ***1996***, *37*, 7521.

SECTION 378: OXIDES - ALKYNES

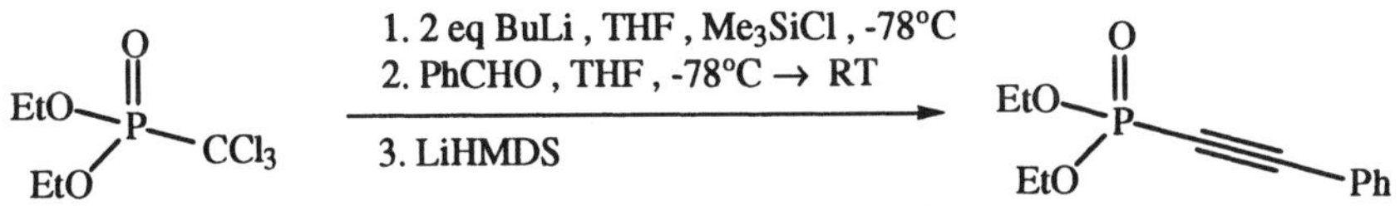
1. 2 eq BuLi, THF, Me3SiCl, -78°C
2. PhCHO, THF, -78°C → RT
3. LiHMDS

89%

Dizière, R.; Savignac, P. *Tetrahedron Lett.*, ***1996***, *36*, 1783.

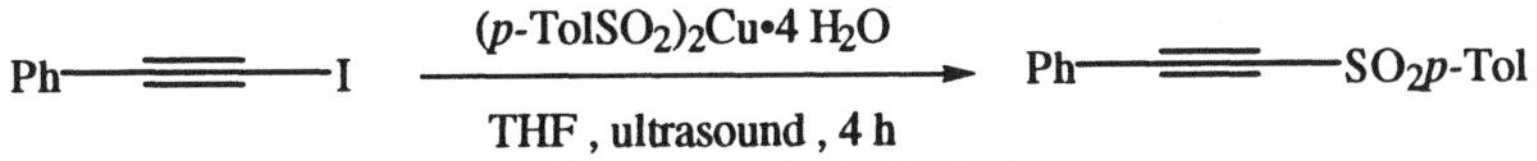
(p-TolSO2)2Cu•4 H2O
THF, ultrasound, 4 h

77%

Suzuki, H.; Abe, H. *Tetrahedron Lett.*, ***1996***, *37*, 3717.

SECTION 379: OXIDES - ACID DERIVATIVES

NO ADDITIONAL EXAMPLES

SECTION 380: OXIDES - ALCOHOLS, THIOLS

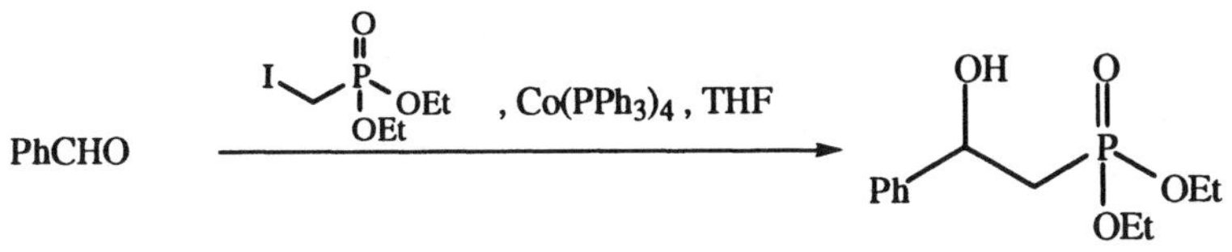
PhCHO
, Co(PPh3)4, THF

81%

Orsini, F. *Tetrahedron Lett.*, ***1998***, *39*, 1425.

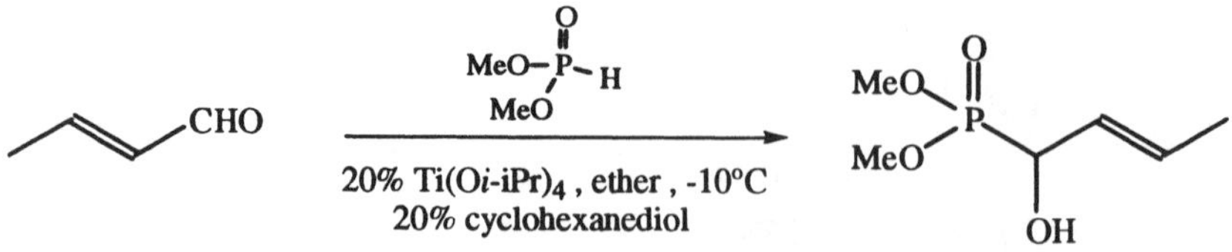
CHO
MeO—P—H
MeO
20% Ti(O*i*-iPr)$_4$, ether , -10°C
20% cyclohexanediol
MeO
MeO
P
O
OH

72% (65% ee , *R*)

Groaning, M.D.; Rowe, B.J.; Spilling, C.D. *Tetrahedron Lett.*, ***1998***, *39*, 5485.

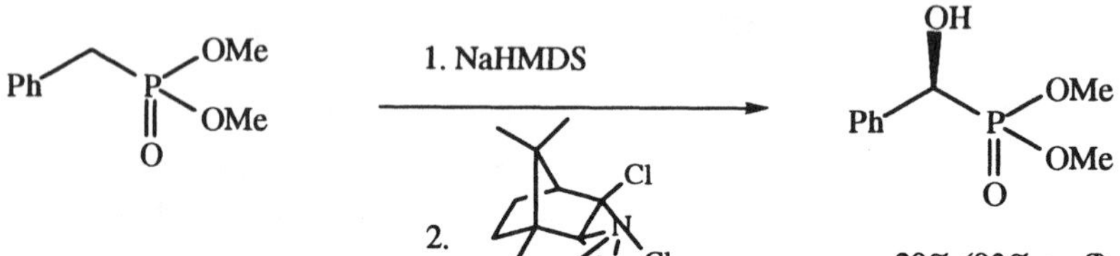
Ph
OMe
OMe
P
O
1. NaHMDS
2.
Cl
N
Cl
O
S
O$_2$
OH
Ph
OMe
OMe

20% (93% e , *S*)

Pogatchnik, D.M.; Wiemer, D.F. *Tetrahedron Lett.*, ***1997***, *38*, 3495.

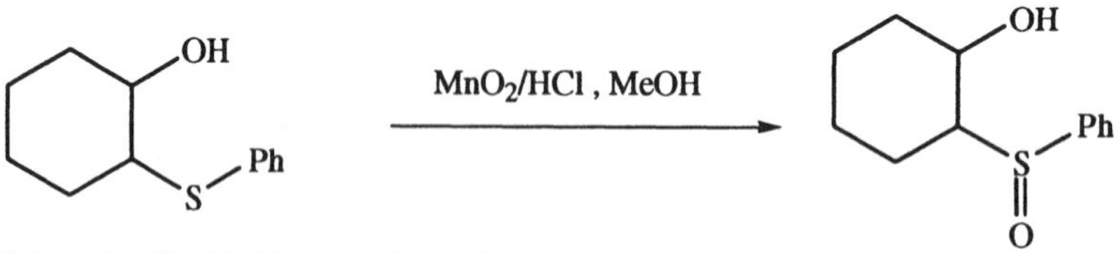
OH
S
Ph
MnO$_2$/HCl , MeOH
OH
Ph
S
O

81%

Gabbi, C.; Ghelfi, F.; Grandi, R. *Synth. Commun.*, ***1997***, *27*, 2857.

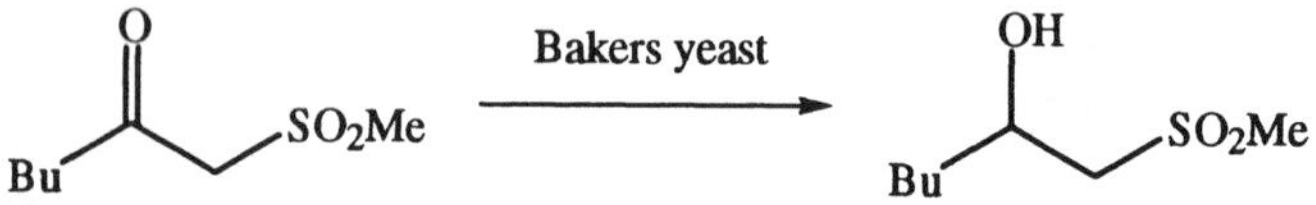
O
Bu
SO$_2$Me
Bakers yeast
OH
Bu
SO$_2$Me

12% (71% ee , *S*)

Maguire, A.R.; Lowney, D.G. *J. Chem. Soc., Perkin Trans. 1*, ***1997***, 235.

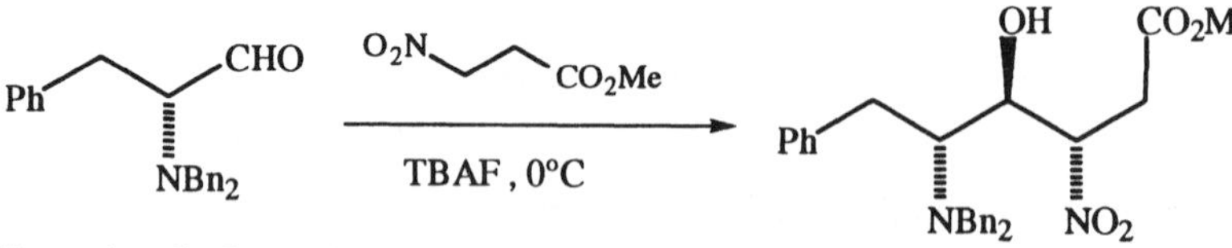
Ph
CHO
NBn$_2$
O$_2$N
CO$_2$Me
TBAF , 0°C
OH
CO$_2$Me
Ph
NBn$_2$
NO$_2$

66% (49:1)

Hanessian, S.; Devasthale, P.V. *Tetrahedron Lett.*, ***1996***, *37*, 987.

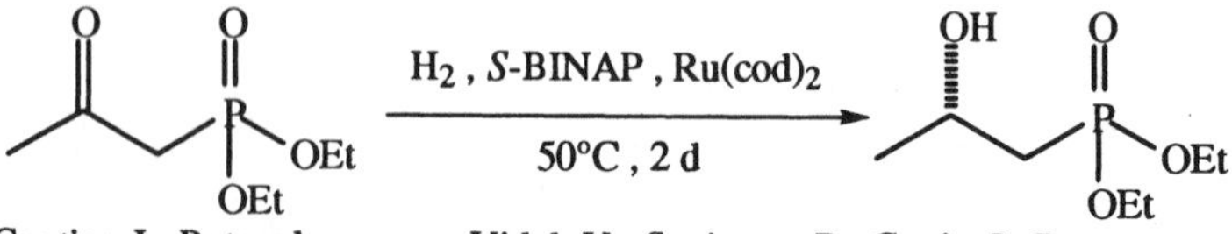
O
O
P
OEt
OEt
H$_2$, *S*-BINAP , Ru(cod)$_2$
50°C , 2 d
OH
O
P
OEt
OEt

quant (94% ee)

Gautier, I.; Ratovelomanana–Vidal, V.; Savignac, P.; Genêt, J.–P. *Tetrahedron Lett.*, ***1996***, *37*, 7721.

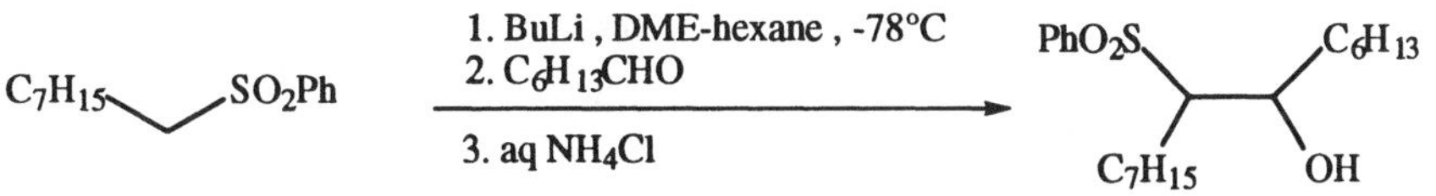

yields variable with 2° sulfones 95%

Hart, D.J.; Wu, W.–L. *Tetrahedron Lett.*, ***1996***, *37*, 5283.

SECTION 381 OXIDES - ALDEHYDES

NO ADDITIONAL EXAMPLES

SECTION 382: OXIDES - AMIDES

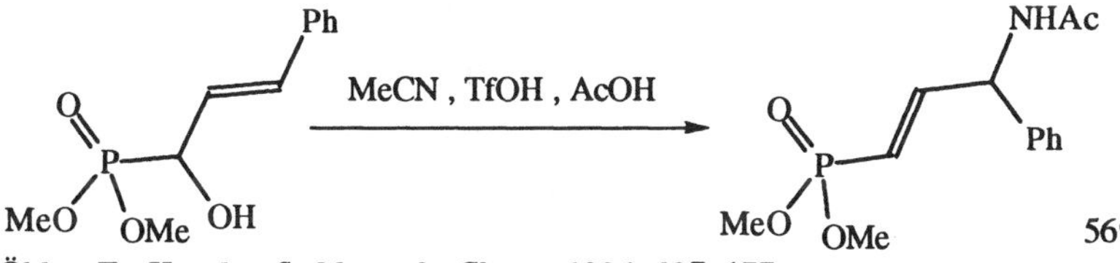

56%

Öhler, E.; Kanzler, S. *Monatsh. Chem.*, ***1996***, *127*, 177.

SECTION 383: OXIDES - AMINES

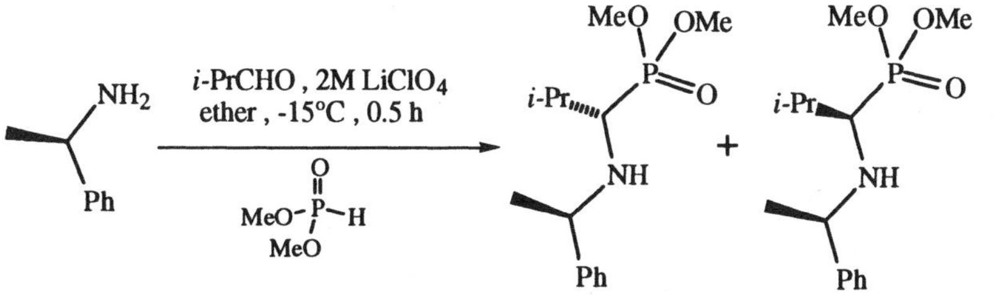

(79 : 21) 95%

Heydari, A.; Karimian, A.; Ipaktschi, J. *Tetrahedron Lett.*, ***1998***, *39*, 6729.

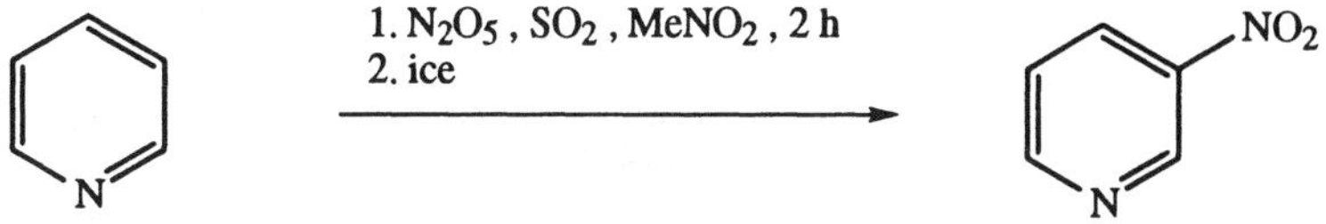

58%

Armnestad. B.; Bakke, J.M.; Hegbom, I.; Ranes, E. *Acta Chem. Scand. B*, ***1996***, *50*, 556.

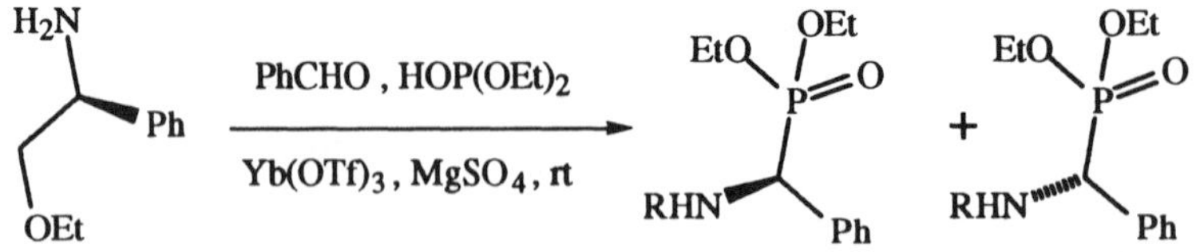

(78 : 22) 95%

Qian, C.; Huang, T. *J. Org. Chem.*, ***1998***, *63*, 4125.

SECTION 384: OXIDES - ESTERS

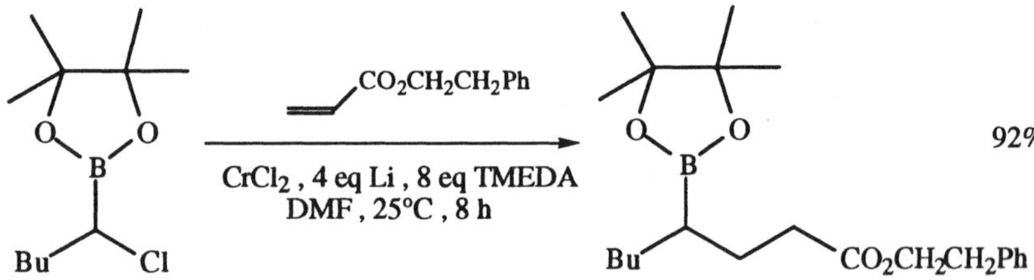

92%

Takai, K.; Shinomiya, N.; Ohta, M. *SynLett.*, ***1998***, 253.

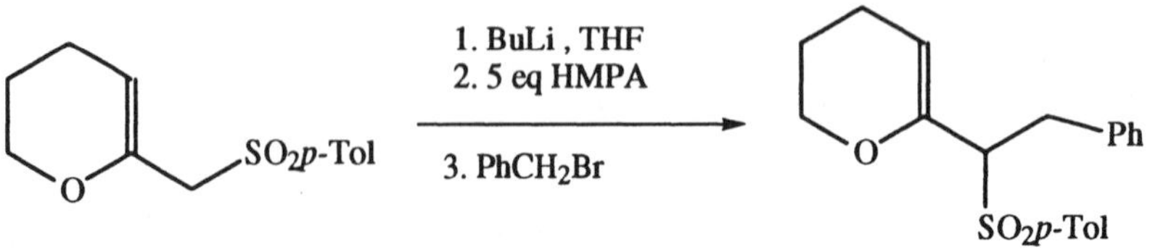

78%

Edwards, G.L.; Sinclair, D.J.; Wasiowych, C.D. *SynLett.*, ***1997***, 1285.

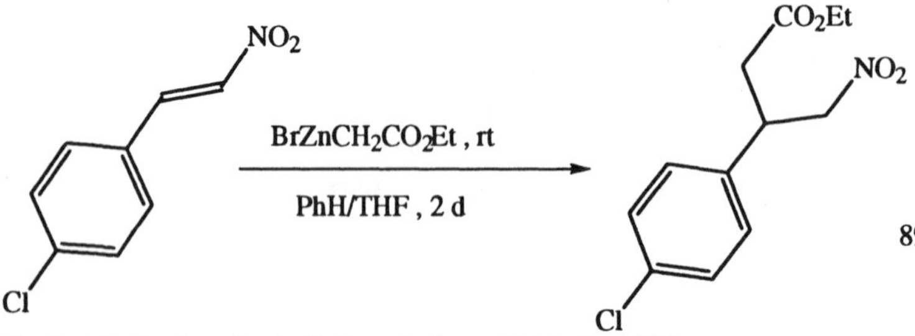

89%

Menicagli, R.; Samritani, S. *Tetrahedron*, ***1996***, *52*, 1425.

SECTION 385: OXIDES - ETHERS, EPOXIDES, THIOETHERS

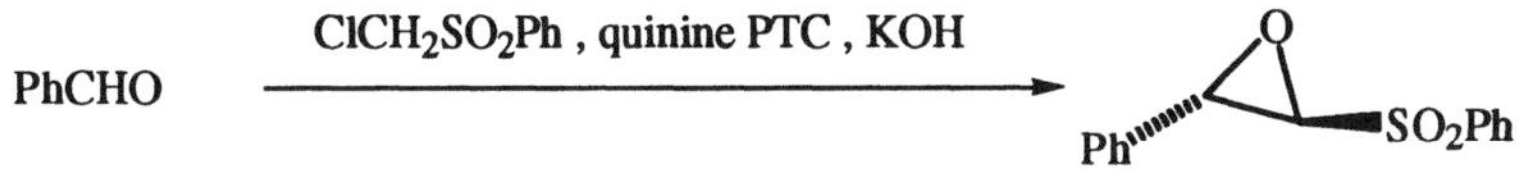
PhCHO
$ClCH_2SO_2Ph$, quinine PTC , KOH
O
Ph
SO_2Ph

85% (69% ee)

Arai, S.; Ishida, T.; Shioiri, T. *Tetrahedron Lett.*, ***1998***, *39*, 8299.

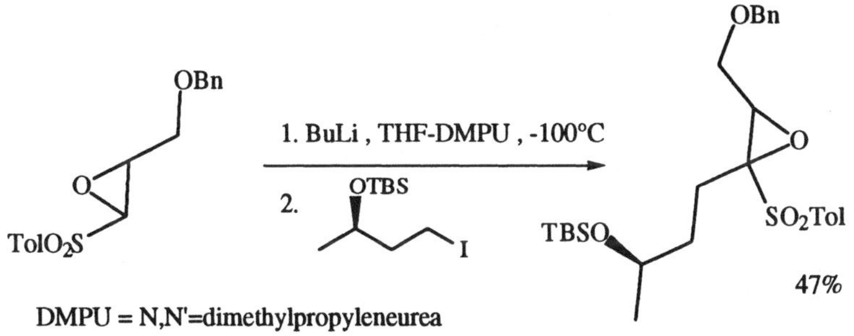
OBn
O
$TolO_2S$
1. BuLi , THF-DMPU , -100°C
2.
OTBS
I
OBn
O
SO_2Tol
TBSO

47%

DMPU = N,N'=dimethylpropyleneurea

Mori, Y.; Yaegashi, K.; Iwase, K.; Yamamori, Y.; Furukawa, H. *Tetrahedron Lett.*, ***1996***, *37*, 2605.

SECTION 386: OXIDES - HALIDES, SULFONATES

NO ADDITIONAL EXAMPLES

SECTION 387: OXIDES - KETONES

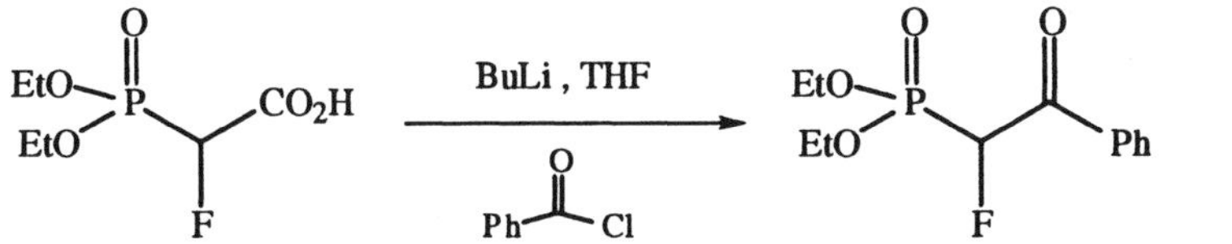
O
EtO
P
EtO
CO_2H
F
BuLi , THF
O
Ph
Cl
O
O
EtO
P
EtO
Ph
F

72%

Kim, D.Y.; Choi, Y.J. *Synth. Commun.*, ***1998***, *28*, 1481.

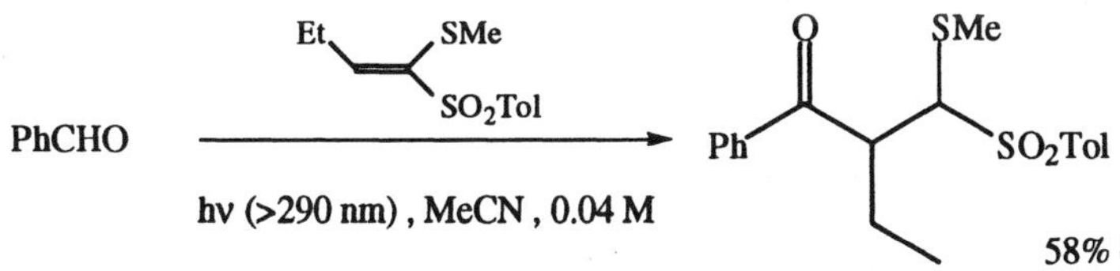
Et
SMe
SO_2Tol
PhCHO
hν (>290 nm) , MeCN , 0.04 M
O
SMe
Ph
SO_2Tol

58% (61:39 dr)

Ogura, K.; Arai, T.; Kayano, A.; Akazume, M. *Tetrahedron Lett.*, ***1998***, *39*, 9051.

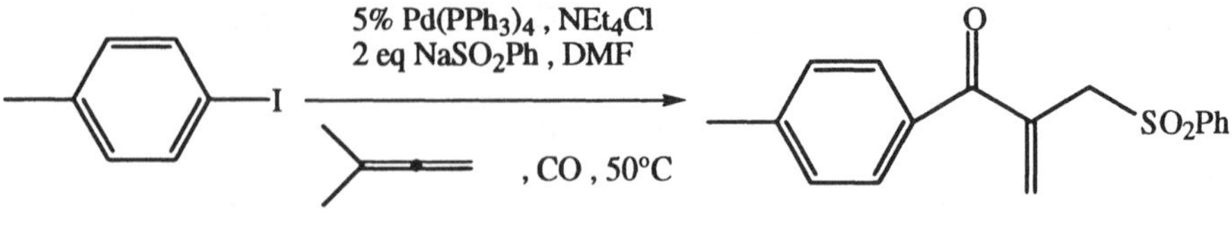
5% Pd(PPh3)4 , NEt4Cl
2 eq NaSO2Ph , DMF
I
, CO , 50°C
O
SO2Ph

61%

Grigg, R.; Brown, S.; Sridharan, V.; Uttley, M.D. *Tetrahedron Lett.*, ***1997***, *38*, 5031.

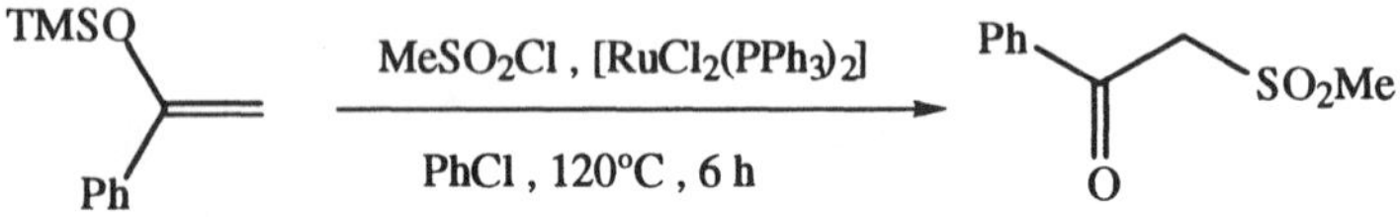
TMSO
Ph
MeSO2Cl , [RuCl2(PPh3)2]
PhCl , 120°C , 6 h
Ph
SO2Me
O

86%

Kamigata, N.; Udodaira, K.; Shimizu, T. *J. Chem. Soc., Perkin Trans. 1*, ***1997***, 783.

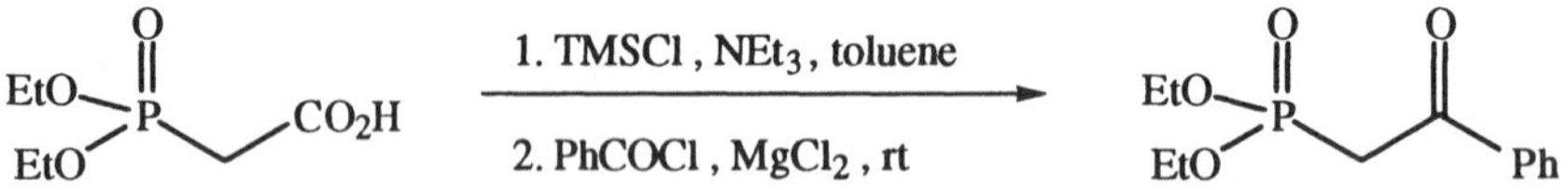
O
EtO
P
EtO
CO2H
1. TMSCl , NEt3 , toluene
2. PhCOCl , MgCl2 , rt
O
O
EtO
P
EtO
Ph

92%

Kim, D.Y.; Kong, M.S.; Lee, K. *J. Chem. Soc., Perkin Trans. 1*, ***1997***, 1361.

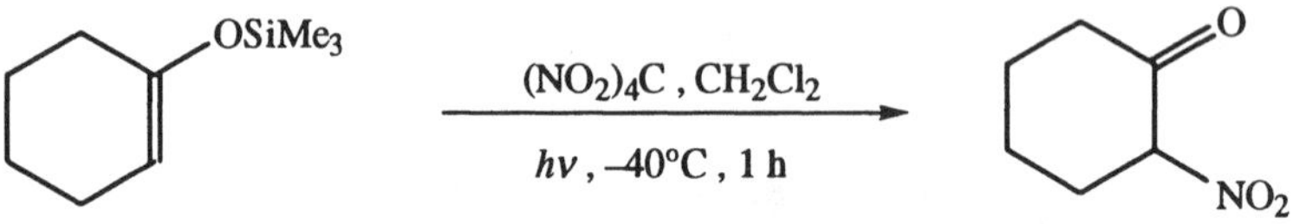
OSiMe3
(NO2)4C , CH2Cl2
hv , –40°C , 1 h
O
NO2

73%

Rathore, R.; Kochi, J.K. *J. Org. Chem.*, ***1996***, *61*, 627.

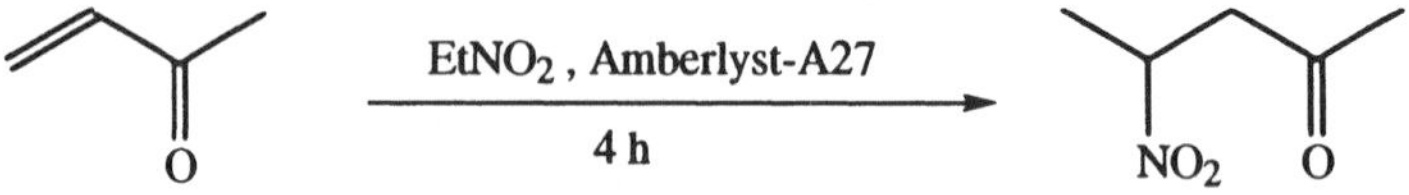
O
EtNO2 , Amberlyst-A27
4 h
NO2
O

76%

Ballini, R.; Marziali, P.; Mozzicafreddo, A. *J. Org. Chem.*, ***1996***, *61*, 3209.

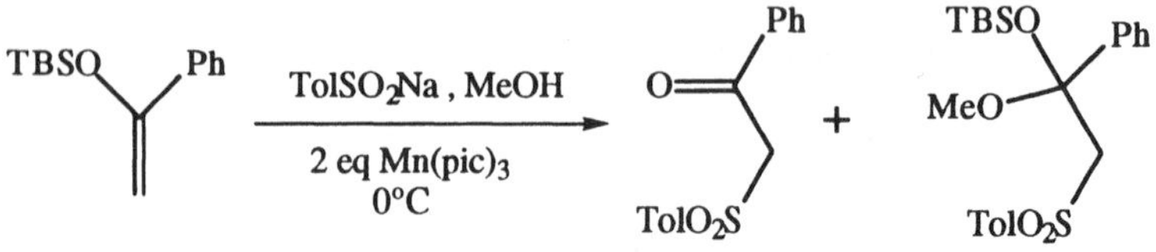
TBSO
Ph
TolSO2Na , MeOH
2 eq Mn(pic)3
0°C
Ph
O
TolO2S
+
TBSO
Ph
MeO
TolO2S

14% 74%

Mochizuki, T.; Hayakawa, S.; Narasaka, K. *Bull. Chem. Soc. Jpn.*, ***1996***, *69*, 2317.

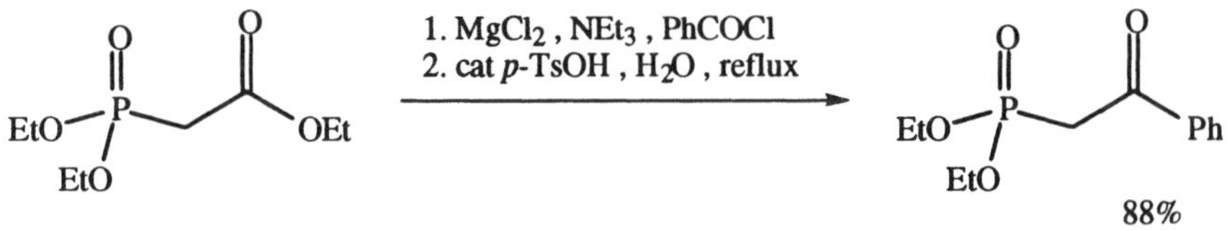
1. $MgCl_2$, NEt_3 , PhCOCl
2. cat p-TsOH , H_2O , reflux
EtO
OEt
Ph

88%

Kim, D.Y.; Kong, M.S.; Kim, T.H. *Synth. Commun.*, ***1996***, *26*, 2487.

SECTION 388: OXIDES - NITRILES

NO ADDITIONAL EXAMPLES

SECTION 389: OXIDES - ALKENES

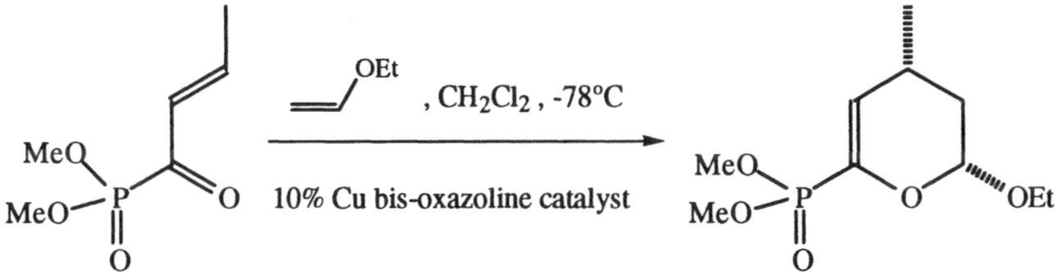
OEt
, CH_2Cl_2 , -78°C
10% Cu bis-oxazoline catalyst
MeO
MeO

89% (99% ee) (99:1 *endo:exo*)

Evans, D.A.; Johnson, J.S. *J. Am. Chem. Soc.*, ***1998***, *120*, 4895.

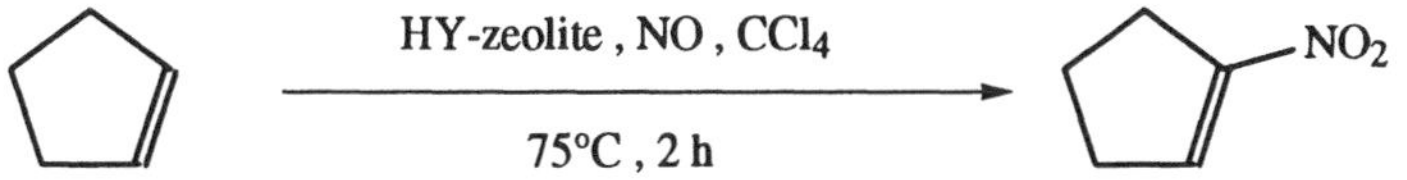
HY-zeolite , NO , CCl_4
75°C , 2 h
NO_2

78%

Sreekumar, R.; Padmakumar, R.; Rugmini, P. *Tetrahedron Lett.*, ***1998***, *39*, 2695.

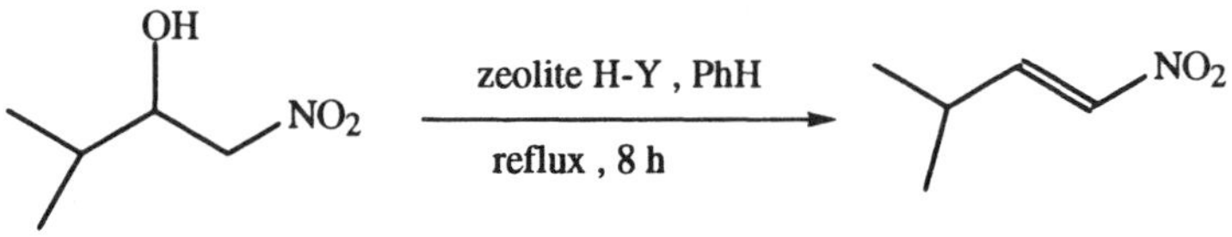
OH
NO_2
zeolite H-Y , PhH
reflux , 8 h

73%

Anbazhagan, M.; Kumaran, G.; Sasidharan, M. *J. Chem. Res. (S)*, ***1997***, 336.

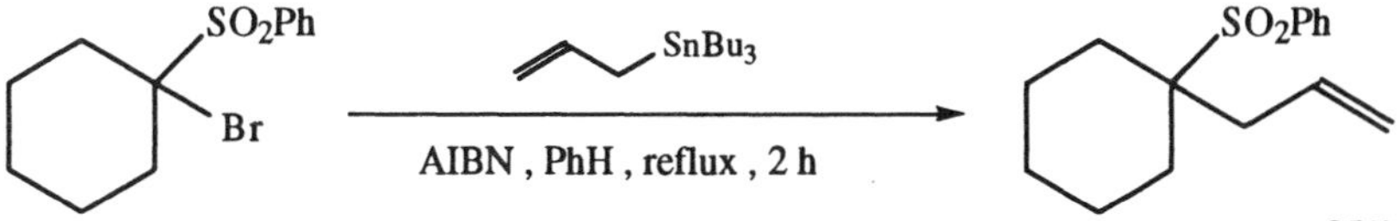
SO_2Ph
Br
$SnBu_3$
AIBN , PhH , reflux , 2 h

85%

Giardinà, A.; Giovannini, R.; Petrini, M. *Tetrahedron Lett.*, ***1997***, *38*, 1995.

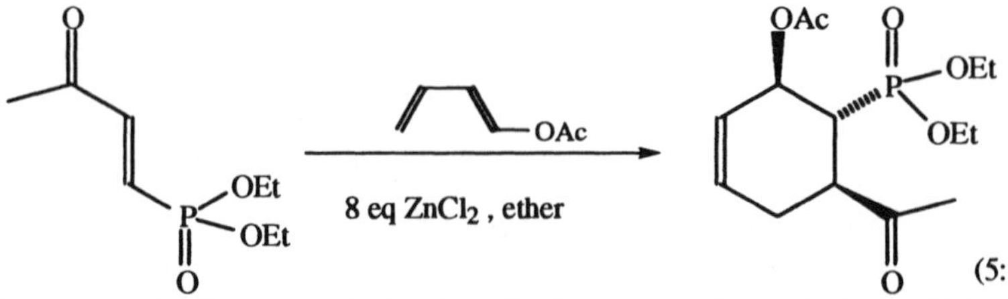
OAc
OEt
8 eq $ZnCl_2$, ether
OAc
OEt
OEt
OEt
OEt

(5:1) 85%

McClure, C.K.; Herzog, K.J.; Bruch, M.D. *Tetrahedron Lett.*, ***1996***, *37*, 2153.

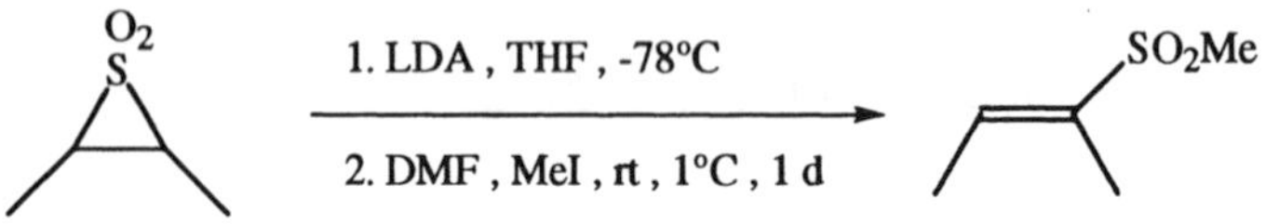
1. LDA , THF , -78°C
2. DMF , MeI , rt , 1°C , 1 d
SO_2Me

40% (>20:1 *E:Z*)

Dishington, A.P.; Mucciioli, A.B.; Simpkins, N.S. *SynLett.*, ***1996***, 27.

$$C_6H_{13}CHO \xrightarrow{MeNO_2\text{, Envirocat EPZG, 100°C, 20 min}} C_6H_{13}CH{=}CHNO_2$$

94% (*E* only)

Bandgar, B.P.; Zirange, M.B.; Wadgaonkar, P.P. *SynLett.*, ***1996***, 149.

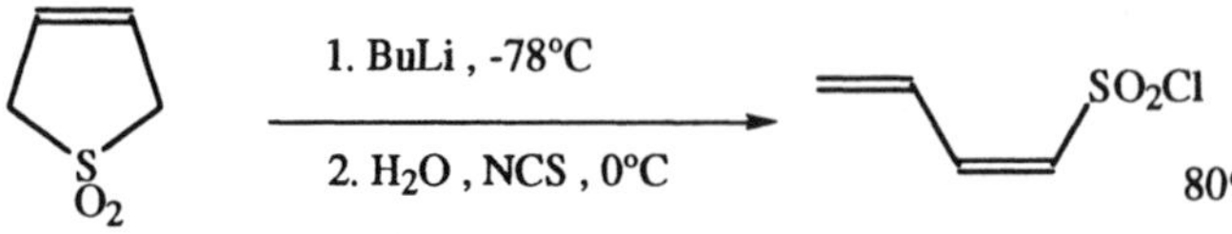
1. BuLi , -78°C
2. H_2O , NCS , 0°C
SO_2Cl

80% (78:22 *Z:E*)

Lee, Y.S.; Ryu, E.K.; Yun, K.-Y.; Kim, Y.H. *SynLett.*, ***1996***, 247.

SECTION 390: OXIDES - OXIDES

NO ADDITIONAL EXAMPLES